Powell's Turtles: Perspectives & Research.

OP / 9.98   NDJ

Science & Natural History   117547

TURTLES
Perspectives and Research

TURTLES
Perspectives and Research

MARION HARLESS
HENRY MORLOCK

Robert E. Krieger Publishing Company
Malabar, Florida

Original Edition 1979
Reprint Edition 1989

Printed and Published by
ROBERT E. KRIEGER PUBLISHING COMPANY, INC.
KRIEGER DRIVE
MALABAR, FLORIDA 32950

Copyright©1979 by John Wiley & Sons, Inc.
Reprinted by Arrangement

All rights reserved. No part of this book may be reproduced in any form or by any means, electronic or mechanical, including information storage and retrieval systems without permission in writing from the publisher.
No liability is assumed with respect to the use of the information contained herein.
Printed in the United States of America.

Library of Congress Cataloging-in-Publication Data
Turtles: perspectives and research / [edited by] Marion Harless, Henry Morlock.
    p. cm.
  Reprint. Originally published: New York: Wiley, c1979.
  Bibliography: p.
  Includes index.
  ISBN 0-89464-319-3
    1. Turtles. I. Harless, Marion. II. Morlock, Henry.
[QL666.C5T87 1989]
597.92--dc19   88-12933
                                                                             CIP

10 9 8 7 6 5 4 3

To those who further our knowledge of Turtles without endangering any of the earth's populations of Turtles

Contributors

Walter Auffenberg
Department of Natural Science
The Florida State Museum
University of Florida
Gainesville, Florida

R. Bruce Bury
National Fish and
Wildlife Laboratory
U.S. Fish and Wildlife Service
Fort Collins, Colorado

Stephen D. Busack
Museum of Vertebrate Zoology
University of California
Berkeley, California

H. Robert Bustard
United Nations Development
Programme
New Delhi, India

Howard W. Campbell
Gainesville Station
National Fish and Wildlife Service
U.S. Fish and Wildlife Service
Gainesville, Florida

Charles C. Carpenter
Department of Zoology
University of Oklahoma
Norman, Oklahoma

David W. Ehrenfeld
Department of Horticulture and
Forestry
Cook College
Rutgers University
New Brunswick, New Jersey

Michael A. Ewert
Department of Zoology
University of Indiana
Bloomington, Indiana

Eugene V. Gourley
Department of Biology
Radford College
Radford, Virginia

Terry E. Graham
Department of Biology
Worcester State College
Worcester, Massachusetts

A. M. Granda
Institute for Neuroscience and Behavior
University of Delaware
Newark, Delaware

Marion Harless
P. O. Box 253
Elkins, West Virginia

Victor H. Hutchison
Department of Zoology
University of Oklahoma
Norman, Oklahoma

John B. Iverson
Department of Biology
Earlham College
Richmond, Indiana

Crawford G. Jackson, Jr.
San Diego Natural History Museum
San Diego, California

Donald C. Jackson
Division of Biological and Medical Sciences
Brown University
Providence, Rhode Island

John Klicka
Department of Biology
Wisconsin State University
Oshkosh, Wisconsin

Warren K. Legler
Department of Physiology
University of Kansas Medical Center
Kansas City, Kansas

I. Y. Mahmoud
Department of Biology
Wisconsin State University
Oshkosh, Wisconsin

Marion L. Manton
Department of Biological Sciences
Macquarie University
Sydney, Australia

James H. Maxwell
Institute for Neuroscience and Behavior
University of Delaware
Newark, Delaware

Edward O. Moll
Department of Zoology
Eastern Illinois University
Charleston, Illinois

Henry Morlock
Department of Psychology
SUNY College of Arts and Sciences
Plattsburgh, New York

Michael V. Plummer
Department of Biology
Hardy College
Searcy, Arkansas

Alice S. Powers
Department of Psychology
Bryn Mawr College
Bryn Mawr, Pennsylvania

Peter C. H. Pritchard
Florida Audubon Society
Maitland, Florida

CONTRIBUTORS

Anton J. Reiner
Department of Psychiatry and
Behavioral Science
Health Sciences Center
SUNY at Stony Brook
Stony Brook, New York

Thomas R. Scott, Jr.
Department of Psychology and
Institute for Neuroscience and
Behavior
University of Delaware
Newark, Delaware

Irwin M. Spigel
Department of Psychology
Erindale College
University of Toronto
Mississauga, Ontario, Canada

Warren F. Walker, Jr.
Department of Biology
Oberlin College
Oberlin, Ohio

Preface

Turtles have adapted to oceans, forests, swamps, and deserts with a success that is as close to permanency as we find in the vertebrate world. The variation in adaptations suggests that there exists no representative or composite turtle to be used as a model for all turtles. The stereotyped notion of "the turtle" should become conceptually extinct with the publication of this book. Some chapters focus on a single species, not as a representative of all chelonians, but as the species with which the investigator is most familiar. Frequently similar kinds of information are lacking for other species, for in reality we have very little knowledge of most turtles. We may succeed in eliminating the habitats of some species before we learn of their present environmental relationships or of their possible behavioral and physiological adaptations to changing environments. Recent actions by interest groups and governments may save some of the species that seemed certain candidates for extinction in this century; some may not survive.

Upon merging independently prepared bibliographies of turtle research several years ago, it appeared to us that many investigators were unaware of the techniques and findings of their colleagues. The purposes of this book are to bring together the diverse approaches to studying turtles, to describe useful techniques, to present and interpret data, and to indicate directions for future research and thought.

A single scientific name for each genus and species is used throughout the book, even though some authors preferred a different taxonomic designation.

Biochemistry, parasitology, gross anatomy, morphology, and paleontology are included peripherally, not because of lack of interest in these areas or denial of their importance, but because of desire to restrict topics to those most clearly related to behavior, physiology, and ecology.

The research methods are not meant as final solutions to the problems associated with data collection, nor are the research results of subsequent chapters meant to be the permanent pronouncements they may appear to be. Future research may benefit from the suggestions in these chapters. More important, future research will benefit from the ideas not in these chapters, but generated by readers who synthesize these many thoughts in terms of their own experiences.

We thank Mindy Rogers Bean and Jill Mason Pritchard for their help in assembling and preparing the original bibliographies. We are also thankful to Mindy for her typing of the voluminous correspondence associated with the beginning of this project. And we are very grateful to Mary M. Conway, Life Science Editor of Wiley-Interscience, for her patience and many helpful suggestions.

MARION HARLESS
HENRY MORLOCK

Elkins, West Virginia
Plattsburgh, New York
August 1978

Contents

1 Taxonomy, Evolution, and Zoogeography
  Peter C. H. Pritchard   1

METHODS

2 Collecting and Marking   *Michael V. Plummer*   45
3 Telemetry   *Warren K. Legler*   61
4 Life History Techniques   *Terry E. Graham*   73
5 Photographic Analysis   *Charles C. Carpenter*   97
6 Laboratory Maintenance   *Howard W. Campbell* and *Stephen D. Busack*   109
7 Anesthesia and Surgery   *James H. Maxwell*   127

VITAL FUNCTIONS

8 Cardiovascular System   *Crawford G. Jackson, Jr.*   155
9 Respiration   *Donald C. Jackson*   165
10 The Central Nervous System   *Alice S. Powers* and *Anton J. Reiner*   193
11 Thermoregulation   *Victor H. Hutchison*   207
12 Feeding, Drinking, and Excretion   *I. Y. Mahmoud* and *John Klicka*   229

SENSORY PROCESSES

13 Eyes and Their Sensitivity to Light of Differing Wavelengths
  A. M. Granda   247
14 The Chemical Senses   *Thomas R. Scott, Jr.*   267
15 Olfaction and Behavior   *Marion L. Manton*   289

REPRODUCTION AND DEVELOPMENT

16	Reproductive Cycles and Adaptations *Edward O. Moll*	305
17	The Embryo and Its Egg: Development and Natural History *Michael A. Ewert*	333

BEHAVIOR

18	Behavior Associated with Nesting *David W. Ehrenfeld*	417
19	Locomotion *Warren F. Walker, Jr.*	435
20	Learning *Henry Morlock*	455
21	Social Behavior *Marion Harless*	475
22	Emotional Reactivity *Irwin M. Spigel*	493
23	Rhythms *Eugene V. Gourley*	509

POPULATION DYNAMICS

24	Population Dynamics of Sea Turtles *H. Robert Bustard*	523
25	Demography of Terrestrial Turtles *Walter Auffenberg* and *John B. Iverson*	541
26	Population Ecology of Freshwater Turtles *R. Bruce Bury*	571

Bibliography 603

Index 669

TURTLES
Perspectives and Research

CHAPTER 1
Taxonomy, Evolution, and Zoogeography

PETER C. H. PRITCHARD

TAXONOMY

Turtles, being poikilothermous, laying cleidoic eggs, and having a typically scaled integument, are unquestionably reptiles. Within the class Reptilia, turtles are usually classified with the earliest or "stem" reptiles of the order Cotylosauria in the subclass Anapsida. This designation alludes to the absence of true temporal fossae in the turtle skull, in contrast to the other reptilian subclasses in which one or two temporal fossae are present on each side of the skull. Only a few living turtles have fully roofed-over skulls; these include the Cheloniidae, Dermochelyidae, Platysternidae, and to some extent *Pseudemydura umbrina* and *Peltocephalus tracaxa*, essentially the forms in which the head is incompletely retractile or nonretractile. More typically, the skull roof is strongly emarginated from behind, leaving a relatively narrow quadratojugal arch. In a few genera (e.g., *Heosemys*) this arch may disappear completely. In one family, the Chelidae, the skull roof is emarginate from below, leaving a narrow parietosquamosal arch, which also disappears completely in one genus (*Chelodina*). Such emarginations are not considered true temporal fossae, even though they serve the same function of allowing the bunched jaw muscles to expand outward when the mouth is forcibly closed.

The earliest attempts to classify the different kinds of turtles, apart from the initial monogeneric arrangement of Linnaeus, relied on habitat differences; Brongniart in 1805 classified the land turtles as *Testudo*, the freshwater forms as *Emys*, and the marine forms as *Chelonia*. Other early classifications separated the turtles on the basis of whether the digits were fused together or separate, the presence or absence of a plastral hinge, or the retractibility of the extremities. Modern classifications usually focus on the plane of retraction of the neck, since this difference not only is reflected in the cervical vertebrae, but also shows a strong correlation with pelvic structure. This classification establishes the Cryptodira, a group of 10 living families in which the neck is vertically retractile, and the Pleurodira, which include the two families of sidenecked turtles. Some authorities make the initial taxonomic division of the testudines into the Athecae and the Thecophora, the former containing only the highly divergent *Dermochelys*. Many extinct turtles do not fit into either the Cryptodira or the Pleurodira and are placed in a separate suborder, the Amphichelydia, in which the neck is nonretractile and the neck vertebrae primitive, lacking ginglymes.

Zangerl (1969) proposed a new arrangement of the higher divisions of the testudines, characterized by four "levels of organization," or suborders, the Amphichelydia, Mesochelydia, Metachelydia, and Neochelydia.

The Amphichelydia of Zangerl is used in a restrictive sense, to include only the very early turtles of the families Proganochelyidae, Proterochersidae, and Kallokibotiidae. This suborder is characterized by very broad vertebral scutes, unstable and variable numbers of shell bones and scutes, extensive series of supramarginals and inframarginals, two pairs of mesoplastra, and intergular, caudal, and intercaudal scutes.

The Mesochelydia include most of the families traditionally placed in the Amphichelydia, as well as the living Chelidae and Pelomedusidae. In these turtles the scute patterns have become more stabilized, though supramarginals and inframarginals are still usually present. The living side-necks, especially the Chelidae, have undergone considerable modification of the scute mosaic, but still retain an "archaic complexion" and, in the Pelomedusidae, such primitive features as mesoplastra.

The diverse and successful suborder Metachelydia includes all the living and extinct marine turtle families, the soft-shells, snapping turtles, mud turtles, Carettochelyidae and Dermatemyidae, and the extinct Aperotemporalidae, Sinemyidae, and Plesiochelyidae. At this level, the costal scutes are stabilized at four pairs (except in *Caretta* and *Lepidochelys*), supramarginals are present only in *Macroclemys*, and the vertebral scutes are relatively narrow. Scutes disappear completely in the Trionychidae, and nearly in the Carettochelyidae. *Kinosternon* and *Sternotherus* lack the entoplastron, while the Trionychidae lack peripheral bones but have neomorphic "preplastra," no entoplastron, and fused epiplastra. The Dermochelyidae lack scutes and neural and peripheral bones, and have ribs that remain free throughout life.

The Neochelydia, which lack supramarginals, inframarginals, intergulars, and similar features, and have relatively narrow vertebral scutes and a repeated tendency toward shell kinesis, are very successful at present in terms of number of species, though only two families, the Testudinidae and the Emydidae, are recognized, and these are often combined into one.

Zangerl's intelligent and sophisticated though also controversial classification can be criticized because of the extreme diversity of turtles included in the Mesochelydia and Metachelydia. Other arrangements of turtle families are given by Romer (1956), Mlynarski (1969), Ckhikvadze (1970), and Williams (1950). All are worthy of consideration, and no attempt is made here to extol the merits of one at the expense of the others.

Turtle taxonomy has traditionally been based almost entirely on osteological and external morphological criteria. This is inevitable in a group in which the fossil forms assume such great importance. However in recent years several efforts have been made to supplement the traditional osteological approach with biochemical or anatomical comparisons in an attempt to provide a wider spectrum of criteria for the classification of the living forms. Notable among these efforts have been the chromatographic and electrophoretic studies of serum proteins of Frair (1964, 1972), the comparisons of fatty acid compositions and ratios of Ackman, Hooper, and Frair (1971), and the studies of comparative penis morphology of Zug (1966).

Frair's studies have generally confirmed traditional views of turtle classification but have provided some interesting insights; for example, not only are the Trionychidae peculiar anatomically, they also have a highly distinctive serum electrophoresis pattern. On serological grounds, *Dermatemys* appears to be closely related to *Staurotypus*, and there is indeed a fossil form (*Xenochelys*) that is a morphological intermediate. The snapping turtles (Chelydridae), on the other hand, appear to be more closely related to the Emydidae than to the Kinosternidae, and this apparent relationship agrees with the anatomical studies of McDowell (1964). The marine turtles, even the divergent *Dermochelys*, have similar serum patterns. McDowell's separation of the Emydidae into Batagurinae and Emydinae is confirmed by a difference in serum patterns between the two groups. *Platysternon* appears to be relatively close to the emydids and quite distinct from the kinosternids and chelydrids. Within the genus *Kinosternon* the small North American species give a different pattern from the larger tropical forms.

The fatty acid studies of Ackman, Hooper, and Frair (1971) indicate that the living marine species are distinct from all others in having *trans*-6-hexadecanoic acid in their depot fat, that *Dermochelys* is exceptional in having 9.5% lauric acid, and the *Dermatemys* can be differentiated from the other freshwater species by details of fatty acids indicative of an herbivorous diet.

Zug's studies of the penis show a close relationship between the cheloniid and dermochelyid sea turtles; a distinction between the Emydidae and the Batagurinae; a close relationship between *Chrysemys* and *Pseudemys*, and between *Deirochelys*, *Emydoidea*, and *Malaclemys*; a close relationship between *Dermatemys* and the kinosternids, and between the carettochelyids and the trionychids. The tortoises (Testudinidae) are distinct from, though related to, the Emydidae in penial morphology, while the Platysternidae appear to be closely related to the Emydidae. The penis of the chelydrids conforms Frair's findings of a relationship between the snapping turtles and the Testudinidae and Emydidae. Within the genus *Sternotherus*, a study of penial morphology would place *S. carinatus* apart from *S. minor* and *S. odoratus*.

EVOLUTION

It was held for many years that a small fossil reptile from the middle Permian of Welt Vreeden (Beaufort West), South Africa, known as *Eunotosaurus africanus*, represented the "missing link" between the cotylosaurs and the turtles. *Eunotosaurus* had a broadened body with eight pairs of expanded ribs, and this, together with the appropriate geological horizon, was considered by Watson (1914) to suggest strongly that *Eunotosaurus* was the ancestral chelonian. Recent authors, notably Parsons and Williams (1961), have argued otherwise; the ribs of *Eunotosaurus*, though expanded, are leaf shaped, thus differ from the pleural bones of true turtles, which initially develop mesially near the vertebral column and have lateral flanges. These flanges progress outward as the turtle matures, so that they eventually form jagged sutures throughout their lengths with their neighbors. Moreover, the fossils of *Eunotosaurus* showed no signs of either plastron or abdominal ribs, and recent studies by Cox of the University of London have shown that *Eunotosaurus* has no dermal armor and no other features suggesting relationship with the turtles. Some critical parts of *Eunotosaurus*, including the skull roof, the neck, and the feet, remain unknown. Since no fossil turtles are found outside Europe earlier than the Jurassic, it would seem reasonable to postulate an origin for the group in what is now Europe.

The oldest unmistakable chelonians, the Proganochelyidae, are from the Triassic of Germany. The closely related *Triassochelys dux* and *Proganochelys quenstedi* had such primitive features as teeth in both jaws and palate, and numerous marginals and supramarginals. However the peculiarly universal arrangement of four pairs of costal scutes and five vertebrals was already present.

One can only speculate how the Proganochelyidae evolved from the primitive reptile stem. Deraniyagala (1930, 1939) postulated an evolutionary "armadillo" stage and illustrated the "saurotestudinate," the "missing link" in turtle evolution. As Mlynarski (1956) pointed out, there is no fossil evidence to support this hypothesis.

Carroll (1969) remarked that there are only two certainties regarding the origin of turtles: (1) that they did not evolve from any group that had already developed a lateral or dorsal temporal opening, and (2) that they did not evolve from any form in which the palate was fused to the braincase. This means that the turtle lineage was already separate in the early Permian. Gregory (1946) and Olson (1947) both assumed that turtles evolved from the diadectomorph cotylosaurs, a group that flourished from the Carboniferous to the Triassic. These primitive reptiles had a complete, imperforate temporal region and a well-defined otic notch, both features shared with the turtles. On the other hand, the later diadectomorphs had the palate solidly attached to the braincase, and it is therefore assumed that the turtles had already become separate by the time this modification occurred.

It is most probable that the earliest known turtles of the family Proganochelyidae were marsh dwelling; the fully aquatic preference of most modern turtles is presumably a secondary modification, since the scaled integument and shelled, terrestrial eggs of turtles and other reptiles are essentially adaptations that permitted these animals to escape from dependence on an aquatic medium. Nevertheless, the broad, shell-encased body inevitably makes turtles slow on land, with little chance of escape from strong-jawed predators, and the move back to the freshwater medium was a rather early development for the group; the vast majority of both modern and extinct turtle genera have or had a freshwater habitat.

One of the first truly terrestrial turtles was the Upper Cretaceous dermatemyid *Zangerlia testudinimorpha*, a relatively large species from the Lower Nemegt Beds of Mongolia (Mlynarski, 1972). The assumed terrestrial habitat is postulated on several grounds: the deep shell, the very strong dorsal sulci (which suggest thick, sculptured scutes such as are typical of many terrestrial tortoises today), the shortened phalanges, and the extensive fusion of the carpal elements. On the other hand, several other features, such as the straight humerus, the somewhat reduced plastron, and the extensive superficial sculpturing of the shell (similar to that of *Carettochelys* or *Tretosternon*) show the influence of recent aquatic ancestors.

The Upper Cretaceous meiolaniid *Niolamia argentina* may have been another Mesozoic terrestrial form. However, the true tortoises of the family Testudinidae did not appear in the fossil record until the mid-Eocene; presumably there were primitive tortoises in existence in the Paleocene. They perhaps reached their greatest abundance and diversity in the Pliocene. Tortoises evolved from primitive emydid turtles, and the emydids have subsequently given rise to several other terrestrial lines (e.g., *Terrapene, Pyxidea, Rhinoclemys*).

Turtles entered the marine environment at an early stage, and indeed some isolated cervical vertebrae from the early Triassic of Germany, which have been described under the generic name Chelyzöon, may conceivably have belonged to a marine turtle, according to Staesch (in Goode, 1967b). The Lower Cretaceous *Desmemys* type of the subfamily Desmemydinae of the family Pleurosternidae, was a marine form, whereas the extinct Toxochelyidae and Protostegidae and the living Cheloniidae and Dermochelyidae are comprised of entirely marine species. The amphichelid family Thalassemyidae, from the Upper Jurassic to the Upper Cretaceous of Europe and Asia, was also marine. Since neither the Desmemydinae nor the Thalassemyidae were ancestral to the Chelonioidea, it appears that turtles have independently evolved marine lineages at least three times. In addition, some of the early side-necks of the Pelomedusidae were marine, and a few of the modern emydids show a preference for estuarine or brackish situations.

Related to the Proganochelyidae, and distinguished by the absence of fusion between the pelvis and the plastron, is *Proterochersis robusta*, the sole species of the family Proterochersidae, and also from the Triassic of Europe. This species had a carapace length of about 30 cm, two pairs of mesoplastra, caudal and intercaudal scutes, very broad vertebral scutes, and a somewhat cruciform plastron. Its vaulted shell suggests a largely terrestrial life.

ZOOGEOGRAPHY

Chelydridae

The **family Chelydridae**, although represented only by two living species, **constitutes an important** element in the modern New World turtle fauna. **The more widespread species is the common snapping

Figure 1   Florida snapping turtle (*Chelydra serpentina osceola*).

turtle, *Chelydra serpentina*, of which two temperate subspecies (*C. s. serpentina* and *C. s. osceola*) and two tropical subspecies (*C. s. rossignonni* and *C. s. acutirostris*) exist. The species is easily recognized by the flat, posteriorly serrated shell, the very long, tuberculate tail, the big head with somewhat dorsally located eyes, the long neck, and the very reduced, cruciform plastron. Large adults have shells about 35 cm long. The Floridian and Mexican races (*C. s. osceola* and *C. s. rossignonni*) are morphologically very similar, for example, in having elongate soft tubercles on the neck. These races are thought to have been separated by Pleistocene cold conditions forcing the population south on the two sides of the Gulf of Mexico, with the subsequent void later being invaded by *C. s. serpentina*, which is now found over most of the eastern and central United States as well as southern Canada (Feuer, 1966). There is a hiatus in the range of the species in south Texas and throughout Tamaulipas. The Mexican and South American races intergrade in Honduras; *C. s. acutirostris* reaches as far south as the Gulf of Guayaquil, Ecuador, but does not cross the Andes. The Florida race, *C. s. osceola*, reaches the largest size (record shell length: 47 cm), though an artificially fattened member of the nominate race has reached a weight of 39 kg.

The alligator snapping turtle, *Macroclemys temmincki*, has its principal breeding areas in the Okefenokee Swamp in southern Georgia, in northern Florida, and in Louisiana. However on occasion individuals may be found up the Mississippi as far as Kansas, Indiana, and Illinois. Some males have shells well over 65 cm long and weights exceeding 100 kg. *Macroclemys* is distinguished from *Chelydra* by its more pointed triangular head, laterally placed eyes, vermiform "fishing lure" on the tongue, supernumerary scutes on the plastron, three strong dorsal keels throughout life, and (a curious, very primitive feature) four or so supramarginal scutes on each side of the carapace.

The fossil history of *Chelydra* extends as far back as the Paleocene in North America, and two fossil species of *Macroclemys*—*M. schmidti* and *M. auffenbergi*, from the Miocene and Pliocene, respectively—have been described. In view of the unmistakable American distribution of the family, it is surprising that the Chelydridae were once widespread in the Old World. Fossil *Chelydra* are known from the Oligocene to Pleistocene of Europe, and numerous species of the genera *Chelydrops* and *Chelydropsis* have been described, mostly from very fragmentary material. Fossil New World genera include the diminutive *Acherontemys*, from the Miocene of Washington, and *Hoplochelys*, with seven species from the early Tertiary.

Kinosternidae

Four genera of the family Kinosternidae are currently recognized. Two of them (*Kinosternon* and *Sternotherus*) are so closely related that they are barely separable, and two others (*Staurotypus* and *Claudius*) show substantial differences not only from *Kinosternon* and *Sternotherus*, but also from each other, and are often included in a separate subfamily (Staurotypinae). *Kinosternon* is recognized by the relatively large plastron whose front and rear lobes hinge freely on a fixed midsection composed of the hyo- and hypoplastral bones. The entoplastron is missing, and the gular scute is single. *Kinosternon* species or "mud turtles" have a collective range extending from Long Island, New York, to Bolivia and northern Argentina; they reach their greatest diversity in Mexico and Central America. The two eastern United States species (the widespread *K. subrubrum* and the Floridian *K. bauri*) are both very small, being adult at around 10 cm; their south-central and southwestern relatives (*K. flavescens* and *K. sonoriense*, respectively) are slightly larger. In certain of the tropical species (*K. herrerai* from northeastern Mexico, *K. angustipons* from Caribbean Costa Rica, and *K. dunni* from coastal Colombia) the plastron is highly reduced so that the soft parts are exposed when the plastral lobes are raised. In certain others (e.g., *K. leucostomum* from Veracruz to Ecuador, *K. scorpioides cruentatum* from Tamaulipas to Panama, *K. abaxillare* from central Chiapas, and *K. acutum* from the base of the Yucatan peninsula) the plastral lobes are so large that the shell can be closed as tightly as that of a box turtle. The shell is almost closable in the Mexican *K. integrum* and the Yucatan *K. creaseri*, but somewhat less so in the South American *K. s. scorpioides* and in the northern and central Mexican *K. hirtipes*.

Many of the tropical forms of *Kinosternon* have tricarinate shells, but the keels tend to disappear with age. The young of most of the species are virtually indistinguishable.

Kinosternon has very limited ability to reach islands, and is absent from the Antilles, with the predictable exception of Trinidad. However a population that has been named *Kinosternon scorpioides albogulare* reaches an amazing population density on the island of San Andrés, in the western Caribbean. A population of the *K. scorpioides* complex is also abundant on the island of Coiba, off the Pacific coast of Panama.

The fossil record sheds little light on the evolution of the genus. One fossil species, *K. arizonense*, has been described from the Pliocene and Pleistocene of Arizona and northern Mexico.

Sternotherus differs from *Kinosternon* in having a reduced, somewhat angular plastron, with little mobility in the hind lobe, and with a relatively long interpectoral seam. The plastral seams are often invaded by soft tissue, particularly in adult males. The gular is small and frequently absent. The genus is virtually confined to the United States; a widespread species, *S. odoratus*, extends from Maine to Texas and to Sauz, Chihuahua, while the large *S. carinatus* is found in the lower Mississippi Valley. The three subspecies of *S. minor* (*S. m. minor*, *S. m. peltifer*, and *S. m. depressus*) are found in northern Florida, south Georgia, and southeastern Alabama; northern and western Alabama and adjacent parts of Mississippi, Georgia, and Tennessee; and Walker County, Alabama, respectively. *S. m. depressus* is an extremely flat turtle whose habits have not been studied. The various species are highly aquatic, and although sometimes found basking, they do not often wander far from water.

Staurotypus, with one species in the Caribbean lowlands of southern Mexico and northern Central America, and another on the Pacific side, reaches a larger size than *Kinosternon* (up to 38 cm in *S. triporcatus*, about 25 cm in *S. salvini*). The three dorsal keels are retained throughout life and indeed become stronger in large adults. The plastron is small and cruciform, with narrow but strong bridges, a large, triangular entoplastron, and a single (anterior) hinge. Only four pairs of plastral scutes are present. The plastral structure is thus quite distinct from that of *Kinosternon*, and subfamilial status, at least, appears to be justified.

The fossil species *Xenochelys formosa*, from the Oligocene of South Dakota, shows characteristics of both *Staurotypus* and *Dermatemys*.

The small *Claudius angustatus* is found from Veracruz to northern

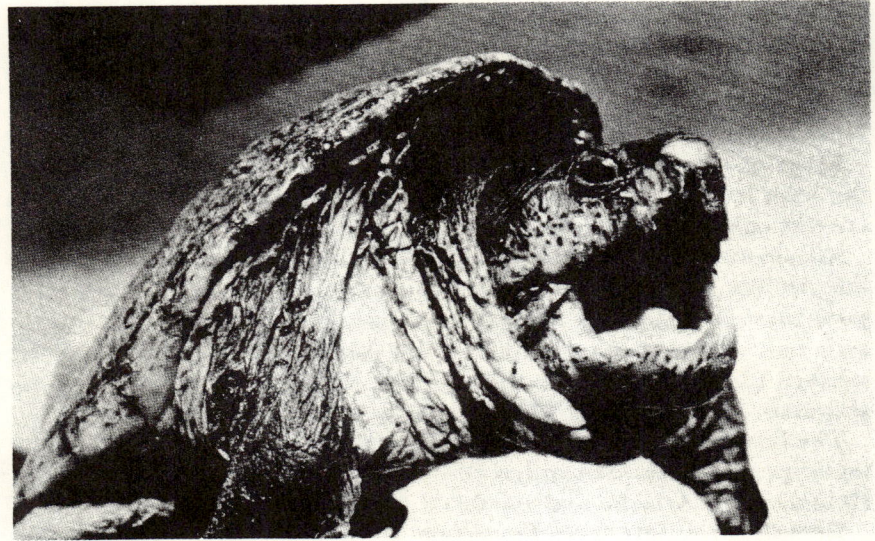

Figure 2 Mexican narrow-bridged mud turtle (*Claudius angustatus*).

Guatemala. It has a thin shell with only traces of lateral keels, a very large head with lateral cusps on the upper jaw, and a long neck. The plastron has only four pairs of scutes and is unique in being composed of only seven bones—the hyo- and hypoplastra are fused into single structures. An entoplastron is present, and there are no plastral hinges. The bridge is exceedingly thin, and the plastron does not have a sutural contact with the carapace. Several characters, including the narrow, unossified bridges, the unhinged, cruciform plastron, the thin carapace, the large head, long neck, and free dorsal rib ends suggest a relationship with the snapping turtles (Chelydridae).

Dermatemyidae

The family Dermatemyidae includes a single living species, *Dermatemys mawi*, which is a rather large, vegetarian river turtle from southern Mexico, northern Guatemala, and Belize. In general form, with its streamlined shell whose seams become indistinct with maturity, as well as its projecting, conical snout, *D. mawi* bears a superficial resemblance to the Asiatic batagurs and their relatives of the family Emydidae. However the presence of a complete series of inframarginal scutes, a single gular, and the reduction of the posterior neural bones suggest that the actual

Figure 3   Mexican river turtle (*Dermatemys mawi*).

relationship lies with the Kinosternidae; the fossil *Xenochelys* appears to be structurally intermediate between the two families. The head of *D. mawi* is rather small, but the jaw surfaces bear a very complex arrangement of pseudo-"teeth" of various shapes and sizes.

Various fossil dermatemyids have been described from the Cretaceous to Miocene of Asia, Europe, and North America; however the non-American forms are somewhat questionable. Among the New World representatives, there was an early tendency toward deviant numbers of plastral scutes, and certain genera, such as *Basilemys* and *Adocus*, also developed exceedingly sinuous, asymmetrical plastral seams. *Adocus* was a low-shelled form with elongate posterior marginal scutes, whereas *Agomphus* had a steeply humped shell and very short marginals.

Testudinidae

The Testudinidae are a group of moderately successful turtles whose most conspicuous characteristic is their terrestrial, usually somewhat xeric habitat. The family is further characterized by club-shaped front feet and columnar hind feet with no independently movable digits, a typically domed carapace, the absence of inframarginal scutes, completely retractile extremities, a short tail, and heavily scaled, impermeable integument of the anterior surfaces of the front limbs, thighs, and soles of the hind feet. The nuchal scute may or may not be present, and in most tortoises the neural bones show a pattern, at least in part of the series, of alternating octagons and squares, the pleural bones showing a corresponding alternation of broad and narrow proximal and distal ends.

Modern tortoises reach their greatest diversity in Africa south of the Sahara, where the widespread genus *Geochelone* is represented by two large species (*G. pardalis* and *G. sulcata*) of the nominate subgenus, as well as by the endemic African genera *Homopus, Psammobates, Chersina, Kinixys,* and *Malacochersus*. *Homopus* includes four species (*H. areolatus, H. femoralis, H. boulengeri,* and *H. signatus*), all of which have relatively circumscribed ranges. Two of them (*H. signatus* and *H. boulengeri*) may form a natural subgroup, characterized by fused carpals, five claws on each forelimb, single inguinal scutes, and the typical presence of only 11 pairs of marginal scutes, though currently taxonomic recognition of such a split is withheld. *Psammobates* has three species (*P. geometricus, P. oculifer,* and *P. tentorius*), the first being almost extinct, and the last being one of the most variable and taxonomically confusing of all turtle species. At present, however, only three races (*P. t. tentorius, P. t. trimeni,* and *P. t. verroxi*) are considered to be geographically definable, thus valid (but see Hewitt, 1933 and 1934, and Archer, 1967b, for a different interpretation). Members of these South African endemic genera are among the smallest of all tortoises, and most of the forms do not reach a carapace length of 12 cm. Structurally they are fairly primitive in that the neural bones are persistently hexagonal. The monotypic *Chersina* is characterized by a single, elongate, spadelike gular scute (the single gular is otherwise found only in *Cylindraspis* and in *G. yniphora* among the Testudinidae), the elongate anterior marginals, the alternating octagonal-tetragonal middle neural bones, and the barely ridged or nonridged alveolar surfaces.

The genus *Kinixys* is also endemic to the African continent, with forest-living species (*K. erosa* and *K. homeana*) in West Africa and a widespread savanna and scrub-living species, *K. belliana*, which reaches as far south as South Africa and has been recorded from Madagascar. Many subspecies of *K. belliana* have been described, but all are controversial, though *K. b. nogueyi*, having only four claws on the front feet and a definable range in West Africa, appears to be more definite than most. The genus is identified by the unique feature of a dorsal hinge, which extends from the middle of the third vertebral scute, down between the second and third costal scutes, and terminates in a cartilaginous infusion between the seventh and eighth marginals. The hinge thus offers protec-

Figure 4   Madagascar tortoise (*Acinixys planicauda*).

tion only to the rear part of the body, though its functions may be more diverse than pure protection (e.g., it may assist in oviposition and respiration). Only the adults are hinged; the young are recognizable by the relative proportions of the third and fourth costals (the latter being the larger), and by the alignment of the seam between the third and fourth costals and the eighth and ninth marginals.

Malacochersus, with the single species *M. tornieri* (again very variable), is another African endemic. This species, commonly known as the pancake tortoise, is found in Tanzania, and despite its unspecialized, relatively robust skull, has a highly fenestrated and greatly depressed shell even as an adult, this being correlated with the characteristic habit of concealment in split boulders and under rocks.

Pyxis and *Acinixys* are monotypic genera endemic to Madagascar. *Pyxis* is unique in having an anterior plastral hinge. It also shows no alveolar ridges, but does have alternating octagonal and quadrilateral neurals. *Acinixys* has an unhinged shell, triturating alveolar surfaces, and hexagonal neurals. Neither species exceeds a shell length of about 15 cm.

Geochelone is the only tortoise genus found in tropical southeast Asia. The subgenus *Geochelone* is represented by two species, *G. elegans* and *G. platynota*, both of which are much smaller than their African counterparts and have a decorative starred pattern on the carapace scutes—a pattern that shows up repeatedly throughout the family. The other Asiatic subgenera are *Manouria* and *Indotestudo*, each with two species. *Manouria* is probably the most primitive of living tortoises, and, indeed, Auffenberg (1971) has recently asserted that some of the earliest known (Eocene) tortoises of the subgenus *Hadrianus* are indistinguishable from *Manouria*. The primitiveness of the modern species is most obviously displayed by their paired supracaudal scutes, depressed, emydidlike shells, and lack of a thickened posteriorly recessed epiplastral lip; and it is perhaps significant that the preferred habitat is much more moist than is typical of tortoises. The two species of *Indotestudo* (*G. travancorica* and *G. elongata*) have elongate, narrow shells and are differentiated from each other by coloration and the presence in *G. elongata* of a nuchal scute. The zoogeographically inexplicable "*Geochelone forsteni*" on Celebes is probably based on a long-established population of *G. travancorica* (Auffenberg, personal communication).

Indotestudo is very closely related to the subgenus *Chelonoidis* which is widely distributed in South America. *Geochelone carbonaria* narrowly enters Central America as well as being present, probably as introduced populations, on several isolated West Indian islands. *G. carbonaria* and *G. denticulata* have a complex pattern of partially sympatric ranges in northern and central South America, with most interesting habitat differences in some parts of the range and a tendency toward more vivid and distinctive colors in areas of sympatry. *G. chilensis*, from the pampas of northern Argentina, is a much smaller species that nevertheless appears to be the closest living relative of the Galapagos giant tortoises. *G. petersi* and *G. donosobarrosi*, described by Freiberg (1973), are thought by Auffenberg (personal communication) to represent sexual and geographical variants of *G. chilensis*.

Tortoises have an unexpected, though readily explicable, ability to cross oceanic barriers and reach both offshore and remote islands. They float well in water, and although poor swimmers, they can survive in water for considerable lengths of time, drifting passively with the current. On a micro scale, certain islands off the Gulf coast of Florida have abundant *Gopherus* populations, and such major islands as Trinidad and Borneo are also inhabited by tortoises identical to those found on the adjacent mainland. Madagascar has several species quite distinct from African forms, and one possibly introduced African species (*Kinixys belliana*). The largest, most spectacular tortoises are those found on rather remote oceanic islands, including the Galapagos and Aldabra, and, until very recently, numerous other BIOT islands, the Seychelles, Comores, and Mascarenes. Primordially, and to some extent even today, these insular giant tortoises were limited only by food or nesting space;

thus equilibrium population densities were much higher than those reached by continental forms.

Only certain islands were suitable for access and colonization by tortoises. A candidate island must be in the tropics, since giant tortoises cannot tolerate prolonged cold spells, being poikilothermous and too big to burrow. The island must be suitably distant from the mainland—if it is too close, the island will have too diverse a fauna, probably including predatory mammals; if it is too far, the chances of a tortoise making a live landfall become remote. There must also be current patterns that could transplant tortoises from the mainland to the island. All the larger Galapagos islands, except for those that have no central elevation, thus no humid zone, once had giant tortoises. On the largest, highest islands, the tortoises reached an enormous size, since lush vegetation was always available. The ultimate population-limiting factor was probably nesting space in the hot lowlands, which are composed principally of lava, with areas of soil suitable for nesting rather scarce. On the smaller, arid islands the tortoises were food limited; thus they not only matured and stopped growing at a smaller size, but also evolved certain characteristics, such as long legs and an open-fronted, "saddlebacked" shell, which enabled them to extend their vertical feeding range. Very dense tortoise populations are maintained on low-lying Aldabra, because it is situated in a warm rather than a cold ocean current, thus has adequate rainfall for growth of abundant vegetation.

The Galapagos tortoises, *Geochelone elephantopus*, are placed in the subgenus *Chelonoidis*, together with the mainland South American species. About 14 subspecies are recognized, of which *G. e. porteri, G. e. vandenburghi, G. e. elephantopus*, and *G. e. guntheri* are completely dome shaped; *G. e. ephippium, G. e. hoodensis, G. e. phantastica*, and *G. e. abingdoni* strongly saddlebacked; *G. e. becki* and *G. e. chathamensis* variable but often saddlebacked; and *G. e. darwini* and *G. e. microphyes* intermediate in shape. The extinct race *G. e. galapagoensis* was also saddlebacked. The Aldabra tortoises are now all included in *Geochelone gigantea*, which is placed in the subgenus *Aldabrachelys* largely on the basis of the unusual skull shape, with a bulbous supranasal area. The extinct Mascarene tortoises (*G. indica* on Reunion, *G. vosmaeri* and *G. peltastes* on Rodriguez, and on Mauritius, *G. inepta, G. leptocnemis*, and others that are probably synonyms of those two species) are placed in the subgenus *Cylindraspis*, defined by the combination of no nuchal scute and a single gular.

Two species of *Geochelone* occur on Madagascar: *G. radiata* on the south and southwestern coastal lowlands and *G. yniphora* near Baly Bay on the west coast. Both are placed in the subgenus *Asterochelys*, defined by a combination of features including the broad pectoral scutes, present nuchal scute, and alternately octagonal and tetragonal neurals. *G. radiata* has a starred pattern, and the highly domed carapace reaches a length of about 38 cm. *G. yniphora* is even larger, has an overall yellowish-tan color, and is remarkable for the single, enormously elongate gular scute. The gigantic Pleistocene *G. abrupta*, also from Madagascar, is thought to be possibly ancestral to *G. radiata*. Auffenberg (1974) provisionally placed it in the subgenus *Aldabrachelys*. Until only a few thousand years ago, Madagascar also had a giant tortoise of the subgenus *Aldabrachelys*, *G. grandidieri*; this was an exceedingly large species with a depressed and very thick carapace whose bony remains are still occasionally found in dry caves.

Tortoises were once widespread and diverse in North America, but various Pleistocene and recent climatic events appear to have caused extensive extinction and retreat. Today only one genus (*Gopherus*) is found in North America, with four species whose ranges are all well separated today, although analysis of courtship patterns suggests that sympatry once occurred (Auffenberg, 1965). *G. polyphemus*, of the extreme southeastern United States, appears to be rather closely related to the Mexican "giant" gopher, *G. flavomarginatus*, a species that has reached a late stage of retreat and is now restricted to a remote internally drained basin in northwestern Mexico, centered around the area where the states of Durango, Coahuila, and Chihuahua come together. The California desert tortoise, *G. agassizi*, and the small Berlandier's tortoise, *G. berlandieri*, from south Texas and northeastern Mexico, share the characteristic of a rather narrow head and appear to be closely related. The genus *Gopherus* is characterized by the specialization of the front feet for excavation of burrows, the median ridge on the premaxilla, and the frequent, though very variable, prolongation of the gular area into a projecting fork (paralleled by *Megalochelys*, an extinct subgenus of *Geochelone*).

The genus *Testudo* (in earlier days an inclusive term for all turtles) is now restricted to certain circum-Mediterranean and Middle Eastern species, *T. hermanni* and *T. marginata* in southern Europe, *T. graeca* in southeastern and southwestern Europe, North Africa, and the Middle East; *T. horsfieldi* in Afghanistan, Pakistan, and parts of the Soviet Union, and *T. kleinmanni* in Egypt. The genus is differentiated from *Geochelone* by the presence of supranasal scales, a nuchal scute, a prootic generally concealed dorsally by the parietals, and a hind plastral lobe that is often slightly movable. *T. kleinmanni*, a dwarf desert species, is placed in a separate subgenus *Pseudotestudo* because the maxilla is unridged and the quadrate does not enclose the stapes (Loveridge and Williams, 1957). *T. horsfieldi* is also divergent in some respects and has recently been placed in the revived genus *Agrionemys* (Mlynarski, 1966). In his recent revision Auffenberg (1974) reduced *Agrionemys* to subgeneric rank.

Emydidae

The family Emydidae includes the majority of the freshwater turtles of the Northern Hemisphere. It is a modern group, characterized by such features as the absence of mesoplastra, inframarginals, and intergulars; and it is closely related to the Testudinidae, but differs in never having elephantine hind feet and (except for a few species) in being adapted for a freshwater rather than a terrestrial habitat. The Emydidae reach their greatest diversity in the eastern United States and in southeast Asia, but the few species found in the Mediterranean area, including North Africa, may on occasion reach great abundance; and two genera are reasonably widespread in Central and South America. The family shows

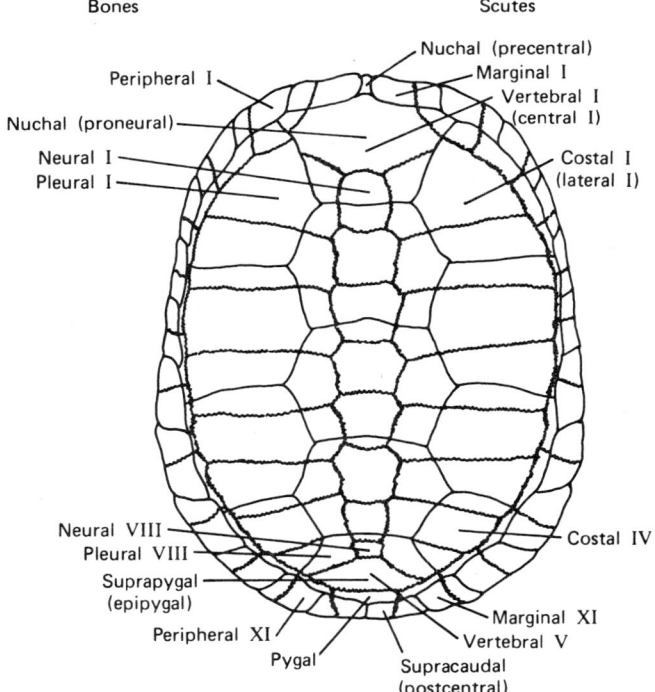

Figure 5 Carapace (divested of scutes) of an emydid turtle (*Emys orbicularis*), with nomenclature of bones and scutes.

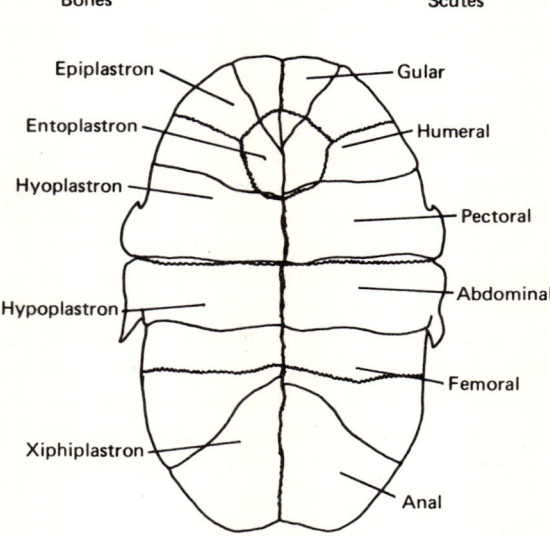

Figure 6 Plastron of *E. orbicularis*.

little ability to reach oceanic islands, but various forms of *Pseudemys* exist on the Greater Antilles, as well as on two of the Bahamas and the Cayman Islands. The family is absent in Australia and in subsaharan Africa. Until recently, three genera (*Emys*, *Clemmys*, and *Geoemyda*) were considered to have both New and Old World representatives, but the necessity of splitting these inclusive genera has now been demonstrated (Loveridge and Williams, 1957; McDowell, 1964); and the Old World and the New World forms are now placed in separate subfamilies, the Batagurinae and the Emydinae, respectively. The subfamilies are differentiated by features in the rear of the skull, the jaw, and the cervical vertebrae. However the correlation of subfamily with hemisphere is not perfect, the Old World *Emys* being considered an emydine, and the New World *Rhinoclemys* a batagurine.

The most widespread genus in the New World is *Pseudemys*, a group of rather generalized terrapins characterized by the foreshortened head and linear skin patterns. Four species groups are recognized, the so-called *scripta*, *floridana*, *concinna*, and *rubriventris* groups, each being named after the principal species in the group. The *scripta* group is characterized by the rounded, rather than flat, ventral surface of the lower jaw, the presence of only three phalanges in the fifth toe, and the absence of an anterior cusp in the middle ridge of the upper alveolar surface. The *floridana* subspecies are larger (up to about 38 cm carapace length) and have a deep bridge; they lack thickened epiplastral lip and a plastral pattern, and, unlike *scripta*, do not become melanistic. The *concinna* subspecies differ from *floridana* in having a dark, linear plastral pattern, a more complex dorsal pattern, and a lower, more streamlined carapace. The *rubriventris* species have a notched and cusped upper jaw margin, as well as a marked tendency toward reddish pigmentation in both carapace and plastron. The distinctions between the groups hold true in most areas, but break down in some (e.g., *P. concinna texana* has a jaw cusp like that of *rubriventris*, whereas some of the western representatives of the *scripta* series, such as *P. s. hiltoni* and *P. s. nebulosa*, have a carapace outline more typical of *concinna*). The *scripta* group is the most widespread, being found over most of the eastern United States, as well as along both coasts of Mexico and Central America, southern Baja California, the Greater Antilles, and extreme northern Colombia and Venezuela. *P. scripta*, or a close relative *P. dorbignyi*, is also found in certain isolated areas of Brazil—in the State of Maranhão on the northern coast, and in the extreme southern part of the country, the range here extending into parts of Uruguay, Paraguay, and northern Argentina. Subspecific designations of the southernmost populations have been discussed by Freiberg (1971).

The subspecies of *floridana* and *concinna* are restricted to the southern United States, while the three species of the *rubriventris* group, *P. rubriventris*, *P. nelsoni*, and *P. alabamensis*, have well-separated ranges in northeastern United States, peninsular Florida, and southern Alabama, respectively. An isolated relictual population in Plymouth County, Massachusetts, although formerly thought to be distinct (Babcock, 1937), is now known to be inseparable from *P. rubriventris* (Conant, 1951; Graham, 1969).

Closely related to *Pseudemys* is *Chrysemys*, whose single species *C. picta*, with four well-defined subspecies, is widespread in the United States (except for the extreme southeast and southwest) and southern Canada, with at least one population of the subspecies *C. p. belli* in Chihuahua, Mexico (Stebbins 1966). Historically, the cooters and sliders of the genus *Pseudemys* and the painted turtles of the genus *Chrysemys* have been repeatedly lumped, then split again into separate genera. Both generic names were initially proposed by J. E. Gray, who was perhaps the ultimate "splitter" in cheloniological history; the name *Chrysemys* dates from 1844, thus has priority over *Pseudemys*, which was first proposed in 1855 for *"Testudo" concinna*. Boulenger in 1889 combined the two groups under *Chrysemys* (Boulenger, 1966), though in the following years several authors took the opposite view (e.g., Baur, 1893; Jordan, 1899). Carr in his numerous publications on the group used *Pseudemys* uniformly for the cooters and sliders (Carr, 1935, 1937a, 1937b, 1938a, 1938b, 1938c, 1942a, 1942c, 1952; Carr and Crenshaw, 1957). However McDowell (1964) argued on osteological grounds *Chrysemys* was no more different from *Pseudemys scripta* than *P. scripta* was from *P. floridana, P. rubriventris*, and their relatives. Thus all three groups would have to be subsumed within the genus *Chrysemys*, or each would have to be recognized generically (with *Trachemys* for *P. scripta*). McDowell chose the former course. This has since been followed by certain authors (e.g., Ernst and Barbour, 1972; Weaver and Rose, 1967), though not by others (e.g., Moll and Legler, 1971; Williams, 1970).

There are clearly valid arguments for both points of view. On osteological grounds alone the two genera appear to be very close, and some contend that only osteological characters are valid at the generic level. On the other hand, Moll and Legler (1971) wrote: "We eschew this [McDowell's] classification at both generic and specific levels because it is based purely on osteology and fails to consider color, markings, and soft anatomy, all of which are important in the taxonomy of emydid turtles (whether or not the characters are preserved in the fossil record)." My feeling is that on balance of all arguments, the genera should be kept separate; a change of name of such familiar animals, where established usage has separated the genera for so many years, should be made only when arguments in favor of discontinuing the separation overwhelm those for keeping it, which in this case they certainly do not. Even in osteological characters *Chrysemys* does show some differences from *Pseudemys*—for example, in the larger posterior palatine foramen, and in the unconstricted first central scute, the narrower triturating surfaces, the unsculptured shell, and the unserrated hind margin of the carapace. Moreover, although the difference between a typical *P. s. scripta* and a typical *P. f. floridana* is very marked, the difference breaks down greatly when one compares one of the large tropical members of the *scripta* group (e.g., *cataspila* or *hiltoni*) with such members of McDowell's subgenus *Pseudemys* as *P. alabamensis* or *P. concinna*. Moreover, were it not for the existence of a relatively large and *Pseudemys*-like western subspecies of *Chrysemys* (which, living as it does in the part of the range of *C. picta* that is not inhabited by any *Pseudemys*, may reasonably be said to represent parallel evolution), one feels that the separation of the genera would be universally accepted. Somehow, *Chrysemys* stays aloof throughout its range from the taxonomic nightmare that is the genus *Pseudemys*, whose different "species" and "subgenera" show such puzzling parallelism and even hybridization; such aloofness ought to be recognized taxonomically.

The map turtles of the genus *Graptemys* are distinguished by the series of vertebral spines, to which the alternative vernacular name of "sawbacks" alludes. These turtles, unlike *Pseudemys* and *Chrysemys*, do not make overland migrations even during very rainy weather, and consequently the various species have evolved in isolation in separate river systems. Two species groups may be discerned: the "broad-head" and the "narrow-head" groups, the former including the widespread *G. geographica*, the Mississippi Valley *G. kohni*, the Chipola–Apalachicola–Chattahoochee–Flint River *G. barbouri*, and the Alabama–Escambia–Pearl River *G. pulchra*; the latter group includes the Pearl River *G. oculifera*, the Pascagoula River *G. flavimaculata*, the Black Warrior–Tombigbee–Alabama River *G. nigrinoda*, the Guadalupe River *G. caglei*, the Colorado River *G. versa*, the Mississippi Valley *G. p. pseudogeographica* and *G. p. ouachitensis*, and the Sabine River *G. p. sabinensis*. The broad-headed forms show the most extreme sexual dimorphism known among turtles, females reaching up to three times the carapace length of males. The broad head is developed only in the mature females, which have massive jaw muscles and strong alveolar surfaces for eating shelled molluscs, their exclusive food; the males, which usually are found in shallower water or nearer the surface, feed principally on insects.

Closely related to *Graptemys*, but differing in having a spotted rather than striped pattern on the limbs and blunt dorsal tubercles, are the diamondback terrapins of the genus *Malaclemys*. Seven subspecies of the single species *M. terrapin* extend from Cape Cod to Texas and have a linear range along the wetlands and intracoastal waterways of the eastern United States; these turtles are uniquely adapted for life in brackish water.

The isolated genus *Deirochelys*, with the single species *D. reticularia*, shows some similarities in pattern to *Pseudemys* and *Chrysemys*, but has a peculiarly elongate, flattened head and an exceedingly long neck; massive development of the longissimus dorsi muscles has caused a corresponding alteration of the dorsal rib ends, which are free of the carapace and bowed downward for about a third of their length. This genus is confined to the southeastern United States. The northern species *Emydoidea blandingi* is probably the closest living relative of *Deirochelys*; it has a somewhat similar carapace shape (though with shallower bridges), as well as a rather long neck and similar skull, but the color pattern, being principally speckled, is entirely different, and the plastron develops a moderately effective hinge.

The genus *Clemmys* includes four species: three in the northeastern United States and one in the far west. *C. insculpta*, is primarily terrestrial;

whereas *C. muhlenbergi* is found in bogs in the northern part of the range and in mountain streams in the south. *C. marmorata* and *C. guttata* are pond dwellers. Structurally, *Clemmys* is a rather generalized genus. The top of the head is covered with undivided skin, and the hexagonal neural bones have their shorter sides flanking the front. The digits are more or less webbed; and the tail is very long in the hatchling, but only moderately long in adults. *C. muhlenbergi* and *C. guttata* are among the smallest of all turtles, reaching adult lengths of about 10 cm.

Box turtles of the genus *Terrapene* are plentiful in the eastern and south central United States and are also found in certain isolated areas of Mexico as far south as the Yucatan Peninsula. With the exception of the tertiarily aquatic *T. coahuila* from the Basin of Cuatro Cienegas, Coahuila, Mexico, the box turtles are terrestrial forms; the habitat of some species, including *T. ornata* and *T. nelsoni*, is distinctly xeric. The most obvious feature of the genus is the plastral hinge, which allows the anterior and posterior shell openings of all but the youngest specimens to be closed completely. In contrast to the mud turtles (*Kinosternon*), the plastron in *Terrapene* has no immovable middle section. Box turtles are long-lived but small, most forms seldom exceeding a carapace length of about 15 cm; the carapace sutures disappear completely soon after adult size is reached, giving a very strong structure, but one with no further capacity for growth. *T. carolina*, the common or eastern box turtle, is one of the most variable turtles known; the four subspecies show a tremendous range in shell shape and coloration.

The remaining New World genus is *Rhinoclemys*, formerly called *Geoemyda*, whose seven species range from northwestern Mexico south to Ecuador and Para, Brazil. They are distinguished from *Terrapene* by their larger size and the absence of the plastral hinge (though the bridge tends to be slightly flexible), and from *Clemmys* by the hexagonal neural bones whose shorter sides flank the rear. Two species, the large *R. funerea* from the Caribbean lowlands of Costa Rica, and *R. nasuta* from Pacific Colombia, are highly aquatic; *R. punctularia* and *R. pulcherrima* are semiterrestrial; and *R. rubida*, *R. annulata*, and *R. areolata* are purely terrestrial. The habitat reflects in the degree of webbing of feet and the cornification of the integument. The terrestrial forms often show dorsal growth annuli as do true tortoises. The shell varies from flat to highly domed, and the subspecies of *R. pulcherrima* and of *R. rubida* are distinguished partly by the degree of elevation of the shell. Only one species, *R. punctularia*, reaches east of the Andes, and it is restricted to the northern coastal area.

The circum-Mediterranean emydids are referable to the genera *Emys* and *Mauremys*. *Emys orbicularis*, the single species of the first genus, is found in Central and Southern Europe, North Africa, and the more humid parts of the Middle East. It is a rather generalized, small species, which develops a slightly hinged plastron with maturity. It is similar to *T. carolina* in having enormously variable coloration, though there is usually at least some light speckling on a dark, often black, background. *Mauremys* include *M. leprosa* and *M. caspica* in the Mediterranean area, *M. mutica* in China, and *M. japonica* in Japan. These turtles are all rather generalized pond and stream dwellers; *M. leprosa* and *M. caspica* reach their maximum abundance in small streams and other bodies of water in highly arid parts of North Africa and Asia Minor. The plastron is rigid in *Mauremys*, and the adult length is usually about 18 to 20 cm.

The remaining emydid genera are all from southeast Asia, where the family may well have originated; although also well established in North America, the genus-to-species ratio there is much lower, which suggests a more recent adaptive radiation in the New World. The genus *Sacalia*, with one or possibly two species in China and Hainan Island, was formerly considered to be inseparable from *Mauremys*, but significant skull differences have now been demonstrated (McDowell, 1964). The remaining Asiatic emydids may be classified into usually giant, herbivorous river turtles with typically developed plastral buttresses, which enclose lateral expansions of the lungs; semiterrestrial forms ecologically akin to the box turtles and wood turtle of the United States; and medium-sized pond turtles, often with broad heads and strong, mollusc-crushing jaws. The first group includes the genera *Batagur, Hardella, Orlitia, Callagur, Kachuga*, and *Annamemys*. The first four genera are monotypic and reach very large adult sizes, often of the order of 60 cm. They are distinguished by the structure of the alveolar surfaces and by various features of the neural bones. All tend to be vegetarian when mature, to have projecting, conical snouts, and to nest on sandbanks in large rivers. *Callagur*, however, has a marked tendency to invade brackish water.

Kachuga, with seven species, is closely related to *Callagur*, but includes not only several very large species but also two or three quite small ones, whose decorative patterns and high, tectiform, tuberculate shells recall the American *Graptemys*. *Annamemys*, with its single species, *A. annamensis*, is a small, poorly known turtle from Vietnam, which is included in this group because of its very strong plastral buttresses.

The second group includes the genera *Melanochelys, Hieremys, Cuora, Notochelys, Geoemyda, Heosemys, Pyxidea*, and *Cyclemys*. *Melanochelys* includes two species of three-keeled, semiaquatic forms, *M. tricarinata* and *M. trijuga*. The former, with a variety of subspecies that extend as far south as Ceylon, is distinguished partially by head pattern and partially by very varying adult size. *Hieremys* includes a single, rather large species,

Figure 7 Asiatic emydid turtle (*Callagur borneoensis*).

H. annandalei, from Thailand, distinguished from its relatives such as *Geoemyda* by the absence of the median ridge on the upper alveolar surface. *Cuora* includes several species of small, somewhat aquatic Asiatic box turtles, very similar to *Terrapene* in superficial aspects, but differing in various skull features and in the failure of the shell bones to fuse with maturity. Most have relatively domed shells, but the population of the widespread *C. amboinensis* in the Philippines is rather flat shelled. *Notochelys*, with the single species *N. platynota*, reaches a shell length of about 35 cm and develops a feeble midplastral hinge with maturity; the distinguishing feature is the constant presence of a small, square, extra central scute between the fourth and last centrals. In other turtles, this is present only as an occasional anomaly.

Notochelys is clearly closely related to *Cyclemys*, which in turn is so close to *Heosemys*, *Geoemyda*, and *Pyxidea* that the generic distinctions may not be even valid if proper comparisons of specimens of all ages and both sexes are conducted. *Cyclemys dentata*, widespread in southeast Asia, is a rather flattened turtle with a broad shell up to about 23 cm long. No evidence of a plastral hinge is present in young or even half-grown specimens, but a reasonably functional hinge develops in adults. *Geoemyda* formerly included about 16 species of both New and Old World emydids, but McDowell's revision of the family demonstrated that it is proper to include only the type species, *G. spengleri*. This is a rather poorly known form, apparently rare in much of its range, with a tricarinate, distinctly flat-topped, posteriorly serrated carapace. Mature *G. spengleri* develop a plastral hinge (cf. Smith, 1931), as does *Pyxidea mouhoti* which although somewhat similar morphologically is easily distinguished by the much larger size and steeper, more elevated shell. *Heosemys*, with several species including the 40 cm *H. grandis* and the spine-shelled *H. spinosa*, is considered distinct in that it lacks entirely the postorbital bar in the skull, thereby any "cheekbone" between the eye and the ear. However Smith (1931) found that large series of specimens showed every degree of devolution, beginning with a slight thinning of the bar and ending with its complete disappearance; thus he doubted the taxonomic usefulness of *Heosemys* at the generic level.

The third group includes the genera *Ocadia*, *Morenia*, *Geoclemys*, *Chinemys*, *Malayemys*, and *Siebenrockiella*. All except *Morenia* and *Chinemys* are monotypic. *Ocadia sinensis* is a primitive pond-dwelling species from China, Hainan, and Taiwan. It is superficially somewhat similar to *Pseudemys*, even having numerous greenish longitudinal lines on the neck. *Morenia* is distinguished from *Ocadia* first by the failure of the humeropectoral sulcus to intersect the entoplastron, second by the more posteriorly located internal choanae, and third by the divided skin on the rear of the head. The two species, *M. ocellata* and *M. petersi*, are found in Burma, Tenasserim, and Bangladesh, and are poorly known. *Geoclemys hamiltoni* is a fairly large species with a thick, tricarinate shell and a large, yellow-spotted head with massive crushing jaw surfaces. It is found in the Ganges and Indus rivers of Pakistan and India. Japanese specimens of the well-known *Chinemys reevesi* are considerably larger and darker than those from China. The genus is distinguished from *Geoclemys* by the longer tail, more posteriorly situated entoplastron, and more anteriorly located choanae. Two other species, *C. kwangtungensis* and *C. megalocephala*, are poorly known and localized inhabitants of mainland China. *Malayemys subtrijuga*, from Vietnam, Thailand, and Malaysia, has a tricarinate, smooth-margined shell, and a large head with projecting snout and strong white lines on the sides. *Siebenrockiella crassicollis* is a widespread species in southeast Asia; it loses the lateral keels on the carapace with maturity, but has the large head and strong jaws of *Malayemys*. It is mostly black, with a few ill-defined white spots on the head.

Although fossil emydids are known from the Paleocene of North America and are moderately plentiful from the Eocene onward, the record does not shed a great deal of light on the evolution of the family. Known fossil genera include *Broilia*, *Palaeochelys*, and *Dithyrosternum* from the Oligocene of Europe, and *Gyremys* and *Palaeotheca* from the Upper Cretaceous and Eocene of North America, but these—especially the last two—are all known from less than adequate material. The fossil genera *Ptychogaster* and *Geiselemys* are sometimes included in a separate family, the Ptychogastridae, and the genera *Clemmydopsis, Shansiemys,* and *Sakya* are sometimes included in the family Sakyidae (Mlynarski, 1969; Ckhikvadze, 1973).

Platysternidae

The family Platysternidae contains a single living species, the Asiatic big-headed turtle, *Platysternon megacephalum*, of which two or possibly three races live in Thailand, Burma, and China (including Hainan). Despite the presence of several very distinctive features (e.g., a completely roofed-over skull, a continuous series of inframarginal scutes on the bridge, and a long tail like a snapping turtle), the family is thought to be related to the Testudinidae and the Emydidae, indeed, some authors subsume the big-headed turtle as a subfamily within the Testudinidae.

Platysternon reaches a carapace length of about 18 cm and has such a large, heavy head and flattened, often concave carapace that the center of gravity is well forward, and the animal has remarkable climbing abilities. The head is nonretractile but dorsally protected by a single large scute; the front of the carapace bears a wide nuchal indentation to accommodate the large head. The jaws are very strong and massive, but the limbs are relatively weak.

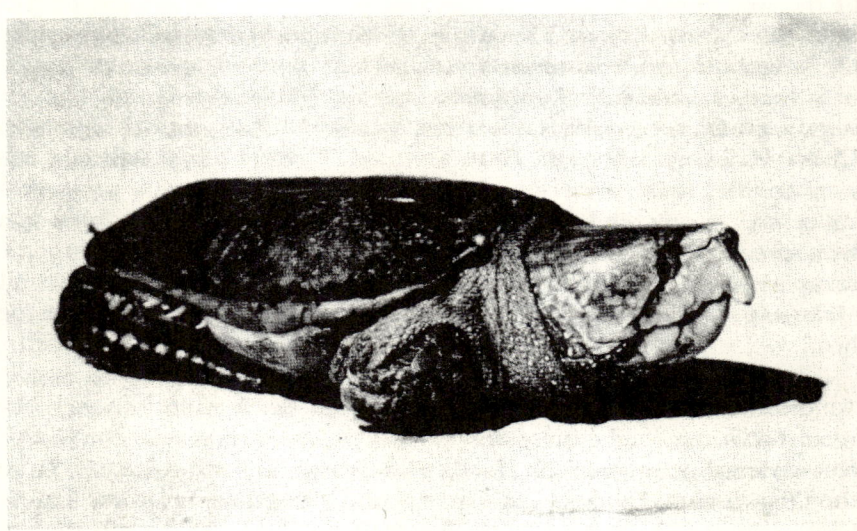

Figure 8 Asiatic big-headed turtle (*Platysternon megacephalum megacephalum*).

Fossil Platysternidae include *Macrocephalochelys pontica*, known only from a large incomplete skull from the Pliocene of the Ukraine, and *Planiplastron tatarinovi*, from the Oligocene of Central Kazakhstan. Neither of these gives any significant information on the evolution of the family, and it is possible that *M. pontica* may not belong to this family, since the type specimen lacks the diagnostic posterior part of the skull.

Trionychidae

The soft-shelled turtles of the family Trionychidae are among the most divergent and distinctive of all turtles. In these species, the scutes are replaced by a continuous layer of soft skin, and the peripheral bones are lost (except for possible rudiments in *Lissemys*); thus the sides and rear part of the dorsal shell (better called the dorsal "disk") are quite flexible, the free rib ends being simply embedded in the cartilaginous material of the shell margin. The scalation of the soft parts, too, is almost completely eliminated; there are only three claws on each foot; and the snout is elongate, terminating in a snorkellike breathing tube. In common with other turtles equipped with such a tube (*Chelus*, *Carettochelys*), the premaxilla of these animals is single. The sharp jaws are disguised by deceptively soft lips, and the neck is exceedingly long, but so completely retractile that when pulled in completely the eighth cervical and the first dorsal vertebrae actually lie belly to belly. Soft-shells range in adult lengths from less than 23 cm to more than 1 m, all the largest species being from the Old World. In adult trionychids, the bony part of the dorsal disk forms a superficial bony callosity with a characteristically sculptured surface. In most of the New World forms, only seven pairs of pleural bones are present. Patterns of plastral callosities, whose arrangement is highly characteristic for each species, form as the turtle matures.

Living soft-shells are divided into two subfamilies, the Trionychinae and the Lissemydinae. The latter are distinguished by the presence of soft, semicircular flaps, which allow the retracted hind feet to be totally concealed. The front part of the plastron also is sufficiently flexible to afford closure of the anterior shell opening; though the value of this protection would seem to be lessened by the soft, boneless nature of the flexible parts.

The entoplastron is missing in trionychids, but the epiplastra are fused into a single boomerang-shaped bone; and neomorphic structures called "preplastra" are present anteriorly (Williams and McDowell, 1952).

The genera of the Trionychinae are distinguished by the placement of the eyes; in *Trionyx* the eyes are relatively far back, so that the postorbital bar is narrower than the orbital diameter. In *Pelochelys*, whose single

Figure 9 Florida soft-shelled turtle (*Trionyx ferox*).

species, *P. bibroni*, is a seagoing Asiatic species that is among the largest of the soft-shells, the eyes are distinctly further forward, and the head is somewhat rounded, with a short proboscis. In *Chitra*, also a monotypic Asiatic genus, the eyes are situated at the extreme front of the skull. *Trionyx* is the only genus that reaches the New World; there are four species in North America, with a variety of subspecies whose collective range extends over much of the United States and parts of northern Mexico. In the Old World *Trionyx* species occur in southeast Asia, and there is also one species in the Middle East and a single very large and widespread species in Africa; populations of the latter extend down the Nile to the Mediterranean, with thriving, purely marine populations off the coast of Turkey (Hathaway, personal communication).

About 180 fossil *Trionyx* species have been described, but most are known only from small fragments that shed little light on the evolution of the family. Perhaps the most ancient trionychid known is *Sinaspideretes wimani*, possibly of Upper Jurassic age, which had a rather deep carapace lacking peripheral bones but with discernible traces of scute sulci. Better known is *Plastomenus*, from a variety of Cretaceous and Eocene North American forms, as well as a single species from the Eocene of Kazakhstan. In this genus the plastral bones were large and almost or completely contiguous; the entoplastron was usually absent and the epiplastra were fused, but preplastra were absent.

Figure 10   New Guinea plateless river turtle (*Carettochelys insculpta*).

Carettochelyidae

Today the family Carettochelyidae includes a single species, *Carettochelys insculpta*, from southern New Guinea. *C. insculpta* is a fairly large, highly aquatic river turtle that shows certain similarities to the Trionychidae; the shell lacks scutes, being covered instead with soft skin, while all surface bones including the top of the head are strongly sculptured, and the snout ends in a blunt proboscis or breathing tube. As with all turtles exhibiting such a nasal development, the premaxillary bone is single. The limbs are flipperlike, similar to those of sea turtles though somewhat more flexible. The neck is not as long as in the Trionychidae, and the very thick bony shell is complete; thus there is none of the flexible carapace margin or soft anterior plastral area such as is shown by trionychids. The skull is short and blunt, with the nasal opening wider than high, the eyes well separated on the sides of the head, and the supraoccipital process equipped with exceedingly broad flanges. The jaw surfaces lack ridges and appear to be adapted for crushing hard-shelled organisms. *C. insculpta* is a fluviatile species that has recently been discovered in the Daly River of Northern Australia (Cogger, 1970; Peters, 1970). Carr (personal communication) found *Carettochelys* to be a common theme of aboriginal rock carvings and believes that the species may have become much more restricted in its Australian distribution since the introduction of the water buffalo, whose mud-churning activities tend to destroy suitable nesting areas.

Fossil carettochelyids are known from deposits of Cretaceous to Oligocene age of Europe, Asia, and North America; fossils of *Carettochelys* itself are known from the Miocene of New Guinea. Extinct genera differ from the living one in having the atlas not fused to the odontoid and in retaining some of the scute divisions in the shell; they are placed in a separate subfamily, the Anosteirinae. Young specimens of *C. insculpta* may show indistinct indications of central scutes (Zangerl, 1959), but these disappear as the turtle grows.

Sea Turtles

Seven species of sea turtles are alive today: the leatherback (*Dermochelys coriacea*), the green turtle (*Chelonia mydas*), the hawksbill (*Eretmochelys imbricata*), the loggerhead (*Caretta caretta*), the flatback (*Chelonia depressa*), the olive ridley (*Lepidochelys olivacea*), and Kemp's ridley (*Lepidochelys kempi*). The leatherback has a totally unique shell structure composed of a thick layer of cartilage and a dorsal continuous sheet of small mosaic bones; there are free ribs, but no neurals, peripherals, or entoplastron,

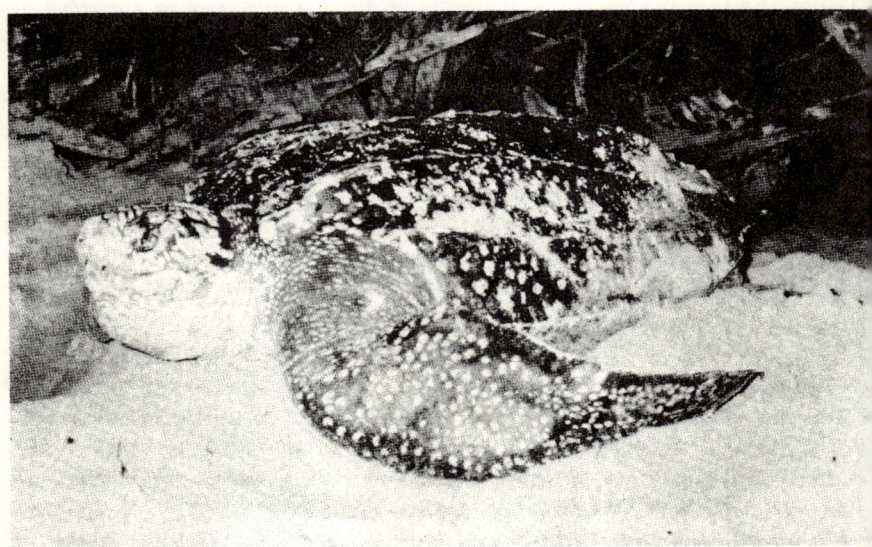

Figure 11 Leatherback sea turtle (*Dermochelys coriacea*).

Figure 12 East Pacific hawksbill turtle (*Eretmochelys imbricata bissa*).

and the plastral bones are highly reduced. These and numerous other essentially embryonic features make this species clearly very different from the others, and it occupies its own family (Dermochelyidae); some even accord it the status of a suborder (Athecae). The other living sea turtles form the Cheloniidae, distinguished by the paddlelike flippers, nonretractile neck, roofed-over skull, and complete series of inframarginals. Subfamilies within the Cheloniidae are controversial, but the best arrangement would place the genus *Chelonia* in a tribe (Chelonini) apart from the others (Carettini) for reasons discussed by Carr (1942b), although *Eretmochelys*, which shares with *Chelonia* the characteristic of only four pairs of costal scutes (the others have five or more), is considered by some to belong in the Chelonini.

The present zoogeography of the sea turtles gives little clue to their areas of origin. The green turtle is an essentially circumtropical species, which typically nests in colonial fashion on either mainland or island beaches, the feeding grounds often being more than 1000 miles from the breeding grounds. The hawksbill has a roughly similar overall range, but tends to nest individually rather than in colonies and is less given to long migrations. The East Pacific populations of both species are darker in color and tend to reach a smaller adult size than the Atlantic populations, but doubtless there are several undescribed subspecies of both in different parts of the world.

The loggerhead has an "antitropical" distribution, the principal nesting grounds being in the eastern United States, South Africa, southern Madagascar, Turkey, Japan, and Queensland (Australia); however nesting also occurs in the tropical western Caribbean, in Cuba and Colombia, and on certain small islands. The species is extremely rare on oceanic islands and throughout the entire East Pacific area. No subspecies have been convincingly described. Migrations of several hundred miles have been reported (Hughes, 1971, 1974; Bustard and Limpus, 1971), with the wintering grounds generally being nearer the tropics than the nesting grounds.

The olive ridley has a very wide, emphatically tropical range in the Indian, East Pacific, and Eastern Atlantic oceans, with breeding colonies also being established in the Guianas. Postnesting migration in this species tends to be a "fanning out" in all directions rather than a definite shift to a different area such as occurs with the green turtle (Pritchard, 1973). Nesting may take place either singly, in small colonies, or in enormous, simultaneous, sometimes diurnal aggregations, up to 100,000 strong, known as *arribadas* (Pritchard, 1969b; Hughes and Richard, 1974). The major *arribada* sites are in the East Pacific (Mexico and Costa Rica), but there is a huge one in Orissa, India, and a lesser one in western Surinam.

The flatback turtle is not known to occur outside northern Australia, where there are important nesting grounds in Northern Territory and Queensland (Bustard, 1973). Kemp's ridley has an even smaller breeding range, almost the entire world's population nesting, usually in diurnal *arribadas*, on a stretch of beach in the Mexican State of Tamaulipas, between Barra Coma and Barra San Vicente (Hildebrand, 1963; Carr, 1963a). Postnesting dispersal takes the turtles principally to the Mississippi Mouth and to Campeche, but a few reach as far as Key West, Florida (Chavez, 1968; Pritchard and Marquez, 1973). Young Kemp's ridleys are most often caught off the Atlantic coast of the United States, and a few have been found in the British Isles and Europe (Brongersma, 1972).

Fossil cheloniids include the rather specialized *Pachychelys* from the Pliocene; *Euclastes, Kurobechelys, Peritresius, Syllomus,* and *Procolpochelys* from the Miocene; *Glarichelys, Chelyopsis, Carolinochelys,* and *Islamichelys* from the Oligocene; and *Eochelone, Erquelinnesia, Gafsachelys,* and *Euclastes* from the Eocene. In addition, various fossil fragments attributable to *Chelonia* and *Caretta* are found from the Upper Cretaceous onward.

In addition to the Cheloniidae and Dermochelyidae, three entirely extinct families of sea turtles are recognized: the Thalassemyidae, the Toxochelyidae, and Protostegidae. The Thalassemyidae were the earliest known sea turtles, being recorded from the late Jurassic and Cretaceous of Europe. They are currently classified with the Amphichelydia and may have been descended from the Plesiochelyidae. The Thalassemyidae had open fontanelles between the rib ends and the sides of the plastron, and like modern sea turtles, the carapace and plastron were not fused at the bridges. In many species the neural bones were reduced in number and size. The entoplastron and even the epiplastra appear to be absent in some species. The limbs were imperfectly adapted for the marine environment, and the family was obviously relatively new to the sea.

The other two extinct families are best known from the Upper Cretaceous of the United States; they included some bizarre and huge species that swam in the great Niobrara Sea that once covered what is now the central United States. The Toxochelyidae, probably derived from the Thalassemyidae, varied in their degree of aquatic specialization; some forms apparently were near-shore dwellers while others were probably pelagic. They differed from the modern Cheloniidae in having cruciform plastra; and although the fore flippers were paddle shaped, the hind feet were essentially chelydriform. The carapace was usually oval or circular, and both carapace and plastron showed lateral fontanelles. The most primitive genera (*Toxochelys, Thinochelys,* and *Porthochelys*), characterized by the absence of a secondary palate, as well as by anteriorly narrow triturating surfaces, small nasal bones, flat neurals, smooth shell margin, and small central and lateral plastral fontanelles, constitute the subfamily Toxochelyinae. The more modern Osteopyginae, to which only *Osteopygis* can be unequivocally assigned, have a secondary palate, lack nasal bones, have a broad mandibular symphysis, at most only slightly serrated peripheral bones, and unkeeled neurals. The most specialized toxochelids (*Lophochelys, Ctenochelys,* and *Prionochelys*), which constitute the subfamily Lophychelyinae, have broad triturating surfaces, the beginnings of a secondary palate, small nasal bones, and a highly serrated median keel, sometimes with separate epineural bones on the tips of the tubercles. Large fontanelles are always present in both carapace and plastron.

The Protostegidae are defined by the elongate skull, with roofed temporal area, hooked beak, highly fenestrated carapace, absence of a secondary palate, presence of nasal bones, and reduced xiphiplastra. Epiplastra were sometimes absent, and there were sometimes nine pairs of pleural bones. The more primitive genera (*Chelosphargis* and *Calcarichelys*), constituting the subfamily Chelospharginae, were relatively small and had comparatively well-ossified shells. Lateral fontanelles were moderate, and the pygal bone was very narrow. *Calcarichelys* had a uniquely serrated median keel with every other neural bone showing a high, triangular lateral profile. The more advanced genera *Protostega* and *Archelon* were much larger; *Archelon ischyros*, the largest sea turtle definitely known, reached a total length of about 3 m. In both of these genera the carapace was exceedingly fenestrated, with an essentially embryonic bony structure.

Collins (1970) has postulated that *Cimochelys benstedi*, from the Lower Cretaceous of England, should be considered a protostegid.

Pelomedusidae

The Pelomedusidae are a family of primitive turtles whose representatives are found today in much of tropical South America east of the Andes as well as in Africa and Madagascar. The family is defined by the lateral plane of retraction of the neck, with correlated muscular development and lateral processes on the cervical vertebrae; the pelvis fused to both carapace and plastron; the presence of a pair of mesoplastra and a single intergular scute; the absence of the nuchal scute; and the skull roof that is emarginate, if at all, largely from behind. Five rather well-defined genera are generally recognized: *Podocnemis*, in South America, in which the shell is well ossified, the mesoplastra small and widely separated, the skull roof only slightly emarginate, and the hind feet each equipped with four claws; *Peltocephalus*, from South America, with a fully roofed skull, hooked jaws, and laterally placed orbits; *Erymnochelys*, from Madagascar, with similarities to both *Podocnemis* and *Peltocephalus*; *Pelomedusa*, from Africa and Madagascar, with a poorly ossified shell, a large midplastral fontanelle that persists until late in life, as well as small, widely separated mesoplastra, a fully emarginate skull roof, and five claws on each hind foot; *Pelusios*, from Africa, Madagascar, and a few outlying islands, with a well-ossified shell, contiguous mesoplastra that make up the posterior border of a well-defined plastral hinge, a fully emarginate skull roof, and five claws on each hind foot. Because of the rigid bridge and fusion of the pelvis to the shell, only the anterior lobe of *Pelusios* is movable, in contrast to hinged cryptodires, such as *Terrapene* and *Kinosternon*.

Podocnemis is an ancient and formerly very widely distributed genus, one of the three living reptile genera already in existence during the Cretaceous (the other two are *Trionyx* and *Crocodylus*). However recent studies suggest that the geological distribution of *Podocnemis* given by Zangerl (1948) was too broad, the genus being known from the Eocene of Spain, the Upper Cretaceous to Recent of South America, and the Eocene to Pleistocene of Africa, but not from North America, Asia, or the Cretaceous and Miocene of Europe. It appears probable that the genus was displaced from Africa by competition with *Pelomedusa* and *Pelusios*. It appears that these two latter genera have only recently reached Madagascar, the populations there being indistinguishable from those of East Africa, and the related Malagasy species, *Erymnochelys madagascariensis*, still survives.

Living *Podocnemis* have a nearly complete series of neural bones, but some of the extinct forms, including *P. venezuelensis* from the Pliocene of Venezuela and *"Eusarkia" rotundiformis* (which may be a *Podocnemis*) from the Eocene of Tunisia, have completely eliminated the neural bones, a development paralleled by most of the living Australian (and some of the South American) chelids. Roger Wood (personal communication) suggests that *P. venezuelensis* more properly belongs to a new genus, which he will describe shortly.

Podocnemis has six living South American species, *P. expansa, P. sextuberculata, P. erythrocephala, P. unifilis, P. vogli,* and *P. lewyana. Peltocephalus tracaxa* is distinguished from *Podocnemis* by several features, including the hooked, snapping-turtle-like jaws, the almost completely roofed skull, and the larger size of the male, compared to the female, as well as by features of the blood sera, karyotypes, and nesting habits. *Erymnochelys madagascariensis* is superficially somewhat similar to *P. tracaxa*, but has a very small intergular scute that only partially separates the gulars. The adult size of the living species of *Podocnemis* varies greatly; *P. erythrocephala, P. vogli,* and *P. sextuberculata* are mature at a shell length of 30 cm or less, but *P. expansa* can reach a carapace length of almost 90 cm.

Pelomedusa is widespread in Africa and Madagascar, but only a single species, *P. subrufa*, is recognized. *Pelusios*, however, is a complex genus whose systematics are still controversial. Loveridge (1941), after examining large series of specimens, asserted that there were only four valid species, *P. adansoni, P. gabonensis, P. subniger,* and *P. sinuatus*. These may be separated into two species groups, *P. adansoni* and *P. gabonensis* with a long anterior plastral lobe and relatively posteriorly located hinge, and *P. subniger* and *P. sinuatus* with the hinge much further forward. However Laurent (1965) adds the following species: *P. carinatus, P. castaneus, P. nanus, P. bechuanicus, P. niger,* and *P. williamsi*; *P. castaneus* has four subspecies and *P. williamsi* two. Roger Wood (personal communication) places *P. adansoni, P. gabonensis, P. nanus,* and *P. niger* in one group, and *P. sinuatus* and *P. subniger* in the other; he considers *P. bechuanicus, P. carinatus, P. castaneus,* and *P. williamsi* all to be synonyms of *P. subniger*. Fossil *Pelusios* are known from the Miocene and Pliocene of Egypt and from the Pliocene of Chad and of Kenya.

Figure 13 Gabon side-necked mud turtle (*Pelusios gabonensis*).

Extinct pelomedusid genera include *Polysternon* from the Eocene to Oligocene of France, *Paralichelys* from the Oligocene and Pliocene of France and Spain, *Stereogenys* from the Eocene and Oligocene of Egypt, *Platycheloides* from the Cretaceous of Malawi, *Carteremys* from the Upper Tertiary of India, *Elochelys* from the Upper Cretaceous of France, and *Shweboemys* from the Eocene to Pliocene of Egypt, Pakistan, and Burma. The three species of *Shweboemys* were somewhat similar to *Podocnemis*, but had tectiform shells and a broad secondary palate with a narrow median cleft. The newly described *Stupendemys geographicus*, from the Miocene of Venezuela, appears to be the largest known turtle of all time.

Chelidae

The other side-neck family, the Chelidae, is today represented by several genera in the South American and Australian regions; the fossil record does not extend earlier than the Pliocene, and no fossil chelids are known from outside the present range of the family. The Chelidae are distinguished from the Pelomedusidae by the absence of mesoplastra and by the presence of a parietosquamosal arch in the skull instead of a quadratojugal arch; some of the species have conspicuously elongate, snakelike necks. A marked tendency in this family is the reduction of the neural bones, which form a complete series only in *Chelus* and *Hydromedusa* and are usually completely absent in *Platemys* and in the Australasian genera.

Probably the best-known of the South American species is the matamata turtle, *Chelus fimbriatus*, a sluggish, highly aquatic animal well described by Freiberg (1971) as having "un aspecto de desarrapado atrabiliario." It has an exceedingly rough, tuberculate shell up to 40 cm long in adult specimens. The head of the matamata is greatly flattened and adorned with numerous tassels and irregular appendages; the mouth is extremely wide, the eyes tiny and placed near the tip of the snout, which ends in an elongate breathing tube.

The two species of *Hydromedusa, H. maximiliani* and *H. tectifera*, are recognized by their extraordinarily long necks and the unique position of the nuchal scute, which is laterally expanded and is excluded from contact with the anterior margin of the carapace by the marginal scutes. These turtles are found in southern Brazil, Uruguay, and northern Argentina.

The genus *Platemys* includes several small species with a characteristic groove down the middle of the carapace. *P. platycephala* is widespread in the Amazonian and Guianas regions, while *P. spixi* and *P. radiolata* (which may be subspecifically related) occur in southern Brazil, and *P. pallidipectoris* in Argentina.

Figure 14   Amazonian toad-headed turtle [*Phrynops (Batrachemys) nasutus*].

The other South American chelids are currently included in the genus *Phrynops*, a complex group divisible into at least three subgenera. The subgenus *Mesoclemmys* is characterized by small adult size and narrow head; the single species, *P. gibba*, is found in the Guianas, Venezuela, Trinidad, and northern Brazil. In the subgenus *Phrynops* the head is very flat, the eyes are dorsally located, and the parietal bones form a definitely flattened surface. It includes three species (*P. geoffroanus, P. hilari,* and *P. rufipes*), which together extend from Colombia through Brazil to Uruguay. *P. hilari* is the largest, reaching a shell length of nearly 50 cm. *Batrachemys*, in which the head is very broad and rounded and the parietals form a very narrow dorsal surface, includes the widespread Amazonian and Guianas species *P. nasutus. P. dahli* is localized in northern Colombia, *P. tuberculatus* (possibly a synonym of *P. nasutus*) in northeastern Brazil, and *P. vanderhaegei* in Paraguay. The species *P. hogei, P. wagleri, P. lutzi,* and *P. boulengeri* are of uncertain validity, and further work is necessary before their subgeneric, or even generic, allocation can be decided (see Luederwaldt, 1926; Mertens, 1967).

In Australia, perhaps the best-known genus is *Chelodina*, containing six species. The neck is as long as that of *Hydromedusa*, but the genera are easily distinguished by the unusual nuchal scute of *Hydromedusa* and the usually absent neural bones and large intergular scute of *Chelodina*. This intergular scute completely separates the humerals and partially separates the pectorals. Three species groups of *Chelodina* are recognized: *C. longicollis, C. novaeguineae,* and *C. steindachneri*, in which the plastron is broad; *C. rugosa, C. expansa, C. parkeri,* and *C. siebenrocki*, in which the plastron is narrow; and *C. oblonga*, in which five to eight neural bones are present. The species are distinguished by differences in shell shape: for example, *C. expansa* is very wide and expanded posteriorly, *C. rugosa, C. parkeri* and *C. siebenrocki* are pear shaped, *C. oblonga* is very narrow, *C. steindachneri* is discoidal, *C. longicollis* is somewhat expanded posteriorly, and *C. novaeguineae* is oval. *C. longicollis* is the best-known and most widely distributed species; *C. expansa* is the largest, reaching a length of nearly 45 cm. One species or another is found in most of the areas of Australia where there is permanent water; *C. siebenrocki, C. parkeri* and *C. novaeguineae* reach New Guinea as well, and the latter is also known from Roti Island, southwest of Timor.

Of the nine or ten described species of the genus *Emydura*, only three are considered valid by Goode (1967b); *E. macquarri* from the Murray River drainage, *E. kreffti* from coastal New South Wales to Queensland, and *E. australis* in the extreme north. However the New Guinea forms *E. subglobosa* and *E. albertisi* are probably valid, though possibly only as subspecies of *E. australis*. The genus is differentiated from *Chelodina* by

ZOOGEOGRAPHY

Table 1 Families and Genera of Living Turtles

Family	Genera
Chelydridae (snapping turtles)	*Chelydra, Macroclemys*
Kinosternidae (mud and musk turtles)	*Kinosternon, Sternotherus, Staurotypus, Claudius*
Dermatemyidae (Central American river turtle)	*Dermatemys*
Testudinidae (tortoises)	*Geochelone, Homopus, Psammobates, Chersina, Kinixys, Malacochersus, Pyxis, Acinixys, Gopherus, Testudo*
Emydidae (pond turtles, etc.)	*Pseudemys, Chrysemys, Graptemys, Malaclemys, Deirochelys, Clemmys, Terrapene, Rhinoclemys, Emys, Mauremys, Sacalia, Batagur, Hardella, Orlitia, Callagur, Kachuga, Annamemys, Melanochelys, Hieremys, Cuora, Notochelys, Geoemyda, Heosemys, Pyxidea, Cyclemys, Ocadia, Morenia, Geoclemmys, Chinemys, Malayemys, Siebenrockiella*
Platysternidae (big-headed turtle)	*Platysternon*
Trionychidae (soft-shelled turtles)	*Trionyx, Pelochelys, Chitra, Cycloderma, Cyclanorbis, Lissemys*
Carettochelyidae (plateless river turtle)	*Carettochelys*
Dermochelyidae (leatherback sea turtle)	*Dermochelys*
Cheloniidae (sea turtles)	*Chelonia, Caretta, Eretmochelys, Lepidochelys*
Pelomedusidae (side-necked turtles)	*Podocnemis, Pelusios, Pelomedusa, Peltocephalus, Erymnochelys*
Chelidae (side-necked turtles)	*Chelus, Hydromedusa, Platemys, Phrynops, Chelodina, Emydura, Elseya, Pseudemydura*

the much shorter neck and the small intergular scute, which completely separates the gular but only partially separates the humerals. The northern species have conspicuously enlarged heads and strong jaws apparently adapted for crushing molluscs.

In the related genus *Elseya*, the crown of the head is covered with a single enlarged scale instead of numerous small ones. The three species

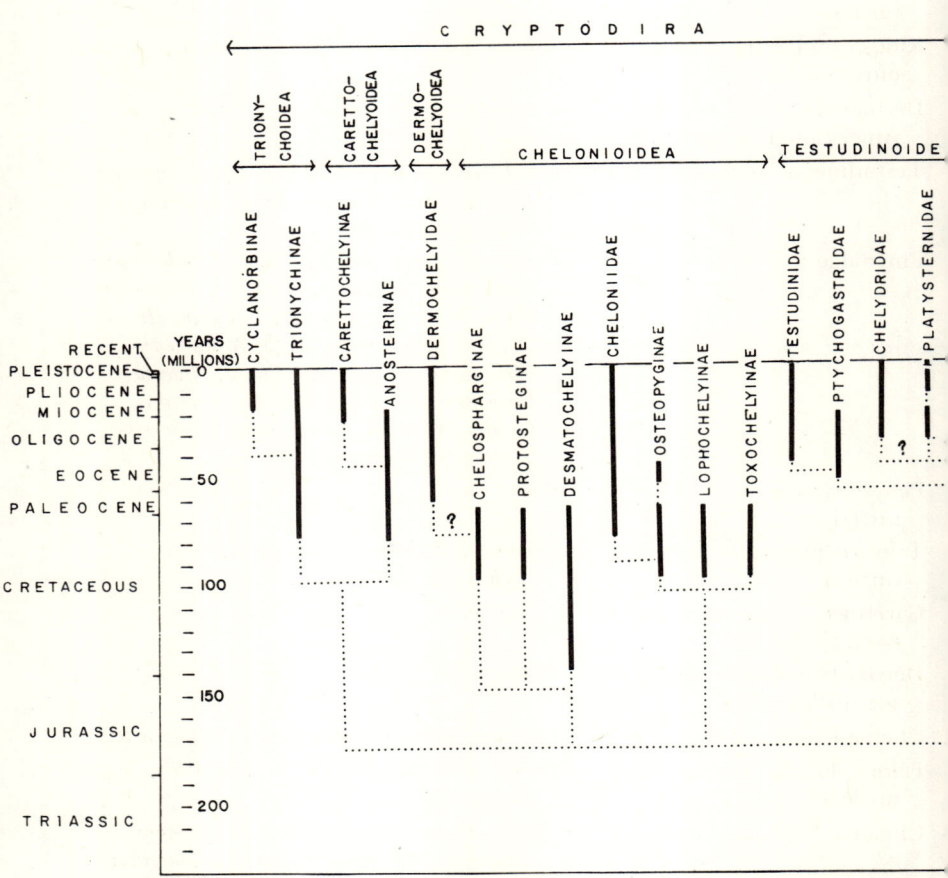

Figure 15 Phylogeny of the testudines, hypothetical in places.

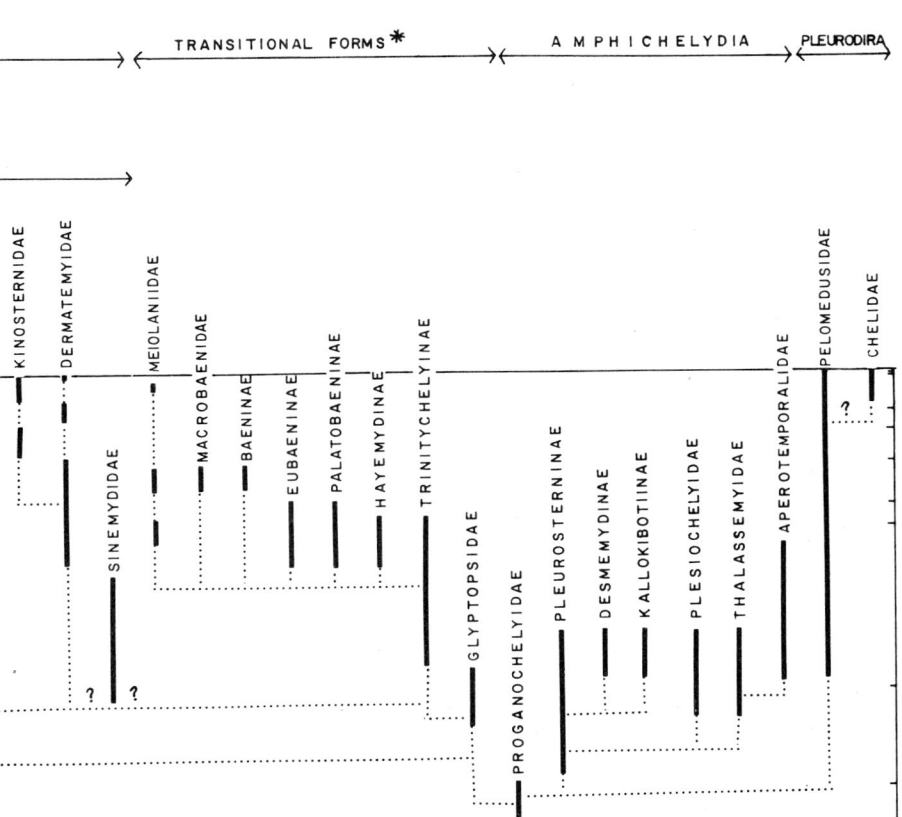

*CONSIDERED TO BE AMPHICHELIDS BY ROMER (1956) BUT CRYPTODIRES BY GAFFNEY (1972)

of *Elseya* (*E. dentata*, *E. novaeguineae*, and *E. latisternum*) are aggressive creatures sometimes known as "Australian snapping turtles." The northern species, *E. dentata*, may reach a length as great as 45 cm.

Pseudemydura umbrina is a diminutive chelid with an exceedingly limited range near Perth, Western Australia. It has an almost completely roofed-over skull and a very large intergular scute, which completely separates the gulars and the humerals and partially separates the pectorals.

Other Extinct Families

Numerous fossil turtle families exist in addition to those mentioned elsewhere in this chapter. Among them are the Baenidae, Macrobaenidae, Eubaenidae, Meiolaniidae, Kallokibotiidae, Neurankylidae, Aperotemporalidae, Plesiochelyidae, Thalassemyidae, Desmatochelyidae, Sinemyidae, and Chelycarapookidae. Nearly all these are placed in the extinct suborder Amphichelydia by most authors, although a recent revision by Gaffney (1972b) assigned several of them to the Cryptodira. They have little relevance to a discussion of living forms; their relationships are illustrated in Figure 15. Most became extinct before the end of the Eocene, but one gigantic form with a horned skull, *Meiolania oweni* of the Meiolaniidae, survived on Lord Howe Island until the uppermost Pleistocene.

METHODS

CHAPTER 2
Collecting and Marking

MICHAEL V. PLUMMER

The scarcity of extensive series of turtles in most herpetological collections is indicative of the difficulties involved in collecting these reptiles. Turtles are bulky animals. Not only do they take up large amounts of space in museums and live exhibits, but they also are difficult to process and transport in numbers in the field. Moreover, especially for aquatic turtles, the equipment involved in collecting them also is bulky and cumbersome. Despite these drawbacks, turtles are collected easily once they are located, and unlike most other reptiles, aquatic turtles are trapped easily. With the renewed interest in field studies of ecology and behavior, innovative and practical methods of collecting and marking turtles have been developed and refined.

More scientists are studying turtles now than ever before, and it must be remembered that some turtles, particularly the tortoises, suffer from the inroads of man. It is the duty of every investigator to practice conservation in every study. Immediate needs must be weighed continually against the irreversible course of extinction.

COLLECTION METHODS

Hand Collecting

Hand collecting takes about as many forms as there are collectors. One cruises an area and looks for turtles or turtle sign; when a turtle is found, it is collected by hand or with some appropriate device.

45

Generally, terrestrial turtles are simply picked up after being located visually. One's luck in locating them can be improved by first selecting suitable turtle habitat. Legler (1960a) reported that *Terrapene ornata* was found most often along fences, in ravines, and in and around piles of cattle dung, where it foraged for invertebrates that inhabit the droppings. He found that by riding a horse through his study area he more easily found turtles because his vision was less obstructed and he could cover the area more quickly.

Carr (1952) mentioned that hunters frequently find their dogs pointing box turtles (*Terrapene* spp.). Schwartz and Schwartz (1974) also noticed this phenomenon and made use of it in an ecological study. Their trained Laborador retrievers located and retrieved more than 90% of the total of 3832 box turtle captures. The dogs worked close to their masters, permitting the exact location of each capture to be plotted. The dogs commonly found turtles under debris, and in the fall they often dug turtles out from their hibernacula, 6 to 7 in. below the surface.

The giant Galapagos tortoises (*Geochelone elephantopus*) leave many characteristic signs (Pritchard, personal communication). In suitable habitat these turtles make permanent, well-defined trails. Slight excavations made where they sleep, droppings, and bent-over vegetation in new trails are other useful signs.

Individuals of *Gopherus polyphemus* can be dug out of their burrows, but it is more expedient to wait at the burrow and capture the tortoise as it enters or leaves. Woodbury and Hardy (1948) used a steel rod with a sharp hook to remove *G. agassizi* from their burrows, but this technique depends on the tortoise being within reach of the rod, a condition that is not always met.

During periods of hot, dry weather, many species of terrestrial turtles are attracted to water; thus a systematic search of pools, streams, and surrounding areas often yields specimens when searches of usual shelter and feeding sites do not.

Even aquatic turtles can sometimes be collected on land. Many freshwater species wander on land, especially during the spring and fall (Cagle, 1944b), and these can be collected easily by hand, often in great numbers. Sexton (1959) constructed "retaining fences" 10 in. high and 100 m long around a pond to detain temporarily turtles that were moving overland and to increase his chances of encountering the turtles.

Carpenter (1955) described "sounding" as a collecting technique: a metal rod with a blunt end is thrust into potential turtle-containing sites. By probing debris and masses of leaves, he found many turtles of five species and even located turtles hibernating 6 to 10 in. below the surface. The collector soon learns the sound of the rod hitting a turtle as opposed

COLLECTION METHODS

to a rock or log. Bishop and Schoonmacher (1921) located hibernating *Clemmys insculpta* by sounding into muskrat burrows and into mud of stream bottoms. Dobie (1971) used a sounding rod with a hook on one end to locate and remove *Macroclemys temmincki* from their underwater retreats. Smith (1947) collected *Trionyx spiniferus* by probing sand bars at the water's edge. The rate of capture by this technique may be enhanced greatly by first locating concentrations of soft-shells by cruising in a boat, then probing those areas of concentrations. Small isolated sandbars well separated from other burrowing substrate tend to have more burrowed turtles per unit than do large, extensive sandbars or mud flats.

The persistent burrowers (e.g., those belonging to the genera *Chelydra*, *Kinosternon*, and *Trionyx*) occasionally rest their heads on the substrate with the remainder of their bodies buried; such animals can be spotted visually in clear water. These turtles can also be located by walking in shallow water and looking for a small depression in the mud or sand; the depression results from the turtle retracting its head periodically after obtaining air. At times a swirl of mud is created by the withdrawal of the head.

In vegetation-choked areas of lakes and small streams a garden rake is useful. Walk slowly along the shore and look for heads or carapaces protruding above the surface, or for disturbances made by foraging or fleeing turtles; then quickly rake the vegetation-entrapped turtle onto shore. Goin (1942) described a specialized dredge for raking turtles out of water hyacinths.

"Noodling" or "muddling" involves feeling around with the hands or feet in the mud bottoms of shallow waters until a turtle is located. Cagle (1950) found this technique to be extremely productive in late summer and winter when turtles were locally concentrated in areas of receding water.

Gibbons (1968b) captured *Emydoidea blandingi* by first spotting the bright yellow chin as the head protruded from the water, then rowing toward it in a boat. The turtles seldom moved after they had ducked their heads under, and were easily collected. This is a productive technique with many turtles, whether capturing with the hands or with a dip net.

In clear water, turtles may be captured by swimming after them. Bider and Hoek (1971) designed a floating blind to facilitate capture. "Water goggling" is a technique originally described by Marchand (1945b) for use in clear Florida streams. The collector wears a face mask and scans underwater while being towed by holding onto a handle on a boat. When he sights a turtle, he releases the handle and a short, rapid swim results in a capture. This method has the advantage of covering large areas quickly and is relatively efficient in capturing individuals of the genus *Graptemys*, which are extremely wary and difficult to capture by any other method. Chaney and Smith (1950) located basking aggregations of map turtles by day and collected them at night when the turtles were found resting on vegetation just below the surface.

Electric shockers have been used with some success (Dobie, 1971; Harris, 1965). Because most reptiles sink to the bottom after being shocked, Gunning and Lewis (1957) found that shockers are best used in shallow, clear water. In every instance in the laboratory they found that unretrieved individuals made a rapid recovery. Other methods utilized in fisheries research may also be applied to turtle collecting, for example, using seines and trammel nets. Carr (1952) found seining to be most effective in areas where large numbers of young are concentrated; mature turtles are more likely to elude the seines. Moll and Legler (1971) took more adult male *Pseudemys scripta* with trammel nets than with any of the six other methods they used. Several aberrant methods that yield specimens are not recommended because turtles and other organisms may be killled in the process: poisoning with rotenone may yield specimens of the genus *Trionyx*, which are more reliant on pharyngeal respiration than are other turtles. Even dynamiting may yield turtles (Webb, 1962). Shooting is unsatisfactory because most wounded turtles are not recovered.

The congregating habit of nesting females on beaches in some freshwater and marine species offers a unique opportunity for collecting. Roze (1964) and Carr (1967b) described techniques for *Podocnemis expansa* and *Chelonia mydas*, respectively.

Sea turtles are captured commercially with large mesh nets set across channels and sloughs in shallow grass flats (Carr, 1952). These nets, buoyed by floats, often are used in conjunction with wooden decoys that attract rutting males, therefore increasing efficiency of capture (Carr, 1952). During the breeding season, sea turtles are easy prey to various types of harpoons and spears as the animals drift preoccupied in copulation or asleep on the surface (Parsons, 1962). Carr (1952) described a more direct method of wrestling the turtle into boats or onto the beach by strong swimmers. Apparently, an important factor in maneuvering these large turtles is keeping the forward edge of the shell tilted upward to prevent the animal from diving.

One of the most unusual methods of capturing sea turtles involves the use of a remora (*Echeneis*). This fish is attached to a long line and allowed to swim among congregating turtles until it attaches itself to the carapace of a turtle, which is then pulled in. Parsons (1962) gave an excellent account of the historical development of this method.

Most freshwater turtles may be taken on baited hooks. Set lines, throw lines, float lines, trot lines, and hooks and lines of fishermen often catch turtles. Usually these methods are secondary to other methods of capture, but Dobie (1971) obtained 90% of his *Macroclemys temmincki* from trot lines of commercial fishermen. Strong recurved hooks set firmly on favorite basking sites may also snag turtles (Carr, 1952; Webb, 1962).

Trapping

Baited Traps. The basic tool of capture of most freshwater species is the baited funnel trap with its many variations (e.g., hoop nets). This device is simply a cylindrical or rectangular frame covered with cotton or nylon netting or wire mesh; an inverted funnel with a horizontally flattened opening projects into the body of the trap (Figure 1). The turtle enters the trap through the funnel and once in, cannot escape. Size, shape, number of funnels, and construction materials vary among collectors. Legler (1960b) described a basic, collapsible hoop net using four hoops of aluminum tubing with a cylinder of treated cotton netting. Other collectors have covered rectangular wooden frames with wire mesh. These traps are sturdy but are very bulky and are not collapsible. To capture *Trionyx muticus*, I use rectangular traps 8 × 20 × 24 in. with a single 8 in. throat, made entirely of 1 in. mesh chicken wire. Traps set in shallow water usually catch more turtles than those set in deep water. Not only is shallow water usually better turtle habitat, but a turtle captured in a trap in which he can readily obtain air is not so prone to activity; thus the animal is less likely to escape or to scare away other turtles (Lagler, 1943b). Inasmuch as water must be deep enough to cover

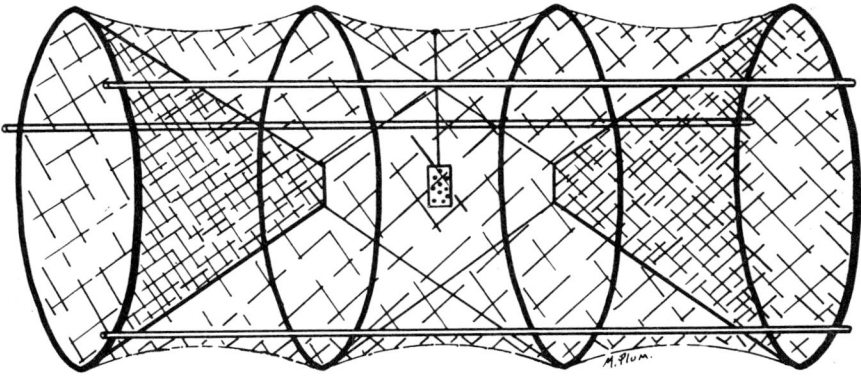

Figure 1 A baited hoop net.

the throat of the trap, rectangular traps have the advantage of lying flush with the substrate; thus they can be set in shallower water than the conventional hoop nets. The size of the vertical opening of the throat apparently is not critical. Turtles do not seem to be hindered by having to push through the low opening (Legler, 1960b). A low vertical opening is preferable because it decreases the chances of escape. Individuals of *Chelydra serpentina*, with body thicknesses exceeding 4 in. frequently push through a 1 × 20 in. throat opening designed for softshells. In running water, traps should be set with the funnel side facing downstream so turtles may follow the bait scent into the trap. Traps always should rest on the bottom in shallow water and should be firmly anchored with a line. Generally, efforts to trap turtles in deep water with floating funnel traps have been unsuccessful.

Fyke nets have also been borrowed from fisheries. These are large hoop nets with two drift fences that extend diagonally outward from the single throat opening. These traps probably do not increase capture rate over that of conventional hoop nets (Lagler, 1943b), and they are much more bulky and difficult to set.

A complete list of the baits that have been used to attract turtles would be quite extensive. Some of the more popular and productive baits include chopped fresh or canned fish, chicken entrails, and other fresh meat.

The type of bait to be used should be chosen in consideration of the type of turtle to be trapped, the relative ease of acquiring and storing bait, and the particular trapping situation at hand. An omnivorous species such as *Chelydra serpentina* would be much more apt to be attracted by bananas than would the purely carnivorous *Trionyx muticus*. Attracting large herbivorous turtles is a perennial problem, yet to be solved. Fortunately, however, some of the purely herbivorous types, such as adult *Pseudemys scripta* in the Rio Grande (Legler, 1960b), are attracted by meat baits. Clark and Gibbons (1969) characterized adult *P. scripta* as "opportunistic carnivores" who feed on meat when it is readily available. Some turtles, such as those belonging to the genus *Graptemys*, tend to be unresponsive to any bait, although they occasionally enter traps.

Collecting in remote areas often necessitates carrying bait and keeping it for extended periods. Legler (1960b) pointed out that canned sardines are a good general purpose bait where fresh bait cannot be maintained. If it does not catch turtles, it will catch fish, which in turn can be used as bait. It is important to be open-minded and opportunistic in any collecting endeavor. A collector who refuses to change established baits even when collecting is sparse may collect far fewer turtles than if

COLLECTION METHODS 51

he had experimented with several different kinds of bait on each population in order to determine the most attractive. Lagler (1943b) noted that *Trionyx spiniferus* often are attracted by such atypical bait as unspoiled watermelon rind.

Although it is widely held that putrid baits are best for attracting turtles, data reveal that fresh bait is by far the most productive (Lagler, 1943b; Breckenridge, 1955; Tinkle, 1958a; Legler, 1960b; Webb, 1962; Ernst, 1965). Lagler (1943b), in a summary of collecting methods, stated that the freshness of any bait seems to be an important factor in determining its value as an attractant. Legler (1960b) found that putrid or partly decayed bait was much less effective than was fresh bait. Ernst (1965) compared six different baits in attracting three species. Only with *Chelydra serpentina* was putrid bait more likely to be attractive. Only an hour after replacing day-old fish in an empty trap with fresh fish, I found that the trap contained 14 *Trionyx muticus*.

The last example illustrates another important point. Baits seem to be the most effective in the first few hours of immersion in water. Although Lagler (1943b) felt that checking intervals of less than 12 hr would disturb turtles and lower the yield, Legler (1960b) found that traps that were checked and rebaited at 1 or 2 hr intervals had a higher capture rate than did traps left for much longer periods. If bait cannot be changed, it will continue to attract turtles for several days, especially if the water is not too warm, although efficiency decreases. Bait should be enclosed in a container to prevent the first turtle that enters from eating it all. Quarter-inch hardware cloth folded into small boxes or 35 mm film canisters with holes punched in them make satisfactory containers.

Nonbaited Traps. Various types of nonbaited traps have been used. These traps have the advantage of not requiring the constant attention of the collector in renewing the bait, yet always remaining operative. Generally, these traps take advantage of the turtles' natural behaviors, which lead them into the trap. Sexton (1959) placed large funnel traps made of chicken wire at the mouths of the inlet and outlet streams of his study pond. A drift fence channeled any turtle attempting to leave the pond via the streams into the traps. He found that *Chrysemys picta* tended to migrate out of the pond in the spring in large numbers, and these traps were efficient in their capture. Gibbons (1970d) utilized a series of pitfall traps in conjunction with drift fences placed around a pond to collect aquatic turtles that wandered on land. Of 199 adult and juvenile turtles seen on land near the pond, only 13 were not caught in these traps. Gibbons (1968c) constructed cylinders of wire mesh, 3 ft in diameter, and 5 ft high, and stood them on end near shore. A 1 ft square door was cut near the bottom with the free edges of the mesh extending inward. Each trap was accompanied by a drift fence. Any turtle moving along the shoreline was channeled into the trap. The tall cylinder left room for fluctuations in water level to prevent drowning the captured turtles. Other investigators have used unbaited funnel traps set in areas of turtle concentrations for random swim-in capture.

Legler (1960a) set unbaited small-mammal live traps along ravines and rock fences ("natural drift fences") and collected a few *Terrapene ornata* in them. Auffenberg (personal communication) uses two methods for trapping gopher tortoises. One involves placing a one-way hinged door of clear Plexiglas over the mouth of the burrow. This allows the turtle to leave but not reenter, and increases the chances of finding it near the burrow opening. Alternatively, Auffenberg sinks a small wooden box or large can immediately in front of the burrow mouth with the edge of the container flush with the surface of the ground. Ideally, upon leaving the burrow, the turtle falls into the pit and remains there until collected.

The various attempts to trap terrestrial tortoises have generally been unsuccessful in terms of time and energy spent by the collector. For most species in most situations, collecting by hand remains the most effective method, after maximizing one's chances of encountering tortoises. Often a collector can have his needs most quickly filled by soliciting specimens from local residents.

The basking habit characteristic of many aquatic turtles can be used to the advantage of the collector. Basking traps in all forms of sophistication have been used extensively. One of the simplest involves attaching a wire mesh basket to the side of a sunning station with the top of the basket flush with the surface of the water. Upon leaving the basking site, the turtles plunge into the water and into the wire basket. Since the top is flush with the surface, the turtles may leave at will and natural movements are not greatly disturbed. This may be a desirable feature in some population studies. Capture rate may be greatly enhanced if the collector makes a sudden appearance from the side opposite the basket. Carr (1952) related that 20 to 30 *Pseudemys floridana suwanniensis* may be taken at one time. Turtles that bask on land may be captured as they rush from their basking sites to the water by suddenly erecting a previously placed collapsed net with a pull cord that extends to a blind (Robinson and Murphy, 1975). Breckenridge (1944) suggests sinking a barrel into water until the rim is near the surface. A cover slightly smaller than the opening is suspended by two nails fixed on opposite sides of the lid. This pivotal lid dumps any turtle climbing onto the barrel. This trap may also have bait placed near the center of the lid. A major disadvantage is that

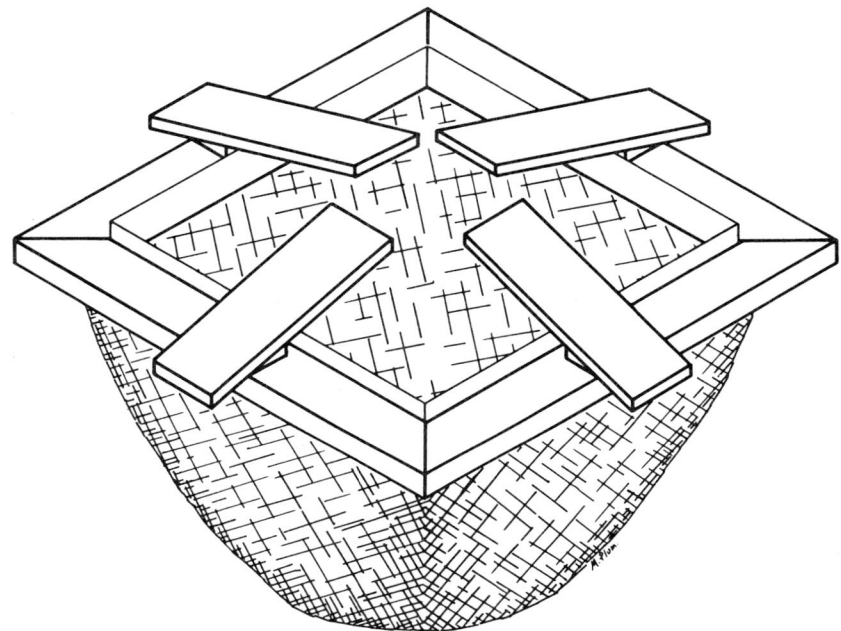

Figure 2 Basic design of a floating basking trap. Turtles basking on the inclined boards plunge into the bag upon leaving their basking station.

this trap does not compensate for changes in water level, which characterize most bodies of water.

Floating basking traps are not only easier to set up but are also immediately responsive to water level fluctuations. Cagle (1950) described a basic floating basking trap that is commonly used along the Mississippi River by commercial collectors. It consists of a square wood frame from which a net or wire bag is suspended in the middle. Boards angled up from the outer edge toward the center of the trap permit turtles to crawl out to bask but not to crawl out once they have plunged into the bag (Figure 2). Some modifications of this basic basking trap include using other types of material for the inclined board (e.g., wire mesh, closely spaced nails driven into the frame at an angle) and replacing the inclined board with single or multiple movable treadles (see Lagler, 1943; Breckenridge, 1955; Ream and Ream, 1966).

Braid (1974) described an innovative basking trap modified from similar traps designed to snare birds of prey. Called the "bal-chatri" trap, the device consists of a wire mesh base cut to a desired size and shape so that it may be molded around a basking log or other suitable site. Erect slip nooses, tied from monofilament fishing line, cover the mesh and snare turtles by the head, feet, or tail when they crawl out to bask. His version of the trap did not capture juveniles.

It is important to recognize that turtles seek out their preferred microhabitat in a pond, lake, or stream. Thus if a collector is seeking a particular species, he must know about the ecological conditions preferred by that species. Generally surveys of turtle species in ponds and streams have shown that species ratios vary greatly between different parts of the same water system (Cagle and Chaney, 1950; Tinkle, 1959b). Cagle and Chaney (1950) and Cagle (1950) noted that even within a very localized habitat, catches may vary with type of bait, method of setting traps, water temperature and depth, and the behavioral patterns of different size classes and sexes. Thus during periods of sexual activity, a mature female that is first in the trap may lure many males to the same trap, whereas if a large aggressive male enters first, he may exclude all others. The many variables influencing total catch in trapping make any systematic study of trapping efficiencies difficult.

There also may be a problem of obtaining an unbiased sample from a population. Ream and Ream (1966) compared five different techniques of capturing turtles and found that no single method was adequate for obtaining an unbiased sample. Each method seemed to capture large numbers of a particular size or sex and lesser numbers of the other classes. This phenomenon is common in population studies, and it undoubtedly has contributed to the unbalanced sex ratios often reported, thus influencing population composition data. To solve this problem, some investigators have stressed the use of diversity in collecting techniques (Mossimann and Bider, 1960; Ream and Ream, 1966). Others have attempted to develop unbiased sampling techniques (Bider and Hoek, 1971).

Collecting methods have been and will continue to be studied with regard to improving their efficiency and practicality. The methods described and reviewed above have proved themselves over time and should serve as a springboard toward innovative ideas.

One point is crucial. Check local wildlife and fisheries laws before collecting. Some states require a scientific collecting permit and/or a fishing license to take turtles. Hoop nets and similar traps are often illegal for catching fish and must be clearly marked as turtle traps. Czajka and Nickerson (1974) provide regulations governing collecting of reptiles in all states. However the provisional collector should consult the appropriate state wildlife agency before collecting, because laws change rapidly and an increasing number of states are passing laws designed to protect nongame species.

MARKING

Field studies often require the recognition of groups or individuals in order to trace their movements, ontogenetic development, roles in behavioral interactions, and so on. The various methods of marking have been devised in an attempt to fulfill these needs. There are three basic requirements of a good mark. First, it should be relatively permanent. This is especially important in such a long-lived animal as a turtle. However in many species there are large size differences between hatchlings and adults, and it is difficult to develop a mark that will not become outgrown or otherwise obliterated. Second, a mark should be clearly visible and capable of being accurately read by the investigator. A numbered tag offers little chance of misreading. A system of notches or toe clips, however, leaves more chance for misreading, especially when several hundred individuals are involved. Last, the mark should hinder the turtle as little as possible. Gibbons (1968c) pointed out that one basic assumption that is essential in population studies is that marking has no influence on an individual's chance of remaining in the population. Any marking system should be analyzed to determine the extent to which it might hinder the animal in its locomotor or other physical movements, in its susceptibility to predation, or in its sexual or social interactions.

Shell Mutilation

The shell has been the target for most turtle marking systems. Woodbury (1956), in a symposium on marking animals in ecological studies, provided a good review of the methods used up until that time. Some investigators carve numbers into the carapace. Nichols (quoted by Cagle, 1939) carved numbers in the carapace of *Terrapene carolina* and found that the numbers remained distinct for 10 to 15 years. However this method is time-consuming and laborious, and cannot be used for species that molt their outer plates. Many workers have experimented with notching the edge of the marginal laminae of the carapace. Cagle (1939) developed a system that could be used to identify up to 2516 individual turtles by notching the laminae using four separate marks. By marking a plastral scute or clipping a toe, the series could be used over again and theoretically, an almost infinite number of possible combinations could

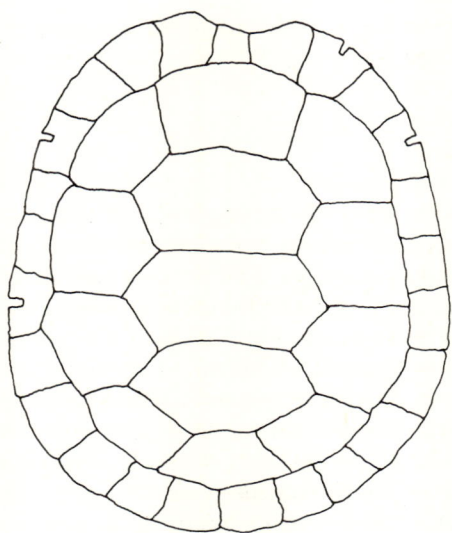

Figure 3 Carapace of an emydine turtle (*Deirochelys reticularia*, KU 3400, Juv). Marginal notches have been added to illustrate an individual marked 4,7-2,4.

be attained. The notching is done by a small file or hacksaw, or with a penknife or scissors on young turtles whose shells have not completely ossified. For speed and for application of a clean-cut mark, Cagle (1944b) substituted a small electric grinder to apply the marks. The marginals are numbered from anterior to posterior on each side and the identifying number of a turtle is the series of numbers representing the specific marginals notched. Thus 4,7-2,4 would be an individual whose fourth and seventh left marginals and second and fourth right marginals are notched (Figure 3). Cagle's system or modification of it is widely accepted and is used by many modern workers. Modifications include utilization of more or less than four notches, drilling holes in the marginals instead of notching, and using different ways of numbering the marginal laminae. Although adult turtles will usually retain the marks for years, it is often necessary to remark young turtles when they are recaptured, since the comparatively rapid growth of the shell may have occluded the marks. Certain techniques may be better suited to particular situations. For this reason the notching system may never be standardized, nor need it be.

The Trionychidae may be marked similarly in spite of their lack of laminae. By viewing the carapace dorsally and imagining the face of a clock superimposed on it, one can cut notches on the edge of the carapace at desired locations to represent numbers corresponding to hours of the clock. Breckenridge (1955) used a leather punch to mark *Trionyx ferox* and found that although the holes in the vascular carapace healed quickly, they were still discernible at 3 to 5 years as shallow sinuses. While working with a population of *T. muticus*, I have recognized individuals marked 3 years previously by Henry S. Fitch, an earlier worker with the same population. The marks were originally cut as notches with a penknife. Although the notches eventually heal and fill out to the margin of the carapace, a V-shaped piece of scar tissue remains. This scar tissue is most easily seen when viewed on the whitish ventral surface.

Problems may arise in differentiating between intentionally applied marks and wounds acquired by an animal in the field. Generally, wounds are irregular in shape in contrast to the smooth outline of marks. This is also a reason for marking more than one scute. Rarely will an accidentally "marked" turtle fit a combination of special marks or be the size of the turtle indicated in the worker's records.

Woodbury (1948) described the manufacture and use of an electric tattooing device that could be used on the clear plain ventral surface of members of the Trionychidae. Breckenridge (1955) used a tattooing device to inject ink into the plastron of *T. ferox*. Conceivably this method could be affected by growth if hatchlings were tattooed. The pigment granules might be dispersed to such an extent that it would be impossible to interpret the numbers.

Woodbury and Hardy (1948) used branding to mark *Gopherus agassizi*. They discovered that if the burn was too deep and affected the underlying dermal tissue, a slow regeneration process was initiated that completely obliterated the brand within a few years. This regeneration could be avoided by using a white-hot wire and burning the shell very quickly. By branding various combinations of plates, a large number of individuals could be marked uniquely. Weary (1969) described the use of an electric branding outfit that can be used in the laboratory. Clark (1971) described a technique in which numbers are formed by bending Hoskins Chromel "A" resistance wire (Malin and Co., Cleveland, Ohio). By heating the bent wire with a small propane torch, numbered brands can be conveniently applied in the field. Clark found that the brand was clearly legible 36 days after marking. No infection or alteration by growth and healing processes marred the brand during this short time. Clark stated that the wire must be left in contact long enough to produce a lasting and legible mark, but not long enough to penetrate the dermal layer and initiate regenerative processes.

Tag Marking

The application of various types of numbered tags has also been popular in turtle studies. Unfortunately, this method is particularly subject to the drawbacks of growth in large species. Carr (1967b) pointed out that it is difficult to devise a single tag that will identify both a 3 oz hatchling and the 300 lb adult turtle. For this reason, tags may be used with the most confidence on small species or on adults of large species. There have been some recent efforts to tag hatchling sea turtles by injecting certain materials into the body and later analyzing tissue samples. For example, Forbes (1972) injected europium chloride in a citrate complex. Turtles that had been marked were identifiable by subjecting tissue samples to neutron activation analysis. Such a sophisticated procedure may be necessary to provide lifelong marks in large species; however it does not solve the problem of providing unique marks to individual turtles.

Carr (1967b) experimented with oval monel metal plates wired to the posterior edge of the carapace of female sea turtles. He was frustrated by the quick loss (often less than 2 weeks) of the tag caused by the vigorous courtship and mating activities of rutting males. Carr has had the most success with the application of metal tags of the type used in animal husbandry for attachment to the ears of livestock. The ear tags are attached to the posterior edge of the front flipper and meet all the basic requirements discussed above. Tagging remains a current topic in the *Marine Turtle Newsletter* (N. Mrosovsky, Ed., University of Toronto). Metal tags attached to the carapace have also been used on freshwater and terrestrial species with some success (Wickham, 1922; Pearse, 1923a; Bogert, 1937; Miller, 1955).

Gaymer (1973) marked giant Aldabran tortoises (*Geoghelone gigantea* by placing numbered, ¾ in. titanium disks into shallow depressions cut in the carapace. The disks were fixed in place by a metal-resin adhesive. Fifty-nine of 150 marked tortoises were relocated after one year; only 2 disks had been lost.

The various brands of "pop rivet" tools available in most hardware stores offer some promise in tag marking. Small aluminum rivets with back-up plates may be applied to the shell through a small hole drilled in the carapace. The back-up plate provides a convenient place to stamp an identifying number. Breckenridge (1955) marked 39 *Trionyx ferox* using brass rivets but never recaptured a single turtle with a rivet in place, although some were recaptured showing rivet holes. I have experienced a 50% loss rate of rivets in as little as 2 months in *T. muticus*. The tissue surrounding the rivet apparently becomes necrotic, and the rivet falls out. However the use of pop rivets in hard-shelled species could be rewarding because the rivets are quickly applied, provide a visual identification number, and protrude very little from either side of the shell. Pough (1970) provides a similar marking method with a device designed to attach buttons by the use of small plastic bulbs. The bulbs are attached through holes drilled in the marginal scutes. Although Pough gave no data pertaining to the retainment of bulbs, he suggested that such plastic bulbs applied to a hatchling would be clamped in place as the shell grew and retained throughout life, particularly if applied to individuals of small species.

Other Marking Techniques

Application of paint to the shell has been used in some short-term studies. Woodbury and Hardy (1948) painted different combinations of scutes in various colors to mark individuals of *Gopherus agassizi*. They found that the paint wore thin after a year or two, although one red-marked tortoise was positively identified 8 years later. Miller (1955) wrote numbers in India ink on the shell of *G. agassizi* and covered them with clear lacquer. He later improved his method by carving numbers into the shell, filling them with pigment, and then covering them with lacquer. Painting has also been used on aquatic species (Tinkle, 1958a; Moll and Legler, 1971). Moll and Legler found that the numbers on hatchling *Pseudemys scripta* remained legible for 1 to 2 months or until the scutes were shed. Since many factors tend to obliterate paint marks this method is, at best, useful only under specific conditions (e.g., short-term behavioral studies) and should be used in conjunction with other more stable methods.

In order to study movements, Carr (1967b) marked *Chelonia mydas* by attaching brightly colored Styrofoam floats to a line that was tied to the carapace of the turtle. He felt that this did not appreciably affect turtle activity. Because of the curvature of the earth's surface, these floats were not visible for more than 1 mile. Therefore Carr attached a helium-filled balloon to the float by a 20 ft line; this increased visibility to about 8 miles. A sausage-shaped balloon offers less resistance to air than does a spherical balloon, thus is less likely to be blown down by winds (Carr, 1967b).

In studying short-term movements of *Trionyx muticus*, I have used as floats small (ca. ½ × 6 in.) sausage-shaped balloons, inflated with 1 atm of pressure. By using different colored balloons, each marked with specific numbers of rings (using a permanent felt-tipped pen), numerous individuals may be identified at a distance through binoculars. The balloon is attached to the posterior edge of the carapace by a fine, stiff wire. Stiff wire has less chance of becoming fouled on obstructions and, if it does become fouled, readily pulls out of the carapace because of the pressure exerted by the turtle.

Davis and Sartor (1975) used a similar method, attaching a wooden dowel in a hole drilled in the nuchal scute and color coding the end of the dowel with paint or plastic tape. Obviously, these two methods are practical only in comparatively obstruction-free waters.

"Turtle trailing," originally used by Breder (1927), is a technique in which a spool of thread is attached to a terrestrial turtle, the free end anchored at some specific spot, and the path of the turtle determined by following the unwinding thread. The original construction of the device was literally a trailer that the turtle pulled. Stickel (1950) improved the technique by mounting the entire device on the carapace, and Emlen (1969) further modified and improved the device in his work on terrestrial homing in *Chrysemys picta*. Both Breder and Stickel used the method in studying home range and movement in *Terrapene carolina*, and the method has become almost standard practice in field work with turtles of this genus (e.g., Legler, 1960a; Dolbeer, 1969; Metcalf and Metcalf, 1970; Reagan, 1974; Schwartz and Schwartz, 1974).

Bennett, Gibbons, and Franson (1970) tagged three species of aquatic turtles with radioactive tantalum-182 in order to study distance from and depth of burrowing from temporarily dried-up ponds. The apparent relative insensitivity of some turtles to gamma radiation (Atland, Highman, and Wood, 1951; Cosgrove and Davis, 1965) and the ease with which small radioactive tags may be inserted into holes drilled in the carapace make this type of "remote sensing" an ideal method for locating turtles when short distances are involved. For longer distances the various types of biotelemetry (see Chapter 3) may be extremely productive.

ACKNOWLEDGMENT

I thank William E. Duellman, Henry S. Fitch, Alan H. Savitzky, and Linda Trueb for reading the manuscript and making numerous suggestions for its improvement. Mr. Savitzky called several pertinent papers to my attention. Dr. Trueb assisted with the illustrations. Thanks go to my wife, Sharon, for typing the manuscript.

CHAPTER 3
Telemetry

WARREN K. LEGLER

Biotelemetry, the transmission of biological information over a distance without direct connections involving wires, makes use of either radio waves or ultrasonic vibrations as carriers of the information. This chapter is devoted almost exclusively to the use of radio waves, since a majority of applications make use of this mode of transmission. These applications fall into three fairly distinct categories of distance between the transmitter and the receiver of the radio waves carrying the information. The early applications, which still form an important category, involved transmission over only a few centimeters or meters, usually in an indoor laboratory setting. Later, transmission of information over distances up to several hundreds or thousands of meters was employed, most often in an outdoor setting, with the organism under study being free to move about at will in its natural environment. Finally, transmission distances of many kilometers, even thousands of kilometers, were involved in some applications. Probably the best known examples of biotelemetry involving these extremes of distance are to be found in the space flight programs, where information regarding the physiological processes of the astronauts is transmitted from spacecraft in earth orbit or even from the moon. The types of information that have been transmitted, too numerous to recount in detail here, include temperature, heart rate, various pressures and flow rates, electroencephalograms, and internal chemical parameters such as pH.

Another type of information that may be obtained, though by a somewhat indirect route, is the location of the organism carrying the transmitter. In this type of application the transmitter serves as a beacon that may be localized in space either by a process of triangulation using receivers spatially separated from each other or by moving a single receiver toward the signal source by noting how the strength of the signal varies as the receiver is moved relative to it.

Several factors make the use of radiotelemetric methods attractive First, they enable relatively easy access to data from internal body cavities. The radio waves travel through the intervening body tissues to a receiving antenna close by, possibly even surrounding the subject. Another advantage conferred by these methods is the freedom from encumbrance from trailing wires and other apparatus that would otherwise be required to transmit signals, even though the source of the signals might be on the surface of the subject and would require no particular discomfort or danger to the subject to pick them up. An example of such a case would be the transmission of an electrocardiogram using surface electrodes.

Finally, and this is an extremely important advantage, the use of a beacon transmitter allows an investigator to follow an individual organism over its entire free-ranging movement in its natural habitat, without the necessity for maintaining continuous close contact over a long period of time. Thus the advantages that radiotelemetry brings to the conduct of studies of the natural history of an organism are enormous.

Radiotelemetry makes possible the intensive study of a few individual members of a species, with each individual being rendered observable essentially at the investigator's will. The classical methods of field biology, largely dependent on live trapping or other catch-as-catch-can methods, seldom allow multiple observations of a specific individual. As an example of the classical methods, consider Fitch's (1963) study of the blue racer, *Coluber constrictor*, carried out over a period of 14 years. He recorded, by live trapping, 1020 individuals for a total of 1688 times. Only 67 (6.6%) were seen more than three times; 679 were seen only once. On the average, a given trap caught a blue racer once in about every 2 months. Fitch and Shirer (1971) force-fed transmitters to eight species of snakes, which retained them for varying periods before disgorging them. Extensive data on movement patterns and other behavioral parameters were obtained. It is particularly significant that of 170 records of five male blue racers, nearly 25% involved finding the snake climbing in a tree. Clearly, this climbing behavior could not have been deduced from the earlier study.

TRANSMITTERS

In applications involving transmission of information other than simply the position of the individual organism, a characteristic of the transmitted radio wave must be changed in some way to code or represent the information. Two of the characteristics that can be changed, or modulated, to carry information are the amplitude and the frequency of the transmitted radio wave. Both types of modulation have wide application in biotelemetry.

In amplitude modulation (AM) an increase in the amplitude of the information signal gives an increase in the amplitude of the transmitted radio wave. Through the use of demodulator circuits in the receiver, the information signal may be recovered. In the ideal case this demodulated signal exactly matches the signal that served to modulate the radio wave carrier at the transmitter end of the communication link.

Frequency modulation (FM) circuits cause the frequency of the radio wave to vary in synchrony with the amplitude of the information signal. Usually, an increase in signal magnitude gives rise to an increase in the frequency of the radio wave. Here too, demodulation circuits in the receiver allow accurate recovery of the information signal. A good discussion of modulation methods, in the biotelemetry context, was given by Mackay (1970).

Pulse modulation, quite closely related to amplitude modulation, is binary in that the radio wave is alternately on, at some constant amplitude, then completely shut off. The length of each of the pulses, during which the radio carrier wave is on, can be made to represent the information being transmitted. For example, a pulse duration of 10 msec represents a temperature of 50°C, a duration of 20 msec represents 30°C, and so on. Pauley (1971) has given analyses of such a modulation scheme.

Mackay (1970) provides a very complete discussion of a number of systems devised for multiplexing, that is, the use of a single radiofrequency carrier to transmit more than one channel of information. To cite one example, Pauley (1971) has described a system for transmission of both temperature and heart rate. In this system each heart beat triggers the transmitter for a period of time that depends on the temperature. Pauley supplies details of the transmitter circuit and also some receiver circuitry specially designed to recover the heart rate and temperature data from the transmitted signal.

Biotelemetry transmitter circuits range in complexity from those involving only a few components to those with component counts running to several dozen. One of the simplest, but still a very useful circuit, appears in Figure 1. This circuit has been described by Shirer and Downhower (1968) as an adaptation of an earlier circuit of Cochran and Lord (1963) to include a pulse modulation scheme of Rawson and Hartline (1964). This transmitter operates in a pulsatile fashion, the transmitter being on only a small fraction of the time, usually 10% or

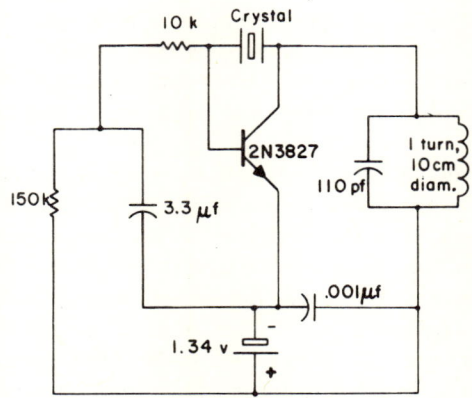

Figure 1 Tracking transmitter schematic.

less. The primary advantage of this type of operation is the saving in battery power that can be realized. In practice, it is found that the pulsatile nature of the transmission does not interfere with interpretation of the signal for tracking and location purposes. The other significant feature of this design is the use of a quartz crystal to provide a very constant and dependable frequency of oscillation. This feature is a virtual necessity if a number of individual organisms carrying transmitters will be in a given area at the same time. Crystal control of the frequency allows easy distinguishing of one transmitter from another. Precise knowledge of the frequency is also important when it is necessary to operate near the limits of the range capabilities of the transmitter-receiver combination.

Lin and Ko (1968) have presented a very careful and complete discussion of the operation of another widely used oscillator, based on the well-known Hartley circuit. The circuit is pulse modulated and lends itself quite well to the transmission of temperature information, though it has been used in many other applications as well.

When working with small animals, the physical size and weight of the transmitter must be kept as small as possible, and the circuit components must be carefully arranged to require a minimum of space. Figure 2 illustrates a component configuration that has been found to be useful for work with turtles. In this circuit, based on Figure 1, the components are soldered together using their own leads as far as possible. The inductor coil in the resonant tank circuit consists of a single loop of wire, which is attached to the turtle around the periphery of the carapace. The rest of the components, except for the battery, are encapsulated in a

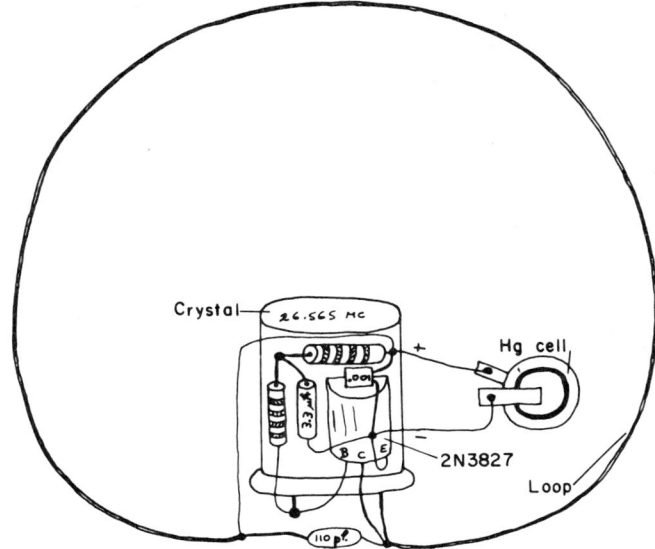

Figure 2   Transmitter component arrangement.

clear epoxy resin, with projecting leads to attach to the small mercury battery. This allows the battery to be replaced easily. If the transmitter is to be considered expendable at the end of the life of the battery, the battery may also be encapsulated with the rest of the components. A liquid silicone rubber material, Silastic, may be used to form a protective coating around the whole assembly and serves as an adhesive to attach the transmitter to the turtle at the rear of the carapace. This coating affords both mechanical protection and also protection from water and other environmental hazards. If batteries are to be changed periodically, the rubber may be cut away to allow replacement, followed by resealing with fresh rubber, which will bond to that encapsulating the rest of the unit. Mackay (1970) discusses the properties of the sealant materials.

Verts (1963) and Cochran and Lord (1963) also provide valuable discussions of the procedures involved in constructing and checking the operation of transmitters.

RECEIVERS AND ANTENNAS

Receivers for biotelemetry applications have not received nearly as much attention as have transmitters. A radio receiver, whether for AM or FM modulated signals, is a complex electronic device requiring a good deal of care and expertise in its design.

Commercially available receivers are satisfactory for many applications. Small and inexpensive battery powered receivers are available for FM signals in the 88–108 MHz band set aside for this broadcast mode. For AM signals the situation is a little more difficult. Since the regular AM broadcast band (530–1600 kHz) is virtually useless for biotelemetry work, unmodified broadcast band receivers are of no value. A number of communications receivers, mostly intended for use by radio amateurs, are attractive for fixed site operations. These receivers, designed as they are for reception of Morse code transmissions, are well suited for reception of pulse modulated biotelemetry signals. There are also some receivers now commercially available for biotelemetry use.

Cochran and Nelson (1963) described an 11 channel portable receiver that operates at about 26.6 MHz. Their paper includes complete instructions for constructing the receiver, though the project should probably not be attempted by the inexperienced. Shirer and Downhower (1968) describe a modification to a portable broadcast band receiver kit to shift its frequency of reception to the 26.6 MHz band. This receiver has proved very successful in field use, demonstrating excellent tuning stability under all conditions. The receiver described by Cochran and Nelson becomes almost unusable in cold weather because of rapid shift of the tuning frequency.

Mackay (1970) gives details of a phaselock loop receiver designed by Keith Brocklesby and Richard Barwick. Though quite complex, this type of receiver can tune very precisely to a transmitter signal, rejecting input at all frequencies outside a very narrow band centered on the transmitter frequency. This narrow band tuning underlies the high sensitivity of the phaselock receiver. The phaselock principle is widely employed in space communication systems, where signals must be sent over long distances with a minimum of power.

A receiving antenna with directional characteristics must be employed to locate a transmitter. Shirer and Downhower (1968) discuss several types, of which the most important type is probably the circular loop antenna. This antenna possesses a figure-8 sensitivity pattern, where the received signal goes to zero when the plane of the loop is perpendicular to the line of sight to the transmitter (Figure 3). However this leaves ambiguous the matter of which of two diametrically opposed directions is the correct one. This ambiguity can be resolved by employing a small sense antenna in conjunction with the loop (Figure 3). The resulting pattern allows determination of the correct bearing to the transmitter.

Legler (1969) carried out an experimental study of a number of

FIELD OPERATIONS

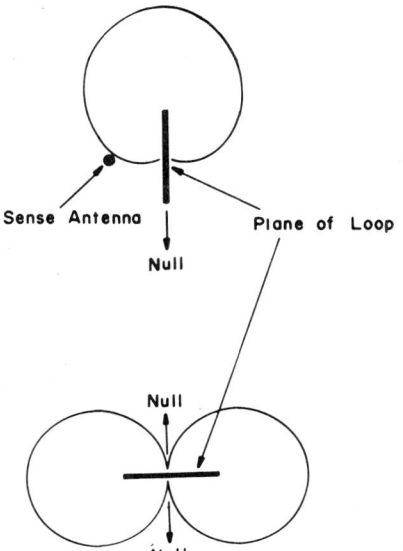

Figure 3 Loop antenna patterns.

factors involved in the propagation of 27 MHz biotelemetry signals over short distances, up to about 500 m. This study examined the effects of terrain, vegetation, size and height of transmitting antennas, and polarization of the transmitted signal.

FIELD OPERATIONS

Biotelemetry studies of turtles ranging freely in their natural environment have been described by Legler (1969). The process of tracking down an individual turtle begins with determination of an initial bearing to the animal by means of a receiver tuned to the frequency of the animal's transmitter. The use of antennas with directional characteristics has been mentioned already. Sometimes this initial bearing is determined through the use of a more sensitive receiver and antenna combination located at a central site. The investigator then moves in the direction of the animal until the signal can be picked up with a smaller portable receiver. The directional characteristics of the portable antenna then are used to guide the investigator to the general area of the animal.

Often the directional clues provided by the antenna pattern become somewhat confused when the receiver comes to within 25 m or so of the transmitter. At this point, a shift to a different mode of direction finding may be helpful. The strength of the received signal is used as a clue. If the signal becomes stronger as the investigator moves, he is heading toward the animal. If the signal fades, the path needs correction. Use of this direction-finding scheme can usually lead the investigator fairly directly to within a couple of meters of the animal. Legler (1969), tracking ornate box turtles, *Terrapene o. ornata*, often found that the position determination had to be accurate to within a fraction of a meter to be able to find a turtle dug in under leaves or in long grass. The transmitter and receiver equipment already described allows an experienced tracker to locate a turtle to this positional accuracy.

Triangulation by means of multiple bearing lines from several receiver locations is another possible means of position determination. Cochran, Warner, Tester, and Keuchle (1965) have described a system with rotating directional antennas located on two towers, spaced a half-mile apart, allowing position determination over a 4500 acre area. Outputs from these antennas are fed to 52 pairs of receivers, allowing simultaneous position determinations of the same number of separate transmitters. Bearing determinations have been shown to be accurate to within ±0.5°. The most favorable geometry, where the bearing lines from the two antennas intersect in a right angle at a point equidistant from the antennas, would localize the animal to a square having about 10 m on a side.

Reception of data of other types, transmitted by means of frequency or amplitude modulation, might not require finding the animal at all; that is, reception of a signal strong enough to allow satisfactory recovery of the data is sometimes sufficient. In other cases visual observation of the animal is needed to properly interpret the data recovered from the telemetry signal. Legler (1971) has given observations of the change in heart rate of a box turtle as it warmed up during a period of basking in the morning sunlight. The knowledge of what the animal was doing during the gathering of the heart rate data was nearly as important as the data.

Recovery of complex telemetered data in the field is a difficult problem, and with it is associated the need to make equipment easily portable and simple to operate. One factor making it easier to develop portable receiving equipment that is capable of very complex functions is the increasing availability of inexpensive integrated circuits. Pauley's (1971) circuitry to recover the temperature information from his heart rate–temperature telemetry system makes extensive use of integrated circuit modules and is easily carried in the field with the radio receiver.

APPLICATIONS

A complete bibliography of the field of biotelemetry would run more than 2000 entries. Mackay (1970) gives a bibliography of about 1500 items, including a number of medically oriented papers. Other bibliographical compilations, though somewhat less recent, are by Barwick and Fullager (1967) and Schladweiler and Ball (1968). The latter work emphasizes studies of wild animals under natural conditions.

Slater (1963) edited the proceedings of a biotelemetry conference held at the American Museum of Natural History. The papers presented there, though now somewhat dated, still represent a very useful introduction to the field. Brander and Cochran (1971) presented additional material.

R. S. Mackay, an energetic and innovative practitioner of biotelemetry for a number of years, has described measurements of deep body temperatures in the Galapagos tortoise and the marine iguana (Mackay, 1964) and measurements of pressures in the gastrointestinal tract of the monitor lizard, *Varanus flavescens*, and the iguana *Ctenosaura pectinata* (Mackay, 1968).

There has been little published application of biotelemetry to studies of turtles. Mackay's studies of the Galapagos tortoise have been mentioned. Carr (1962, 1965, 1967b) has tried, though apparently without much success, to use radiotracking methods in following the migratory movements of the green sea turtle, *Chelonia mydas*. These studies are very demanding of the transmitting and receiving equipment because of the aquatic environment and the great distances involved.

Legler (1969) studied the movements of a small population of ornate box turtles, *Terrapene o. ornata*, ranging freely in a natural environment including both deciduous woods and open grassy areas. Several individuals were studied for nearly a year. Though they varied widely in choice of summer habitat, all the turtles moved to a wooded area for hibernation. In several cases this move to the hibernation site was a precipitous event, some turtles moving several hundred feet to a place of hibernation between one day and the next.

Another turtle was equipped with a special transmitter that was keyed on by each beat of the animal's heart (Legler, 1971). Stainless steel screws were introduced through small holes drilled through the plastron to serve as pickup electrodes. This animal, out of doors in a large pen in an open grassy area, was studied for a month. Figure 4 shows the change in heart rate of this turtle as he warmed up during a morning period of basking in the sun. At the point indicated, after about an hour motion-

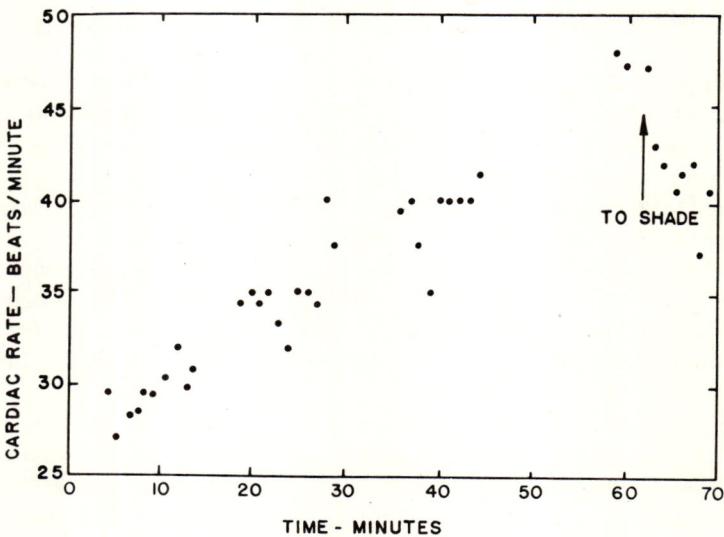

Figure 4 Cardiac rate during basking.

less in the sun, the turtle abruptly left this spot and moved into the shade nearby. Figure 5 summarizes a month's observations of heart rate as a function of the ambient air temperature. The points were all obtained under what were thought to be thermal equilibrium conditions, with the turtle sitting quietly in shade and exhibiting a nearly constant heart rate.

Schwartz and Schwartz (1974) used radiotracking of a small number of animals during a very extensive study of the three-toed box turtle, *T. carolina triunguis*, in central Missouri. Their studies extended over a period of 9 years and included observations of more than 1000 individual turtles. They found that radiotracking did permit obtaining considerable data that otherwise would have been lost. They describe the activities of one individual before and after winter hibernation. The radio transmitter allowed the investigators to find the turtle even though dug in under several inches of vegetation and soil on a number of occasions before finally staying down through the winter months.

Using ultrasound in vegetation-free water, Moll and Legler (1971) were able to receive transmission from *Pseudemys scripta* 1500 m distant. Dense aquatic vegetation reduced the transmitting distance to 100 m or less. With traditional marking, it frequently happened that the far-ranging adults were never observed again; ultrasound tracking permitted data collection on movement, home ranges, and homing behavior.

Bury (1972) used radiotracking methods in a study of the Pacific

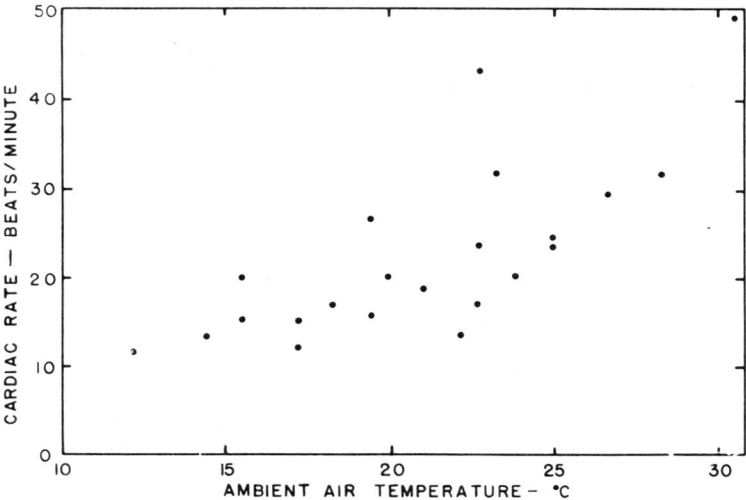

Figure 5 Heart rate versus temperature.

pond turtle, *Clemmys marmorata*, in northern California. He found that his transmitters could be picked up 150 m away even when submerged in 1 m of water. At a depth of 0.5 m, the transmission could be picked up at a distance of 500 m. He also made use of temperature-sensitive transmitters placed in the stomachs of the turtles in studying deep body temperatures during basking.

Robert L. Meeks (personal communication) encountered no transmission difficulties in detecting snapping turtles under 6 in. of mud in several feet of water; signals from such animals could be picked up a quarter-mile away. With these large animals, Meeks was able to use two size "D" mercury batteries (1.35 V, 14,000 mA · hr each) on his transmitters. The batteries were sealed in "Scotchcast" electrical resin for long-term use. Eighteen turtles were followed, and most of the transmitters were still sending signals that could be picked up by the handheld receiver at the end of 12 months. The 0.5 wavelength loop antenna covered the range of 30.10 to 30.25 MHz.

Application of biotelemetry methods to nearly all species of turtles should be a fairly straightforward task. In addition to obtaining the location of the animal, the temperature and heart rate are easily within reach (Pauley, 1971). Satisfactory pickup of the electrical potentials from the heart does, however, require a bit more than simply introducing electrodes through holes in the plastron. Using red-eared sliders, *P.*

scripta elegans, Pauley (personal communication) successfully implanted electrodes close to the heart by surgically removing a circular section of the plastron with a hole saw, suturing the electrodes to the cardiac muscle tissues, and replacing the circular plastron section with epoxy cement. The turtles tolerated this procedure quite well (see chapter 7).

FINAL COMMENTS

In beginning biotelemetry work with turtles it is probably best to start with beacon transmitters similar to that shown in Figure 1, using a portable receiver with a loop antenna. Ranges of 100 to 200 m can be reliably obtained in most conditions. Experience gained in conducting studies of this sort can lead to more advanced work with systems for obtaining physiological data. Temperature is by far the easiest quantity to obtain and probably represents a natural next step. Heart rate is a reasonable thing to attempt, provided the necessary surgery can be done. Internal pressures and flow measurements might be possible with the larger species. Surgery would be required here also, and the increased bulk of the complex circuitry would require a larger animal to carry it.

CHAPTER 4
Life History Techniques

TERRY E. GRAHAM

Life history investigations of turtles have not been refined to the same degree as those of many other vertebrates. According to Fitch (1949), herpetologists were preoccupied with the identification and naming of animals. The unfortunate consequence was that some of our most skilled turtle biologists were little concerned with the ecology of these animals and had little, if any, knowledge of such life history basics as growth, feeding habits, and population dynamics. During the past 30 years interest in turtle life history has increased considerably.

A great deal of our knowledge has been based on information derived from chance observations made in the field at opportune places and times. Such data are useful, and serve as a starting point for more thorough investigation. Fitch (1949) felt that a life history study should be intensive and prolonged over several years, depending on the longevity, developmental rate, and abundance of the species in question, but such long-term studies are usually neither feasible nor practical. The life history study should be pursued on an undisturbed area where the species is reasonably accessible and abundant. Today it is not uncommon to find a field study precipitously halted or seriously jeopardized by the construction of a housing project, shopping center, or other "improvement." State and federal parks and wildlife sanctuaries are undoubtedly among the safest places in which to initiate a study. Not only do they frequently provide habitat that best approximates the pristine, but they also stand as areas that are less vulnerable to imprudent human intervention.

Prerequisites to life history study include habitat analysis and identification of plant and animal species found in the study area. Topographic maps and climatic data are invaluable. Before the study is initiated the researcher should become thoroughly familiar with the literature on the species; such preliminary literature searching often yields clues to which avenues of approach to the study may be potentially fruitful.

Fitch (1949) outlined an approach to ecological life history studies that might easily be applied to turtles. Cagle's (1953) ambitious and admittedly elaborate outline for a reptile life history covered most of the major areas of concern, including extensive coverage of areas most investigators up until now have considered very little. He suggested that the following major questions be answered:

1 What are the morphological characteristics of the population to be studied? Are the data taken only from individuals of the genus, species, or subspecies intended?
2 What is the geographic range?
3 What is the age and sex composition of a local population?
4 What is the density of the population?
5 What is the potential reproductive capacity? What is its relation to the realized reproductive performance? What are the best measures of natality?
6 What major factors control the relationship between the number of surviving young and the number of eggs produced by females?
7 What are the characteristics of the young? Do they show any characteristic behavior traits? How is the behavior pattern related to growth and survival?
8 What are the characteristics of the growth curve of individuals in the local population?
9 What is the annual cycle of activity, and what factors exert major influence on this cycle?
10 What is the diel cycle of activity?
11 What are the food habits, and how are they related to growth and survival?
12 Does this form show any characteristic and genetically limited group behavior patterns?

This outline is presented here in skeleton form, and the reader is strongly urged to study the source (Cagle, 1953) carefully. Cagle conceded that a researcher could not hope to cover all the questions his outline posed. He felt, however, that an awareness of these questions might stimulate the researcher to make observations that might otherwise be omitted.

Many of Cagle's questions can be answered simply and directly by familiar methods. Knowledge of specialized techniques, however, is an asset in the analysis of species identification, age and sex determination, reproductive potential, survivorship, population size, movements, growth, and food habits. A discussion of selected techniques follows.

TECHNIQUES

Species Identification

The precise identification of the turtle is obviously necessary. Although many investigators are undoubtedly familiar with procedures for such definition, it may be helpful to mention a process by which the identity of unfamiliar species may be established.

A chief tool of the taxonomist is the zoological key for identification of a particular animal group (e.g., family, genus, species). Keys to turtles most often employ characters related to the head, limbs, or shell. Carr (1952), Ernst and Barbour (1972), Conant (1975), Stebbins (1966), and Cochran and Goin (1970) diagram and describe most of the reliable taxonomic characters used in turtle identification. Keys to North American species are found in Carr (1952), Ernst and Barbour (1972), and Blair, Blair, Brodkorb, Cagle, and Moore (1957); keys to foreign species are included in the works of Wermuth and Mertens (1961) and Pritchard (1967).

Immature turtles are often difficult to identify positively from a key; they should be carefully compared with the photographs and species accounts found in the references just mentioned. There may be exceptions to any key because turtles show individual variation and most keys are designed to permit identification through use of average characters. Some characters may vary from the given range of measurement. Occasionally some specimens appear with characters intermediate in expression between those of two closely related taxa; sometimes an entire population may demonstrate such intermediate expression. The correct identification of such intergrade individuals requires knowledge of the characters used to differentiate the two taxa.

Age Determination

Almost inevitably, anyone examining a turtle will sooner or later inquire or ponder about its age. Flower (1925) suggested that turtles have the greatest longevity of any vertebrate. According to "Rubner's hypothesis" (Oliver, 1955), turtles belonging to northern populations of a species found in both northern and southern latitudes will exhibit greater life spans than their counterparts of more southerly locales. The southern populations are active for a greater part of the calendar year and in essence "live more" during this period than do the northern populations. The low metabolic rate characteristic of a hibernating turtle might therefore help to extend the animal's life span. The turtle shell has also been suggested as having great survival value in warding off predators and protecting turtles from the hazardous effects of fire and environmental radiation. Carr (1952) generalized that selected turtles do attain considerable age and that they grow and reach maturity faster than is normally supposed.

The methodology of age determination is generally limited to the four techniques that follow.

Deck (1927), Price (1951), Oliver (1954), Graham and Hutchison (1969), and Latham and Schlauch (1969) presented discussions of the longevity of the box turtle, *Terrapene carolina*, based on animals carrying date/initial inscriptions, many of them authenticated. For the most part, however, this technique has very limited application; obviously it cannot be used with uninscribed animals, and these are the rule rather than the exception in natural populations.

The second technique utilizes records kept on captive animals (Flower, 1925; Conant and Hudson, 1949), which may live longer than their counterparts in natural populations.

A third method involves counting the growth rings or "annuli" on the scutes (laminae) of the shell. These growth rings indicate periods in which little or no growth occurred. Between these rings lie growth zones that represent periods of uninterrupted growth. Moll and Legler (1971) illustrated the fashion in which these growth rings arise in *Pseudemys scripta* (Figure 1). The ring-counting method relies heavily on the normally valid (for north temperate species) assumption that only one growth ring is formed annually. Growth rings fade with age in many species so that often only the most recent ones are visible. Turtles, such as *Chrysemys picta marginata*, which experience ecdysis (scute-shedding), tend to lose their earlier annuli rather quickly (Sexton, 1959a), whereas those such as *Terrapene ornata* (Legler, 1960a), which apparently do not shed their scutes, bear complete ring records. Many species possess minor (accessory) rings in addition to major (primary) growth rings. The major rings are deeper and represent annual growth cessation concurrent with hibernation, whereas the minor annuli are shallow and are formed in variable numbers during brief growth arrests within a single growing season. When all the major growth rings are clear, they form an accurate age record. But it should be kept in mind that when a turtle

TECHNIQUES

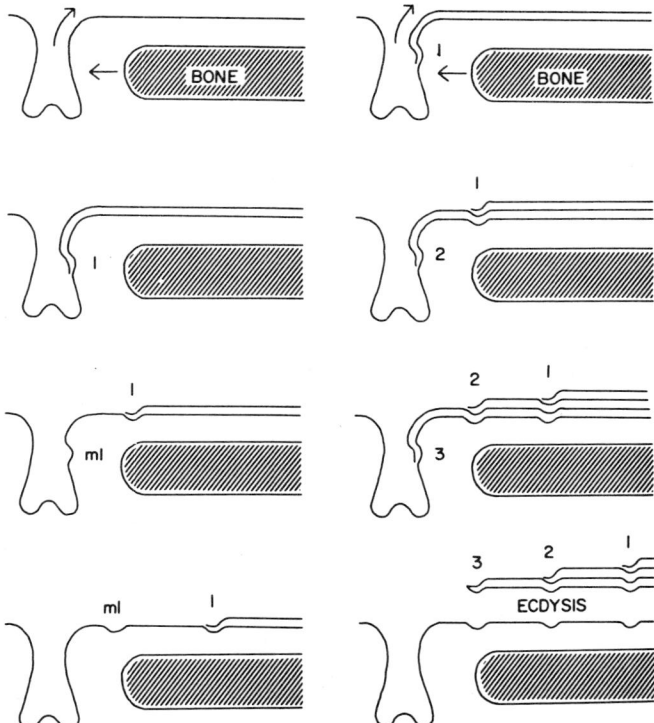

Figure 1 Growth of left abdominal lamina of a young *Pseudemys scripta* illustrated from sections. The ventral surface of the plastron appears at the top of each cut and the interabdominal seam is at the left. Arrows show direction of growth and shading indicates bone. The left column (from the top) shows how major (l) and minor (ml) growth rings are formed. The right column shows formation of major growth rings (1-3). Multiple layers of epidermis are shed together and the grooves remaining after ecdysis are only impressions of growth rings. (From Moll and Legler, 1971.)

attains full growth, or grows irregularly, these rings lose validity as age indicators. Miller (1955) and Woodbury and Hardy (1948) concluded that the growth rings of *Gopherus agassizi* are unsuitable for age estimation. The neotropical slider, *P. scripta*, forms several major growth rings each year (Moll and Legler, 1971). Thus for this species ring counting would produce erroneous estimates of age.

Recognizing that scute-shedders lose rings with time, Sexton (1959a) proposed an alternative technique of age estimation for the midland painted turtle, *C. p. marginata*. He showed that it was possible to accurately estimate the age at which discernible growth rings had been

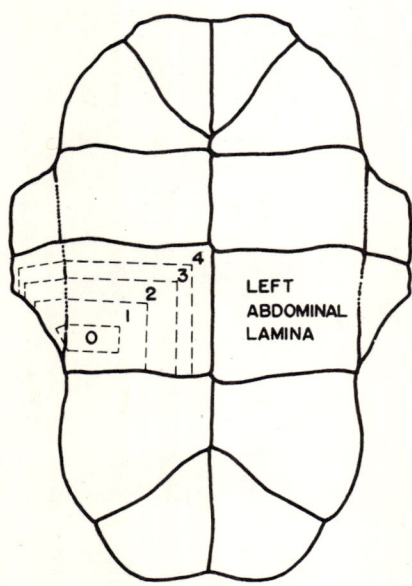

Figure 2 Diagram of the plastron of *Chrysemys picta* showing growth rings on right abdominal lamina only. This turtle is in the fourth season of growth after emergence from the nest. (From Sexton, 1959a.)

formed, even when earlier growth rings were lacking or obliterated. His estimates were based on turtles having complete sets of rings. Using the means and standard deviations of these measurements, he determined ages of growth rings in animals with incomplete sets. He used the growth rings on the right abdominal lamina only (Figure 2), and listed the medial lengths of these growth rings (Table 1).

Table 1 The Medial Border Lengths of Growth Rings on the Right Abdominal Lamina of Female *Chrysemys picta marginata*[a]

Ring of Growing Season	Range of Ring (cm)	Mean Ring Length (cm)	Standard Deviation	Number of Animals Measured
1	0.9–1.4	1.14	0.12	62
2	1.2–1.8	1.54	0.14	62
3	1.5–2.3	1.86	0.17	62
4	1.8–2.5	2.11	0.18	49
5	2.0–2.75	2.34	0.17	26
6	2.1–2.9	2.56	0.17	41
7	2.45–3.1	2.78	0.19	23
8	2.7–3.4	3.01	0.21	15

[a]Modified from Sexton (1959a).

TECHNIQUES

To illustrate Sexton's method, consider that a female *C. p. marginata* has three visible growth rings on her right abdominal lamina. Their medial border lengths are 2.6, 2.8, and 3.0 cm. If we compare these values with those for females in Table 1, we can identify the rings as growth rings 6 through 8. Though it is not always this simple to identify the rings, the ages of all turtles could be estimated to within a year, which is far better than being able to state only that an animal is over 3 years old or is in some other broad age category. The more visible annuli present, the greater the accuracy of this method. Large adults with few clear rings can be aged only tentatively, but at least the approximate age range can be established. The use of this technique (Sexton, 1959a) is based on the following assumptions:

1 There is a detectable growth increment each year.
2 Only one major growth ring is added each year.
3 Major growth rings are not lost.
4 The medial border length of a major growth ring does not change after its formation.

It should be obvious from the foregoing discussion that some of these assumptions may not be met in certain populations. Caution should always be exercised to ensure that the assumptions are valid for any given population *before* this technique is employed.

Sexton (1959a) concluded that since growth rate varies from place to place, his age-size data were valid only for the population he studied and that separate age-size tables should be constructed for any turtle population that is to be studied demographically. Moll and Legler (1971) prepared such a table and used it to estimate the ages of sliders in which some growth rings had been lost.

A fourth turtle aging procedure, suggested by Mattox (1935), relies on annular rings in the long bones of the appendages. He demonstrated a correlation between the number of rings in polished sections of the humerus and femur of *C. p. marginata* and the size of the individual. From this he felt it tenable to correlate these rings with turtle age. Such correlation has not yet been demonstrated. Dobie (1971) found little correlation between bone annulus count and the number of growth rings on the shell scutes of the alligator snapping turtle, *Macroclemys temmincki*. He used vertebrae and mandibles, however, and since Mattox (1935) reported poor results with mandibles, this negative finding is not surprising.

Sex Determination

Analysis of turtle population structure is based on the division of the population into age and sex classes. This procedure allows for definition

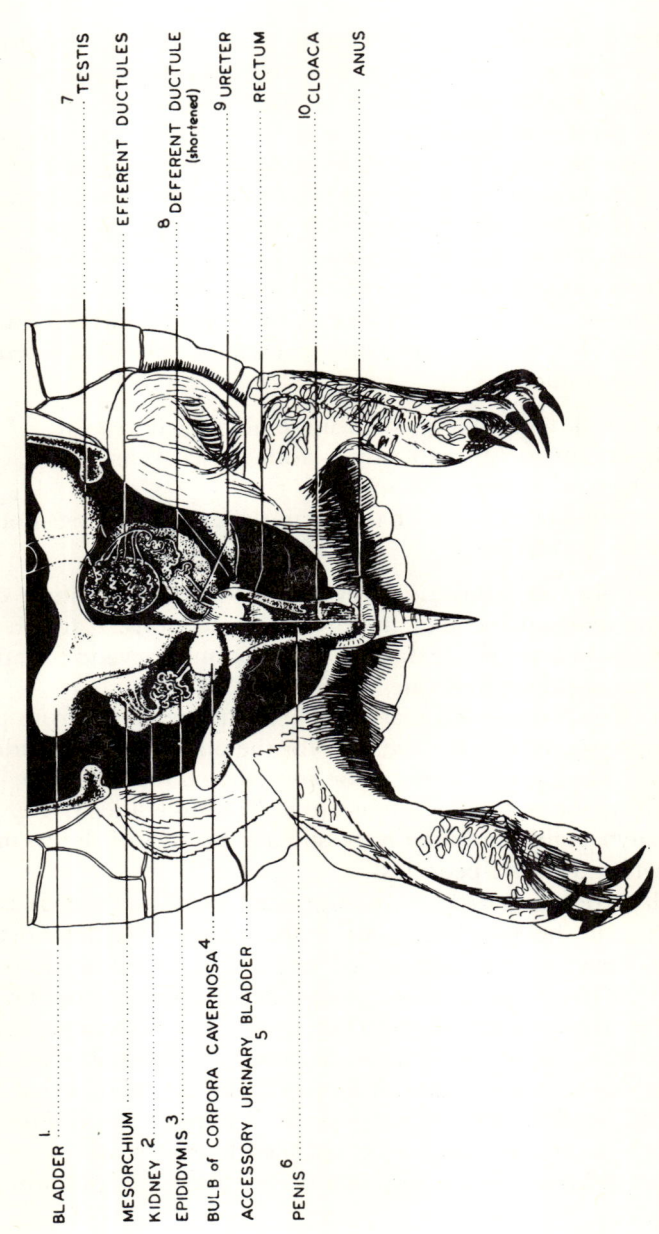

Figure 3 Ventral view of urogenital systems of male (top) and female (bottom) *Chrysemys picta*. (Adapted from Ashley, 1955.)

Table 2 Sex Characters in Selected Turtles[a]

Species	Criteria for Sexing[b]
Chelydra serpentina	PREANAL TAIL LENGTH: greater than 86% of posterior plastral lobe length; *less than 86% of posterior plastral lobe*. ANAL OPENING: posterior to rear carapacial margin; *anterior to rear carapacial margin*
Sternotherus odoratus	INNER HINDLEG SURFACE: has a small patch of tilted scales; *is relatively smooth*. TAIL: long and stout, ending in a blunt nail; *short and slender, lacks terminal nail*. SKIN AREA BETWEEN PLASTRAL SEAMS: broad; *narrow*
Clemmys insculpta	PREANAL TAIL LENGTH: twice as long as that of female; *anal opening anterior to carapacial margin*. PLASTRON: concave; *flat*. POSTERIOR PLASTRAL MARGIN: *more nearly approaches rear carapacial margin*. FORELIMB SCALES: prominent; *reduced*
Terrapene carolina	IRIS: red; *yellow-brown*. POSTERIOR PLASTRAL LOBE: concave; *flat or a bit convex*. TAIL: long and stout; *short and tapering narrowly*. HINDLIMB NAILS: short, strong, curved; *long, straight, more slender*. PLASTRON LENGTH: usually exceeding 140 mm; *usually less than 140 mm*
Malaclemys terrapin	CARAPACE LENGTH: 100–140 mm; *150–230 mm*. ANAL OPENING: posterior to rear carapacial margin, *anterior to rear carapacial margin*. HEAD SHAPE: dorsal outline narrow, more pointed; *dorsal outline blunt, broader*. CARAPACE SHAPE: more pointed behind; *more rounded behind*
Graptemys barbouri	CARAPACE LENGTH: 90–130 mm; *170–260 mm*. ANAL OPENING: posterior to rear carapacial margin; *anterior to rear carapacial margin*. TAIL: long and thick; *short and slender*
Chrysemys picta	FORECLAWS: elongated; *considerably shorter*. ANAL OPENING: posterior to rear carapacial margin; *anterior to rear carapacial margin*. CARAPACE HEIGHT: depressed; *more domed*. OVERALL SIZE: smaller in all shell dimensions than the female

TECHNIQUES

Table 2 *(Continued)*

Species	Criteria for Sexing[b]
Deirochelys reticularia	TAIL: long and thick; *short and slender*. SIZE: *females larger than males*. ANAL OPENING: posterior to carapacial margin; *anterior to carapacial margin*
Gopherus agassizi	GULAR PROJECTIONS: longer; *shorter*. CLAWS: massive; *lighter*. PLASTRON: concave; *flattened*. TAIL: thick and long; *short and slender*. OVERALL SIZE: males larger than females
Chelonia mydas	CARAPACE: tapering more posteriorly; *more rounded posteriorly*. REAR PLASTRAL LOBE: narrow; *wide*. TAIL: vertically prehensile, tipped with a heavy nail, extends far beyond rear of carapace; *barely reaches rear edge of carapace*
Trionyx spiniferus	CARAPACIAL PATTERN: retain the juvenile dark ocelli or white dots; *lose juvenile pattern and develop mottled lichenlike markings on the carapace*. CARAPACE LENGTH: 127–216 mm; *165–457 mm*. TAIL: long and thick with anal opening near tip; *short and slender with anal opening farther from the tip*

[a]Adapted from Carr (1952); Ernst and Barbour (1972).
[b]Female characters are italicized.

of breeding females, yearlings, and older age classes. With this information we can evaluate turtle production and survival, which are essential aspects of an analysis of population dynamics.

Ashley (1955) presented detailed instructions for the laboratory dissection of the reproductive system of turtles (genus *Chrysemys*). Figure 3 illustrates ventral views of both male and female urogenital systems taken from his manual. According to Ashley, the reproductive anatomy is essentially the same for all chelonian species. Sexing by examination of the internally located gonads is normally certain, even in fairly young specimens. The drawback to this procedure is that the animals must be sacrificed, then inspected after dissection.

The use of secondary sex features, rather than dissection, is advisable whenever valid secondary sex characters have been established. Among the general examples of sexual dimorphism are the longer, stouter male tail, the more flexible and often prehensile male tail, a specialized scale on the tip of the male tail, a concave male plastron, elongated nails on the forefeet of some males, secondary melanism in some old males, an enlarged in-curved nail on the flippers of some marine males, and size dimorphism in some species (in some species males are larger, whereas in some others they are significantly smaller). Table 2 is a compilation of criteria for sex determination of selected species.

It is unfortunate that acceptable means are not available for externally sexing juvenile turtles. It is possible that thorough analyses of preserved specimens in existing collections might reveal additional minor external morphological differences correlated with anatomical sex but not dependent on sexual maturity. Perhaps some aspect of blood morphology shows sexual dimorphism, or a sexing technique involving microscopic scanning of the cloaca might be developed. Research in this area is greatly needed.

Reproductive Potential

Data on reproductive potential may be found in Chapter 16, but illustrative studies and definitions are included here.

Annual reproductive potential is represented by the total number of eggs a female turtle is capable of laying in a year (Gibbons, 1968d). An approximation of the total egg number laid each year can be estimated for an entire population if the mean annual reproductive potential is multiplied by the number of mature females estimated for the population (Gibbons, 1968c). Figure 4 illustrates the potential rate of increase of a pair of *Pseudemys scripta* from two different habitats (Gibbons, 1970b). Faster growth, hence earlier maturity, and larger body size, hence larger clutch size, may be responsible for the theoretical increase in birthrate predicted for the Par Pond population living in warmer waters of a nuclear reactor reservoir. The assumptions accompanying the figure may not be valid, but they do allow us to compare the potential reproductive rates of the two populations.

If a turtle nests only once each year and clutch size can be determined, it is a relatively simple matter to determine reproductive potential. Multiple clutches are not uncommon, however, and are the rule for some species (see Chapter 16). In some other species there is latitudinal variation, with southern populations producing more than one clutch each year. Other species consistently produce no more than one clutch annually, but may show considerable latitudinal variation in number of

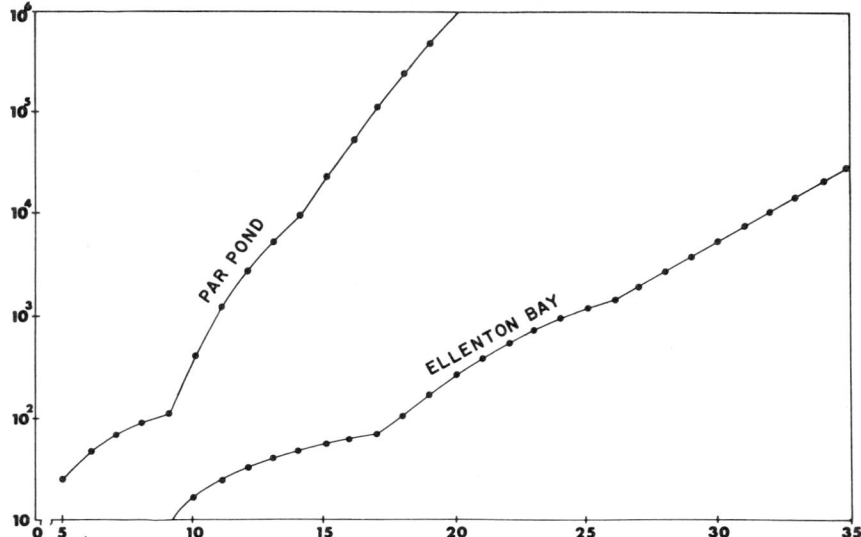

Figure 4 Potential rates (abscissa, years; ordinate, total number) of increase of a pair of *P. scripta* from two different habitats in South Carolina based on five assumptions: (1) average clutch sizes for Ellenton Bay and for Par Pond are 3.99 and 11.62, respectively, (2) individuals from both populations average two clutches per year, (3) age at maturity is 9 years for Ellenton Bay females and 5 years for those from Par Pond, (4) sex ratios are 1:1, and (5) mortality is zero for all age classes in both populations. (From Gibbons, 1970b.)

eggs per clutch, making it difficult to generalize from one population to a second at a different latitude.

Tinkle (1961) indicated that as a general rule maximum reproductive potential can be calculated by adding the number of large yolk-filled follicles that could be ovulated in one season to the number of eggs already present in the oviduct and/or the number of ovulation scars in the ovaries. In his work with *Sternotherus odoratus*, which lays a single clutch throughout its range, Tinkle demonstrated that reproductive potential increases as samples are taken farther north. For *S. odoratus* taken early in the year, reproductive potential was calculated solely by counting the enlarged, potentially ovulatory follicles. During the ovulation-nesting season reproductive potential was estimated by adding the number of oviducal eggs or corpora lutea to the number of remaining large follicles. Tinkle stated that postreproductive females did not provide reliable information because the corpora lutea, which would have indicated the number of eggs ovulated, had disappeared, and the follicles representing the next year's egg complement were not yet enlarged. He concluded that there is probably no means by which the reproductive potential can be calculated exactly; however the suggested counts of follicles, lutea, and eggs will give an estimate of maximum egg production for purposes of comparison.

Tinkle (1961) used the term "reproductive potential" in the sense of "partial potential reproductive capacity" of Allee, Emerson, Park, Park, and Schmidt (1949), who wrote: "partial potential is the maximum reproduction possible for the species population under a given set of conditions" (Allee et al., 1949, p. 272). This is opposed to "absolute potential," which is the maximum reproduction possible for a population under ideally optimal ecological conditions.

Survivorship

Deevey (1947) said that if one studies a cohort of individuals born together, survivorship of any age class in the population is derived by subtracting deaths of the previous age interval from the number alive at the beginning of that previous interval. To calculate survivorship in any population, the initial population size or the number of individuals at age 0 must be known (Organ, 1961). Usually the number of hatchlings produced by a stable population is taken as the number of individuals at age 0.

When the number of survivors is plotted on the ordinate and the age interval on the abscissa, the resulting curve is called a survivorship curve. Such curves are valuable in furthering our understanding of population dynamics. They reveal the age interval at which the animal is most vulnerable. Alteration of mortality during such critical intervals could conceivably have a profound influence on the future density and structure of the population. Our knowledge of the actual survival of early age classes of turtles is meager, however.

Hirth and Schaffer (1974) calculated theoretical minimum survival rates of hatchling *Chelonia mydas* necessary to maintain a stable, constant size population. Since many of the assumptions necessary to their calculations were theoretical and admittedly unrealistic, the minimum survival rates necessary to maintain stable populations in their examples were probably underestimated. Nonetheless their approach serves as an excellent basis for modeling hatchling survival.

Gibbons (1968c) estimated that 2% of the eggs in a Michigan population of *Chrysemys picta* developed into juveniles, which were then recruited into the population. He obtained this estimate by first approximating the total number of eggs laid by a population in a year, then calculating what percent of this egg total produced juveniles. To accomplish the latter he first estimated the number of juveniles in the population using a mark-release-recapture procedure. Gibbons suggested that the survival pattern from the egg stage to maturity was similar to the type IV curve of survivorship (Slobodkin, 1961).

Another approach to estimation of survival rate was taken by Yahner (1974). He approximated survival rates for different weight classes of *Terrapene carolina* by calculating the percentage recovery of animals originally found in 1968–1969, then recovered in 1972. One drawback to this approach is its reliance on the assumption that all previously captured individuals are still present in the sampling area and have a 1.00 probability of being caught.

Population Size

Although a very rough idea of population size can sometimes be derived by hand collecting and basking censuses, more refined procedures are required to obtain a precise population estimate. Many techniques are available for the estimation of population size. An excellent coverage of these methods, including their mechanics, application, and suitability was presented by Overton (1971). The estimation method selected must be carefully chosen for the particular species, time, and study area; not all techniques are applicable to turtle research.

All methods of turtle population estimation make two fundamental assumptions that must be taken into account (Davis, 1963):

1 Recruitment (including natality) and mortality (including emigration) during the period of data collection are negligible, or a correction for these effects can be made.
2 All individuals in the population have an equal (or known) likelihood of being counted. This means that individuals must not become trap shy or trap addicted in live-trapping studies; they must mix randomly in the population after release; they must not group according to size or sex. In preparing a population estimate, it is vital that the investigator determine whether these assumptions are satisfactorily met; if they are not met, the estimate will be erroneous.

Of the general approaches to determination of population size, the two most applicable with turtles are a true census and a sampling estimate.

The true census approach is probably the most obvious method for determination of population size. Unfortunately this technique is not normally practical because it is rarely possible to locate all individuals in a natural population. One could directly count turtles in random sample plots and extrapolate these to the entire study area. This approach is not recommended, however, unless the animals are known to be randomly distributed throughout the entire area. But random distribution is rare in natural populations: it occurs only where the environment is uniform and the animals have no tendency to aggregate. Some degree of clumping is undoubtedly the most common type of dispersion (internal distribution pattern) found in nature. Odum (1971) stated: "Determination of the type of distribution, of the degree of clumping (if any), and of the size and permanence of groups are necessary if a real understanding of the nature of the population is to be obtained, and especially if density is to be correctly measured." Direct counts of turtles can be made in areas where they are concentrated (e.g., in hibernacula or on nesting grounds). Direct counts can also sometimes be made in confined situations or in water bodies that are enclosed and then drained.

Probably the most important of the sampling estimate methods are the mark-recapture techniques. Methods of turtle capture and marking are discussed in Chapter 2. The formula one chooses for the estimation must be consistent with the properties of the capture and marking technique used (Overton, 1971). If the investigator is cognizant of the assumptions on which a particular method is based, he can critically evaluate the specific requirements of his capture and marking effort. As discussed in Chapter 2, many types of trapping lack the capacity to reflect turtle populations in a truly representative fashion. Ernst (1971c) indicated that if several collecting methods are used simultaneously, the bias caused by any one method can be overcome. Another way to circumvent capture bias in population estimation is to use a different capture technique for each capture trial (Overton, 1971).

The mark-recapture method is based on initially capturing, marking, releasing, and later recapturing animals. Ideally all the marking and releasing should be done at one time and the recapturing at another time. The interval between the release and recapture should not be too great; a few days is usually sufficient. Assumptions basic to mark-recapture methods are as follows:

1 Turtles do not lose their marks.
2 There is no appreciable recruitment (births or immigration) during the study period.
3 There is no difference in the mortality of marked and unmarked animals.
4 Marked and unmarked turtles have an equal likelihood of being captured.

While the method is straightforward, the habits of a species often do not fulfill the assumptions (Davis, 1963). Assumption 1 is usually valid provided a permanent mark has been applied. Assumption 2 may be true for short periods, particularly in physically isolated populations during nonbreeding times. Assumption 3 is most often true, but should be carefully checked. Assumption 4 is difficult to assess in that trap shyness (or addiction), territorial behavior, and nonrandom population dispersion all negate it.

Tinkle (1958) pointed out some shortcomings of the mark-recapture technique he applied to river populations of turtles. At one site population estimation was precluded because only 2 of 36 marked animals were recaptured. Marked turtles may have left the area or become more wary; continual disturbance of the area may have frightened turtles away; or the number of turtles in the area may have been so great that the chances of recapture were remote. Tinkle concluded, however, that with refinement of the mark-recapture technique, reliable estimates of river turtle population size might be obtained. He used the modified Lincoln Index which is based on the very simple ratio:

$$N/M = n/m \quad \text{or} \quad N = Mn/m$$

where the population total N is shown to be related to the number marked and released M in the same way that the total captured later n is related to the number marked recaptured m. The population estimate is calculated from the equation by substitution of M, the original number marked and released, n, the total captured during the census period, and m, the number of marked animals recaptured during the census period (Davis, 1963).

Moll and Legler (1971) used a version of the Lincoln Index to obtain a rough estimate of population size in *Pseudemys scripta* from Panama. Ernst (1971c) used the original form of the Lincoln Index to estimate population size of *Chrysemys picta* in Pennsylvania. Ernst indicated that few hatchlings and juveniles were collected, probably because of their secretive habits, and that this caused a low Lincoln Index estimation.

Gibbons (1968c) used the Lincoln Index to estimate the number of *C. picta* in different age, sex, and size classes. He averaged the estimates for six collecting periods to get his final estimate for each category. Because the assumption of random mixing of marked individuals into the population cannot be met in natural populations, he collected more intensely in certain places within the study area; thus more marked turtles were released in some places than in others. To minimize the error caused by nonrandom redistribution, he later collected representatively in all parts

Table 3 The Schnabel Method of Population Estimation Based on Data for New Hampshire *Chrysemys picta*[a]

Day	Number Trapped, $u + r$	Marked Animals in Area, m	$m(u + r)$	$\Sigma m(u + r)$	Recaptures, r	Σr	Estimated Population, P_s
1	2	—	—	—	—	—	—
2	4	2	8	8	0	—	—
3	1	6	6	14	0	—	—
4	2	7	14	28	0	—	—
5	2	9	18	46	1	1	46
6	2	10	20	66	0	1	66
7	5	12	60	126	2	3	42
8	2	15	30	156	1	4	39
9	5	16	80	236	0	4	59
10	7	21	147	383	2	6	64
11	4	26	104	487	0	6	81
12	2	30	60	547	0	6	91
13	3	32	96	643	1	7	92
14	2	34	68	711	1	8	89
15	3	35	105	816	2	10	82
16	3	36	108	924	0	10	92
17	7	39	273	1197	4	14	86
18	6	42	252	1449	0	14	104
19	8	48	384	1833	3	17	108
20	5	53	265	2098	0	17	123
21	8	58	464	2562	6	23	111
22	10	60	600	3162	7	30	105
23	7	63	441	3603	6	36	100
24	8	64	512	4115	4	40	103

[a]Graham (unpublished).

of the study area. For further discussion of the Lincoln Index method, see Hayne (1949).

A variation of the Lincoln Index approach involves accumulation of the captures and recaptures over a period of time. There are several ways in which this can be done (see Overton, 1971). With the Schnabel (1938) method, animals are captured, marked, and released daily. For each day a record is kept of all animals caught, the number of recaptures, and the number of animals marked. The actual method of calculation is basically the same as that of the Lincoln Index, but in the Schnabel method M gets progressively larger. Population estimates can be calculated by this technique on a daily basis. The Schnabel formula is given as follows:

$$P = \frac{\Sigma m(u + r)}{\Sigma r}$$

where P is the population estimate, m the number captured, marked, and released, r the number recaptured each day, and u the number of unmarked individuals captured each day. The summation in the formula is over the number of days/trials. Table 3 illustrates the mechanics of the method.

One drawback of the Schnabel method is that since calculation of the standard error is quite complicated, the confidence limits of the estimate cannot be determined easily (for details see Ricker, 1958). According to Davis (1963), the Schnabel method smooths the data by accumulation and makes the estimates quite uniform. When Table 3 is examined, the question remains, What is the best way to average the estimates? A rather complicated answer was given by Overton (1971) and is not discussed here. A further question remains: When can one stop sampling? Robson and Regier (1964) discussed this question and presented a useful graph for determination of appropriate sample size under various conditions.

In conclusion, most estimation techniques have low precision; that is, the variation of the estimates is high. In spite of this, low precision estimates are better than none at all. As Gibbons (1968c) reminds us, an estimate is a rough approximation of population size, not a calculated absolute.

Movements

In some instances it is possible to make continual direct observation or capture-release-recapture studies on individual turtles and thereby ascertain their spatial and temporal movements, but this routine can be monotonous and time-consuming in the extreme. In addition, direct observation is seldom either practical or possible. A few basic methods, together with the more lengthy approaches previously mentioned, can be applied to the study of home range, annual and diel activity, homing, rate of travel, migration, emigration, and immigration.

Some studies have employed the capture-release-recapture method to analyze aquatic turtle movement (Cagle, 1944b; Pearse, 1923a; Sexton, 1959b; Williams, 1952). The relatively simple approach used by these workers consisted of capture; marking; recording the exact location of capture; release; and one or more subsequent recaptures (with the location of each recapture recorded exactly). When the recorded locations are plotted on a map or grid system, a picture of the pattern of movements is constructed.

Other methods of movement analysis include the use of thread-trailing devices (e.g., Breder, 1927; Emlen, 1969; Legler, 1960; Stickel, 1950), sonic telemetry tags (e.g., Moll and Legler, 1971), radioactive pin tags (e.g., Bennett, Gibbons, and Franson, 1970); and drift fence traps (e.g., Sexton, 1959; Gibbons, 1970d). The last technique seems to be particularly valuable in the study of juvenile turtles. Details and additional studies may be found in Chapters 2 and 3.

Growth

Estimation and absolute detemination are the two fundamental kinds of analysis of growth. The steps for absolute determination are as follows:

1 Capture, measure, and mark.
2 Release.
3 Recapture and remeasure to directly determine the growth increment, if any.

Normally the measurement taken is the total length of carapace or plastron (Figure 5), using a vernier dial caliper accurate to 0.1 mm. This

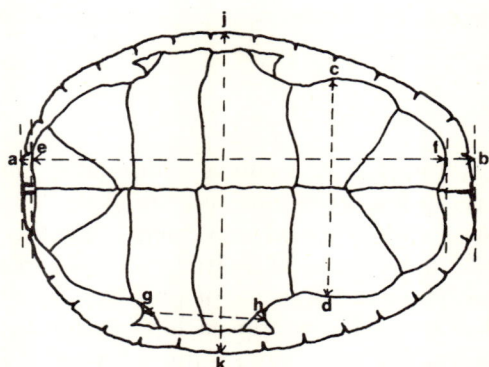

Figure 5 Method of measuring a turtle shell: carapace length, a–b; carapace width, j–k; plastron length, e–f; plastral hind lobe width, c–d; length of bridge, g–h.

method of growth resolution has been employed, for example, on *Trionyx ferox* (Breckenridge, 1955), *Pseudemys scripta* (Cagle, 1946; Moll and Legler, 1971), *Malaclemys terrapin* (Hildebrand, 1932), *Chrysemys picta* (Pearse, 1923a; Sexton, 1965), and *Sternotherus odoratus* (Risley, 1933b). Of course absolute growth measurement can be performed on captive as well as wild animals but, as with age data, the results must be carefully considered, since they reflect growth under artificial conditions. Measurement of absolute growth is the most exact method of assessing growth; that is, it lacks much of the error inherent in growth estimation techniques. Drawbacks to the method include the amount of time consumed in the process, failure to recapture some initially measured animals, and the difference in growth intervals between release and recapture for each turtle, usually found if all the individuals are not released on the same day and recaptured on a single subsequent day. Otherwise, growth comparison of individuals is quite difficult.

The procedure of growth rate estimation in turtles was first demonstrated by Sergeev (1937), who used a formula widely employed by ichthyologists. This technique used growth rings, plastral laminae lengths, and plastron length computed by summing the median border lengths of all plastral laminae on one side of the animal.

Cagle (1946, 1954) stated that the relative lengths of the abdominal laminae and the plastron remain approximately the same throughout life in *P. scripta* and *C. picta*. In addition, he suggested the use of plastral laminae in growth studies because plastral growth rings are formed more in approximation to a plane than are carapacial growth rings. If the formula adapted by Sergeev (1937) is to be used, the assumption that no ontogenetic change in the ratio of the selected plastral lamina to overall plastron length must hold. Moll and Legler (1971), for example, found that in *P. scripta* from Panama the abdominal lamina got relatively larger with age; therefore they utilized a modification of Sergeev's formula.

In the summation method (Cagle, 1946) medial border lengths of growth rings on all right (or left) plastral laminae are summed for a given age class (growth ring number) to estimate plastron length. Figure 6 illustrates the method of mensuration required. Since the summation method is not based on the assumption of ontogenetic constancy in the ratio of lamina length to plastron length, it is generally a more reliable method of growth estimation than that of Sergeev (1937). Growth estimation by summation has been performed on *P. scripta* (Cagle, 1946), *Chelydra serpentina* (Gibbons, 1968a), *Clemmys guttata* (Graham, 1970), and *P. rubriventris* (Graham, 1971b).

In conclusion, some turtle species are well suited for growth study

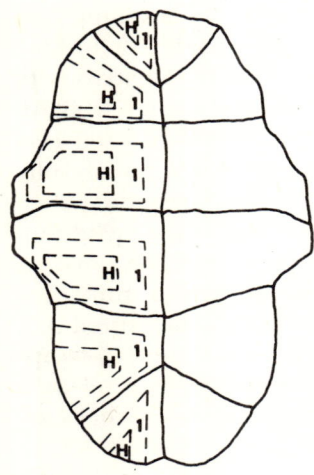

Figure 6 Summation method of estimating plastron growth. The lengths of the median borders of all growth rings of the same year class are summed to estimate plastron length at hatching and the end of each growing season thereafter. The median border of each ring is that portion parallel to the longitudinal axis of the shell. The individual shown here is in its second growing season. H indicates median borders of hatchling rings, and 1 shows median borders of rings formed at the end of the first growing season.

because both their age (from growth rings) and their size at a given age (from plastral measurement) can be accurately established.

Food Habits

The food a turtle eats depends not only on what the individual prefers but also on what is available. The person intending to analyze turtle food habits should attempt to get some idea of food preferences and availability before beginning the detailed investigation. Field observation of feeding turtles is probably the best way to acquire this preliminary information, and after it is completed animals can be collected and removed to the laboratory for study. In addition, preliminary observation of feeding habits (where, when, and how feeding occurs) is valuable. For example, if one knows when an animal feeds, specimens can be collected and examined during or just after that time. In this way the digestion of the food will be minimal, and this will allow for an easier identification of each food item. The laboratory work is normally time-consuming and the investigator needs a wealth of technical information to properly identify the food items found. Counts of specific food items, the percentage of each type of food in the diet, and other measurements are routinely taken (Korschgen, 1971).

The three basic ways to sample ingested food items from turtles are: (1) sacrifice of the animal and removal of its digestive tract, (2) examination of fecal material, and (3) back-flushing material from the stomachs of live animals. Each approach has advantages. Method 1 probably yields the best quantitative results, but it is becoming increasingly undesirable to sacrifice animals, as this method requires. Folkerts (1968) suggested that examination of turtle feces has the advantages that the animals can be marked and released for further studies, space and preservatives are saved, and collection of fecal material is less time-consuming than dissection. However, items that leave no trace in the feces are overlooked, meaningful volumetric analysis is precluded, and fecal remains of food items are often difficult to identify. The back-flushing method has been used by Legler (1977) on unanesthetized turtles. The basis of the technique is to displace the stomach contents out through the esophagus with water injected from a continuous pump syringe into the stomach by way of a flexible tube inserted through the mouth and esophagus. With this procedure animals need not be sacrificed; moreover, they can be released after sampling and subsequently resampled and rapid examination (20 per hour) is possible (Legler, 1977). The only apparent drawback to this method is that one cannot be certain that the entire stomach contents have been flushed out, but this could be tested initially by immediate dissection of a few back-flushed animals. With refinement, this useful technique should receive wide acceptance; anesthesia may be incorporated (see Chapter 7).

Regardless of the method used to obtain the food sample, the next step is the actual quantitative and qualitative study of the food material. Probably the most frequently used combination of quantitative techniques includes the numerical method, the frequency of occurrence method, and the volumetric method. When used separately, each approach has its drawbacks; when they are combined these shortcomings are minimized.

Data should demonstrate what foods are eaten, the number of individuals of each kind of food eaten, the frequency with which each category of food is eaten, and the portion of the total comprised by each kind (Lagler, 1956). The study of food habits produces more meaningful results from large samples. This suggests that it is best to obtain an exhaustive sample at the outset. However common sense should dictate that, particularly where animals are to be sacrificed, the sample should be no larger than is absolutely necessary.

This rather brief coverage of life history techniques is by no means a complete review of all possible methods, but it is hoped that the details of the selected techniques reviewed here will serve as a starting point for students of turtle life history and as a brief survey for turtle researchers not familiar with field techniques.

CHAPTER 5
Photographic Analysis

CHARLES C. CARPENTER

I hope that this chapter will be of assistance to others in their attempts to use certain photographic techniques in the study of reptile behavior. I do not profess to be an expert on the subject of photography; rather I have used cameras, primarily motion picture cameras, as research tools and have found the results to be useful in many ways.

CAMERAS

Though a still camera can record events, the motion picture camera, in addition, can capture sequences of actions, stop actions, measure the speed of actions and events (time-motion), record simultaneous events and sequences, and isolate acts and act systems (Russel, Meade, and Hayes, 1954).

The studies that can be aided by motion pictures and some of the types of question that can find possible answers through the use of motion pictures are many. Functional morphology is one of these areas; the motion picture can record locomotion and gait (Gans, 1962; Urban, 1965; Walker, 1971a), appendage use (Synder, 1952), trunk involvement (Bustard, 1969), the use of specific organs (Ulinski, 1972), and feeding movements (Frazzetta, 1962), and it can give insight into locomotor adaptations (Barclay, 1946; Ireland and Gans, 1972; Synder, 1952). The ontogenetic changes of such actions can be recorded and compared, along with changes due to season, temperature, social position, and other factors. I have used this technique to record postures, postural changes, and movement sequences through time to compare many species of lizards (Carpenter, 1961, 1962, 1963, 1965, 1966a, 1966b, 1967, 1969; Carpenter and Grubitz, 1961; Carpenter and Mosley, 1967; Carpenter, Badham, and Kimble, 1970; Ferguson, 1971; Gorman, 1968; Jensen, 1971; Kastle, 1963, 1965; Parcher, 1974). Courtship sequences can be analyzed (Ferguson, 1970; Weaver, 1970). Color and color changes can be documented. With the proper equipment, one can record and correlate particular movements with particular sounds (Conant, 1973; Whitson, 1971).

Still cameras, especially the available 35 mm reflex varieties, have the advantage of ease in handling, versatility in speed under poor light, and close working distance; moreover, they can be equipped with powerful and fast telephoto lenses, which may be necessary for obtaining adequate pictures in the field. However only intermittent events can be captured, and actions occurring between shots are missed (Evans, 1953).

The motion picture cameras generally used for animal behavior work are the 16, 8, and super-8 mm reflex types that permit viewing the animal as it is being photographed. The 16 mm equipment is larger and heavier, and the film is more expensive, but it offers advantages over the smaller sizes in presenting a larger picture with greater resolution, an important feature in detailed analysis of behavior.

A motion picture camera should be mounted on a good steady tripod if the resulting film is to be used for detailed analysis, for when pictures are taken by hand holding the camera, the inevitable jiggle can make qualitative analysis very difficult.

The older, and still available, spring-wind movie cameras usually have the disadvantage of needing to be wound at least twice for each full roll of film, and this may cause an interruption and missing of a behavior sequence. Fortunately many motion picture cameras are equipped with power drives, that is, a motor to drive the mechanism that works the shutter and the film reels and indirectly controls film speed. Some have these power packs built into the camera. The power packs are usually rechargeable, and this necessitates a power source for recharging after a certain number of hours of use. The better power-driven movie cameras are equipped with a tachometer that permits the operator to see the speed of the drive (frames per second). The cameras also have frame counters that can be used in making records of particular sequences. Motion picture cameras can be equipped with excellent zoom lenses, which generally eliminate the need for another lens and add great versatility in moving in and out for close up details and broad views of the action being photographed.

FILM

The speed (frames per second) at which the film is taken will determine the amount of blur obtained. If the animal moves very rapidly and the film is taken at 16 mm/sec (usually the slowest speed for taking acceptable movies), blurred motion will be encountered when the film is projected frame by frame. Thus for very rapid motion, taking pictures at faster speeds is required for detailed analysis of the object. Slow motion (on projection) is accomplished by taking the movies at rates of speed of 32 frames per second and greater. The faster you take your film, the more you can slow down action, but the faster you use up your film, the more light is necessary and usually a more expensive camera, and film costs are higher because the film is used up faster. But this is the only way that certain behaviors can be captured for detailed analysis (Frazzetta, 1962; Urban, 1965).

The choice of film depends on the particular conditions, and relevant information may be obtained from standard sources of photographic information. For behavior work, I prefer color film because the color patterns are often important in interpreting behavior sequences and present a truer picture of an event. Filming in bright sunlight, I have used a color film with a slow speed, since faster color films may be too fast for so much light.

In some behavior work it has proved helpful to simulate behavior patterns by producing an animated film. By the use of the animation technique, one can select the actions to be emphasized, or one can add an animated graph of the action as it proceeds. The techniques of animation are not easy or inexpensive, but once mastered, they provide a unique method for presenting comparative behavior materials (Carpenter and Mosley, 1967).

VIDEOTAPE

A very useful tool is a closed-circuit television system. This system includes a power zoom lens television camera mounted on a pan and tilt base. The pan and tilt permit the observer to follow moving animals with the television camera. The camera, zoom lens, and pan and tilt are all mounted on one cart and manipulated from a control module mounted on another cart. All can easily be moved for use at different observation areas in the laboratory. The television monitor is jacked into the television camera as well as a ½ in. videotape recorder. Two variations of this system are (a) having separate monitors in different laboratories (some television cameras have small monitors built into the camera) and (*b*) using a remote arrangement (i.e., the control module and monitor in my office, with the camera, pan, and tilt in an environmental chamber). With the latter setup, the observer can monitor behavior remotely in a closed chamber with no disturbance to the animals other than the camera. The videotape recorder is kept close at hand to switch on and record activity if appropriate.

Videotape does not give the resolution for analysis offered by motion pictures and it is less reliable as a method for measuring time sequences. The television system initially is much more expensive to obtain; once purchased, however, it is cheaper to operate in that the videotapes can be used, erased, and used repeatedly (Davis and Jackson, 1970; Jackson and Davis, 1972a, 1972b).

Thus the motion picture camera is used to get permanent, high resolution records for detailed analysis, whereas the videotape is used to capture repeated sequences of behavior for quantitative comparison of sequences and events.

Television cameras are very sensitive to light, and we have observed behavior of rats under a 100 W red light bulb on the monitor and recorded this behavior on fast motion picture film by taking movies of the action on the monitor screen (Carpenter and Stalling, 1968). The latter would not be necessary if a videotape recorder were available.

RECORDS

It is very important to develop and maintain a good system of recording data for each film. These data should include (when known) data, time of day, place, species, sex, age, individual animal, and if known or important, social position. If possible, record the context of the pictures being taken (courtship, aggression, feeding, etc.). If such data are not recorded when known, much information of value may be lost, and later use of the film for other unsuspected purposes may be impaired.

As films begin to accumulate, it becomes necessary to establish a catalog to locate a particular film and to know what is present on each film without projecting it. I have tried two methods of cataloguing. The first method involved cutting and splicing together, in chronological sequence, all of the film pertaining to one species or to one type of behavior. In doing this it is very important to keep the information records straight and in sequence with the film sequence. Splicing film involves considerable time in editing the film for cutting.

The second method, which I now use, is to keep each roll of film intact and numbered chronologically by date taken, some rolls often including shots of a number of species. I keep a looseleaf catalog of species, and under each species, the roll of film and footage where the shots of each species can be found. With the availability of the Vanguard Motion Analyzer, I now review each roll of film as it is returned from being developed, and record the sequence of frame numbers pertaining to each species and type of behavior. Thus I can return to any particular frame on any particular film roll, for further observation. Intact behaviors can be analyzed and reanalyzed as additional data are collected. Interpretations made originally may change as the same animals are filmed at a later time or as additional members of the same or related species are studied.

STORAGE OF FILM

Do not permit stored film to get hot (85°F and above) or cold (below 60°F). I use old refrigerator cabinets for storage, installing appropriate shelves or racks, and filing the films by species or by number. Storing each roll in a can protects them all from dust.

I find it best to keep my films in 100 or 50 ft reels and store them in this manner, unless it is a longer or composed film for a certain purpose. Large reels, 500 ft or more, are cumbersome both for storing and for locating desired footage for viewing.

Some investigators insist that a duplicate film be made on each initial processing before the original is ever viewed. The original film is stored as the record of the original data and as a source of additional duplicates in case the work copy is damaged or worn during analysis and repeated viewings. Especially valuable films should be duplicated as a matter of course. Duplication of films originally put together by splicing eliminates the problem of film splice breakage during presentation. A local television station may be willing to produce duplicates; otherwise the duplicates will cost as much or more than the original processing.

Behavioral films may be submitted to the American Archive at Pennsylvania State University, where they will be judged for inclusion in the collection of *Encyclopedia Cinematographica* and, if accepted, made available to other investigators. Several short films on tortoise and sea turtle behavior sequences are currently listed.

STAGING

Space may be a very important factor in permitting animals to perform certain behaviors in a natural way; close confinement inhibits many behaviors. Know the behavioral states of the animal, for they may vary circumdielly as well as seasonally and with reproductive state, health, and captivity.

When an observer is well acquainted with a species, he learns to recognize initial actions or steps in behavioral sequences, which precede the action to be photographed. Anticipating actions is of primary importance when attempting to obtain motion pictures, otherwise much costly film is wasted on unproductive shots.

There are many techniques that can be used in staging, and these depend on the animal. A neutral contrasting background in an obstruction-free area is desirable so that the details of behavior are readily discernible. Record the size of the animal so that a scale can be determined for later measurements and proportional measurements of movements—moving a zoom lens will vary the size of observed objects.

Take pictures from different angles, or a predetermined angle, so that the resulting film will show the details to best advantage (Urban, 1965). Sometimes by using mirrors, two or more angles of the same action can be obtained from the same picture.

Frequently I have found it necessary to have all the animals individually marked for later recognition on the film. The type of mark used varies with the size, structure, and type of animal, whether color or black and white film is used, and whether color, patterns, or symbols are used as marks. These marks should be coded so that all individuals and sexes can readily be determined, and they should be placed on the animal in a position that will show up in the picture. If it is not possible to use marked individuals, be sure to keep accurate records with the film of the individuals in a particular photograph. For certain types of behavior studies it may be beneficial to place a discernible mark (point of reference) at some special point on the animal, so that this particular spot can be followed as it changes position or posture during a behavioral action.

Members of some species do not perform with the observer in full view, but most species quickly habituate to captivity and perform well in view of the observer who uses a regular place for observing. For those that appear to be disturbed by the prolonged presence of an observer, a simple blind can be erected with viewing ports and a camera port. An adaptation of such a blind, or a large observation chamber, is to provide the port with a one-way mirror, the observer working from a darkened area behind it. We have taken successful still photographs and motion pictures through a one-way mirror using fast film, both indoors and outdoors.

ANALYSIS

The value of photographs, and particularly motion pictures, depends on the quality of the pictures, the clearness of detail, the size of the object,

ANALYSIS

the speed of the action, the angle of the shots, and the knowledge of the individual and the context of its actions. With good quality films, there are many techniques that can be used to extract different types of data.

For motion picture analysis, the conventional movie projector permits only one type of projection, which is difficult to use for study other than repeated viewing of an event. Special projectors are now available (sometimes called analysis projectors, or time-motion projectors) especially for 16 mm films, which permit single frame advance and viewing and are equipped with extrapowerful blowers to keep the film from melting. Some of these machines have frame counters, and the direction of the film movement can be reversed. Such projectors can be equipped with lenses of different lengths, which permit projecting images of different sizes on a screen. The single frame advance machine also permits the prolonged observation of one frame of a movie film for the desired length of time, having the effect of stopping motion for analysis.

When I first started time-motion analysis of behavior (Carpenter and Grubitz, 1961), I did not have a movie camera available. What I used was a simple stopwatch and a split-timer stopwatch; the latter makes possible the timing of two successive events or actions simultaneously, under the control of the observer. This method is certainly of value if no other technique is available, but success depends on previous knowledge of a behavior pattern and the parts to be timed (Carpenter, 1961, 1962, 1963).

When I acquired a versatile 16 mm movie camera and was able to obtain film taken or exposed at a known speed (frames per second), the stopwatch method was abandoned, for the speed of an action (time-motion) could be calculated from the number of frames occupied on the film by the action. On a conventional projector (without stop motion or frame counter) this is cumbersome; with a film editor it is feasible. Using an analysis projector with single frame advance and a frame counter, it is a simple operation. In all these techniques the film must be taken at a known speed.

Once good motion pictures of an action or behavior sequence are available, it is also possible to measure movement change. To do this, I made a grid by painting horizontal lines across a screen (a large poster board or painted plywood) at 1 in. intervals, producing a vertical scale that was then numbered. When the motion picture was projected onto this screen, and held as a single frame, the size of the animal could be varied by the distance between the projector and the grid screen, the largest size image of the animal giving the most accurate results. Starting with the film frame that immediately preceded the action, I chose an obvious point on my animal (top of head, eye, etc.) and recorded its

Figure 1 Sketch of Vanguard Motion Analyzer designed for the time-motion analysis of 16 mm motion picture film.

position as the appropriate number from the vertical scale grid line. As each frame was advanced, and if the animal action moved through the vertical scale, the chosen point of reference would move, and for each frame, the numerical position of the reference point on the grid scale was recorded. When the sequence of numbers for the entire action is recorded, the numbers can be plotted against the known film speed (frames per second) to produce a graphical representation of the action through time on a vertical scale. For displaying lizards I have called this the display-action-pattern graph (Carpenter, 1967, 1969; Carpenter, Badham, and Kimble, 1970). It is very important that the projector, screen, and frame projection be steady for accuracy.

ANALYSIS

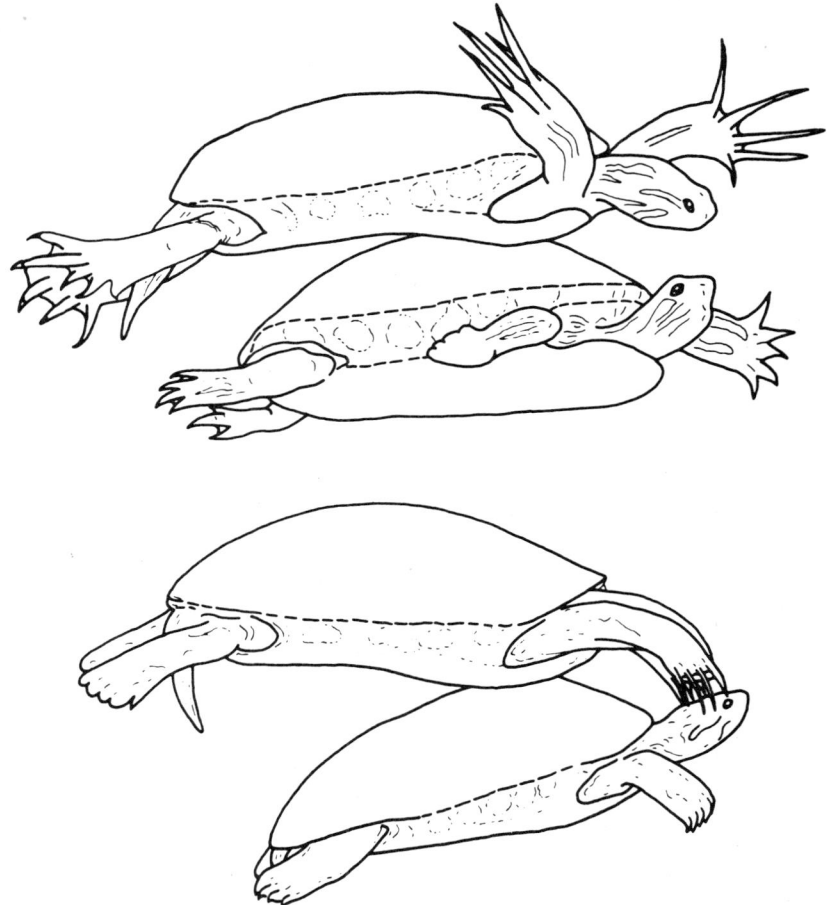

Figure 2 Two postures during courtship of a female *Pseudemys nelsoni* by a male *P. floridana hieroglyphica*, Dallas Zoo, January 7, 1974. Drawing taken from projected 16 mm motion picture.

A single machine is now available that greatly simplifies and increases the reliability of such measurements and also adds other dimensions to the use of movie films. Although quite expensive, possibly it may be borrowed from or shared with physical education or sports departments. The machine is a Vanguard Motion Analyzer (Vanguard Instrument Corporation, Long Island, N.Y.) and consists of a large boxlike base with a large ground glass view screen on which the image is reflected from a mirror receiving the image from the projector head that attaches to the top of the base (Figure 1). The head contains the projector

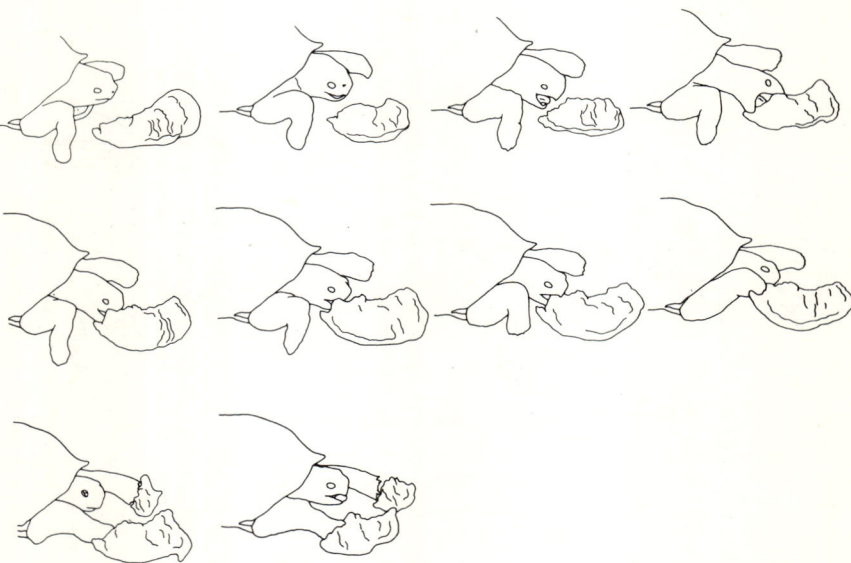

Figure 3 Head and foot movement sequence of a *Terrapene ornata* biting off a piece of apple, University of Oklahoma Biological Station, July 1972. Drawing taken from projected 16 mm motion picture sequence of 80 frames (5 sec duration at 16 frames per second).

apparatus for moving the film, magnifying the image, and counting film frames (a resettable counter), the projector lamp, and a powerful blower for cooling. The expensive projector bulbs have a use-life of about 50 hr. The great value of the projector head is that it is a very precise instrument, always projecting the film onto the view screen in exactly the same position, reducing or eliminating errors in measurement, which are inherent in the previous method using the analysis projector and grid screen. The base also has a single frame advance button, a switch that controls forward or backward movement of the film, and a rheostat that permits the film to be advanced at slow to fast speeds. The view screen has a horizontal grid that is attached to a circular frame surrounding the view area, the latter marked with a 360° (vernier) rotatable scale. The vertical and horizontal coordinates are each equipped with a movable fine wire that produces a line across the view screen. Each coordinate has its own control wheel. Thus each coordinate can be moved up or down or from side to side, respectively. Each coordinate is connected to a numerical readout window that produces numbers from 00.00 to 99.99. These scales can be set, or reset, as desired. The projector head can be rotated so that the angle of the image on the projected frame can

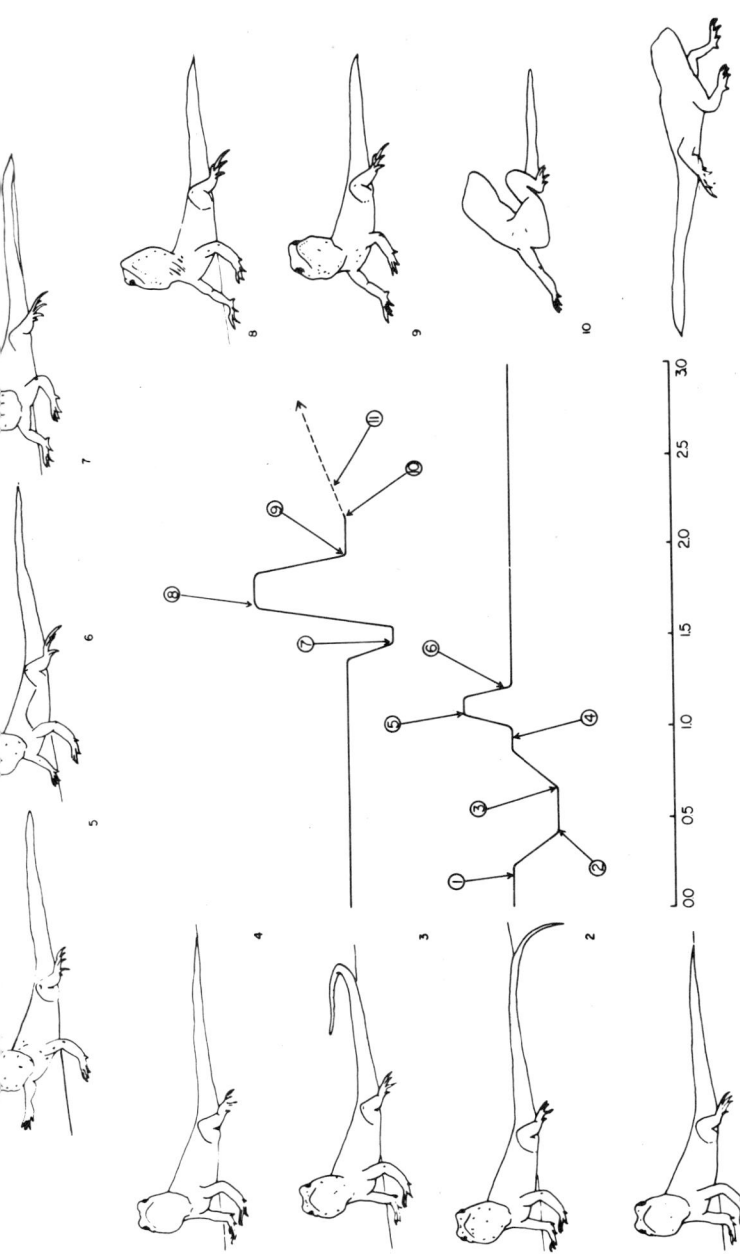

Figure 4 Display-action pattern of an aggressive male agamid lizard *Amphibolurus muricatus* from Australia. Drawing taken from projected 16 mm motion picture sequence. Four types of action are indicated: 2 and 3, tail twitch; 5 and 6, leg wave; 7, 8, and 9, push-up-head-bob; 10 and 11, lizard immediately shifts posture and jumps away. The time sequence is indicated in seconds. This action represents action through 34 frames of film taken at 16 frames per second. Circled numbers with arrows indicate the position on the display-action-pattern graph relating to the respective numbered lizard posture drawings.

be aligned appropriately. Such versatility permits sophisticated measurements on the horizontal and vertical planes and angles of these planes. The frame counter and single frame advance permit accurate measurements of speed of performance when speed of film (frames per second) is known (Ferguson, 1971; Purdue and Carpenter, 1972a, 1972b). I am sure there are possibilities for film study and measurements with this machine that have not yet been devised or exploited.

DRAWINGS FROM MOTION PICTURES

Motion pictures can be used to produce accurate drawings or sketches of particular postures or position, or to produce accurate sequential sketches of behavioral actions (Figures 2, 3, 4). Using the analysis projector with single frame advance, the frame(s) showing the posture or position desired is projected onto a sheet of paper, taped to a wall, in the size desired (size is affected by the position of paper or length of lens). With the frame held by stop-motion, the projected image desired is traced on the paper. For a sequence of postures or position in a movement pattern, one sketch is made; the film is advanced frame by frame until the next desired posture or position is present, and it is sketched. This procedure is followed until the sequential sketches of the desired behavior are obtained. Such a technique can be used to advantage to capture the movement positions in interactions between two or more individuals (Blanc and Carpenter, 1969; Carpenter, 1969).

CHAPTER 6
Laboratory Maintenance

HOWARD W. CAMPBELL and STEPHEN D. BUSACK

The care of turtles in the laboratory can be relatively simple and straightforward if the planning and design of the facilities and the husbandry program take into account the turtles' specific requirements and the goals of the research. All manner of behavior, ecological preferences, and requirements exist within the order Chelonia and husbandry programs must be attuned to these variations if healthy, physiologically and psychologically "normal" animals are desired for research. This chapter outlines the areas of greatest concern and provides some guidelines for the uninitiated, but in the final analysis, any researcher will be well repaid by first learning about the natural biology of the species he plans to work with.

HOUSING

Housing requirements vary widely depending on the species to be maintained, the numbers of animals to be held, and the purpose of the facility. Studies of behavior presumed to be normal in the natural habitat should be undertaken only under conditions that are naturalistic both physically and in terms of normal population densities. All facilities should be constructed to satisfy the basic requirements of the species for adequate space of the proper nature (land, water, or a proper mix of the two), temperature, photoperiod, and cleanliness. The most efficient approach to solving this problem, especially if a diverse series of species will be maintained over any period of time, is to construct a facility that offers a mosaic of physical parameters within which the individuals can move to satisfy their immediate needs. When two or more species are to be housed together, serious consideration must be given to their compatibility, both behaviorally and in regard to their environmental requirements. The aggressive *Chelydra serpentina*, for example, should not be kept with other species, and fully aquatic and semiaquatic species should not be mixed with terrestrial species unless the holding facilities are constructed to accommodate both habitats.

Some of these points may appear to be self-evident, but too often they are *not* evident in laboratories of universities and other research institutions nor even in the homes of captive turtle enthusiasts. We feel that the reliability and replicability of research will improve as these simple biological concepts become more widely acknowledged.

The construction of naturalistic experimental housing for behavioral studies is beyond the scope of this chapter. Investigators who wish to approach this problem should first become familiar with the species through reference to a general text (e.g., Carr, 1952; Ernst and Barbour, 1972; Pope, 1939; Pritchard, 1967), research articles when available, or consultation with a specialist in turtle ecology.

In the laboratory situation where the primary desire is simply to maintain a number of animals in good health over a short period, the problems are less complex. Simple holding pens should allow the turtles a period to recover from the generally unnatural and physically debilitating conditions in transport and at the dealer's shop. Two weeks, preferably three, should be allowed for acclimatization. Thus, depending on the turnover rate, several sets of facilities may be needed to maintain a constant supply of experimental animals.

Terrestrial species can be maintained in containers that provide adequate space, have sides over which the animals cannot climb, and can be easily cleaned. Cages can be constructed of plywood, wire of any mesh sufficiently small to restrain the species (and folded inward or otherwise altered at the top to prevent climbing out), concrete blocks, and so on. Large glass aquaria also make satisfactory cages, but these can be difficult to clean. Rough concrete substrates should be avoided. They are not only difficult to clean, but also abrasive to the turtles' plastrons and feet.

Various materials have been used for litter in terrestrial turtle cages. Each has some merits and some disadvantages, but no litter appears to be necessary if the cage floor is situated so that it can be periodically washed down. In our experience, wood and bark chips, sawdust, sand, peat and sphagnum moss, vermiculite, and other particulate materials all have a

HOUSING

tendency to adhere to food or to be carried or pushed into the water supply. The resulting accidental ingestion is a potential source of health problems.

Pritchard (1967) recommended a minimum cage space of 33 in.^{2} for each linear inch of turtle shell. We regard 15 in.^{2} of cage space per inch of turtle as minimal for laboratory purposes when frequent cleaning is possible; 20 in.^{2} per inch of turtle would be more adequate.

Cage layout for terrestrial turtles should provide a localized source of heat at one end, away from food and water supplies. The heat source should be positioned so as to prevent contact or close approximation by the turtles. If 120 W infrared heat lamps are used, they should be suspended 3 to 4 ft above the cage, and their area of influence should not extend throughout the enclosure. The turtles should have adequate room to move out of the heated area so that body temperature can be behaviorally regulated. Ordinary incandescent bulbs of 60 to 100 W can also be utilized, in banks of multiple bulbs if the cage size so indicates. The placement should be adjusted so that there is always 6 or 8 in. between the bulb and the top of the turtle's carapace. If large numbers of turtles are to be maintained in one enclosure or cage, care should be taken to provide an adequate area of high thermal availability to allow all the animals to thermoregulate simultaneously. Social order may modify basking, with subordinate individuals forced into atypical activity patterns or denied adequate access to heat.

Other heat sources such as warm water pipes and heating strips of various design and brand name can be effectively incorporated into the cage design.

In addition to heat lamps, it is desirable to provide some sources of ultraviolet light, such as Vita Lite™ or GroLux™, for indoor cages. Normal laboratory lighting systems will suffice for light, but it should be used in conjunction with an efficient "artificial sunlight" system (Bartlett, 1971) to ensure adequate vitamin D metabolism. Excessive ultraviolet light, however, may cause problems (Wallach, 1970).

Housing for aquatic species is somewhat more elaborate because of the necessity of providing both aquatic and land areas and the difficulties of cleaning the water. Cullum and Justus (1973) described a most elaborate and satisfactory physical system that allows for excellent control of water quality, temperature, and light, and requires little modification for most purposes. For a laboratory seriously involved in long-term studies, the expense and effort of setting up this system will be well repaid. Less committed programs, however, can still maintain healthy turtles.

As for terrestrial cages, the basic requirements of aquatic turtle tanks are ease of cleaning, adequate space (including a dry area for almost all but the marine species), and proper heat and light sources. Almost any material will serve for the containers. Glass aquaria are good, but expensive and often difficult to clean. Soapstone sinks serve well, as do fiberglass-coated plywood boxes if these are elevated and fitted with a drain to facilitate cleaning. The stainless steel tanks and sinks often available in the laboratory are good and easy to clean, but noisy when numbers of large turtles are banging around in them. As with terrestrial cages, rough surfaces such as unfinished concrete should be avoided, especially when maintaining thin-skinned species of the family Trionychidae.

Cage size is of particular importance, for overcrowding in aquatic cages will compound problems in maintaining water cleanliness. Adequate dry land area is also important to facilitate basking and drying. Bury and Wolfheim (1973) have shown that turtles may move about out of the water to enhance drying after preferred temperature levels have been reached, and overcrowding will function to restrict this potentially important activity. Pritchard (1967) recommended 144 in.^{2} of wet tank space per 4 linear in. of turtle. If adequate filters are used or if a frequent cleaning schedule is maintained, 100 in.^{2} of wet tank space per 4 in. of turtle should be sufficient.

Light and heat provisions for aquatic turtles are equivalent to those for terrestrial species. Water temperatures should be maintained by appropriate water heaters.

Certain species will do well over any extended period of captivity only when provided with a saltwater environment; this is especially true of the sea turtles and the diamondback terrapins, *Malaclemys terrapin* (Campbell, 1967; Frye, 1973). If such species are being held for a very brief period, they can be kept in fresh water; however, it is recommended that a level teaspoon of table salt be added per pint of water to prevent fungal infections (Allen and Littleford, 1955). For longer periods of maintenance, a seawater system is needed. Natural seawater is optimal, but any of the various commercial seawater mixtures is adequate. Spoczynska (1970, 1971) used an artificial medium having a specific gravity of 1.025 successfully with both leatherback and green turtles. Numerous books on this subject (e.g., Straughan, 1970) are available from most tropical fish shops, and the materials recommended are generally available.

Kohler (1972a–1972c) lists and reviews the generally available books on reptile care. Part II (1972b) deals specifically with turtles, but much of value may be found in books reviewed in other sections. Also see Murphy (1969, 1972, 1973a).

TEMPERATURE

Proper temperature is a foremost concern in the maintenance of captive reptiles. Without a proper thermal regime, no amount of attention to other variables will produce healthy specimens.

Turtles vary rather widely in their thermal preferences, but most if not all species require temperatures higher than the average laboratory air temperature of approximately 22°C. Brattstrom (1965) compiled records of body temperature for a number of species, and Boyer (1965), Fitch (1956), and Moll and Legler (1971), among others, discuss the natural temperature relations of species frequently found in laboratories. (Also see Chapter 11.)

The minimum ambient temperature to which turtles should be exposed if continuous normal activity is desired is 21°C. For most effective laboratory maintenance, temperatures between 24 and 30°C should be provided. Water temperatures should also be above 21°C, preferably at 23 to 26°C.

Kaplan (1957a) suggested that specimens can be stored at 5 to 6°C for long durations with only periodic warming exposures for feeding. He noted certain physiological parameters that appeared to remain stable through cold storage periods. We recommend against such a procedure except, possibly, for certain specialized experimental purposes where physiologically normal, active animals are not needed. Without even considering health problems with undigested food and the humane aspects of the issue, we point out that cold storage is not likely to provide specimens of normal responsiveness without a 2 or 3 week acclimation period.

DAY LENGTH

Day length and temperature are closely related for most of the temperate zone species commonly used in the laboratory. The importance of photoperiod to the regulation of basic hormonal states (see Chapter 16) should be considered by workers whose results may be influenced by variation in hormonal state. Day length control can be achieved by the use of readily available timers. Turtles should not be disturbed during their nocturnal cycle. In many institutions the janitorial staff can be scheduled to clean during the day; if not, it may be necessary for the research staff to assume the janitorial duties in the animal room.

For observational studies of behavior, it is desirable to have stepping switches programmed to simulate dawn and dusk, with heat lamps operating for only a portion of the daily light cycle. For simple maintenance of experimental animals for short periods, one timing system will suffice. If one photoperiod is selected and all experimental animals are given the opportunity to adjust to it, a greater uniformity in physiological and psychological data can be achieved. Photoperiods of 10 to 12 hr of light with 14 or 12 hr of dark have been satisfactory in our work, but others can be used as appropriate to the purpose of the research. (Chapter 23 gives more information on photoperiod.)

FEEDING

The proper diet in sufficient quantity will eliminate many common ailments and problems associated with keeping turtles in captivity. All turtles require essentially the same nutrients normally supplied in packaged foods for mice and rats. To compensate for captive conditions, however, larger quantities of calcium, copper, and vitamin D are also required, especially for young, growing specimens.

Aquatic species will obtain sufficient protein and vitamins if fed a mixed diet of raw meats such as chicken or beef liver and heart, chopped whole fish, aquarium snails, crayfish, mealworms and other insects, live minnows, goldfish, guppies, and fresh shrimp. If foods such as these are difficult to acquire or maintain, commercial preparations are available that supply the necessary requirements. Many of the soft, noncereal, canned dog and cat foods supply protein and other nutrients (Anonymous, 1972). Apparently little has been written on the effectiveness of the commercially prepared fish and turtle balanced diets such as Trout Chow™ (Purina) and Turtle-Ettes™ (Longlife). These carefully prepared, high-protein, nutrient-rich feeds should not be confused with the pet trade's "ant egg" or dried insect turtle foods, which are inadequate or worse for maintenance. Waltrous (1971) and Crombie (personal communication) assert that excellent growth and health can be achieved with the high-protein products, and our own experience supports this view. Certainly, if supplementary feedings of a variety of vegetables and other plant material are provided, the commercial products appear to be fully adequate for a basic diet. Nonfibrous greens are best, and celery tops, *Elodea*, duckweed, watercress, and other water plants make good supplements. Cabbage, spinach, and kale, may also be used. "Iceberg" lettuce contains few nutrients (Anonymous, 1973), but other greener lettuces may be fed.

Terrestrial species may be fed mealworms, earthworms, raw meat, hard-boiled eggs, dog and cat foods, and canned baby foods supple-

FEEDING

mented with dry grass, raw or cooked vegetables, fresh fruit, and dandelion greens and flowers (Pritchard, 1967; Frye, 1973). Some tortoise fanciers recommend an occasional feeding of dried apricots soaked in vitamin water as a means of increasing essential vitamin intake.

Turtle diets should be regularly augmented with calcium for shell growth and hardness (Frye, 1973). For aquatic species, an additive-free block of plaster of paris may be placed in the tank to provide calcium as it dissolves. Also, turtles will chase cuttlebone around and bite off chunks, which also serves to keep beaks in trim. Dissolved or powdered calcium, crushed eggshell, animal bonemeal, and bone phosphates may be added to the foods of all turtles to provide extra calcium.

Several vitamin preparations are available to complement even the best of balanced diets. Cod liver oil applied directly to food will supply vitamins A and D when natural sunlight is not available (Burke, 1970). Liquid baby vitamin compounds, ABDEC™, and AVITRON™, serve as acceptable supplements and may be added directly to the tank water. Several drops to 5 gal of water will be sufficient. KIT-VITE™ multiple vitamins, available in paste form, do not dissolve in water and may be applied directly to food. Frye (1973) recommended as a general procedure that the diet be supplemented with a balanced preparation that provides calcium as well as vitamins A, B-complex, C, D, E, and K.

Turtles should be fed regularly, and each particular laboratory situation will dictate the most appropriate mode. If the number of aquatic turtles is large and an adequate filtration system is available, food may simply be placed in the tank two or three times a week. To avoid fouling the water, excess food should be removed at the end of an hour or two. When dealing with a small number of animals in individual aquaria, the water should be changed after feeding to prevent bacterial buildup. To avoid thermal shock when changing the water in this manner, ensure that temperature changes are not abrupt (Pritchard, 1967). With terrestrial species any excess fruits, vegetables, meats, or moistened commercial foods should be removed after all animals have had an opportunity to eat their fill.

Animals fresh from the field, as well as commercially obtained stock, may require a few days to a few weeks to acclimate before they will feed. Heavy and light feeding cycles are normal and are not necessarily indicative of problems. An animal may suddenly feed less than it did; unless an illness is apparent, however, the individual's appetite will generally increase again after a few days. Do not resort to force feeding immediately. If the fasting continues, place the turtle overnight or for 24 hr in a shallow bath of water enriched with a multiple vitamin–mineral preparation at higher than usual concentration; then offer a preferred food item. If temperature and day length are controlled and incipient hibernation is not indicated, prolonged nonfeeding is an indication of a possibly serious disorder. The specimen's overall health should be evaluated carefully, and force feeding by intraesophageal intubation with a dietary supplement rich in the B-complex vitamins and vitamin A should then be initiated.

On occasion, newly acquired specimens are received in a severely starved and dehydrated condition. If they do not respond to the vitamin-mineral bath within 24 hr, they should be force fed rather than risk losing them. Frye (1973) suggested intraesophageal administration of special nutrient-replacement products such as Petkalorie™ (Haver-Lockhart) and Nutrical™ (EVSCO), perhaps mixed with pureed infant meat diets. He also recommended isolation and quiet as almost as important as the nutritional replacement for specimens in this condition.

Hatchlings require a high-protein diet rich in calcium and should be fed daily to stimulate proper growth and behavior. All food should be cut small enough to enable the smallest turtle to feed without choking.

The scope of this chapter is limited and specific recommendations for food preferences for the world's turtles would require too much space, since each genus and, in some cases, each species has its own preferences. Most species used in laboratory experiments belong to the genera *Chrysemys, Pseudemys*, and *Terrapene*, however; and they will all do well if the previous guidelines are followed. *Kinosternon* and *Graptemys* from the New World and *Malayemys* from southeast Asia, are known to eat large amounts of snails but may accept a more easily procured diet. Unless specifically indicated by the research design, exotic species should be avoided because natural history data are unavailable for many and, lacking these data, adequate laboratory provisions are difficult to devise. Additionally, each turtle, as with individual dogs, rats, and people, may have its own preferences. With experimentation and patience, however, an adequate diet can be found for most species and most individuals. Species-specific patterns are also present in feeding behaviors, and it is best to explore the ecology of the experimental animal in advance of setting up captive colonies (Campbell and Campbell, 1972).

DISEASE

The majority of the disease problems encountered with captive turtles result from improper husbandry, sanitation, and general hygienic conditions, either in the laboratory or by introduction of infected animals subjected to substandard conditions in dealers' shops—the "maladaptation syndrome" of Cowan (1968). Frye (1973) recommended 4 to 6 weeks isolation before introducing new specimens into the laboratory population. This duration may be excessive for laboratories in which a relatively rapid turnover of specimens is common, but it should be required whenever valuable or irreplaceable specimens (endangered species, etc.) are maintained. New arrivals should be carefully inspected for lesions, unnatural swellings, evidence of exudate from the nostrils, and any other external symptoms of disease, and treated as indicated by the symptomatology.

Hunt (1957) determined the causes of death for more than 200 autopsied turtles and reviewed the pertinent literature. He found that about 75% of these deaths occurred through diseases of the alimentary and respiratory systems. *Entamoeba* sp. was cited as the chief agent in the alimentary disorders, and lobar pneumonia, resulting from overcrowding during transport, was the primary respiratory problem. Murphy (1973b) reviewed the proposed treatments and reported his own success with the drug Tylocine® administered either in food (as Tylan) or with a catheter and syringe at a dosage of 25 mg/kg of soft-tissue weight daily.

Frye (1973) illustrated the symptoms of respiratory disease: failure in appetite, torpor, nasal discharge, open-mouth breathing, and wheezing are common, and aquatic species may develop a loss of equilibrium while swimming. He recommended Ampicillin® or Chloromycetin® plus ascorbic acid injected intramuscularly in the rear of the thigh. Dosage for Ampicillin® was given as 3 to 6 mg/kg; for Chloromycetin®, 10 to 15 mg/kg, in divided doses, was recommended.

Marcus (1971) noted that a variety of microorganisms may cause pulmonary infection. Doyle and Moreland (1968, 1969) and Reichenback-Klinke and Elkan (1965) studied mycobacterial pulmonary infections. These appear resistant to treatment by antibiotics (Wallach, 1969). Turtles have also been reported to be subject to a variety of pulmonary mycotic infections, including species of *Aspergillus* and of *Mucor* (Marcus, 1971; Wallach, 1969).

With such a variety of pathogens attacking the pulmonary systems of captive turtles, the first response of the laboratory manager to symptoms of pulmonary infection should be to isolate the causative agent so that the appropriate remedial action can be taken. Pulmonary infections should pose no problem in laboratory stock if new specimens are properly isolated and if the laboratory facilities provide for proper thermoregulation, diet, and sanitation, as well as adequate space.

Although it has been suggested that turtles are vulnerable to viral infections (Smith and Coates, 1938; Tamm, 1952), there are few data to support virus as a major problem in captive populations (Marcus, 1971;

Murphy, 1973b). This is a research area that should be highly rewarding to a laboratory with proper facilities.

A wide variety of gram-negative bacilli have been implicated in turtle disease, including *Aeromonas, Bartonella, Citrobacter, Clostridium, Mycobacterium, Proteus, Pseudomonas, Salmonella* and *Serratia*. Marcus (1971) considered the gram-negative bacilli to be the primary cause of morbidity and mortality in captive turtles. Murphy (1973b) has provided an excellent review of the literature on these diseases and their treatment.

Ulcerative stomatitis, which produces lesions in the oral mucosa with ultimate progression to bone necrosis and osteomyelitis ("mouth-rot," in the terminology of reptile hobbyists) is occasionally reported in turtles (Frye, 1973). A number of gram-negative bacilli have been suggested as the responsible pathogens (Wallach, 1969) and *Aeromonas hydrophila* isolated from infections has been used to infect healthy snakes (Page, 1961, 1966). Improper diet, especially the lack of vitamin C, has been suggested as a contributing factor in this disease (Wallach, 1969). Injury to the oral mucosa provides entry to the pathogen. Wallach recommended topical application of thimerosal, hydrogen peroxide, or sulfonamide for mild cases. Advanced cases require careful cleaning and debridement before application. This local treatment should be supplemented by a single dose of systemic antibiotic and a dietary supplement of 10 to 30 mg of ascorbic acid for 10 days (Wallach, 1969).

Marcus (1971) suggested disinfection of infected cages with nontoxic compounds such as benzalkonium chloride or sodium hypochlorite solution. He further suggested local treatment with benzalkonium chloride, in addition to the antiseptics mentioned by Wallach (1969), and suggested Tetracycline as the antibiotic of choice.

Cleansing of the affected area with hydrogen peroxide and Betadine® solution, with administration of Ampicillin® (3–6mg/kg), multi-B complex vitamins, and ascorbic acid was the treatment of choice of Frye (1973).

Kaplan (1957b) first described the condition known as septicemic cutaneous ulcerative disease (SCUD) in turtles, caused by *Citrobacter freundi*. Symptoms of this disease include general paralysis, lethargy, anorexia, skin ulceration, and a film over the eyes. Infection of an entire turtle colony is possible through contaminated water, and turtles of the genera *Chrysemys, Emys,* and *Pseudemys* have been infected (Murphy, 1973b).

Kaplan (1958b) successfully treated this disease using chloramphenicol with an initial dosage of 6 mg/100 g of soft-tissue weight, followed by 3 mg/100 g given twice a day for 7 days. Jackson and Fulton (1970) used chloramphenicol in tank water, 250 mg/20 gal, added two or three a times a week just before feeding, along with several hours of ultraviolet radiation each day. Frye (1973) recommended daily injections of chloromycetin succinate (chloramphenicol) at 10 to 15 mg/kg in divided dosages, except where dehydration or impaired renal or hepatic function is indicated.

Bacterial infections are most generally apparent as gastroenteritis, accompanied by anorexia, constipation or diarrhea, emesis, and frequently, pneumonia (Wallach, 1969). Sodium sulfamethazine, 1 oz/gal, for 10 days (Wallach, 1969), and oral administration of chlortetracycline bisulfate at 100 mg/lb of body weight (Gray, Davis, and McCarten, 1966) are given as treatments for this condition. *Pseudomonas aeruginosa* is the most frequently identified pathogen in this disease (Gray et al., 1966). Other treatments for the *P. aeruginosa* infections are reviewed by Murphy (1973b).

Sodium sulfamethazine at the above mentioned dosage is recommended as a standard treatment for all newly acquired turtles before they are introduced into the laboratory stock. Chloromycetin® at half or one-quarter the recommended mammalian dose for a 10 day period has also been used (Dieterich, 1967).

Dobbs (1973) recommended Neomycin sulfate, with homatropine methylbromide, when gastroenteritis is indicated by hemorrhaging, discolored or mucoid stools, regurgitation, or loss of appetite. The dosage suggested is 110 mg/kg of soft-tissue weight, administered orally. Dosage for this drug (with polymixin sulfate; Daribiotic®, Beecham-Massengill) administered intramuscularly or intravenously, was given as 10 mg/kg by Frye (1973). He did not recommend use of this medication when impaired renal or hepatic function or dehydration are indicated.

The occurrence of numerous serotypes of *Salmonella* and *Arizona* in captive turtles has received considerable attention. The presence of these bacteria in a laboratory holding facility certainly constitutes a potential public health problem of some magnitude. Jackson, Jackson, and Fulton (1969) recorded the presence of *Citrobacter* and *Edwardsiella*, in addition to *Salmonella* and *Arizona*, in a series of wild and captive turtles; McCoy and Seidler (1973) identified seven potential pathogens in fecal swabs from captive *Pseudemys scripta elegans*. Jackson and Jackson (1971) found an infection rate of 12.1% overall for *Salmonella* and *Arizona* in a sample from a series of zoological parks. Marcus (1971), Murphy (1973b), and Wallach (1969) briefly review the literature on the occurrence of the pathogenic forms.

Salmonella and *Arizona* do not, in general, appear to constitute a problem for a laboratory colony of turtles other than the potential threat to the laboratory workers (Frye, 1973; Marcus, 1971). Healthy turtles seldom show any symptomatology of *Salmonella* or *Arizona* infection, but any decline in the health of an individual could be expected to initiate enteritis (Taylor, 1969). Septicemia with necrotic foci in the liver and other viscera may result from *Salmonella* infection (Marcus, 1971); and McNeil and Hinshaw (1946) attributed the death of two Galapagos tortoises to massive infections of *S. newport* and *S. sandiego*.

Treatment of infected individuals with antibiotics effective against gram-negative bacteria, such as chloramphenicol, was suggested by Murphy (1973b), and Frye (1973) also suggested this or Neomycin sulfate or Furoxone® (Eaton) given orally at a dosage of 25 to 40 mg/kg. Marcus (1971) pointed out that it is extremely difficult to eliminate the carrier state for *Salmonella* and that, in man, cholecystectomy may be required. Laboratory staff may best respond to the threat of salmonellosis by careful personal hygiene and, except where suspicious symptomatology exists, no specific effort is indicated to eradicate the pathogens, other than scrupulous tank cleanliness.

Abscesses of various origins are occasionally encountered. Both pure and mixed cultures of *Aeromonas* spp., *Citrobacter* spp., *Enterobacter* spp., *Escherichia* spp., *Mycobacteria* spp., *Peptostreptococcus* spp., *Proteus morganii*, *P. rettgeri*, *Pseudomonas* spp., *Salmonella marina*, and *Serratia* have been isolated from abscesses (Frye, 1973). Treatment involves drainage or thorough removal of debris from the abscess, flushing the cavity with 3% hydrogen peroxide and Betadine® solution, and packing with Furacin® (Eaton) if possible (Frye, 1973).

Disease of the shell is not an uncommon problem. Algae, fungi, and bacteria may all be implicated in such cases. Many aquatic species normally harbor growths of algae on the shell. Under natural conditions this association appears to cause no harm to the turtle, and various authors have considered this to be a symbiotic relationship that benefits the turtle by providing a degree of camouflage. In the laboratory the algal growth may become excessive, and damage to the shell and even death may occur (Hunt, 1957). Algal problems can be prevented if clean water and adequate basking facilities are provided. Treatment consists of washing with iodine solution or 1% copper sulfate solution (Marcus, 1971).

Fungal infections appear as pale areas on the shell or as eroded shell or elevated plates. Specimens suspected of harboring fungal infections should be isolated. Infected specimens should be kept out of the water and dry, except for necessary feeding, and the local areas should be cleaned and painted with Betadine® solution twice daily (Frye, 1973).

PARASITES

A wide diversity of ecto- and endoparasites have been described. Murphy (1973c, 1973d), Telford (1967, 1971), and Hunsaker (1966) have reviewed the various genera and species in detail.

The great majority of species of parasites are not generally a problem in turtles that are well cared for. Telford (1971) argued that, in nature, most parasites exist in a balance with their hosts, becoming a serious health problem only under the unusual conditions and stress of captivity. Additionally, the possibility of cross-infection between host species of unfamiliar parasite species may cause some problems for mixed laboratory stock. A parasite species that exists in its normal host with no untoward health effects may prove devastating when introduced into an unnatural host species.

Ectoparasites generally found on turtles are leeches of several genera (Murphy, 1973d), ticks, especially *Ornithodoros* sp. and *Amblyomma* sp., and mites, notably the common snake mite, *Ophionyssus natricis*. These parasites have been implicated in the transmission of certain blood parasites, and the snake mite appears to be a mechanical vector of hemorrhagic septicemia, at least in snakes (Camin, 1948).

Control of these ectoparasites is straightforward, and their elimination will prevent any cross-transmission of blood parasites because nearly all helminth and protozoan blood parasites require a hematophagous vector, usually an arthropod, for transmission (Telford, 1971).

Leeches may be removed with forceps and the attachment area painted with any standard topical antiseptic.

New specimens can be deticked by painting each tick with a drop of alcohol, then removing it carefully with forceps. Afterward the area should be treated with an antibiotic. Mites can be removed from new specimens by soaking them overnight in warm water. Murphy (1975) recommended a solution of Ortho Dibrom 8-E® at 2 to 4 cc/gal of water or Diazinon 25-E® at one part 25% concentrate to 240 or 480 parts water.

Once specimens are clean, mites and other arthropods should be no problem, but Vapona No-Pest Strips™ can be hung in the cage room (King, 1971). Small strips can also be hung in individual cages. In general, insecticides should be avoided in reptile quarters. Murphy (1973d) suggested that Baygon 1.5 emulsifiable insecticide (*o*-isopropoxyphenyl methylcarbamate) is safe if *not* sprayed on surfaces that come in direct contact with the specimens. The use of silica aerogel for the control of mites, although once popular, is suspected of responsibility for deleterious pulmonary effects and is not recommended (Frye, 1973).

As noted previously, blood parasites generally cause no problems in the laboratory situation. If it is desirable to evaluate the blood parasitic burden in a turtle colony, the techniques outlined by Telford (1967) and Benbrook and Sloss (1961) are recommended.

Protozoan infections in the intestine may occasionally cause problems, especially when the specimens are in generally poor health otherwise (Reichenback-Klinke and Elkan, 1965). If bloody, mucoid, or otherwise discolored or unusual stools indicate trouble, the specimen(s) should be isolated, the cage(s) thoroughly cleaned and, if feasible and desirable, an attempt should be made to determine the causative organism, using the techniques of Telford (1967) or Benbrook and Sloss (1961).

When an obvious pathogen cannot be isolated, Murphy (1975) recommended the following general treatments:

1 Daily oral doses of Biosol® (Upjohn) Neomycin sulfate at 10 mg/kg until the condition is eliminated.
2 Same, in drinking water at 1 tsp/gal.
3 Seven days of oral dosage with Daribiotic® (Massengill) at 22 mg/kg.
4 If regurgitation occurs, a tetracycline product every 3 days at 25 mg/kg.

If trichomoniasis is indicated, the use of sulfonamides may eliminate the problem. Infected animals should be isolated and treated with sodium sulfamethazine in the drinking water for 10 days at the dosage of 1 oz/gal (Wallach, 1969). Murphy (1975) also suggested Emtryl® (Salisbury Labs) in aqueous solution, administered intraesophageally at 106 mg/100 g of body weight.

Amoebiasis is frequently encountered in laboratory turtle colonies. Several amoeba species in the genera *Entamoeba* and *Endolimax* have been isolated, but *Entamoeba invadens* appears to be the causative agent in most cases. Frye (1973) recommended 0.5 mg/kg intramuscular dosages of Emetine® hydrochloride (Eli Lilly) daily for 10 days. Hydration of the specimen must be maintained to prevent buildup of excessive serum levels. Murphy (1975) recommended a dosage of 40 mg/kg daily for 7 days with this drug. EnteroViform® has been used successfully for treatment in snakes at a dosage of 0.3 g/kg (Backhaus, 1963). Lomatil® (Searle) was used successfully to treat amoebiasis in a Galapagos tortoise (Dobbs, 1973) and was recommended at an oral dosage of 1.25 mg/kg by Murphy (1975).

The most effective control of amoebiasis is a prophylactic treatment of all incoming turtles. Ratcliffe (1966) reported on the usage of Tetracycline hydrochloride for this purpose for snakes and turtles. For general use of tetracycline hydrochloride, Dobbs (1973) recommended a dose of 20 mg/lb of body weight of Tetrachel-Vet® (Rachell), administered orally.

Coccidian infections in turtle colonies can be controlled by the use of sodium sulfamethazine in the drinking water at the dosage of 1 oz/gal for 10 days (Wallach, 1969). Lehmann (1972) recommended a treatment of Bayrene® (sulfamethoxydiazine). This is administered at 80 mg/kg on the first day and at a rate of 40 mg/kg for 4 more days. Dobbs (1973) suggested the use of sulfa compounds and Amprolium® for coccidian infections, but did not discuss dosage levels. Frye (1973) suggested sulfadimethoxine or sulfamethazine orally at a dosage of 0.5 g/kg on the first day and 0.25 g/kg for 3 additional days. He cautioned against the use of these drugs when dehydration or impaired renal or hepatic function is indicated.

The most frequently obvious parasites are worms of a diversity of species (see Murphy, 1973d, for a review of the recorded species). The presence of worms in the stool does not necessarily indicate a health problem in the colony. As mentioned, captive conditions and cross-infections may heighten problems with parasites, and a general treatment of all new specimens should be considered. Murphy (1973d), however, argued that the potential toxicity of the anhelmintic drugs precluded their use except where a specific problem is encountered. The decision to administer any of the more toxic anhelmintics should be made only on the basis of the health of the captive colony, not on the basis of human aversion to the sight of worms in the stools of the turtles.

Nematode infections can be controlled in turtle colonies when necessary with Piperazine (diethylenediamide) using an oral dosage of 50 mg/lb of body weight twice at 1 week intervals (Hunsaker, 1966; Murphy, 1975). Dobbs (1973) suggested 25 mg/lb at 1 week intervals. The common salts of Piperazine are said to be relatively harmless to reptiles (Murphy, 1973d) and are frequently used as a prophylaxis. Standard routes of administration of this and other vermifuges are on the food, in the drinking water, or, preferably, by intraesophageal administration.

Acute nematode infections can be treated with thiabendazole (Thibenzole®, Merck) administered intraesophageally or on the food. Backhaus (1965) recommended a dosage of 100 mg/kg at biweekly intervals, whereas Dobbs (1973) recommended 18 mg/100 g of body weight. Frye (1973) pointed out that since this drug is hygroscopic, it must be properly premixed with water to prevent gastrointestinal problems or fluid or electrolyte shifts. Atgard V® (Shell) is recommended when necessary for *Ascaris* and *Ancyclostoma*, and for flukes (Strongyloids) (Dobbs, 1973). This is a potent drug, however, and care should be exercised in its administration. Wallach (1969) recommended that its pellets be seasoned for 3 days in water to reduce the potency. It can then be administered directly into the stomach or with food. The dosage suggested by Vandeford (1968) was 12.5 mg/kg of soft body weight for 2 days. Murphy (1975) recommended 27.5 mg/kg of body weight for 2 days. Dobbs (1973) warned that underdosing may cause the parasites to migrate out of the intestine into the body cavity, causing the death of the turtle.

Tetrachlorethyline has been used for helminths in snakes (Nelson, 1950), and Santonin® was reported as an anhelmintic in tortoises, used in conjunction with castor oil (Nöel-Hume and Nöel-Hume, 1954).

Several genera of cestodes are known to parasitize reptiles. Di-N-butyl–tin oxide administered orally at a dosage of 35.0 mg/kg of soft-tissue weight was suggested by Wallach (1969) for control of both cestodes and acanthocephalan worms. Lehmann (1972) reported on the use of Niclosamide (Yomesan®, Chemagro) and Piperazine-Niclosamide (Manosil®). The latter drug resulted in fatalities with double doses, but Niclosamide appeared to be safe at a dosage of 150 mg/kg of body weight. For positive control of cestodes, the stool should be checked after 4 to 6 weeks and the dose repeated if necessary. Scoloban® (Bunamidine hydrochloride) at a dosage of 25 to 50 mg/kg of body weight is suggested by Frye (1973), but only when no heart condition is suspected. A frequency of once every 2 to 3 weeks if necessary was recommended. Murphy (1975) also recommended Diphanethane 70® at standard dog dosage.

On occasion miasis appears, involving the larvae of various flesh- and screwworms, such as *Sarcophaga* sp. and *Cuterebra* sp. Death may result from the action of these larvae (Rainey, 1953; Rosskopf, 1975), and they should be removed when noted in captive specimens. The infected area can be opened with a scalpel, the larvae manually removed with forceps, and the area cleaned and treated with an antibiotic (Murphy, 1973d).

TRAUMATIC INJURY

If turtles are properly housed and cared for, and if inappropriate species mixes are avoided, there should be little need to cope with traumatic injury in laboratory stocks. Accidents will happen, however, and both intra- and interspecific aggression may result in injury, even among or between generally compatible species.

In the wild, turtles seem to be able to recover from wounds that appear to be lethal. Loss of limbs, eyes, and massive shattering of the shell are all survived, at least on occasion. Captive specimens often also survive such injury if the quarters are sufficiently clean, and the victim is in good health otherwise. The most appropriate immediate response to severe injury appears to be to isolate the individual in a scrupulously clean environment, apply an antibiotic, and watch. If there is a sustained loss of appetite and a decrease in alertness and usual activity, an experienced veterinarian should be consulted.

The only unusual attribute of turtles as far as treatment of their injuries is concerned is their shell. Injury to limbs and other soft parts can be treated as would comparable injuries to other reptiles, but damage to the shell requires specialized techniques. In treating injuries of this type the wound should first be cleaned. If the tissue is necrotic, Kymar® (Armour-Baldwin) should be applied (Murphy, 1975). Frye (1973) recommended the use of Daribiotic® (Beecham-Massengill) in a tepid Ringer's solution as a general cleansing agent. After the loose tissue has been removed and shell fragments debrided, he suggested that all exposed edges be treated with Betadine® solution. When this has dried, either in air or assisted by a hairdryer, the shell can be restored using a fiberglass fabric impregnated with epoxy resin. Frye (1973) noted that the heat of polymerization caused no problem in his operations using Devco 5-Minute Epoxy™ (Devcon).

CHAPTER 7
Anesthesia and Surgery

JAMES H. MAXWELL

Although little research has been devoted entirely to anesthesia in turtles and none to surgery, a substantial amount of information does exist in the form of methodological notes in research articles and scattered references in related material. Data are available on a number of genera and a variety of anesthetics.[1] Usually the number of animals involved in each case is small, the situation unique, and the data limited; but collectively, the information is enlightening and valuable.

ANESTHESIA

Anesthetizing turtles remains a difficult task; their sluggish blood flow and low metabolism result in long delays in the induction of anesthesia, prolonged deep anesthesia, and lengthy recovery periods. Their ability to voluntarily suspend respiration for long periods makes it difficult to administer inhalation anesthetics and makes the observation of respiratory behavior a nearly useless diagnostic tool. These problems have prompted a widespread search by those using turtles for an effective anesthetic procedure. This section reviews most of these efforts. Others, armed with this information, now will be spared the frustration and waste involved in unknowingly using an inefficient technique, and perhaps still others will be challenged to attempt new and possibly more successful ones.

[1] A list of drug manufacturers and products related to this chapter is provided in the appendix.

Injectable Anesthetics

Pentobarbital Sodium. The most commonly used anesthetic is the barbiturate, pentobarbital sodium (Nembutal, Abbott Laboratories). Its great popularity as an anesthetic for laboratory animals in general (Strobel and Wollman, 1969) has led to attempts to assay its effectiveness in turtles (Kaplan and Taylor, 1957; Young and Kaplan, 1960; Hunt, 1964). There has been a great range of doses employed, each investigator having determined empirically, apparently, a serviceable level (see Table 1). The significant differences in dosages determined in this way may reflect the pharmacological nature of the drug itself. There is a direct relationship between the dosage employed and the plane of anesthesia achieved and some degree of reaction to noxious stimulation is observed under all but the most massive doses (Adriani, 1955; Hall, 1971). This quality may be partially responsible for the variety of dosage levels reported in the literature because, in each case, the subjects were exposed to different surgical procedures and, one may presume, different degrees of noxious stimulation.

Other factors possibly contributing to the variance in dosage levels used are species, routes of administration, and generally unreported or unknown factors, such as body temperature, sex, and age.

The efforts of Kaplan and Taylor (1957) produced the first systematic work devoted solely to the study of anesthesia in turtles, specifically adult *Pseudemys scripta*, the species most widely used in laboratories. Their study includes observations on such details as corneal reflex, muscle tone, gross movements, pupil diameter, heart rate, rectal temperature, and, of primary interest here, average dosages, induction times, and durations of deep anesthesia. Deep anesthesia was defined in this study in terms of muscle relaxation, absence of movement, and lack of response to unspecified noxious stimuli. No data were provided concerning excessive or fatal dosage levels. Three routes of administration were employed: I.P. (intraperitoneal), I.V. (intravenous), and I.C. (intracardiac). Surprising, perhaps, is the finding that the average effective dosage was similar with all three routes: about 16 mg/kg. Other characteristics were more route dependent. I.C. injections, made with a long needle inserted between the neck and forelimb, produced the most rapid induction, about 30 min, while I.P. and I.V. injections averaged 65 and 53 min, respectively. The shortest period of anesthesia was for the I.C. route. At 3 hr after the onset of deep anesthesia, 55% of the 38 turtles receiving I.C. injections were still deeply anesthetized. No animals remained deeply anesthetized longer than 10 hr.

Our experience suggests that a slightly higher dosage level for Nembutal is required than that reported by Kaplan and Taylor. We have implanted stainless steel jaw screws in over 40 *P. scripta elegans* under Nembutal anesthesia and have found that a suitable dosage is about 20 mg/kg, a value below the average of all the studies listed in Table 1 (Granda, Maxwell, and Zwick, 1972).

The Nembutal in our research is generally diluted with turtle Ringer's solution at a ratio of 1 part of Nembutal to 4 parts of dilutant to assure a widespread distribution of the drug. Turtle Ringer's solution contains 8.04 g of sodium chloride, 0.22 g of potassium chloride, and 0.25 g of calcium chloride plus a trace of sodium carbonate in each 1000 cc of distilled water (Rogers, 1938).

Both I.P. and I.M. routes were used to administer Nembutal with about equal success, although the I.P. route produced somewhat more rapid induction. In either case, induction averaged slightly over 1 hr. The action of Nembutal was variable even though such factors as classification (*P. s. elegans*), weight (approximately 1.2 kg), sex (female), temperature (24–27°C), and surgical procedure were controlled. Some specimens reached a surgical level of anesthesia within 45 min, while others did not become deeply anesthetized even after several hours. When the turtle was not deeply anesthetized within 2 hr, a supplemental 10 mg was given, usually with the result that the surgical level was reached within another hour.

Since turtles do not catabolize barbiturates rapidly (Dessauer, 1970), recovery from Nembutal is a slow process. A turtle receiving 20 mg/kg typically remains deeply anesthetized for several hours. A drowsy period lasting several more hours follows. During this period the turtle usually exhibits cycles of undirected locomotion and somnolence and should not be allowed in deep water. The animals that require more than the initial dose typically take longer to recover, even though they originally appear to be more resistant to the drug. Some of the turtles that receive large total amounts of Nembutal (40 mg/kg) take days to recover, if they recover at all. We find a dosage exceeding 40 mg/kg to be fatal in about 50% of the cases and one of more than 50 or 60 mg/kg to be almost always fatal. In contrast, Young and Kaplan (1960) report that nearly 50% of *P. scripta* receiving only 20 mg/kg alone by I.C. injection failed to recover. Perhaps there is an inherent safety factor associated with the slower acting routes where the drug diffuses into the blood more slowly. In support of this hypothesis is the finding by Hartse and Rechtschaffen (1974) that the mortality rate decreased when multiple small doses given over a period of time were substituted for a single large injection (also see Flanigan, Knight, Hartse, and Rechtschaffen, 1974; and Flanigan, 1974).

Table 1 Doses of Nembutal Used with Turtles

Reference	Subjects	Dosage	Route	Type of Surgery
Bantli (1974)	*Pseudemys scripta elegans*	20 mg/kg	I.P./or subcutaneous	Craniotomy
Belekhova and Kosareva (1971)	*Emys orbicularis* *Testudo horsfieldi*	5–20 mg/kg	—	Craniotomy
Butler and Knox (1970)	*Chrysemys picta belli* (150–800 g in weight)	6.0 mg/animal	I.M.	Hypophysectomy
Flanigan et al. (1974)[a]	*Terrapene carolina bauri* *Terrapene c. carolina* *Terrapene c. major* (498–830 g in weight)	20–35 mg/animal (small multiple doses)	I.P.	Implant screws in skull and carapace
Gandal (1958)	*P. s. elegans* *T. c. carolina*	6.0 mg/kg	Subcutaneous	Plastron puncture
Granda et al. (1972)	*P. s. elegans*	20 mg/kg	I.M.	Implant jaw screws
Gulick and Zwick (1966)	*P. scripta*	35 mg/kg	I.P.	Craniotomy
Hall and Ebner (1970)	*P. scripta*	37 mg/kg	—	Craniotomy

Hartse and Rechtschaffen (1974)	*Geochelone carbonaria*	27–98.2 mg/kg (small multiple doses)	I.P.	Implant screws in skull and carapace
Heisey (1970)	*P. s. elegans* *C. picta*	66 mg/kg	Oral	Craniotomy
Hunt (1964)	*Testudo g. graeca*	18 mg/kg	I.P.	—
Kaplan and Taylor (1957)[b]	*P. scripta*	16 mg/kg 17.5 mg/kg 15.5 mg/kg	I.P. I.V. I.C.	— — —
Karamian et al. (1966)	*E. orbicularis*	10–20 mg/kg	I.V.	Craniotomy
Morlock (1972)	*P. s. elegans* *C. p. picta*	25 mg/kg 40 mg/kg	I.P.	Craniotomy
Patterson and Gulick (1966)	*Pseudemys*	30 mg/kg	I.P.	Implant jaw screws
Zagorulko (1968)	*E. orbicularis* *T. horsfieldi*	10–15 mg/kg	I.V.	Craniotomy

[a]See also Flanigan (1974).
[b]See also Kaplan (1958a), Young and Kaplan (1960), Kaplan (1969).

Unlike Hall and Ebner (1970b), Young and Kaplan (1960) advise against warming experimental animals to speed recovery from Nembutal anesthesia. Although they did find a 25% reduction in recovery time when *P. scripta* were adapted to 30°C, they also observed that the combination of heat and pentobarbital excessively raised the animals' heart rates (also see Hutton, Boyer, Williams, and Campbell, 1960).

Nembutal may not be equally effective for all species. Hunt (1964) reported that *Testudo g. graeca* required 18.2 mg/kg, a dosage near that required by *P. scripta*. But, Morlock (1972) found that *Chrysemys picta* required 40 mg/kg, whereas *P. scripta* could be anesthetized with only 25 mg/kg. The validity of this difference is reinforced by the fact that animals of both genera were subjected to identical surgical procedures, although it should be noted that only a few animals of each species were used.

Except for the studies of Kaplan and Taylor (1957), Young and Kaplan (1960), and Hunt (1964), none of the research summarized in Table 1 was devoted specifically to the study of anesthesia, and the dosage levels are not necessarily presented as guidelines for specific classifications of turtles.

Nembutal Mixtures. Nembutal has been used successfully in combination with other drugs, for example, with gallamine triethiodide, a muscle relaxant, in *P. scripta* (Walsh, Houk, Atluri, and Mugnaini, 1972; Bantli, 1974) (see Tables 1 and 2).

Belekhova and Kosareva (1971) employed a Chloralose-Nembutal mixture; α-Chloralose (Sigma Chemical Co.) is a glucose and chloral compound. Equi-Thesin was a commercially prepared mixture similar to the Chloralose-Nembutal compound that, when injected I.P. at a dosage of 1.5 ml/kg, had an effect on *P. scripta* similar to a 20 mg/kg dose of Nembutal. Unfortunately Equi-Thesin is no longer manufactured. If required, 500 ml of a suitable substitute for Equi-Thesin could be made up from the following formula: chloral hydrate, 42.5 mg; pentobarbital sodium, 9.7 mg; magnesium sulfate, 21.2 mg; propylene glycol, 44.3% w/v; alcohol, 11.5%; and distilled water, q.s.

A combination of drugs used by Young and Kaplan (1960) with *P. scripta* appears to be particularly effective. They found that when an I.M. injection of 10 mg/kg of the tranquilizer chlorpromazine hydrochloride (Thorazine, Smith, Kline and French Laboratories) was followed in 10 min by an I.C. injection of 10 mg/kg of Nembutal, anesthesia was achieved within an average of 15 min from the first injection. Deep anesthesia persisted for about 3 hr. The induction times and durations of anesthesia were more reliable than with Nembutal alone. Without chlorpromazine, some turtles failed to reach a surgical stage. This rapid induction technique is not recommended where speed is not essential because of the technical difficulties and the potential danger associated with I.C. injections (Calderwood, 1971), and the authors caution that the anesthetic must be injected slowly.

Urethane. Ethyl carbamate (Urethane, Aldrich Chemical Co.), the ethyl ester of carbaminic acid, has been used, but not often (Kaplan and Taylor, 1957; Patterson, Evering, and McNall, 1968; Zagorulko, 1968). In the most comprehensive examination given this drug as used with *P. scripta*, average effective dosages ranged from 1.7 g/kg (I.V.) to 2.8 g/kg (I.P.), and the range of induction times was similarly great: 2.1 (I.C.) to 4.5 hr (oral) (Kaplan and Taylor, 1957). Nearly all the turtles were deeply anesthetized for at least 10 hr. The unreliability of dosage level and effect has led those experienced with Urethane to advise against its use (Kaplan and Taylor, 1957; Hunt, 1964; Kaplan, 1969). In addition, it has a proved carcinogenic potential in some laboratory animals and it is potentially harmful to those handling it (Strobel and Wollman, 1969; Hall, 1971).

Etorphine. Etorphine (M.99, American Cyanamid), a morphine derivative with potent analgesic properties, has been used in veterinary practice with I.M. injections at a dosage of 0.3 to 2.7 mg/kg. Galapagos tortoises (*Geochelone elephantopus*) weighing 45.4 to 62.7 kg have been successfully anesthetized with 10 to 15 mg. Induction times were less than 20 min, and anesthesia persisted for 0.5 to 3 hr (Wallach, 1969).

Tribromoethanol. Tribromoethanol (Avertin, Squibb and Sons) was widely used before barbiturates became popular, but it is not currently employed routinely (Hall, 1971; Soma, 1971). Tribromoethanol has been used with oral (Mosby and Cantner, 1955; Brown, 1969; Heisey, 1970), rectal, and I.P. (Hunt, 1964) routes of administration. Hunt reports average induction times of about 53 min, with deep anesthesia lasting from 1 to 2 hr for a dosage of 250 mg/kg administered I.P. The same dosage was used for oral administration by Heisey. Hunt recommended a rectal infusion of 400 mg/kg for turtles in poor condition.

Tricaine Methanesulfonate. Tricaine methanesulfonate (TMS) (formerly M.S. 222, Sandoz) is available as ethyl *m*-Aminobenzoate, methanesulfonic acid salt (Aldrich Chemical Co.), or Finquel (Ayerst Laboratories). Satisfactory anesthesia has been achieved in turtles following I.M. or I.P. injections of 0.6 to 0.75 g/kg (Granda, Matsumiya, and

Sterling, 1965; Hertzler and Hayes, 1969; Granda, Maxwell, and Zwick, 1972). Deep anesthesia lasts from 1 to several hours. Recent evidence suggests that TMS is more effective if buffered in the region of pH 7–8 than if used in the commercially available acidic form (Ohr, 1976).

Inhalation Anesthetics

The primary difficulty associated with inhalation anesthetics is the problem of administration. The strong odor, characteristic of those agents, appears to be obnoxious to turtles, and they react to the vapors with long periods of suspended respiration (apnea). Often rather drastic measures have to be taken to get the vapors into the animals' lungs. Furthermore, anesthetic agents typically have a relaxing effect on the respiratory muscles of reptiles (Brazenor and Kaye, 1953; Betz, 1962), and since volatile drugs are principally eliminated by respiration, some form of artificial ventilation must be employed to effect a reasonably rapid recovery.

Diethyl Ether. Diethyl ether, commonly known simply as ether (Squibb and Sons), has been administered to *P. scripta* (Kaplan and Taylor, 1957; Boycott and Guillery, 1962) and *T. graeca* (Hunt, 1964; Walker and Berger, 1973) with mixed results. Ether was recommended by Kaplan and Taylor, who found the average induction time to be about 37 min. A similar value was reported by Boycott and Guillery.

Kaplan and Taylor used a tight-fitting mask to administer the ether. A high initial concentration with minimum air was most effective. Presumably the investigators ceased administration when deep anesthesia was reached: no maintenance concentration was reported. Anethesia was more lengthy than with Nembutal, and nearly 40% of the animals were still deeply anesthetized at 10 hr. In contrast, Boycott and Guillery describe the recovery of their animals as "rapid," with the turtles eating within 24 hr.

Hunt noted that most tortoises are able to resist anesthesia for hours by apnea when merely confined in an anesthetic chamber. He also described the great difficulty often encountered in fitting a mask to a tortoise and intimated that possible harm could come to the animal in the struggle. Notwithstanding the difficulties, he found that with a mask, induction usually took 30 to 50 min.

There are other negative factors to be considered with this agent. Ether is flammable and, in the right proportions to air, explosive. It should be used only under rigid control and away from spark or flame. Ether is also irritating to skin and mucous membranes and can cause death if the subject accidentally aspirates it (Hall, 1971). Under the right conditions ether might be used effectively, but it certainly should not be used casually.

Halothane and Methoxyflurane. The fluorinated hydrocarbon halothane (Fluothane, Ayerst Laboratories) has been used to anesthetize *P. scripta* (Calderwood, 1971) and *T. graeca* (Gans and Hughes, 1967; Rosenberg, 1970, 1974).

Calderwood's induction concentration was 3% by volume mixed with oxygen, delivered to a small-animal anesthesia chamber by a vaporizer. Induction was difficult because of a long period of apnea. The addition of 5 or 10% carbon dioxide to the gas mixture resulted in increased respiration volume and decreased induction times. It was suggested that induction might be achieved more easily if the halothane were delivered through an endotracheal cannula. Calderwood inserted a tube through a hole in a bite block into the glottis, which was first sprayed with a local analgesic. Postinduction anesthesia was maintained with a 1.5% by volume concentration of halothane.

Gans and Hughes (1967) and Rosenberg (1970) administered a mixture of halothane and carbon dioxide to *T. graeca* in desiccators. Rosenberg decerebrated his animals under halothane and later paralyzed them with tubocurarine (1–2 mg). They were respirated by bronchial cannulae at a rate of 3 cycles/min with a tidal volume of 10 to 12 ml.

Rosenberg (1974) modified his procedure to achieve a more rapid induction of anesthesia. A hole was quickly drilled through the carapace over one lung and halothane vapor was injected directly into the exposed organ. Later, after the anesthetized subject was decerebrated, another hole was drilled to the contralateral lung, the trachea was occluded at the neck, and the lungs were ventilated with an artificial respiration pump. Still later, the tortoise was paralyzed with an initial dose of 5 mg of tubocurarine. Electrical recordings from the spinal cord were stable for up to 13 days when the animal was stored between recording sessions at 3 to 6°C.

We had no success with attempts to anesthetize turtles by enclosing them in a desiccator containing a gauze pad soaked with methoxyflurane (Penthrane, Abbott Laboratories). Methoxyflurane has a relatively low vapor pressure, and at 23°C it is possible to obtain only a 4% concentration even with a vaporizer (Hall, 1971). Because of this property methoxyflurane is not recommended as an induction agent, though it might be useful as a maintenance anesthetic after induction has been obtained by other means.

Neither halothane nor methoxyflurane is explosive or flammable at clinical concentrations and room temperature, and both can be used with electric cautery. The vapors of both agents are nonirritating and do not cause excessive mucus secretion as ether does; in liquid form, however, they could be lethal if aspirated.

Nonchemical Anesthesia

Hypothermia. Anesthesia through hypothermia can be induced by placing the turtle in a refrigeration unit (Spigel and Ellis, 1966; White and Ross, 1966) or by immersing it in crushed ice (Butler and Knox, 1970; Spigel, Ramsay, and Seggie, 1970; Granda, Maxwell, and Zwick, 1972).

Hypothermia does not offer rapid induction: it may require several hours to achieve the desired depth of anesthesia. In addition, prolonged surgical procedures must be carried out while the animal remains chilled, as on a tray of ice (Butler and Knox, 1970), and with supplements of local analgesics (Calderwood, 1971). The depth of anesthesia is not great (Granda, Maxwell, and Zwick, 1972), and the turtle usually reacts sluggishly to noxious stimuli (Musacchia and Sievers, 1956). Another undesirable side effect is the occasional respiratory infection that follows prolonged or repeated chilling. Lethargy, bubbling nose, and difficulty in breathing are the most obvious symptoms. Chloramphenicol (Chloromycetin, Parke, Davis & Co.), administered either orally or by injection, is effective in combatting the infection, and a satisfactory treatment for bacterial diseases is detailed by Doyle and Moreland (1969). (Also see Chapter 6.)

Hypothermia does offer the benefit of rapid recovery upon rewarming to normal temperatures, and there is the further advantage that it is a relatively safe procedure.

Local Analgesia

Local analgesics are particularly valuable with turtles because they permit the use of a lower level of anesthesia during surgery, and recovery periods are shorter.

Lidocaine. Lidocaine (Xylocaine, Astra Pharmaceutical Products, Inc.) is a valuable analgesic for use with turtles. Chemically, lidocaine is diethylaminoacetyl-2,6-xylidine hydrochloride, a very stable compound that can be stored and resterilized indefinitely. Lidocaine has great tissue-penetrating power and, as a consequence, has a rapid onset of effect (Hall, 1971).

Dantzler and Schmidt-Nielsen (1966) used lidocaine in a 2% solution to produce analgesia in the hindlimbs of unanesthetized but restrained animals (*P. scripta* and *Gopherus agassizi*) before surgically exposing the femoral vein. Once the exposure was made, the vein was cannulated with a polyethylene tube, PE 50, blood samples were taken, and injections given. Robin, Vester, Murdaugh, and Miller (1964) and Haning and Thompson (1965) reported using a similar technique. In addition, Robin et al. (1964) were able to take blood samples over a period of several days by keeping the catheter tied in place and filled with sodium heparin (Upjohn Co.) diluted 1:500 with saline.

We have used lidocaine routinely to infiltrate the temporal muscles of the head and to bathe the surgical wounds incurred during craniotomies performed on curarized turtles. It is nonirritating, and Hall (1971) reports that concentrations as high as 8% have been applied to the corneas of rabbits with no reaction. Lidocaine is also effective on membranous tissue and can be sprayed on the glottis prior to entubation.

Xylocaine Ointment (lidocaine in polyethylene and propylene glycols, Astra Pharmaceutical Products, Inc.) is especially useful on surgical wounds. It adheres well to the tissue, prevents drying, and maintains analgesia for long periods.

Procaine Hydrochloride. Another local analgesic of possible value is procaine hydrochloride (Sherry Pharmaceutical Co., Inc.; Novocain, Winthrop Laboratories). The analgesic effects of this drug are similar to those of lidocaine, although procaine has poorer penetrating power and is therefore slower acting and not effective when applied as a surface analgesic.

Muscle Relaxants

The most widely used muscle relaxants have no anesthetic or analgesic properties, but they are valuable by themselves and as adjuncts to true anesthetics.

Tubocurarine. Curare is extracted from the plant *Chondrodendron tomentosum*. The active ingredient is the alkaloid D-tubocurarine, a quaternary ammonium base (Adriani, 1955, 1970). Tubocurarine is known as a competitive neuromuscular blocking agent because it adsorbs to the acetylcholine receptors at the postsynaptic membrane, thereby blocking the transmitter action of acetylcholine. Tubocurarine is not a depressant of the central nervous system and does not pass the blood-brain barrier (Adriani, 1962; Alexander, 1969; Tavernor, 1971).

Tubocurarine is not absorbed significantly through unbroken skin and is not effective taken orally except in massive doses. It is thought either that the drug is excreted from the alimentary tract much faster than it is absorbed (Alexander, 1969) or that it is destroyed as it passes through the mucous membranes of the gastrointestinal tract (Adriani, 1962). The drug is absorbed from muscular sites after hypodermic injections, but paralysis occurs much more rapidly if the route of administration is intravenous. High rates of blood flow will hasten the induction time and quicken recovery as well. Any condition that reduces blood flow, such as vasoconstriction caused by reduced temperatures, will delay both induction and recovery (Adriani, 1962). Poikilothermic animals that under normal circumstances have a relatively sluggish blood flow, succumb to paralysis and recover from it more slowly than warm-blooded animals.

The paralyzing effects of tubocurarine do not occur uniformly throughout the entire striated muscle system, and there is a minimal effect on smooth muscle tissue (Adriani, 1970), although Strobel and Wollman (1969, p. 1392) report a possible "atropine-like" effect on smooth muscle responses to acetylcholine. Generally, in mammals (Adriani, 1970; Tavernor, 1971) the stage of induction or recovery can be determined roughly by noting the pattern of paralysis of the subject, but this is not possible in turtles. It is improbable that the musculature of turtles is homologous with that of mammals or that its susceptibility to curarization follows the mammalian pattern. For example, in the turtle eye, the ciliary group and the sphincter of the iris are composed of striated muscles (Gillett, 1923) which, unlike the smooth mammalian intraocular muscles, probably behave pharmacologically like skeletal muscle in the presence of a curariform agent. The observation that a turtle is not perceptibly breathing does not conclusively indicate that the turtle is completely paralyzed. A better indicator is the motor response to a noxious stimulus such as a toe pinch or a light tap on the head. In turtles, by the time such a response has been abolished, breathing has completely ceased.

The ability of turtles to go without breathing for long periods (Belkin, 1963) does have an advantage in that it relieves the urgency, experienced when mammals are used, of initiating the artificial respiration procedures required by the use of muscle relaxants. One can wait until the turtle is immobilized and has a relaxed glottis before attempting to insert the endotracheal cannula. No preanesthesia is required with turtles as it is with animals having a closer dependence on oxygen.

Anticholinesterase drugs are routinely used as antidotes for tubocurarine (Adriani, 1955, 1962, 1970; Lumb, 1963; Alexander,

1969; Tavernor, 1971). These drugs, such as neostigmine methylsulfate (Prostigmin, Roche Laboratories) and edrophonium (Tensilon, Roche Laboratories) are thought to inhibit acetycholine hydrolysis, thus increasing the amount of available acetylcholine. Anticholinesterase drugs are short acting, and if recurarization occurs, additional doses of the anticurariform agent must be administered.

It must be remembered that tubocurarine is merely a muscle relaxant and in no way prevents the appreciation of external stimuli. Surgery may be performed on a curarized subject without any struggling because the subject is unable to resist. Humane considerations dictate, however, that some anesthetic or analgesic agent be administered concurrently with the relaxant so that the turtle's comfort is reasonably assured.

Tubocurarine in Practice. The use of muscle relaxants with turtles is not discussed in the standard veterinary anesthesia texts. The only published information comes from scientific studies that required immobilized turtles and the use of the paralytic agent was a secondary concern. Published data, summarized in Table 2, show that tubocurarine chloride has been the most widely used immobilizing agent for turtles. The variance in dosage level probably stems from the lack of information available to the experimenters and from the fact that the dosage level for muscle relaxants in turtles is not very critical. Large doses are apparently not toxic to turtles, and small doses will usually cause paralysis if given enough time. The one study that reported the recovery of the turtles at the end of the experiments is also one that employed a relatively high initial dosage (Granda, 1962). These turtles were given sufficient prostigmin methylsulfate at the end of each experimental run to effect recovery. No harmful effects of the large tubocurarine injection were noted.

Experiments in this laboratory use paralyzed *Pseudemys scripta elegans* as subjects, and some unpublished information is available on the techniques employed to immobilize them. Tubocurarine chloride, 15 mg/ml (Abbott Laboratories), is used as the relaxant. Each milliliter of the drug is diluted with turtle Ringer's solution to a volume of 10 cc before I.M. injection at a dosage of 8.0 to 9.0 mg/kg. Multiple small-volume injections are made in the hindlimbs of the turtles to ensure a wide distribution, and consequently more rapid absorption. Injections at this dosage produce complete paralysis within 25 to 50 min at 24 to 27°C, as determined by a lack of response to a sharp tap on the muscles of the neck. Smaller doses can be used when there is less need for rapid paralysis and shorter durations of immobilization are required. Paralysis can be achieved by the administration of 3.0 to 4.0 mg/kg of tubocurarine chlo-

Table 2 Doses of Relaxants Used with Turtles

Reference	Subjects	Dosage	Route
Armington (1954)	*Pseudemys scripta*	8 mg/kg tubocurarine	—
Bantli (1974)	*P. s. elegans*	2 mg/kg initial dose gallamine	I.V.
Granda (1962)	*P. s. scripta*	30–60 mg/animal tubocurarine	I.M.
Granda and O'Shea (1972)	*Chelonia m. mydas* (200–400 g)	3 mg/450 g tubocurarine	I.M.
Iwase and Lisenby (1965)	*P. s. elegans*	2–3 mg/kg gallamine diluted 4:1 in Ringer's	I.V.
Robbins (1972)	*P. s. elegans* (2–4 kg)	15 mg/animal tubocurarine	I.M.
Rosenberg (1974)	*Testudo hermanni* *T. graeca*	5 mg/kg initial dose tubocurarine	I.M.
Skoloda (1973)	*P. s. elegans* (0.6–3.8 kg)	7–15 mg/animal tubocurarine	I.M.
Strejckova and Servit (1973)	*T. graeca* *Emys orbicularis* (0.8–1.0 kg)	5 mg/kg tubocurarine	—
Volanschi and Servit (1969)	*T. graeca* (150–300 g)	0.6–1 mg/animal tubocurarine	I.P.
Walsh et al. (1972)	*P. s. elegans*	1–2 mg/kg gallamine	I.P.

ride injected in the manner described above. Induction time will be on the order of 1 hr or more.

Soon after complete paralysis is observed, the turtle is placed on a mechanical respiration device. Here, the mouth is opened and the tongue is depressed, thereby exposing the glottis (Figure 1). The muscles of the laryngeal prominence are relaxed, and it is a simple matter to insert a pliable vinyl cannula (2.0 mm O.D.) several millimeters into the trachea. The cannula is secured with adhesive tape to the lower jaw, and the short, exposed end is inserted into a plastic "Y" connector. The length of the exposed end is kept as short as possible to keep the tidal air volume at a minimum. One side of the "Y" connector leads, by way of a plastic tube, to a small animal respirator (Model No. 607, Harvard

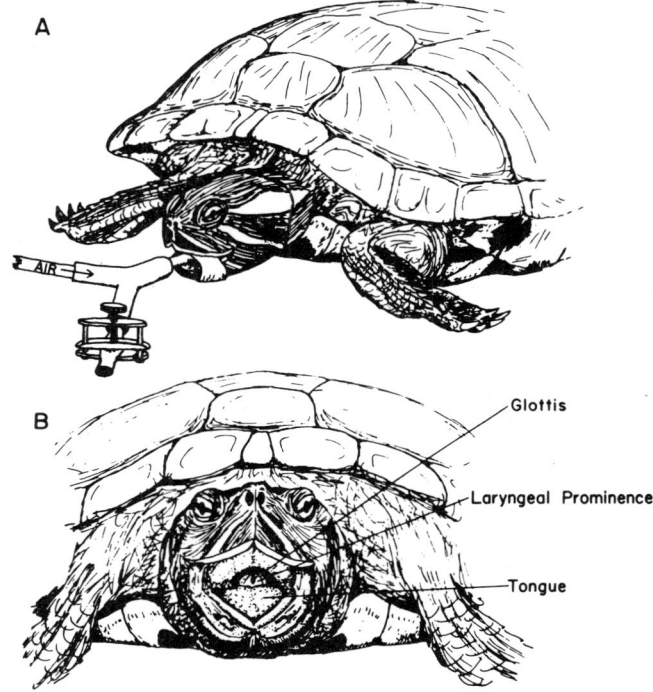

Figure 1 (a) A simple arrangement for an endotracheal cannula including a small tube clamp that regulates air pressure to the lungs and allows exhaled air to escape. (b) The mouth and larynx of a *Pseudemys scripta*.

Apparatus Co.) set to deliver 150 cc of air at a rate of 5 cycles/min. The other leg of the "Y" connector is supplied with a 3.0 cm section of rubber tubing and an adjustable tube clamp. The rubber tube allows excess air to be bled off. The clamp is adjusted so that there is a just-noticeable movement of the tissue lying between the turtle's forelimbs and neck during the positive respiration stroke. No negative pressure is applied. Exhalation is accomplished through the slow, natural collapse of the turtle's lungs. The expirate escapes easily through the tube clamp assembly.

Of course, whenever small animals are ventilated by a small cannula there is a danger of the cannula becoming clogged with exhaled mucus. We have injected 1.0 ml of a 10% solution of atropine sulfate (A. J. Buck and Sons) I.P. on the assumption that this would prevent the formation of excess mucus. No direct evidence is available concerning the effectiveness of this precaution, although there have been very few instances of cannula obstruction with its use. It might be noted, however, that atropine may produce undesirable neurological side effects in turtles (see Hartse and Rechtschaffen, 1974), and one author cautions against its use specifically with small intubated animals because atropine tends to thicken tracheobronchial secretions, which then heighten the probability of obstruction (Soma, 1971).

Recovery from Paralysis. The duration of paralysis varies greatly among individual turtles, and the length of time any particular specimen will remain immobilized is relatively unpredictable. Turtles receiving single doses of 8.0 to 9.0 mg/kg of tubocurarine chloride have remained paralyzed from 8 to 36 hr or more. One might expect an animal receiving a reduced dosage to recover somewhat more rapidly. In our laboratory paralyzed turtles have been maintained, in apparent good health, for several days. Maintenance requires only continued artificial respiration, protection against desiccation, and periodic small doses, 3.0 mg/kg, of the relaxant (J. E. Fulbrook, personal communication). Occasional small muscular reactions observed days after the initial injection indicate that recovery might be expected even after such prolonged paralysis. Generally, however, when recovery is desired the animal is given multiple I.M. injections totaling 1.0 mg of Prostigmin diluted with Ringer's solution to 0.12 mg/ml. The turtle is then confined in a dry container and observed. Usually, unless the dose of tubocurarine was light or the animal was near recovery naturally, 1.0 mg supplements of the antidote are required. Resumption of artificial respiration may be necessary if normal breathing ceases for a prolonged period and, of course, the turtle should not be allowed in deep water until it has been observed to recover completely, which may take several hours.

Turtles can and do recover the power to move about, but we are not sure whether they are completely sound even when that ability has returned. One specimen of *P. scripta* was observed to remain in a permanent trancelike state. This animal kept its neck completely extended except when greatly disturbed and continually tried to climb the walls of the home tank. Other animals seem to recover to apparently normal behavior and health.

Gallamine Triethiodide. There are several muscle relaxants available, but only gallamine triethiodide (Flaxedil, American Cyanamid Co.) has been used to any extent with turtles.

Flaxedil is similar in action to tubocurarine but, because of its relative impotency, larger doses are employed. We find I.M. injections of 30 to

40 mg/kg generally produce paralysis in *P. scripta* in 1 hr. In contrast, Walsh et al. (1972) and Bantli (1974) report using only about 2.0 mg/kg for *P. scripta*. The subjects in both these studies were presedated or preanesthetized with Nembutal, which probably served to potentiate the relaxation effect. That combination is particularly appealing because of the small dosages involved.

Iwase and Lisenby (1965) used Flaxedil to prevent movement in *P. scripta* using a dosage of only 2 to 3 mg/kg diluted 1:4 with Ringer's solution. Their procedure of injecting the drug directly into the jugular vein is probably the reason such a low dosage could be used. Such routes always seem to be more effective, although they are technically more difficult to use.

Euthanasia

It may be necessary to kill a turtle as an essential part of an experiment or as a way to alleviate the suffering of an incurably diseased or injured animal. The act should be performed by a fully trained person in a manner that is as humane as possible.

In selecting the proper method from the many available, several factors should be weighed: the procedure should not interfere with the interpretation of experimental data; the procedure should bring about a death that is as rapid, painless, and as esthetically acceptable as possible; finally, the procedure should be economical and safe for the operator (Breazile and Kitchell, 1969).

The simplest and most esthetically acceptable killing procedure is to administer an overdose of one of the injectable anesthetics. In most laboratory animals an I.V. route would be used, but since this requires some minor surgery in turtles, with their horny skin, an I.P. route is acceptable. A dose of 100 mg/kg of pentobarbital should be fatal. When extreme rapidity is required or when it is undesirable to contaminate the tissue with chemicals, turtles can be decapitated with a guillotine device of the type manufactured by the Harvard Apparatus Company.

SURGERY

Individual experimental or medical operations may require different degrees of surgical intervention, but all require that the turtle be suitably immobilized by physical restraint, anesthetization, or paralysis. In all cases the use of local analgesics is recommended.

Craniotomies

The turtle's head must be secured in a way that permits access to the area to be exposed. Usually the most satisfactory way to secure the head is to hold it with blunt ear bars that fit snugly into the depressions formed by the tympanic membranes found behind the angle of the jaws. The head is then held level by some form of clamp on the snout. Simple and satisfactory holders can be easily constructed (Figure 2); more sophisticated devices are available commercially (David Kopf Instruments, Tujunga, Calif.).

Once the head is secured, the site of the operation is infiltrated with a local analgesic and the skin is dissected from the area of the skull to be removed. There is a large degree of variation in turtle skull morphology, even among animals of the same genus and size. The relationship between the external skeletal landmarks such as cranial sutures and the neural structures is not strict; therefore fairly large areas of bone are usually removed so that particular neural structures can be seen. Figure 3 shows the approximate relationship between major skeletal and neural structures in *P. scripta* (also see Ashley, 1962; Gaffney, 1972a).

If caudal neural structures, such as the tectum or cerebellum, are to be exposed, it is necessary to dissect the temporal muscle away from the temporal fossa. This can be best accomplished with a blunt, flat instrument, such as a scalpel handle. The muscles are first injected with an analgesic, then gently separated from the bone without cutting. The muscles on both sides can be held out of the way with a single small spreader or with cotton packing.

An alternative approach is to remove the temporal muscle entirely. This is a difficult procedure and is not recommended unless the widest possible exposure is required. If this course is followed, it is necessary to cauterize and remove the underlying superficial temporal artery and its branches (Bantli, 1974).

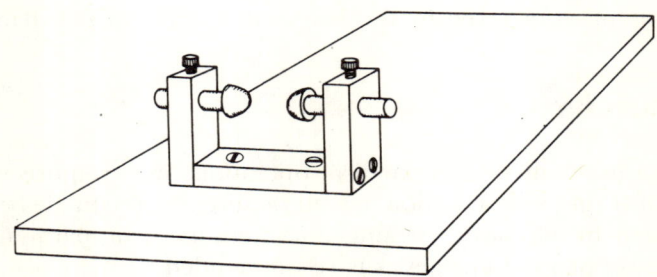

Figure 2 A simple, easily constructed head holder for turtles.

SURGERY

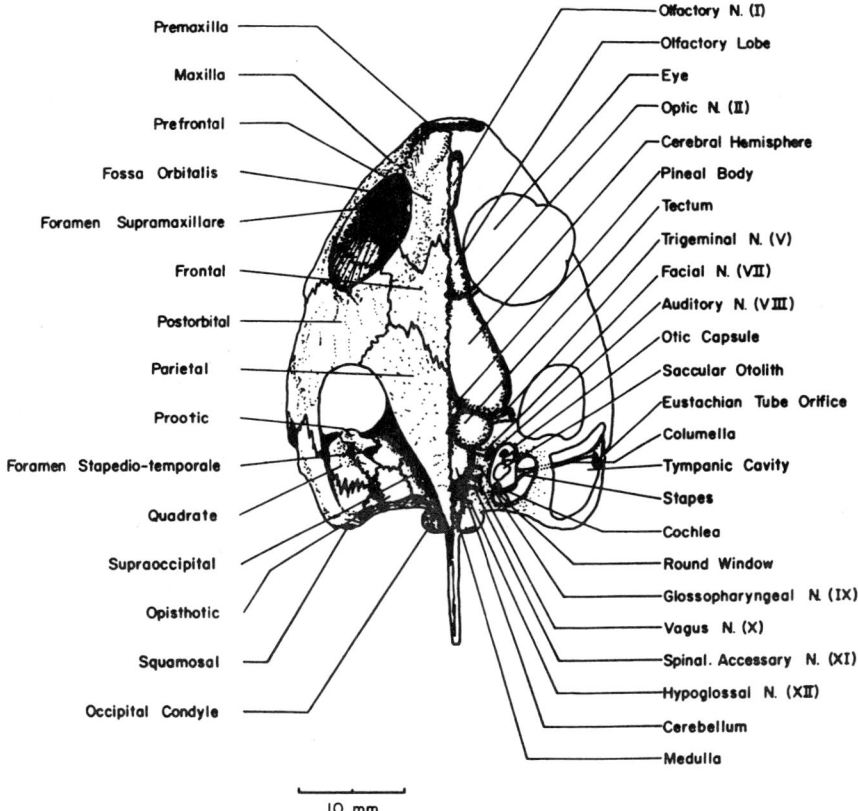

Figure 3 A dorsal view of the skull of a *P. scripta* with the right side removed to show the underlying neural structures as well as the auditory and vestibular apparatuses.

Bone material is most easily removed with an air-turbine dental drill. This high-speed instrument can "brush" away the skull over the desired area until the remaining bone is thin enough to pick away with probe and forceps. Exposures can also be made with either an ordinary (i.e. slower-speed) dental drill, trephine, or rongeurs, but these procedures are usually more difficult. With whatever method employed, care must be taken not to disturb the superior sagittal sinus found dorsal to the brain over the midline. The sinus extends from the most posterior portion of the brain anterior to the olfactory lobes. Rupture of this vessel results in severe bleeding that not only obscures the field of view but also may result in the turtle's death.

Once the desired exposure is made, the dura mater is slit and either removed or reflected out of the way. This must be done carefully to avoid rupturing the small blood vessels that trace the surface of the brain and are often in close association with the dura. Any vessels that cannot be reflected with the dura should be gently separated from that membrane and either left intact or, alternatively, cauterized and removed. The very thin pia mater does not hinder most experimental procedures such as ablation, electrical recording and stimulation, or chemical application; nor does it impede or damage delicate microelectrodes.

If the cranial opening is to be closed, the cavity is packed with absorbable sponge (Gelfoam, Upjohn Co.). Small openings need only have the skin, kept moist with Ringer's solution or Xylocaine Ointment, sutured back in place. Large openings should be covered with dental acrylic, and the skin need not be replaced. In both cases it is recommended that the surface be coated with collodion to protect against moisture.

Hypophysectomy

The pituitary gland (hypophysis) is encased in a bony cavity ventral and posterior to the optic chiasm. It can best be reached using a ventral approach through the roof of the mouth.

The anesthetized turtle is held ventral side up, with the shell and head firmly secured. The lower jaw is held open with a small retractor and the basisphenoid bone, located far back in the roof of the mouth, is exposed under local analgesia. A hole is drilled through the basisphenoid bone and the sella turcica, which encapsulates the pituitary (see Ashley, 1962, his fig. 10; also Gaffney, 1972a, his fig. 2) is peeled back. The pituitary can be snipped off and the stalk cauterized. The cavity is filled with Gelfoam and sealed with dental acrylic (Butler and Knox, 1970). The ventral approach to the brain permits access to neural tissue as well as the hypophysis and might be useful where structures such as the cranial nerves are not easily reached using a dorsal approach.

Ear Surgery

Though lacking an external ear, turtles do possess internal auditory structures accessible by a dorsal approach through the temporal fossa or by a lateral approach through the squamosal bone (see Figure 3; also see Wever and Vernon, 1956b, their fig. 1).

The sole auditory structure visible externally is the tympanic membrane, which appears as a round, slightly depressed area posterior to the angle of the jaw. Just medial to the tympanic membrane lies the air-filled tympanic cavity with its single columella, the homologue of the mammalian ossicular chain. The posterior portion of the tympanic cavity can be entered without disturbing the auditory apparatus by drilling laterally through the squamosal bone at a point approximately 4 mm posterior to the margin of the tympanic membrane (see Figure 3). The inner auditory structures as well as the vestibular mechanisms, all housed in the otic capsule, are best approached through the temporal fossa. The temporal muscle is removed in the manner described earlier and the bone overlying the otic capsule removed. The bones encasing the auditory and vestibular apparatuses are the prootic, quadrate, opisthotic, and squamosal bones though, as Figure 3 shows, a dorsal approach will also involve removal of some portion of the supraoccipital, which constitutes the posterior-medial floor of the temporal fossa (Patterson, Evering, and McNall, 1968).

Jaw Screw Implant

The conditioned-avoidance technique developed by Granda, Matsumiya, and Stirling (1965) to obtain sensory thresholds requires the position of the turtle's head to be controlled and monitored closely. One way to accomplish this task is to use a control line attached in some way to the turtle's head, and the simplest manner of attachment is by means of a small loop held to the jaw with screws and dental acrylic (Granda, Maxwell, and Zwick, 1972; Maxwell and Granda, 1975).

Implanting jaw screws is a relatively simple matter. Once the turtle has been anesthetized, the head is secured in a simple holder of the type illustrated in Figure 2. The mouth is held open with a bite bar pulled tightly back into the angle of the jaw. The toothlike cusps found on both sides of the jaw (see Figure 4) are ground off with a dental drill to make a flat surface. Two small holes are drilled through the mandible at a point slightly posterior to where the cusps were. In our experience the best tool to use for drilling holes is a small hand-held drill (Moto-tool, Dremel Manufacturing Co.) capable of holding a bit at least 5 cm long of a size just smaller than the screws employed. Self-tapping, flat-headed, stainless steel, 0-80 screws about 6 mm long are first sterilized, then screwed in tight (Figure 4). Screws of this size will not seriously weaken the jaw of a 20 cm turtle. Once the screws are in place, a loop of stainless steel wire is fastened around the screw tips and sealed in place with dental acrylic. The acrylic is formed in generous amounts around the screw tips and over the smooth horny lower surface of the jaw. Acrylic is also used to seal around and over the screw heads on the upper surface of the jaw (Figure 4).

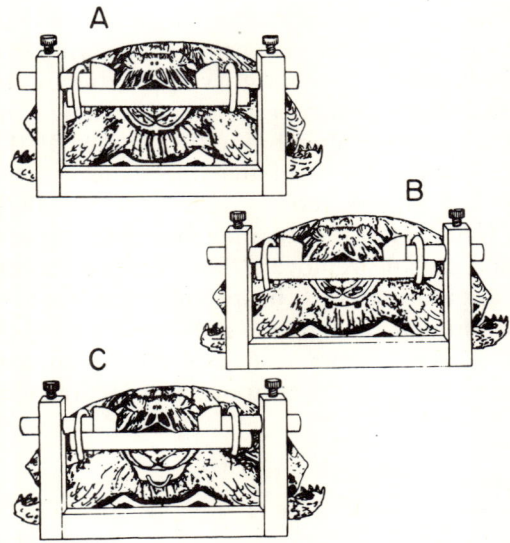

Figure 4 Illustrations of how a small loop can be attached to a turtle's jaw by means of jaw screws. Note the toothlike cusps on the inner surface of the jaw in (*a*) and the liberal use of dental acrylic in (*c*).

The jaw screw implant usually remains in good condition for well over a year of experimental use, and the turtles show no change in their eating pattern. Eventually, however, the screws work loose. When this happens, the old screws can be removed, the holes cleaned out and new screws implanted. If a turtle is to be taken out of an experiment but not sacrificed, as might be the case with a well-trained subject, the screws can be removed and the jaw tissue will regenerate completely without special care in about a year.

Thoracic and Abdominal Surgery

It is beyond the scope of this work to describe the numerous operations that may be performed either for experimental or veterinary purposes. Adrenalectomy and cardiac puncture techniques are described in detail, so that these can serve as models for other thoracic or abdominal procedures.

Adrenalectomy. The adrenal glands can be reached most easily with a dorsal approach through the posterior portion of the carapace.

Butler and Knox (1970) employed hypothermia anesthesia because that technique results in a greatly reduced blood flow to the internal organs. A ¾ in. diameter trephine was used to drill an opening in the carapace at the intersection of the fourth vertebral shield with the third and fourth costal shields. The oblique abdominis muscle was cut and held apart with hemostats to expose the kidney and the nearby testis. The testis was displaced, and the adrenal gland, found on the anterior ventral portion of the kidney, was removed by cautery. Gelfoam was applied to the wound on the kidney to retard bleeding and the testis was returned to its normal position ventral to the kidney. The oblique abdominis was sutured with silk thread and the cavity filled with Gelfoam. The disk of bone and shell removed with the trephine was replaced and held with a resin cement. After the procedure was completed bilaterally, the shell was bound with tape for 24 hr to prevent movement of the fore and hindlimbs and to minimize postoperative bleeding.

It may be wise to use male turtles for experimental adrenalectomy procedures whenever possible. In the female the ovaries are found in the same anatomical location as the male testes, and the oviducts are often filled with developing eggs. The resulting crowding may make the adrenalectomy difficult.

Cardiac Puncture. As with adrenalectomy, hypothermia is a good method to produce anesthesia before performing a cardiac puncture. The turtle is immobile, there is reduced blood flow, and, most important, the heart is still for several seconds between contractions (Hutton, et al., 1960), thereby easing the chore of puncturing the organ.

The heart may be exposed by drilling through the anterior-median corner of the right abdominal shield of the plastron (Zangerl, 1969). Depending on the requirements of individual procedures, different sized openings can be made: small openings should be made with small drill bits and larger openings with a trephine or the type of hole saw used with hand drills. If the hole is made carefully, the shell and bone material can be removed without puncturing the peritoneal membrane. In acute procedures, such as a perfusion, the peritoneal membrane and the pericardium can be dissected away for a clear approach to the heart. In chronic or veterinary procedures, these membranes can be carefully slit and held aside with hemostats or, alternatively, left completely intact.

To sample blood, a 22-gauge needle attached to a heparinized syringe is quickly thrust into the heart and the required amount of blood withdrawn. If a trephine or hole saw is used, the disk of shell and bone can be placed back in the hole and sealed as in the adrenalectomy. If multiple blood samples are to be taken, the opening can be plugged with a closely fitting stopper and sealed with wax or tape. Turtles can survive numerous bloodlettings, living without special attention for months or years in their laboratory tanks (Altland and Parker, 1955; Musacchia and Sievers, 1956; Rapatz and Musacchia, 1957; Hutton and Goodnight, 1957; Hutton, 1960; Dessauer, 1970).

Other methods of cardiac puncture do not involve making a hole through the plastron. The heart may be approached with a long needle directed either through the soft tissue between the neck and forelimb or through the rear leg pocket (Dessauer, 1970). One can also push an 18-gauge needle directly through the plastron at the point where the pectoral and abdominal shields intersect on the midline. The needle is inserted with a rotating motion, using a fair amount of pressure, posterodorsal at an angle 20° from the vertical. The heart is pierced at a depth of 0.5 to 1.0 in. (Gandal, 1958; also see Lyman, 1945; Frair, 1963; Belkin, 1964). These two latter methods are quicker than the hole drilling procedure, but neither affords a view of the heart and there is, therefore, a heightened risk of injury to that organ.

Aside from cardiac puncture, various other bloodletting techniques are used with turtles. Blood samples can be obtained from the femoral vein (Robin et al., 1964; Haning and Thompson, 1965; Danzler and Schmidt-Nielsen, 1966), the jugular vein (Lopes, 1955), or the carotid artery (Crenshaw, 1962). Blood in microliter amounts can be repeatedly obtained from the retroorbital sinuses using disposable heparinized pipettes with polished tips (Frair, 1963). In this technique, the pipette tip is gently guided in next to the eyeball until it ruptures the fine blood vessels lining the orbit (also see Riley, 1960, and the section on local analgesia above).

The anatomical location of the other internal organs can be determined by consulting the laboratory guide of Ashley (1962).

DISCUSSION

It is apparent that many people have been engaged in finding ways to immobilize turtles when their efforts could have been more profitably directed elsewhere. What is needed most in the field of turtle anesthesiology is a comprehensive investigation of a variety of anesthetics and muscle relaxants. Ideally, such a study would vary such parameters as sex, age, species, weight, and body temperature and would arrive at a standardized technique that might bring order to a chaotic situation.

Most desperately needed in the field of surgery, aside from a standard anesthetic technique, are stereotaxic atlases for the brains of the most commonly used turtle species. At present it is necessary to remove a large area of the skull, exposing gross neural landmarks, before finer neural structures can be approached. Even following this procedure there are relatively few deep-lying brain structures that can be reached with confidence and consistency. Stereotaxic atlases for turtles of the quality now available for mammalian and avian brains would be of inestimable value to those engaged in the study of comparative neurology.

ACKNOWLEDGMENT

The research for this chapter was supported by grant EY 01540 from the National Eye Institute, National Institutes of Health.

APPENDIX A LIST OF DRUG MANUFACTURERS AND THEIR ASSOCIATED PRODUCTS

Abbott Laboratories
North Chicago, Ill. 60064

Pentobarbital sodium
   (Nembutal)
Tubocurarine chloride
Methoxyflurane
   (Penthrane)

A. J. Buck and Sons, Inc.
Cockeysville, Md. 21030

Atropine sulfate

Aldrich Chemical Co., Inc.
Milwaukee, Wis. 53233

Ethyl carbamate
   (Urethane)
Ethyl m-aminobenzoate, methanesulfonic acid salt (Tricaine methanesulfonate)

American Cyanamid Co.
Pearl River, N.Y. 10905

Etorphine (M. 99)
Gallamine triethiodide
   (Flaxedil)

Astra Pharmaceutical Products, Inc.
7 Neponset Street
Worcester, Mass. 01606

Lidocaine (Xylocaine and Xylocaine Ointment)

Ayerst Laboratories, Inc.
Veterinary Medical Division
685 Third Avenue
New York, N.Y. 10017

Halothane (Fluothane)
Tricaine methanesulfonate (Finquel)

Parke, Davis and Co.
P.O. Box 118
Detroit, Mich. 48232

Chloramphenicol
(Chloromycetin)

Roche Laboratories
Division of Hoffmann-LaRoche, Inc.
Nutley, N.J. 07110

Neostigmine methylsulfate
(Prostigmin)
Edrophonium (Tensilon)

Sherry Pharmaceutical Co., Inc.
Long Island City, N.Y.

Procaine hydrochloride

Sigma Chemical Co.
P.O. Box 14508
St. Louis, Mo., 63178

α-Chloralose

Smith, Kline and French Laboratories
1500 Spring Garden Street
Philadelphia, Pa. 19101

Chlorpromazine hydro-
chloride (Thorazine)

Squibb and Sons
P.O. Box 4000
Princeton, N.J. 08540

Ether

Upjohn Company
7000 Portage Road
Kalamazoo, Mich. 49001

Gelfoam
Sodium heparin

Winthrop Laboratories
90 Park Avenue
New York, N.Y. 10016

Procaine hydrochloride
(Novocain)

VITAL
FUNCTIONS

CHAPTER 8
Cardiovascular System

CRAWFORD G. JACKSON, JR.

Although the cardiovascular system of turtles has been studied by a number of investigators over a long period of time, only a very few species have been studied in detail. Among those contributing to the current status of our knowledge are Bojanus (1819), Burne (1905), O'Donoghue (1917, 1918), Goodrich (1919), Kimball (1923), Thomson (1932), Schepers (1939), Kaushiva (1940), Mathur (1946), Taylor (1952), Dunlop (1955), McDowell (1961), Girgis (1961b, 1962a, 1962b), Grignon and Grignon (1962), Juhasz-Nagy et al. (1963), and Burda (1965). The basic architecture will probably be little modified by future studies, although the few forms examined do indicate that slight taxonomic (as well as the expected individual) variations are present within the order.

Fundamentally, the cardiovascular system consists of a heart having three distinctly different chambers, an arterial system, and a venous system that includes a well-defined renal portal system.

THE HEART

The heart is essentially reptilian, with the exception of a few morphological and functional characteristics that may be chelonian attributes.

The atria are relatively thin walled (with right wall slightly thicker), possess ridges (musculi pectinati) or trabeculae internally, and are unequal in size, the cavity of the right being noticeably larger in all forms studied thus far. The atria are completely separated from each other by the interatrial septum (= interauricular septum or septum auriculorum). The sinus venosus is a relatively large chamber attached to the dorsal surface of the atria. It is usually located more in association with the right atrium, into whose cavity it always discharges deoxygenated blood through the slitlike sinuatrial opening (= sinuauricular aperture), which is guarded by a valve composed of two thin flaps.

The wall of the sinus venosus is muscular, but it is somewhat thinner than that of the atria. Blood enters the sinus venosus through four veins: (1) left precaval (= descending vena cava or anterior vena cava), (2) right precaval (= descending vena cava or anterior vena cava), (3) postcaval (= ascending vena cava or posterior vena cava), and (4) left hepatic. Minor variations of the entrance point of these veins into the sinus venosus undoubtedly occur.

The left atrium receives oxygenated blood through two pulmonary veins, the openings of which are separate (except in some Trionychidae), but usually close and unguarded by valves.

The ventricle is the most muscular chamber of the heart and has an internal wall characterized by numerous somewhat spongy trabeculae, although definite trabeculae carneae cordis or musculi papillares may be lacking. The volume of its cavity is relatively small, reflecting the low stroke volume of the heart. The interior of the ventricle is subdivided into two major portions, a feature of great physiological significance. The names cavum grandum (= cavum dorsale or cavum magnum) and cavum parvum (= cavum pulmonale or cavum ventrale) have been applied to these portions, which are in continuity anatomically but are functionally separated by a muscular ridge, the interventricular septum. The cavum grandum can be considered to be further subdivided into a cavum arteriosum (= cavum sinistrum) on the left side and a cavum venosum (= cavum dextrum) on the right. Blood from the left atrium passes through the left atrioventricular valve into the cavum arteriosum, and blood from the right atrium enters the cavum venosum after passing through the right atrioventricular valve. Both right and left aortic arches originate in the cavum venosum, conducting blood from that cavity to the systemic circulation of the body. The pulmonary circulation begins in the cavum pulmonale, out of which the pulmonary trunk arises to transport blood to the lungs via the pulmonary arteries. The cavum venosum and cavum anteriosum are connected by a passageway that White (1959) recognized as the "interventricular canal." This canal is bounded anteriorly near the base of the ventricle by the membranous origins of the single-flapped right and left atrioventricular valves. These valves probably serve dual and reciprocal functions in (1) preventing backflow of blood from the ventricles to the atria during ventricular systole, and (2) during ventricular diastole, blocking exchange of blood between the cavum arteriosum and the cavum venosum.

THE HEART

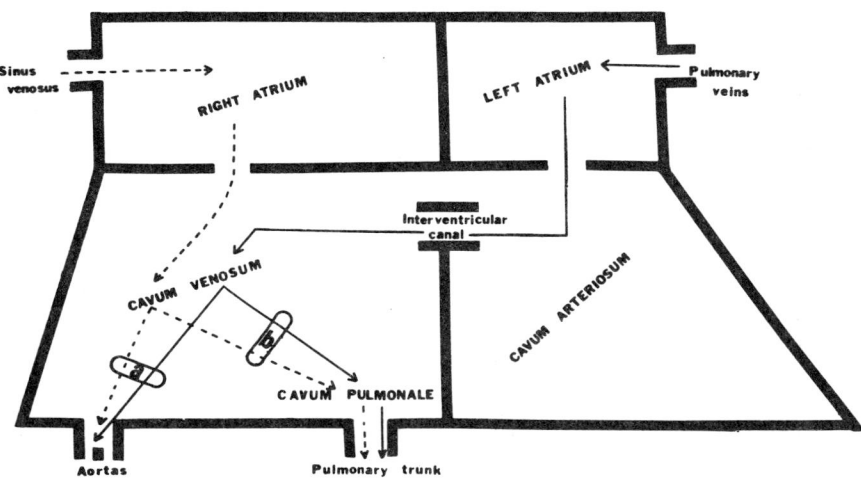

Figure 1 Schematic of chelonian heart and pathways of blood flow. Dashed lines indicate relatively deoxygenated blood as returned to the heart from the venous system. Solid lines indicate relatively oxygenated blood as returned to the heart after transiting the capillary beds of the lungs. Systemic pathway is indicated by **a**, pulmonary pathway by **b**. Figure prepared by Marguerite M. Jackson.

As a result of the investigations of a number of workers (Mathur, 1944, 1946; Girgis, 1961b; White, 1959, 1968; Steggerda and Essex, 1957; White and Ross, 1966; Millen et al., 1964) it appears that the course of blood through the chelonian heart is as shown in Figure 1. Relatively deoxygenated blood returning to the right atrium passes into the cavum venosum of the ventricle. From here, it follows two routes, one leading out of the heart by way of the right and left aortas, and one leading through the cavum pulmonale of the ventricle to exit the heart by way of the pulmonary trunk. Blood returning to the left atrium is relatively oxygenated (unless the animal is hypoxic) and passes into the cavum arteriosum of the ventricle. From here it enters the cavum venosum, where it is presented with the same two exit routes as the deoxygenated blood coming from the right atrium. The relative importance of pathways *a* and *b* in Figure 1 depends on the physiological state of the turtle. White and Ross (1966) found that when breathing, *Pseudemys scripta* routed 60% of its heart output into the pulmonary circuit (pathway *b* of Figure 1) and 40% into the systemic circuit (pathway *a* of Figure 1). The same investigators (White and Ross, 1965) were able to demonstrate with various techniques the existence of a "right-to-left shunt" and a "left-to-right shunt," that is, an increased favoring of pathway *a* at the expense of pathway *b*, and vice versa, for apneic turtles versus breathing turtles, respectively.

Although slight morphological variations of the interventricular septum remain to be described, functionally this muscular ridge acts as a directing baffle that channels incoming blood toward the pulmonary and systemic routes. Mixing of blood within the ventricle is further inhibited mechanically by the presence of trabeculae.

ARTERIAL SYSTEM

Blood exiting the heart must do so through three vessels: the pulmonary trunk (arising out of the cavum pulmonale), the left aorta or left systemic arch, and the right aorta or right systemic arch (the latter two arising out of the cavum venosum). The pulmonary trunk (= common pulmonary arch) extends anteriorly for a short distance, then bifurcates into right and left pulmonary arteries carrying blood to their respective lungs. The left aorta extends dorsolaterally, giving off either two or three major arteries before it joins posteriorly with the right aorta to form the dorsal aorta. The presence or absence of a shunt (the ductus arteriosus or ductus Botalli) between each pulmonary artery and its corresponding aortic arch appears to be variable. Bojanus (1819) describes both a right and left ductus Botalli in *Testudo europaea* (now *Emys orbicularis*). Gegenbaur (1901) claimed that small patent ducts were present in *Chelydra serpentina*. Where patent ducts do not exist, it seems that at least a fibrous vestige, the ligamentum arteriosus, remains.

The first major branching of the left aorta is usually a common gastric artery followed by a coelic artery, although Girgis (1962b) found them to have a common short origin (the "gastro-coelic artery") but dividing almost immediately in *Trionyx triunguis*. The third (usually) major branch from the left aorta is the anterior (= superior of some authors) mesenteric artery.

The common gastric artery bifurcates at the stomach to form the dorsal (= posterior) gastric artery (supplying the greater curvature region of the stomach) and the ventral (= anterior) gastric artery (supplying the lesser curvature region).

The coeliac artery divides into the anterior pancreaticoduodenal artery (supplying blood to the pylorus, liver, pancreas, and duodenum through many variable branches) and the posterior pancreaticoduodenal artery (also supplying the liver, pancreas, and duodenum). The cystic artery is an extension of the posterior pancreaticoduodenal artery and sends blood to the muscular wall of the gallbladder.

ARTERIAL SYSTEM

The anterior (= superior) mesenteric artery courses posteriorly in the mesenteries breaking up into a fanlike arrangement that penetrates various parts of the small intestine. The large intestine is served by the posterior (= inferior) mesenteric artery, a prominent branch of the anterior mesenteric artery.

In contrast to the left aorta, which usually has three major branches, the right aorta has only one, the brachiocephalic (= innominate) artery. This major vessel arises from the left aorta immediately after the latter leaves the ventricle. The right aorta then curves sharply to the right, coursing posteriorly and mesially to join with the left aorta posteriorly and form the dorsal aorta. Near its beginning the brachiocephalic gives rise to the common coronary artery unless the origin is from the base of the right subclavian artery. The common coronary artery quickly bifurcates into a ventral coronary artery (which runs along the ventral aspect of the ventricle) and a dorsal coronary artery (running on the dorsal aspect of the ventricle). The latter artery divides into right and left atrial (= auricular) branches going, respectively, to the posterior walls of the right and left atria. The brachiocephalic artery after a very short straight course bifurcates into the right and left subclavian arteries, then almost immediately into what are often called right and left common carotid arteries (or simply, "carotid" arteries, which has led to some confusion).

The subclavians with their branches supply arterial blood to most of the anterior part of the body with exception of anterior portions of the neck, and the head. The distribution pattern is complex, and confusion exists because the few species studied in detail are inadequate to assess the extent of the variability of patterns due to taxonomic and individual differences. Another problem encountered is the difficulty in determining the synonymies existing between the vessels described by various workers.

The thyroid (= thyroideal) arteries are among the first given off by the subclavians. The esophageal artery ramifies over the ventral surface of the esophagus while sending branches to the ventral cervical muscles and to the trachea.

Each subclavian extends anterolaterally, bifurcating into an ascending axillary artery and a descending axillary artery. The ascending axillary curves anteromedially to branch repeatedly in the scapular region. One of the first branches (usually) is the musculocutaneous artery, which sends branches to the dorsal body wall and to the skin of the posterolateral sides of the neck and head. The common vertebral artery and its branches supply blood to the deeper areas at the base of the neck, passing through the intervertebral foramina. The dorsal brachial artery runs from the ascending axillary in an anterolateral direction to supply (via its smaller branches) the dorsal area of the brachium and antebrachium. It eventually extends into the forefoot as the manus artery.

Immediately distal to the origin of the dorsal brachial, the common descending intercostal (= vertebral artery of some authors) artery arises. It runs caudally alongside the dorsal vertebrae in the channel created by the arching rib heads. Medially, it sends branches to the spinal cord while laterally it sends regular branches, the transverse intercostal (= intercostal of some authors) arteries toward the rim of the carapace.

Distal to the common descending intercostal, the prominent marginocostal artery arises to run laterally and caudally near the rim of the carapace. It branches variably, sometimes terminating by anastomosing with the epigastric artery or branches of it. Some of its lateral branches may terminate in the plastron or the skin adjacent to it. Medially, the marginocostal may establish connections to the common descending intercostal via the transverse intercostals.

The major vessel, the descending axillary artery, arises near the concentration of pectoral musculature and quickly sends a dorsal branch (= internal circumflex artery) into the pectoral muscles dorsal to the coracoid bone. Distally, the descending axillary bifurcates into an anterior branch (= external circumflex humeral artery) and a posterior branch (= thoracic artery). The anterior branch extends anteriorly to the ventral surface of the antebrachium, where it becomes the ventral brachial artery. The posterior branch runs into the mass of the pectoral musculature.

The most complex situation encountered among the branches of the brachiocephalic is that of the carotid circulation. McDowell (1961) has demonstrated taxonomic differences in this circulation, but much remains to be clarified. An additional problem is present regarding the nomenclature and the great number of vessel names and their homologies and synonymies. The right and left common carotid arteries arise from the brachiocephalic and run prominently along the ventral side of the neck toward the head. The first branching may be two (occasionally three) small arteries that run to the corresponding bronchus. These are the bronchial arteries. Distal to them arises the thymic artery (= thymus gland artery), which supplies blood to the thymus gland. A variable number of arteries, mostly of very small caliber, extend from the common carotid to the spinal cord by way of the intervertebral foramina, either ramifying into arterioles in the dura mater or piercing it to anastomose with arteries immediately dorsal or ventral to the cord. The most posterior of this group, and usually the largest, is the spinal artery (= ramus anastomoticus), which runs caudally on the dorsolateral aspect of the neck while sending small superficial branches to the cervical muscles and deeper ones to the cervical portion of the spinal cord. A number of small cervical arteries run medially and supply blood variously to the spinal cord plus adjacent muscles and connective tissue. Near the head a short branch, the hyoidean artery courses ventrolaterally sending small branches to the hyoid musculature, the posterolateral muscles and overlying skin of the head.

The intracranial arterial circulation is quite complex and beyond the scope of this treatment.

BRANCHES OF THE DORSAL AORTA

The dorsal aorta, formed from the junction of the right and left aortas, runs caudally just right of the midline of the body. Near its origin, a small, variable number of diminutive arteries ramify dorsally into the overlying body musculature. The next branches of the dorsal aorta supply blood to kidneys, adrenal glands, and gonads. The number of pairs of vessels and their distribution are known to vary both taxonomically and individually. In most chelonians, two or three pairs of renal arteries run laterally to supply blood to the kidneys and closely associated adrenal tissue, while one or two pairs of genital or gonadal arteries extend to the gonads. In males these may be termed spermatic or testicular arteries; in females, ovarian arteries. Girgis (1961b) found in *Trionyx triunguis* that separate renal and genital arteries did not exist. Kaushiva (1940) reported that the same situation obtained in *Lissemys punctata*. In these cases, adrenorenal-genital arteries are recognized and usually consist of two pairs, although bilateral asymmetry in numbers of vessels was reported by Girgis (1961b).

The next branching from the dorsal aorta consists of the large right and left epigastric arteries. Each epigastric artery arises dorsally of the kidney, then runs laterally toward the periphery of the shell. Shortly after its origin, it may give off a small recurrent intercostal artery that runs dorsally and medially toward the vertebral column before branching into anterior and posterior segments associated with the vertebrae. Considerably distal to the origin of the recurrent intercostal, the epigastric artery divides into an anterior branch running along the curve of the carapace and giving off various branches, among which are the lipoidal artery to fat masses and plastral artery to the bony plastron. Anteriorly, the anterior branch terminates near or anastomoses with the marginocostal artery. The posterior branch always supplies the pelvic musculature, and in *T. triunguis* (Girgis, 1961b) even gives rise to the major arteries (femoral and sciatic) of the upper and lower hindleg, respectively. In most chelonians, however, it appears that these major limb arteries are not branches of the epigastric but instead arise from the stem of the common iliac artery.

In forms where the external iliac and internal iliac arteries arise from the dorsal aorta by a common trunk, the circulation pattern is as follows: the common iliac artery arises just posterior to the origin of the epigastric. Almost immediately, the common iliac bifurcates into an internal iliac artery and an external iliac artery. The internal iliac sends small branches into the pelvic musculature and the bony carapace (carapacial arteries), the accessory bladders (vesicle arteries), and distal portions of the oviducts. One large branch, the hemorrhoidal artery, is present to supply blood to the cloacal wall.

The external iliac artery bifurcates, sending branches to the pelvic musculature, then continuing ventrally into the thigh as the femoral artery. The dorsal branch extends into the lower leg as its main artery, the sciatic. Here it branches into digital arteries to the toes.

The dorsal aorta proceeds posteriorly into the tail after giving off all its paired branches. The terminal portion of the dorsal aorta is the caudal (= middle sacral) artery.

THE VENOUS SYSTEM

The venous system (in addition to hypophyseal and hepatic portal systems) contains a renal portal system by which blood collected from capillaries in the tail, hind limbs, and some portions of the pelvis may be routed to the kidneys, where it is distributed through venous capillary networks around the convoluted tubules.

The venous system shows more individual variability than the arterial system, although it remains less thoroughly studied than the latter system.

Two of the most prominent paired veins of the pleuroperitoneal cavity are the abdominal (= ventral abdominal) veins. They extend anteriorly from the pelvic region (where they are usually united by a transverse abdominal vein) to the liver, where they terminate. Just before turning dorsally to enter the liver, each abdominal receives a pectoral vein from the pectoral girdle musculature, and a pericardial vein from the surface of the pericardial sac. Posteriorly in the bladder region, each abdominal receives a small vesical (= vesicle) vein. The pelvic girdle musculature is drained by the right and left pelvic veins, which may be unequal in diameter, with the left often the larger. Each pelvic usually enters its corresponding abdominal just posterior to the entrance of the vesical vein. The medial thigh muscles are drained by the crural vein, the

THE VENOUS SYSTEM

dorsolateral thigh muscles by the femoral vein. Both may unite, then enter the abdominal. The external iliac vein is very short and receives several small tributaries from the abdominal wall before entering the abdominal vein. The lipoidal (= adipose) vein drains fat masses in the groin and enters the abdominal vein near the crural. The epigastric vein is prominent, curving with the carapace and emptying into the external iliac (or occasionally the renal portal). The caudal vein drains the side of the tail, and a cloacal vein drains the wall of the cloaca. Both join to form the ischiadic vein, which empties into the renal portal vein.

The Renal Portal System

Near the confluence of the epigastric and external iliac veins, a large vessel, the renal portal vein, extends anteriorly and dorsally. Just before reaching the kidney, each renal portal may receive one or more small carapacial veins as well as a few tiny tributaries draining the dorsal muscle masses in the region of the kidney. At about the midpoint of the kidney, the renal portal bifurcates into an anterior branch, the vertebral vein, and a posterior branch, the hypogastric vein. The vertebral vein extends anteriorly along the vertebrae, running in the channel created by the rib heads. Along the sutures between the pleural (costal) bones, intercostal veins coming from the marginocostal vein connect with the vertebral vein. The hypogastric receives tributaries that drain, in part, the bladder, the cloacal wall, and the secondary sex organs.

The Hepatic Portal System

The hepatic portal system originates as capillary beds in the walls of the digestive viscera (stomach, small intestine, pancreas, gallbladder) and spleen. Blood drains from these organs to a second set of capillary beds, ramifying in the substance of the liver.

The hepatic portal vein extends transversely, embedded in the dorsal wall of the liver, sending numerous variable branches into the tissue of the liver lobes. As noted earlier, it receives blood from the right and left abdominal veins. On its left side, the hepatic portal receives numerous (usually five or more) small gastric veins that drain the wall of the stomach. The adjacent pancreas is drained by two or more anterior pancreatic veins entering the hepatic portal at about its midpoint, and two or more posterior pancreatic veins that enter the portal vein near its right terminus. Near the entrance of the posterior pancreatics, several small cystic veins enter the portal, draining blood from the wall of the gallbladder and cystic duct. Slightly posterior, the duodenal vein, draining the duodenum, enters the portal. On the right side of the pleuroperitoneal cavity, several small splenic veins drain blood from the spleen into the hepatic portal vein. Most posteriorly, the mesenteric veins, numerous but variable in number, drain the walls of the small (in part) and large intestines. They may anastomose into one or two trunks before entering the hepatic portal near the spleen.

The Systemic Veins

As stated previously, blood enters the right atrium through four great vessels: the left and right precavals, the postcaval, and the left hepatic vein. These vessels are the systemic veins.

Each precaval vein collects blood through four main tributaries, each of which has its own secondary tributary drainage. The most medial tributary is the thyroscapular (= thyreoscapular) vein, which not only drains the thyroid gland but, by way of other venous channels, drains the subscapular musculature. Anterolateral to the confluence of the thyroscapular and the precaval, the usually larger internal jugular vein empties into the precaval, carrying blood from the head and neck region. It makes an anastomosis cranially with the external jugular vein, and also, through its tributary esophageal vein's plexus, brings blood from the walls of the esophagus. The largest of the four precaval tributaries is the subclavian vein. Each subclavian is formed from the confluence of the external jugular vein running proximally from the head and neck, and the brachial vein from the forelimb. As the external jugular courses toward the heart, it receives several vertebral veins that drain blood from the vertebral areas. The brachial becomes the axillary vein upon entering the body cavity. The fourth tributary, the scapular vein, is the most lateral and dorsal of the series and drains the musculature of the scapular region.

The postcaval vein just before entering the heart receives a variable number of hepatic veins which drain blood from liver capillary beds into the postcaval. At its posterior extremity, the postcaval is formed from the confluence of the right and left renal veins which drain the kidneys. Also in this region, one or more pairs of gonadal or genital veins may appear and join the postcaval, bringing blood from the gonads or gonoducts.

The left hepatic vein drains blood from the left lobe of the liver into the sinus venosus from which it enters the right atrium.

The Pulmonary Veins

The right and left pulmonary veins transport oxygenated blood from the capillary beds of their respective lungs to the left atrium.

CHAPTER 9
Respiration

DONALD C. JACKSON

Turtles are a diverse group found in strikingly different habitats, but much of this chapter deals with a few species of freshwater turtles because physiologists have generally studied readily available animals and also have preferred to study species on which background data are available. The slider or red-eared turtle, *Pseudemys scripta elegans* has been studied extensively in many laboratories and is prominent in this discussion.

RESPIRATORY MECHANICS AND BUOYANCY CONTROL

Like other reptiles, turtles generate airflow into the lungs by a suction pump mechanism; the pressure within the lungs is lowered below atmospheric pressure to cause inspiratory flow. Verification of this has come through experiments in which pressures inside the lungs and body cavity of turtles were measured during normal breathing (McCutcheon, 1943; Gans and Hughes, 1967; Gaunt and Gans, 1969).

The mechanism for lung ventilation, though similar in principle to that in other reptiles, is drastically modified structurally by the rigid shell of turtles. It is unwise to generalize because the rigidity and extent of the shell are properties that differ quite markedly among turtles; however in the turtles that have been studied, a fairly consistent pattern emerges that probably applies to most other species. The pressure changes within the lungs are generated by muscles that expand or retract the anterior

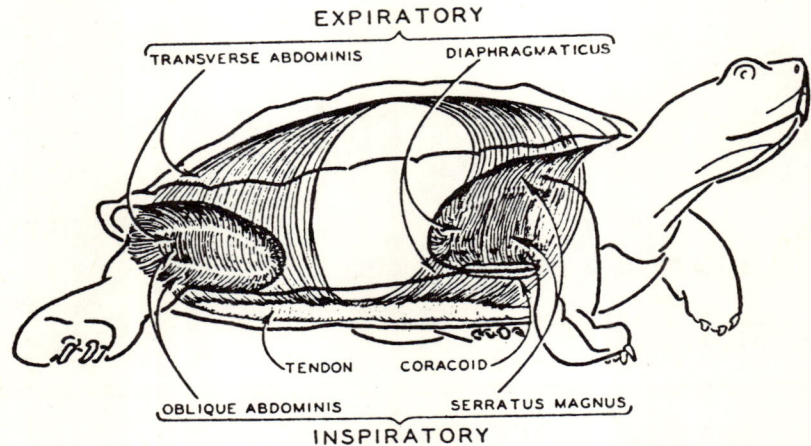

Figure 1 Schematic diagram of the respiratory muscles of *Malaclemys terrapin*. (From McCutcheon, 1943.) By permission of The University of Chicago Press.

and posterior limb pockets. Figure 1 illustrates the organization of the major respiratory muscles of *Malaclemys terrapin*. The action of these muscles alters the pressure in the entire body cavity and, as revealed by the measurements of Gans and Hughes (1967) on *Testudo graeca*, the instantaneous pressure is nearly uniform throughout the body. The ribs are fused to the shell, and there is no true diaphragm separating thoracic and visceral compartments. In certain species [e.g. *Lissemys punctata*: Shah (1962)] presumably primitive phylogenetically, gas flow is apparently powered by a striated muscle sheath that invests the lung. Contraction of this muscle, the *muscularis striatum pulmonale*, forces air out of the system.

The relative importance of the various respiratory muscles apparently is highly variable among turtles (Shah, 1962) and even in the same species, depending on its situation. An interesting example of the latter point was reported by Gaunt and Gans (1969) for the snapping turtle, *Chelydra serpentina*, which has a considerably reduced plastron. When on land, its soft exposed surface tissues permit considerable sagging of the visceral structures under the force of gravity. The sagging viscera exert tension on the dorsally placed lungs, and this tension tends to inflate the lungs. On land, therefore, inspiration is largely a passive process involving little muscle activity, while expiration is always an active event (Figure 2). When *C. serpentina* is submerged, its viscera are supported by the water, and the lungs are subjected to the hydrostatic pressure exerted by the water; expiration is more or less a passive event (depending on the

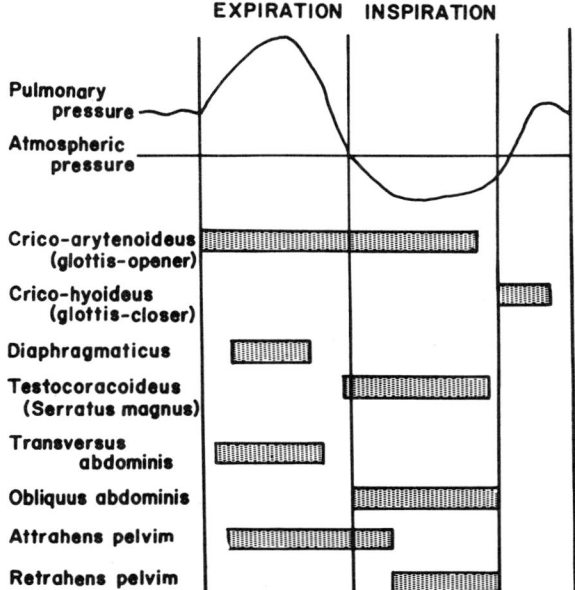

Figure 2 The relationship between the electrical activity of the major respiratory muscles of *Chelydra serpentina* and pulmonary pressure changes during a respiratory cycle. (From Gaunt and Gans, 1969.)

depth), while inspiration requires active muscular contraction. *C. serpentina* is economical with its own resources and, where possible, utilizes external forces to move air into or out of its lungs. In *Testudo graeca*, a tortoise with a very complete shell, both phases of gas flow require muscular activity. In some turtles the pectoral girdle is involved in respiration and, during breathing movements, the forelimbs can be observed to move in and out. These limb movements may be an essential component of the mechanism (Wolf, 1933), or they may serve merely an accessory role (Gans and Hughes, 1967).

The organization of the respiratory pump has two functional consequences that are of considerable importance to the buoyancy control mechanism of certain freshwater species. First, the body volume of turtles must remain rather uniform, except when tidal volume displacements occur during actual breathing. Volume changes (from normal) would cause bulging or retraction of the limb pocket tissue, thus disturbing the action of the respiratory muscles. The shape and the orientation of these soft tissues are probably essential for the efficient performance of the respiratory pump. It is of interest that the body weight of fasted

Sternotherus minor did not differ significantly from the body weight of paired, fed animals after 90 days (Belkin, 1965). Presumably, the body substance lost during the fast was replaced by water to maintain body volume. The second consequence to buoyancy control of the organization of the respiratory pump involves the cloaca. Since the forces exerted by the respiratory muscles are felt throughout the body cavity, these forces can just as easily induce water flow in and out of the cloaca of a submerged turtle as airflow in and out of the lungs of a breathing turtle. Whether air or water flow occurs depends on which sphincters are open during respiratory muscle activity. Cloacal water flow in *Emys orbicularis* was observed by Lüdicke (1936), and I have seen cloacal flow, concurrent with tracheal airflow, in *Pseudemys scripta elegans* in experiments (unpublished) in which the cloaca was cannulated. In both species, the cloacal flow appeared to pass in and out of the cloacal bursae (Smith and James, 1958), which are strategically located lateral to the cloaca against the inner surface of the posterior limb pockets. It is unlikely that cloacal water flow occurs normally when a freshwater turtle breathes air, since this would contribute little to gas exchange; but the water flow may be important when compensatory body volume changes are required, such as those involved in buoyancy control.

The ability of aquatic turtles to control their buoyancy is revealed by their nearly effortless movements in deep water. Their specific gravity is obviously close to that of water despite the heavy shells, which in some freshwater species account for 30% of the animal's total body weight in air (Hall, 1924; Jackson, 1969). In *P. s. elegans*, the body elements that are heavier than water together exert a sinking force of about 14 g for each 100 ml of body volume. Of this 14 g, the shell accounts for about 75%, or 10.5 g. To achieve neutral buoyancy, there must be a buoyant force equal to the sinking force. In *P. scripta,* as most probably in other species, the buoyant force is provided mainly by the lung air, which consequently must occupy some 14% of the body volume in a neutrally buoyant animal. It is of interest to note that soft-shelled turtles, Trionychidae, with their low density shells, have very small lung volumes compared to the Emydid species (Agassiz, 1857).

The buoyancy mechanism was studied experimentally in the sea turtle *Caretta caretta* (Jacobs, 1939) and in the freshwater *P. s. elegans* (Jackson, 1969; 1971b). In both studies, external weights or flotation devices were attached to the shells of the turtles to experimentally alter their buoyancy and the turtles were able, within limits, to reestablish normal buoyancy by adjusting their resting lung volumes. Jacobs also noted that *C. caretta* could selectively alter the volume of the anterior portion of the lung with respect to the posterior portion, or the right lung, for example, with respect to the left lung.

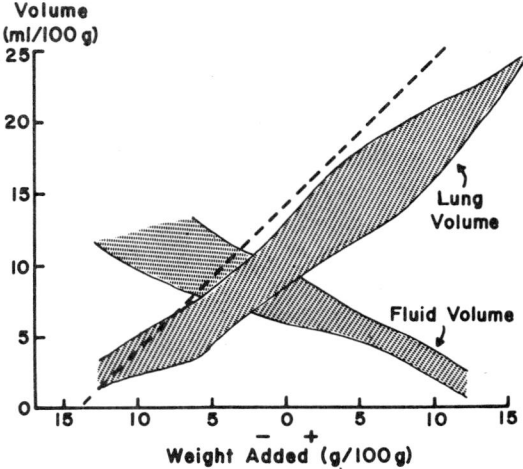

Figure 3 Lung volumes and fluid volumes (urinary bladder and cloacal bursae) of turtles (*Pseudemys scripta elegans*) following buoyancy adjustments to attached weights (+) or floats (−). Dashed line describes lung volumes necessary to maintain neutral buoyancy with the changing loads. (Modified from Jackson, 1971.)

I was particularly interested in how the lung volume adjustment could occur within the rigid shell of *P. s. elegans* (Figure 3). A compensatory volume change as discussed above did occur; the fluid volume contained in the urinary bladder and cloacal bursae was found to change in inverse fashion to the lung volume change, so that total body volume remained essentially unchanged (Figure 3). This suggests that the bladder and cloacal bursae, as they function in this mechanism, are analogous to the ballast tanks of a submarine. They lose equivalent amounts of fluid when lung volume is increased and they take on water when lung volume shrinks.

An important mechanical result of this compensation was that the pressure in the lung of *P. scripta* was maintained close to atmospheric pressure over the entire range of resting lung volumes. Without the accompanying change in fluid volume, the increases or decreases in lung air would have caused serious pressure changes in the lungs and body cavity, as well as causing the unfavorable bulging or retracting at the limb pockets. When the fluid volume adjustment occurred, the lung pressure, tested by experimentally changing lung volume on paralyzed animals, changed very little, and these changes approximated those observed on lungs removed from the shell (Figure 4). The volume-pressure relationship, or compliance, of excised lungs of *P. s. elegans* is very steep compared to mammalian lungs; that is, the gas volume can be

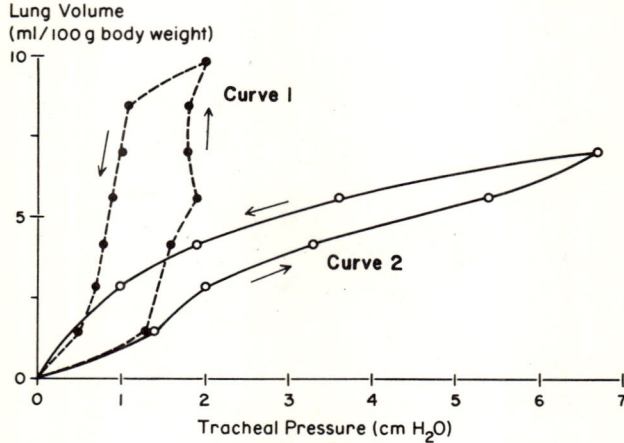

Figure 4 Total respiratory pressure-volume (compliance) curves measured on paralyzed but otherwise intact turtles (*P. s. elegans*). To obtain curve 1, the inflation and deflation of the lungs was accompanied at each step by equal but opposite changes in urinary bladder volume. To obtain curve 2, the lungs alone were inflated and deflated. (Modified from Jackson, 1971b.)

changed with only a small change in pressure. The specific compliance (compliance normalized to resting lung volume) of the lungs of 10 specimens of *P. scripta* averaged 1.7 (cm $H_2O)^{-1}$ compared to various mammalian lungs, which had specific compliances of about 0.2 (cm $H_2O)^{-1}$ (Crosfil and Widdicombe, 1961). With its unique respiratory structure and its highly compliant lungs, *P. s. elegans* is able to vary its resting lung volume from about 6 to 20 ml/100 g of body weight with pressure changes of less than 1 cm H_2O.

VENTILATION AND GAS EXCHANGE

The phylogenetic development of lung function is considerably advanced in the reptiles, since essentially all the gas exchange required for normal activity can occur in the lungs. The overall exchange process is complex and includes the movement of gas in and out of the lungs, the exchange of oxygen and carbon dioxide between the alveolar gas and the blood, and the transport of the respiratory gases in the blood to and from the tissues of the body.

Lung Ventilation

The normal respiratory cycle of turtles begins with an expiration, is immediately followed by an inspiration, and concludes with a compression of the lung air (McCutcheon, 1943). In the final compression phase the glottis is closed and no airflow occurs. This end inspiratory basic pattern is similar to that of other reptiles but differs from the mammalian pattern, which is end expiratory.

The rhythm of breathing appears to differ in terrestrial and aquatic turtles. Terrestrial turtles, such as *Testudo graeca*, breathe in a repetitive manner, although the interval between individual cycles is variable (Gans and Hughes, 1967). Freshwater turtles, on the other hand, characteristically take a rapid series of breaths, then become apneic for a variable period of time before taking another series of breaths (Figure 5). This episodic breathing pattern has been observed in many species, including *Malaclemys terrapin* (McCutcheon, 1943), *Chrysemys picta dorsalis* (Randall, Stullken, and Hiestand, 1944), *Pseudemys concinna* (Belkin, 1964), *P. s.*

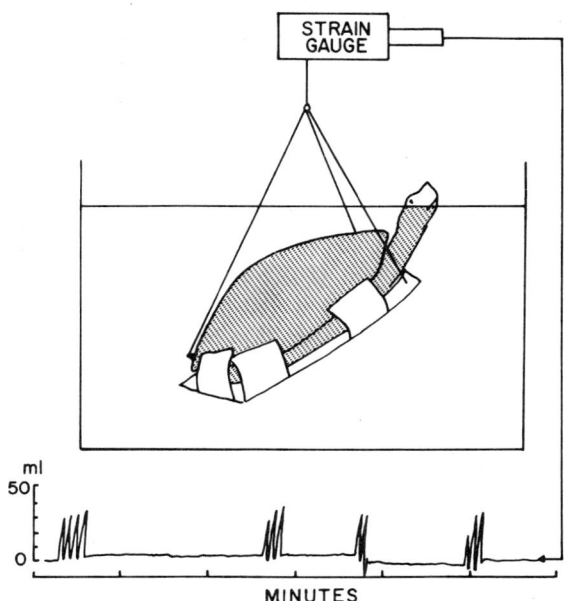

Figure 5 Episodic breathing pattern characteristic of freshwater turtles as measured by the buoyancy technique (Jackson, 1971). In this record, the upward deflection is expiration, which is the event normally initiating a breathing cycle. The turtle is submerged in water but can elevate its head above the surface to breathe.

elegans (Frankel, Spitzer, Blaine, and Schoener, 1969; Jackson, 1971a), *Chelydra serpentina* (Gaunt and Gans, 1969), and *Chelus fimbriatus* (Lenfant, Johansen, Petersen, and Schmidt-Nielsen, 1970). Episodic breathing persists even when the animals are out of water, but it most likely derives from the typical aquatic behavior of these turtles. Many activities are normally conducted underwater, but at intervals the turtle swims to the surface and breathes. Although during an episode of breathing these turtles breathe at a rapid rate, their overall rate of breathing is rather low, on the order of 1 or 2 breaths per minute. This rate, as well as the duration of the apneic interval, is highly variable and must depend primarily on the metabolic rate of the animal. In unrestrained turtles, the overall breathing rate was low and the voluntary apneic intervals were long, lasting in some instances for 1 or 2 hr (Belkin, 1964; Lenfant et al., 1970).

Tidal volumes and respiratory minute volumes have been measured infrequently. In *P. s. elegans*, tidal volume averaged 18.5 ml/kg at 20°C, but was somewhat lower at 10 and 30°C (Jackson, 1971a). Compared to other reptiles and mammals, this is a large tidal volume, but *P. scripta* also breathes at a much slower rate than these other animals. Respiratory minute volume in *P. s. elegans* was about 20 to 25 ml/kg · min. This particular variable varied directly with metabolic rate in this species but, surprisingly, was independent of experimental temperature.

Gas Transfer in the Lungs

The lungs are partitioned into gas exchange units or alveoli. The degree of partitioning differs among species and also within the lungs of a given turtle. In the red-eared turtle partitioning increases the internal gas exchanging surface area to nearly 8 times the area of the outer surface of the lungs (Perry, 1972). The partitioning decreases posteriorly, but even in the most caudal region the lungs are perfused by blood and considerably partitioned so that gas exchange should occur in this region. Wolf (1933) had suggested that little or no exchange occurs caudally, but that air from this region passes through the forward gas exchange areas during expiration much as the gas from the caudal air sacs of birds exits by way of the gas exchanging parabronchi. Physiological experiments are required to resolve the details of regional gas exchange.

Little is known about the gas exchange between alveolar gas and pulmonary capillaries in turtles. Several observations, both anatomical and physiological, suggest that the exchange process is less efficient than in mammalian lungs. First, the alveolar surface area in *P. s. elegans* is only some 0.2 m^2/kg of body weight compared to about 1 m^2/kg in man

(Perry, 1972). The alveolar dimensions are quite variable, but average some 1 to 2 mm in diameter in various species (Tenney and Tenney, 1970). A second factor is the organization of the pulmonary capillaries in the interalveolar septa. Unlike mammalian lungs, where a single capillary network is interposed between adjacent alveoli, a double layer occurs in the septa of *P. s. elegans*, presumably limiting gas exchange to the roughly 50% of the capillary surface that faces the alveoli (Perry, 1972).

The possibility exists that the respiratory gases, particularly oxygen, may not reach equilibrium between alveolar gas and capillary blood during the passage of blood through the lungs. However a variable fraction of the oxygenated arterial blood is recycled through the pulmonary circulation (a left-to-right shunt) because the left and right ventricular outflows from the heart are incompletely separated (Steggerda and Essex, 1957). Although this recycling means added work for the heart, it may enhance the oxygen saturation of the arterial blood that is pumped into the systemic circulation. Again, physiological studies are needed in which simultaneous measurements are made of the oxygen pressure of alveolar air and of pulmonary venous blood to establish the completeness of oxygen exchange in the pulmonary capillaries. This will not be simple, since alveolar gas composition may be quite variable in different regions of the lung because of ventilation-perfusion inequalities.

A right-to-left shunt (or lung bypass) also exists in *C. serpentina* (Steggerda and Essex, 1957), *P. scripta* (White and Ross, 1966) and *Chelus fimbriatus* (Lenfant et al., 1970). Lenfant et al. demonstrated the impact of this shunt on gas exchange by simultaneously measuring P_{O_2} values in the alveolar gas and arterial blood (from the carotid artery). An alveolar-arterial P_{O_2} gradient of some 15 to 25 mm Hg was observed. A smaller gradient for P_{CO_2} was found that generally was in the direction of the alveolar gas but, paradoxically, reversed toward the end of the apneic phase. Lenfant et al. concluded that the right-to-left shunt gradually increased during apnea, which supports previous findings by Millen et al. (1964) and White and Ross (1966). It appears that equilibration between the alveolar gas and the pulmonary capillaries is reasonably complete, but that the final arterial values are altered by admixture with mixed venous blood (right-to-left shunt).

Blood Gas Transport

The basic features of blood gas transport in freshwater turtles were described 50 years ago by Southworth and Redfield (1926). They noted two features of the respiratory properties of the blood of *Pseudemys concinna* that distinguished it from mammalian blood: (1) low hematocrit, which averaged some 20% in their sample, and (2) the unusually high carbon dioxide content of the turtle's blood, which was about 80 vol % (36 mM/liter). In terms of the oxygen and carbon dioxide dissociation curves, nothing particularly unusual was noted other than the effects attributable to the low hemoglobin, which decreased both the oxygen capacity and the carbon dioxide buffering capacity of the blood (compared to mammalian blood). These observations have been confirmed and extended to many other chelonian species, and some attempts have been made to interpret the findings in terms of the natural history of the various species.

Smith (1929) found serum carbon dioxide concentrations exceeding 40 mM/liter in several freshwater turtles and values of 30 mM/liter or more in a marine species. Smith also reported the remarkably high carbon dioxide content of the pericardial and peritoneal fluids of the freshwater forms and suggested that these fluids, together with the blood, help adapt these turtles to prolonged diving and the resultant metabolic acidosis. A recent study (Jackson and Silverblatt, 1974) confirmed the role of the blood carbon dioxide in this regard, but failed to substantiate Smith's proposed role for the other fluids rich in carbon dioxide.

Oxyhemoglobin dissociation curves have now been made on many turtles. McCutcheon (1947) reported that the blood of marine species had less affinity for oxygen than the blood of freshwater species which, in turn, was less than that of terrestrial species. His explanation was that marine turtles, with their reduced shells, are more active, therefore require the higher unloading oxygen tensions that low affinity blood affords. Sullivan and Riggs (1967) tentatively concluded, in contrast, that the blood of active underwater swimmers from freshwater families (Chelydridae, Kinosternidae, and Trionychidae) has a higher affinity for oxygen than does the blood of more terrestrial species. Sullivan and Riggs confirmed McCutcheon's finding that the blood of marine turtles has a low affinity, however. Lenfant et al. (1970), based on their study of the sluggish freshwater matamata, *Chelus fimbriatus*, suggested that the low affinity for oxygen (thus high unloading capacity for oxygen) observed by them may be advantageous to this species during its long periods of breath holding. The low affinity of the blood of other diving turtles also may serve to enhance oxygen extraction from the blood during apnea. A related finding is that the Bohr shift is large in *C. fimbriatus* (Lenfant et al., 1970) and, since this shift facilitates the unloading of oxygen during acidosis, it also may be an adaptive feature for a diving species.

CONTROL OF BREATHING

For a complex organism, respiratory gas exchange must be accomplished in a way that preserves essential features of the internal environment. Oxygen must be supplied to the cells, not merely in adequate amounts, but at pressures that are compatible with the normal functioning of the cells. Carbon dioxide loss must also be carefully controlled, since the partial pressure of carbon dioxide (P_{CO_2}) in the body fluids is an important determinant of the acid-base status. For most air-breathing vertebrates, it is the regulation of carbon dioxide that is dominant in the control of respiratory gas exchange. It appears that the regulation of P_{CO_2} and pH normally governs breathing in at least some turtles as well, and that oxygen uptake is usually accomplished simply as a consequence. Factors influencing respiration and respiratory control include environmental parameters, such as temperature and ambient pressure, as well as behavioral or physiological ones, such as activity level, diving, hibernation, and variable homeostatic requirements.

Metabolic Rate

For an air-breathing animal, pulmonary ventilation must be matched in some consistent fashion to metabolism. Oxygen supply and excretion of carbon dioxide must match oxygen consumption and carbon dioxide production, respectively, so that the internal levels of these gases remain stable. Measurement of breathing volumes of *P. s. elegans* by the buoyancy principle (Figure 5) necessitated some restraint on the animal; but even under these quasi-resting conditions, the oxygen consumption ($\dot{V}_{O_2}$) of a series of turtles at the same temperature varied by as much as four- to fivefold. In a highly predictable manner, these changes in $\dot{V}_{O_2}$ were accompanied by proportionate changes in respiratory minute volume ($\dot{V}_E$), so that the ratio, $\dot{V}_E/\dot{V}_{O_2}$, was relatively unchanged (Figure 6). Expressing ventilation by this ratio, $\dot{V}_E/\dot{V}_{O_2}$, emphasizes the regulatory function of breathing, since it is the relationship between ventilation and metabolic rate that determines alveolar gas composition (with allowance made for the dead space and the respiratory exchange ratio).

The regularity of the turtles' breathing control belies one's initial assumption, based on casual observations, that a turtle's respiration is erratic, thus poorly controlled. It probably is less precise than mammalian control in the sense that the controlled variables of the system, P_{O_2}, P_{CO_2}, and pH, are permitted to fluctuate over a wider range. This is illustrated by the serial measurements by Lenfant et al. (1970) of alveolar

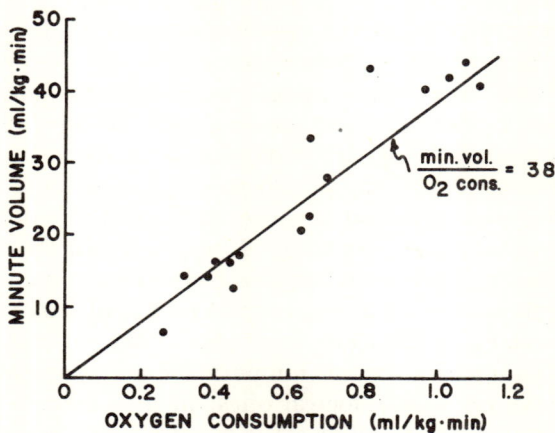

Figure 6 The relationship between respiratory minute volume and oxygen consumption of *P. s. elegans* at 20°C. (Adapted from Jackson, 1971a.)

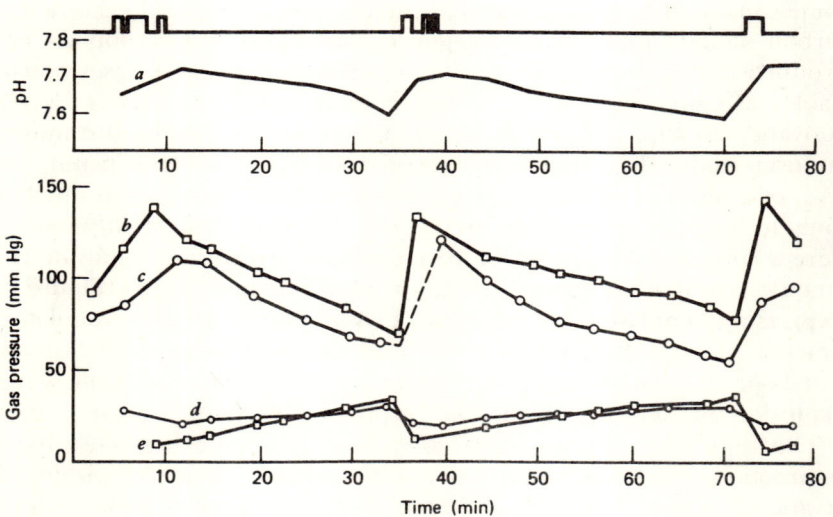

Figure 7 Time course of partial pressure of oxygen and carbon dioxide in arterial blood and alveolar gas of *Chelus fimbriatus*. Three successive breathing sequences are depicted by the vertical bars at the top. The curves represent the following variables: *a*, arterial pH; *G*, alveolar P_{O_2}; *c*, arterial P_{O_2}; *d*, arterial P_{CO_2}; *e*, alveolar P_{CO_2}. (From Lenfant et al., 1970.) By permission of ASP Medical and Biological Press, Amsterdam, and C. Lenfant.

and arterial P_{O_2} and P_{CO_2} and arterial pH over several successive breathing episodes in *Chelus fimbriatus* (Figure 7). Control can be revealed only by observing the system for an adequate time period. Each data point in Figure 7 is the mean value for a test period lasting approximately one hour.

Temperature

A turtle's body temperatures usually approximate the temperature of its local environment during the steady state (see Chapter 11). What effect does temperature change have on the respiratory control of a turtle? One might guess (and data from other reptiles support the guess) that since an increase in temperature speeds up the metabolic rate, it should also speed up the ventilation. As noted, an increase in metabolic rate at a given temperature promotes increased ventilation, thus preserves normal blood gas homeostasis.

However simultaneous measurements of oxygen consumption and respiratory minute volume on *P. s. elegans* revealed no change in the mean resting ventilation volumes of turtles at temperatures between 10 and 30°C, despite a large rise in metabolic rate over this range (Figure 8). This indicates that an increase in metabolic rate induced by higher temperature does not have the same stimulating influence on respiration

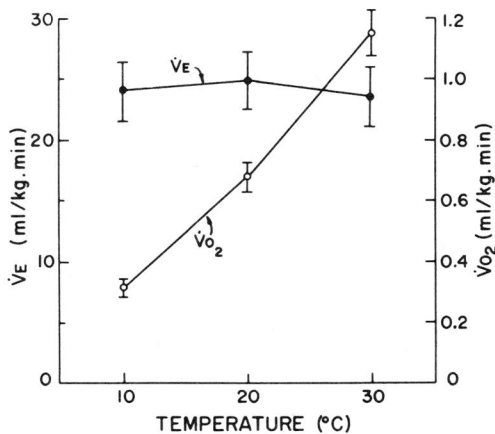

Figure 8 The relationship between respiratory minute volume $\dot{V}_E$ and temperature and between oxygen consumption $\dot{V}_{O_2}$ and temperature of turtles (*P. s. elegans*) breathing air. (Data from Jackson, 1971a; Jackson, et al., 1974). By permission of ASP Medical and Biological Press, Amsterdam.

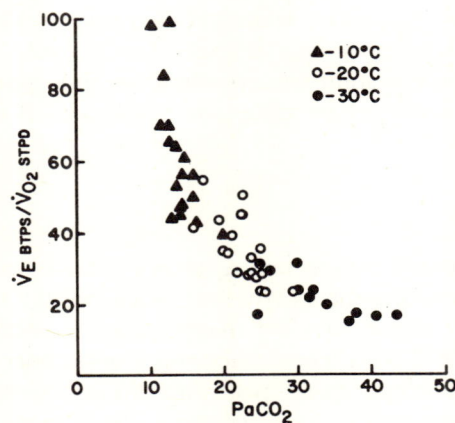

Figure 9 The dependence of arterial P_{CO_2} on the ventilation-metabolism ratio $\dot{V}_E/\dot{V}_{O_2}$ in *P. s. elegans* breathing air at different temperatures. (From Jackson, Palmer, and Meadow, 1974.) By permission of ASP Medical and Biological Press, Amsterdam.

as does a rise in metabolism when it occurs at a constant temperature. A consistent relationship between $\dot{V}_E$ and $\dot{V}_{O_2}$ exists at each temperature studied, but the ratio $\dot{V}_E/\dot{V}_{O_2}$ differs.

The dependence of alveolar gas composition on the $\dot{V}_E/\dot{V}_{O_2}$ ratio makes it obvious that *P. s. elegans* is not preserving the same blood-gas homeostasis when its temperature changes, since $\dot{V}_E/\dot{V}_{O_2}$ is dependent on temperature. Simultaneous measurements of $\dot{V}_E/\dot{V}_{O_2}$ and arterial P_{CO_2} and pH have documented this (Figure 9). Arterial P_{CO_2} varied directly with body temperature in accordance with the fall in the $\dot{V}_E/\dot{V}_{O_2}$ with temperature. Arterial pH varied in an inverse manner with temperature. According to the terminology employed in respiratory physiology, these changes denote hypoventilation at high temperature or, alternatively, hyperventilation at low temperature. However such descriptive terms are inappropriate in this case, since the observed changes in the ventilation state of *P. scripta* at different temperatures do not reflect a breakdown of normal control, but rather are required for proper acid-base balance.

Inspired Gas Composition

Insight into the nature of respiratory control can be gained by causing an animal to breathe a gas mixture containing more carbon dioxide or less oxygen than normal, and observing its respiratory reaction. Most air-breathing animals increase breathing activity when either change is sufficiently severe, although it is usual for there to be somewhat more sensitivity to changes in carbon dioxide. One would predict a priori that turtles would be rather insensitive to either an increase in carbon dioxide or a decrease in oxygen, since many, particularly the diving species, are remarkably tolerant to prolonged breath-holding. Also, the normal control of the respiratory variables, at least as illustrated by *C. fimbriatus* (Figure 7), permits rather wide fluctuations of the blood values of these variables. One might predict, however, that body temperature would have a pronounced effect on the response of turtles to these gases, since temperature influences normal breathing as just described. With these considerations in mind, experiments were undertaken in my laboratory to test the respiratory response of *P. s. elegans* to carbon dioxide and low oxygen gases at various temperatures. Background information indicated that various species do indeed respond to increased carbon dioxide and decreased oxygen by increased breathing activity, but quantitative evaluation of sensitivity to these gases was lacking. Species studied earlier included *Chrysemys picta dorsalis* (Randall, Stullken, and Hiestand, 1944), *Terrapene c. carolina* (Altland and Parker, 1955), *P. scripta* (Frankel et al., 1969), and *C. fimbriatus* (Lenfant et al., 1970).

Inspired gases rich in carbon dioxide induced hyperventilation in *P. s. elegans* (Jackson, Palmer, and Meadow, 1974). There was detectable hyperventilation with 2% carbon dioxide, and when 6% carbon dioxide was breathed, ventilation averaged 10 times the normal air-breathing volume. This degree of responsiveness to carbon dioxide, comparable in relative terms to the response reported for mammals, exceeded expectations, since *P. scripta* is noted for its tolerance to long periods of diving and accompanying hypercapnea. Also unexpected was the temperature independence observed; the relative magnitude of the response to carbon dioxide was identical at 10, 20, and 30°C. However the carbon dioxide response curves, while similar at each of these temperatures (Figure 10), do not coincide, since as we showed earlier, both P_{CO_2} and $\dot{V}_E/\dot{V}_{O_2}$ are temperature-dependent variables. Nonetheless, these results suggest that the respiratory control system of *P. scripta* is equally sensitive to changes in P_{CO_2} throughout this temperature range.

The result when gases low in oxygen were breathed was quite different. The response to hypoxia was highly temperature dependent; the higher the temperature, the greater was the ventilatory sensitivity and response (Jackson, 1973). To understand this temperature dependency, recall that at 10°C an air-breathing turtle has a very high ventilation ratio ($\dot{V}_E/\dot{V}_{O_2}$) and is, in effect, hyperventilating. On the basis of acid-base homeostasis, as was said earlier, this is not true hyperventilation, but rather normal ventilation for 10°C. But from the standpoint of oxygen

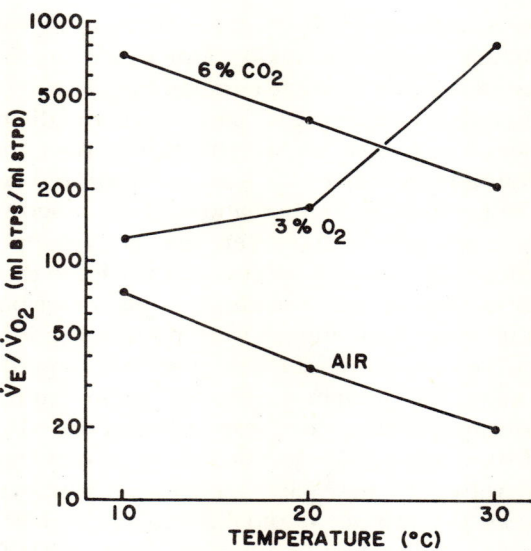

Figure 10 The ventilation ratio $\dot{V}_E/\dot{V}_{O_2}$ of *P. s. elegans* at various temperatures breathing air, 3% oxygen (in nitrogen), or 6% carbon dioxide (in air). Note that ventilation is plotted logarithmically. Points are mean values taken from Jackson (1973) and Jackson et al. 1974. By permission of ASP Medical and Biological Press, Amsterdam.

intake, the turtle *is* hyperventilating at 10°C and, to such an extent, very little ventilatory response occurred even when 3% oxygen was breathed. At higher temperatures, on the other hand, the trend was toward hypoventilation (again with respect to oxygen acquisition), so that at 30°C the turtles exhibited a pronounced ventilatory response when they breathed 3% oxygen and even responded slightly to 10% oxygen.

The strikingly different effects of temperature on the ventilatory response to these two gases is illustrated in Figure 10. The proportionate increase in ventilation at each temperature when 6% carbon dioxide was breathed may be interpreted as follows: the normal, air-breathing ventilation of the animal is altered by temperature to satisfy acid-base requirements. To achieve this, the turtle regulated its blood P_{CO_2} at a different value at each temperature. Apparently, however, sensitivity to carbon dioxide is shifted with temperature, so that a given change in blood P_{CO_2} (induced by breathing carbon dioxide) stimulates an equivalent respiratory response, whatever the temperature (Figure 11). Thus when the turtles breathed 6% carbon dioxide, their ventilation (V_E/V_{O_2}) increased by about tenfold at 10, 20, and 30°C.

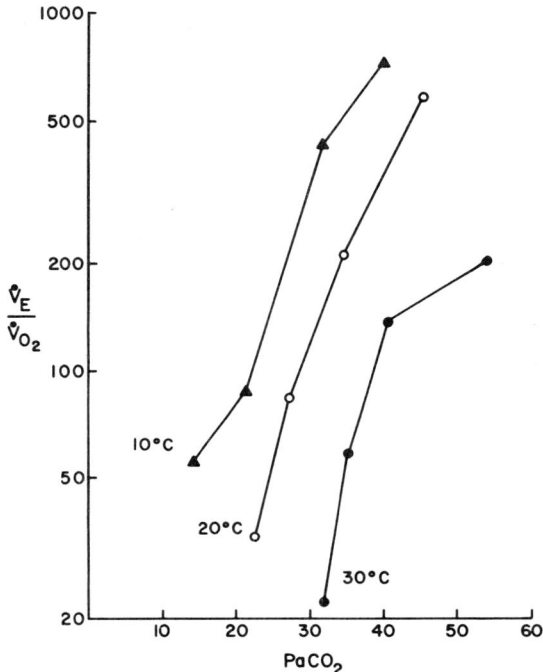

Figure 11 The dependence of ventilation ($\dot{V}_E/\dot{V}_{O_2}$) on arterial P_{CO_2} of *P. s. elegans* breathing air and various gases containing carbon dioxide at different temperatures. (Modified from Jackson et al., 1974.) By permission of ASP Medical and Biological Press, Amsterdam.

The temperature-dependent response to low inspired oxygen may be interpreted as follows: the shift with temperature in the normal, air-breathing ventilation (for acid-base regulation) inevitably affects the acquisition of oxygen from the inspired air. As the temperature rises and the relative ventilation ($\dot{V}_E/\dot{V}_{O_2}$) falls, a greater fraction of the oxygen must be extracted from the inspired air to satisfy the metabolic needs. From the standpoint of oxygen uptake, therefore, the turtle is hypoventilating at 30°C (relative to 10°C), whereas in terms of acid-base regulation it is breathing as it should. This relative hypoventilation at high temperature reduces the "safety margin" for oxygen uptake and can explain, at least in part, the greater sensitivity (or vulnerability) of this species to reduced oxygen at 30°C. The acid-base and oxygen requirements of breathing seem to be coming into conflict as the temperature of the animal is increased. The possibility exists that at some upper critical temperature, the simultaneous satisfaction of both these homeostatic requirements can no longer occur even when air is breathed. A fundamental conflict such as this could be a factor that limits tolerance to high temperature.

It is apparent from these observations that a close relationship exists between temperature, ventilation, and acid-base balance. The important homeostatic function involved is clearly the regulation of acid-base balance at different body temperatures. Robin (1962) observed that the arterial pH of *P. s. elegans* varied inversely with body temperature, and Howell, Baumgardner, Bondi, and Rahn (1970) observed a similar effect on *Chelydra serpentina*. In other poikilothermic species, blood pH has been found to decrease by about 0.015 unit per degree celsius increase in body temperature. According to Rahn (1967) this pH change is necessary to maintain a constant OH^-/H^+ ratio (or relative alkalinity) in the blood.

To achieve this temperature-related pH regulation, the physiological mechanisms responsible for pH regulation must respond to temperature in the proper way. In *P. s. elegans*, ventilation apparently adjusts when temperature changes; whether this is so for other chelonian species remains to be established.

DIVING

A notable aspect of chelonian respiratory physiology is the ability of many turtles to survive for quite long periods without breathing. As described earlier, the normal voluntary breathing pattern of some freshwater species includes apneic intervals lasting for 1 or even 2 hr at 20 to 25°C (Belkin, 1964; Lenfant et al., 1970). During these voluntary dives, the turtles were relatively inactive and were probably depleting their internal oxygen stores at a slow rate. Belkin (1964) calculated that specimens of *Pseudemys concinna* under resting conditions could readily tolerate dives lasting 2 or 3 hr before reaching critically low internal oxygen tensions. The large lungs of this and related species are the chief reservoir of oxygen, with the blood-bound oxygen accounting for most of the balance. According to the data of Lenfant et al. (1970), cited in Figure 7, the arterial P_{O_2} of *Chelus fimbriatus* was still above 50 mm Hg after 30 to 40 min of voluntary breath holding; low metabolic rate, coupled with rather plentiful oxygen stores, permits long periods of apnea before stressful hypoxia ensues. During this time underwater activities may be carried out, supported by aerobic metabolism.

But what if the aerobic time limit is exceeded? For most diving vertebrates, whose brain and heart tissues cannot function even temporarily in the absence of oxygen, the physiological adaptations to apnea are restricted to mechanisms for conserving the available oxygen. Many turtles, on the other hand, can continue to function long after internal oxygen stores are exhausted, even in an anaerobic environment. Numerous studies (e.g., Johlin and Moreland, 1933; Robin, Vester, Murdaugh, and Millen, 1964; Belkin, 1963) document the unusual tolerance of aquatic chelonians to oxygen deprivation. Clearly, the continuous dependence on oxygen of certain vital organs is not a limiting factor in diving turtles.

Anaerobic glycolysis evidently provides metabolic energy during prolonged anoxia. Johlin and Moreland (1933) found that lactic acid, an end product of anaerobic glycolysis, increased twentyfold in the blood of a freshwater turtle as a consequence of a 24 hr exposure to nitrogen. Similar increases in blood lactate have been noted in *P. scripta* (Robin et al., 1964; Jackson and Silverblatt, 1974). A convincing demonstration of the critical importance of the anaerobic pathway to the anoxic turtle was provided by Belkin (1962), who observed that when musk turtles, *Sternotherus minor*, were injected with iodoacetic acid (a specific inhibitor of glycolysis), they drowned much sooner than noninjected controls.

To further characterize diving metabolism, the total metabolic rate of *P. s. elegans* was measured by direct calorimetry during diving and correlated with blood and lung oxygen levels (Jackson, 1968). The dives (which lasted 4 hr at 24°C) were subdivided into three metabolic phases (Figure 12). During phase I, which lasted about 20 min, metabolism was unchanged from the predive level. Oxygen levels were falling rapidly, but still exceeded critically low levels. In phase II, also lasting about 20 min, metabolic rate fell rapidly to about 35% of the initial value, while available blood and lung oxygen was depleted. In phase III, which continued for the balance of the 4 hr dive, metabolism slowly declined to about 15% of the initial rate, and blood and lung oxygen remained low and unchanged. When iodoacetic acid was given to these turtles in the calorimeter before diving (repeating Belkin's experiment), death occurred after about 60 min of diving; phases I and II, the aerobic phases, were unchanged, but phase III, the anaerobic phase, was abolished.

There is no indication that turtles can survive indefinitely in the anoxic state; in fact, many of the studies cited terminated with the death of the animal. However the ability to support all their vital processes by anaerobic metabolism for even a limited time is a most useful adaptation that enables turtles to remain underwater for a long time. The time they can survive depends primarily on their metabolic rate which, in turn, depends on their activity level and body temperature. Improved survival

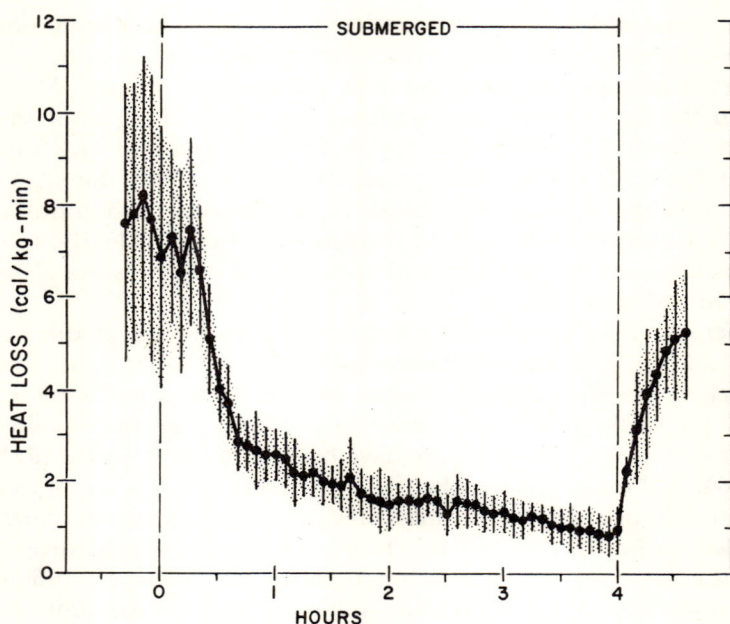

Figure 12 Calorimetric heat loss of *P. s. elegans* before, during, and after 4 hr experimental dives at 24°C. Solid lines connect mean values (10 animals) calculated over 5 min intervals. Vertical lines represent standard deviations above and below the mean for each interval. (From Jackson, 1968.) By permission of The American Physiological Society.

at low temperatures was clearly revealed by Musacchia (1959) on *Chrysemys picta*. The conditions for survival become most favorable during winter hibernation when metabolic rate is minimal; turtles survive for weeks or even months in situations where lung breathing is impossible. It is still not clear, though, whether survival during hibernation depends on anaerobic processes or on low level aerobic processes supported by the uptake of oxygen from the local environment.

Direct oxygen extraction from water has been reported in many species of turtles and, in some, it contributes significantly to underwater survival at ambient temperatures of 20 to 25°C. In the soft-shelled turtles, *Trionyx spiniferus asper* (Dunson, 1960) and *T. triunguis* (Girgis, 1961), aquatic breathing is particularly prominent, although Girgis contended that only a low level of metabolic activity could be sustained by this avenue of exchange. In *P. scripta*, extrapulmonary oxygen uptake is of little consequence in this midtemperature range (Jackson and Schmidt-Nielsen, 1966). Belkin (1968) compared oxygen uptake and

Table 1 Mean Values of Oxygen Consumption ($\dot{V}_{O_2}$) and Survival Time of Two Species of Turtles Submerged in Water Equilibrated at Different P_{O_2} Values[a]

P_{O_2}	Pseudemys scripta		Sternotherus minor	
	$\dot{V}_{O_2}$[b] (mM/kg · hr)	Survival Time (hr)	$\dot{V}_{O_2}$[c] (mM/kg · hr)	Survival Time (hr)
0	—	20	—	13
154	0.05	23	0.36	59
738	0.25	28	1.05	>5000

[a]From Belkin (1968).
[b]Air-breathing $\dot{V}_{O_2}$ = 1.26.
[c]Air-breathing $\dot{V}_{O_2}$ = 1.10.

survival in *S. minor* and *P. scripta*, submerged in water at 22°C equilibrated with nitrogen, air, or 100% oxygen (Table 1). In nitrogen-equilibrated water, *P. scripta* survived longer, attesting to the anaerobic capabilities of this species. Surprisingly, however, the addition of oxygen to the water, even 100% oxygen, did not significantly improve the survival time of *P. scripta*. *S. minor*, in contrast, was greatly aided by the dissolved oxygen, and in 100% oxygen-equilibrated water could apparently survive indefinitely without breathing. Root (1949) measured extrapulmonary oxygen uptake in a related species, *S. odoratus*, but discounted its significance, since it amounted to only one-eighth of the air-breathing oxygen uptake. Belkin (1968), however, considered that Root's evaluation was misleading, since the air-breathing oxygen uptake values were considerably above the minimum or resting level. With true resting values, the relative importance of aquatic breathing would have been much more evident. In Belkin's study, *S. minor* was unable to maintain a normal oxygen uptake by water breathing alone unless the water P_{O_2} was abnormally high. This is probably true of other species as well at temperatures above 20°C, but at lower temperatures, since metabolic demands are reduced and oxygen solubility in water is enhanced, normal oxidative metabolism may be possible in a nonbreathing, submerged turtle. More studies at low temperature are needed to substantiate this and to clarify the metabolic state of naturally hibernating turtles.

Most of the oxygen directly absorbed from the water passes into the body through the skin, even though the skin would seem ill-suited for this function in most turtles. Particular importance in aquatic breathing was originally ascribed to the buccopharyngeal epithelium by Gage and Gage (1886), based on their study of various soft-shelled turtles. Recent studies, however, have shown that this route accounts for only about a third of the total oxygen uptake (Root, 1949; Girgis, 1961). It has been suggested that there are species differences with respect to buccal pumping and its supposed function (Hansen, 1941; McCutcheon, 1943; Belkin, 1968). The cloacal bursae have also been implicated as underwater gas exchange structures in the species in which they are present. Submerged turtles do move water in and out of the cloacal bursae, as noted by Lüdicke (1936) in *Emys orbicularis*, but the respiratory function of this behavior is unproved.

During diving, the homeostatic functions of breathing are disturbed and a consistent pattern of changes occurs in the blood. These include elevated blood P_{CO_2}, increased acidity, increased lactic acid concentration, and reduced blood P_{O_2}. The severity of these effects in turn depends on the duration of the dive, the metabolic rate of the animal, and the contribution of extrapulmonary oxygen and carbon dioxide exchange. Freshwater turtles are able to survive extreme changes in these variables (Johlin and Moreland, 1933; Robin et al.,1964). Not so well understood in the overall adaptation to apneic diving is the ability to

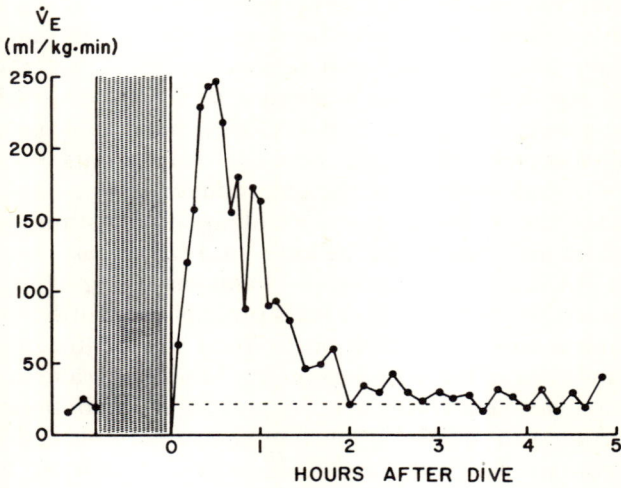

Figure 13 Respiratory minute volume $\dot{V}_E$ of a turtle (*P. s. elegans*) before and after a 3 hr dive at 24°C. The time scale of the diving period (shaded area) is compressed. The dashed line is the mean predive $\dot{V}_E$. (From Jackson and Silverblatt, 1974.) By permission of The American Physiological Society.

restore rapidly and accurately normal homeostasis once breathing is resumed. This problem was recently studied in *P. s. elegans* (Jackson and Silverblatt, 1974). Standardized dives lasting 3 or 4 hr at 24°C resulted in anoxia plus severe respiratory and metabolic (lactic) acidosis. When permitted to breathe, the experimental animals gradually (over the course of about 30 min) increased their pulmonary ventilation to a peak response that averaged 9 times the predive ventilation (Figure 13). By hyperventilating, the turtles were able to restore blood oxygen rapidly to above normal and, within 2 hr to correct blood pH to normal (Figure 14). Many more hours were required, however, to correct completely the metabolic acidosis.

Since the blood P_{O_2} was above normal and the blood P_{CO_2} was below normal within 30 min after the dive in these turtles, much of the hyperventilatory response must be attributed solely to changes in the acidity of the blood, or at a central receptor site such as in the brainstem.

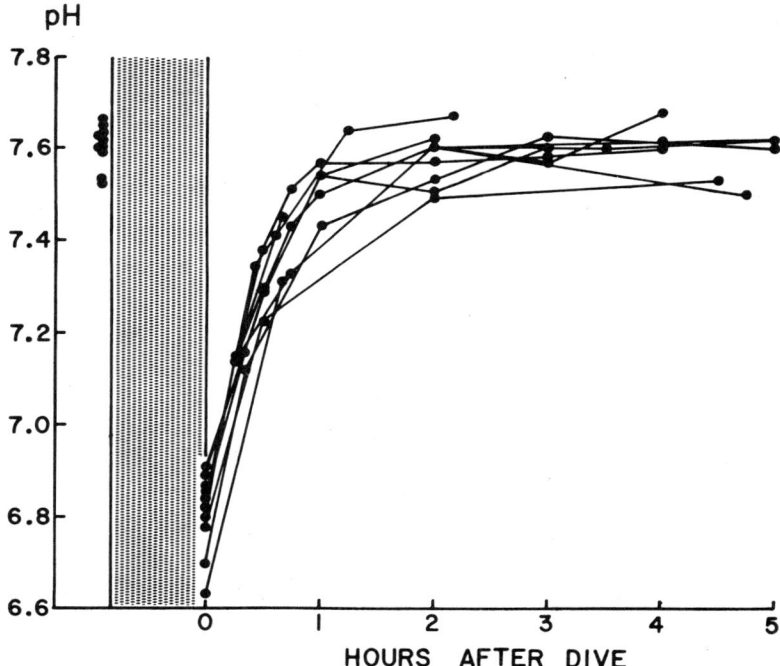

Figure 14 Arterial pH of 10 *P. s. elegans* before and after 3 to 4 hr dives at 24°C. The measurements at the end of the dive (time 0) were made before the resumption of breathing. (From Jackson and Silverblatt, 1974.) By permission of The American Physiological Society.

The intensity of this postapneic ventilation, together with the previously described sensitivity of the same species to induced changes in P_{O_2} and P_{CO_2}, seem to conflict with the turtles' assumed tolerance to such changes during voluntary diving. In many vertebrate divers, in fact, the ventilatory sensitivity to inspired carbon dioxide has been reported to be less than in nondivers (Andersen, 1966). My view is that the respiratory system of *P. scripta* is relatively insensitive to changes in the controlled variables during diving, perhaps because of an inhibitory influence acting on the respiratory center, whereas when the turtle is breathing, this inhibition is released and the full effect of the stimulus can be realized. Such a mechanism could permit wide fluctuations in the blood gas and acid-base variables during apnea, but could induce rapid restoration back to normal upon the initiation of breathing. Such an inhibitory mechanism, however, exists only hypothetically.

PROBLEMS AND PROSPECTS

It is clear that our knowledge of the respiratory physiology of turtles is incomplete. For some areas more sophisticated methods must be employed, and observations must be extended beyond the few common species to sample the diverse adaptations that doubtless exist in other chelonian groups. Following are specific examples of possible directions for future work based on these deficiencies.

Gas Exchange

Aside from the results of Lenfant et al. (1970), almost no information has been obtained on gas exchange at the alveolar-blood interface. Furthermore, this lone study employed the South American species, *Chelus fimbriatus*, for which few other data are available. Certain basic data are lacking, such as dead space volume and diffusion capacity.

Of great importance in gas exchange is the distribution of gas and capillary blood flow in the lung, or ventilation-perfusion ratio ($\dot{V}/\dot{Q}$ ratio). Ideally, the proper relationship between these two flows should exist in the whole lung and in each alveolar unit. Significant regional deviations from this ideal ratio disturb the normal gas exchange function of the lungs, particularly with respect to oxygen uptake. Nothing is known concerning $\dot{V}/\dot{Q}$ matching in turtle lungs, although the apparent reduction in capillary density in caudal lung units of certain species (Wolf, 1933) suggests that it could be important. Failure to achieve adequate gas exchange could result from such $\dot{V}/\dot{Q}$ inequalities together

PROBLEMS AND PROSPECTS

with possibly incomplete diffusion equilibrium and shunting of blood from the right to the left heart. The effects of temperature on these processes would be worth studying, since ventilation and cardiac output are probably affected differently, and the effects of temperature on shunting and diffusion exchange are unknown.

Apparently nothing is known concerning the factor or factors that control shunting of blood except that hypoxia or some other factor related to apnea probably induces bypass shunting (Lenfant et al., 1970; Johansen et al., 1970). This phenomenon may be the evolutionary basis for the pulmonary flow restrictions in mammals, which occur in localized regions of the lung where ventilation is inadequate. Thus the dramatic effect in turtles may be a useful model for studying this important mammalian reflex. A related problem is how the diving turtles make use of the large oxygen depot in the lungs if, during a dive, the lungs are bypassed by the blood.

Control of Breathing

Major gaps exist in our understanding of respiratory control concerning the receptor structures responding to changes in the regulated variables and concerning the central nervous system organization which integrates control. Research on these topics should be carried out on species such as *P. scripta*, for which essential background data are available. Tentative anatomical and physiological characterization of carotid body chemoreceptors were made by Frankel et al. (1969) in *P. scripta*, but the structures they examined were diffuse and variable. Comparative evidence favors the location of such structures in the region of the aortic arches.

The central nervous integrating mechanism, or respiratory center, was localized in the medullary region of the brainstem by Lumsden (1924). He found that sectioning of the brain above the fourth ventricle had no effect on breathing in a tortoise (species not given), whereas sectioning below the fourth ventricle abolished breathing. The close similarity between the respiratory control system in turtles and mammals suggest that chemoreceptors exist in the brainstem region. In mammals, central chemoreception is associated with the composition of the cerebrospinal fluid and the brain extracellular fluid, particularly with respect to the hydrogen ion concentration. According to Heisey (1970) *P. s. elegans* and *C. picta* possess larger volumes of cerebrospinal fluid per unit brain weight than do mammals. This property, coupled with the marked respiratory sensitivity of *P. s. elegans* to carbon dioxide, may make this species a potentially valuable model for studying medullary chemoreception. Initially, however, the existence and location of central chemoreceptors must be established.

Temperature

Thermal change is certainly the most prominent environmental variable affecting respiration in ectothermic animals. My measurement on *P. s. elegans* (Jackson, 1971a) is best described as a decrease in the ventilation to metabolic rate ratio ($\dot{V}_E/\dot{V}_{O_2}$) with temperature. However a study on the box turtle, *Terrepene carolina* (Altland and Parker, 1955) and several studies on lizards (e.g.: Dawson and Bartholomew, 1958; Templeton and Dawson, 1963; Bennett, 1973) do not reveal the same relationship. In contrast, the findings for other lizards were similar to those for *P. s. elegans* (Nielsen, 1961; Giordano and Jackson, 1973). Careful investigations that consider all the functions of breathing must be carried out to resolve this issue, assuming that there is a single solution.

Breathing serves both the purposes of oxygen supply and acid-base homeostasis and, as discussed earlier, these functions may conflict at high temperature. Furthermore the respiratory system may serve the added function of evaporative heat loss (see Chapter 11). How is this function incorporated by a mechanism already serving two independent (and possibly conflicting) functions? Measurements of ventilation, metabolic rate (oxygen consumption plus carbon dioxide production), as well as blood acid-base and blood gas values, should be made in representative turtles from different thermal environments at or near their upper temperature limit.

At the other extreme, very low temperature, important problems in chelonian respiratory physiology remain unresolved. Hibernating turtles often bury themselves in the mud at the bottom of a pond, and it is difficult to imagine how adequate convective gas transport through such a substrate could occur.

Extrapulmonary oxygen exchange occurs to a significant extent in many species of aquatic turtles, but its primary importance may be at temperatures lower than those usually studied. At low temperature, the metabolic demand for oxygen is reduced, while the available oxygen (dissolved in the water) is probably increased. For these reasons, some turtles may rely on extrapulmonary surfaces for a greater proportion of their gas exchange at low temperature, such as is the case with many amphibians (Whitford and Hutchison, 1963). There is little doubt that carbon dioxide loss through nonpulmonary surfaces is of far greater importance quantitatively than is oxygen uptake, because of the higher solubility of carbon dioxide in water; but carbon dioxide loss by this route has not been studied in turtles. Because of the primary importance of carbon dioxide in respiratory control and acid-base homeostasis, this important topic deserves further investigation.

Comparative Studies

Descriptive comparative studies in some areas, such as lung architecture, respiratory musculature, and blood oxygen transport, need to be followed up by more physiological investigations. Excellent models are the thorough descriptions of respiratory muscles and their functions in *Testudo graeca* (Gans and Hughes, 1967) and in *Chelydra serpentina* (Gaunt and Gans, 1969). This type of study is needed on marine turtles and other species to clarify the function of a well-developed muscle sheath enveloping the lungs. Possible functions in various turtles include expiratory flow activation, gas mixing within the lungs, buoyancy control, and the maintenance of proper gas volume distribution in the lungs.

Other contrasts between turtles that could be profitably studied include respiratory control in diving versus nondiving species, ventilation and acid-base relationship in desert versus aquatic species, and a correlation between the degree of alveolar development (see Perry, 1972, for review), and gas exchange capacity in various species. Many of the special adaptations of different species must be reflected in some aspect of their respiratory physiology.

CHAPTER 10
The Central Nervous System

ALICE S. POWERS and ANTON REINER

Improvements in existing experimental neuroanatomical techniques (e.g., the Nauta, Nauta-Gygax, and Fink-Heimer methods for the selective staining of degenerating fibers and terminals) and the development of new techniques (autoradiographic and horseradish peroxidase methods for tracing fiber connections) have served to renew interest in the comparative neuroanatomy of the brain. This chapter integrates recent findings with the older descriptive work to provide a more current perspective on the organization of turtle brains.

The interested reader may consult Ariens-Kappers, Huber, and Crosby (1936) for a useful review of ideas about central nervous system (CNS) organization in turtles before the advent of recent experimental techniques. In addition, extensive descriptions of the brains of turtles may be found in the following works: Johnston (1915) on the telencephalon of *Terrapene carolina*; Goldby and Gamble (1957) on the telencephalon of various reptiles; Northcutt (1970) on the telencephalon of *Chrysemys picta*; Papez (1935) on the diencephalon (and to some extent the mesencephalon) of *Chelonia mydas*; Tuge (1932) on the brainstem of *Pseudemys scripta elegans*; Cruce and Nieuwenhuys (1974) on the brainstem of *Testudo hermanni*; and Baumann (1966) on the entire CNS of *C. picta*. A stereotaxic atlas of the forebrain and midbrain of *C. picta* has recently been constructed (Powers and Reiner, in manuscript).

GROSS MORPHOLOGY

Figure 1 depicts a turtle brain, as seen from ventral, dorsal, and lateral aspects. The forebrain, midbrain, and hindbrain are arranged in a linear

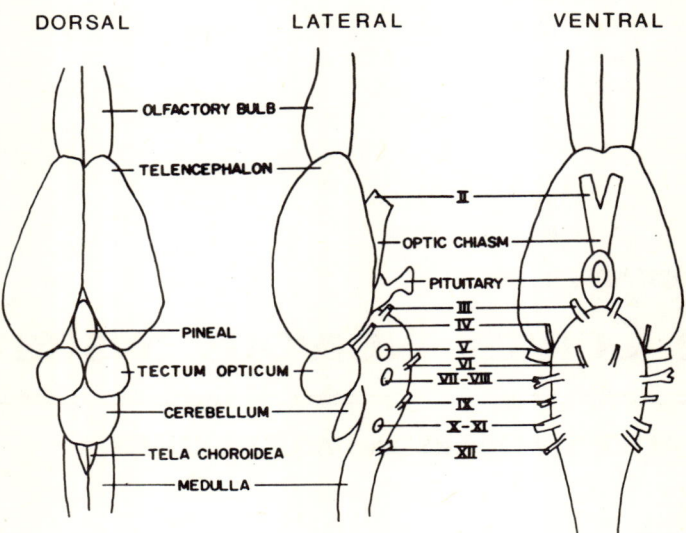

Figure 1 Dorsal, lateral, and ventral views of turtle brain.

fashion along the anterior-posterior axis, without any cephalic or cervical flexures. The cerebral hemispheres are present as a pair of conspicuous rostral enlargements. The cerebral hemispheres in turtles are larger and better developed than in amphibians, but are considerably smaller and not as well differentiated as in birds and mammals. A pair of olfactory bulbs form anterior extensions of the cerebral hemispheres and in turn connect with the olfactory sacs by way of the olfactory nerve. Dorsally, the cerebral hemispheres obscure the diencephalon from view. The hemispheres are separated along the midline by a longitudinal fissure, which widens at the caudal border of the telencephalon. A choroidal sac, which forms the roof of the diencephalon, is present in this widening. The pineal body, or epiphysis, is attached to this sac. From the lateral and dorsal aspects, a pair of prominent swellings are evident posterior to the cerebral hemispheres. These are the tecta, and they form the roof of the midbrain. Caudal to the tecta, the cerebellum is visible as an unpaired flap of tissue extending from the brainstem over the fourth ventricle. The fourth ventricle is covered by the tela choroidea, which is involved in the secretion of cerebrospinal fluid. From the ventral and lateral aspects, the optic tracts and chiasm are visible beneath the forebrain. Caudal to the chiasm, the pituitary body is visible. Twelve pairs of cranial nerves are present. The spinal cord of turtles is externally distinguished by the great enlargement of cervical and lumbar segments over thoracic and sacral segments. Trunk musculature in turtles is largely absent; thus spinal segments associated with trunk musculature are greatly reduced. The brain and spinal cord are sheathed in two meninges, a dura mater externally, and a leptomeninx that intimately covers the brain. Both a subdural and an epidural space are present in turtles. The epidural space contains a great amount of tissue, in which large epimeningeal veins are present.

INTERNAL STRUCTURE AND CONNECTIONS

The Telencephalon

The most prominent features of the cerebral hemispheres are a well-differentiated basal mass of tissue (consisting of such structures as the amygdaloid nuclei, the paleostriatum augmentatum, the globus pallidus, and the dorsal ventricular ridge) and a band of cortical tissue overlying the ventricle. Traditionally, it was thought that the predominant input to both basal and cortical regions of the telencephalon was olfactory (Johnston, 1915, 1916, 1923; Ariens-Kappers et al., 1936). However Gamble (1956) demonstrated that input from the olfactory bulbs to the cerebral hemispheres was quite restricted: only the anterior olfactory nucleus, the medial, dorsomedial, and lateral cortices, and the nucleus of the lateral olfactory tract were found to receive a projection (Figure 2).[1] Furthermore, other researchers (Hall and Ebner, 1970a, 1970b; Parent, 1976) have demonstrated a substantial input to the cerebral hemispheres from the dorsal thalamus. Based on such findings about the connections of the telencephalon, and based on work on the histochemistry of the telencephalon, it now seems reasonable to recognize four general subdivisions of the telencephalon: (1) olfactory, consisting of the olfactory bulbs and piriform cortex; (2) limbic, consisting of amygdaloid and septal areas and medial and dorsomedial cortices; (3) internal striatum, consisting of globus pallidus and paleostriatum augmentatum; and (4) external striatum, which includes the dorsal cortex, pallial thickening, dorsal ventricular ridge, and core nucleus (see Figure 2). This schema for subdividing the vertebrate telencephalon is that of Nauta and Karten (1970).

The foregoing subdivision of the telencephalon departs from traditional ideas in that older viewpoints did not recognize the internal and

[1] A list of the abbreviations used in Figures 2 to 6 appears at the end of the chapter.

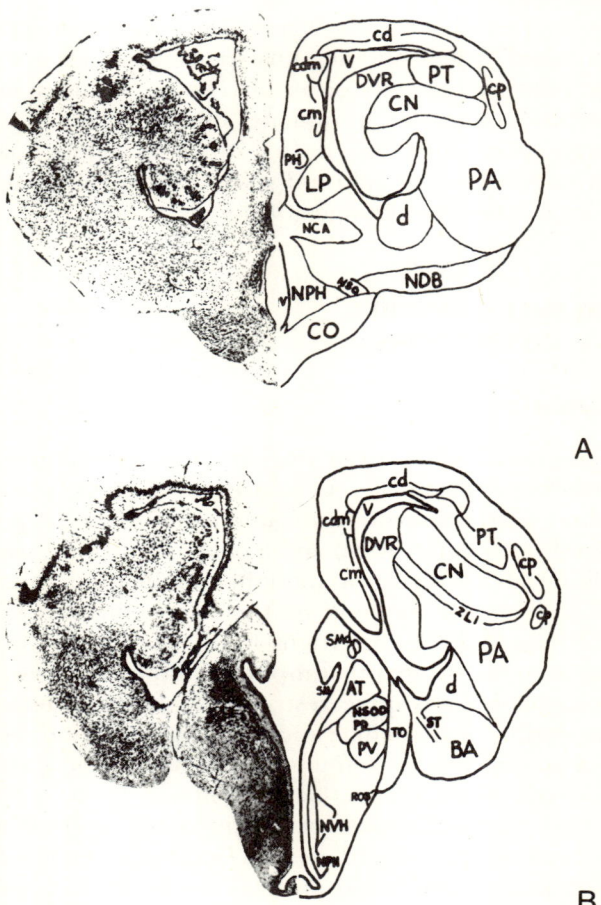

Figure 2 Photomicrograph and drawings of transverse sections through the forebrain of *Chrysemys picta*. (*a*) Taken through the cerebral hemispheres at a rostral level. (*b*) Taken at a more caudal level through the hemispheres. The rostral diencephalon is evident ventrally and medially.

external striatum as distinct regions. The internal and external striatum together were thought to be comparable to the basal ganglia of mammals (Johnston, 1923). Histochemically, however, the internal and external striatum appear to be distinct subdivisions, with only the internal striatum being comparable to the basal ganglia of mammals (Baker-Cohen, 1968). The external striatum appears to be more comparable to the neocortex of mammals. This histochemical similarity of the external striatum to the neocortex is supported by data on the connections of the telencephalon, since it has been shown that the external striatum is the recipient of extensive input from the dorsal thalamus, as is the neocortex in mammals. Hall and Ebner (1970b) have shown that the dorsal cortex receives a projection from the nucleus geniculatus lateralis, pars dorsalis, of the thalamus and that the core nucleus and the dorsal ventricular ridge receive a projection from nucleus rotundus of the thalamus—both thalamic nuclei receive visual input (see section on the diencephalon). There is reason to believe that somatic information also reaches the telencephalon, since Orrego (1961) found evoked potentials to somatic stimulation in the anteromedial dorsal cortex. Auditory information, too, may reach the telencephalon, since nucleus reuniens of the thalamus, a presumed homolog of the medial geniculate nucleus in mammals (Papez, 1936), projects to the dorsal ventricular ridge (Parent, 1976). Such an auditory pathway has been demonstrated in alligators and lizards (Pritz, 1974b; Foster and Peele, 1975). In turtles, however, the precise terminal field within the dorsal ventricular ridge has not been delineated.

Little is known with certitude about the connections of the other three telencephalic subdivisions. The efferent projections of the internal striatum are presumed to pass by way of the ventral peduncle of the lateral forebrain bundle to the mesencephalic tegmentum (Ariens-Kappers et al., 1936). Parent (1976) has shown that the substantia nigra of the tegmentum projects to the paleostriatum augmentatum (Figure 2). In an earlier study, Parent (1973) showed that the substantia nigra contained a large number of catecholaminergic cell bodies, whereas the paleostriatum augmentatum contained a large number of catecholaminergic nerve endings.

The connections of the olfactory bulb were described earlier. In addition, a stria medullaris has been described and is said to provide olfactohabenular connections. Little is known of the connections of the limbic portion of the telencephalon. A medial forebrain bundle, providing a connection of the septum and medial and dorsomedial cortex with the hypothalamus and tegmentum, and a stria terminalis, providing a connection between amygdaloid and hypothalamic regions, have been described but not shown experimentally (Goldby and Gamble, 1957).

The Diencephalon

Four diencephalic subdivisions are generally recognized: (1) epithalamus, (2) dorsal thalamus, (3) subthalamus, and (4) hypothalamus and hypophysis. The epithalamus consists of the habenulae (noted earlier as possibly receiving olfactory input by way of the stria medullaris) and the pineal body. The dorsal thalamus consists of a number of well-defined nuclear groups. Traditionally, it was believed that in terms of afferent and efferent connections, the dorsal thalamus of turtles was associated mainly with the tectum rather than with the telencephalon (Ariens-Kappers et al., 1936). However current evidence indicates that the dorsal thalamus has a substantial projection to the external striatum of the telencephalon. Thus the dorsal thalamus of turtles would appear to be comparable to that in mammals.

The most prominent thalamic nucleus in turtles is nucleus rotundus (Figure 3). Nucleus rotundus is apparently involved in visual functions, since it receives input from the superficial (visual) portion of the optic tectum (Foster and Hall, 1975), and since all rotundal units studied respond to visual stimulation (Belekhova and Kosareva, 1971). A second dorsal thalamic nucleus, the nucleus geniculatus lateralis, pars dorsalis, receives optic tract terminations (Kosareva, 1967; Knapp and Kang, 1968a, 1968b; Hall and Ebner, 1970a). The efferents of these two nuclei have been described. A third thalamic nucleus, the nucleus geniculatus lateralis, pars ventralis, also receives optic tract terminations, but the efferents of this nucleus are unknown. Ebbesson (1969) has described spinal projections to a rostral and lateral zone of the thalamus. Nucleus

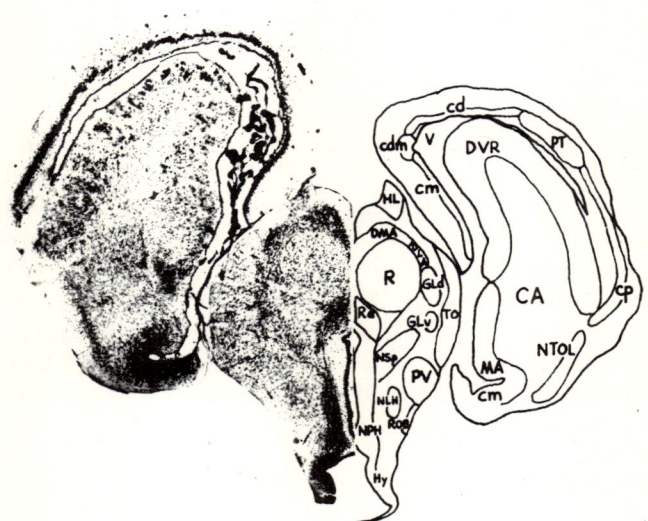

Figure 3 Photomicrograph and drawing of a transverse section at a middiencephalic level. Note the prominence of nucleus rotundus of the dorsal thalamus. The cerebral hemispheres are still evident laterally.

reuniens presumably receives (auditory) input from the torus semicircularis of the midbrain, since this is the case in alligators and lizards (Pritz, 1974a; Foster and Peele, 1975). Nucleus reuniens and two additional thalamic nuclei, dorsomedialis anterior and dorsolateralis anterior (Figure 3), have been described as projecting to the external striatum of the telencephalon (Parent, 1976). However the terminal fields of these projections within the telencephalon are unclear.

The subthalamus is thought to be connected with the internal striatum of the telencephalon by way of the ventral peduncle of the lateral forebrain bundle (Ariens-Kappers et al., 1936). There has been no experimental work on the subthalamus. Crosby (1969) has presented a description of the hypothalamic and hypophyseal regions of the turtle brain. Parent and Poitras (1974) have described the catecholaminergic innervation of this region.

The Mesodiencephalon

The mesodiencephalon (also called the pretectum) is marked by the presence of the posterior commissure (Curwen and Miller, 1939). Three pretectal nuclei, lentiformis mesencephali, geniculatus pretectalis, and area pretectalis, receive retinal input (Knapp and Kang, 1968a, 1968b; Bass, 1976). In addition, nucleus opticus tegmenti (also called the nucleus of the basal optic root), located in the basolateral tegmentum, receives retinal input (Figure 4). Reiner and Karten (1978) have shown that both nucleus opticus tegmenti and nucleus geniculatus pretectalis project to the cerebellum. Two additional pretectal nuclei, the dorsal pretectal nucleus and the ventral pretectal nucleus, receive tectal projections (Hall and Ebner, 1970a; Foster and Hall, 1975).

The Mesencephalon

Two general regions of the mesencephalon are recognizable, the tectum (overlying the tectal ventricles) and the tegmentum (comprising the basal region of the mesencephalon). The tectum is a layered structure; the six layers recognized here are (1) stratum opticum, (2) stratum fibrosum et griseum superficiale, (3) stratum griseum centrale, (4) stratum album centrale, (5) stratum griseum periventriculare, and (6) stratum album periventriculare (Figure 5). Optic tract terminations occur in the two superficialmost layers, stratum opticum and stratum fibrosum et griseum superficiale (Knapp and Kang, 1968a, 1968b). The stratum griseum centrale receives input from the cells of the overlying layers (mainly the stratum fibrosum et griseum superficiale) and projects to

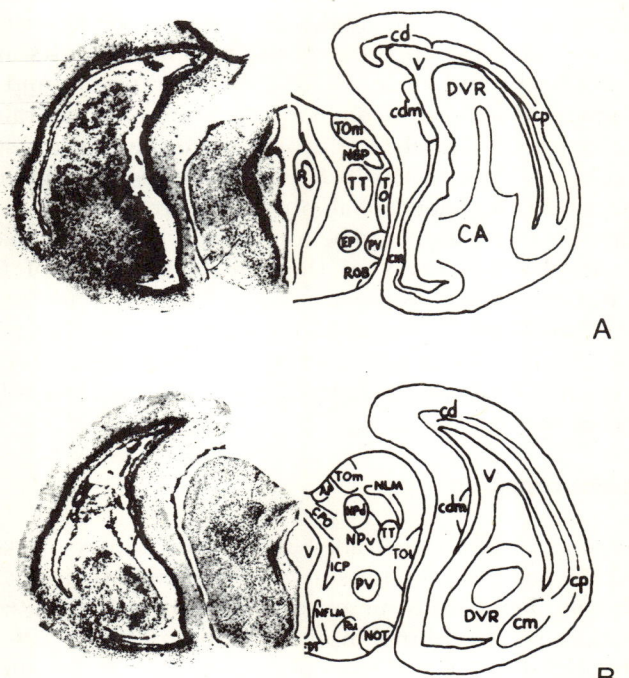

Figure 4 Photomicrograph and drawings of sections taken through (a) rostral and (b) midpretectal levels. The cerebral hemispheres are still evident laterally.

nucleus rotundus (Foster and Hall, 1975). The fibers of this tectorotundal pathway traverse the stratum album centrale, enter the tectothalamic tract, and course rostrally to nucleus rotundus. The stratum griseum periventriculare gives rise to descending ipsilateral and contralateral pathways that emerge from the tectum by way of the stratum album periventriculare. These descending pathways terminate in reticular areas of the brainstem (Foster and Hall, 1975). Other than the retina, the only known sources of tectal afferents are the spinal cord (Ebbesson, 1969) and nucleus isthmi, pars magnocellularis (Foster and Hall, 1975). The input from the spinal cord ends in the stratum griseum periventriculare.

It is clear that in alligators and lizards auditory projections from the cochlear nuclei of the brainstem terminate in a structure of the caudal midbrain called torus semicircularis (Pritz, 1974a; Foster and Peele, 1975). A torus is present in the midbrain of turtles also, and although experimental data are lacking, it seems likely that the torus of turtles also receives a projection from the cochlear nuclei.

INTERNAL STRUCTURE AND CONNECTIONS 201

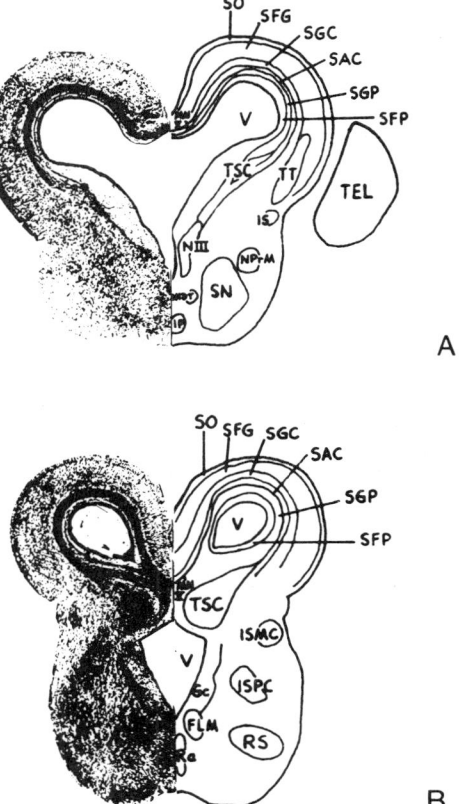

Figure 5 Photomicrograph and drawings of sections taken through (a) midtectal and (b) caudal tectal levels. The torus semicircularis, the presumptive homolog of the inferior colliculus of mammals, is present along the midline in (b).

As noted, there is a substantia nigra in the ventrolateral tegmentum. Medially, a nucleus ruber has been described, but its afferents have not been determined. Ten Donkelaar (1976a, 1976b) has described a rubral projection to the spinal cord. Nucleus profundus mesencephali, found dorsal to substantia nigra, has been said to receive telencephalic input (Johnston, 1915), but this suggestion requires experimental scrutiny.

The Cerebellum

The cerebellum has been described in detail by Larsell (1932, 1967). Three layers are present: a molecular, a Purkinje, and a granule cell

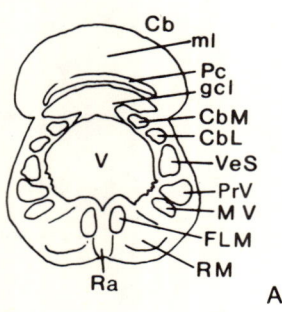

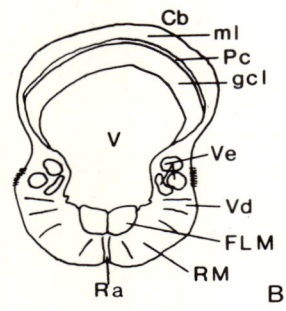

Figure 6 Drawings of transverse sections taken at (a) rostral and (b) midcerebellar levels, illustrating the location of the deep cerebellar nuclei and several brainstem nuclei.

layer (Figure 6). The organization of the cellular elements within these layers seems comparable to that in other amniotes. Medial and lateral deep cerebellar nuclei have been described (Larsell, 1967), and cerebellar output is presumed to pass by way of them to the brainstem. Of cerebellar afferents, ventral and dorsal spinocerebellar tracts have been demonstrated experimentally (Ebbesson, 1969), and a purported tectocerebellar projection has been shown not to exist (Foster and Hall, 1975). As noted previously, there is input to the cerebellum from the nucleus opticus tegmenti and the nucleus geniculatus pretectalis; both receive retinal input. Finally, Larsell (1967) has described vestibular input, both primary fibers from the vestibular ganglion and secondary fibers from the vestibular nuclei of the brainstem, to the cerebellum.

The Brainstem

Cruce and Nieuwenhuys (1974) have presented a detailed treatment of the brainstem. In brief, cranial nerve nuclei are organized along the same four functional columns as in other vertebrates: somatic motor, visceral motor, visceral sensory, and somatic sensory, with the somatic motor column extending through the mesencephalon. There is no pontine enlargement in turtles. A reticular core is present throughout the brainstem. At the spinobulbar junction, dorsal column nuclei are present.

The Spinal Cord

Dorsal and ventral horns are well developed and divide the spinal cord into dorsal, lateral, and ventral funiculi. Collaterals of primary afferents pass by way of the dorsal funiculus and terminate in the dorsal column nuclei. Ascending fibers of the lateral funiculus terminate throughout the reticular core of the brainstem and some reach the tectum as well as the dorsal thalamus (Ebbesson, 1969). Dorsal and ventral spinocerebellar tracts ascend in the dorsolateral and ventrolateral cord, respectively. The descending pathways of the cord were described by ten Donkelaar (1976a, 1976b). Descending fibers course in the ventral and lateral funiculi. Spinal pathways arise from nucleus interstitialis of the tegmentum (also called the nucleus of the medial longitudinal fasciculus), the vestibular nuclei, and the reticular formation; all terminate on the medial portion of the ventral horn. A rubrospinal pathway ends in the intermediate zone of the spinal gray.

OVERVIEW

Earlier treatments of the brains of nonmammals stated that the tectum was the highest correlation center for sensory input and that the forebrain was primarily olfactory. With more recent experimental neuroanatomical techniques, it has become clear that sensory input reaches the diencephalon and telencephalon by discrete pathways. The tectum is a center that appears to be primarily involved in vision (e.g., Hayes and Hertzler, 1967; Hertzler and Hayes, 1967), rather than in the integration of input from various exteroceptive modalities.

ABBREVIATIONS

AP	Area pretectalis
AT	Area triangularis
BA	Nucleus basalis amygdalae

CA	Nucleus centralis amygdalae
Cb	Cerebellum
CbM	Nucleus cerebellaris medius
CbL	Nucleus cerebellaris lateralis
cd	Cortex dorsalis
cdm	Cortex dorsomedialis
cm	Cortex medialis
CN	Core nucleus of the dorsal ventricular ridge
CO	Chiasma opticum
cp	Cortex pyriformis
CPo	Commissura posterior
d	Area d
DLA	Nucleus dorsolateralis anterior
DMA	Nucleus dorsomedialis anterior
DT	Decussatio tegmenti
DVR	Dorsal ventricular ridge
EP	Nucleus entopeduncularis
FLM	Fasciculus longitudinalis medialis
Gc	Griseum centrale
gcl	Granule cell layer of the cerebellum
GLd	Nucleus geniculatus lateralis, pars dorsalis
GLv	Nucleus geniculatus lateralis, pars ventralis
HL	Nucleus habenularis lateralis
Hy	Hypophysis
ICP	Nucleus interstitialis commissuralis posterioris
IP	Nucleus interpeduncularis
IS	Nucleus isthmi
ISMC	Nucleus isthmi, pars magnocellularis
ISPC	Nucleus isthmi, pars parvocellularis
LP	Lobus parolfactorius
ml	Molecular layer of the cerebellum
MA	Nucleus medialis amygdalae
MNV	Nucleus mesencephalicus nervi trigemini
MV	Nucleus motorius nervi trigemini
NCA	Nucleus commissuralis anterioris
NDB	Nucleus fasciculi diagonalis Brocae
NFLM	Nucleus fasciculi longitudinalis medialis
NGP	Nucleus geniculatus pretectalis
NIDT	Nucleus interstitialis decussationis tegmenti
NLH	Nucleus lateralis hypothalami
NLM	Nucleus lentiformis mesencephali
NOT	Nucleus opticus tegmenti
NPd	Nucleus pretectalis dorsalis

ABBREVIATIONS

NPv	Nucleus pretectalis ventralis
NPH	Nucleus periventricularis hypothalami
NPrM	Nucleus profundus mesencephali
NSO	Nucleus supraopticus
NSOD	Nucleus decussationis supraopticae dorsalis
NSp	Nucleus suprapeduncularis
NTOL	Nucleus tracti olfactorii lateralis
NVH	Nucleus ventralis hypothalami
PA	Paleostriatum augmentatum
Pc	Purkinje cell layer of the cerebellum
PD	Pedunculus dorsalis fasciculi prosencephali lateralis
PH	Primordium hippocampi
PrV	Nucleus sensorius principalis nervi trigemini
PT	Pallial thickening
PV	Pedunculus ventralis fasciculi prosencephali lateralis
R	Nucleus rotundus
Ra	Nucleus raphes
Re	Nucleus reuniens
RM	Nucleus reticularis medius
ROB	Radix opticum basalis
RS	Nucleus reticularis superior
Ru	Nucleus ruber
SAC	Stratum album centrale
SFG	Stratum fibrosum et griseum superficiale
SFP	Stratum fibrosum periventriculare
SGC	Stratum griseum centrale
SGP	Stratum griseum periventriculare
SM	Sulcus medius
SMd	Stria medullaris
SN	Substantia nigra
SO	Stratum opticum
ST	Stria terminalis
TEL	Telencephalon
TO	Tractus opticus
TOl	Tractus opticus, pars lateralis
TOm	Tractus opticus, pars medialis
TSC	Torus semicircularis
TT	Tractus tectothalamicus
V	Ventriculus
Vd	Nucleus descendens nervi trigemini
Ve	Nucleus vestibularis
VeS	Nucleus vestibularis superior
ZLI	Zona limitans interstriatica

CHAPTER 11
Thermoregulation

VICTOR H. HUTCHISON

Contrary to the textbook aphorism that poikilotherms have body temperatures of the surrounding environment, many "cold-blooded" animals are relatively independent, within limits, of ambient temperatures (T_a) in nature. Indeed, it has been argued that there is no fundamental difference between ectothermic (external source of heat) and endothermic (internal source of heat) animals, since the two types vary only in the degree of their ability to maintain thermal homeostasis (Cloudsley-Thompson, 1969). To date, however, no reptile has been shown to possess the highly sophisticated physiological mechanisms for long-term high rates of endogenous heat production; the lack of effective insulation in reptiles is another major difference.

The exertion of some control over body temperature (T_b), which is especially evident during periods of activity, is achieved through combinations of various degrees of behavioral and physiological regulation, as well as through the utilization of favorable microclimates. The mass of many turtles is great enough to prevent rapid changes in T_b through heat exchange with the environment. Thus the larger land tortoises are found in the tropics and subtropics; the larger aquatic turtles tend to avoid temperature extremes encountered by terrestrial reptiles.

BODY TEMPERATURES

The earliest recorded observation on body temperatures of turtles was probably made in 1872 by Walbaum, who found that T_b varied only 1 or

2 degrees from that of the environment (Baldwin, 1925a). Few additional studies on the T_b of turtles were made until relatively recent times, and chelonians have received far less attention than other reptiles. Many investigators were stimulated to accumulate field data on body temperatures of reptiles by the initial studies of Cowles and Bogert (1944) and Colbert, Cowles, and Bogert (1946). Reviews of the literature on reptile body temperatures and thermoregulation have been provided by Brattstrom (1965), Cloudsley-Thompson (1971), Mayhew (1968), and Templeton (1970). Definitions and discussions of the terms and techniques used in the study of reptilian thermal relations have been reviewed by Brattstrom (1970) and Templeton (1970). Figure 1 summarizes the various energy exchanges between a turtle and its environment.

Table 1 is a compilation of body temperature data from field and laboratory studies and is expanded from a listing by Brattstrom (1965). The *voluntary minimum* is the T_b at which an animal voluntarily emerges or retreats; it usually represents the lowest recorded T_b of an active individual. The *voluntary maximum* is the T_b at which an animal retreats to a lower environmental temperature (shade, water, burrow); this usually represents the highest T_b recorded of an active individual. The *preferred temperature* (also sometimes called the optimum, eccritic, or mean activity temperature) is the often narrow range of T_b at which animals are active, or the T_b resulting from selection in a laboratory thermal gradient. The preferred temperature often represents the mean of all recorded body temperatures of active animals in the *normal activity range* (the temperatures between the voluntary minimum and voluntary maximum). Since the T_b selected by a reptile under natural conditions may not be the same as that selected in a thermal gradient under controlled laboratory conditions, Licht, Dawson, Shoemaker, and Main (1966) limited "eccritic temperature" to field measurements and "preferred temperature" to laboratory gradients. It should be emphasized that present data do not show clearly whether the T_b of turtles is maintained over a fairly small range around an eccritic temperature, as is true of many lizards.

The validity of T_b values taken in the field and used to calculate voluntary maxima and minima or eccritic temperatures may be highly questionable. Only when coupled with a knowledge of the prior thermal history and behavior of the animal during the preceding few hours can T_b have much significance. For example, metal cans filled with water may exhibit temperature distributions very similar to those of reptiles in the field (Heath, 1964). Turtles may be active at a given time at nonpreferred temperatures when behavioral thermoregulation is made impossible by ambient physical conditions or because of the necessity to escape from predators. The optimum temperature for some particular func-

Table 1 Temperature (°C) Relations of Turtles[a]

Species	Voluntary Minimum	Voluntary Maximum	Mean Preferred or Average Activity	Lethal or Critical Thermal Maximum Acclimated	Lethal or Critical Thermal Maximum Mean	Reference[b]
Chelydridae						
Chelydra serpentina	5	24.5	—	20	(39.46)	5, (15)
Macroclemmys temmincki	18	—	—	20	(39.35)	1, (15)
Kinosternidae						
Sternotherus carinatus	14	34	33.29	20–24	40.25	19
S. odoratus	10	34	20.7–22.4*	20.	(41.03)	12,* 19, (15)
			21–26	20–24	41.60	12
S. minor	—	—	—	20	40.4	15
Kinsternon bauri	—	—	—	20	40.6	15
K. flavescens	18	32	25.06	20–24	43.25	19
K. subrubrum	—	—	—	20–24	40.95	19
Emydidae						
Clemmys guttata	8–10	—	21.1–24.9*	20	(41.98)	7, 12,* (15)
C. insculpta	—	—	—	20	41.3	15
C. marmorata	9	(34)	(24–32)	—	(40.0)	5, (6)
Terrapene carolina	—	—	25.6	20	(41.9–43.0)	5, (15)
T. ornata	13	35.9	28–30	?	(40)	16, (10)
Malaclemmys terrapin	—	—	—	20	(41.8–42.5)	15
Graptemys pseudogeographica	—	39	32.7	20	(41.0)	4, (15)
Chrysemys picta	8	27.8–32	20.7–22.4*	—	—	9, 15*
Pseudemys c. concinna	—	—	—	20	41.8	15

Table 1 (Continued)

Species	Voluntary Minimum	Voluntary Maximum	Mean Preferred or Average Activity	Lethal or Critical Thermal Maximum Acclimated	Lethal or Critical Thermal Maximum Mean	Reference[b]
P. floridana	—	—	—	20	40.85	15
P. rubriventris	—	—	—	20	39.36	15
P. nelsoni	—	—	—	20	40.42	15
					43–44	21
P. scripta	10	37	25.5*	20	(41.0–41.69)	5,* 7, (15)
Deirochelys reticularia	25.5	25.6	25.5	20	41.3	15
Emydoidea blandingi	—	—	—	20	(39.5)	(15), 5
Rhinoclemys pulcherrima	—	—	—	20	43.5	15
Testudinidae						
Gopherus agassizi	15.0	37.8	30.6	20	(43.1)	5, (15)
G. berlandieri	—	—	—	20	40.4	15
G. polyphemus	—	—	34–35	20	(43.9)	3, (15)
Testudo hermanni	16.0	34.0	25–30	?	39–42	8
T. horsfieldi	—	34–35 (30–34)	—	—	—	1, 22
Geochelone gigantea	—	—	27–30	—	—	14
G. elephantopus	—	—	28–33	—	—	18
Cheloniidae						
Chelonia mydas	18.0	30.0	29.7–(29.9)	—	—	(13), 17
Eretmochelys imbricata	—	—	28.5–29.0	—	—	13
Lepidochelys olivacea	—	—	28.7	—	—	20

Dermochelydae						
Dermochelys coriacea	12.2	—	(30.6)–32.0	—	—	11, (20)
Trionychidae						
Trionyx spiniferus	—	35.7	32.7	20	(41.05)	4, (15)
T. ferox	—	—	—	20	39.0	15
Chelidae						
Chelodina longicollis	—	—	—	20	30–44	24

[a] See text for definitions and warnings on too strict application of these data. The asterisks and parentheses refer to the corresponding citations in the reference column.

[b] *References:*

1. Allen and Neill (1950).
2. Andreev (1948).
3. Bogert and Cowles (1947).
4. Boyer (1965).
5. Brattstrom (1965).
6. Bury (1972).
7. Cagle (1950).
8. Cherchi (1956, 1960).
9. Ernst (1972).
10. Fitch (1956).
11. Friar, Ackman, and Mrosovsky (1972).
12. Graham (1972).
13. Hirth (1962).
14. Honegger (1967).
15. Hutchison, Vinegar, and Kosh (1966).
16. Legler (1960).
17. McGinnis (1968).
18. MacKay (1964).
19. Mahmoud (1969).
20. Mrosovsky and Pritchard (1971).
21. Pritchard and Greenwood (1968).
22. Sergeyev (1939).
23. Sexton (1959).
24. Webb and Witten (1973).

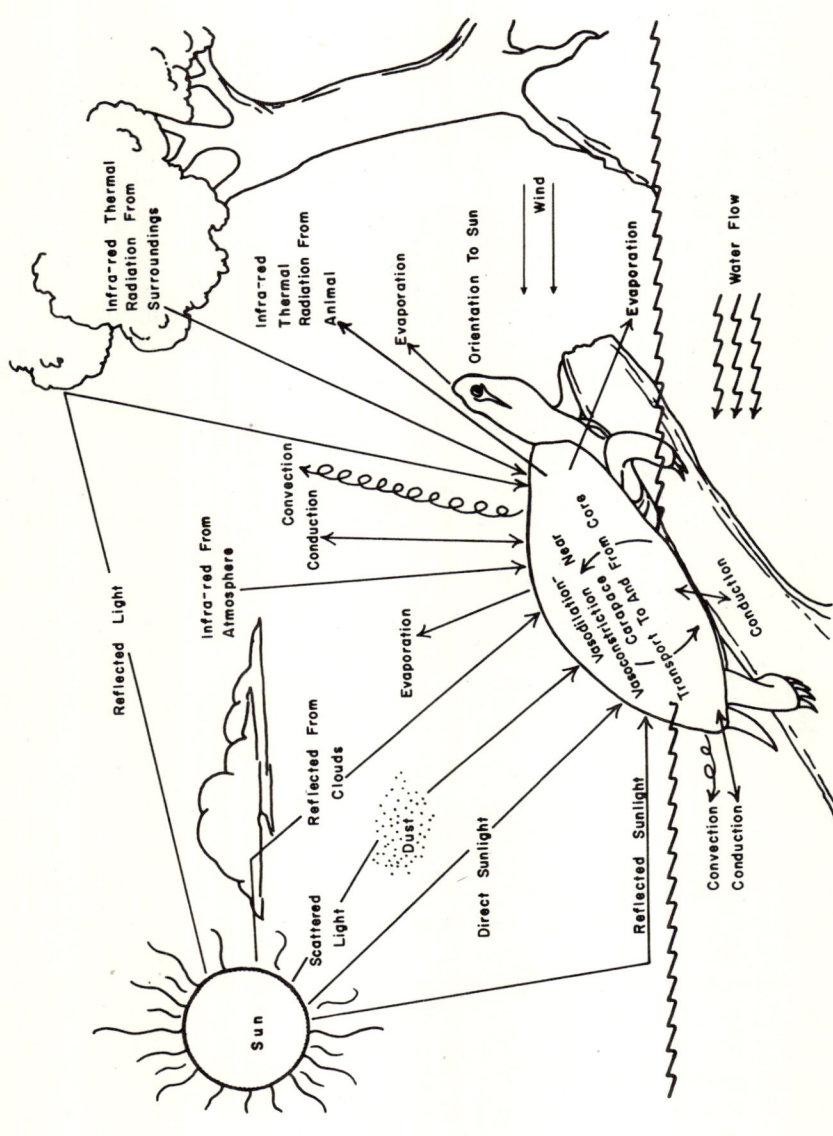

Figure 1 Energy exchange of a turtle in a natural environment. The animal is on a basking site with its hind quarters in the water.

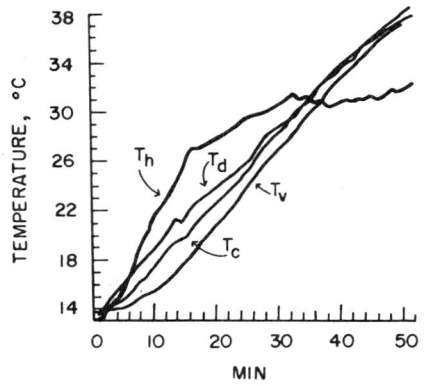

Figure 2 Temperatures that resulted from radiant heating in a 308 g *Chelodina longicollis*: T_h, head temperature; T_d, dorsal temperature; T_c, cloacal temperature; T_v, ventral temperatures. (From Webb and Johnson, 1972.)

tions may not coincide with the "preferred temperature." For example, the optimum temperature for feeding in aquatic turtles may be lower than that for digestion (Kenyon, 1925; Chesley, 1934; Anderson and Wilbur, 1948). Thus during a daily cycle of activity, an ectotherm may have a wide range of T_b's, some of which may not be optimal for certain physiological processes. Thermal optima may vary with the physiological state of the individual (Bustard, 1967).

Traditionally reptilian body temperatures have been measured from the cloaca, with rapid-reading mercury or thermistor thermometers. The more recent development of biotelemetric techniques has provided opportunities for the continuous monitoring of T_b under natural conditions (MacKay, 1964; McGinnis and Voigt, 1971). With implantation of two or more telemetry devices in different parts of the body, notable differences in T_b within individuals were observed in snakes by Dill (1972). Similar differences of T_b in different portions of the body certainly occur in turtles, as indicated by laboratory studies where the temperatures of the brain and the posterior body differed significantly during heating (Webb and Johnson, 1972; Webb and Witten, 1973). These findings (Figure 2) cast further doubt on the validity of acutely measured cloacal temperatures in the field. Future studies should include not only measurements of long-term changes in T_b and T_a, as well as continuous observations of behavior, but also continuous recordings of brain and body cavity temperatures.

BEHAVIORAL THERMOREGULATION

Thermoregulatory behavior has been well documented in many ectotherms and has been defined as "a repertoire of movements and postures which used under appropriate natural conditions, results in a relatively constant and frequently high temperature" (Heath, 1970). Such behavior allows turtles to regulate T_b (within limits) by activities such as emergence from shelter, retreat, basking, orientation with respect to wind or angle of incidence of solar radiation, changes in posture, and selection of different temperatures in water or substrate. Thus by utilization of extrinsic sources of heat and by control of the rates of heat gain and loss, some turtles are able to elevate T_b to some "preferred" level appreciably higher than that of the immediate surrounding environment; the range of behaviorally regulated T_b during activity is close to the more physiologically regulated T_b of endotherms.

Although few detailed investigations have been made on behavioral thermoregulation in turtles, several valuable studies have high heuristic value. Sexton (1959), Boyer (1965), and Ernst (1972) studied thermal relations in *Chrysemys picta*, and Moll and Legler (1971) investigated these relationships in the same species in Panama. Bury (1972) examined the thermal ecology of *Clemmys marmorata*, including an intensive analysis of basking behavior. Mahmoud (1969) included useful information on thermoregulation in his study of the ecology of Kinosternidae. Similar investigations have been made on *Gopherus agassizi* (McGinnis and Voigt, 1971), *G. berlandieri* (Voigt and Johnson, 1976), *Geochelone elephantopus* (MacKay, 1964), *Sternotherus odoratus* (Edgren and Edgren, 1955). More limited studies have been made on *Geochelone sulcata* (Cloudsley-Thompson, 1970), *G. gigantea* (Honegger, 1967), and *Testudo horsfieldi* (Andreev, 1948).

The principal method by which many turtles regulate T_b is movement in and out of sun and shade or water. Such basking behavior is most highly developed in the Emydidae (Boyer, 1965). Among marine forms only *Chelonia mydas* is known to bask out of water (Balazs, 1974). The larger body size of many turtles places restrictions on heliothermism. Strong insolation through a low air mass can warm 1 cc of water about 0.5°C in 1 min (Fry, 1967). Thus an animal more than a few centimeters thick would warm fairly slowly (Figure 3).

The angle of incidence of sunlight on the surface of a basking turtle, though due to a combination of factors, is determined largely by the orientation and shape of the body and the angle of elevation of the sun. Experiments on water-filled models made from aluminum and painted black showed that the flattened *Trionyx ferox* shape heated faster than a high-roofed *Sternotherus carinatus* model. When the angle (25°) of incidence of light was changed from anterior to lateral, the *S. carinatus* model had a slightly greater rate of heat gain. In live animal experiments, the peaked carapace of *S. odoratus* allowed for the slowest heat

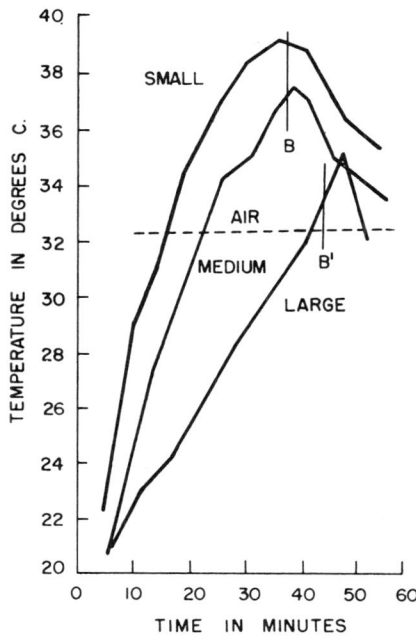

Figure 3 Effect of body size on heat gain in *Pseudemys scripta elegans*. Turtles (72–1800 g) were placed in direct midday sunlight and were removed at B and B'. (Modified from Boyer, 1965.)

gain among several species studied. Although the flattened shape of *T. ferox* is the most favorable for radiation absorption, heat gains of this species were similar to those of emydids of the same weight with much more rounded carapaces; the lighter color and leathery texture of the carapace, as well as a possible greater rate of water loss, can apparently offset the factor of shape (Boyer, 1965). The reflectivity of the integument has been shown to be an important component of the rate of heat gain in lizards (Hutchison and Larimer, 1960; Norris, 1967), but apparently no comparable studies have been made on turtles. Light "intensity" has been listed as the most important physical factor in determining the T_b of a basking turtle at a given time (Boyer, 1965), but more correctly it would be the "total incident solar energy."

The substrate temperature of a basking site depends primarily on the specific heat of the material, but Boyer (1965) found no evidence that thermal conductance was an important factor in basking site selection. The reflectivity of the area surrounding the turtle can also influence heat exchange, but this variable has not been considered in previous studies.

Numerous authors have shown that air and water temperatures are important in initiating or terminating basking. Wind can also be an important factor in basking behavior through the increased rate of water loss and cooling from exposed turtles; fewer basking turtles are seen on windy days (Boyer, 1965).

Other factors that may influence the orientation of basking turtles include the degree of crowding on the basking site, the size of the site itself, and the drive in some turtles (especially emydids) to attain high elevations, possibly to attain better fields of vision to detect intruders.

Basking sometimes occurs with part of the body remaining in the water and resting on submerged portions of the basking site, by floating at the surface, or by completely submerging in shallow water (Cagle, 1944b, 1950; Boyer, 1965; Moll and Legler, 1971; Pope, 1939). Turtles have often been seen swimming or resting in water beneath ice; Sexton (1959) found that *Chrysemys picta* swimming just under the ice had a T_b about 2°C higher than the surrounding medium and suggested that absorption of the longer wavelengths of solar radiation might account for the increase. Gibbons (1967) also explained this behavior as attempted thermoregulation. However similar small differences between T_b and T_a have been noted in turtles in water under laboratory conditions, and it was suggested that the differences were due to physiological regulation (Baldwin, 1925b; Ernst, 1972). Ernst (1972) has compiled data on the relationships of activity to T_b in *C. picta* (Table 2). Similar data for the same species in another geographical area were described by Sexton (1959).

Terrestrial turtles can maintain a fairly narrow zone of temperature by movement in and out of shade or a burrow. The desert tortoise

Table 2   Relationship of Activity to Cloacal Temperatures (°C) of Painted Turtles, *Chrysemys picta*[a]

Activity	N	Range	Cloacal Temperature Mean	Standard Deviation
Basking	8	22.5–29.0	25.2	2.43
Feeding	127	14.0–26.0	19.4	2.80
Moving in water	254	8.0–26.1	17.7	3.73
Moving on land	33	22.2–26.8	24.5	2.90
Dormant in water (high temperature)	75	15.8–23.1	17.5	2.33
Dormant in water (low temperature)	22	4.5–8.0	6.2	2.71

[a] From Ernst (1972).

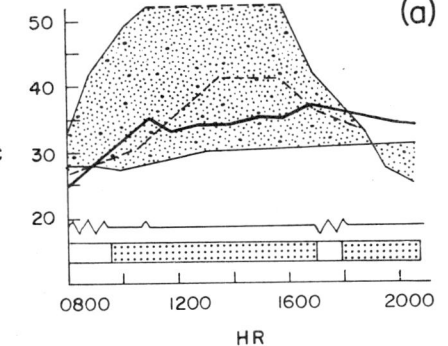

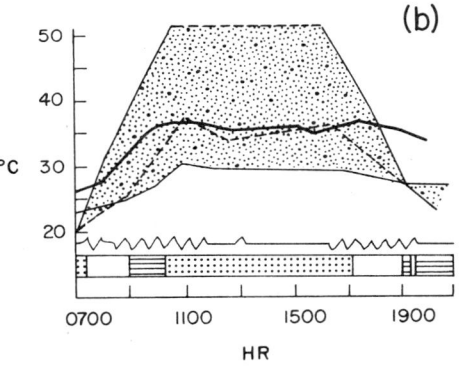

Figure 4 Deep body temperatures measured with radiotelemeters in a free-ranging desert tortoise (*Gopherus agassizi*) which (*a*) retreated to a previously excavated burrow during midmorning and (*b*) began digging a new burrow during midmorning of the next day. Heavy solid line is deep body temperature; dashed line, ground surface temperature in shade; stippled area, range of ambient temperature available to animal with upper limit (upper dashed line) showing the ground surface temperature in the sun and lower limit (lower solid line) the burrow air temperature. Sawtooth lines show periods of activity. Horizontal bar: stippled, tortoise in burrow; horizontal lines, in shade; open bar, in sunlight. (From McGinnis and Voigt, 1971.)

(*Gopherus agassizi*), for example, retreats into its burrow at midday to the only place near ground level where there is a T_a below the lethal range. A retreat to the burrow at night can extend the higher body temperature into midevening (Figure 4). The evening retreat to the burrow lessens during midsummer, with the result that a lower body temperature at the onset of morning activity permits a longer period of activity at the surface during the season of rising T_a. The carapace surface acts as a buffer against the solar radiation (Rose, 1969) and aids in keeping the deep body temperature as much as 10°C cooler than the surface (McGinnis and Voigt, 1971); the same is true for *Geochelone elephantopus* (MacKay, 1964). In spring and autumn *G. agassizi* may be active during most of the midday period (Woodbury and Hardy, 1948). Similar thermoregulatory behavior has been reported for *Geochelone gigantea* (Honegger, 1967), *Testudo horsfieldi* (Andreev, 1948), and *G. sulcata* (Cloudsley-Thompson, 1970).

Basking may serve other than a thermoregulatory function or aid to digestion. Drying may be necessary to maintain the health of the integument by keeping the skin and shell free from parasites (Cagle, 1950). Algal growths on aquatic and semiaquatic turtles can cause severe deterioration of the shell and underlying tissues (Neil and Allen, 1954; Edgren and Tiffany, 1953). Basking may also aid in the production of vitamin D (Pritchard and Greenwood, 1968).

Photoperiod may play an important role in basking behavior (Cloudsley-Thompson, 1970), as it does in temperature selection for some species in laboratory thermal gradients.

The experimental approach to thermoregulatory behavior has not been widely used with turtles, but it offers an excellent opportunity to supplement and understand observations in the field. The use of experimental thermal gradients where T_b from animals with known thermal histories (acclimation) can be monitored continuously under controlled photoperiod may provide information that is unavailable from field studies.

Graham (1972) continuously monitored with thermocouples the deep body temperatures of three turtle species in a laboratory gradient. Before release in the gradient, the turtles were acclimatized to various combinations of temperature (15, 25, and a daily cycle of 15 to 25°C) and photoperiod (LD8:16—i.e., 8 hr of light alternating with 16 hr of darkness—and LD16:8). Graham found that mean preferred temperatures of *Clemmys guttata* and *Chrysemys picta* were significantly higher for acclimation to the longer photophase (period of light) at a given temperature. In *S. odoratus*, the differences were not significant, although the mean preferred temperatures were higher under the longer photophase (Figure 5). The observed lack of a significant effect of photoperiod in *S. odoratus* may reflect the tendency of this species to be less thermoregulatory in its behavior. The adaptive significance of an increased thermal preference with an increased daily photophase length may lie in the ultimate effect of selection for higher temperature on thermal resistance. As discussed below, thermal tolerance was enhanced when *C. picta* were acclimated at LD16:8 compared to those at LD8:16 (Hutchison and Kosh, 1965).

The operant conditioning technique for studying behavioral thermoregulation in reptiles described by Regal (1970) offers a new approach that should be tried with turtles.

THERMAL TOLERANCE

The lethal limits of temperature have been determined either by exposing animals to a constant test temperature and determining the time to

THERMAL TOLERANCE

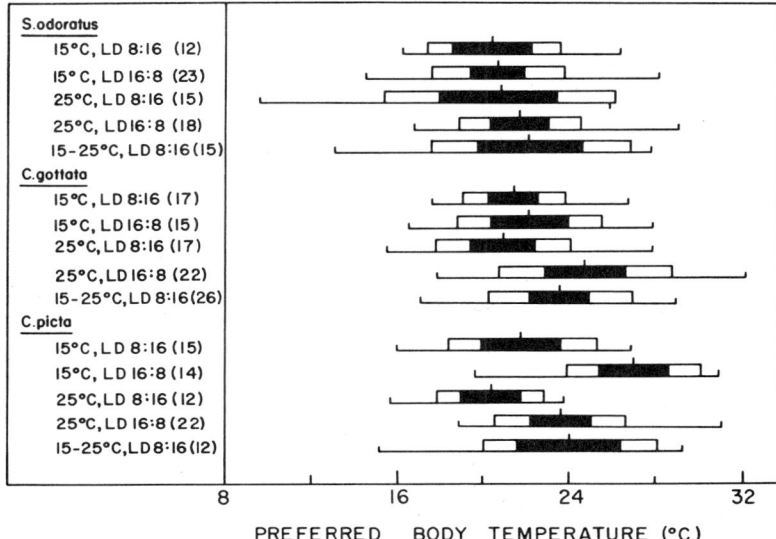

Figure 5 Preferred body temperatures of *Chrysemys picta*, *Clemmys guttata*, and *Sternotherus odoratus* determined in a thermal gradient tank after the animals had been acclimatized for at least 2 weeks to the conditions listed for each run. In each sample the horizontal line indicates the range; the long vertical line, the mean preferred temperature (MPT); one black and one white rectangle combined on either side of the mean indicate 1 standard deviation; one black rectangle on either side of the mean, 2 standard errors. The number in parentheses for each condition indicates the total number of animals examined for a given sample (= number of days of measurement). These data result from accumulation of the 48 half-hour temperature records per animal day. (From Graham, 1972.)

death or by heating or cooling animals from a nonlethal level at a rate of about 1°C per minute until some physiological end point is reached. The first method has been extensively used with fishes. As discussed by Fry (1967, p. 381), "both methods are amenable to statistical analysis and both have their purposes." The method of exposure to constant test temperatures is probably more valuable for analysis of *physiological* function, but it requires much time and equipment, and large sample sizes. The method of slow heating or cooling has been widely applied by herpetologists and is known as the "critical thermal maximum or minimum" (CTMax or CTMin; often "CTM" when applied only to upper limits). The term was introduced by Cowles and Bogert (1944) and was defined as "the thermal point at which locomotory activity becomes disorganized and the animal loses its ability to escape from conditions that will promptly lead to its death." This definition has been broadened to incorporate statistical variation (Lowe and Vance, 1955) and standardized conditions of testing (Hutchison, 1961). The end point used for determining the CTMax is often the onset of spasms (OS), although the loss of the righting response (LR) is also used. Mahoney and Hutchison (1969), Hutchison, Vinegar, and Kosh (1966), and Brattstrom (1970) have compared the use of OS and LR as end points. The CTMax or CTMin offers an immediate *ecological* index, since many lower vertebrates may reach this upper lethal limit from acute fluctuations in the environmental temperatures above their limits of tolerance.

As with measurements of thermoregulatory behavior, the determination of CTMax or CTMin must be made on animals with known thermal histories, since acclimation (physiological adjustment to one environmental factor) to temperature or acclimatization (adjustment to two or more environmental factors) to combinations of temperature and photoperiod, for example, can significantly change the end points (Hutchison, 1961).

Values for the CTMax of turtles are given in Table 1. These limiting temperatures are correlated with both habitat and distribution; the lowest values are found in the aquatic chelydrids ($\bar{x} = 39.4°C$) and trionychids ($\bar{x} = 40.0°C$), intermediate values in the semiaquatic emydids ($\bar{x} = 41.6°C$), and the highest ($\bar{x} = 43.4°C$) in the terrestrial testudinids. Although no significant geographic variation in CTMax was found in the box turtle, *Terrapene carolina*, or the stinkpot, *S. odoratus*, widely separated populations of painted turtles, *C. picta*, differed significantly and were correlated with the zoogeography of the species (Hutchison, et al., 1966).

The sensitivity to high temperature not only is a function of acclimation or acclimatization but may also follow a distinct daily rhythm. Kosh and Hutchison (1968) showed that *C. picta* acclimatized to long days (LD16:8) and moderate temperatures (20°C) had appreciably more thermal tolerance in later morning (0930 hours), early (1330) or late afternoon (1730), and evening (2130) than at late night (0130) or early morning (0530). Turtles acclimatized to short days (LD8:16) and lower temperatures (10°C) showed a similar pattern. However a third group of turtles exposed to "unnatural" combinations of moderate temperature (20°C) and short photoperiod (LD16:8) showed a phase shift in the rhythm such that they were most tolerant of high temperatures at times of the day (late night or early morning) that would not be adaptive in nature (Figure 6). Because the time of day at which animals are tested for thermal tolerance is important, some previous data may need to be reexamined.

Seasonal variation in thermal tolerance and thermal selection has been demonstrated in most temperate zone ectotherms examined, but apparently there are only scattered field observations on turtles.

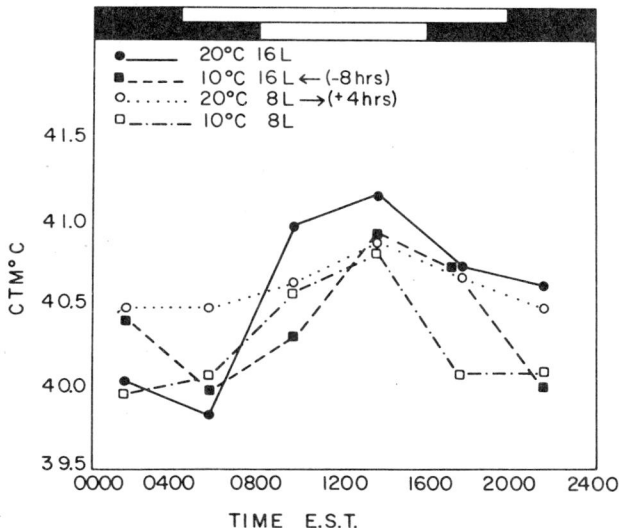

Figure 6 Variations of CTMax of *Chrysemys picta* acclimatized to combinations of different temperatures and photoperiods: 16L, 16 hr of light alternating with 8 hr of darkness (LD16 : 8); 8L, 8 hr of light alternating with 16 hr of darkness (LD8 : 16). Note similarity of amplitude responses (or oscillatory patterns) when certain curves are shifted relative to the LD16 : 8 curve at 20°C. Arrows and figures in parentheses at upper left designate direction and amount of time these curves were phase shifted. (From Kosh and Hutchison, 1968.)

Webb and Johnson (1972) and Webb and Witten (1973) have shown that significant differences may exist in head-body temperatures during heating (Figure 7) of the Australian long-necked turtle, *Chelodina longicollis*, and they question the validity of cloacal temperatue as an endpoint value for CTMax. The brain temperature may be the most important to measure, since the spasms used as an end point occur simultaneously throughout the body and are phenomena of the central nervous system. Also, the head temperature at the OS, in contrast to temperatures elsewhere in the body, does not vary with body size. Webb and Witten (1973) also noted temporal variations in the CTMax, similar to those observed by Kosh and Hutchison (1968), with the use of head temperature but not body temperature.

Except for an occasional anecdotal observation in the field, virtually no data apparently exist for the CTMin of turtles, and there are few data for other reptiles. This paucity of data reflects the greater difficulty in finding a physiological end point at low temperatures that meets the criteria of the definition. Perhaps the sudden change observed in the electrocardiograms of snakes at low temperatures (Landreth, 1972) may

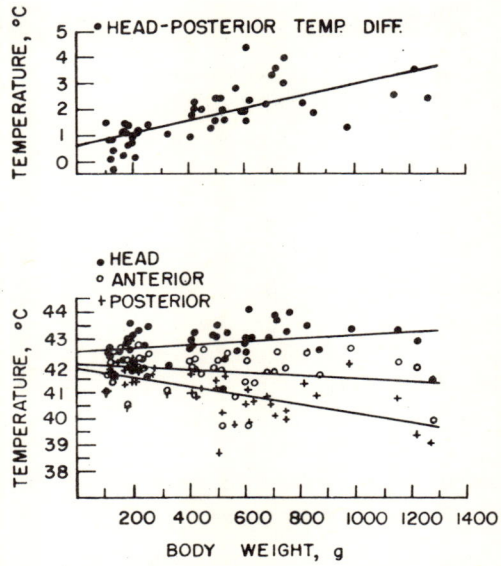

Figure 7 Head and anterior and posterior body temperatures at the CTMax in *Chelodina longicollis* (lower) and the head-posterior body temperature gradient at the CTMax (upper) of turtles heated in water. (From Webb and Witten, 1973.)

prove useful as an end point. Data from other ectotherms suggest that turtles can withstand an acute T_b of at least $-2°C$, provided ice crystals are not formed within the internal tissues. The activity of some turtles under ice in winter, described previously, suggests that the CTMin and factors influencing it would be a fruitful area for investigation.

Although the state of hydration has a significant effect on thermal tolerance in many ectotherms, it has not been widely studied in turtles. The loss of water in terrestrial or basking turtles may play a role in increasing tolerance to high temperature, as it does in amphibians (Hutchison, 1961).

BRUMATION ("HIBERNATION") AND ESTIVATION

True "hibernation," at least as the term is used for endotherms, is probably inappropriate as applied to the behavior of reptiles. No clear physiological differences have been shown between reptiles in cold torpor (low body temperature) and those in so-called hibernation. True hibernation involves not only a low body temperature, but also numerous other changes in the endocrine system, blood volume and composition, heart rate, cellular function, and so on. Similarly, estivation in reptiles may be only a limited "dormancy" during summer periods of high temperatures, as hibernation is winter dormancy in reptiles. Mayhew (1965) has suggested that the term *brumation* (from *bruma*, the Latin word for winter) be used for the winter dormancy of ectotherms and that *hibernation* be reserved for endotherms. Whittow (1973) supports the view that hibernation cannot occur in reptiles, since they are not able to execute the required cardiovascular changes and do not possess brown fat or shivering mechanisms required for rapid emergence from winter torpidity.

Voigt and Johnson (1976) used radiotelemetry to study summer estivation and thermoregulation in *Gopherus berlandieri*. In contrast to *G. agassizi* (McGinnis and Voigt, 1971) the Texas tortoise engages in extensive preemergent basking and does not emerge from estivation until the body temperature has reached minimal voluntary limits; this behavior allows it to avoid predation by escaping to shelter when disturbed rather than depending on a passive defense.

Brumation in turtles has been widely studied in the field (*Pseudemys scripta*, Cagle, 1950; *Terrapene carolina*, Carpenter, 1957; *Malaclemys terrapin*, Lawler and Musick, 1972; *Kinosternon subrubrum*, Bennett, 1972; *Clemmys guttata*, Netting, 1936; *Sternotherus odoratus*, *S. carinatus*, and *Kinosternon flavescens*, Mahmoud, 1969; *Chrysemys picta*, Ernst, 1972). These and other workers have described the sites of winter dormancy and water or air temperatures at which the animals retreat to burrows, stumpholes, mud bottoms of ponds, and so on. Such studies are valuable in increasing our understanding of the natural history of turtles, but many more detailed investigations are needed, especially on the T_b and T_a of dormant animals at the time of retreat and emergence. Almost nothing is known about the physiological changes that accompany brumation or cold torpor in ectothermic vertebrates. *C. picta* in cold torpor at $-2°C$ to $+6°C$ for 1 to 3 months showed a significant hemodilution due to removal of blood cells; no significant changes were seen in plasma proteins or specific gravity of the blood (Musacchia and Sievers, 1956). Observations on captive turtles indicate that brumation and estivation are not inherent seasonal rhythms but may be due to environmental conditions of temperature, moisture, photoperiod, and food supply. However an underlying seasonal cycle of sensitivity to these environmental factors may be involved, and this possibility should be investigated.

PHYSIOLOGICAL THERMOREGULATION

Behavioral changes that result in controlled body temperatures of ectotherms are accompanied by significant internal functional changes. Perhaps the most obvious physiological changes involved in thermoregulation are alterations in circulatory flow and in the pattern of distribution of the blood.

As an external heat source (e.g., solar radiation) diminishes, the T_b of a turtle may decline more slowly than one would predict from the mass-surface area relationship of the animal. On the other hand, the increase of T_b during heating may be greater than expected from purely physical relationships (Table 3). These differences in the rates of heating and cooling are achieved by an increase of conductance to the body surface by vasodilation or a decrease by vasoconstriction, respectively. Small elevations in core temperature of reptiles may also result in major changes in the magnitude and distribution of cardiac output (Baker and White, 1970). *Pseudemys floridana*, *Chelydra serpentina*, and *Geochelone carbonaria*, when heated from 10 to 30°C or cooled from 30 to 10°C in water, heated 25% faster than they cooled; in air the aquatic species' rates of cooling were exceeded by rates of heating due to evaporative water loss. Both cutaneous and carapace blood flows were increased by heating and decreased by cooling. Such vascular responses to temperature change represent alterations in functional insulation and thereby may play an important role in thermoregulatory capacity (Weathers and White, 1971; Lucey, 1974).

Spray and May (1972) showed that two basking turtles, *Chrysemys picta* and *Pseudemys scripta*, heated significantly faster than they cooled; two terrestrial species, *Gopherus polyphemus* and *Terrapene carolina*, cooled significantly faster than they heated. The rate of heat exchange in dead basking turtles was the same as the cooling rate in live animals, whereas the heat exchange of the two terrestrial species when dead was similar to their heating rates when alive (Table 3). Heating rates of *Gopherus agassizi* in the field may be up to 10 times faster than cooling rates; under controlled laboratory conditions the two rates were equal (Voigt, 1975), but in *G. berlandieri* heating was faster than cooling (Voigt and Johnson, 1977). Thus three different patterns have been observed in three species of the same genus. Spray (1972), using ballistocardiographic techniques, found a weight shift during heating and cooling of *C. picta* and *Terrapene ornata* that occurred in opposite directions; he attributed the results to shifts in blood volume toward or away from the carapace in response to heating or cooling. These results are further evidence for a significant role of the active control of conductance, which is utilized in different directions by turtles from two diverse habitats.

Table 3 Summary of Heat Exchange Ratios for Turtles[a]

Species	Weight (g)	Air-flow (cm/sec)	(Alive) cooling/ heating	(Dead) cooling/ heating	Reference[d]
Geochelone carbonaria	349		0.71	—	1
Pseudemys floridana	925–3740	Water	0.88	—	1
P. floridana	3909	51	1.02	—	1
P. floridana[b]	925–3740	Water	0.73	—	1
Chelydra serpentina	1120–3700	Water	0.69	—	1
C. serpentina	1136–4045	51	1.00	—	1
C. serpentina[b]	1120–3700	Water	0.65	—	1
Chrysemys picta	575–640	20	0.830	—	2
C. picta	230	0	0.803	1.000	2
C. picta[c]	217	0	0.007	—	2
C. picta	480	20	—	1.119	2
Pseudemys scripta	288	Water	0.745	—	2
P. scripta	383–560	20	0.774	—	2
P. scripta	288	0	0.757	—	2
P. floridana	480–595	Water	0.421	—	2
Gopherus polyphemus	737–945	20	0.434	—	2
G. polyphemus	1781,2814	20	—	1.100	2
Terrapene carolina	550,570	20	2.162	—	2
T. carolina	454,434	20	—	1.060	2

[a]From Spray and May (1972).
[b]Injected with atropine.
[c]Deafferented.
[d]References:
1. Weathers and White (1971).
2. Spray and May (1972).

All reptiles have an obligatory loss of heat by evaporation of water from body surfaces at all temperatures (Bentley and Schmidt-Nielsen, 1966), but the augmentation of this water loss at high temperatures is thermoregulatory. At high temperatures turtles may augment water loss, thus lower T_b, by salivating, panting, and by urinating on the legs. Baldwin (1925a) found that *T. carolina* showed rapid respiration, frothing of the mouth, and the accumulation of moisture on the head and around the eyes when the temperature was raised above 27°C. Riedesel et al. (1971), working with the same species, found that salivation began when T_b reached 32.3 to 40.4°C. Frothy saliva from the mouth and nostrils ran down the head and neck, and was smeared onto the front legs by wiping of the mouth. Urination occurred erratically, but usually preceded salivation. The urine was rubbed onto the back legs. Salivation in *T. ornata* is copious at high temperatures and is an integrated response influenced by both T_a and T_b (Riedesel, 1973; Sturbaum and Riedesel, 1974). Salivation at high temperatures has also been reported in *Geochelone elongata* (Swindells and Brown, 1964) and *G. sulcata* (Cloudsley-Thompson, 1968b). The large bladder in turtles may serve to store urine for emergency thermoregulation (Riedesel et al., 1971).

In the spurred tortoise, *Testudo graeca*, at elevated temperatures heat from evaporative cooling from water released through eye moistening, panting, urination, and salivation may exceed the metabolic heat production. Thermoregulatory salivation began at 39.5±0.4°C irrespective of body size, but panting rates were lower in turtles with higher weights (Cloudsley-Thompson, 1974).

The extent of panting, salivation, and urination as thermoregulatory functions among turtles needs further study. The control and efficiency of these augmentations to water loss also require more investigation.

Strong circumstantial evidence exists for the partial "endothermy" of leatherback turtles (*Dermochelys coriacea*); Friar, Ackman, and Mrosovsky (1972) found a deep T_b 18°C higher than the water temperature from which the turtle was taken. The sightings of leatherbacks in cold northern waters led Pritchard (1969a) to suggest that this species might be endothermic. Greer, Lazell, and Wright (1973) have presented good anatomical evidence for a countercurrent heat exchanger in the front and rear flippers. This structure would greatly reduce the loss of endothermally derived heat and would aid in maintaining a T_b higher than T_a, but it is not yet known whether these turtles can increase metabolic heat production under certain environmental conditions, as in the case of the incubating Indian python (Vinegar, Hutchison, and Dowling, 1970). The subepidermal fat layer in these turtles could serve as an insulating layer, suggesting that members of this species may be facultative endotherms. Certainly these recent observations suggest that the remarkable physiological thermoregulation of leatherbacks offers great opportunity for further study.

The mechanisms of behavioral and physiological thermoregulation reviewed above imply the existence of a fairly precise sensitivity to temperature, as well as an effective control.

Heath (1965, 1970) adapted Hardy's (1961) models of thermoregulation to poikilotherms, especially "on-off" regulators and "proportional controllers." The *on-off control* has a thermostat that controls an effector output in an all-or-none fashion. In turtles the onset of panting or salivation is an example of this type of control. Two such systems with different temperature settings may be coupled to result in thermoregulation, such as in the emergence and retreat of a basking turtle. When the set-point temperature exceeds the high value, retreat occurs. After the turtle cools below the low set point, it again emerges. Between the high and low set points is a refractory range where no such shuttling behavior is elicited.

The large temperature oscillations of an on-off control system are avoided by *proportional control*, where the effector output is continuously proportioned to the deviation of T_b from the control level. The postural changes and changes in orientation to the sun in basking turtles may act as a proportional controller. A recently emerged turtle may orient its body to receive solar radiation at the maximum angle of incidence and expose a maximum surface area to the sun by extending the neck and limbs. As the T_b approaches the set point, the orientation and posture may change to reduce radiative heat gain.

In *rate control* the effector output varies with the rate of temperature change. The differences in heating and cooling rates of turtles and the associated vasomotor changes that result in alterations of blood flow to the carapacial areas may be examples of rate control.

The evidence for thermosensitive receptors in reptiles has implicated the hypothalamus, since Rodbard, Samson, and Ferguson (1950) showed a temperature-sensitive area, manifested by changes in blood pressure, in the midportion of the cerebrum at the level of the third ventricle in *Pseudemys scripta elegans*. Heath, Gasdorf, and Northcutt (1968) have specifically localized the area to the anterior hypothalamus. Work on other reptiles has also suggested that the hypothalamus contains thermoreceptors involved in thermoregulatory control, but Hammel, Caldwell, and Abrams (1967) have shown that thermoregulatory responses in lizards may depend on T_b as well as hypothalamic temperature. Spray and May (1972) offered the first indication that peripheral receptors function in thermoregulatory control in reptiles. An individual *C. picta* was deafferented by the cutting of the dorsal roots of the thoracic nerves supplying the sensory supply of the carapace. The ability of the turtle to alter its rate of heat transfer was completely eliminated, since the heating and cooling rates occurred at almost the same heating rates as other turtles in air (Table 3).

FUTURE STUDIES

Although there is a paucity of studies on thermoregulation in turtles, research activity in this area has been increasing. Patterns of temperature regulation in reptiles may serve as models of evolutionary development of thermal regulation in birds and mammals. "Synthesis of such interactants as microclimate, predators, vulnerability, stage of development, behavior and its central organization and the underlying physiological adjustments, should yield greater perspective concerning reptilian thermoregulation and the evolution of endothermy" (White, 1973). The evolutionary divergence of chelonians from other extant reptilian groups should be borne in mind in any comparative studies. However features held in "common between reptiles, birds and mammals can be presumed present in the reptilian ancestors of each group" (Heath, 1970).

Future investigations on reptilian thermoregulation will most likely see a much increased use of turtles. The hardiness of these animals, along with the comparative ease of laboratory maintenance and their current availability and diversity, make them most suitable for study. More important, however, many significant questions on the evolutionary adaptations for thermoregulation remain to be asked, and answered.

ACKNOWLEDGMENTS

Margaret Kohl provided bibliographic assistance and Ginna Davidson helped with the figures.

CHAPTER 12
Feeding, Drinking, and Excretion

I. Y. MAHMOUD and JOHN KLICKA

FEEDING

Feeding Adaptations

In chelonians the horny jaws are used in capturing prey. The broad tongue aids in manipulation to a certain degree. The hyoid apparatus, attached to both the sternum and the lower jaw by several muscles, facilitates swallowing. The broadness of the basihyal aids in the pumping action of the throat during swallowing.

Between the braincase and the outer shell of the skull, the temporal roof is generally emarginated, providing the region behind the orbits with two large temporal spaces for muscle attachments. Each of these spaces is in direct communication anteriorly with the orbit and the subtemporal vacuity. The powerful adductor mandibulae (or capiti mandibularis) muscles that close the jaws originate mainly from the temporal spaces (also from the supraoccipital crest if present) and descend anteriorly through the subtemporal vacuity to insert on the mandible. The emargination or "scalloping out" of the temporal space may result in loss of contact between squamosal and parietal bones or even between the squamosal and orbital bones. In other cases the emargination leaves only a thin cheek bar behind the orbit; in extreme cases, as in *Chelodina*, even the cheek bar is gone, leaving the entire temporal space wide open. The degree of emargination may accommodate a larger muscle mass, which then provides greater force for crushing. A comparative myological study relating to emargination would be valuable.

The small, weak depressor mandibulae, which originates laterally from the posterior part of the skull, inserts on the back end of the lower jaw. Having the jaw-closing muscles much stronger than the jaw opening muscles is an important evolutionary development in chelonians, since the head is the main instrument for capturing, cutting, and crushing.

Turtles possess certain jaw features as adaptations for dealing with particular foods. The jaws of the highly carnivorous soft-shell turtles, for example, are very sharp for cutting prey. The upper jaw of the green turtle has vertical ridges, and the lower jaw is strongly serrated as an adaptation for grazing. The hawksbill, which is not as heavy a grazer, has a smooth or slightly serrated lower jaw and the upper jaw lacks strongly elevated ridges. The serrated jaws of the desert tortoise, *Gopherus agassizi*, are used for plant shredding. Bellairs (1970) pointed out similarly serrated beaks in *Batagur baska* and *Geochelone denticulata*. He also noted the hooked beaks of *Macroclemys temmincki* and *Dermochelys coriacea*, which may aid in capturing prey. Bleakney (1963) suggested that the backward projecting spines lining the esophagus in *D. coriacea* might aid in swallowing its jellyfish prey.

Members of the genus *Sternotherus* change their feeding habits from insectivorous in juveniles to molluscivorous in adults, and the musculature of the jaw becomes larger and broader with adulthood. According to Ernst and Barbour (1972), *Graptemys barbouri* and *G. pulchra* also develop enlarged jaw musculature, apparently useful for crushing. Male *G. pulchra*, however, have weaker jaw muscles and probably rely more on insect prey. In the red-bellied turtle, *Pseudemys rubriventris*, the median ridges on the crushing surfaces on the lower jaws are tuberculate, an adaptation for feeding on aquatic plants (Ernst and Barbour, 1972). Similar jaw adaptations exist in the cooters *Pseudemys floridana* and *P. concinna*.

Smith (1961) proposed that the "choanal rakers," which extend across the slitlike choana in the green turtle, function as strainers during feeding. When water enters the mouth during feeding, the pressure of the tongue presumably presses on the roof of the mouth, forcing excess water out through the choanal slits. Although no myological studies have been conducted to substantiate Smith's hypothesis, a positive correlation exists ontogenetically between the presence of the rakers and plant consumption.

Sexton (1959) stated that *Chrysemys picta marginata* hunts its food by striking its head into aquatic vegetation; when small animals move out, the turtle pursues them. We have seen mud turtles making swift and apparently random movements through mats of aquatic plants, disturbing and feeding on small prey such as tadpoles.

Moll and Legler (1971) observed that *Pseudemys scripta* swam slowly, pushing floating vegetation with their forelegs and biting indiscriminately at it. Adults tore leaves of *Elodea* and *Najas* from the stems and swallowed them whole. As is true for most species, large items were held by the jaws and torn by the front claws.

Members of the genera *Kinosternon*, *Chrysemys*, *Pseudemys*, *Graptemys*, and *Trionyx*, and the marine turtles, rely on muscular coordination and speed to capture fast-moving prey.

Aquatic turtles may hunt their food leisurely. Members of *Chelydra* and *Sternotherus* frequently walk or swim very slowly near the bottom with the neck fully extended; from time to time the head is flexed downward, apparently smelling certain objects.

Turtles of the genera *Chelydra*, *Macroclemys*, and *Chelus* all engage in rather passive methods of catching prey. Newman (1906a) described the snapping turtle lying on the bottom with open mouth. Its vermiform tongue moves about, and any fish approaching the "lure" is quickly snapped up. The alligator snapping turtle's wiggling, bifurcated "fish lure" changes from grey to red when it is moving, perhaps because of vasodilation controlled by a nervous mechanism. The side-necked matamata has fleshy processes around the head; small fishes nibble at these filaments which sway with the movement of the water (Carr, 1963b). When the turtle opens its mouth, distending the pharynx floor, the prey are swept in with the onrushing water, which then escapes from the sides of the mouth.

Turtle collectors from the Lemberger Company, Oshkosh, Wisconsin, informed us that the primary food source of the omnivorous snapping turtles during early spring is frog eggs. The turtles use their jaws and front claws to break the gelatinous masses into smaller ones, then swallow them. We have observed snapping turtles, with their necks extended, swim rapidly to the surface, grab ducklings by the legs, drag them to the bottom, and proceed to tear the prey apart. Turtle collectors have also observed this method of predation by snappers on hatchlings of common terns and blackbirds.

Belkin and Gans (1968) described an unusual feeding technique in the pelomedusid turtle, *Podocnemis unifilis*. This species feeds on fine, floating food by skimming the water's surface. "Neustophagia" was also seen in *Chrysemys picta* as communicated to them by another observer. Belkin and Gans noted that the closely related *P. expansa* did not exhibit neustophagia. It was concluded that since *P. expansa* lives in a permanently stable aquatic habitat with an abundant food supply, there is no need for neustophagia. On the other hand, *P. unifilis* inhabits unstable aquatic habitats such as oxbows and swamp pools. These habitats frequently are richly supplied with minute floating foods such as algae, small insects, and crustaceans, but poorly inhabited by larger prey.

We observed what appeared to be neustophagia in yellow mud turtles *Kinosternon f. flavescens* in western Oklahoma shortly after they emerged from hibernation in temporary pools and mudholes. Water skimming was seen more frequently when there was an abundance of minute crustaceans emerging after spring rains. Thousands of concostracod shells were present in the feces of collected turtles. Neustophagia is perhaps an adaptive behavior necessary for survival in this species. When feeding behavior in the four kinosternid species of Oklahoma was studied in captivity, only *K. f. flavescens* showed occasional neustophagia; the others that inhabit permanent aquatic habitats did not engage in surface skimming.

Some turtles use their jaws to grasp the food, then direct it toward the throat by utilizing the inertia of the object. Gans (1969) called this method "inertial feeding." The jaws open to release the prey, and the head shifts to a new position while the prey remains inert. Muscles of the head and neck must be well coordinated and strong enough to enable the animal to regrasp the object after the shift in position has been attained. Both aquatic and terrestrial turtles shift to a preferred position to grasp prey. We have noticed this over a long period of time, particularly in a captive box turtle. The rapid plucking motion of a box turtle feeding on a garter snake as described by Hutchison and Vinegar (1963) is used when box turtles feed on earthworms; slower, smoother, engulfing motions are made when feeding on grubs, fruits, and so on.

Dietary Preferences and Changes

Techniques used for determining feeding preferences of turtles in natural settings are given in Chapter 4. Research such as that reported by Lagler (1943a) includes stomach analyses of food items for numerous species in one locale, and accounts of single species (e.g., Legler, 1960a) detail food acquisition and factors influencing preferences. Numerous anecdotal descriptions exist, particularly in the early herpetological literature. Summary statements of available information on food preferences are generally included with the species descriptions given by Carr (1952), Pritchard (1967), and Ernst and Barbour (1972). No attempt is made here to catalog dietary preferences of the world's turtles, but general trends in dietary changes are presented, with a few studies cited as examples.

Age and Size Factors

With regard to food preferences, age is not readily separable from size in turtles that have not yet reached maturity. Similarly, in species with unequal sizes of the two sexes at maturity, any dietary difference between the sexes is likely to be correlated with size of the animals.

Most of the chelonians studied show dietary changes with age. Young are chiefly carnivorous, but as they grow older and larger they become predominantly omnivorous or herbivorous.

Marchand (1942) examined the stomach contents of *Chrysemys picta dorsalis* and found a reversal of food preference as turtles reached maturity. About 88% of the adult foods consisted of plant material, and 12% included insects and amphipods. In juveniles, animal and plant foods accounted for 85 and 10%, respectively. The slider *Pseudemys scripta elegans* also showed a reversal in feeding habits from predominantly carnivorous in young to predominantly herbivorous in adults (Marchand, 1942; Clark and Gibbons, 1969). Juveniles fed on small insects, which have the high calcium content necessary for turtle shell growth. It is possible to assume that dietary requirements change with age partially because of changes in physiological demands. The availability of food is another factor influencing a change in feeding. As turtles grow, they require more food; larger and abundant food items can be acquired with less time and energy. As mentioned, small foods may be eaten if they are extremely abundant.

Although the kinosternid turtles of Oklahoma were found to be omnivorous, turtles under 5 cm carapace length fed mainly on small insects, algae, and carrion (Mahmoud, 1968).

Following observations of green turtle hatchlings in captivity, Carr (1967b) reported that the young turtles must be pelagic, relying on floating or close-to-the-surface food, including small invertebrates. When green turtles weigh 10 lb, they begin making the transition to grazing. Their growth becomes very rapid, and little energy is expended in foraging on the unlimited food supply.

Seasonal Factors

Since most turtles are opportunistic feeders, some species change their feeding habits according to the availability of food items in different seasons. During spring and summer when berries are abundant, wood turtles feed on them, but later these turtles invade streams and feed on aquatic insects (Carr, 1952). Several investigators (e.g., Grant, 1960; Auffenberg and Weaver, 1969) have demonstrated that the Texas tortoise, *Gopherus berlandieri*, prefers red and green foods over blue. Preference shifts from green during the spring to red during August and September. Such color preference is directly related to the natural color changes as the red fruits of *Opuntia* mature during that time of the year.

Pearse (1923b) reported that for the western painted turtle during early summer the major diet was insects (60.7%) followed by plants (32%). In September the plant items made up 55%, and the insects were 36.2%. Such a reversal is attributed to the abundance of insects during earlier summer; and as the insect population declined, the turtles depended more on plants as food.

In the kinosternid turtles of Oklahoma, feeding habits in summer were primarily carnivorous, in the winter herbivorous (Mahmoud, 1968a).

Few data are available on the optimum temperatures at which chelonians will accept food. When the body temperature is near the upper or lower limits of the normal activity range, animals usually do not feed. Obtaining data on the optimal feeding temperatures would be extremely useful in the study of behavior and ecology of a species. Moll and Legler (1971) recorded the cloacal temperatures of 108 sliders feeding or foraging and found a range from 25.9 to 34.4°C. Freshwater and terrestrial turtles generally feed when the temperature of the body is between 20 and 32°C. This range can be even wider, especially in northern latitudes.

Leatherback turtles, *Dermochelys coriacea*, are found feeding on jellyfish along the coast of Canada and New England at the end of the summer in water of about 12°C. This is some 15°C below the temperature of the water where they normally nest in Florida. In this species, however, the body temperature may be more than 15°C above the ambient temperature (Friar, et al., 1972).

Feeding Behavior in Captivity

In Chapter 15 Manton presents information on the role of olfaction in feeding, and Burghardt (1970) has reviewed studies of food preference and discrimination. Morlock discusses the topic in regard to laboratory studies of learning in Chapter 20. In general feeding has not been widely studied in laboratory situations; chemoreception has not been well separated from visual cues, nor has chemoreception been clearly delineated into the roles of taste and smell in feeding. Studies in captivity frequently do not take into account behavior of the animals in their natural environment. Given the lack of natural surroundings, degree of crowding, limited movements, and so on, the study in captivity may still shed some light on some aspects of feeding behavior.

FEEDING

Crowding in captivity, for example, may contribute to the establishment of a social hierarchy that otherwise would not exist in nature. Boice (1970) reported a "feeding-order hierarchy" in captive box turtles (*Terrapene c. carolina*). We have seen aggressive behavior in captive kinosternid turtles. "Dominants" were usually the first to attack and to ingest the food placed in the tank; some time later the others fed on what was left over. When a small amount of food was introduced, biting, shoving, and pushing with forelimbs and body were very common. Turtles holding food with the mouth were frequently chased by others, with the food either dropped or grabbed. The turtles with food usually swam to a corner facing the tank wall and swallowed rapidly. In some cases the turtle with the food was chased by several others trying to grab the food. This occurred even when the tank contained an abundance of food. Later, when the other turtles became aware of the remaining food, the chasing behavior diminished. We have observed similar chasing, biting, shoving, and pushing behavior in emydids such as the western painted turtle and the red-eared turtle.

In one comparative study (Mahmoud, 1968a), two species of mud turtles (*Kinosternum subrubrum hippocrepis* and *K. f. flavescens*) and two species of musk turtles (*Sternotherus odoratus* and *S. c. carinatus*) were tested for preference for size and type of prey, preference for dead or living prey, ability to capture fast- and slow-moving prey, and chemoreception. The chemoreception trial used liver; prey used for other trials included juvenile garter snakes, june bugs, dipteran larvae, tadpoles, frogs, and snails. Adaptations to the different natural habitats of the turtles were reflected in responses on the laboratory trials.

The Establishment of Food Preferences in Hatchlings

Recent studies indicate that hatchlings given food for the first time may develop a preference for the initial food. Mahmoud and Lavenda (1969) have demonstrated that food preference for the initial diet was only temporary in three groups of individually housed hatchling red-eared turtles. The 30 day initial diet of minced *Tenebrio molitor*, ground beef, or ground fish was followed by preference testing with the one familiar and two unfamiliar foods on 3 consecutive days. Groups were then fed one of the other foods for 8 days and once again tested with all three foods for 3 days. The third food item was then fed each group for 8 days, and the final 3 days of preference tests were completed. The turtles showed a distinct preference for the original diet for the first feeding test, and this preference was maintained after the second feeding test, but not after the third. The establishment of food preferences for the original food was indeed transitory and was completely broken down after a short period. The 30 day exposure to the initial food might be below the threshold value to establish a prolonged food preference, but the data do not indicate the existence of a threshold.

Burghardt and Hess (1966) indicated a food preference for the initial food in snapping turtle hatchlings after exposure to the initial diet for only 1 week. Burghardt (1967) suggested the existence of an innate preference for a particular food. It is possible that the behavior observed here does not involve an all-or-none phenomenon, that longer initial exposure might require a longer alternate feeding time for the eradication of food preference. The time required to abolish a previously established food preference may be a measure of the strength of that preference. Unfortunately, no extensive studies along these lines have been conducted. It is conceivable that the strength of food preference for initial diet might be related to the predominant food in the natural habitat.

Food Reserves in Hatchlings

Mahmoud and Klicka (1971) have demonstrated that keeping unfed snapping turtle hatchlings at 21°C for 135 days has virtually no effect on liver glycogen; in fact, if they are kept at 4°C for the last 60 days of the 135 day fast, the liver glycogen concentration increases slightly. This may be of adaptive value in northern latitudes, since similar temperature variations are experienced under natural conditions. Thus if the hatchlings winter in the nest, as is often the case in the northern latitudes (Carr 1952), they would still have a large glycogen energy reserve to draw on in the spring, when they face the strenuous task of emerging. It has been suggested that in poikilothermic vertebrates glycogen is used to supply energy during periods of intense muscular activity, while protein and/or fat are the primary energy sources during fasting (Lovern, 1940; Stimpson, 1965). In this regard we have found that although glycogen stores are not depleted in the unfed hatchling, the yolk sac that is carried over from the egg is completely resorbed during the first 30 days of fasting (21°C), indicating the importance of this rich lipid reserve to the newly hatched turtle. In the colder climates the energy reserve of the yolk probably enables the fasting hatchlings to winter in the nest, whereas in warmer regions it allows the awkward, newly hatched animal a period of time for adjustment and orientation to its environment before beginning the pursuit of prey.

DRINKING

Relatively few physiological and behavioral studies of drinking in chelonians exist. Thus the discussion in this section is based on fragmentary information and on our personal observations.

Tortoises and land turtles drink water periodically when it is available to them. They usually consume a large amount of water at one time. In captivity, gopher tortoises are capable of increasing their body weight between 41 and 43% by drinking (Miller, 1932). They accomplish this by drinking slowly and steadily, sometimes for an hour or so. We have noticed that captive box turtles also drink copiously at one time. The large amount of water taken in may be stored in the bladder and used gradually as needed. If so, this behavior is indeed a valuable one, since surface water is not always available in terrestrial habitats.

Land turtles and tortoises possess certain physiological and behavioral adaptations to meet water shortages. For example, cutaneous water loss in chelonians has been shown to be inversely related to aridity of habitat (Schmidt-Nielsen and Bentley, 1966; Bentley and Schmidt-Nielsen, 1966). Metabolic rate in reptiles, and particularly in chelonians, is much lower than in homeotherms (Schmidt-Nielsen, 1964); accordingly, reptilian water loss is also less.

Tortoises and land turtles extend their necks and immerse their lower jaws during drinking to allow the water to enter the mouth passively. When the mouth cavity is filled, the jaws close, and the throat muscles force the water into the esophagus. Afterward the mouth is open again, and the process of water filling is repeated.

Since feeding usually takes place under water, aquatic turtles may drink large amounts of water to facilitate swallowing during food intake. Freshwater turtles are primarily ureotelic, thus require large amounts of water to get rid of the urea and ammonia. We have sacrificed individuals of several species of freshwater turtles for various research purposes and frequently have found the bladder fully distended with fluid. Whether the fluid accumulation in the bladder has anything to do with the ureotelic method of excretion remains unknown.

Land turtles and tortoises obtain a great deal of fluid from their plant diet. The importance of water provided as a by-product of oxidative metabolism in desert chelonians remains unknown.

Auffenberg (1963) described an unusual method of drinking employed by some land tortoises from areas of low rainfall in South Africa. When he simulated rainfall on several captive specimens of *Psammobates tentorius trimeni*, he noticed that almost all the specimens immediately assumed the following position: the legs were fully extended, the gular plate of the plastron was pressed against the surface, the front limbs were fully extended anteriorly, and the external surface of the manus was pressed against the sides of the head. When this position was assumed, the carapace was tilted toward the front, and the water falling on it flowed downward and forward. The recurved edges of the posterior marginal scutes of the carapace and bridge serve as a "gutter" when water flows anteriorly toward the head. The water is then conveyed from the shell to the head by way of the external surface of the forearm.

Auffenberg also reported similar observations on captive specimens of *Kinixys homeana* from South Africa, except since the marginal scutes in this species are more recurved than in *Psammobates*, the "gutter" is more effective in water transport. The drinking habit of *Homopus areolatus* from South Africa is similar to that of *K. homeana* except that the posterior part of the carapace is not raised as high as in *K. homeana*. Auffenberg noted that *H. areolatus* poke their heads into the loose soil, presumably to drink water; other tortoises, *Gopherus agassizi*, *G. polyphemus*, and *Chersine angulata*, behave similarly.

EXCRETION

Nitrogen Excretion

Although the majority of reptiles excrete uric acid as the major nitrogenous waste product, this is by no means the case in chelonians. Reports in the early literature indicate that the major excretory product of the turtle kidney is more often urea (Magnus and Muller, 1835; Schiff, 1835; Marchand, 1845; Lewis, 1918; Wiley and Lewis, 1927; Clementi, 1929) than uric acid (Mills, 1886; Munzel, 1938). Some controversy has arisen on this point inasmuch as independent investigations have implicated each compound as the predominant waste product within a single species (Clementi, 1929; Munzel, 1938; Moyle, 1949). Undoubtedly some of this confusion has resulted from differences in the experimental techniques employed. Some researchers have sampled urine taken directly from the bladder, and others have used samples of voided urine; however the two may differ because of length of time spent in the bladder and differential reabsorption of water, urea, and ammonia by the bladder epithelium. More important in generating some of this controversy, perhaps, has been the limited attention paid to the prior history of the animal with regard to diet, availability of water, activity, season, and ambient temperature. Future research along these lines must take such factors into consideration to generate consistent results.

Generally speaking, the mode of nitrogen excretion in chelonians appears to be related mostly to availability of water. Some aquatic species such as *Chelonia mydas* are predominantly ammonotelic (Khalil, 1947), whereas semiaquatic species such as *Pseudemys scripta* and *Chelodina longicollis* are ureotelic (Dantzler and Schmidt-Nielsen, 1966; Rogers, 1966), and desert species excrete primarily uric acid (Moyle, 1949; Baze and Horne, 1970). Moyle (1949) studied the pattern of nitrogen excretion in eight species ranging in habitat from aquatic to arid. His findings led him to group these animals into three categories: (1) aquatic and semiaquatic turtles, which excrete ammonia and urea in roughly equal amounts, (2) terrestrial species, which are restricted to moist environments and excrete primarily urea, and (3) terrestrial species, which live in arid environments and are predominantly uricotelic. However considerable variability in the mode of nitrogen excretion exists within a species, and, indeed, in some species such as *Testudo kleinmanni* a single individual can vacillate between ureotelism and uricotelism (Khalil and Haggag, 1955). What induces such shifting in waste product metabolism in this species remains obscure and deserves further investigation.

In contrast to the situation in most ophidians and lacertilians, it is becoming increasingly apparent that chelonians, including those living in xeric habitats, have retained the capacity for ureogenesis. Biochemical studies indicate that virtually all turtles examined to date have all the enzymes of the ornithine cycle present in their livers at one time or another (Florkin, 1949; Brown and Cohen, 1969; Mora, Martuscelli, Ortiz-Pineda, and Soberon, 1965; Baze and Horne, 1970). The finding by Baze and Horne (1970) that the ornithine cycle enzyme arginosuccinate synthetase is normally absent from the liver of the uricotelic desert species *Gopherus berlandieri* but can be induced by such environmental pressures as fasting and dehydration, is of considerable interest. This is contrary to what one might expect, since a greater degree of water economy can be predicted from a uricotelic habit than a ureotelic one. It could be, however, that this is a mechanism brought into play not for the purpose of excreting waste nitrogen as urea instead of uric acid, but rather to increase the osmotic concentration of the tissue fluids in an attempt to retain as much water as possible in a manner similar to that of elasmobranchs. Uric acid, which is insoluble, cannot serve this function and could remain the major nitrogenous waste voided. Along these same lines Gilles-Baillen (1970) has demonstrated that in the euryhaline species *Malaclemys terrapin centrata*, adaptation to 100% seawater is accomplished, in part, by increasing the blood urea concentration, and Dantzler and Schmidt-Nielsen (1966) have found that the increase in blood osmolality in the dehydrated desert tortoise *G. agassizi* is principally due to an increase in urea concentration. Such findings suggest an important role for urea in turtle osmoregulation and could explain why the orinithine cycle is active in predominantly uricotelic species.

Water and Electrolyte Balance

Inasmuch as the nephron of the reptilian kidney lacks the loop of Henle, a blood hyperosmotic urine cannot be produced by way of a countercurrent mechanism as it is in the mammal (Schmidt-Nielsen, 1964). Although the ability to excrete nitrogenous wastes as uric acid represents a very important water conservation measure because the relative insolubility of this compound reduces the osmotic work of the tubule or bladder in water reabsorption, the fact remains that the dehydrated or salt-loaded turtle has no renal mechanism for decreasing blood osmolality by elimination of excess salt. In marine species such as *Chelonia mydas* and *Caretta caretta* and in the brackish water *Malaclemys terrapin*, elimination of excess salt is apparently accomplished through an extrarenal route, the lacrimal gland of the eye orbit (Schmidt-Nielsen and Fange, 1958; Abel and Ellis, 1966; Dunson, 1970; Cowan, 1971). Thus such organisms might be capable of obtaining osmotically free water by drinking seawater and eliminating sodium by way of the salt gland (Holmes et al., 1963); however Bentley, Bretz, and Schmidt-Nielsen (1967) could find no evidence that *M. terrapin* ever drank seawater when it was kept in this medium. The orbital salt gland may serve as an excretory route for the relatively large amounts of the potentially toxic K^+ ingested in the herbivorous diet of *C. mydas* (Holmes and McBean, 1964). Excretion of both Na^+ and K^+ by the orbital salt glands of marine turtles may be under the control of adrenal steroid hormones, and the possibility exists that salt loading by drinking seawater triggers the release of these hormones, which then stimulate both renal and extrarenal excretion of the dietary K^+ (Holmes et al., 1963).

There is currently little evidence for extrarenal salt excretion in either freshwater or terrestrial species, and in one desert species, *Gopherus agassizi*, attempts to demonstrate orbital salt glands by salt loading have failed (Schmidt-Nielsen and Dawson, 1964). Cowan (1969) did a histological study on the lacrimal and Harderian glands of several freshwater emydids after subjecting the turtles to various salt concentrations and found no indication that these glands are active in salt excretion. This is not surprising, since freshwater species generally are faced with a prob-

EXCRETION

lem opposite that of marine turtles (i.e., excessive water intake and ion loss). The problem of excessive water intake in these freshwater forms is easily solved by the production of a copious blood hypoosmotic urine. Although the chelonian kidney is not capable of filtering as great a volume of fluid as the mammalian organ, owing to a relative deficiency in both size and number of glomeruli, it stands considerably ahead of most other reptilian kidneys in this respect (Cordier, 1928; Marshall and Smith, 1930). Undoubtedly some of the ion loss experienced by freshwater turtles is made up from the ingested food, but some of these species also alleviate this loss by active absorption of ions from the aqueous environment through the pharyngeal lining, cloacal bursae (accessory bladders), and cloaca (Dunson, 1967b). Dunson and Weymouth (1965), for example, have demonstrated that the soft-shell turtle *Trionyx spiniferus* can actively absorb Na^+ from an aqueous medium containing as little as 5 μmole of Na^+ per liter, and since fresh water in nature generally ranges between 300 and 500 μmole of Na^+ per liter, a large safety factor appears to exist here. The involvement in this process of accessory bladders may explain why these organs are found in freshwater turtles but not in others.

The handling of electrolytes by the renal tubules of chelonians has not yet been adequately investigated either quantitatively or qualitatively. The turtle kidney cannot put out a hyperosmotic urine, but this, of course, does not mean that it exerts no control over electrolyte balance. In the euryhaline *Malaclemys terrapin* the urine osmolality increases fivefold when the animal is moved from fresh water to seawater, but it still does not become hyperosmotic to plasma (Bentley et al., 1967). Recent work by Butler (1972a) implies that some tubular reabsorption of Na^+ occurs in *Chrysemys picta belli*, and this may be controlled to some extent by adrenal steroids (Butler and Knox, 1970; Butler, 1972b). However Trobec and Stanley (1971) have found that Na^+ balance is not under endocrine control in this species during the winter months. Spigel (Chapter 22) reviews the effects of experimental manipulation of environmental stimuli on excretory electrolytes and the possible involvement of the pituitary-adrenal axis. Active secretion of uric acid by the distal tubule has been demonstrated in *Pseudemys scripta* and *Gopherus agassizi* (Dantzler and Schmidt-Nielsen, 1966), but whether secretion of any ions such as K^+ or H^+ occurs here as it does in the distal tubule of the mammal is unknown. More attention has been paid to the active reabsorption of electrolytes by the chelonian urinary bladder because of the potential this structure has in being used as a research tool to study ion fluxes. *In vitro* studies have demonstrated active uptake of Na^+ from the mucosal to serosal surface in urinary bladders from *P. scripta* (Klahr and

Bricker, 1964; Bricker and Klahr, 1966; Brodsky and Schilb, 1966; Wyssbrod, 1969) and *Testudo graeca* (Bentley, 1962). It is interesting to note the difference in energy dependency of the Na^+ transport mechanism in these two species. Complete dependency on oxidative metabolism was found in *T. graeca* (Bentley, 1962), but only the energy generated by anaerobic metabolism was utilized in the *P. scripta* transport scheme (Klahr and Bricker, 1964). There are indications that Cl^- (Brodsky and Schilb, 1963) and H^+ (Green, Steinmetz, and Frazier, 1970) or HCO_3^- (Schilb and Brodsky, 1972) are also actively transported across turtle bladder epithelium.

Renal responses to water deprivation and salt loading appear to differ appreciably between freshwater turtles and desert tortoises. According to Dantzler and Schmidt-Nielsen (1966), the kidney nephron in both *P. scripta* and *Gopherus agassizi* behaves in an all-or-none manner and is not continuously variable as in mammals. Thus a decrease in glomerular filtration rate in these turtles is brought about by a reduction in the number of functioning nephrons rather than a reduction in amount filtered by each nephron. However the authors found that when blood osmolality begins to rise in the freshwater *P. scripta*, the permeability of the renal tubules also increases (hence water reabsorption occurs), just as it does in mammals. On this basis *P. scripta* can produce a ureteral urine that ranges from blood hypoosmotic to blood isoosmotic, depending on the state of hydration. In addition, as the blood osmolality increases, the glomerular filtration rate decreases, and cessation of all glomerular function occurs when blood osmolality increases by only 20 milliosmols. In contrast, in the desert tortoise, *G. agassizi*, tubular permeability remains low and constant regardless of changes in blood osmolality, and glomerular filtration rate does not begin to decrease until blood osmolality increases by more than 50 milliosmols. In this species renal shutdown does not occur until blood osmolality is increased by more than 100 milliosmols. Such kidney function results in the continuous output of a ureteral urine that is blood hypoosmotic. The authors point out that this may be of importance in uricotelic species, since concentrating the urine in the kidney tubule might result in precipitation there of uric acid, which could cause tubular damage and blockage. A much better solution would be to have water reabsorption occur at the level of the bladder that would allow the uric acid to precipitate out in the bladder and avoid obstruction of kidney function. Indeed, Dantzler and Schmidt-Nielsen have shown that although the ureteral urine is hypoosmotic in *G. agassizi*, samples taken directly from the bladder are always isoosmotic. They have found the bladder in this species to be extremely permeable to water, ions, and small molecules like urea, and when the hypoosmotic ureteral urine enters the bladder it rapidly equilibrates with the blood and becomes isoosmotic. State of hydration of the animal does not alter bladder permeability in the tortoise, although it appears to do so in *P. scripta*. This suggests that the tortoise might experience a problem in voiding excess water, should it become well-hydrated. However Dantzler and Schmidt-Nielsen found that in the water-loaded tortoise the ureteral urine bypasses the bladder and is spontaneously passed to the outside, thereby enabling the animal to get rid of the excess water by way of a hypoosmotic urine. The bypass mechanism appears to involve a strong muscle sphincter in the bladder neck. The ureters enter the bladder neck above the sphincter; thus when the sphincter is closed, the ureteral urine flows into the cloaca and out of the body. More information is needed regarding the control of this sphincter mechanism.

In amphibians neurohypophyseal hormones govern glomerular filtration rate and tubular and bladder permeability. Whether similar control exists in chelonians has not yet been adequately studied, but preliminary investigations suggest that such hormones regulate these parameters in freshwater turtles and not in desert species (Bentley, 1962; Dantzler and Schmidt-Nielsen, 1966; Butler, 1972a, 1972b).

REMARKS ON FUTURE RESEARCH

Most workers have ignored the methods and techniques by which turtles obtain their food and have concentrated on the type of food consumed. There are certainly many unique morphological and behavioral feeding devices that must be explored in gaining an understanding of the overall picture of feeding behavior.

An investigation of comparative feeding habits among hatchlings of different species relative to water depth, temperature, and other environmental factors, food preference, onset of feeding after hatching, and so on, would be very interesting.

To our surprise, we found no study of the physiology of digestion in chelonians. Elucidations of drinking behavior and the physiological mechanisms that initiate drinking would be significant contributions, especially in the terrestrial species. At the present there is only fragmentary work in the area of excretion and electrolyte balance. A comparative study of excretion in species from various habitats would give us a better understanding of excretory mechanisms relative to physiological demands.

SENSORY PROCESSES

CHAPTER 13
Eyes and Their Sensitivity to Light of Differing Wavelengths

A. M. GRANDA

MORPHOLOGY AND ANATOMY OF THE EYEBALL

Turtles possess rather remarkable visual systems, and it is one of those comforting egocentricities to suppose that however remarkable they may be, visual systems evolved to their full glory in "higher" human primates. There is little in mammals, however, to match the rich diversity of gross anatomy, retinal structure, variety of photopigments, and neural connections distributed among the fishes, amphibia, reptiles, and birds (Walls, 1942; Duke-Elder, 1958; Bridges, 1965, 1972; Crescitelli, 1972; Ali, 1974).

Several sound principles are shared in common across all vertebrate classes. These virtues include the standard three tunics of *sclera, choroid,* and *retina,* the latter completely inverted in contradistinction to the invertebrates. The eyeball itself forms a dark chamber filled with clear aqueous and vitreous humors under pressure. The anterior portion of the sclera is transparent and forms a permanent corneal lens of specific refractive power depending on habitat. Within the eye is a variable lens, whose machinery of accommodation is especially interesting in terms of its evolutionary development (see below). The entire organ is protected by a bony orbital cavity in which it is embedded, further cushioned by fascia and muscle. In its orbit the eye is moved by extraocular muscles innervated by cranial nerves.

The special reptilian adaptations of the vertebrate eye derive largely from the fact that reptiles were the first class to completely adapt to life on dry land. Accommodation, for example, is obtained by deformation of the lens through striated ciliary muscles cantilevering from a ring of scleral ossicles rather than by a to-and-fro cameralike movement of the lens as it is in fishes (Duke-Elder, 1958). The lens is unusually soft and transparent, lending itself to that purpose. There is also a ventral transversalis muscle, which orients the lens nasally for convergence in binocular vision.

In some ways turtle eyes have primitive features, consistent with the claim that turtles are one of the most ancient of surviving reptiles; in other ways, their eyes have rather sophisticated adaptations. For example, turtle eyeballs, though variable in size, are small compared to those of birds and primates. There is little change in curvature in the limbus region of the sclera and cornea, a feature that is developed in birds to a marked degree (Duke-Elder, 1958). Embedded within the thick, well-developed corneal epithelium, overlapping in several layers, are the scleral ossicles—really thin, flat, bony plates that vary in number from 6 to 15 depending on species (Duke-Elder, 1958). The scleral ossicles play a critical role in the accommodative process, for the ossicular ring acts to stiffen the sclerocorneal sulcus (see Figure 1), which otherwise would be envaginated by the intraocular pressure during the process.

The pupil is round and somewhat responsive to light, the effect being confounded by the act of accommodation. The powerful sphincter muscle for accommodation is formed by the ciliary body separating to approach the lens and the annular pad or *Ringwulst* of the lens itself. The ciliary body has processes consisting of tall finlike structures (Figure 2), which make contact with the lens at an equatorial thickening, the annular pad. The elements of the pad are radially distributed and act to transmit the strain of the ciliary process directly to the very pliable lens. In freshwater turtles the lens is more flat than in sea turtles (Figure 3), but the greater curvature is very much expected in sea animals, where corneal refraction is obviated by immersion in seawater and the lens must assume the entire burden of accommodation. In room light, *Pseudemys scripta* has a pupillary opening of about 1.5 mm in diameter (Baylor and Fettiplace, 1975). The pupil does respond to the drug pilocarpine, but very little, if at all, to atropine (Northmore, personal communication).

The iris in freshwater species is beautifully colored in reds, greens, yellows, and browns, probably for camouflage as in birds and fishes. In sea and terrestrial turtles the coloration is not marked, but there is an interesting sexual dimorphism in the terrestrial *Terrapene carolina*: the iris of the male is red and that of the female brown (Duke-Elder, 1958).

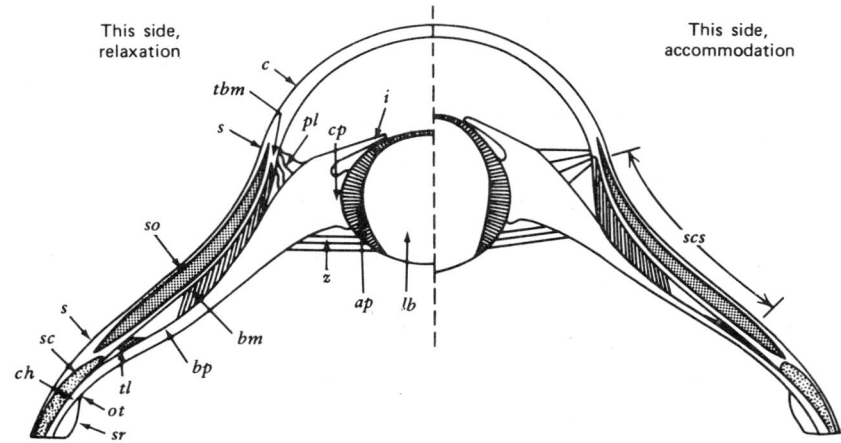

Figure 1 Composite diagram of accommodation in reptiles: *ap*, annular pad of lens; *bm*, Brücke's muscle; *bp*, base plate of ciliary body; *c*, cornea; *ch*, choroid; *cp*, ciliary process; *i*, iris; *lb*, lens body; *ot*, ora terminalis; *pl*, pectinate ligament; *s*, sclera; *sc*, scleral cartilage; *scs*, sclerocorneal sulcus; *so*, scleral ossicle; *sr*, sensory retina; *tbm*, tendon of Brücke's muscle; *tl*, tenacular ligament; *z*, zonule. (Taken from Walls, 1942, p. 275, with the kind permission of the publisher.)

The blood supply, as in most reptiles, is indirect and operates mostly by diffusion to supply nutrients for metabolism. The retina is accordingly avascular, an advantage in avoiding extraneous nonvisual structures in the reception of light, but of little help where the metabolic requirements are great. In lizards, for example, there is a conus papillarus to supplement the vascularization. In turtles, except for an embryological attempt at a glial conus on the optic disk in some species, adults show nothing on the disc at all. The surface of the disk smoothly continues into that of the surrounding retina.

The Retina

In the retina of *Pseudemys scripta*, single cones are predominant and form three distinct types, containing separate visual pigments (Liebman and Granda, 1971). The single cones and principal members of double cones contain oil droplets that are brilliantly colored, ranging from rich reds to pale yellows. The cones are 30 to 40 μm, with a 2 μm base diameter tapering to 1 μm at the outer segment tip. The oil droplets are graded in size, from the large red ones, 5.8 to 7.1 μm in diameter, to orange-yellow, 3 to 5.8 μm in diameter, to the small clear ones, 2.4 to 3.7 μm in diameter (Granda and Haden, 1970).

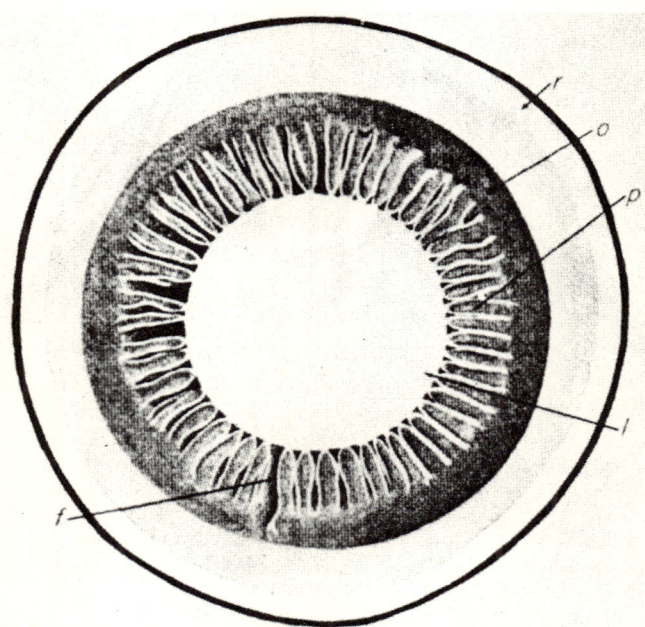

Figure 2 Anterior segment of the left eye of *Emys orbicularis*, seen from behind, showing junction of ciliary processes with lens capsule: f, unclosed portion of embryonic fissure; l, lens; o, orbicularis ciliaris; p, ciliary processes; r, retina. (Taken from Walls, 1942, p. 277, with the kind permission of the publisher.)

Double cones containing a bona fide principal receptor of single cone morphology are intimately allied with a fat accessory member without an oil droplet. The accessory cone contains a "rod" pigment (Liebman and Granda, 1971). The oil droplets are either yellow or clear in the principal members. Their contained photopigment appears to be red sensitive (Richter and Simon 1974). There are rods, too, once thought to be absent in turtle eyes (Schultze, 1866); these structures lack oil droplets and have a large rectangular morphology (3 μm in diameter × 40 μm long). Rods contain a visual pigment that absorbs light in the same spectral region as does the visual pigment in accessory cones. The photoreceptors, both rods and cones, connect to second-order neurons, the bipolar cells, and the horizontal cells. Horizontal cells consist of luminosity cells (L cells) of two types that differ in morphology and receptive field size, and also chromaticity cells (C cells) that are hyperpolarized by some wavelengths of light and depolarized by others.

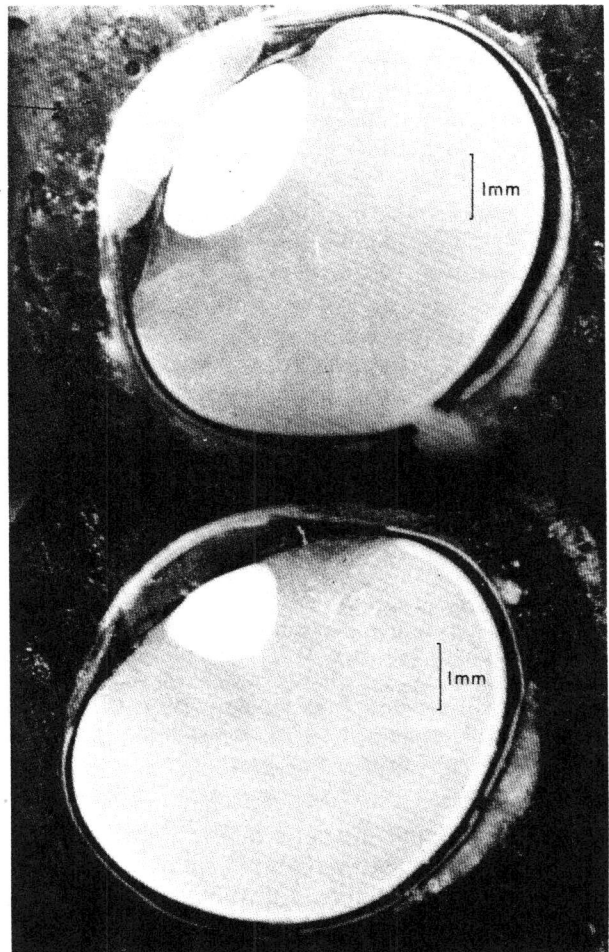

Figure 3 Frozen hemisection of eyeball and lens. (*Top*) *Pseudemys*: note shape of lens. (*Bottom*) Sea turtle: the lens is more strongly curved. (The photographs were kindly provided by Dr. D. P. M. Northmore.)

In addition to the photoreceptors, bipolar and horizontal cells, there are two other cell types to complete the retinal architecture. Ganglion cells receive their signals mainly from bipolar cells and communicate thus from the photoreceptor-bipolar sequence by way of long axons largely to the optic tectum and to the thalamic complex. Amacrine cells are also present and are thought to function as receivers of centrifugal

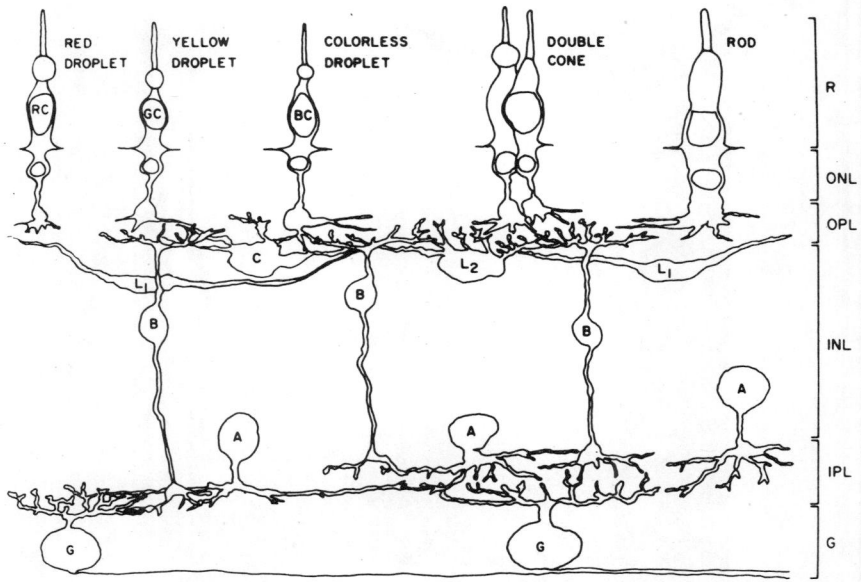

Figure 4 Diagrammatic arrangement of retinal cells in the eye of *P..scripta*; only cones have oil droplets, accessory cones and rods do not: RC, red cone; GC, green cone; BC, blue cone; L_1, luminosity horizontal cell, type 1; L_2, luminosity horizontal cell, type 2; C, chromaticity horizontal cell; B, bipolar cell; A, amacrine cell; G, ganglion cell; R, receptor layer; ONL, outer nuclear layer; OPL, outer plexiform layer; INL, inner nuclear layer; IPL, inner plexiform layer. The principal cone of the double cone pair is thought to be red sensitive with an orange droplet; see text.

signals from the isthmooptic nucleus in the diencephalon. Amacrine cells make synaptic contact with ganglion cells and may influence them (Marchiafava, 1976). Figure 4 shows all these cell types in stylized format.

SENSITIVITY TO COLOR

This discussion describes evidence from photochemistry to identify visual pigments, from electrophysiology to identify the contributions of visual elements in the pathway where possible, and from psychophysical experiments that define visual sensitivity in expressed behavior.

Photochemistry

For light to be detected it must be absorbed, and the light-absorbing substances contained in the outer segments of the photoreceptors are the visual pigments. All visual pigments are *carotenoids*, derived from a common structural plan of subunits linked by single and double bonds called *conjugated chains*. The part of the pigment molecule that absorbs light is the prosthetic *chromophore* that forms an integral part of a lipoprotein complex or *opsin*. The opsin determines the characteristic absorption of light for a particular photopigment, but the chromophore-opsin complex determines the photochemical activity associated with vision.

In the dark-adapted eye, deprived of light absorption, the visual pigment exists in an 11-*cis* form of the aldehyde of vitamin A (Hubbard and Wald, 1952a, 1952b). The action of light is to isomerize the chemical structure from the cis form to an all-trans form. This cis-trans isomerization, a change in spatial configuration, is critical to vision for each structure has its own chemical and physical properties.

All known visual pigments are based on vitamins A_1 and A_2, now also referred to as retinol and 3-dehydroretinol (Figure 5). The maximum absorbance of vitamin A_1 pigments are in the neighborhood of 500 nm, whereas those based on vitamin A_2 pigments are shifted to the longer wavelengths.

There is a broad assignment of vitamin A_1 pigments to creatures confined to land environments and to those in the sea. Freshwater fishes, on the other hand, have a maximum absorbance shifted to about 540 nm. Wald (1939a, 1939b, 1941) found that these red-shifted photopigments were based on vitamin A_2. The division is in keeping with an overall classification of visual systems proposed by him, where $retinal_1$ combines with rod and cone opsins to give pigments with maximum absorption at 500 and 562 nm, respectively, for rhodopsin and iodopsin, whereas $retinal_2$, in like combination, has maximal absorption at 522 and 620 nm for porphyropsin and cyanopsin (Wald, 1958, 1959; Wald, Brown, and Smith, 1953).

It is recognized now that this simple classification does not hold, and the visual pigments derived from both $retinal_1$ and $retinal_2$ have broad and overlapping spectral regions. In freshwater fishes both systems are often present in a single eye (Bridges, 1965, 1972).

Photopigments

Turtles are distributed over several environments: land, sea, and fresh waters. The green turtle, *Chelonia mydas mydas*, belongs to the marine family Cheloniidae, and the red-eared turtle, *Pseudemys scripta elegans*, is a member of the freshwater family Emydidae. For these two species, Liebman and Granda (1971) have isolated differing sets of photopigments that conform to the vitamin A_1–vitamin A_2 environmental dis-

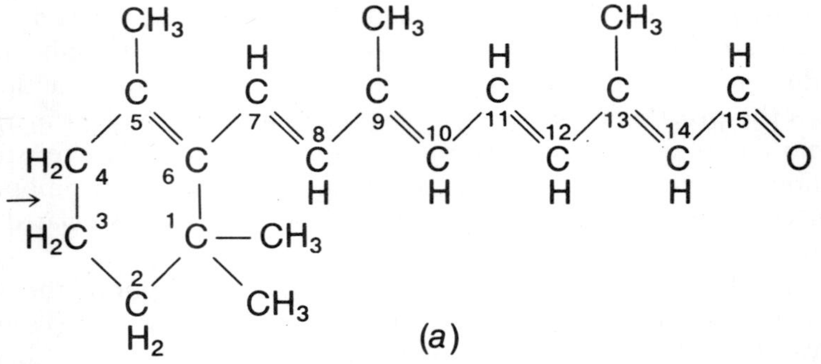

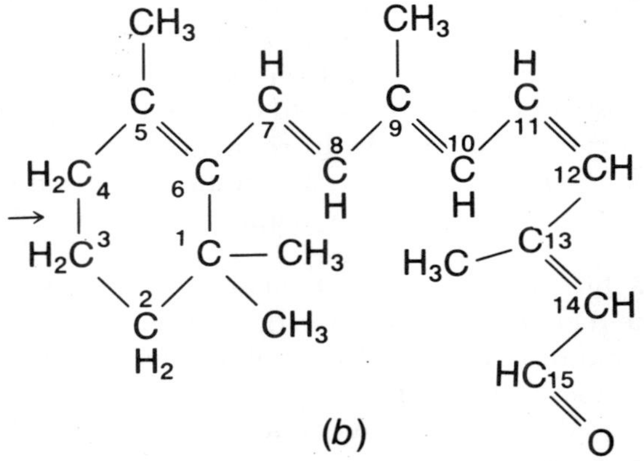

Figure 5 Chemical structures of (a) all-trans and (b) 11-cis retinal, the aldehyde form of vitamin A_1. A similar arrangement exists for 3-dehydroretinal, the aldehyde form of vitamin A_2. The two substances shown differ by the presence of a double bond in the ring indicated by arrows.

tribution. *C. mydas* possess retinal$_1$ pigments, and *P. scripta* possess retinal$_2$ pigments. Based on microspectrophotometric measurements of the visual pigments *in situ* (Liebman, 1972), particular photopigments were identified in individual photoreceptors. Both species have predominantly cone retinas, although rods, too, are present. *C. mydas* rods, measuring 3 × 40 μm, easily seen in most preparations, contained a visual pigment absorbing maximally at 500 to 505 nm. In general morphology the rod outer segments were large and rectangular, in contrast to the cones, where the outer segments were small and tapering, often with oil droplets present. The cones contained three photopigments housed in separate receptors absorbing maximally at 440, 502, and 562 nm (see Figure 6).

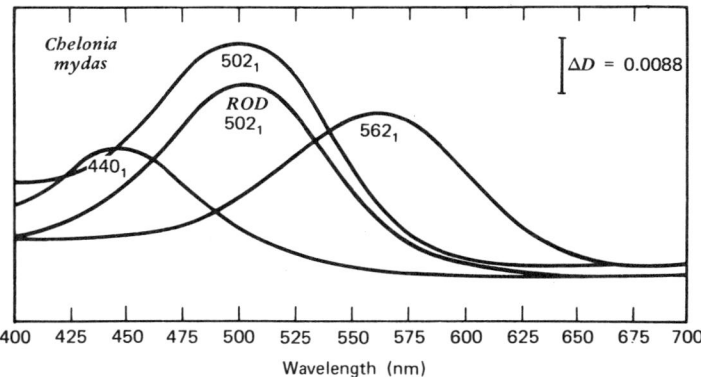

Figure 6 Smoothed density spectra for *C. mydas*. (After Liebman and Granda, 1971.)

A similar distribution existed for *P. scripta*, except that the absorption maxima were shifted to the longer wavelengths. The rod outer segments, similar in appearance to *C. mydas* rods, contained a photopigment absorbing maximally at 518 nm. The cones, with outer segments 30 to 40 μm long, and tapering from a 2 μm diameter base to a 1 μm terminal diameter, contained photopigments with absorption maxima of 450, 518, and 620 nm (Figure 7).

The pigments are appropriate to the prevailing light conditions. The ocean environment in which *C. mydas* spend the bulk of their lives transmits maximally at 470 to 475 nm (Jerlov, 1968). In fresh waters,

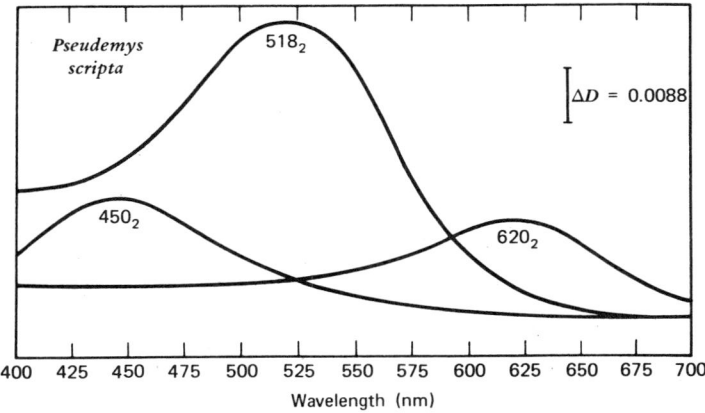

Figure 7 Smoothed density spectra for *P. scripta*. (After Liebman and Granda, 1971.)

contaminated by fine light-scattering particles and dissolved organic materials, transmission maxima are shifted to beyond 600 nm within very little depth (Clarke and James, 1939). Both species appear to capitalize on the available aquatic light in their particular environments.

Oil Droplets

Light reaching the outer segments of the cone photoreceptor must first traverse oil droplets situated at the distal border of the inner segments (Figure 4). Richly colored, they are a conspicuous feature of fresh, unstained retina. The oil droplets in *C. mydas* are orange and yellow, with a smaller grouping of colorless droplets. *P. scripta* has droplets colored a deep red, another group shading from orange through pale yellow, and a third group of clear droplets.

The droplets can be exceedingly dense. The large red droplets of *P. scripta* show a range of peak absorbance in excess of 50 logarithmic units, an extremely high optical density (Liebman and Granda, 1975). The red color is thought to be the carotenoid *astaxanthin* (Wald and Zussman, 1938; Karrer and Jucker, 1950). The characteristic spectral shape for *astaxanthin* also holds for the orange oil droplet, apparently the same substance but at lower concentrations, 6 to 9 logarithmic units (Liebman and Granda, 1975). The yellow droplet is made up of a different carotenoid, perhaps *zeaxanthin* or *xanthophyll*, with a range of 12 to 18 logarithmic units. The clear droplets absorb scarcely at all and are virtually flat across the measured spectrum.

C. mydas has no red oil droplets (Granda and Haden, 1970). Unlike *P. scripta*, the orange and yellow droplets present show almost, but not quite, identical carotenoid structure. It is not certain what these carotenoids may be. They are present in much more dilute concentrations—4 to 7 logarithmic units for orange, 3 to 4 logarithmic units for yellow—ranges not nearly as extreme as in *P. scripta*.

Since the time of Krause (1863), the oil droplets have been postulated as the basis for a color discriminating system. The different transmission bands of the colored oil droplets were thought to act to differentiate the light falling on separate cones for color discrimination. The system is simple in that it requires but a single cone pigment. But since there are three cone pigments, that theory obviously cannot hold, and the question of their possible use and significance remains. They apparently act as sharp cutoff filters, transmitting on the long wavelength side and absorbing heavily in the short wavelengths (Figure 8).

One could very well expect the transmission of the oil droplet to be in agreement with the maximum light-absorbing capabilities of the photo-

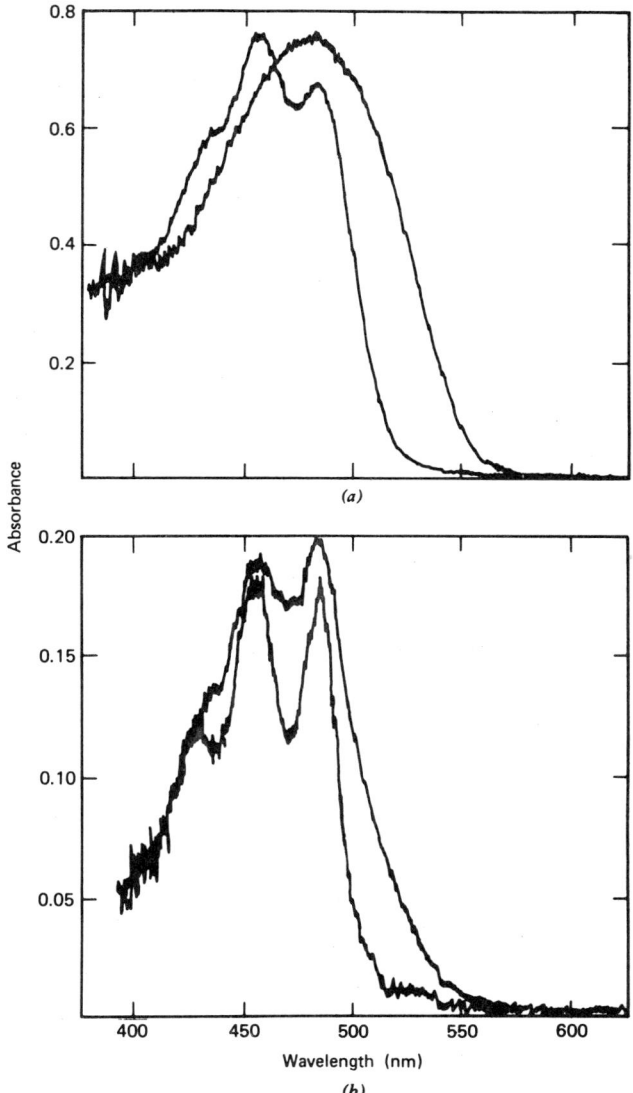

Figure 8 Resolved oil droplets after dilution in mineral oil. (a) *P. scripta* three-fingered spectrum of yellow droplet (left-hand trace), singly peaked spectrum of red and orange droplets (right-hand trace). (b) *C. mydas* sharply fingered spectrum of yellow droplet (lower trace), orange droplet (upper trace). (From Liebman and Granda, 1975.)

pigment to which they are conjoined. Single cones containing visual pigment 620_2 (identified as maximum absorption wavelength based on vitamin A_2) have red or orange droplets, those containing visual pigment 518_2 have yellow droplets, and those containing visual pigment 450_2 have a clear or colorless droplet (Granda and Liebman, unpublished data).

The combinations act to shift the maximum absorption peak of the photopigment to longer wavelengths; for example, the red-sensitive photopigment absorption of 620 nm in *P. scripta* is shifted to 640 nm by a red oil droplet, the green-sensitive photopigment of 518 nm is shifted to 550 or 560 nm by a yellow oil droplet, and the blue-sensitive photopigment of 450 nm, which is coupled with a clear oil droplet, is very little affected. A similar long wavelength shift occurs in *C. mydas*.

The entire red shift was supposed by Walls and Judd (1933) to enhance contrast by absorbing short wavelength light to a greater degree. The portion of the visible light contributing to scatter and glare is much reduced in this fashion, since blue light scatters more than red.

Electrophysiology of the Eye

The spectral sensitivity of *P. scripta*, derived from electrical measurements of the entire eye, shows maximal sensitivity in the red region of the visible spectrum (Armington, 1954; Deane, Enroth-Cugell, Gongaware, Neyland, and Forbes, 1958; Granda, 1962; Granda and Stirling, 1965, 1966). The electroretinograms from which the sensitivity measurements are derived are mass potentials developed across the eye in response to light. The potentials change in amplitude and wave form as a function of wavelength and intensity. The stimulus intensity required to produce a criterion level of response at each wavelength plots the spectral sensitivity of the eye. In the dark-adapted eye of *P. scripta* the most sensitive region is in the longer wavelengths, at 640 to 650 nm (Figure 9). There is also a well-defined hump at 570 to 580 nm and little evidence of much activity below 480 nm (Granda, 1962). With light adaptation, no shift of peak sensitivity occurs.

The lights required to elicit well-defined electroretinograms are intense, very much above absolute threshold. In an effort to use lower criterion values signifying dim light stimulation, Granda and Stirling (1966) summated wave forms on a digital computer to extract small responses from the noise level of the recording system. The animals showed great sensitivity to red light in the region 640 to 660 nm, and also near 580 nm, as previously reported for *P. scripta*. Of particular interest was a well-defined blue-green process below 500 nm. The results could be

SENSITIVITY TO COLOR

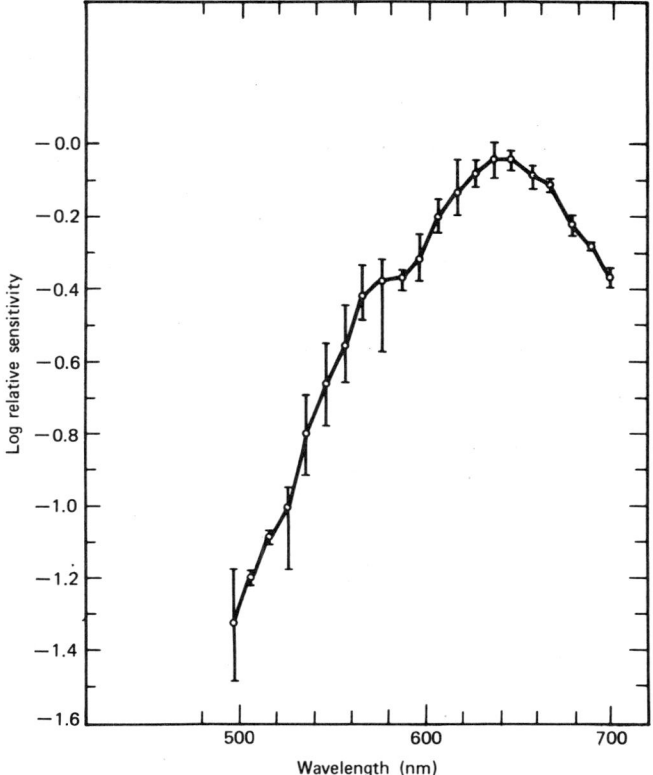

Figure 9 Mean spectral sensitivity for *P. scripta* determined from electroretinograms under dark-adapted conditions. The range for each point among three animals is indicated by vertical lines. (From Granda, 1962.)

accounted for by assuming that the red-sensitive receptors were plentiful and low in sensitivity, while the blue-sensitive were few and high in sensitivity. In dim light fewer of the red-sensitive receptors would be excited, thus uncovering the smaller number of blue-sensitive receptors. At more intense stimulus lights, the large number of red-sensitive receptors activated would simply mask any contribution from the blue end.

Recording intracellularly from individual cones, Baylor and Hodgkin (1973) reported that the largest number of responses came from red-sensitive cones, those from green-sensitive cones less often, and those from the blue cones or rods rarely. The spectral sensitivity curves from 17 cells (8 with maxima at 630 nm, 6 at 550 nm and 3 at 460 nm) appear in Figure 10. The groupings agree well with the absorption maxima of

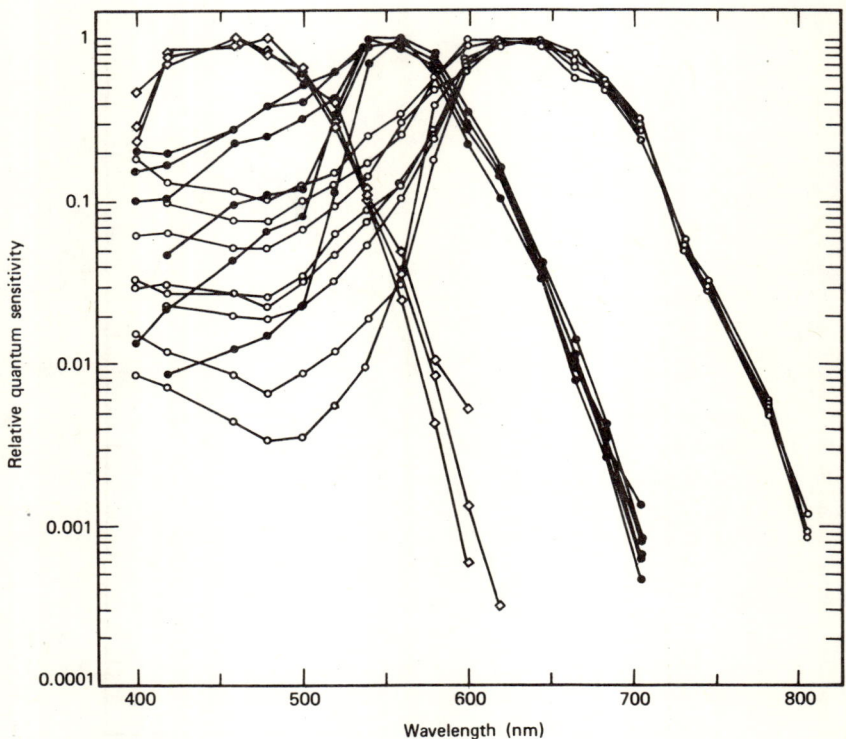

Figure 10 Spectral sensitivities of cones determined from intracellular recordings in photoreceptors. The ordinate is relative quantum sensitivity, the abscissa wavelength. Collected results from 17 cells. (Taken from Baylor and Hodgkin, 1973, with the kind permission of the authors.)

the photopigments conjoined with individual oil droplets (Liebman and Granda, 1971; Liebman, 1972).

The frequencies of various classes of oil droplets can provide a clue to the relative frequency of receptor types present if oil droplet and photopigment are uniquely paired. A distribution of 38% for red oil droplets, 47% for orange-yellow oil droplets, and 15% for clear oil droplets found by Granda and Haden (1970) agrees well with the figures reported by Peiponen (1964) and Baylor and Fettiplace (1975).

Baylor and Fettiplace were successful in separating orange from yellow droplets, for the orange ones were always located in the principal members of double cones; the accessory members contained no oil droplets. The photopigment of the principal cone was red sensitive, the other green sensitive and coupled to it (Richter and Simon, 1974). The orange oil droplets are diluted red oil droplets (Liebman and Granda, 1975), and both red and orange oil droplets are paired with red-sensitive cones, which is in keeping with the finding. To arrive at a distribution of photoreceptors, the orange and red oil droplet percentages of Baylor and Fettiplace (1975) were combined to yield 45% for red-sensitive cones, 38% for green-sensitive cones, and 17% for blue-sensitive cones. The green-sensitive cones in double cone arrangements, where there is no oil droplet, have been included with the regular green-sensitive cones that have yellow oil droplets. With this analysis, red-sensitive cones predominate in the retina of *P. scripta*, and it is little wonder that their responses dominate the spectral sensitivity of the eye.

In *C. mydas*, as might be expected from the photopigment absorption, the entire spectral sensitivity curve is shifted to the shorter wavelengths, for the eye is based on vitamin A_1, not on vitamin A_2, as in *P. scripta*. Granda and O'Shea (1972) recorded electroretinograms from the eye of *C. mydas* and found spectral sensitivity to peak at 520 nm with secondary peaks at 600 and 450 to 460 nm (Figure 11). Compared to *P. scripta*, the red sensitivity for *C. mydas* was displaced about 50 nm toward the short

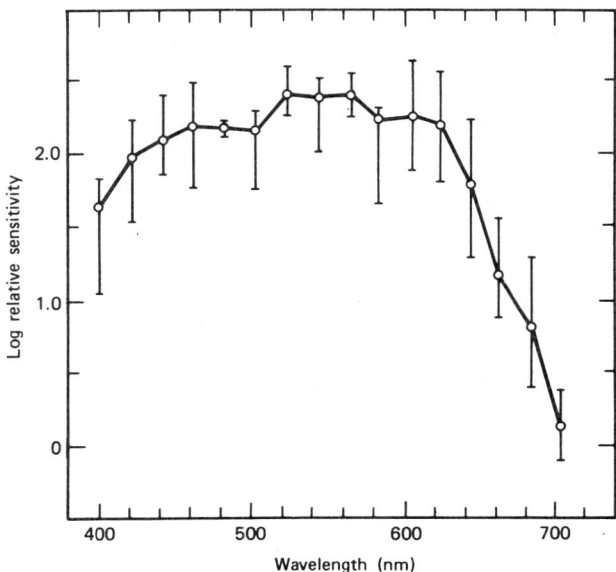

Figure 11 Mean spectral sensitivity curve for *C. mydas* determined from electroretinograms under dark-adapted conditions. The range for each point among four animals is indicated by vertical lines. (From Granda and O'Shea, 1972.)

wavelengths. At 500 nm, sensitivity in *P. scripta* had fallen almost 1.4 logarithmic units from the peak value of 640 to 650 nm. *C. mydas*, on the other hand, was sensitive in the blue spectral region; the spectral curve showed a peak very clearly reflecting the entire shift of sensitivity to the shorter wavelengths.

The discrepancy between absorption spectra of the photopigments and the electrically determined, spectral sensitivity can be accounted for by the presence of colored oil droplets, as in *P. scripta*. Where the oil droplets are colored and sufficiently dense (Liebman and Granda, 1975), they shift the peak absorption of the pigments to longer wavelengths. The distribution of oil droplets in *C. mydas* is 36% for orange droplets, 43% for yellow droplets, and 21% for clear droplets (Granda and Haden, 1970). If these droplets are paired with the photopigments absorbing maximally at 562, 502, and 440 nm (Liebman and Granda, 1971), their frequencies predict closely the relative peak sensitivities shown in Figure 11.

Psychophysics of Color

Turtles can learn to press keys and solve mazes to obtain rewards of food and water and also to avoid electric shock (see Chapter 20). Applied to vision, these techniques allow the determination of threshold measures that depend on an integrated signal routed to the brain.

Working with *Mauremys caspica* and food reward, Wojtusiak (1933) reported three maxima in the spectral sensitivity curves: a prominent peak in the red region of the spectrum near 634 nm, and secondary peaks near 500 nm and in the very short wavelengths. Armington (1954) also used a situation involving food. In a thoroughly dark-adapted *P. scripta* he showed a peak sensitivity of 525 nm for the one animal investigated, with a small shoulder in the long wavelengths. The function was different from the red-dominated electroretinographic curves also determined under dark-adapted conditions.

Sokol and Muntz (1966) found maximal sensitivity at 420 to 500 nm after their *Chrysemys picta* were adapted to the dark for 30 min. Under light-adapted conditions they found essentially the same results (Muntz and Sokol, 1967). Graf (1967), on the other hand, found a sensitivity function that peaked between 625 and 640 nm in the same species under light-adapted conditions (Figure 12). In this study turtles were required to match steady lights with flickering lights at various wavelengths and intensities.

All these turtles are freshwater species. They all showed spectral

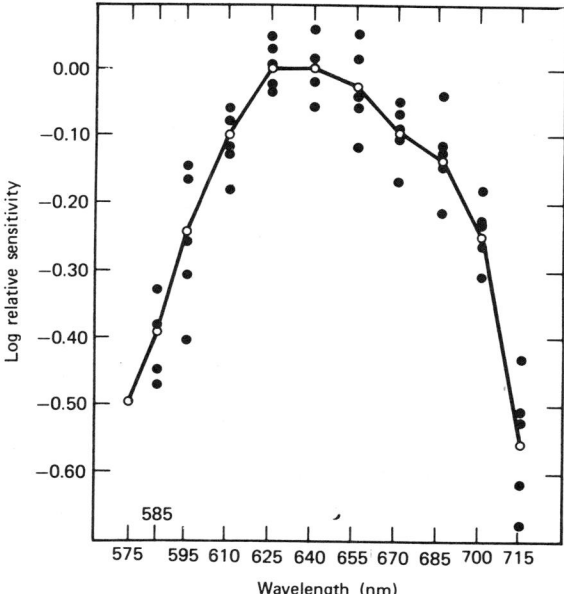

Figure 12 Mean relative spectral sensitivity curve (open circles) for five animals (*Chrysemys picta*), obtained by behavioral responses to flicker photometry. The filled circles are individual data points. (Taken from Graf, 1967, with the kind permission of the author.)

functions that were sensitive to red light under light-adapted conditions with strong test lights, and blue light sensitivity with weak stimulus lights and dark adaptation.

This conclusion for freshwater turtles was corroborated by Granda, Maxwell, and Zwick (1972), who studied the temporal course of dark adaptation in *P. scripta*, using conditioned avoidance to electric shock. Dark adaptation curves for red light reached asymptote in about 90 sec in the dark. Curves for shorter wavelengths, however, took longer to reach asymptote and showed discontinuities indicative of more than one receptor system operating (Figure 13). Spectral sensitivity curves obtained at times in the dark ranging from 10 sec to 45 min showed a shift in peak sensitivity from the red (633 nm) early in dark adaptation, to the blue (466 nm), when the turtle was well dark adapted (Figure 14).

With this technique we used chromatic adaptation to isolate the receptor mechanisms. Spectral sensitivity curves were determined against red,

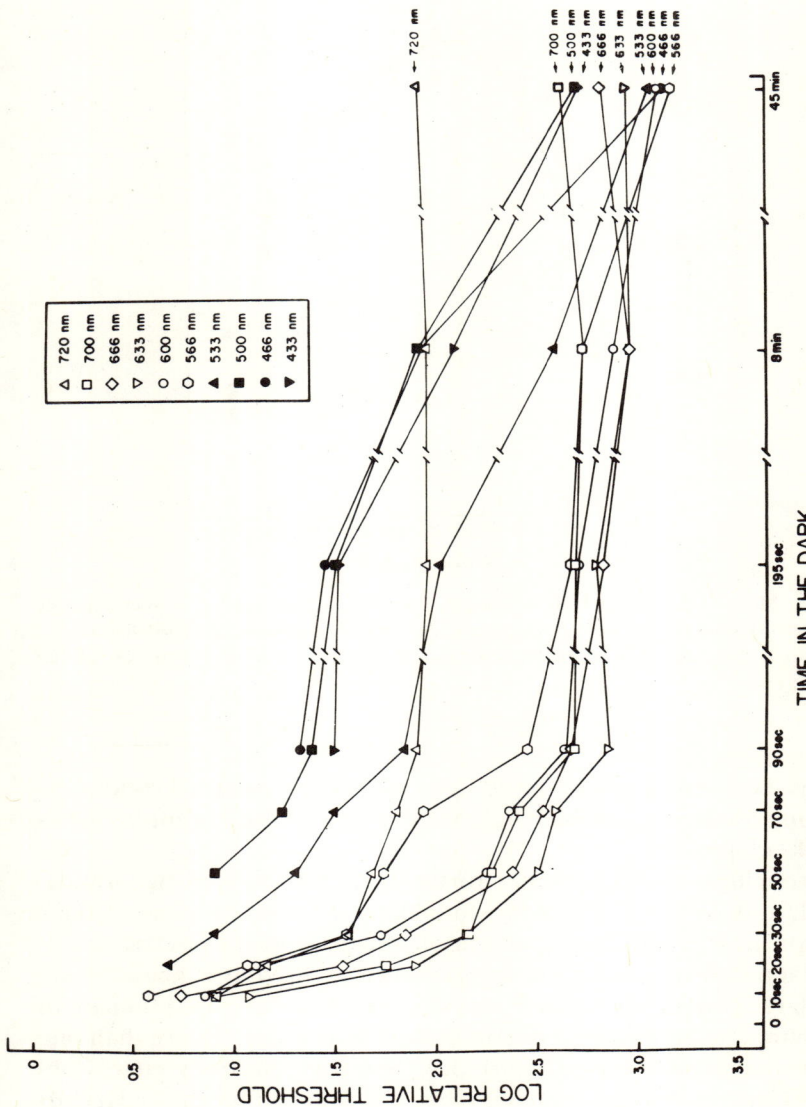

Figure 13 Dark adaptation functions for one subject (*P. scripta*) measured with test lights ranging from 433 to 720 nm. The values plotted on the abscissa are continuous to 90 sec; interruptions at indicated points thereafter permit a compression of the time scale. (From Granda, Maxwell, and Zwick, 1972.)

CONCLUSION

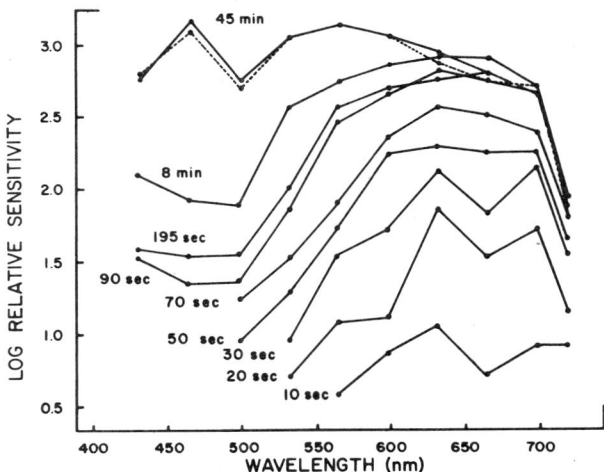

Figure 14 Spectral sensitivity functions for one subject (*P. scripta*). The number accompanying each curve indicates the time in the dark at which the curve was defined. (From Granda et al., 1972.)

green, and blue backgrounds equated for energy. Threshold values so derived show significant decreases in sensitivity appropriate to the spectral regions to which they were adapted. Theoretical spectral sensitivities of three underlying receptor mechanisms were then derived to account for the results. The peak sensitivities of the theoretical mechanisms compared rather well with the known absorption maxima of the three classes of photopigments found by microspectrophotometry (Liebman and Granda, 1971).

CONCLUSION

A variety of techniques can be brought to bear on the single problem of sensitivity to color, thereby providing valid assurance of its function. Photochemistry, electrophysiology, and conditioned behavior all have contributed insights to visual activity. The problem is made more fascinating and complex by the distribution of species in many habitats that make different demands on the visual apparatus because of variations in lighting conditions.

ACKNOWLEDGMENTS

I am indebted to J. H. Maxwell and J. E. Fulbrook for assistance with the figures and their renderings, and to Mrs. C. Groot for careful typing and editing of the manuscript. The work was supported by grant EY 01540 from the National Eye Institute, U.S. Public Health Service.

CHAPTER 14
The Chemical Senses

THOMAS R. SCOTT, JR.

OLFACTION

When comparative neurology was introduced in the late nineteenth century, olfactory systems were viewed as the common thread running through all vertebrate species, possibly holding the key to forebrain evolution. In this context, the nasal anatomy of turtles represents a primitive stage of that development. Sensory epithelium of the accessory olfactory organ (Jacobson's organ, vomeronasal organ) is less localized than in any other amniote. The accessory olfactory bulb, which receives fibers largely from Jacobson's organ, is exceptionally robust and well defined. There is no pronounced concha (an extension of the lateral wall into the nasal cavity) as there is in more recently evolved vertebrates. These unique characteristics suggest that the evolutionary line leading to turtles diverged from the main trunk of reptilian evolution before the branches leading to other reptiles, birds, and mammals (Parsons, 1959).

That olfaction is critically important is testified to by the rich variety of behavior it supposedly mediates and by the fact that in *Pseudemys scripta* fully one-third of the 15 mm telencephalon is occupied by the olfactory bulbs.

Gross Structure of the Nose

The nose consists of a pair of membranous sacs that open through the external nares (nostrils) at the anterior end and through the internal nares into the palate at the posterior end. This particularly simple nasal structure has a short, tubelike vestibulum extending from the external nares to the nasal cavity and a nasopharyngeal duct leading from the cavity to the internal nares and palate. At the posterior end of this duct lies the opening to the oral cavity, the choana. This point is marked in some species by a lateral flap with or without papillae, which may be vestigial, or by a lateral ridge. Turtles that show a more advanced development tend toward the simple ridge or lack any choanal structure (Parsons, 1968).

The nasal cavity (cavum nasi proprium) is divisible into two regions. Ventrally lies the intermediate region or pars respiratoria to which the vestibulum and nasopharyngeal ducts attach. Posterodorsally is the olfactory region (pars olfactoria) a lateromedially compressed hemisphere lined with sensory epithelium. The dorsal margin of the intermediate region opens wide into the olfactory area, allowing easy communication between them. The boundary is marked by medial and lateral ridges. There are three principal variations on this general scheme across different families. (1) In Emydidae, the dorsal half of the oval nasal cavity (the olfactory region) is lined with sensory epithelium and Bowman's glands. The ventral half has three parallel bands of sensory epithelium coursing along the lateral, ventral, and medial walls (Jacobson's organ) and lacks Bowman's glands. (2) In Testudinidae, the olfactory region presents a convex bulge into the nasal cavity on its lateral aspect. The sensory epithelium of Jacobson's organ is restricted to the medial wall of the intermediate region. (3) In Cheloniidae, there is only a small olfactory region in the posterodorsal quarter of the cavity. The large intermediate region displays only two pockets of sensory epithelium, one located anteroventrally and one anterodorsally. The Trionychidae are similar, but the epithelium of Jacobson's organ lies in a complex pattern of grooves.

The functional significance of Jacobson's organ is not clear, though it does mediate certain olfactory responses and, in some reptiles, plays a part in feeding, courtship, and aggregation (Nobel and Clausen, 1936; Nobel, 1937; Wilde, 1938). It is a tubelike structure in turtles, opening into the hard palate and partially lined with sensory epithelium. Receptors in the epithelium give rise to the medial trunk of the olfactory nerve, which courses to the accessory olfactory bulb (posterodorsal one-third of the bulb). The lateral trunk is composed of axons originating in olfactory epithelium and terminates in the main olfactory bulb.

The entire nasal cavity receives a rich blood supply. Jacobson's organ is especially well vascularized, and changes in blood pressure have been suggested as a mechanism by which odorous substances could be inhaled or expelled (Allison, 1953; Shibuya, 1964). Sea turtles have the ability to close the vestibular passage to the nasal cavity while resting or sleeping underwater. Closure apparently results from an increased blood flow in the plexus beneath the vestibular epithelium, causing protrusions from the dorsolateral and ventromedial walls to meet. Even though the major variation in nasal cavity volume is controlled through the vascular system, there is some evidence for smooth muscle involvement as well (Tucker, 1971).

In addition to the olfactory nerve, the nasal cavity is served by the trigeminal nerve and also receives autonomic innervation from the sphenopalatine nerve and from nervus terminalis, a ganglionic fascicle of fibers coursing from the olfactory peduncle to the nose. The latter is apparently also autonomic, possibly supplying blood vessels and other structures that support the olfactory and vomeronasal regions (Larsell, 1918).

Receptor Anatomy

Transduction of olfactory information seems to be effected by at least three structures: (1) specialized receptors located in olfactory epithelium lining the olfactory region of the nasal cavity, (2) receptors in sensory epithelium of the vomeronasal organ, located primarily on the medial wall of the intermediate region, and (3) free nerve endings of trigeminal axons, which are poorly defined anatomically.

Olfactory Epithelium. Olfactory epithelium in all turtles is typically yellow-brown material, 0.4 to 1.0 mm thick. It consists of three cell types: receptors, supporting cells, and basal cells, each arranged in its own layer. A bipolar neuron receptor sends a dendritic process toward the mucosal surface and an axon toward the olfactory bulb. The spindle-shaped soma lies in the basal one-third of the epithelium, below the level of its supporting cells. Its length is proportional to the epithelial thickness, but all somas show a similar diameter of 5 to 8 μm. Its dendritic "olfactory rod" is rather long (20–90 μm) and thin (1 μm) and presents a slight enlargement at the external limiting membrane where it branches into 8 to 12 fine hairs or cilia. These are quite thin (0.1 μm) and extend approximately 25 μm into the mucous coat. Each cilium in cross section exhibits nine delicate paired filaments (50 Å diameter) running axially and situated around a centrally placed pair. Individual filaments are of variable length (Graziadei, 1971). Olfactory cilia are motile in turtles, but their movement is slow and disorganized in contrast to the strong synchronous beat of cilia in the respiratory area. This, along with electrophysiological considerations, raises the question of whether the cilia are true receptor elements or whether their asynchronous flailing simply serves to retain and blend mucus on the mucosal surface.[1]

The axons of receptor neurons form the olfactory nerve. They are quite thin (0.1–0.4 μm) and unmyelinated, resulting in very slow conduction rates of approximately 0.15 m/sec. The axons project through the basal cells and processes of sustentacular cells. They are then bound into bundles of at least several hundred by large Schwann cells and proceed with neither synapse nor collateral to the glomeruli of the olfactory bulb.

Supporting, or sustentacular, cells are relatively massive cells that surround receptive neurons. Distally, the supporting cells have a seven- or eight-sided projection with an olfactory rod lying at each angle. They have no cilia. The cytoplasm is distinguishable by its fine granules of yellow and light brown pigment. The basal portion, extending among the bodies of receptor cells, is compressed by them and is often forked at the basement membrane. These cells release a granular secretion that mixes with the mucus from Bowman's glands on the mucosal surface.

Supporting cells had long been thought to physically isolate individual receptors, but transmission electron micrographs show several points of direct communication among receptors (Graziadei and Pierantoni, 1969). It is not known why such contacts exist, though mutual excitatory and inhibitory connections of this sort are commonly employed in information processing. The contacts do not result from simple overcrowding.

The small basal cells form a single underlying row along the basement membrane.

The olfactory epithelial surface is bathed in a sheet of watery secretions that emanates from Bowman's glands. The glands themselves are 100 μm or more beneath the mucosal surface, in the basal layer of cells. They are coextensive with olfactory epithelium, disappearing immediately upon the transition to respiratory epithelium. The cell bodies are large and triangular or columnar, having simple or slightly branched tubes with short ducts opening to the surface of the epithelium at regular intervals. Their cytoplasm has masses of yellow pigment granules scattered throughout, and there is some evidence that this pigment is necessary for olfaction to proceed normally (Allison, 1953). The terminal knobs of receptor neurons are constantly bathed in the secretions of Bowman's glands such that any odorous molecule must be dissolved in mucus before stimulating a receptor hair.

---

[1]Recordings finding no difference in olfactory-evoked responses with the cilia intact and disrupted also raise this question (Graziadei and Tucker, 1970).

Vomeronasal Epithelium. Although clearly olfactory in function, Jacobson's organ is derived embryologically from ectodermal placode, separate from the neural crest. Vomeronasal epithelium is pale grey, in contrast to the rich yellow that characterizes olfactory epithelium. The sensory epithelium is clearly delineated from respiratory epithelium by several characteristics. First, it usually occupies the medial aspect of the intermediate region. Second, the epithelium suddenly thickens when vomeronasal receptors are encountered. In addition, the motile respiratory cilia disappear and the epithelium takes on the characteristic laminar appearance seen in olfactory epithelium, owing to the presence of supporting, sensory, and basal cells. Vomeronasal glands, specifically related to this epithelium and analogous to Bowman's glands, also emerge. In fact, vomeronasal epithelium is structurally quite similar to that of the main olfactory area. Bipolar receptors are surrounded by supporting cells that permit interreceptor contact at the various levels outlined above. Vomeronasal axons project through comparable epithelial layers to form the medial branch of the olfactory nerve, which terminates in the accessory olfactory bulb. The primary difference between the two receptor types lies in their exposed apical endings. Whereas olfactory receptors terminate in a rounded protrusion from which cilia emerge, vomeronasal receptors exhibit a tapering, cone-shaped projection with branched microvilli throughout. They lack cilia, which do not seem to be necessary for the transduction of olfactory information. It has been suggested that the cilia of olfactory receptors serve as a dense web that retains mucus, since the dorsal position of the olfactory epithelium would cause the watery mucus to separate out under the influence of gravity. The ventrally located vomeronasal receptors would not encounter this problem. Cilia might also serve, with microvilli, to increase the surface area of the receptive dendrites, which are simply bare nerve endings.

The Olfactory Nerve

The olfactory nerve is composed of first-order axons and in *Pseudemys scripta*, extends approximately 6 mm from the mucosa to the olfactory bulb. It has two separate branches: (1) the medial olfactory nerve innervates vomeronasal epithelium plus olfactory epithelium from the medial wall of the olfactory region and terminates in the accessory olfactory bulb, (2) the lateral olfactory nerve serves the lateral and dorsal walls of olfactory mucosa and synapses in the main olfactory bulb. Nerve fibers are quite uniformly in the 0.1 to 0.4 μm range.

Electrophysiological studies have shown that upper parts of the epithelium project to the dorsal part of the bulb, while lower epithelium cells send axons to the ventral bulb. Electrical stimulation to the dorsal aspect of the nerve evokes surface-negative responses in the dorsal olfactory bulb, but also surface-positive responses in the ventral bulb. Stimulating the ventral section of the nerve gives precisely the opposite response, suggesting that both areas receive topographic projections (Orrego, 1961a). Retrograde degeneration experiments support this conclusion. In addition, the mediolateral topography inherent in the separation of the medial and lateral branches of the nerve is obvious. An anteroposterior topography between epithelium and bulb has been demonstrated in rabbits (Adrian, 1950). In a system noted for breadth of neural sensitivity and the attendant patterns of sensory input, there is, nevertheless, a certain amount of topographic organization.

The Olfactory Bulb

The olfactory bulb is a rather constant structure in vertebrates, both in its brain location and its cellular composition. Still, turtles, in comparison to most other reptiles, have large, well-defined olfactory bulbs and particularly robust accessory bulbs. These structures are just anterior to the main mass of the forebrain and are connected to it by thick fiber bundles.

Axons composing the olfactory nerve pass through the ethmoid bone and spread over the surface of the bulb. Within the nerve, fibers are intertwined in complex patterns, yet there are no collaterals nor arborizations until the glomeruli are reached. Here all first-order neurons come into synaptic contact with the dendrites of mitral and tufted cells, the only synapse encountered between the olfactory stimulus and the cerebral cortex. Glomeruli are well-defined spherical masses, each of which is circumscribed and apparently functionally autonomous. Complete isolation is avoided by periglomerular cells whose somas lie among the glomeruli and whose axons provide communication among them. Further interaction is provided by collaterals from the primary dendrites of mitral and tufted cells that might contact several glomeruli. A mitral cell typically has two primary dendrites that, along with some accessory dendrites, extend to the glomeruli. This arrangement is an intermediate step between amphibians, whose mitral cells have multiple dendritic processes, and mammals, where each mitral cell has but one primary dendrite.

It is estimated that 15,000 olfactory nerve fibers converge on each glomerulus (Orrego, 1961b) which, on the postsynaptic side, might receive dendrites from only 10 to 15 mitral cells. This enormous amount of convergence affords an opportunity for the spatial summation that is thought to be a critically important aspect of information transfer in the olfactory system.

Although the olfactory bulb is structurally complex, it presents an orderly arrangement of layers, often alternating between neuronal projections and somas. At the surface is seen the incoming layer of densely packed primary axons as they separate out from their well-defined bundles. Just beneath them is the glomerular layer composed of the arborization of primary axons and the dendritic tufts of mitral cells with which they synapse. Here also are interposed the periglomerular cells which, combined with certain displaced mitral somas, group themselves along the internal glomerular border (Crosby and Humphrey, 1939), foreshadowing the better defined external granular cell layer seen in mammals. A third layer, the external molecular or plexiform layer, is composed mainly of the apical dendrites of mitral cells. In addition, tufted cells are found which make synaptic connections with mitral cells and the branching apical dendrites of granular cells whose somas are deep within the bulb. A handful of displaced mitral cell bodies completes the composition of this plexus. The first three bulbar layers, then, are presynaptic (axonal), synaptic (axodendritic), and postsynaptic (dendritic). Mitral cells form the fourth layer. These are the critical sensory neurons whose axons compose the olfactory tracts. They are the largest cells in the bulb and are roughly triangular. The layer is only two or three neurons thick (10–40 μm). Granular cell dendrites also pass through this region. Next is the internal molecular or plexiform layer, composed mainly of mitral cell axons as they form the lateral olfactory tract. Here are also found dendrites and axons of granular cells, whose displaced somas scatter sparsely through this layer, making it poorly defined in places. Still, the overriding feature of this layer is densely packed second-order axons. Most centrally placed is the internal granular cell layer, where most granular cell somas are located. These are round or oval and are grouped in broken strands, often only one cell thick, arranged concentrically around the ventricle. Synapses on mitral cells appear to form part of an increasingly elaborate intrabulbar association system whose development accompanies the increased anatomical specificity noted earlier. From amphibians to reptiles to mammals, as their number of primary apical dendrites decreases, mitral cells receive a progressively more restricted input from the epithelium, and a reduced opportunity for spatial summation. The complex system of recurrent collaterals involving feedback loops from mitral and granular cells could serve to decrease olfactory threshold while still increasing anatomical specificity. One manifestation of this system is that a mitral cell's spontaneous activity is considerably greater even as its input becomes more restricted in mammals (Allison, 1953). Turtles represent an intermediate stage in this trend.

The accessory olfactory bulb, which occupies a dorsal position, is quite well developed and very similar to the main bulb (Goldby, 1934). In several cases it has cell layers continuous with, and not easily distinguishable from, those of the main bulb.

From the olfactory bulb two pathways project to the telencephalon. One is the well-defined lateral olfactory tract composed of large, heavily myelinated axons that originate in the medially placed mitral cells. The other is a diffuse system collectively termed the medial olfactory tract. This includes axons from more lateral mitral cells plus those from deep granular cells. The plexus of mitral cell axons bifurcates at the posterior end of the bulb into lateral and medial tracts. The medial tract has fewer, less dense, lightly myelinated or unmyelinated fibers, some of which synapse with ipsilateral deep telencephalic structures while others cross through the anterior commissure to form a contralateral diffuse projection system.

Central Connections

Second-Order Neurons. The cell masses of the olfactory bulbs grade gradually into telencephalic structures. Moving posteriorly, the mitral cell layer thins out and disappears in the ventral half of the bulb and is replaced by scattered cells. Concurrently, the granular cells disintegrate, and undifferentiated grey of the anterior olfactory nucleus is substituted for it. At this point the only distinguishing feature is a mass of densely packed neurons, including some pyramidal cells, on the ventrolateral aspect that forms the lateral part of the nucleus. Farther posterior, behind the accessory bulb, the anterior olfactory nucleus encircles the ventricle completely while the dense lateral part becomes continuous with piriform cortex. The lateral anterior olfactory nucleus is sometimes simply called an anterior projection of piriform cortex.

The dorsal region of the nucleus becomes more distinct, though never well differentiated, and transforms into general cortex. The amorphous medial anterior nucleus grades into anterior hippocampus dorsally and into the septal region ventrally, both near the cephalic tip of the olfactory tubercle. Similarly the undifferentiated ventral mass is continuous with the rostral tip of paleostriatum dorsally and with the olfactory tubercle ventrally. The gradations from olfactory bulb through the anterior nucleus and into telencephalon vary slightly among species. The description offered here is for the freshwater species *Emys orbicularis*.

OLFACTION

The differences noted in other turtles principally involve the anteroposterior extent of each area and the defined level at which a transition is made. As a general rule, the size of the anterior olfactory nucleus is directly proportional to the amount of forward specialization of the cortex. Both these factors indicate a large anterior nucleus in turtles.

Turtle cortex ranges from 200 to 500 μm in thickness and includes three main areas of olfactory input. Medially, a small area of dorsal hippocampus responds to electrical stimulation of the olfactory nerves; however most of hippocampus, as well as septum, gives no response (Orrego, 1961a). This is somewhat enigmatic, since these areas are commonly believed to play a major olfactory role in lower vertebrates. Hippocampal responses are apparently controlled through the medial olfactory tract whose fibers course along its surface before running toward (and probably through) the hippocampal commissure. Dorsally, general cortex, particularly the posterior sections, shows clear evoked responses to olfactory nerve stimulation. This activity is mediated primarily through lateral olfactory tract axons. Lower amplitude activity can be recorded over much of general cortex, though at the very posterior pole there is no response. Laterally, piriform cortex receives the bulk of olfactory information by way of the lateral olfactory tract (Figure 1). This is an ample region, clearly separated from the rest of the forebrain. Its large and multipolar neurons are at some depth below the superficial olfactory tract fibers, and dendrites extend apically to receive their input. Less significant contributions from the lateral olfactory tract go to the anterior olfactory nucleus, the nucleus of the olfactory tract, the olfactory tubercle, paleostriatum (nucleus of the stria terminalis), and parts of the amygdala. The well-differentiated amygdala consists of a large central mass and a lateral region, both of which probably receive olfactory fibers, mainly from the accessory bulb.

Tertiary Connections

Third-order olfactory pathways are not well defined. However two general systems seem to emerge. One is the olfactovisceral system through which olfactory information influences motivation and arousal levels. It includes the hippocampus and its fiber connections to the hypothalamus. Furthermore, Orrego (1961a) contends that spatial separation begins even in the mucosa, where olfactovisceral fibers form a ventral bundle. The second system is termed olfactosomatic and involves piriform and, secondarily, general cortex. Projections from these somatosensory areas activate somatic motor nuclei of the mesencephalon by way of the habenulopenduncular tracts. This system, Orrego believes, originates in

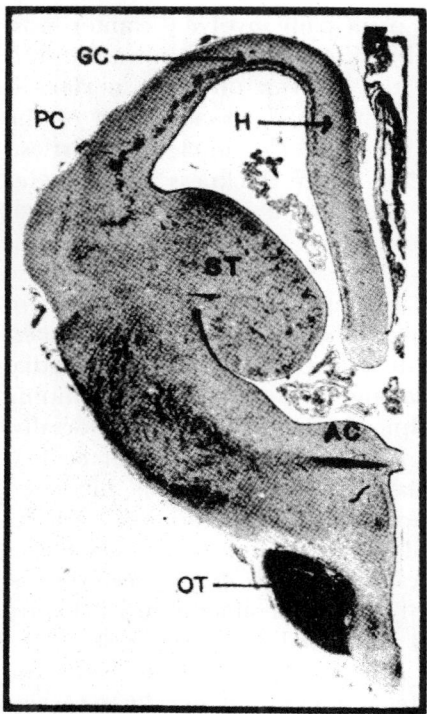

Figure 1 Transverse section of cerebral hemisphere. H, hippocampus; GC, general cortex; PC, pyrifrom cortex; OT, olfactory tract; AC, anterior commissure; ST, striatum. (From Orrego, 1961a.)

a dorsal bundle from the mucosa. Tertiary connections must also include commissural systems through which olfactory fibers can project to contralateral bulb and cortex.

The Contralateral and Efferent Systems

Stimulation of the olfactory bulb evokes a characteristic response from the contralateral bulb. The response is of considerable amplitude, long latency (130 msec), and long duration (100 msec), and seems to represent the activity of contralateral granular cells. The cells that generate the crossing fibers are at the external margin of the internal granular layer, cells that also send axons into the olfactory tracts. In addition, bulbar stimulation evokes activity from the contralateral piriform cortex, apparently by way of the contralateral bulb. At the same time normal olfactory transmission through the contralateral bulb is blocked for at least 300 msec. Piriform cortical responses to contralateral olfactory tract stimulation are not diminished, suggesting that the point of inhibition is in the bulb. Granular cell activity probably stimulates contralateral granular cells, which, in turn, relay impulses to the contralateral cortex while simultaneously inhibiting mitral cells through recurrent collaterals. Crossing fibers project through the anterior commissure.

Interhemispheric connections are profuse, running through the anterior, habenular, and hippocampal commissures.

Efferent connections are not well established, though piriform cortex stimulation evokes a response from both bulbs as well as contralateral cortex and results in contralateral blockage for several hundred milliseconds (Orrego, 1962). Orrego suggested that efferent fibers project from this area of the cortex to each bulb, specifically to contralateral granular neurons that inhibit mitral cells while activating the contralateral piriform cortex. Apparently precautions were not taken by Orrego against the possibility that antidromic stimulation of ipsilateral bulb could cause these events through the afferent connections mentioned earlier.

Receptor Physiology

The nasal cavity in turtles is served by three sensory nerves: olfactory, vomeronasal, and trigeminal. Since all three show chemical sensitivity (although the trigeminal nerve also responds to tactile and temperature stimulation), the entire nasal mucosa, not just olfactory epithelium, should be considered to be a chemoreceptive organ. Nasal capacity, from nares to choana, in *Gopherus polyphemus* is about 0.15 cc. With simple nasal geometry, the estimated mucosal surface area approximates 1 cm^2; the mucosa is about 1 mm thick.

At first glance, the olfactory epithelium appears to be uncomplicated. Olfactory receptors and nerve fibers are part of the same cell, and this also holds true for vomeronasal and trigeminal receptors. Antidromic stimulation shows that a small twig of the olfactory nerve (12–15 μm containing several hundred fine axons) innervates a slightly elongated area of less than 1 mm^2 on the mucosa. However this structural simplicity stands in contrast to the complex electrical potentials observable in various areas, recordings so difficult to interpret that even the origin of the generator potential has yet to be agreed upon.

Basically, evoked neural activity consists of high frequency multiunit or single unit activity (best recorded from the nerve fibers) and slow potentials (from the mucosa).

Slow Potentials. Two types of slow potential are definable. One, first defined in frog mucosa, is evoked only from sensory epithelium and is

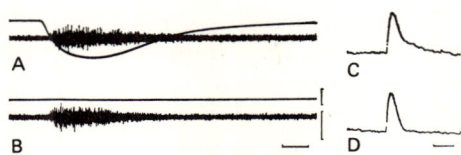

Figure 2 Negative slow potential and nerve twig response. A, produced by a small puff of amylacetate near saturation. B, recorded after absorbent paper put on epithelium for 4 min was removed. Calibration, above, 1 mv; below 40 μv. Time, 0.5 sec. C, D, integrated activity in A, B. Time, 2 sec. (From Shibuya, 1964.)

called the electro-olfactogram (EOG) (Ottoson, 1956). The other is a distinct evoked DC shift in the recording baseline.

Electro-olfactogram. In turtles the EOG was first confirmed in *Gopherus polyphemus* by Shibuya and Shibuya (1963). Upon removing half of the mucous covering of the epithelium, Shibuya (1964) reported that the EOG decreased severely or disappeared, but evoked activity from the olfactory nerve was undiminished (Figure 2). Subsequent research (e.g., Moulton and Biedler, 1967; Takagi and Yajima, 1964, 1965; Tucker and Shibuya, 1965; Nakajima, 1964; Lowenstein, Terzuolo, and Washizu, 1963; Ottoson and Shepard, 1967) has focused on techniques of stimulation and recording and on explanatory aspects of the EOG. The EOG is a slow, monophasic potential evoked from the yellow (receptive) parts of the mucosa by odorized air.

DC Shift. DC shifts are recorded near the mucosal surface and are evoked by chemical and electrical stimulation. Deep within the mucosa,

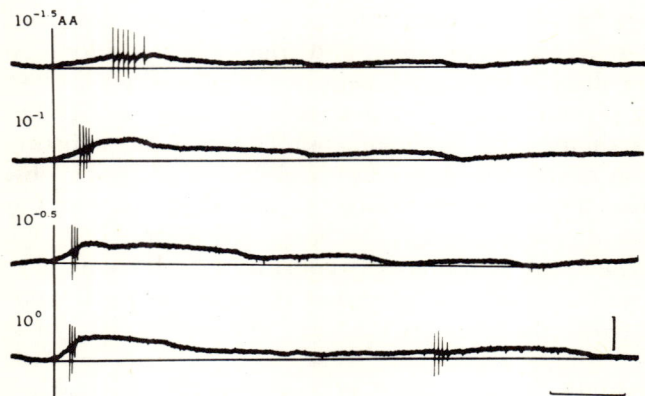

Figure 3 Olfactory cell activity to amylacetate. Vertical line on left indicates beginning of DC shifts. Time, 1 sec. Calibration, 1 mv. (From Shibuya, 1969.)

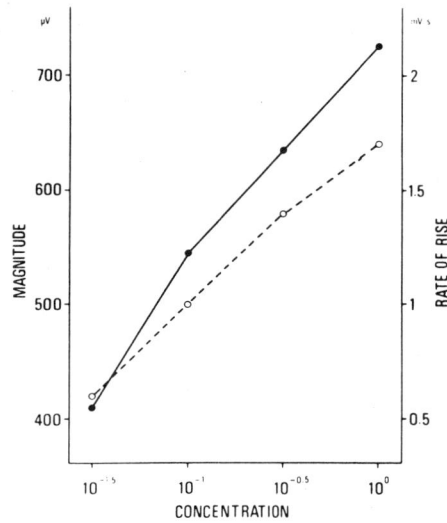

Figure 4 Magnitudes and rates of rise of the DC positive shift in response to amylacetate. (From Shibuya, 1969.)

below the inner limiting membrane, this potential disappears. The amplitude and duration of the shift increase monotonically with stimulus concentration (Figures 3 and 4). Relative to the EOG, the DC shift has a long latency, but this difference diminishes with increasing stimulus intensities (Figure 5).

This is one reason for the suggestion of Shibuya (1969) that DC shifts originate in the receptor cell and represent the true olfactory generator potential.

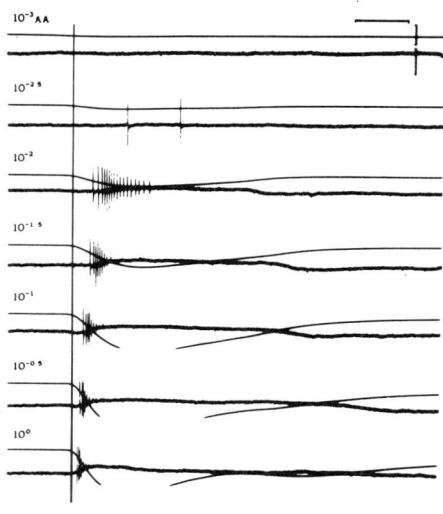

Figure 5 Simultaneously recorded EOGs and olfactory cell activities to amylacetate. Vertical line on left indicates beginnings of EOG. Time, 1 sec. Calibration, 2 mv, 1 mv from above. (From Shibuya, 1969.)

Action Potentials. Until recently most high frequency recordings from the olfactory system of turtles have involved nerve twigs. However in the past decade individual receptor neurons have been isolated, using extremely fine micropipettes. Extracellular action potentials can be recorded as superficially as 200 to 250 μm beneath the mucosal surface. In *Gopherus polyphemus* they are of 2 mV amplitude, mono- or diphasic, and of 3 to 4 msec duration. Compared to the EOG they are initially positive. Successive spikes in a train decrease in amplitude (particularly with high stimulus concentrations) are are easily disrupted by repeated rapid stimulation. Shibuya (1969) recorded a negative potential of 10 to 25 mV upon penetrating a receptor, but was not convinced that this represented a true resting potential value.

Mathews and Tucker (1966) reported on the evoked responses of 40 single units to the application of 30 qualitatively different stimuli in *G. polyphemus*. They found a very low mean spontaneous rate (0–1 spikes/min), which precluded the possibility of inhibitory responses. Maximum firing rates were near 10 spikes/sec. Using low to moderate concentrations, typically 10^{-3} amylacetate,[2] they found some degree of specificity, each neuron being sensitive to from 1 to 14 of the 30 stimuli, with no discernible discharge patterns characteristic of a particular stimulus. Mathews (1972) reported essentially similar results.

In contrast, Shibuya and Shibuya (1963), working with the same species, reported no specificity among single receptors. Using six dissimilar odorants, they found that all units showed only a slight differentiation to this stimulus range. They also reported the common presence of a significant spontaneous rate, from which certain stimuli caused a decrease in firing. Typically, one odorous puff (0.5 ml of 10^{-1} amylacetate) evoked 4 to 15 spikes in a rapid train, successive potentials having severely depressed amplitudes. The number of discharges was roughly proportional to the logarithm of stimulus concentration. With longer periods of stimulation (30 sec) there were clear phasic and tonic components of the response.[3] Barring anesthetic effects,[4] the most obvious source of discrepancy in the results of these two studies is stimulus concentration.

Primary Nerves. Olfactory epithelium is served by the olfactory nerve. The vomeronasal nerve innervates vomeronasal mucosa as well as a

---
[2] $10°$ Amylacetate indicates totally saturated air at 20°C. Thus 10^{-3} is 0.1% saturation.
[3] The phasic components are believed to be the more significant, since olfactory receptors are stimulated intermittently during normal respiration.
[4] Shibuya and Shibuya used ethyl urethane of unspecified concentration. Although Mathews and Tucker do not identify their anesthetic, that laboratory typically employs ethyl urethane.

patch of mucosa in the respiratory tract anterior to olfactory epithelium. These receptors are immediately exposed to air entering the nasal cavity, and could account for the unusually large and rapid vomeronasal responses in turtles. Finally, the ophthalmic and maxillary divisions of the trigeminal nerve terminate in nasal mucosa, probably in free nerve endings. These three nerves do not simply serve redundant functions. Rather, the vomeronasal nerve, which on the whole responds to chemical stimulation as vigorously as does the olfactory nerve, detects stimuli to which the olfactory nerve is rather insensitive. For example, vomeronasal responses exceed those of the olfactory nerve for lower chemicals in a homologous series (methanol, ethanol). The reverse is true for higher members—in this case, from propanol on. Similarly the crossover point in the fatty acid series occurs between isovaleric and n-valeric acids. Since the level of autonomic activity and tongue movements greatly affect the accessibility of its receptors, the vomeronasal nerve is more dependent on these factors than is the olfactory.[5] Another complementary function may exist for the vomeronasal system; it may be especially tied in with emotional and motivational conditions (Moulton and Beidler, 1967).

Trigeminal receptors serve olfactory and pain modalities (Moulton and Beidler, 1967). It appears that the receptors are simply free nerve endings, but whether the same terminal serves both functions is not known. However even a supposedly "pure" olfactory stimulus such as phenyl ethyl alcohol activates fibers in the trigeminal nerve at lower concentrations than are required to stimulate olfactory axons (Tucker, 1963a).

All three nerves display similar effects when certain stimulus parameters are varied. Olfactory responses are little affected by humidity (0–88%), temperature (20–30°C), pH, ionic strength, osmotic pressure, and the nature of the carrier gas (argon, nitrogen, oxygen). The important parameters are quality and concentration of the odorant, its flow rate, and its accessibility to receptive sites (Figure 6). Simply, of what type is the molecule and how readily does it encounter a receptive membrane?

Tucker (1963b) has compared multiunit responses from olfactory, vomeronasal, and trigeminal nerve twigs in *Gopherus polyphemus*. Although all showed broad sensitivity to chemical stimulation, olfactory nerve fibers possessed the greatest range (Figure 7). As stimulus concentration increased, nerve twig potentials grew in ogive fashion, probably reflecting the recruitment of less sensitive fibers (Boeckh, 1969). The low

---

[5]Nasal accessibility is partially determined by reflexes mediated through the trigeminal nerve. By stimulating the cervical sympathetic nerve, the size of the opening to the vestibular chamber and the central olfactory cavity is greatly reduced.

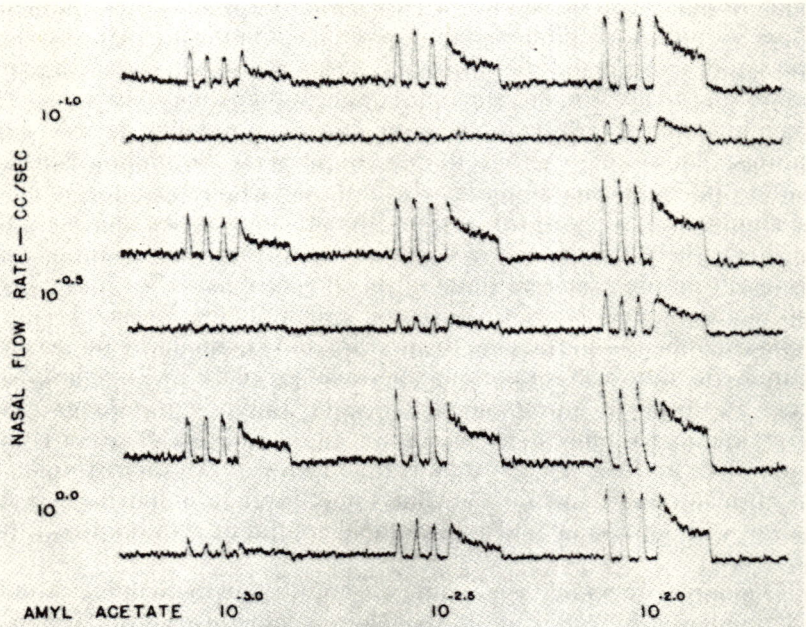

Figure 6 Integrated olfactory responses recorded simultaneously from 2 nerve twigs. The unit of concentration is saturation at 20°C. Odor was on 5 sec of every 20 sec followed by 1 min on. (From Tucker, 1963a.)

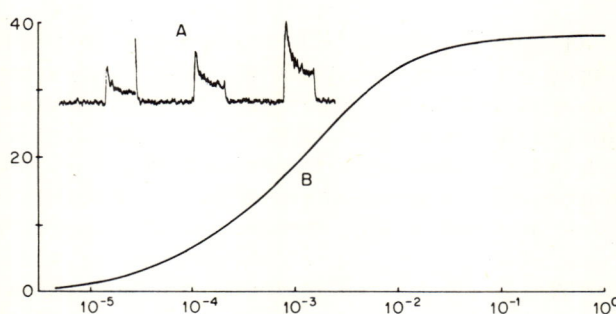

Figure 7 Olfactory nerve twig activity to amylacetate. Concentration unity is saturated air at 20°C. A, time course of integrated activity to stimuli 10^{-3}, $10^{-8/3}$, and $10^{-7/3}$ of unity. B, integrated activity (arbitrary units) in relation to stimulus concentration at receptors. (From Boeckh, 1969.)

conduction velocities (0.15 m/sec) that characterize olfactory axons also pertain to vomeronasal and trigeminal fibers. In the trigeminal nerve, the fibers that are sensitive to chemical stimulation are among the very slowest conducting. Evoked responses in olfactory and vomeronasal nerves typically consisted of a large, reproducible phasic component followed by a more variable tonic response. The amplitude of the phasic activity increased monotonically according to the logarithm of stimulus concentration. In contrast, the tonic component became larger up to a maximum, then decreased in amplitude as the stimulus became more intense. At very high concentrations, the tonic response occasionally even fell beneath the summated spontaneous activity levels, resulting in an excitatory discharge upon stimulus termination. These "on-off" responses have been described in other subjects as well (Takagi and Shibuya, 1959). Reflecting on the source of the generator potential, the parallels between this stimulus-response relationship, and that described for the EOG are impressive.

As a rule, trigeminal sensitivities and response characteristics more closely approximated those of the vomeronasal than of the olfactory nerve. The vomeronasal nerve, however, displayed significant phasic potentials and tended to develop "off" responses more readily than did the olfactory nerve. Trigeminal recordings showed a slow increase to a tonic level commensurate with stimulus intensity and a slow decrease to spontaneous levels upon termination.

Electrical Activity in the Olfactory Bulb

The olfactory bulb, with its laminar construction, and with projecting apical and basal dendrites, is reminiscent of mammalian cortical organization, evoking suggestions that the olfactory bulb represents a primitive form of cortex (Orrego, 1961b).

Bulbar Response Characteristics. Synaptic conduction through the glomeruli is a slow process. Responses in piriform cortex, evoked by pre- and postglomerular stimulation, show latency differences of 8 to 10 msec (25 vs. 15–17 msec) (Orrego, 1961b), a delay in keeping with the great amount of spatial summation required for synaptic transmission here. Gross recordings from the bulb show a characteristic, long-lasting response evoked by olfactory nerve stimulation (Figure 8). This consists of a sharp diphasic potential of only a few milliseconds latency succeeded by a low amplitude, diffuse surface negative wave, then by a much higher voltage wave of 50 to 60 msec duration. Occasionally a fourth peak is observed; it is rather widespread and of variable amplitude. The sharp, short latency peak can best be recorded from anterior bulb where

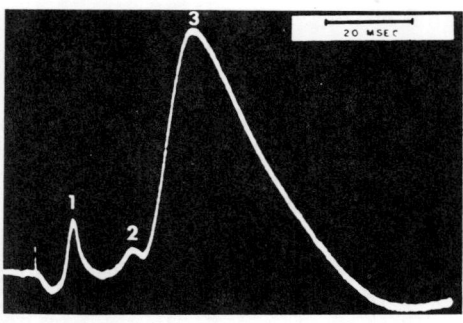

Figure 8 Electrical stimulation of the olfactory nerve. Monopolar recording bulb surface to indifferent electrode in the ventricle. Upward deflection indicates surface negativity. (From Orrego, 1961b.)

primary axons are still dense. Its latency increases in direct proportion to the distance of stimulation, which suggests that this potential represents integrated spike activity from the ascending nerve axons. This notion is reinforced by the presence of di- or triphasic wave forms, which are characteristic of axonal transmission. This peak lasts a full 10 msec, reflecting the very slow conduction speed in these thin fibers.

The second wave also appears to be of presynaptic origin. Conditioning shocks do not block it, as they do later components of the bulbar evoked response, and activation of intrabulbar inhibitory circuits have no effect. The most likely origins of this small potential are the interglomerular presynaptic fibers, though the tips of mitral cell apical dendrites could make some contribution.

The high amplitude, long duration potential that follows reflects the activity in dendrites and somas of mitral cells. This is the primary component of the evoked response, and it can amount to several millivolts extracellularly. It is subject to blockage by the conditioning shocks and inhibitory circuits described below.

When a fourth peak is present, it probably originates in granular cells. These neurons are responsible for most of the bulb's spontaneous potentials and are the source of inhibitory axons that tonically affect mitral cell activity. The excitability cycle of the bulb following olfactory nerve stimulation reflects this inhibitory influence. If stimulation is repeated within 10 msec, temporal summation occurs. But this brief period is followed by a prolonged depression of the third and fourth components of the evoked response, which results in complete refractoriness for 500 msec and a relative refractory state for a full 15 sec (Orrego, 1961b). Moreover, excitability cycles recorded from piriform cortex mimic bulbar patterns almost precisely. The actual mechanism of inhibition is presumed to be similar to that in mammals (Cajal, 1955). Recurrent collaterals branch from the main axon of a mitral cell in the olfactory tract and form excitatory synapses with granular cells. These in turn send axons back to the basal dendrites of mitral cells (not the neuron that originates this loop), effecting recurrent inhibition. However even in a system of small fibers and slow conduction, the period of inhibition is such that some more persistent mechanism could be functioning.

Poststimulus depression is also observed in the glomeruli. Olfactory nerve stimulation elicits activity, then prolonged inhibition from postsynaptic dendrites. Even subthreshold preglomerular stimulation, evoking no spikes in the bulb, inhibits an ordinarily suprathreshold stimulus subsequently applied to the olfactory nerve. It seems that even small depolarizations result in prolonged IPSP's in mitral cell apical dendrites.

Thus inhibition can occur at the glomeruli and at the basal dendrites of mitral cells. The former type apparently results from an extended refractory period, while the latter is attributable to a specialized inhibitory circuit. In concert these influences account for the strong and prolonged bulbar depression following olfactory nerve stimulation (as determined by piriform cortex recordings). Acting alone, the recurrent inhibition from granular cells results in considerably less depression when a postglomerular stimulus is applied. Finally, repetitive olfactory tract stimulation, which bypasses both mechanisms, evokes incremental responses from cortex (Orrego, 1961b).

Activation of granular cells can result from sources other than olfactory tract collaterals. Interbulbar fibers allow communication between granular cells on each side, so that stimulation of one bulb results in blockage of mitral cell conduction in the contralateral bulb. Granular cell stimulation causes an inhibition of approximately 130 msec in contralateral granular cells, followed by prolonged facilitation. Accordingly, low frequency stimulation evokes an incremental response (summing excitations), whereas high frequency stimulation (greater than 10 per second) results in a decremental response (summing inhibitions). Long-term granular facilitation would put a tonic inhibitory influence on mitral activity. Thus the granular cells seem to be a critical determinant of olfactory responses. Whether they are activated by olfactory tract collaterals, by contralateral bulbar stimulation, or by efferent fibers emanating from contralateral piriform cortex, they can block the normal olfactory transmission route. Support for this schema is derived from mammalian studies (Adrian, 1950; Kerr and Hagbarth, 1955). On a gross level, recurrent circuits inhibit responses in neighboring mitral cells, possibly "sharpening" afferent signals.

Single mitral cell activity from the olfactory bulbs of *G. polyphemus* has been studied by Mathews (1972). The major differences between olfactory receptor cells and these second-order neurons are response rates and response types. With regard to the former, mitral cells discharge spontaneously approximately once per second, compared with 0.0 to 0.05 spike/sec from epithelial neurons. Peak activity among mitral cells is 39 spikes/sec, compared with 16 spikes/sec for epithelial cells. Receptor cells give purely facilitatory responses, whereas bulbar neurons respond with inhibition almost as often as with excitation. The introduction of inhibitory responses at the bulbar level has been reported in a wide variety of experimental subjects and probably plays a critical role in sensory coding.

Cortical Olfactory Responses

The functional characteristics of cortical olfactory activity in turtles has never been systematically determined. Most electrical recordings have been designed to delineate anatomical connections, and several findings have already been mentioned. Laterally, the piriform cortex is the primary projection area of olfactory input. From here fibers project to the contralateral piriform cortex, the contralateral bulb, and mesencephalic motor nuclei (an "olfactosomatic" system). Dorsally general cortex also receives major projections and influences the same motor systems. Medially, hippocampus receives rather tenuous projections and influences the hypothalamus as part of an "olfactovisceral" system. Each cortical area sends connections to the homologous contralateral structure. But concerning the breadth of sensitivity of these neurons, their characteristic responses (excitatory, inhibitory, or some combination), levels of spontaneous activity, temporal sequences, and sensory encoding mechanisms, there is no information.

TASTE

Turtles are studied more for being eaten than for eating. There are considerable anecdotal observations and some experimental work on turtles' eating behavior (see Chapter 12). Concerning gustatory anatomy, however, little is known, and about physiology and taste coding, the literature cites nothing.

The gross morphology of the mouth and tongue, at least, have been described. The dorsum of the tongue is marked with long, narrow papillae that terminate in pointed or rounded extremities. The lateral margin of the tongue also possesses papillae, but these are flattened and considerably smaller than those of the dorsum (Bradley, 1971). The presence of taste buds in these papillae is variable. As testimony to the paucity of recent experimentation, Tuckerman's (1892) analysis of the chelonian tongue remains the most widely quoted. He reported the presence of some buds in the tongue of tortoises, but a complete lack in snapping turtles. Within the decade Tucker (personal communication) performed serial sections through the tongue of gopher tortoise and discovered no taste buds. In view of the tortoise's eating habits, however, Tucker remains convinced that it can make gustatory discriminations. Typically a gopher tortoise places its nose near its prospective meal. If the object passes this test, the tortoise next presses the tongue against it. Only then will the animal make a decision to eat.

Other studies suggest that turtles can make gross gustatory discriminations (Burghardt, 1967; Honigman, 1921). Boycott and Guillery (1962) trained *Pseudemys scripta elegans* to eat meat according to olfactory and visual cues, and perhaps to gustatory cues.

In sum, olfaction occupies broad areas of the chelonian nervous system and is represented by numerous sensitive receptors. The inconsequential effects of gustation on behavior, and the lack of well-defined receptors and prominent central sites for taste all indicate that turtles are more concerned with chemical stimuli that emanate from a distance.

ACKNOWLEDGMENT

The author had the support of research grant NS 10405 from the National Institutes of Health during the preparation of this manuscript.

CHAPTER 15
Olfaction and Behavior

MARION L. MANTON

There are many anecdotal accounts of supposed olfactory behavior and, as the preceding chapter shows, turtles are well equipped anatomically for such behavior. It is unlikely that so well-developed a sensory system has no natural function, and experiments to determine the extent of the role of chemoreception in the lives of turtles are sorely needed.

Chemoreception is the most primitive of all the sensory systems and remains the major sense for most invertebrates. It has been replaced by vision as the dominant sense of tetrapods, including turtles. There is no concrete evidence that olfaction is critical in any of the life functions of turtles. On the other hand, sniffing appears to be a standard part of most turtles' behavioral repertoires, and it is normal practice in observing animal behavior to assume that sniffing means that odors are being examined.

The sense of smell is also implied in frequent reports that freshwater turtles are able to locate carrion for food and can be quite competent bait stealers from fishermen even in muddy water. Schmidt and Inger (1957) tell the gruesome story of an elderly man who used a tethered snapping turtle to recover the bodies of people who had drowned. According to lore, turtles are more likely to be attracted and caught when bait begins to decay in traps (Ernst and Barbour, 1972), but data indicate that fresh bait is actually responded to more rapidly by most species (see Chapter 2).

The scant research on utilization of olfactory cues in feeding is reviewed by Burghardt (1970). Definitive studies on the isolation of olfactory, taste, and visual cues have not been carried out.

ODOR PRODUCTION

Though it is not scientifically established that all turtles can smell, there is no doubt that some turtles themselves can be smelly. Musk turtles exude a characteristic odor when distressed, and *Sternotherus odoratus*, otherwise known as stinkpot, has been appropriately named. Freshly caught specimens, when handled, produce a yellowish, volatile liquid with a strong musky odor from two pairs of glands located beneath the anterior and posterior edges of the plastral bridges. One pair opens through axillary pores, the other through inguinal pores. The function of the odor as an alarm signal or for other intraspecific communication has not been thoroughly investigated. Since the liquid is released when the turtle is disturbed, whether alone or with other animals, it may be a protective device to repel predators. Other *Sternotherus* species have, to humans at least, less pungent odors. Other aquatic species, including *Chelydra serpentina*, may also emit a potent musky odor.

Box turtles possess musk glands, and Legler (1960a) reported that adult *Terrapene ornata* may produce odors in very stressful situations. Neil (1948) found that newly hatched box turtles, when disturbed, were able to produce a powerful stench similar to that of *S. odoratus*.

Ehrenfeld and Ehrenfeld (1973) found that the morphology and histochemistry of the paired axillary and inguinal glands are closely similar in both *S. odoratus* and the green turtle (*Chelonia mydas*) despite the wide gap taxonomically. The whitish substance from the glands of the green turtle has only an extremely faint odor to humans, even though the secretion in both species contains a PAS-positive, protein-rich, nonacidic substance resembling a glycoprotein. The difference appears to be related to additional cells in *S. odoratus* containing many small lipid droplets. Similar glands occur in all families of turtles except the Testudinidae (Loveridge and Williams, 1957). These large exocrine glands, each consisting of several lobules surrounded by a thick capsule of striated muscle, have a structure suited to the rapid expulsion of their products. Further study is needed to determine the nature and role of the glandular secretions with respect to the natural history of each turtle group.

OLFACTORY BEHAVIOR

Mahmoud (1967) reported that during the initial stages of courtship in *S. odoratus* and three other species of kinosternids, the male appears to determine sex by approaching a turtle from behind and, with head extended, feeling or smelling the tail of the other animal. If the encounter is with another male, courtship usually ceases, but if a female has been found, courtship continues, with the male moving to the female's side to gently nose at the bridge region in the area of the musk glands.

The courtship of a pair of captive *Pseudemys concinna suwanniensis* was observed by Jackson and Davis (1972). It commenced with the aroused male swimming behind the female with his nose very close to the rear portion of her tail. They suggest the possibility of a chemical releaser from the female's cloaca. Marchand (1944) reported such trailing in the field.

The entirely terrestrial tortoises do not have axillary or inguinal glands, but all four species of the genus *Gopherus* have well-developed integumentary glands beneath the chin in a slightly medial position on each mandibular ramus. They are larger in the males than in the females and are inactive except during the breeding season, when they are often swollen. Electrophoretic and chromatographic analyses (Rose, Drotman, and Weaver, 1969; Rose, 1970) of the aromatic gland secretions show that they contain lipids, proteins, and an esterase. The protein concentration of the female secretion is much lower than that of the male. Rose's behavioral observations suggest that their role is that of an olfactory cue during courtship or aggression.

Auffenberg (1969c) observed that during the courtship of the gopher tortoise (*G. polyphemus*), while the male is occupied with preliminary head bobbing, the female may rub her chin across her forelimbs. Weaver (1970) reported the same behavior in the male. The action probably transfers scent from the chin glands to the front legs. Both sexes possess an enlarged scale near each elbow. When the male changes from head bobbing to biting, the next courtship stage, he often bites the forelegs as well as other regions about the head for quite some time before mounting occurs. Auffenberg (1966b) speculated that the male's head bobbing may serve not only as a visual stimulus but also to spread the scent from his chin glands to act as a chemical stimulus to the female. If this were established, a true reproductive pheromone could be said to exist for at least the gopher tortoises. Woodbury and Hardy (1948) and Weaver (1970), in describing the courtship of the desert tortoise (*G. agassizi*) and the Texas tortoise (*G. berlandieri*), respectively, reported similar head bobbing, followed by biting, once again by the male only.

Further support for the theory that olfaction is used by these tortoises comes from Weaver (1970), who observed controlled confrontations between captive male tortoises. After much head bobbing, the males make contact and sniff each other's head and feet, making lateral wiping motions on the surface of the forelegs for 1 or 2 min. Head bobbing occurs in all *Gopherus* species when they are sniffing objects or food or when two or more tortoises meet. Ernst and Barbour (1972) mentioned that *G. agassizi* interact socially by nodding rapidly at each other and sometimes touching noses before they part. Other circumstantial evidence that they can smell includes the observation by Nichols (1957) that one captive adult tortoise sometimes urinated inside the quarters, after which other tortoises would not sleep there until the area had been completely washed and aired.

Weaver (1970) has made one of the few controlled investigations of olfaction in turtles. After field observations of their courtship and combat, he designed a scent chamber to aid in identifying the cues used by *G. berlandieri* to distinguish sex. Males and females were tested for their responses to cloacal material for each sex as well as to control material. The findings indicated that the males are unable to distinguish between male and female cloacal scent, but females respond more to the cloacal scent of males.

Weaver suggested that females could use this ability to aid in sex recognition, and males might instead use the chemical differences between male and female chin gland secretions for the same purpose. Males respond to females when the chin glands are active in both sexes but only to males when the chin glands are inactive. It would be interesting to repeat this experiment using chin gland extracts. In fact, Weaver's simple apparatus lends itself to a wide range of olfactory discrimination studies.

Weaver sees olfaction as only one process in a series of steps involved in sexual discrimination and suggests that head bobbing in *Gopherus* species may be evolving to serve as a visual signal. Head movements for olfaction alone have a straight, simple forward motion, whereas those during courtship and combat have a strong vertical component, with the neck stretched out and the head held high. Eglis (1962) classified olfactory sniffing movements in tortoises and found them to be of taxonomic value.

The work of Auffenberg (1965) with the partially sympatric South American tortoises, *Geochelone carbonaria* and *G. denticulata*, demonstrates that their head bobbing acts as a species-specific visual signal and that the same type of head movement occurs during olfaction. His observations include the following. During the breeding season males challenge all tortoiselike objects with head bobbing. If identical head bobbing is the response, the tortoise is identified as a conspecific male (visual signal alone). If there is no such response, the male walks to the rear of the other tortoise to sniff the cloacal region, using a similar head movement (visual plus olfactory signal). A conspecific, sexually active female is mounted immediately. Auffenberg believes the basic motor pattern of the olfactory movement gave rise to its use as a visual signal to aid in species and sex recognition.

Head bobbing may be associated with olfaction in tortoises, but throat pumping seems to be characteristic of turtle olfaction under water. Freshwater and marine turtles can often be seen squirting water out their nostrils, with accompanying pulsating movement of the throat. Cahn (1937), while studying pharyngeal respiration in the soft-shelled turtle *Trionyx muticus*, placed grains of carmine in the water close to the nostrils and noted that the carmine flowed into the nostrils as the hyoid apparatus dropped. Root (1949) found that buccopharyngeal respiration is not important in submerged *S. odoratus* and suggested that the frequent underwater throat movements are related to olfaction. The marine loggerhead turtle (*Caretta caretta*) has been observed underwater with its nostrils open and the floor of the mouth moving up and down (Walker, 1959). Similar behavior can be observed in active green turtles, suggesting a mechanism for moving water across chemoreceptors in the mouth or nasal cavities for chemical sampling. When the turtle is resting there are no such movements, and the nostrils are closed by distension of areas of highly vascular connective tissue just within each nostril (Walker, 1959). The main adaptive modification to keep air in and water out of the lungs as the turtles take in and swallow food and pump water through the nostrils is the reflexive closure of the glottis at all times except during expiration and inspiration (McCutcheon, 1943). Throat movements do not occur during breathing periods. McCutcheon recorded their frequency during the presentation of food extracts to the diamondback terrapin, *Malaclemmys terrapin centrata*, and found a significant increase. He suggested that olfaction is the main function, enabling air to be pumped in and out of the nasal cavity instead of relying on passive diffusion of odor molecules.

ODOR DISCRIMINATION

Most aquatic air-breathing vertebrates close their nostrils when submerged and are believed to be unable to smell underwater, although they remain untested in this respect. It is known that some aquatic mammals show extreme reduction of the olfactory organs (Evans and Bastian, 1969). The olfactory anatomy of aquatic turtles, however, closely resembles that of terrestrial turtles. Tucker and Shibuya (1965) recorded from the primary olfactory nerve of *Terrapene carolina*, first with the odors in gaseous phase, then with the same odors in an aqueous phase. The recordings made with the turtle "underwater" are closely similar to those made with the odorants delivered in air. The distinction between a mechanism for smelling underwater rather than, or in addition to, smelling in air is important. The ability to detect an odorous substance dissolved in water is a prerequisite sensory capability if, as has been suggested, some turtles use waterborne chemical cues for orientation in homing or navigation (Koch, Carr, and Ehrenfeld, 1969; Ernst, 1970b).

One of the few behavioral experiments is that of Boycott and Guillery (1962), who used a freshwater species to test olfactory learning. [They and Burghardt (1970) cite an earlier Russian study by Poliakov (1930) claiming olfactory conditioning.] Red-eared turtles, *Pseudemys scripta elegans*, were trained to discriminate between water containing food alone and water containing food plus an olfactory stimulus (amylacetate, vanillin, or eucalyptus oil). The experimental tank was equipped so that a 2 sec electric shock could be delivered, sufficient to cause head retraction. The training consisted of allowing the turtle to eat the food when presented alone, or delivering an electric shock if the turtle attempted to eat in the presence of the olfactory stimulus. The chemical, when present, was added centrally and allowed to diffuse throughout the tank before the turtle was placed in it. Because turtles are air breathing, each trial was made long enough (1 min) to permit sampling both of the water and of the air above it in the tank. Thus the procedure did not differentiate between chemoreception in water versus air. The turtles that learned the discrimination showed a gradual improvement in performance over the 6 week learning period.

The olfactory fibers were sectioned at the point where they enter the olfactory bulb, to control for the possibility that the discrimination might be dependent on taste or general chemical sense. Following postoperative recovery there was a marked initial decrease in performance, implying an olfactory basis for the acquired discrimination. However about 16 weeks after surgery the performance was almost as good as it had been in the preoperative period. Subsequent sacrifice and histological examination showed that some olfactory fibers had survived in all lesions and therefore could have been responsible for the recovery of the discrimination. The greater the lesion, the less the initial postoperative discrimination. Despite problems with the surgical procedure, Boycott and Guillery's experiment does establish that freshwater turtles can be trained to make discriminations based on chemical cues.

A behavioral assay for underwater olfaction in aquatic turtles was devised by Manton, Karr, and Ehrenfeld (1972a). The ability of the green turtle (*Chelonia mydas*) to detect various chemical substances dissolved in water was investigated using operant conditioning techniques (see Chapter 20). In the experimental tank the turtles were trained to press an underwater key; this produced a chemical stimulus that was released intermittently into the water current flowing continuously through the tank. In the presence of this chemical signal, a response to a second key produced food reinforcement. An equal number of controls with water delivery alone were programmed for each experimental session. Once trained, the green turtles maintained a steady base rate of response to the signal production key of an average of 10 per minute, with some turtles making more than 1000 responses in each 1 hr session. This response rate is considerably higher than that which would be expected from freshwater turtles. With their limbs modified into flippers, green turtles move rapidly around the tank. They cannot retract their heads; thus there are no delays through time spent withdrawn in the shell. Because of their high activity level, they are always willing to work for food reinforcement; there is never any problem of periodic lack of interest in food, sometimes encountered in research with other turtles. The entire behavioral sequence of pressing the signal key, swimming over to the food reinforcement key, and pressing it, invariably occurred with the head completely under water.

The results show that *C. mydas* is capable of underwater chemoreception, a rare adaptation for an air-breathing vertebrate (Figure 1). The range of pure chemicals detected includes volatile organic compounds of the type usually smelled by air breathers: alcohols, aldehydes, esters, and tertiary amines. Nonvolatile amino acids were not detected. During the release of the detectable chemicals, the turtles directed their nostrils downward and appeared to be pumping water through the nasal cavities by means of throat movement. Breathing pauses, during which the nostrils were above water, were infrequent and were made only during well-defined breaks in responding. It appears virtually impossible for the turtles to have been using any airborne odor cue.

Since the experiment, as outlined, did not differentiate between chemoreception mediated by olfaction and that mediated by taste, a method for producing temporary anosmia (Alberts and Galef, 1971) was modified for use with these marine animals (Manton et al., 1972b). Olfaction was blocked by intranasal irrigation with zinc sulfate solution. The ability to detect chemicals underwater was lost completely for periods of 1 to 5 days, indicating that *C. mydas* can smell underwater. Recovery of olfaction occurred gradually over a period of several days. Intranasal irrigation with control solutions of saline or magnesium sulfate had no effect on the performance of the chemical discrimination, implicating Zn^{2+} as the critical ion.

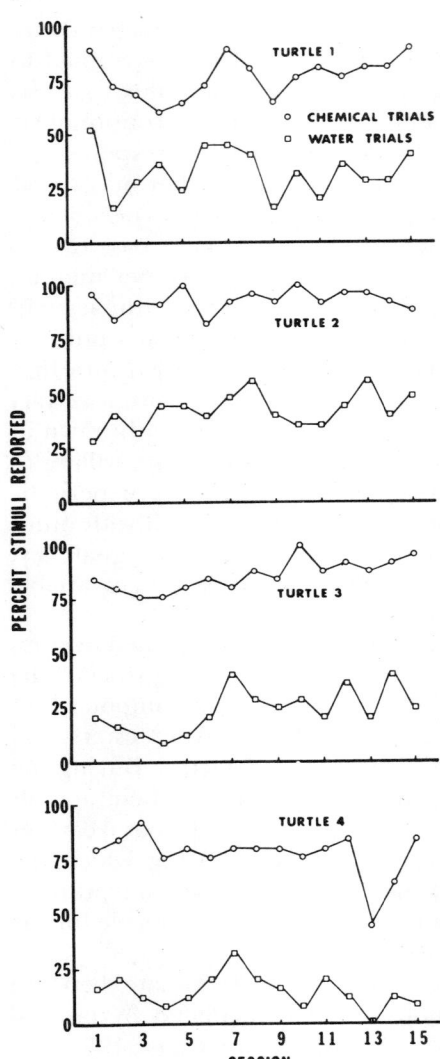

Figure 1 Percentages of correct and false reports for four green turtles during presentation of a chemical stimulus for 15 consecutive sessions: circles, percentage of correct reports after chemical release; squares, percentage of false reports after water release. (From Manton et al., 1972a.)

Anosmia induced by zinc sulfate has several advantages over surgical techniques. First, it acts peripherally, eliminating the problem of interference with central neural connections that occurs with olfactory bulb ablations and olfactory nerve sectioning. Second, it is reversible, thus there is no loss of individuals from the turtle population. Third, the treatment is relatively nontraumatic, and the turtles can be tested the same day as treated, without the necessity for a lengthy recovery period as required after surgery. Fourth, it affords the opportunity for field studies of the role of olfaction in both the open-sea navigation of mature green turtles and the observed specificity of nest-site selection on the breeding beach (Carr and Carr, 1972). In general, the use of zinc sulfate to induce anosmia offers promise of opening up new areas of investigation of the interaction between olfaction and behavior among a wide range of vertebrates; and it may provide an additional tool for the study of olfaction itself. No work has been done on the histology of the olfactory epithelium of the green turtle either before or after the zinc sulfate application; Mulvaney and Heist (1971) present data for rabbits.

The operant procedure described permits reasonable estimations of olfactory acuity to be made. Calculations were based on studies made with an indicator dye added to the tank water in the same way and in the same concentrations as the test chemicals. Detection based on these results occurred at approximate concentrations of 5×10^{-5} M or 5×10^{-6} M depending on the test solution used. Further unpublished tests using progressively lower concentrations of chemicals have since demonstrated detection at an approximate concentration of 5×10^{-9} M. The relatively acute underwater olfaction has not yet been definitely related to any behavior in *C. mydas*. Food location appears to be entirely visual, which is not surprising because as adults they feed on immense beds of turtle grass (*Thalassia, Zostera*) in shallow tropical seas. As far as is known, mating behavior is also initiated by visual cues. Green turtles whose nostrils are blocked by vaseline-soaked cotton plugs are able to make a direct heading for the ocean even though it is out of sight of the beach (Caldwell and Caldwell, 1962). The "sand smelling" response as the female emerges from the sea is the only behavior that invites an olfactory interpretation (Carr and Ogren, 1960). Arriving turtles repeatedly bend their heads downward and touch their noses to the sandy bottom in shallow water near the nesting beach and continue to do so as they move up the beach until they reach dry sand above wave action. It is, however, difficult to imagine what sort of long-lasting recognizable cue could be associated with shifting beach sand. The part played by olfaction in nest-site specificity is currently under investigation. The ability to smell underwater supports the theory that chemical cues may play a role in the open-sea reproductive migration of *C. mydas*.

The problem of sensory adaptation, a diminished response to a steady stimulus, must be accounted for both in the laboratory research and in the open sea. Rate and extent of adaptation is variable and is the basis for the division into phasic (rapidly adapting) and tonic (slowly adapting) receptors. Chemoreceptors in general adapt slowly (Hodgson and

Roeder, 1956; Kleerekoper, 1972); such prolonged behavioral response supports the use of olfaction in long-range orientation. Recovery from olfactory adaptation can be quite rapid (Cain and Engen, 1969). In the open ocean, sensory adaptation could be eliminated every time the turtle's head emerges above water to breathe, or possibly by diving below the shallow surface current to the level of the thermocline. Assuming that adaptation is a feature of the turtle's olfactory system, its airbreathing habit is an obvious advantage, one not shared by salmon. Yet the classic accounts of salmon navigation by chemical cues do not mention the problem of sensory adaptation (Hasler, 1966).

Sensory adaptation was not a problem in the experiment on green turtle olfaction because adaptation typically occurs in the presence of a constant, unchanging stimulus. The procedure resulted in a steadily decreasing stimulus concentration between trials so that most of the chemical was washed out (confirmed by dye tests) by the time of the next trial (1–4 min). A sudden increase in chemical concentration was then satisfactory as a new discriminative stimulus.

ORIENTATION TO ODORS

A major question in all cases of orientation along an odor "trail" in air or water is how the animal steers toward the odor source. It has been demonstrated mathematically and by wind tunnel and dye tests in water (Sutton, 1947; Wright, 1964; Kleerekoper, 1972) that because of turbulence, characteristic of moving fluids, a chemical gradient probably cannot provide useful directional information over more than a very short distance from the odor source. An aquatic animal seeking the source of a waterborne odor is in much the same position as a flying insect following an odor trail in the air.

New ideas have been proposed to replace the earlier hypothesis that animals following a chemical trail can perform incredibly delicate discriminations along a shallow concentration gradient. The prevalent view is that detection of the stimulus triggers movement upstream in a discrete current. Certain insects (Butler, 1970) and fishes (Kleerekoper, 1967; Hodgson and Mathewson, 1971) have been shown to orient in this way. Presumably, if the source is near and conditions remain relatively stable, the animal reaches its goal.

However it is difficult to understand how an animal embedded in a current containing an odor could make course corrections. Reports of male moths locating females across featureless lakes and on night flights are hard to explain. External cues appear to be necessary. Some insects use vision to sense the direction of apparent movement of objects on the ground over which they are flying (Kellog, Frizel, and Wright, 1962), and fish swimming in streams and rivers can use the banks or bottom.

It is not known what type of path (straight, zigzag) is taken upstream by green turtles en route to the nesting beach. The use of radio transmitters with signals reflected from a satellite to onshore receivers should solve the problem. The photographic capabilities of "spy" satellites are apparently more than sufficient to resolve an individual sea turtle moving through the ocean. As pointed out by Koch, Carr, and Ehrenfeld (1969), the most logical course is for the turtle to swim straight upstream under the stimulation of the odor plume emanating from the nest-site goal, using a sun-compass sense as a directional cue. The more information the turtle acts on, the more reliable the outcome is likely to be, and several other backup sources of information were also suggested, such as prevailing wave and wind direction and inspection of the apparent movement of the thermocline.

The nature of the proposed chemical cues for ocean currents is unknown. At least the dilution of the chemical substance as it travels downstream has been eliminated as a problem in one instance. Koch et al. (1969), using available oceanographic data, calculated that a substance added to the South Equatorial current at Ascension Island would undergo surprisingly little dilution (100- to 1000-fold) after traveling about 1200 miles to the vicinity of Brazil. Carr (1972) has called attention to evidence that olfactory cues might be available to turtles migrating to a mainland nesting beach.

Imprinting to Odors

Essential to the navigational hypothesis is the belief that turtles are "imprinted" (Hess, 1964) to the characteristic odor of the water and/or the sand of their birthplace (Carr, 1967a). Olfactory imprinting requires, first, the development of a specific chemical reference mechanism in the nervous system during a sensitive period of hatchling development, and second, the retention of the stored information in the absence of the appropriate chemical cues for 6 to 8 years until the first reproductive migration. Evidence for such a system has been found in salmon (Madison, Scholz, and Hasler, 1972).

That green turtles are capable of retaining an acquired behavior over fairly long intervals was shown after olfactory discrimination tests were discontinued. After almost a year had elapsed, three turtles were returned to the experimental apparatus and tested as previously described, with cinnamaldehyde as the test chemical. The turtles immediately resumed their former rapid responding.

There is time for olfactory imprinting to occur in the green turtle, since the eggs are buried 2½ ft under the sand for approximately 60 days before hatching and the shell and its membrane are permeable to moisture and gases in the sand. The hatchlings emerge from their nest together, having spent perhaps an additional week working their way to the surface; then they travel down the beach and swim actively beyond the surf. The unique quality of the water and/or sand might be permanently registered in their "olfactory brain" at critical stages in their embryonic development or after hatching or after entering the sea. On reaching reproductive maturity, they need only respond, using the sun for the general heading upon encountering the unique chemical to which they were earlier imprinted. A further possibility to increase the likelihood of this event is that their response threshold to the critical olfactory stimulus is lowered by their internal physiological state during the reproductive period. An additional advantage to the theory is that, unlike the cues available in celestial navigation systems, olfactory cues become stronger and less difficult to follow as the goal is approached. Subsequent nest-site specificity to a particular beach cove or section of beach might be explained by the initial imprinting of the developing hatchling to the odor of that site. The survival value conferred by the protected, predator-reduced site would perpetuate this behavior.

However imprinting to any stimulus has never been established for marine turtles. Food imprinting has been studied in freshwater turtles (Burghardt and Hess, 1966; Burghardt, 1967; Mahmoud and Lavenda, 1969), but the original preference was modified after exposure to alternate food experiences. Thus one criterion for imprinting, irreversibility of the response, was not met. This requirement has been questioned, and the reversibility of sexual imprinting has been shown (Klinghammer and Hess, 1964). The entire concept of imprinting needs revision to fit the facts. At least the concept implies that there is a critical stage in early development when exposure to an appropriate stimulus enhances subsequent responding to that stimulus.

SUMMARY AND CONCLUSIONS

There is much to be done to determine the role of olfaction in the lives of turtles. Throughout the literature authors have assumed that a turtle observed touching its nose to an object is actually smelling. They are

SUMMARY AND CONCLUSIONS

probably right, but rigorous testing has been rare. Fifty years of work on orientation and homing have failed to establish the particular guidance system used. The problem can be approached indirectly with controlled tests in the laboratory of turtles' abilities to perceive hypothetical environmental cues. Techniques are available for temporary sensory deprivation, both visual and olfactory, opening up novel approaches in the field and the laboratory. More frequent collaboration with organic chemists is necessary to assist in the analysis of olfactory stimuli. The sophisticated apparatus of experimental psychology is available for discrimination research.

A turtle's environment is as packed with chemical diversity as it is with visual variety. Vision seems to trigger most turtle behavior, and no behavior yet described appears to be dominated by olfaction, although some of the foregoing findings implicate the more primitive sense of olfaction as a backup sensory system. It may yet turn out to be extremely important in social and sexual interaction and as an aid to navigation.

REPRODUCTION AND DEVELOPMENT

CHAPTER 16
Reproductive Cycles and Adaptations

EDWARD O. MOLL

Of the many adaptive features incorporated into the life history of a species, the pattern of reproduction is paramount in respect to ultimate survival. Knowledge of the reproductive pattern, therefore, is basic to many studies concerning evolution and ecology.

The treatment here is chiefly ecological, although certain morphological and physiological considerations could not be excluded. Untested hypotheses and speculations have been mentioned with the hope of provoking additional study. Data and conclusions presented herein have been obtained from a survey of published literature, from my own investigations of turtles in Central America and the central United States, and from various colleagues who have graciously allowed me to cite some of their unpublished observations.

In preparing Figures 5 to 7, I relied heavily on comprehensive regional works such as Ernst and Barbour (1972) and Carr (1952) for the Nearctic; Alvarez del Toro (1960), Freiberg (1967), Moll and Legler (1971), Medem (1960, 1962), and Santos (1955) for the Neotropics; Loveridge (1941) and Loveridge and Williams (1957) for the Ethiopian region; Bourret (1941), Deraniyagala (1939), Pope (1935), and Smith (1931) for the Oriental region; Nikolskii (1915) and Fukada (1965) for the Palearctic; de Rooij (1915) and Goode (1967b) for the Australian region, and Wermuth and Mertens (1961) for the world. Additional examples were added from shorter papers by Auffenberg (1969a), Sexton (1960), Roze (1964), Goode and Russell (1968), Archer (1968b),

Limpus (1971), DeVries (1968), and Medem (1966). For comparisons of size I have used the maximum reported carapace length for turtles and the mean, greatest diameter for eggs; though not necessarily the best indicators of size, these measurements were available for the greatest number of species.

SEXUAL MATURITY

Onset of maturity in most chelonians correlates more with attainment of some minimum size than with age (Cagle, 1950; Ernst, 1971d; Gibbons, 1968d; Legler, 1960a). Male *Sternotherus odoratus* is an apparent exception that matures at various sizes but approximately the same age throughout its range (Tinkle, 1961). Males of many species mature smaller and often earlier than females (e.g., *Chrysemys picta, Graptemys barbouri, Malaclemys terrapin, Melanochelys trijuga, Terrapene ornata, Sternotherus odoratus*, and *Trionyx muticus*). In other species both sexes mature at similar sizes (e.g., *Clemmys guttata, Chelydra serpentina, Kinosternon flavescens, Macroclemys temmincki*, and *Rhinoclemys funerea*). In a few species, females may mature at the smaller size (e.g., *Gopherus agassizi, K. leucostomum*, and *S. minor*). Advantages of the first pattern, such as reduced intersexual competition, rapid maturation of males, and increased reproductive capacity of females, are obvious. Advantages of the latter patterns are less clear but males of certain of these species (e.g., *G. agassizi*) are known to be more competitive and aggressive for females, space, and so on, thus making large size advantageous. Whether this is true for all these species generally remains to be demonstrated.

Latitudinal variation in size and age at maturity has been reported in several species but no general pattern is evident as yet. Two wide-ranging species, *S. odoratus* (Tinkle, 1961; Gibbons, 1970a) and *C. picta* (Christiansen and Moll, 1973; Moll, 1973) mature earlier (except male *S. odoratus*) and at smaller sizes in lower latitudes. In *Pseudemys scripta*, another wide-ranging species, size at maturity varies little within the temperate zone (Cagle, 1950), but tropical populations mature at larger sizes (Moll and Legler, 1971), a reverse of the aforementioned trend. Although the latitudinal trends differ, all three species mature larger and attain maximum size in the portions of their range lacking close relatives. Further investigation of these trends should consider competitive influences (see McNab, 1971) as well as climatic ones.

On the basis of data from squamate reptiles (Tinkle et al., 1970; Fitch, 1970), larger species of chelonians might also be expected to mature later than smaller ones, but no such trend is evident. For example, the large *Chelonia mydas* (Hendrickson, 1958) and the small *S. carinatus* (Tinkle, 1958b) may mature in 4 to 6 years, whereas the medium-sized desert tortoise may require 15 to 20 years (Woodbury and Hardy, 1948). When considering separate populations, factors such as heredity, habitat productivity, and length of growing season may be more important determinants of maturation time than mere size.

MALE REPRODUCTIVE CYCLE

Knowledge of male chelonian reproductive cycles has accumulated slowly. A few papers describing portions of the cycle appeared in the early part of the century (e.g., Allen, 1906; Pellegrini, 1925) but groundwork for relating the cycle to chelonian ecology was best laid in the 1930s. This important decade was marked by Risley's excellent studies of spermatogenesis in *S. odoratus* (1933b, 1933c, 1936, 1938), by the first—and to date the last—experimental attempt to determine environmental timing mechanisms of the male cycle by Burger (1937), and by a brief account of spermatogenesis in *Terrapene carolina* (Hansen, 1938).

Research was renewed in the early 1950s with descriptions of the male cycle in *T. carolina* (Altland, 1951) and *Mauremys leprosa* (Combescot, 1954a, 1954b). Sud (1956) described in detail subcellular changes during spermatogenesis in *Lissemys punctata* but did not discuss the seasonal timing of the cycle. Miller (1959) provided a general review on spermatogenesis in reptiles.

In the 1960s Legler (1960a) included a description of the male cycle in his thorough study of *T. ornata*. Lofts and Boswell (1961) reported the testicular cycles of *M. caspica*, placing special emphasis on seasonal changes in lipids. Gibbons (1968d) described seasonal macroscopic changes in testes of *C. picta*. Reviews were provided by Forbes (1961) and Lofts (1968, 1969).

With the advent of the 1970s interest in male reproductive cycles suddenly accelerated. Almost as many species were studied in the early part of this decade as in all previous ones (*C. picta:* Ernst, 1971d; Christiansen and Moll, 1973; Moll, 1973; *P. scripta:* Moll and Legler, 1971; *S. odoratus:* Spaet, 1973; *S. carinatus, Kinosternon flavescens, K. subrubrum:* Mahmoud and Klicka, 1972; *K. flavescens:* Christiansen and Dunham, 1972; *Macroclemys temmincki:* Dobie, 1971; and *Chelydra serpentina:* White and Murphy, 1973).

Four widely separated families have now been examined (Emydidae, Chelydridae, Trionychidae, and Kinosternidae) and, at least in the temperate zone, the male cycle is concordant. Spermatogenesis begins in the spring, peaks in late summer, and ends in the fall as spermatozoa leave the testes to overwinter in the epididymides. Spring is the peak period for breeding in most species, although fall and continuous breeding may also occur (Hildebrand, 1929; Risley, 1938; Cagle, 1950; Legler, 1960a; Tomko, 1972). Spermatozoa produced in the late summer of one year usually fertilize eggs in the following spring. For comparison with amphibians and with other reptiles, see reviews by Licht (1972a), Lofts (1968), and Miller (1959).

An additional cycle involving the endocrine cells (Leydig and Sertoli cells) complements the spermatogenic cycle. Activity in these cells directly correlates with an increase in cell volume and nuclear diameter along with depletion of lipoidal inclusions (Miller, 1959). Based on these criteria, endocrine cell activity peaks during gametogenic quiescence, rapidly decreases as spermatogenesis is renewed, and is minimal at the peak of spermiogenesis.

As germ cells multiply and accumulate, the testes increase macroscopically in size. Peak enlargement occurs at the height of spermiogenesis (between late July and late September in the majority of temperate-zone species). Testicular regression follows, paralleling the expulsion of spermatozoa into the epididymides. Complementary size fluctuations occur in the accessory tubules. Epididymides attain peak diameters in late fall and early spring as spermatozoa enter, then regress after spring mating. If mating does not occur, "involuntary sperm release" (Mahmoud and Klicka, 1972) may still bring about regression. In addition to the size increase associated with spermatogenesis, testes of some species undergo a second, lesser period of enlargement between brumation (hibernation) and the onset of spermatogenesis (Legler, 1960a; Moll, 1973). Legler (1960a) correlated this increase with proliferation of the sustentacular cytoplasm and the subsequent regression with the period of active mating.

Microscopic details of the temperate-zone cycle can be conveniently considered in five somewhat overlapping phases described below. Examples of cell types mentioned in the account and representative views of each phase appear in Figures 1 and 2, respectively. Although variations do occur (chiefly in timing of the phases), the following characteristics of each phase seem consistent in five representative species I have examined *Pseudemys scripta, Chrysemys picta, Chelydra serpentina, Sternotherus odoratus,* and *Trionyx muticus*) and with those reported in the aforementioned literature.

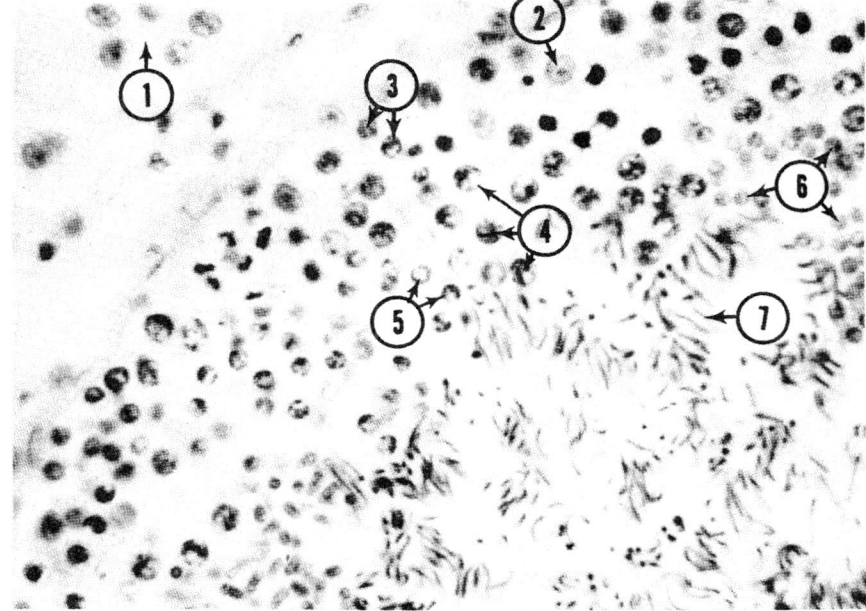

Figure 1 A portion of the interstitium and seminiferous epithelium from the testes of a Panamanian *Kinosternon leucostomum* collected 27 September (×670): 1, island of Leydig cell; 2, Sertoli cell; 3, spermatogonia; 4, primary spermatocytes; 5. secondary spermatocytes; 6, spermatids; 7, cluster of spermatozoa.

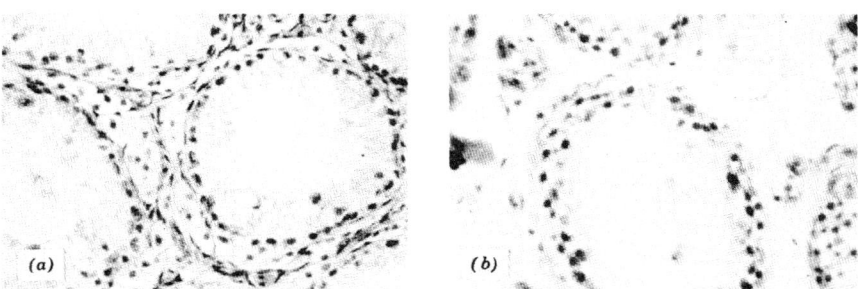

Figure 2 Stages of spermatogenesis in *Chrysemys picta* (×200). (*a*) At end of germinal quiescence; note vacuolated Leydig cells ringing seminiferous tubules (14 April, Coles County, Ill.). (*b*) Gonial proliferation (23 June, Ashland County, Wisc.). (*c*) Peak of spermatocytogenesis, just before spermiogenesis (31 July, Ashland County, Wisc.). (*d*) Spermiogenesis (20 August, Ashland County, Wisc.). (*e*) Spermiation (29 September, Ashland County, Wisc.). (*f*) Postspermatogenic tubule; note debris in Sertoli cytoplasm (7 November, Coles County, Ill.).

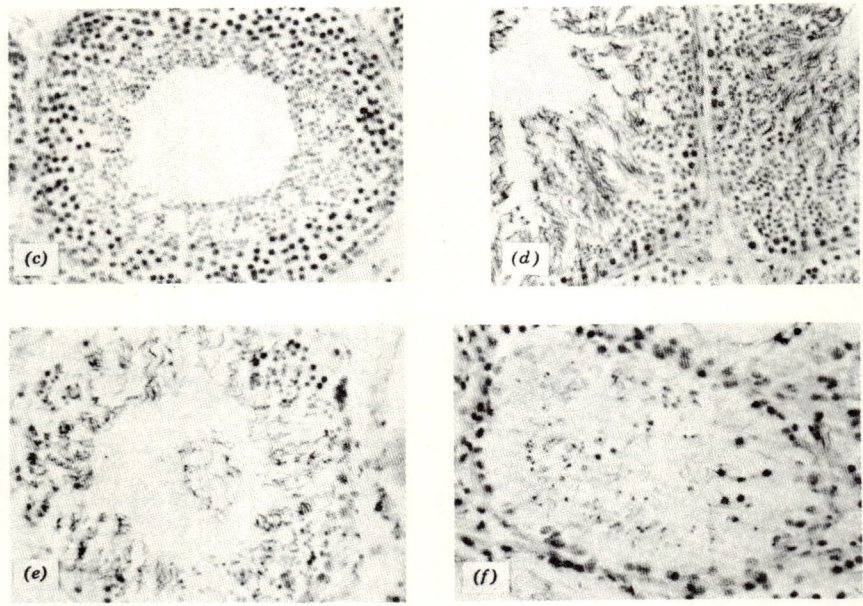

Figure 2 (*Continued*).

Germinal Quiescence

The period of germinal quiescence (Figure 2*a*) occurs between gametogenic cycles and in the temperate zone includes the period of brumation. Seminiferous tubules decrease to minimum diameter as Leydig cells proliferate and enlarge, filling intertubular spaces; the latter event is most pronounced in *C. serpentina*. Within the tubules, debris from the previous cycle often remains within the lumina and enmeshed within the Sertoli cytoplasm. The Sertoli nuclei may migrate to some extent from the basement membrane toward the lumen. The seminiferous epithelium comprises numerous Sertoli cells along with a few inactive spermatogonia. Inter- and intratubular lipids (associated with Leydig and Sertoli cells) attain peak levels at this time.

Gonial Proliferation

As photoperiod and temperatures increase, spermatogonia begin to proliferate through mitosis, surpassing Sertoli cells in number and eventually becoming several rows deep at the basement membrane. Preceding or paralleling gonial proliferation (Figure 2b), the Sertoli cytoplasm (sustentaculum) increases in height, in some cases nearly occluding the tubule lumen. Most leftover debris disappears. Lipoids begin disappearing from intertubular areas.

Spermatocytogenesis

Primary spermatocytes appear and undergo reduction division, first to secondary spermatocytes, then to spermatids, during spermatocytogenesis (Figure 2c). Secondary spermatocytes seem to be very transitory and are usually few in number compared to the other cell types. Diameters of seminiferous tubules increase throughout this phase, the rate of increase accelerating rapidly during the latter part. Intratubular lipoids decrease as this stage progresses. Intertubular lipids show a resurgence in some species (e.g., *C. picta*), only to decrease again during spermiogenesis.

Spermiogenesis

Spermatocytogenesis continues throughout the early part of the spermiogenesis phase (Figure 2d), and most stages of spermatogenesis are abundantly displayed. A key feature of this phase is maturation of spermatids to spermatozoa, which accumulate in featherlike bunches at the distal ends of Sertoli cells, partially occluding the lumen. Seminiferous tubules attain maximum diameters, and Leydig cells, compressed by the expanding tubules, are minimum size, few in number, and contain relatively little or no lipids. Sertoli cells, with most of their nuclei again near the basement membrane, also contain minimal amounts of lipids.

Spermiation

Following the peak of spermiogenesis, mature sperm dissociate from the Sertoli cells and enter the lumen (spermiation, Figure 2e), from which they pass into the epididymides. Spermatocytogenesis followed by spermiogenesis dwindles to an end during this period. Lipids again begin accumulating within the inter- and intratubular tissues. This phase may be completed before or after brumation, depending on the species and its geographic location.

Testicular cycles of different species differ chiefly in respect to timing and length of the different phases. This variation has been noted geographically and ontogenetically. A definite compression of the phases in the cycle is evident at higher latitudes. Latitudinal extremes of a temperate zone species commonly attain the peak period of spermiogenesis at about the same time even though they may begin gonial proliferation as much as two months apart. Spermatogenesis also tends to be prolonged in lower latitudes. In *C. picta* spermatogenesis extends from March into November in Louisiana and from May into October in Wisconsin, even though the peak spermiogenic period in both areas is August (Moll, 1973).

The degree of testicular regression may also vary latitudinally, with testes regressing to relatively smaller sizes in lower latitudes. *C. picta* testes regress to 44% of the maximum size in Louisiana, 52% in Illinois, and 61% in Wisconsin. Shorter activity periods in northern areas may crowd cycles to the extent that insufficient time remains for complete regression before recrudescence (Christiansen and Moll, 1973; Moll, 1973).

Moll and Legler (1971) compared the male cycle in three tropical species, *Pseudemys scripta, Rhinoclemys funerea*, and *Kinosternon leucostomum*. *P. scripta*, a recent tropical invader, retained the basic temperate-zone pattern, with peak Leydig cell activity during gametogenic quiescence. *R. funerea* and *K. leucostomum* demonstrate long-standing adaptation to a tropical existence. Although year-round data are incomplete for these species, a reexamination of gonadal materials discussed by Moll and Legler (1971) suggests deviations from the temperate-zone pattern in that spermatogenesis, though cyclic, has relatively short periods of germinal quiescence, and maximum Leydig cell activity does not necessarily occur during gametogenic quiescence.

Departures from the previously described chelonian patterns include a report of continuous spermatogenesis in *Macroclemys temmincki* (Dobie, 1971) and a lack of cyclic Leydig cell activity in *Sternotherus odoratus* (Risley, 1938). These reports warrant further study. Dobie based his conclusions on the year-round presence of sperm in the accessory tubules of the testes, but turtles with seasonal spermatogenesis may retain spermatozoa within the epididymus throughout the year (Legler, 1960; Moll and Legler, 1971; Christiansen and Dunham, 1972). Risley (1938) observed little seasonal change in Leydig cells and concluded that they were not cyclic. Interestingly, Risley (1933c) had earlier reported that Leydig cells were at maximum size and number during fall, winter, and early spring. In support of the earlier paper, preliminary investigations using frozen sections and lipid-specific stains (Spaet, 1973) indicate that Leydig cells of *S. odoratus* accumulate and lose lipids in a manner similar to that described for other species.

FEMALE REPRODUCTIVE CYCLE

Few papers have considered the entire female cycle. Microscopic aspects of oogenesis were described by Agassiz (1857), Munson (1904), and Altland (1951). Seasonal changes of fine oviduct structures have also been recounted for *Terrapene carolina* (Hansen and Riley, 1941) and *Mauremys leprosa* (Combescot, 1954b, 1955). Seasonal macroscopic changes have been reported for a variety of temperate-zone species: *T. ornata* (Legler, 1960a); *M. leprosa* (Combescot, 1954b); *M. caspica* (Lofts and Boswell, 1961); *Chrysemys picta* (Moll, 1973; Christiansen and Moll, 1973; Ernst, 1971d;. Gibbons, 1968d); *Chelydra serpentina* (White and Murphy, 1973); *Macroclemys temmincki* (Dobie, 1971); *Kinosternon flavescens* (Christiansen and Dunham, 1972; Mahmoud and Klicka, 1972); and *K. subrubrum, Sternotherus carinatus,* and *S. odoratus* (Mahmoud and Klicka, 1972). Female cycles of three tropical species, *P. scripta, R. funerea,* and *K. leucostomum,* were described by Moll and Legler (1971).

Females generally have an annual cycle, but exceptions exist, most notably among the Cheloniidae. The green turtle *Chelonia mydas* typically has a 3 year cycle throughout its range (Carr and Ogren, 1960; Hendrickson, 1958; Harrisson, 1956b). Cycles of 2 and probably 4 years are also evident to a lesser degree, and individuals may shift from one to another (Carr and Carr, 1970). Similar periodicity is indicated for *Caretta caretta* (Caldwell, 1962; Hughes and Mentis, 1967b) and *Eretmochelys imbricata* (Carr, Hirth, and Ogren, 1966). *Lepidochelys* species differ by nesting annually (Pritchard, 1969), although annual nesting occurs in other cheloniids occasionally (Hughes and Mentis, 1967b; Dix and Richardson, 1972). There is little evidence for multiyear cycles among freshwater species, although Gibbons (1969) reported that large, old *Deirochelys reticularia* may shift to a biennial or triennial cycle, and Dobie (1971) suggested that some *Macroclemys temmincki* may nest in alternate years.

The temperate-zone female cycle consists of the four phases or periods described next.

Follicular Enlargement

Enlargement resulting from yolk accumulation (vitellogenesis) begins in late summer or fall and continues (except during hibernation) until completion of the second phase in the spring. Depending on the species, age, and geographic location, completion of enlargement for the first clutch of follicles may occur before or after hibernation. The snapping turtles (*Macroclemys* and *Chelydra*) normally complete follicular develop-

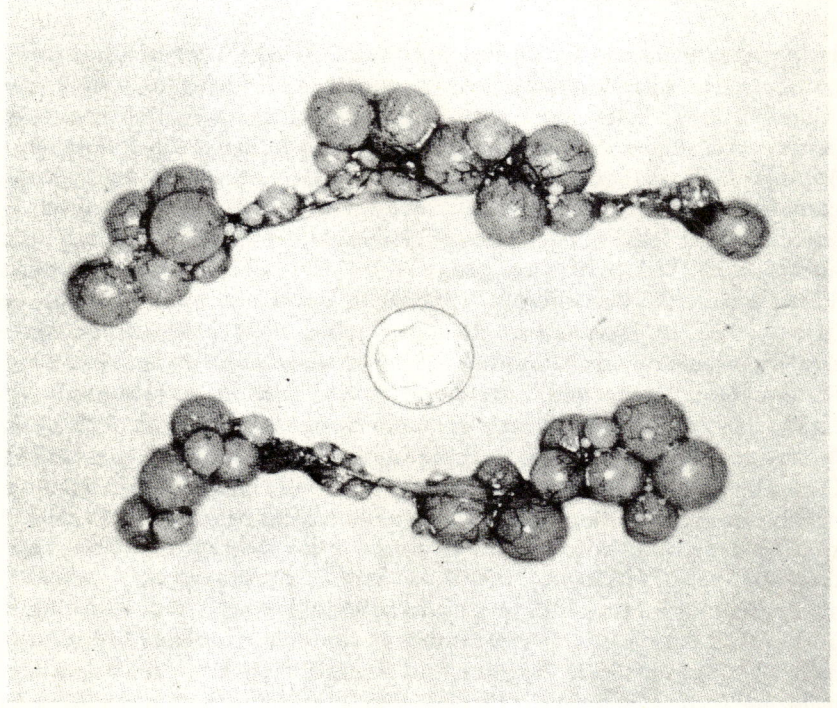

Figure 3 Ovaries of *Pseudemys scripta* (Randolph County, Ill., 14 October). Note enlarging follicles of several size groups. A dime is pictured for scale.

ment before hibernation (Dobie, 1971; White and Murphy, 1973), whereas *Kinosternon* and *Sternotherus* species (Christiansen and Dunham, 1972; Mahmoud and Klicka, 1972) and *Mauremys caspica* (Lofts and Boswell, 1961) complete vitellogenesis just before ovulation. Studies of *Chrysemys picta* (Moll, 1973; Christiansen and Moll, 1973; Ernst, 1971d; Gibbons, 1968d; Powell, 1967) indicate that in the north the first clutch of follicles enlarge to near ovulatory size before hibernation, but farther south most enlargement occurs between spring emergence and the nesting season.

Follicles of turtles laying multiple clutches tend to enlarge in groups (Figure 3). Immediately before the nesting season several size groups are distinguishable. Agassiz (1857) commented on these distinct sets but thought they represented clutches to be laid in subsequent years. The number of groups of enlarged follicles present immediately before the nesting season gives a rough approximation of the number of clutches that may be laid in a single season.

Ovulation and Intrauterine Period

As each set of large follicles is ovulated, progressively smaller sets enlarge to ovulatory size. During an ovulatory period one ovary may be more active than another, though in the next period its counterpart may become more active (Legler, 1960a; Ernst, 1971d).

Ova from one ovary do not necessarily enter the oviduct of the same side but can cross over to its counterpart. This phenomenon—widespread and perhaps universal among turtles—has been reported for species of *Terrapene*, *Chrysemys*, and *Emydoidea*, by Legler (1958), *Sternotherus* by Tinkle (1959a), tropical *Pseudemys* by Moll and Legler (1971), *Kinosternon* and *Sternotherus* by Mahmoud and Klicka (1972), *Macroclemys* by Dobie (1971), and *Chelydra* by White and Murphy (1973). Legler (1958) suggested that equalization of eggs in the oviducts might be an advantage, whereas Tinkle (1959a) and Moll and Legler (1971) considered it more likely that the distribution of oviductal eggs is due to chance.

The intrauterine period is highly variable among species, and the possibility of multiple clutches is not always considered in published reports. Turtles are also capable of retaining eggs for extraordinarily long periods when optimum nesting conditions are lacking (Cagle and Tihen, 1948). A sampling of estimates is as follows: 2½ to 3 weeks in Michigan *Chrysemys picta* (Gibbons, 1968d); 4 to 5 weeks for *C. picta* and 3 weeks for *Clemmys insculpta* in Nova Scotia (Powell, 1967); 2 to 3 weeks for Kansas *Terrapene ornata* (Legler, 1960a); 20 to 35 days for Michigan *Sternotherus odoratus* (Risley, 1933b); and 5 to 8 weeks for Wisconsin *S. odoratus* (Edgren, 1960). Based on reports of time between nestings (Table 1), some of these estimates may be high.

Nesting Period

Oviposition may occur from one to several times during a year depending on such factors as latitude, species, and age (or size). Evidence of multiple clutches in a single season appears in Figure 4. Nesting seasons commonly fall between late April and late July throughout much of the range of north temperate species. In lower latitudes, however, with less intense climatic limitations, nesting often begins earlier in the year and is extended over a greater period. Nesting of Florida species, for example, often occurs as early as February or March; and in *Pseudemys floridana*

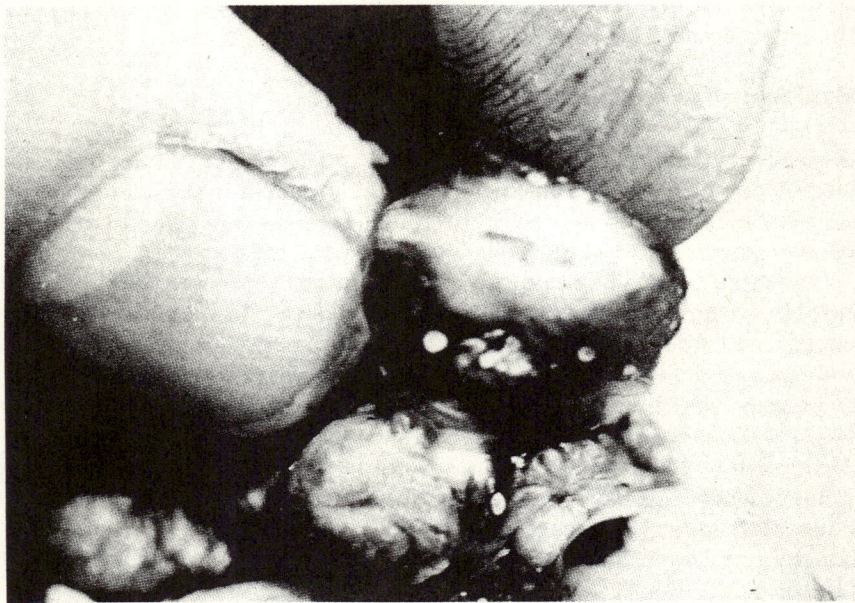

Figure 4 Two of the 73 corpora lutea found on the ovaries of a *Trionyx spiniferus* collected in Coles County, Illinois, on 8 July. The size difference between the corpora lutea indicates that they resulted from ovulation of different clutches. Probably four or five clutches were laid by this female in the nesting season.

peninsularis and *Deirochelys reticularia* (Carr, 1952), nesting may occur throughout the year. A split nesting season has been recorded for at least one species, *Eretmochelys imbricata* (Nakajima, 1920, in Fukada, 1965).

Table 2 lists the species that lay more than one clutch per year. Logically those laying multiple clutches require longer nesting seasons than those laying only one. Times between subsequent nestings are often surprisingly short (see Table 1); periods of renesting within 1 to 3 days, though rare, may occur (Hendrickson, 1958; Pritchard, 1969). Since egg-laying habits frequently become unnatural in captivity, no such reports are cited here other than Mitsukuri (1895) on *Trionyx sinensis*, where turtles were kept under relatively natural conditions.

Clutch size may remain constant with each successive clutch as in *Chrysemys picta* (Gibbons, 1968d) or it may vary. *Chelonia mydas*, for example, lays fewer eggs in the first and last clutches (Carr, 1967b); clutch size may also diminish with each clutch laid (Carr and Hirth, 1962; Pritchard, 1971a). The latter pattern is also characteristic of *Mauremys japonica* (Seo, 1926, in Fukada, 1965), *Terrapene ornata* (Legler, 1960a), and *Caretta caretta* (Caldwell, 1959; Lebuff and Beatty, 1971).

Table 1 Internesting Intervals

Species	Locality	Internesting Period (days)	Reference
Dermochelys coriacea	South Africa	9	Hughes and Mentis (1967b)
	Surinam	10	Pritchard (1969)
Caretta caretta	United States	12–15	Caldwell (1962)
	South Africa	14–16	Hughes and Mentis (1967b)
Chelonia depressa	Australia	13–18	Limpus (1971)
C. mydas	Sarawak	10	Hendrickson (1958)
	Costa Rica	12–14	Carr and Giovannoli (1957), Carr and Ogren (1960)
	Surinam	13–14	Pritchard (1969)
Eretmochelys imbricata	Seychelles	13–15	Hornell (1927)
Lepidochelys kempi	Mexico	20–28	Chavez (1969)
L. olivacea	Surinam	17–30+	Pritchard (1969)
Batagur baska	Burma	ca.14	Maxwell (1911), in Smith (1931)
Chrysemys picta	Illinois	ca.14	Moll (1973)
	New Mexico	ca.14	Christiansen and Moll (1973)
	Michigan	14–21	Gibbons (1968d)
Graptemys barbouri	Georgia	15–24	Wahlquist and Folkerts (1973)
Mauremys japonica	Japan	10–15	Seo, 1926, in Fukada (1965)
Trionyx sinensis	Japan	10–25	Mitsukuri (1895)
Geochelone chilensis	Argentina	ca.15	Auffenberg (1969a)

Table 2 Maximum Number of Clutches per Year Estimated for Various Chelonians[a]

Family and Species	Number	Locality	Reference
Emydidae			
Batagur baska	3	Burma	Maxwell (1911), in Smith (1931)
Chinemys reevesi	3	Japan	Fukada (1965)
Chrysemys picta	4 or 5	Louisiana	Moll (1973)
Clemmys insculpta	1	Nova Scotia	Powell (1967)
Deirochelys reticularia	3	Missouri	Anderson (1965)
Emydoidea blandingi	+	Michigan	Gibbons (1968b)
Graptemys barbouri	4	Georgia	Cagle (1952)
G. geographica	2 or 3	Illinois	Moll (unpubl.)
G. pseudogeographica	3	Oklahoma	Webb (1961)
G. pulchra	6	Alabama	Shealey (1976)
Malaclemys terrapin	5*	North Carolina	Barney (1922), Hildebrand (1929)
Mauremys japonica	3	Japan	Seo (1926), in Fukada (1965)
Melanochelys trijuga	+	Ceylon	Deraniyagala (1939)
Pseudemys dorbignyi	+	Argentina	Freiberg (1967)
P. floridana	+	Florida	Carr (1952)
P. scripta	5 or 6	Panama	Moll and Legler (1971)
Rhinoclemys funerea	4	Central America	Moll and Legler (1971)
R. punctularia	+	Colombia	Medem (1962)
Terrapene carolina	4	Florida	Oliver (1955)
T. ornata	2	Kansas	Legler (1960a)
Testudinidae			
Geochelone chilensis	2	Argentina	Auffenberg (1969a)
G. elegans	3	Ceylon	Deraniyagala (1939)
G. elephantophus	+	Galapagos Islands	Beck, in Lampkin and Turner (1967)
G. pardalis	7*	California	Rowe and Janulaw (1971)
Gopherus agassizi	+*	California	Miller (1955)
G. berlandieri	+	Texas	Auffenberg and Weaver (1969)
Testudo horsfieldi	4	Russia	Sergeev (1941)
Dermochelyidae			
Dermochelys coriacea	3 or 4	Ceylon	Deraniyagala (1939)
Cheloniidae			
Caretta caretta	5	South Africa, Florida	Hughes and Mentis (1967b) Caldwell (1962)

Table 2 (Continued)

Family and Species	Number	Locality	Reference
Chelonia depressus	4	Australia	Limpus (1971)
C. mydas	11	Sarawak	Hendrickson (1958)
Eretmochelys imbricata	5	Japan	Nakajima (1920), in Fukada (1965)
Lepidochelys kempi	3	Tamaulipas, Mexico	Chavez (1969)
L. olivacea	3	Surinam	Pritchard (1969)
Kinosternidae			
Kinosternon bauri	3	Florida	Einem (1956)
K. flavescens	1	New Mexico, Oklahoma	Christiansen and Dunham (1972) Mahmoud and Klicka (1972)
K. leucostomum	+	Central America	Moll and Legler (1971)
K. subrubrum	3	Louisiana	Moll (unpubl.)
Sternotherus carinatus	2	Southeastern United States	Tinkle (1958b)
S. minor	2	Southeastern United States	Tinkle (1958b)
S. odoratus	3	Illinois and Louisiana	Moll (unpubl.)
Chelydridae			
Chelydra serpentina osceola	3	Florida	Ewert (unpubl.)
C. serpentina	1	Tennessee	White and Murphy (1973)
Macroclemys temmincki	1	Louisiana	Dobie (1971)
Trionychidae			
Lissemys punctata	+	Ceylon	Deraniyagala (1939)
Pelochelys bibroni	2	Southern China	Mell (1929), in Pope (1935)
Trionyx muticus	3	Southeastern United States	Webb (1962)
T. sinensis	4*	Japan	Mitsukuri (1895)
T. spiniferus	4 or 5	Illinois	Moll (unpubl.)
Chelidae			
Elseya latisternum	+	Australia	Legler (unpubl.)
Emydura sp.	+	Australia	Legler (unpubl.)
Phrynops dahli	+	Colombia	Medem (1962)

[a]Asterisk "*" denotes data from captives. Plus "+" indicates multiple clutches are probable but the number is unknown.

Latent Period

Between periods of follicular enlargement occurs the latent period, which is so named because ovaries are at minimal size and presumably at minimal activity. Some activity, however, does occur; Altland (1951), Legler (1960a), and Ernst (1971d) reported that new oogonia appear on the germinal ridges at this time. In addition certain follicles left over from previous cycles may undergo atresia during this period and disappear. Christiansen and Moll (1973) also described the occurrence of "minor follicular atresia" in which certain groups of follicles became vascular and less turgid but did not disappear.

The length of the latent period may vary with latitude. In *Chrysemys picta* this period of quiescence becomes progressively shorter with higher latitudes; in northern Wisconsin the period may only last a month or less, whereas it may last 3 months in Louisiana (Moll, 1973).

In the tropics both cyclic and more or less continuous patterns are in evidence. Few have been well studied. Cyclic forms include both recent invaders such as *Pseudemys scripta* (Moll and Legler, 1971) and *Chelydra serpentina acutirostris* (Medem, 1962), along with such long-time tropical residents as the sea turtles and *Rhinoclemys funerea, Podocnemis expansa, Batagur baska,* and *Geochelone carbonaria.* Others seemingly have more continuous periods of reproductive activity, at least as a population: *Lissemys punctata* from Ceylon (Deraniyagala, 1939); *Kinosternon spurrelli (leucostomum), K. dunni, R. punctularia,* and *R. annulata* from Colombia (Medem, 1962), and *K. leucostomum* from Central America (Moll and Legler, 1971).

RELATIONSHIPS OF CYCLES

Inasmuch as reproduction of many cyclic species is timed so that sperm and egg mature simultaneously, the lack of synchronization between the sexual cycles of most turtles is unusual. Males in the temperate zone produce mature sperm as much as half a year before they will be used to fertilize eggs. Furthermore the spermatogenic cycle begins as the ovarian cycle is winding down, and vice versa. Though the advantage is obscure, such a pattern is made possible through the ability of both sexes to store mature sperm over long periods. Females of *Malaclemys terrapin* (Barney, 1922), *Terrapene carolina* (Ewing, 1943), and *Chelydra serpentina* (Smith, 1956) can store sperm for several years, and eggs may be fertilized at least 4 years after insemination.

Baker (1938), Lack (1954), and others have observed that reproductive cycles of most species are timed (through natural selection) so that eggs and young are brought forth at the most favorable period for their survival. Although these favorable conditions are the "ultimate cause" of a particular breeding season, usually another set of factors ("proximate causes") initiates the sexual cycles at the proper time of the year (Baker, 1938). In turtles neither set of factors is well studied, but the sexes are either stimulated by different factors or react to the same factors in an antipodal manner. Because of the sperm storage capability, the "ultimate factors" are presumably more important in respect to the female cycle.

"Proximate causes" in chelonians probably include a combination of endogenous and exogenous factors. With respect to endogenous factors, the close similarity of reproductive cycles and nesting times for most temperate-zone species is suggestive of an inherent rhythm. The strong resemblance of reproductive cycles between temperate and tropical populations of *Pseudemys scripta* (Moll and Legler, 1971) also supports this idea.

Age (size) is another endogenous factor to be considered. Older (larger) males may begin spermatogenesis earlier and continue it longer in the year than younger (smaller) ones (Moll and Legler, 1971; White and Murphy, 1973). Females parallel this situation; older (larger) individuals nest earlier (e.g., Hammer, 1969) and, in species laying multiple clutches, continue nesting later than younger (smaller) members of the population (e.g., Moll and Legler, 1971).

Of the exogenous "proximate causes" the most important factors seem to be temperature, light, and moisture. In a review of the literature, Licht (1972a) suggested that temperature was the most important timing cue for breeding cycles in reptiles and that photoperiod at best had a secondary influence. Some information on turtles supports this idea. Lofts and Boswell (1961) concluded that the spermatogenic cycle of *Mauremys caspica* parallels the temperature cycle more closely than any other environmental factor. Females of most temperate-zone species begin follicular enlargement in the fall when temperatures are declining. If declining temperatures were the causal factor, northern species would be expected to begin follicular enlargement before southern ones. Such a schedule has in fact been demonstrated for *Chrysemys picta*, which begins enlargement as much as 3 months earlier in the north (Moll, 1973). Even among tropical and subtropical species, temperature may influence reproduction. *Chelonia mydas* is distinctly cyclic over most of its range, but nesting occurs year round at Sarawak, in the Seychelles, and in the Gulf of Siam, all areas having minimal annual temperature fluctuation (Hendrickson, 1958; Carr, 1952). Similarly, although *Chelonia depressa* has a restricted nesting season in southern Australia, it nests continuously in tropical northern Australia (Limpus, 1971). The factors above suggest a correlation between reproductive cycles and temperature, but experimental evidence is lacking.

In regard to light, the experiments of Burger (1937) with male *Pseudemys scripta* suggested that increased photoperiod (3–7 hr) in mid-November inhibited the spermatogenic cycle already in progress and activated a new cycle. The design of this early study has been criticized for poorly controlled temperature, small sample sizes, and unhealthy subjects (Licht, 1972b). Photoperiod, nevertheless, is known to affect the nonreproductive physiology of turtles (Hutchinson and Kosh, 1965) and merits careful consideration in future studies.

Working with *Clemmys insculpta*, Brenner (1970) found a direct correlation between ovarian fat content and maturation of the ova and an inverse correlation between the amounts of body and ovarian fat. Photoperiod and temperature influenced the rate of body fat utilization and thus affected ovarian development in the species as well.

Moisture may be an important "proximate factor" in tropical regions having distinct dry and rainy seasons, but this has not been established.

"Ultimate causes" of reproductive seasons, though also poorly studied, tend to be more obvious than "proximate causes." It seems logical, for example, that the spring and early summer nesting seasons of most temperate species allow the most vulnerable period of turtle development to take place during the most favorable season for growth. In the tropical species with cyclic reproduction, cycles usually correlate with wet and dry seasons. *Pseudemys scripta* (Moll and Legler, 1971), *Podocnemis expansa* (Roze, 1964), and *Chelydra serpentina acutirostris* (Medem, 1962) nest in the dry season. The "ultimate causes" of this timing, nevertheless, may differ with the species.

Nesting during the dry season is advantageous to *P. scripta* because eggs can be laid in more open, warmer, drier nest sites with minimal flood danger. Rains commence about the time of hatching, aiding the hatchlings by softening the earth for emergence, by providing more herbaceous growth for cover, and by stimulating their movements toward water (Moll and Legler, 1971). In the case of *Podocnemis expansa* the favored nesting areas are inundated until the dry season (Roze, 1964).

Rhinoclemys funerea of Central America (Moll and Legler, 1971), *Dermatemys mawi* from Chiapas (Alvarez del Toro, 1960), and *Dermochelys coriacea* from Ceylon (Deraniyagala, 1939) all nest in the rainy season. *D. mawi* apparently need high water to move back into flooded arroyos where they nest; no advantages to the other species have been reported.

Finally, competition seems to be an "ultimate cause" worthy of mention. Different nesting seasons may have evolved in certain colonial nesting species to alleviate competition for nest sites. For example, Carr and his students have observed that although *Eretmochelys imbricata* and *Dermochelys coriacea* utilize the same nesting areas in Costa Rica as *Chelonia mydas*, their nesting periods overlap but slightly. Though virtually impossible to prove, this is a likely example of resource partitioning.

CLUTCH SIZE AND REPRODUCTIVE POTENTIAL

At their extreme, chelonians are the most prolific of any amniote group. Some species regularly lay clutches in excess of 100 eggs, and more than 10 clutches may be laid in a year. It is possible then for some individuals to have an annual reproductive potential exceeding 1000 eggs. *Chelonia mydas* holds the record for both maximum clutch size, 226 eggs[1] (Pritchard, 1969), and maximum number of clutches per year, 11 (Hendrickson, 1958). Most turtles are far more conservative. If the sample of 109 species plotted in Figures 5 to 7 is representative, the majority (over 60%) lay clutches averaging less than 10 eggs and less than 10% regularly lay clutches exceeding 30 eggs. Similarly, two or three clutches per nesting season are probably the mode for most species, and 10 or more are a rare exception (Table 2). Undoubtedly many complex factors influence clutch size and reproductive potential.

The relationship between clutch size and body size is generally direct. Large species tend to lay larger clutches than small species (Figure 5). Within a species, clutch size increases with body size and age, although this trend is less evident and may be absent in species laying small clutches. Species that reportedly show this trend are *Chelonia mydas* (Carr and Hirth, 1962; Pritchard, 1969), *Pseudemys scripta* (Cahn, 1937; Cagle, 1950; Gibbons, 1970b; Moll and Legler, 1971), *P. rubriventris* (Smith, 1904), *Chelydra sepentina* (White and Murphy, 1973; Yntema, 1970), *Macroclemys temmincki* (Dobie, 1971), *Malaclemys terrapin* (Barney, 1922; Hildebrand, 1929), *Melanochelys trijuga* (Deraniyagala, 1939), *Kinosternon bauri* (Einem, 1956), *Sternotherus odoratus* (Tinkle, 1961), *Trionyx muticus* (Cahn, 1937; Muller, 1921), *T. spiniferus* (Newman, 1906a), and *T. sinensis* (Mitsukuri, 1905). Populations of *Chrysemys picta* are variable, with some showing a positive correlation (Ernst, 1971d) and some showing none at all (Cagle, 1954; Gibbons and Tinkle, 1969).

The number of clutches per individual probably correlates with size and/or age in local populations of most species, but this has only been

---

[1]Medem (1960) reported finding 350 eggs in a nest of *Podocnemis expansa*, but this could represent more than a single clutch. Roze (1964) reported that an average clutch size is 82 to 86 and that the record number is only 150.

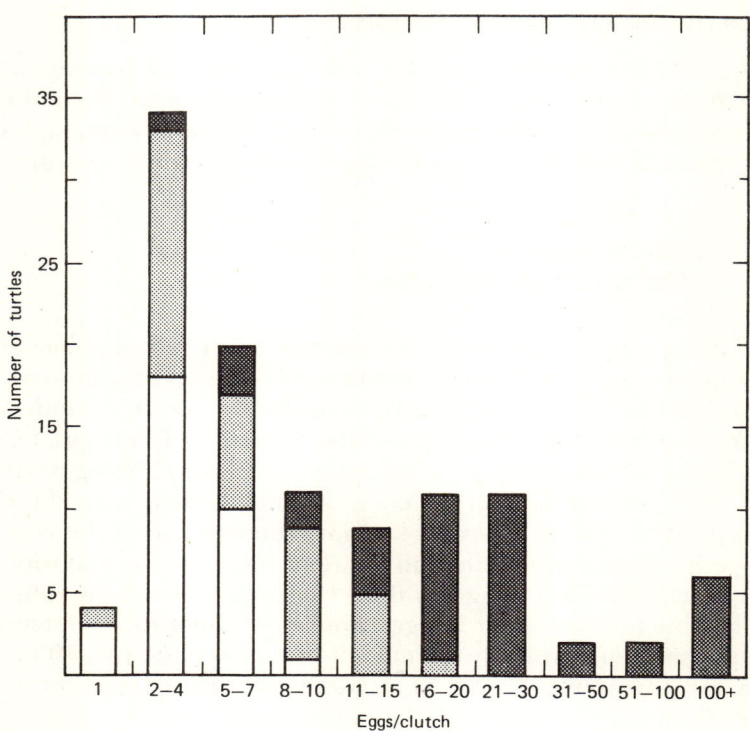

Figure 5 Histogram showing the mean or usual clutch size of 109 chelonians in relation to maximum reported carapace length (CL). Open bars are for small turtles (less than 200 mm CL), dark-stippled bars for large turtles (exceeding 300 mmCL) any light-stippled bars for medium turtles (200-300 mm CL). Chelonians represented in each clutch size category are as follows.

1 egg: *Chersine angulata, Kinosternon angustipons, K. leucostomum, Malacochersus tornieri;* **2 to 4 eggs:** *Chinemys kwantungensis, Chrysemys picta dorsalis, Clemmys guttata, C. muhlenbergi, Cuora amboinensis, C. trifasciata, Cyclemys dentata, Geochelone chilensis, G. elegans, G. elongata, Gopherus berlandieri, Hieremys annandali, Homopus areolatus, H. femoralis, Kinixys belliana, Kinosternon bauri, K. dunni, K. flavescens, K. scorpioides, K. subrubrum, Ocadia sinensis, Phrynops dahli, P. nasuta, Platysternon megacephalum, Psammobates tentorius, Rhinoclemys annulata, R. funerea, R. punctularia, Sacalia bealei, Sternotherus carinatus, S. minor, S. odoratus, Testudo graeca;* **5 to 7 eggs:** *Chinemys reevesi, Chrysemys p. marginata, C. p. picta, Clemmys insculpta, C. marmorata, Gopherus agassizi, G. polyphemus, Graptemys barbouri, G. pulchra, Kachuga smithi, Kinosternon hirtipes, Melanochelys trijuga, Mauremys caspica, M. japonica, Pelusios adansoni, Platemys platycephala, Terrapene carolina, T. ornata, Testudo horsfieldi, Trionyx cartilagineus;* **8 to 10 eggs:** *Chelodina longicollis, Chrysemys picta belli, Deirochelys reticularia, Emydoidea blandingi, Emys orbicularis, Geochelone carbonaria, G. elephantophus, Graptemys pseudogeographica, Kinosternon abaxillare, Malaclemys terrapin, Pseudemys scripta elegans;* **11 to 15 eggs:** *Chelodina expansa, Emydura macquarri, Geochelone denticulata, Graptemys geographica, Lissemys punctata, Pelusios subniger, Podocnemis sextuberculata, Psammobates geometricus, Pseudemys dorbigny;* **16 to 20 eggs:** *Chelus fimbriatus, Cycloderma frenatum, Dermatemys mawi, Geochelone pardalis, Pseudemys concinna, P. floridana, P. scripta* (tropical), *Trionyx ferox, T. muticus, T. sinensis, T. spiniferus;* **21 to 30 eggs:** *Batagur baska, Carettochelys insculpta, Chelydra serpentina acutirostris, C. s. rossignoni, C. s. serpentina, Kachuga trivittata, Macroclemys temmincki,*

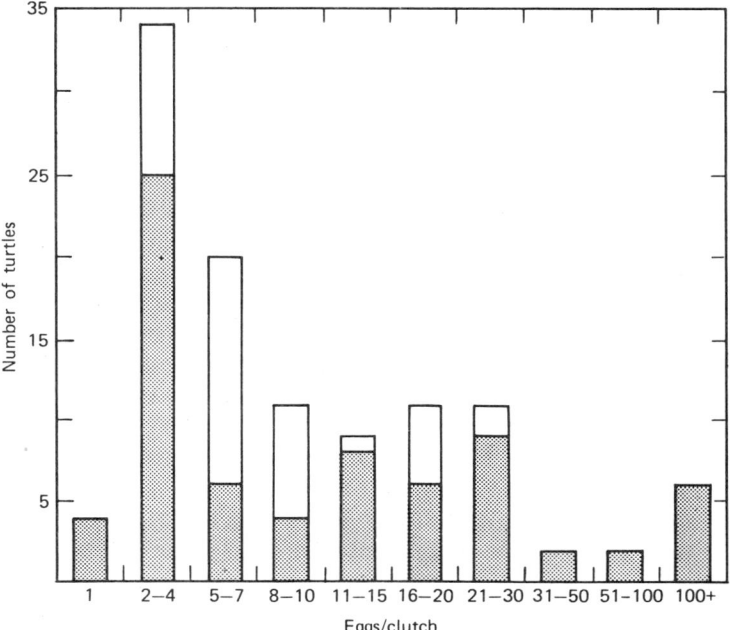

Figure 6 Histogram showing the mean or usual clutch size of 109 chelonian species and subspecies listed in Figure 5 in relation to geographic distribution. **Stippled bars represent** tropical and subtropical species (i.e., having the major portion of the distribution between 30°N and 30°S latitude). Open bars represent temperate species (i.e., having the major portion of the distribution north or south of 30° latitude).

suggested for *T. sinensis* (Mitsukuri, 1895) and *P. scripta* (Gibbons, 1970b; Moll and Legler, 1971). The correlation does not necessarily hold outside of local populations; *C. p. dorsalis*, the smallest painted turtle subspecies, lays more clutches than any of the larger races of the species (Moll, 1973).

Logically, peak reproductive performance of most turtles can be expected during the middle years of life after most growth is completed. Supporting this contention is the report (Barney, 1922) that a stock of *Malaclemys terrapin* raised in captivity increased in reproductive potential through the twenty-fifth year, then declined. The unusually small first clutches of many species may result from one ovary maturing before the

---

Pelochelys bibroni, Podocnemis cayennensis, P. unifilis, Pseudemys rubriventris; **31 to 50 eggs:** *Chelonia depressa, Kachuga dhongoka;* **51 to 100 eggs:** *Peltocephalus tracaxa, Podocnemis expansa;* **more than 100 eggs:** *Caretta caretta, Chelonia mydas, Dermochelys coriacea, Eretmochelys imbricata, Lepidochelys kempi, L. olivacea.*

other, as in *Terrapene ornata* (Legler, 1960a). Correspondingly, senescence may lower reproductive potential as one ovary becomes senile before the other (Legler, 1960a), as older females assume biennial or triennial cycles (Gibbons, 1969) or as females become completely nonreproductive (Cagle, 1944d).

Geographically no absolute trends in clutch size are evident between species (Figure 6). A proportionately large number of tropical chelonians lay small clutches, paralleling a similar trend in birds (Lack, 1968; MacArthur, 1972), but others lay the largest clutches known. Temperate species tend to lay intermediate-sized clutches. Intraspecifically, average clutch size of wide-ranging species tends to increase from lower toward higher latitudes, paralleling the trend in endotherms (Sadlier, 1973; Lack, 1954). *Chrysemys picta* (Christiansen and Moll, 1973; Ernst, 1971d; Gemmell, 1970; Moll, 1973; Powell, 1967), temperate-zone *Pseudemys scripta* (Cagle, 1950), *Terrapene ornata* groups (Milstead and Tinkle, 1966), *Sternotherus odoratus* (Tinkle, 1961), *Trionyx spiniferus*, and *T. muticus* (Webb, 1962) follow this pattern, but tropical *P. scripta* (Moll and Legler, 1971) reverses the trend and lays larger clutches in lower latitudes.

Longer nesting seasons in lower latitudes suggest a general trend for more clutches to be laid in lower latitudes—for example, *C. picta* (Christiansen and Moll, 1973; Moll, 1973), *P. scripta* (Moll and Legler, 1971), and *S. odoratus* (Gibbons, 1970a). If there is a general trend, however, associating latitude with reproductive potential, it is not yet clear. Among painted turtle populations, the reproductive potential of *C. p. belli* from Wisconsin is greater than *C. p. dorsalis* from Louisiana (Moll, 1973) but about the same as *C. p. belli* from New Mexico (Christiansen and Moll, 1973). Populations of tropical *P. scripta* have greater reproductive potential than temperate ones (Moll and Legler, 1971). But on the other hand, northern populations of *S. odoratus* have greater reproductive potential than southern ones (Tinkle, 1961).

Kinds and amount of food can vary reproductive potential indirectly through the effects on growth rate, maturation time, and ultimate size attained or, more directly, by affecting the energy reserves available for reproduction. Some knowledge of these relationships comes to us from turtle farming experiments. Long-term studies of *Malaclemys terrapin* by Barney (1922) and Hildebrand (1929) demonstrated that time to maturity could be reduced 1 or more years by altering the diet and feeding throughout the winter. Haga (1970) observed that annual productivity of *P. scripta* females on a "turtle farm" could be increased by "heavy feeding"; the increase, however, was temporary, and "breeders" treated in this fashion had to be replaced in 1 or 2 years. Under more natural

DISCUSSION

surroundings the increased growth and reproductive potential of *P. scripta* living in heated lakes may be attributable to the increased productivity at lower trophic levels of the food chain (Gibbons, 1970b). Gibbons and Tinkle (1969) also suggested that dietary differences were a primary cause of size and reproductivity variation observed between local populations of *C. picta*.

EGG SIZE

Turtle eggs range in diameter from approximately 20 mm in *Trionyx sinensis* up to the 60 mm spherical eggs of *Dermochelys coriacea* and *Geochelone elephantopus* or the 70 mm oblong eggs of *Batagur baska*, certain *Kachuga*, and *Rhinoclemys*. One of the longest recorded is a 76 × 39 mm egg from *R. funerea* (Moll and Legler, 1971), an awesome effort, considering that carapace length for this species averages less than 250 mm.

Relative egg size correlates inversely with clutch (Figure 7). This has also been reported intraspecially for several species (Risley, 1933b: Allard, 1935: Legler, 1960a: Moll and Legler, 1971). Egg size may also correlate inversely with latitude (Milstead and Tinkle, 1967: Cagle, 1950: Moll and Legler, 1971: Webb 1962). Some correlations between egg size and turtle size have been also reported (Caldwell, 1959: Yntema, 1970: Newman, 1906a). In general, since clutches, they would be expected to lay relatively smaller eggs (see Chapter 17).

DISCUSSION

Clutch size, egg size, and number of clutches may be regarded as an adaptive compromise for survival achieved over millions of years of evolution. Lack (1954, 1968), Williams (1966), Salthe (1969), Tinkle (1969), and Tinkle et al. (1970) have discussed the adaptive significance and interrelationships of these and other factors in reproductive patterns of other vertebrate groups. Many of the same principles probably apply to chelonians, but certain key categories of data such as longevity are lacking for most species. Clutch size may be visualized as a compromise involving the interplay of such factors as the optimum egg size for hatchling survival, and the selective advantages of numbers, both operating within restrictions of the female's body capacity. Similarly the number of clutches may be viewed as a compromise between such advantages as separating clutches in time and space and of liberating the

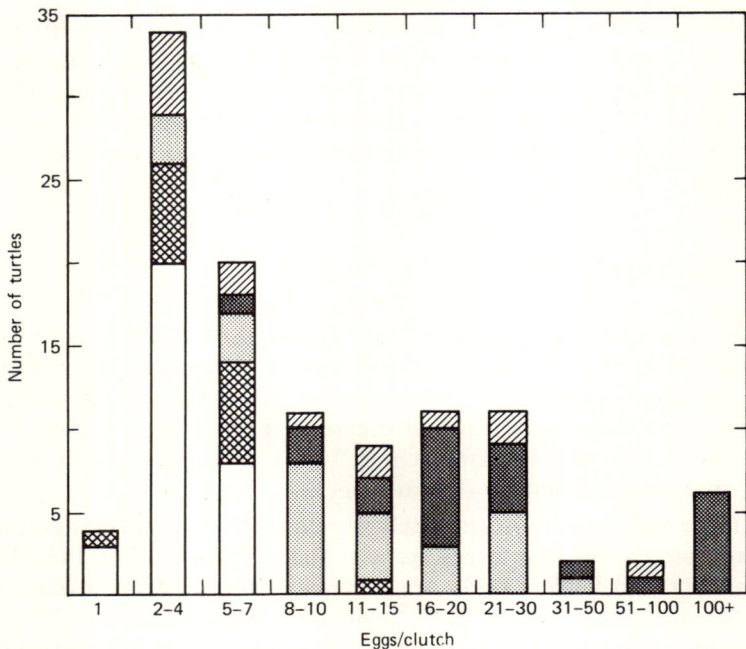

Figure 7 Histogram showing the mean or usual clutch size of 109 chelonian species and subspecies listed in Figure 5 in relation to relative egg length (mean egg length/maximum carapace length of species). Open bars represent species having relative egg lengths of 20% or more: cross hatched bars, 15 to 19%: light-stippled bars, 10 to 14%: and dark-stippled bars, less than 10%. Diagonally striped bars represent species for which data are lacking.

reproductive potential from limitations of female body size, weighed against added risks such as repeated exposure to predation, all operating within such restrictions as length of the favorable nesting season and the energy reserves available for reproduction.

Different species have met these compromises in a variety of ways. Multiple clutches have seemingly enhanced the adaptability of most patterns, thus are common to the majority of chelonians. Other than this feature, reproductive patterns vary considerably, especially in respect to clutch size. Moll and Legler (1971) considered large clutches (10+) as primitive among chelonians and associated small clutches with morphological and ecological specialization.

Chelonian reproductive patterns have seemingly evolved in two diametrically opposed directions, represented at the extremes by the patterns of sea turtles and tropical mud turtles. Characteristics associated with these extreme patterns may be summarized as follows. *Pattern I:* large clutches of relatively small eggs (see, e.g., Figure 8); multiple

DISCUSSION 329

Figure 8 Relative size of egg and plastron in a *Pseudemys scripta* (323 mm plastral length) from Panama, a recent tropical invader.

clutches produced during a well-defined nesting season; communal nesting in well-defined ancestral nesting areas; careful construction of covered nests. *Pattern II:* small clutches of relatively large eggs (see, e.g., Figure 9); acyclic or continuous nesting; solitary nesting with no special nest area; nests poorly constructed or not even attempted.

The success of both patterns is attested to by the continued existence of the species using them. Judging from present-day species, large aquatic genera have tended to evolve toward pattern I (e.g., *Batagur, Chelonia, Dermochelys, Podocnemis, Trionyx*), and smaller, semiaquatic to terrestrial, often morphologically specialized species have tended toward pattern II (e.g., *Cuora, Cyclemys, Kinosternon, Rhinoclemys*). The testudinids, having been liberated from most aquatic ties, are a special case; nevertheless larger, more generalized forms (e.g., many species of *Geochelone*) tend to lay larger clutches, whereas smaller, highly specialized forms (e.g., *Homopus, Kinixys, Malacochersus*) lay small clutches.

The survival stratagems of these two patterns are quite different. In

Figure 9 Relative size of egg and plastron in *Kinosternon leucostomum* (113 mm plastral length) from Panama, a long-time Neotropical resident.

pattern I survival is achieved in the time-honored, most conservative manner—selection for fecundity. All other factors being equal, the turtle laying the most eggs should leave the greatest number of surviving offspring. There are good indications, however, that the extreme of pattern I has become self-defeating. Advances in fecundity and the stereotyping of the pattern have served to greatly intensify density-dependent controls such as predation. Wild dogs and peccaries, which search for prey on the green turtle beaches at Tortuquero, coyotes at the ridley sites in Tamaulipas, and carnivorous birds on the arrau islands in Brazil have become as attuned to the turtles' cycles as the turtles themselves. Add man to the predators, and it is little wonder that turtles possessing such patterns are finding their way to the endangered species list.

If any turtles are to inherit the earth, it may well be those evolving toward pattern II. In a sense pattern II may have been forced on chelonians because of a phylogenetic reduction in body size along certain lineages. Small size is frequently associated with specialization and may be a factor in resource partitioning and coexistence with larger species. Whatever the selective advantages, small size restricts the reproductive capacity of the turtle. The resulting dilemma is that for small turtles to produce as many eggs as their ancestors, the eggs must be relatively small, but, if too small, they negate the advantage of quantity. Judging from the surviving species, egg diameter cannot be decreased much beyond a 20 mm greatest diameter and still be adaptive. A logical course, then, in the evolution of small turtles has been selection for eggs with greater survival capabilities rather than for mere numbers. Large eggs provide the hatchling with a large food reserve and allow it to attain a later stage of development before hatching (Lack, 1968). Furthermore, to animals such as turtles, which abandon their eggs, the characteristics of pattern II probably limit predation to an appreciable degree.

The discussion above is necessarily simplified. Between these extreme patterns lies a wide range of intermediates that utilize certain advantages of each pattern, and no doubt many selective factors other than those discussed are involved. The complexity of the subject becomes evident when tropical and temperate-zone patterns are compared. Species with extreme patterns seem to be best represented in the tropics, and most temperate-zone species have patterns somewhere between. On one hand, the large number of tropical species laying small clutches of relatively large eggs supports the hypotheses of Dobzhansky (1950) and MacArthur and Wilson (1967) that tropical species will tend to produce fewer offspring with greater competitive fitness than will their temperate-zone relatives. But on the other hand the species laying the largest clutches with the relatively smallest eggs are also tropical. I share the view expressed by MacArthur (1972): "Some ingenious observations are needed to shed new light on this old problem."

ADDENDUM

A number of papers have been published since this chapter was written. Without attempting to mention all, the following are recommended to augment this account: Wilbur et al. (1974), Iverson (1977), Lofts and Tsai (1977), Singh (1977), and Swingland (1977).

CHAPTER 17
The Embryo and its Egg: Development and Natural History

MICHAEL A. EWERT

Since there seems never to have been a review of studies of turtle eggs and embryos, this chapter contains considerable review material along with previously unreported data. There is a consensus that the egg stage is the weakest phase in terms of perpetuation of the chelonian life cycle. It is hoped that the ecological bias within the following focus will contribute toward development of a wildlife management oriented conservation outlook for the order.

OBTAINING AND INCUBATING EGGS

Eggs for research come from three major sources: (1) nests of wild populations in their natural environments, (2) layings of captive stocks of researchers, zoos, hobbyists, and commercial turtle farmers, and (3) eggs taken from oviducts of wild-caught females. The first source is perhaps least detrimental to survival of wild populations, since breeding age animals are not removed. The egg supply, however, is dependent on the nesting season, and considerable labor may be expended in locating nests and laying females. Transferring eggs from the field to the laboratory requires special handling and may take several days, depending on transportation and distance. In localities with congeneric sympatry, eggs may not be immediately assignable to species. Finally, the nest eggs have been contaminated by the nest environment and could impose difficulties on experiments demanding aseptic or approximately aseptic conditions.

Eggs obtained from captive nestings usually have involved individuals that were once wild. Two of the most successful breeding programs of species outside their natural ranges have been those of Kirsche (1967) for *Testudo hermanni* and Shaw (1970) for *Malacochersus tornieri*. Each reported viable egg production over a period of several years. Mitsukuri and Ishikawa (1886), Mitsukuri (1891), and Cunningham, Woodward, and Pridgen (1939) obtained large numbers of eggs from commercial turtle farms.

Oviducal eggs have been a major source of material (e.g., Agassiz, 1857; Mahmoud, Hess, and Klicka, 1973). Cagle (1944c) and Yntema (1964) discussed obtaining these eggs, and Yntema provided considerable detail on working with them. The main advantage is that gravid females may be shipped relatively easily from the field without loss of viability to the eggs; moreover, the eggs can be stored in a nearly suspended state of development for at least a week, usually longer, within the female oviducts. The cost of this method has usually been the life of the female parent, since captives tend not to release their eggs until after they are no longer viable. Eggs in females of many species may be detected by palpating with one's fingers the soft tissues anterior to the hind legs, particularly adjacent to the bridge. One must be careful to distinguish between premature eggs, ripe eggs, and lumpy intestines, which may be inflated with compressed vegetable material as typical of *Pseudemys concinna* and *Rhinoclemys pulcherrima*. In some species, notably testudinids, eggs may be relatively remote from the palpable area, rendering palpation ineffective. Taking an X-ray photograph is a possible alternative. Eggs tend to be ripe as soon as their shells have become completely calcified; the status may be detectable through palpation. In kinosternids a few days may pass when the eggs feel hard and ripe yet are not and break within the oviducts if palpated vigorously. If gentle palpation reveals that the eggs seem hard, an additional 1 to 2 weeks at 27°C, depending on the species, will ensure that they are ripe without loss of viability. In contrast, ripe eggs of American *Trionyx* species appear to have a short period of viability within the oviducts, at maximum about a week at room temperature, and should be removed as soon as they appear to be ripe.

Major considerations in laboratory incubation of eggs involve provisions of appropriate temperature and moisture. Although lesser concerns have been expressed over oxygen supply and illumination, I am

OBTAINING AND INCUBATING EGGS

unaware of mortality in laboratory incubations attributable to inadequacies in either. Perhaps too little attention has been devoted to fungal and microbial contamination, although eggs are remarkably resistant to both. New (1966) reviewed aseptic culture techniques used in incubation of many reptilian embryos, and Yntema (1964) elaborated his aseptic methods for *Chelydra serpentina*. Most workers surround their eggs with a natural or seminatural substrate, specifically sand or loam with sand (Coker, 1910; Jayakar and Spurway, 1966; Shaw, 1966, 1970; Goode, 1967a; Kirsche, 1967; Bustard and Greenham, 1968; Bustard et al., 1969; Mahmoud et al., 1973) or peat moss or peat moss mixed with soil (Hausmann, 1968b; Hoessle, 1969; Schmidt, 1970; Zovickian, 1971; Sachsse, 1974). Goode sterilized his incubation media with improved results. Cunningham et al. (1939) tried wet cheese cloth as an artificial substrate, and Lynn and von Brand (1945), Legler (1960a), and Moll and Legler (1971) used moist cotton. Legler (1956) compared sand, sphagnum moss, paper towels, and cotton concluded that moist cotton gave the best results in his work with *Terrapene ornata*. Yntema (1964) kept *C. serpentina* eggs partly submerged in water, and Ehrengart (1971) maintained *Testudo hermanni* eggs on a flat surface in an atmosphere of humid air. Optimum humidity varies with the species; generally, parchment shelled eggs require close to 100% relative humidity, and brittle shelled eggs survive well with less and may actually demand drier conditions if from desert dwelling tortoises.

Most species require conditions for incubation warmer than room temperature, and commercial incubators may be used. Simple provisions involve an incandescent light bulb in a box (e.g., Ehrengart, 1971) or a warmed space in which covered trays of eggs may be placed. Thermal limitations vary with species, but 28 to 30°C probably is safe for all.

Experiences from more than 10 years of incubating eggs of Kinosternidae, Chelydridae, Emydidae, and Trionychidae have left me with some personal preferences and recommendations. If aseptic conditions are not required, eggs from field nests work well and transport easily in a passenger car. Those laid in sand I carry in trays of sand; those from other substrata I place in moist sand or moist paper towels. Eggs a day or less old, which are easily recognized, endure rougher treatment than older eggs. Those in which the vitelline membrane has adhered to the shell should be treated carefully, starting with caution in lifting them from their egg chamber in the nest without rolling them.

Eggs in gravid females of many if not all species may be "forced" out through hormonal induction without sacrificing the female. Yntema (1964) obtained eggs from *C. serpentina* after injecting pituitary extract presumably containing oxytocin as the active egg-forcing substance. I found that as little as 1 unit of synthetic oxytocin per 100 g body weight freed eggs in *Sternotherus odoratus, Chrysemys picta, Staurotypus salvini,* and *Rhinoclemys pulcherrima.* An injection of 2 to 3 units per 100 g is probably not excessive for small turtles. A preparation of 60 mg of dried, lyophilized whole beef pituitary shaken in 10 cc of water and injected in the amount of 1 cc of supernatant per kilogram of turtle into the inguinal region works well. Eggs have appeared between 20 min and 6 hr later, and placing the turtle in warm (30–32°C) as opposed to cold water appears to facilitate laying. Species yielding eggs included *Pelusios sinuatus, Kinosternon bauri, K. sonoriense, Sternotherus minor, Clemmys muhlenbergi, Terrapene carolina, Graptemys geographica, G. oculifera, G. pseudogeographica, Pseudemys scripta, Rhinoclemys areolata,* and *Trionyx muticus.* When provided with suitable nesting substrate, females of *C. serpentina, Clemmys insculpta, S. carinatus, K. leucostomum,* and *K. scorpioides* constructed complete nests containing eggs within a few hours of injection. Two *S. odoratus* yielded premature, unripened eggs. In about 140 females injected, adverse effects of the injected substances have not been evident, although repeated injection of the whole pituitary extract could lead to immune reactions. The method should be especially valuable to investigators working with rare and endangered species, or otherwise precious individuals, and to those interested in nest construction.

For permanent incubation media I have tried sand, cotton, and vermiculite; but sand lost appropriate moisture rapidly, and cotton occasionally fostered microbial infection. A suitable moisture content for starting parchment-shelled eggs is obtained by adding water to *fresh* vermiculite in excess of 5 min adsorption capacity, then draining the excess after 5 min. The initial preparation feels sticky but in about a day loses this stickiness and remains thoroughly damp. A single layer of eggs is covered with 0.5 to 1 cm of vermiculite. Water is added when surface vermiculite feels dry, which is relatively infrequently when egg trays are kept partially covered. There has been no loss of eggs to microbial infection. Vermiculite particles, which tend to cling to the shells of eggs, seem too messy for aseptic incubation.

EXTERNAL FEATURES OF EGGS

Table 1 provides a survey according to family of shape, shell form, and size as determined from the literature, museum specimens, and fresh material personally acquired. Since descriptions of regional faunas do not consistently include egg data and museum collections retain very few eggs, Table 1 represents only 70% of extant species. Many of the data are derived from numerous reports on single species. Works containing more extensive egg data include: for Africa, Anonymous (1937), Loveridge and Williams (1957), Villiers (1958); for southeast Asia, De Rooij (1915), Smith (1931), Mell (1938), Deraniyagala (1939), Bourret (1941), Fukada (1965), Minton (1966); for Australia, Goode (1967b); for Europe, Gadow (1901); for North America, Carr (1952), Ernst and Barbour (1972); for South America, Mondolfi (1955), Medem (1962, 1969), Freiberg (1967), and for Central America, Moll and Legler (1971).

Shape

Species with spherical eggs (Table 1) usually lay more than 10 in a clutch; those with elongate eggs typically produce fewer than 10 at a time. This phenomenon is obvious in a contrast between species with a maximum clutch size of two eggs, which lay elongate eggs exclusively, and species with clutches averaging more than 50 eggs, which lay only spherical eggs. The trend is maintained within the family Testudinidae in relation to the maximum recorded clutch for each species and within the chelid genus *Phrynops* (see Chapter 16). Within the species *Trionyx muticus*, ovate eggs, where the greatest diameter exceeds the least by 3 mm or more, occur more frequently in southern populations (14.2% of 520 eggs in 82 clutches), where clutches typically contain 5 to 7 eggs, than northern (0.7% of 452 eggs in 20 clutches), where 20 egg complements are common. The spherical egg may be an adaptation for packing large numbers of eggs into the oviducts.

Egg shape follows taxonomic lines in the large families of Kinosternidae and Emydidae, where eggs are always elongate, and in Trionychidae, where they are approximately spherical. However egg shape is variable in Neotropical Pleurodira and in Testudinidae, hence imperfect as a diagnostic character.

Elongate eggs that have more complex shapes than the typical ellipsoid include the tapered or "bird egg" shape in *Playsternon megacephalum*, *Cuora trifasciata* (Mell, 1938), *Rhinoclemys areolata* and *Platemys platycephala*. Double ellipsoids with unequal lengths, widths, and depths occur in species of *Kinixys* (Villiers, 1958). Eggs that are flattened on one or more sides or dented irregularly are common in given clutches of species laying parchment-shelled eggs, notably *Pseudemys scripta* (Cagle, 1950), *Chrysemys picta*, and *Graptemys pseudogeographica*.

Size

Variability in egg size occurs in any given species. Examination of 20 to 40 clutches from single populations indicates that the largest viable eggs

Table 1 Characteristics of Eggs According to Family: Shape, General Shell Form, and Extremes in Size as Shown in Species Laying the Largest and Smallest Eggs[a]

Family and Species	Percentage of Species with Egg Data	Shape		Shell Form			Linear Dimensions (mm) (weights in g)	Reference (for individual species)
		Spherical	Ellipsoid	Parchment (pliable)	Hard—Expansible	Brittle		
Pelomedusidae								
(Ethiopian)	70		XXX	XX	X			
Pelusios adansoni	50	X	XXX	X	XX		18–19 × 29.5–33	Villiers (1958)
(Neotropical)			X		X	XX		Goeldi (1898),
Podocnemis expansa		X	X	X			36.5–43 × 44–48	Medem (1969)
Chelidae								
(Australian)	90		XXX			XXX		
Chelodina longicollis	50	XX	XX			XXX	12.5–21.3 × 21.0–33.8	Vestjens (1969)
(Neotropical)			X		X	X	33 × 55	De Rooij (1915)
Chelus fimbriatus		X	X		X	X	34.3 × 37.5	Goeldi (1898)
Dermatemyidae	100		X		X	X		
Dermatemys mawi			X		X	X	33.4–34.8 × 57.2–63.0	Holman (1964)
Kinosternidae	80		XXX			XXX		
Sternotherus odoratus			X		X	X	12.6–18.1 × 21.5–30.7 (2.2–5.6)	this account

Species	%									Size (mm)	Reference
Kinosternon dunni								X	X	25 × 44–45	Medem (1962)
Staurotypus triporcatus								X	X	21.4–26.3 × 35.6–44.6	Legler and Moll (unpubl.)
Chelydridae											
Chelydra s. serpentina	100	XX	X		XX	X				19.9–32 × 19.9–32.4 (7–17.3)	Yntema (1970b), Vogt (unpubl. data), this account
Macroclemys temminicki		X			X		X			33.0–44.0 × 34.0–51.8 (22.9–30.4)	Dobie (1971), this account
Platysternidae											
Platysternon megacephalum	100		X	X				?	?	22–22.5 × 37–37.5	Mell (1938), Mebs (1963)
Emydidae											
(Emydinae)			XXX	XXX		X					
(Batagurinae)			XXX	X		XX			XX		
Emys orbicularis	90		X			X					Gadow (1901)
Chrysemys picta	60		X	X						15 × 25	Ernst (1971b), this account
										14.6–20.7 × 23.1–38.7 (3.2–9.1)	
Clemmys muhlenbergi		X		X						13–16 × 28–36 (4.5–4.7)	Wright (1918), Zovickian (1971), this account
Orlitia borneoensis								X		41 × 79	De Rooij (1915)
Callagur borneoensis			X					X		36.6–44.5 × 71.3–76.3	De Rooij (1915), this account
Testudinidae											
Testudo kleinmanni	80	XX	XX	?				XXX	X	22–23 × 28–30	Flower (1933), Loveridge and Williams (1957)

Table 1 *(Continued)*

Family and Species	Percentage of Species with Egg Data	Shape			Shell Form			Linear Dimensions (mm) (weights in g)	Reference (for individual species)
		Spherical	Ellipsoid	Parchment (pliable)	Hard—Expansible	Brittle			
Psammobates tentorius			X			X		21.3–23.5 × 27.0 × 34.5	Anonymous (1937), Loveridge and Williams (1957)
Geochelone elephantopus		X				X		55.2–56.5 × 59.5–59.7 (107–113)	Shaw (1966)
Carettochelidae	100	X				X			
Carettochelys insculpta		X				X		37–42 × 38–42(30.3)	De Rooij (1915), this account

Trionychidae		50	XXX		
Trionyx sinensis			X	XXX	10–23 × 10–23	Mitsukuri and Ishikawa (1886), this account
Cycloderma frenatum			X	X	28.5–34.1 × 29.6–35.8	Loveridge and Williams (1957)
Cyclanorbis senegalensis			X	X	36 × 36	
Cheloniidae		100	XXX	XXX		
Eretmochelys imbricata			X	X	35–38 × 35–38	Deraniyagala (1939)
Chelonia mydas			X	X	49.0–58.7 × 49.0–58.7	Carr and Hirth (1962)
Dermochelyidae		100	X	X		
Dermochelys coriacea			X	X	49–53 × 51–57 (61.8–84.3)	Deraniyagala (1939)

a"X" indicates presence; for families with several species, "XX" indicates frequent occurrence, "XXX," the predominant mode; "?" indicates that the literature has been vague or contradictory.

Figure 1 Egg size tends to increase with carapace length in the family Testudinidae, size is represented as a linear unit, the cube root of length times width times depth. Each point represents a separate species according to the following list (corresponding sequence on the figure is left to right, then upper to lower): *Homopus boulengeri, H. areolatus, Psammobates oculifer, P. tentorius, H. femoralis, Malacochersus tornieri, Kinixys belliana, Testudo hermanni, Gopherus berlandieri, K. homeana, Geochelone elongata, G. travancorica, G. elegans, G. chilensis, T. horsfieldi, T. graeca, K. erosa, Gopherus polyphemus, G. agassizi, Geochelone radiata, G. emys, G. carbonaria, G. pardalis, G. denticulata, G. sulcata, G. gigantea, G. elephantopus.*

have between 2 and 3 times the mass of the smallest. Variation between subspecies sometimes increases this range. Such intraspecific variation tends to hamper analyses of interspecific variation.

Some of the more interesting interspecific comparisons require data on adult body size (Figure 1, Table 2). Statistics on body size are complicated, however, by apparent effects of local habitat (Gibbons and Tinkle, 1969; Gibbons, 1970b) and regional geographic variation (Moll, 1973), plus the unavailability for most species of parameters of mean body size and size at onset of reproduction. Body size estimates (Table 2) represent large, but not record sized individuals (see Smith, 1931; Wermuth and Mertens, 1961; Boulenger, 1966; and Conant, 1975).

In addition to egg size, Table 2 considers hatchling size. Data were occasionally available for one but not the other, and use of both extends the list. Hatchling size corresponds to egg size when the form of measure is weight. In absence of data on weight, comparisons are based on linear measure, which may be somewhat complicated by varied shapes of eggs and hatchlings. This variation is slight at the familial level. In determination of mean hatchling sizes numerous measurements were taken from birth plates of specimens that had grown since hatching (Cagle, 1946). Estimates for species where the birth plates are lost or distorted through ecdysis were excluded.

EXTERNAL FEATURES OF EGGS

The most obvious relationship between egg size and other factors is that larger species tend to lay larger eggs, as shown for Testudinidae in Figure 1. When analyses are restricted to medium-sized species of given families, however, considerable variation in egg size remains. One source of variability may involve broad climatic regions as given in Table 2. Species that range extensively within wet tropical regions, or tropical rain forests as designated by Richards (1952), are separated from species of other climatic regions. Rain forest species and subspecies produce proportionately larger eggs and hatchlings than species and subspecies of temperate climates or tropical climates with distinct wet and dry seasons. Causal factors relating climate to size have not been determined, but speculation allows for differences in intensities of egg or hatchling predation, types of predators, and lengths of laying seasons. Additional significance of egg size is discussed in Chapter 16.

Shells

The eggshell consists of a mineral layer overlying one or two fibrous organic layers (Young, 1950). Characteristics of these layers, particularly the mineral layer, differ appreciably among species and may be grouped into three categories.

Parchment-shelled eggs indent easily when laid and tend to expand through uptake of water rather early during incubation (e.g., Cunningham and Hurwitz, 1936; Cunningham and Huene, 1938). The shell appears to expand uniformly and assumes a rubbery texture such that eggs bounce when dropped gently. The shell sags when eggs are dehydrated. Toward the end of incubation, the mineral layer tends to slough off as small granules.

Brittle-shelled eggs do not indent easily when laid. The mineral layer will shatter locally if local pressure is applied, and typically, as in kinosternid eggs, such damage includes rupturing of the organic layers, hence exposure of the inner contents. Expansion may occur toward the end of incubation, producing cracks in the outer mineral layer, which fractures into large pieces as the fibrous mineral layers expand. Occasionally, as in *Kinosternon bauri* and *Rhinoclemys areolata*, the entire shell may fracture before pipping, permitting albumin to seep out (Einem, 1956; present account). Uneven moistening during incubation can lead to excessive expansion in a local area and subsequent breaking of the shell, the extraembryonic membranes, and occasionally the yolk sac. Death of the embryo commonly follows. Brittle-shelled eggs may not expand at all during incubation, but if expansion causes the mineral layer to fracture, it tends to slough off in large flakes toward the end, as in eggs of

Table 2 Linear Measurements of Adults, Eggs, and Hatchlings of Selected Species According to Family and Regional Climate of Occurrence

Family, Regional climate, and Species	Carapace Length of Large Adult (mm)	Average or Median Size of Egg (mm)	Size of Hatchling[a] Carapace Length (mm)	Size of Hatchling[a] Plastron Length (mm)
Pelomedusidae				
Subtropical Seasonal				
Pelomedusa subrufa	300	22 × 38	29.7 (5H)	25.2
Pelusios subniger	350	21 × 36	34.9 (8H)	33.5
MEAN	325	22 × 37	32.3	29.4
Tropical Wet				
Pelusios gabonensis	300	—	42.3 (8H)	39.8
Podocnemis sextuberculata	300	—	46.0 (8H)	42.2
MEAN	300		44.2	41.0
Chelidae				
Warm Temperate				
Chelodina expansa	380	25 × 37	37.1 (H)	31
C. longicollis	250	20 × 30	28.8 (H)	28.3
C. oblonga	270	21 × 35[b]	31 (H)[b]	—
Elseya latisternum	280	21 × 32[c,d]	—	—
Emydura macquarri	300	23 × 33	29.3 (H)	—
MEAN	296	22 × 33	31.5	29.7
Tropical Wet				
Chelodina novaeguineae	280	20 × 29[c,d]	—	—
Elseya novaeguineae	300	33 × 55	46.3 (10H)[c]	37.6[c]
Emydura subglobosa	300	18 × 34[c]	28.2 (2H)[c]	23.1[c]
Phrynops geoffroanus	300	32 × 34	45.5 (15H)	38.7
P. gibbus	200	31 × 44	46.8 (4H)	38.6

Table 2 (Continued)

Family, Regional climate, and Species	Carapace Length of Large Adult (mm)	Average or Median Size of Egg (mm)	Size of Hatchling[a]	
			Carapace Length (mm)	Plastron Length (mm)
Platemys				
platycephala	200	28 × 51[c,d]	56.5 (5H)[c]	52.4[c]
P. radiolata	200	—	41.9 (2)	37.6
MEAN	254	27 × 41	44.2	38.0
Kinosternidae				
Warm Temperate—				
Subtropical				
Kinosternon				
hirtipes	150	17 × 28	—	—
K. integrum	160	16 × 30	27.0 (14H)	22.4
K. sonoriense	160	19 × 33	27.0 (1H)	23.5
Sternotherus				
carinatus	150	18 × 32	27.2 (6H)	19.6
MEAN	155	18 × 31	27.2	21.8
Tropical Wet				
Claudius				
angustatus	150	18 × 32[c]	35.4 (4)	24.5
Kinosternon				
angustipons	120	22 × 40	—	—
K. dunni	175	25 × 45	—	—
K. leucostomum	170	20 × 37	33.7 (5H)	26.3
K. scorpioides	180	19 × 35	29.8 (3H)	24.3
MEAN	159	21 × 38	33.1	25.0
Chelydridae				
Temperate—				
Subtropical				
Chelydra serpentina				
serpentina	400	27 × 28	32.1 (154H)	22.6
C. s. osceola	360	27 × 28	32.1 (52H)	22.5
MEAN	380	27 × 28	32.1	22.6
Tropical Wet				
C. s. acutirostris	400	33 × 36	36.6 (1H)	23.7
C. s. rossignoni	360	34 × 35[c]	37.7 (4H)	23.3
MEAN	380	34 × 36	37.0	23.5

Table 2 (*Continued*)

Family, Regional climate, and Species	Carapace Length of Large Adult (mm)	Average or Median Size of Egg (mm)	Size of Hatchling[a]	
			Carapace Length (mm)	Plastron Length (mm)
Emydidae				
Temperate				
Clemmys insculpta	220	23 × 33	35.0 (90H)	29.2
C. marmorata	180	21 × 38	27 (H)	—
Emys orbicularis	200	17 × 30	26.8 (10H)	23.9
Terrapene carolina major	180	22 × 38	33.6 (24H)	30.0
T. coahuila	165	17 × 33[e]	30.4 (2H)[c]	—
Malaclemys terrapin	230	21 × 31	28.8 (H)	26.1
Graptemys barbouri	270	26 × 36	36.6 (41H)	32.9
G. p. pesudo-geographica	250	22 × 35	32.1 (222H)	30.4
Chrysemys picta belli	210	19 × 32	26.7 (87H)	25.9
Pseudemys concinna suwanniensis	400	27 × 42	36.0 (54H)	32.4
P. floridana peninsularis	370	25 × 36	35.8 (21H)	32.5
P. scripta elegans	270	22 × 36	32.3 (71H)	29.8
Deirochelys reticularia	240	23 × 37	33.5 (7H)	30.0
Emydoidea blandingi	240	24 × 39	36.5 (27H)	32.1
Mauremys caspica	240	20 × 35	32.4 (7H)	27.9
M. mutica	180	21 × 38[c]	34.0 (9H)[c]	29.6
Ocadia sinensis	230	25 × 40	34.9 (3)	33.1
Chinemys kwangtungensis	200	27 × 51	—	—
Cuora flavomarginata	170	26 × 48	40.5 (10Y)	38.1
MEAN	234	23 × 37	32.8	32.4
Tropical Seasonal				
Terrapene nelsoni	145	27 × 42	—	—

Table 2 (*Continued*)

Family, Regional climate, and Species	Carapace Length of Large Adult (mm)	Average or Median Size of Egg (mm)	Size of Hatchling[a]	
			Carapace Length (mm)	Plastron Length (mm)
Rhinoclemys				
pulcherrima	200	31 × 52	50.9 (8Y)	45.2
R. rubida	200	25 × 62[cf]	51.9 (5Y)	48.5
Geoclemys				
hamiltoni	330	—	37.2 (6)	34.3
Kachuga				
smithi	230	23 × 44	39.2 (4)	36.7
Kachuga tecta				
tentoria	230	—	38.6 (10Y)	34.3
Pyxidea				
mouhoti	180	25 × 40[c]	34.8 (3)	33.9
MEAN	216	26 × 48	42.1	38.8
Tropical Wet				
Pseudemys				
scripta ornata	350	28 × 42	36.5 (42H)	35.2
Rhinoclemys				
annulata	200	37 × 71[c]	63.9 (9H)	55.2
R. areolata	200	31 × 60[c]	54.7 (11H)	49.9
R. funerea	320	35 × 68	63.4 (12H)	57.7
R. punctularia	270	37 × 71	56.3 (2Y)	52.6
Malayemys				
subtrijuga	210	22 × 44	35.3 (11H)	31.1
Melanochelys				
trijuga thermalis	210	27 × 47	44.1 (4)	39.5
Morenia				
ocellata	210	—	42.2 (1)	37.5
Cuora amboinensis				
Domed biotype	200	32 × 43	47.3 (10Y)	43.0
Depressed biotype			48.8 (11Y)	37.5
C. trifasciata	200	27 × 50	51.7 (1)	49.1
Notochelys				
platynota	320	—	56.2 (14H)[c]	46.5[c]
Cyclemys				
dentata	240	35 × 57	56.2 (12H)	49.5
Heosemys				
spinosa	220	—	63.2 (10Y)	58.1
H. sylvatica	120	—	44.3 (2)	38.5
Geoemyda				
spengleri	130	—	42.2 (4H)	36.4

Table 2 (*Continued*)

Family, Regional climate, and Species	Carapace Length of Large Adult (mm)	Average or Median Size of Egg (mm)	Size of Hatchling[a]	
			Carapace Length (mm)	Plastron Length (mm)
Siebenrockiella crassicollis	180	19 × 45	52.0 (12H)	43.2
MEAN	224	30 × 54	50.4	45.0

[a]Number in parentheses gives sample size when known (for the columns on hatchling size); "H" indicates that the series included one or more actual hatchlings; "Y," that the series included one or more babies, practically hatchlings.
[b]*Source:* Burbidge (1967).
[c]This may be the first published measurement of the developmental stage (egg or hatchling).
[d]*Source:* Legler (unpubl. data).
[e]*Source:* Brown (1974).
[f]*Source:* Friedman (unpubl. data).

Geochelone elephantopus (Shaw, 1966) and species of *Trionyx*. Dehydration of brittle-shelled eggs does not alter the shape of the mineral layer. During early to middevelopment, air pockets may appear interior to the fibrous shell membranes in some species (e.g., in *K. flavescens, Staurotypus salvini, Mauremys mutica, Melanochelys trijuga*) and between the mineral layer and fibrous membranes in others (e.g., *Trionyx* spp.). Late in development fibrous layers separate from the mineral layer in most species.

Eggs with *hard* but *expansible* shells tend to expand and behave like parchment-shelled eggs during incubation. At the time of laying, the major difference is observed inasmuch as the shell does not indent easily. Hence although these eggs behave unlike brittle-shelled eggs during incubation, their initial hard appearance has lead to confusion with brittle-shelled eggs in published accounts.

Parchment-shelled eggs and hard, expansible eggs can occur within different populations of the same species, as for example, with *Chelydra s. serpentina* from Minnesota having the former and *C. s. osceola* from south Florida having the latter. Shell flexibility may even vary between eggs from different females of the same population, suggesting that this distinction is superficial. The rare intermediate state between brittle-shelled and expansible-shelled eggs occurs in eggs of *Chelodina longicollis* where Legler (personal communication) found that the mineral layer begins to fracture into small flakes midway through the incubation period. Premature oviducal eggs of many species appear to be parchment-shelled despite the form they assume when mature.

Affinities of shell form to taxonomic groups is variable (Table 1). In some of the larger groups, exclusively parchment-shelled or brittle-shelled eggs have been reported. In contrast, the Baturinae exhibit all three types as follows: parchment-shelled, *Batagur baska* (Loch, 1950); hard expansible, again *Batagur baska* (Khan bin Momin Khan, 1964) and *Kachuga smithi* (Minton, 1966); and brittle shelled, *Rhinoclemys funerea* (Moll and Legler, 1971) and *Pyxidea mouhoti*. South American members of the genus *Podocnemis* appear to show the same variability: parchment-shelled, *P. expansa* (Medem, 1969); hard expansible, *P. unifilis* (Medem, 1964), and brittle-shelled, *P. vogli* (Pardo, 1969).

There has been some speculation that turtle eggs change from pliable to brittle-shelled during and just after the laying process, since least diameters of hard-shelled eggs are larger than the distance between the carapace and the plastron adjacent to the cloacal orifice of the laying female in *Gopherus berlandieri* (Grant, 1960; Paxson, 1962) and in *Homopus areolatus* (Eglis, 1963). Eggs of *H. areolatus* are said to be soft-shelled (Anonymous, 1937; Rose, 1962) and hard-shelled (Eglis, 1963). Eglis noted that the carapace and plastron were flexible with respect to each other, and Patterson (1970) has demonstrated flexibility of body shell in relation to calcium in the diet of *G. agassizi*, a factor that could fluctuate during egg formation. Eggs of species of *Rhinoclemys* are especially large, yet they always seem to be brittle-shelled, whether laid in water or on land. Superficially, the body shell of *R. areolata* presents a rigid, unhinged aspect. However immediately after one captive specimen laid an egg, my fingers could flex the carapace and plastron at the cloacal orifice. Similarly, egg-laying *R. pulcherrima* and *Clemmys guttata* could be flexed. Other emydids not included among kinetic species (Bramble, 1974) but weakly kinetic are *Heosemys spinosa* and *Sacalia beali*, which appear to have unusually large hatchlings (Table 2; Mell, 1938). In absence of any careful studies of turtle eggshell hardening, it is suggested that females' shells, not the shells of eggs, change in shape during laying.

Table 3 shows the correlation between qualitative aspect and quantitative data on the layered structure of egg shells. In brittle-shelled eggs the mineral layer exceeds 60% of the overall thickness; in parchment-shelled it is 50% or less. Hard expansible eggs overlap these categories with 41 to 73%. There does not appear to be any relationship between absolute shell thickness and shell form.

The mineral layer of parchment-shelled eggs has a nodular aspect because of the presence of a loose formation of crystalline structures; Young (1950) termed these sphaerulites, and they are pictured in detail in Agassiz (1857). During expansion of the shell the nodules separate

Table 3 Some Characteristics of Eggshells[a]

Family and Species	Approximate Size of Egg (g)	Percentage of Shell Thickness in:		Relative Thickness[b]	Actual Shell Thickness (mm)	Shell Type	Surface Aspect
		Fibrous Layers	Mineral Layers				
Pelomedusidae							
Pelusios sinuatus	11	50.2	49.8		0.19–0.24	Parchment	Nodular
Chelidae							
Chelodina expansa	18				0.25–0.40	Brittle	Minutely pitted
Elseya dentata	20				0.5	Brittle	Smooth to irregularly rough
Pseudemydura umbrina[c]	9				0.205	Brittle	
Kinosternidae							
Kinosternon scorpioides	7	16	84		0.35–0.51	Brittle	Smooth (pores near lesser circumference)
K. subrubrum	4	28.7	71.3	2.26		Brittle	Smooth
Sternotherus carinatus	6	17	83		0.28–0.36	Brittle	Minutely pitted
S. odoratus	4	23.2	76.8	1.40	0.17–0.20	Brittle	Smooth

Species	n						
saltvini	11	25	75		0.24–0.38	Brittle	Minutely pitted (pores near lesser circumference)
Chelydridae							
Chelydra serpentina	11	49.4	50.6	2.02	0.20–0.46	Parchment to hard	Nodular
Macroclemys temmincki	27	58.9	41.1	2.51	0.21–0.24	Parchment to hard	Nodular
Emydidae							
Clemmys guttata	6	64.9	35.1	1.45	0.51–0.21	Parchment to hard	Nodular
Terrapene carolina	10	63.5	36.5	1.45		Parchment	Nodular
Graptemys geographica	10	63.6	36.4	1.43		Parchment	Nodular
Chrysemys picta	6	49.1	50.9	1.00	0.12–0.13	Parchment	Nodular
Pseudemys concinna	16	59.7	40.3	1.64		Parchment	Nodular
Emydoidea blandingi	14	45.2	54.8	1.86		Hard	Nodular
Emys orbicularis[a]	6	27	73		0.27–0.28	Parchment to hard	Nearly smooth
Melanochelys trijuga coronata	15	29	71		0.30–0.48	Brittle	Smooth to minutely nodular

Table 3 (Continued)

Family and Species	Approximate Size of Egg (g)	Percentage of Shell Thickness in:		Relative Thickness[b]	Actual Shell Thickness (mm)	Shell Type	Surface Aspect
		Fibrous Layers	Mineral Layers				
Rhinoclemys areolata	34	19	81		0.31–0.43	Brittle	Smooth with pores
Testudinidae							
Geochelone elegans	27	30.8	69.2		0.49–0.54	Brittle	Irregularly rough
G. radiata[c]	37	22.4	77.6		0.59–0.84	Brittle	Smooth
Gopherus polyphemus	40	38.0	62.0	3.02		Brittle	Smooth
Carettochelidae							
Carettochelys insculpta[e]	40				0.4	Brittle	Smooth

Trionychidae					
Trionyx					
euphraticus[c]	8	16.7	83.3		0.23–0.29	Brittle	Smooth
T. ferox	14	27.1	72.9		0.20–0.25	Brittle	Smooth
T. spiniferus	12	27.3	72.7	1.98	0.15–0.19	Brittle	Smooth
Cheloniidae					
Chelonia mydas	52				<0.33	Parchment	
Caretta caretta	35	66.0	34.0	2.52		Parchment	Nodular
Dermochelyidae					
Dermochelys					
coriacea	78	69.0	31.0		0.34–0.37	Parchment	

[a] Unspecified sources of data include Agassiz (1857) and the present author's observations.
[b] Graphic representations from Agassiz (1857) do not agree with current analyses of fresh material. In order to utilize data from Agassiz' extensive research, *Chrysemys picta* has been assigned a value of 1.00 and his other data are presented relative to this.
[c] *Source:* Burbidge (1967).
[d] *Source:* Young (1950).
[e] *Source:* Legler (unpubl. data).

singly or into small groups affixed to the underlying fibrous membranes. In brittle-shelled eggs the mineral layer may manifest a smooth external aspect or a nonnodular roughness, which Young felt might result from long retention in the oviducts but may be characteristic of some species. Deraniyagala (1939) reported that eggs of *Melanochelys trijuga thermalis* and *Geochelone elegans* had crystals in intermeshed series of polygons tangential to the egg. Large pores occur at interstices between crystals, and in *G. elegans* may exude fluid during development. Pores also occur in the mineral layer in the brittle-shelled eggs of *Claudius angustatus* and *Kinosternon scorpioides* where they seem to be most numerous near the least circumference and in *Rhinoclemys areolata* and *R. pulcherrima* where they are broadly distributed. Young found pores in *Emys orbicularis* and *Trionyx euphraticus* eggs, but not in *Testudo graeca* eggs. Young identified sphaerulite crystals in eggs of the same three species as aragonite, a rhombic form of calcium carbonate, found in shells of some invertebrate eggs (Needham, 1963).

Young suggested that there are two fibrous membranes in the shell in *E. orbicularis* and *T. euphraticus*, as is the case in birds (Romanoff and Romanoff, 1949), but only one in *T. graeca*. I applied a tangential shearing force to tear eggshells of several species and found two parts in *Pelusios sinuatus, Elseya latisternum, Macroclemys temmincki, Chrysemys picta, Melanochelys trijuga, Geochelone elegans, Carettochelys insculpta*, and *Trionyx ferox*. However I have never seen the membranes parting from each other naturally during development or hatching. Crystalline nodules are deeply embedded in the outer fibrous layer in *C. picta*. In *Sternotherus odoratus* only a single membrane is distinguishable. An egg from a tank housing several species had a distinct five-ply shell structure of two thick crystalline layers between three thinner and more fibrous layers. This chance discovery indicates that a modest effort might add considerably to the little known about shell morphology.

INCUBATION PERIOD

The Developmental Period After Laying

Embryonic development can be analyzed in stages preceding and following laying. "Incubation period" here refers to the developmental period from laying to hatching. Although the general body of knowledge on chelonian incubation periods is relatively abundant, it is far from exhaustive; quantitative data are frequently lacking; and terminology is not uniform. "Hatching" means climbing out of the eggshell to some workers, emergence from the nest hole to others; but to me it means

INCUBATION PERIOD

primarily pipping of the eggshell. The period from pipping to leaving the eggshell tends to vary from less than 1 day to about 4 days, depending on the species and maturity of the individual baby turtle. Unless stated otherwise, consideration is limited to emergence records of species that begin burrowing to the soil surface immediately after leaving the eggshell, or a few days after pipping.

Data are divided into two groups: (1) those where incubation temperature was controlled or at least monitored under artificial conditions (Table 4), and (2) those where incubation conditions were natural or approximately natural (Table 5). In the latter the most frequent observations are from natural nests; next in frequency are observations from artificial outdoor nests approximating natural nests with regard to species geographic range and egg depth; a few records were obtained in open air buildings in the tropics.

It is generally accepted that temperature accelerates development in poikilotherms and that morphogenesis can be normal over a range of several celsius degrees despite the varying rate. Given the many environments in which turtle eggs incubate, how do developmental rates of different species compare when temperature as a factor is held constant? Might species in certain environments have evolved accelerated rates of development where either the number of consecutive days possible for incubation or the number of days offering warm soil conditions is at a minimum?

Eggs from some regional populations of a given species were incubated at two or more constant temperatures (Table 4). The accelerating effects of higher temperatures on developmental rate are evident and are contrary to statements such as those of Cunningham (1939) and Cagle (1950). There is variation between regional populations of the same species incubated at the same temperature. It is best illustrated by mean incubation times within the rigidly controlled 25 and 30°C series from populations of different latitudes. Among eight species (footnoted in Table 4) and their division into 14 species × temperature groups allowing intraspecific comparison, higher latitudes are associated clearly with more rapid development in 13, possibly reflecting adaptation to cooler climates. In terms of all meaningful comparisons, that is, those involving significantly different latitudes, 44 among 45 support this contention, and only one datum, Tennessee *S. odoratus* at 30°C, is non-supportive (exclude Lynn and von Brand Maryland *C. picta* at 25°C as experimental error). The mechanism accounting for the shortening of high latitude incubation periods could be fixed in the genome of regional populations, or it could be a phenotypically plastic expression of acclimation prior to laying.

Natural incubation periods of northern cool temperate turtles gener-

Table 4 Incubation Periods at Monitored and Often Controlled Temperatures Under Artificial Conditions

Family and Species	Average Incubation Period (days) at Temperature Regime as Stated					Reference[a]
	Below 25°C	25–25.5°C	25–30°C	29.5–30°C	Mostly above 30°C	
Pelomedusidae						
Pelusios subniger				58.0 (4)		
Chelidae						
Chelodina expansa		160		135		Goode and Russell (1968), Legler (unpubl. data)
C. longicollis				75.5		Goode and Russell (1968)
C. oblonga	170 at ca. 23°C					Burbidge (1967)
Elseya dentata				160		Legler (unpubl. data)
E. latisternum				53.5		Legler (unpubl. data)
Emydura macquarri				44.1		Goode and Russell (1968)
Chelus fimbriatus			208 at 28.3–29.5°C			Hausmann (1968b)
Phrynops gibbus			152 at 28–32°C			Medem (1973)

Kinosternidae				
Claudius angustatus	207 at ca. 24°C		150 at ca. 28°C	Hausmann (1968a)
Staurotypus salvini		230.6 (8)		Schmidt (1970)
Kinosternon subrubrum[b]				
Maryland-Virginia		76, 103.6 (16)		
Florida-Louisiana			71.8 (13) at 30–32°C[c]	Lynn and von Brand (1945)
K. bauri	118 at 24.4°C	114.1 (9)	89.8 (8)	Lardie (1975)
		124.7 (61)	93.2 (86)	
K. flavescens		108.6 (23)	94 at 27.2°C	Lardie (1975)
			108.1 (1) at 26–30°C	
			106.1 (17)	
K. leucostomum		174.5 (4)	148.5 (2)	
K. scorpioides		176.6 (11)	90.5 (15)	
Sternotherus carinatus		116.8 (5)	93.5 (2)	
S. minor	141.7 (3) at 22–25°C		89.5 (4)	
S. odoratus[b]	122.4 (54) at 22–25°C			Ewert and Legler (unpubl. data)
Wisconsin		80.5 (152)	69.0 (6) at 27.4°C	
Tennessee		86.7 (39)	59.4 (64)	
			70.3 (18)	

Table 4 *(Continued)*

Family and Species	Average Incubation Period (days) at Temperature Regime as Stated					Reference[a]
	Below 25°C	25–25.5°C	25–30°C	29.5–30°C	Mostly above 30°C	
Chelydridae						
Chelydra serpentina[b]						Lynn and von Brand (1945)
Northern Florida	106.6 (5) at 22–25°C	91.2 (18)		63.1 (12)		Yntema (1968)
Wisconsin		72, 75.1 (34)		60.0 (13)		
New York	140 at 20°C			63.0		
Missouri		82.0 (24)		66.7 (20)		
Arkansas		90.8 (5)		73.0 (5)		
Florida		97.5 (22)	80.0 (13) at 26–30°C	77.6 (18)		
Macroclemys temmincki		113.3 (42)	90.0 (87) at 26–30°C	81.4 (35)		
Emydidae						
Clemmys guttata		70.2 (9)			44.0 (24) at 30–32°C[c]	
C. insculpta Northern Michigan		67.0 (33)		46.1 (14)	40.4 (14) at 30–32°C[c]	

Species / Location					Reference
Connecticut				43.1 (20) at 30–32°C	Zovickian (1971)
C. muhlenbergi	62.3 (3)			43–47 at 26.7–36.6°C	Pieau (1971, 1974a)
Emys orbicularis	77–82		55		
Terrapene carolina[b]					
Indiana	91.2 (15) at 22–25°C		52.6 (47)		
Missouri	87.5 (11) at 22–25°C	60.8 (21) at 27.4°C	53.7 (26)	48.4 (4) at 30–32°C	
Florida	83.9 (10)		56.9 (23)		
T. ornata	125 at 23.9°C	70 at varying 27.8°C	55.0 (17)	59 at varying 32.8°C	Legler (1960a)
	98.3 (10) at 22–25°C	88.8 (20)			
Malaclemys terrapin			61–68		Cunningham (1939)
Graptemys barbouri	88.6 (9)	76.7 (48) at 26–30°C	57.5 (6)		
G. geographica	82.5 (30)		57.1 (20)	51.1 (36) at 30–32°C	
G. kohni	85.5 (4)				
G. oculifera			62.8 (4)		
G. pseudogeographica[b]					
Minnesota	89.3 (18) at 22–25°C	81.0 (57)	52.1 (53)		
Missouri	91.2 (20) at 22–25°C	85.2 (11)	56.5 (20)		
Texas	93.1 (12)				

Table 4 (*Continued*)

Family and Species	Average Incubation Period (days) at Temperature Regime as Stated						Reference[a]
	Below 25°C	25–25.5°C	25–30°C	29.5–30°C		Mostly above 30°C	
Chrysemys picta[b]							
Northern Michigan		66.2 (20)		47.5 (13)			Ream (1967),
Wisconsin	95 at 21–23°C	74, 71.1 (69)		51, 51.0 (18)			Mahmoud et al. (1973)
Connecticut		72.0 (3)			48.7 (20) at 30–32°C[c]		
Maryland		63, 76.0 (12)			49.5 (11) at 30–32°C[c]		Lynn and von Brand (1945)
Tennessee		77.4 (20)	62.0 (5) at 27.4°C	56.3 (17)			
Pseudemys concinna	114 (1) at 22–25°C	93.3 (3)		65.5 (4)			
P. decorata				69.4			Inchaustegui Miranda (1973)
P. floridana		101.6 (13)		68.0 (18)			
P. nelsoni				48.8 (4)[y,d]			Legler and Ewert (unpubl. data)
P. rubriventris		75.2					Graham (1971)
P. scripta[b]							
Kansas-							
Tennessee		93.0 (32)	68.9 (7) at 27.4°C	58.7 (13)			

Texas-Louisiana-Mississippi	112.5 (73) at 22–25°C	100.9 (16)		69, 62.4 (63)	Cagle (1950)
Florida	103 at 24°C	100.8 (24)		63.8 (53)	Moll and Legler (1971)
Panama					Inchaustegui Miranda (1973)
P. stejnegeri				70.3	
Emydoidea blandingi	81.6 (10) at ca. 24°C	71.3 (64)	52.4 (11) at 27.4°C	49.3 (9)	47.4 (45) at 30–32°C
Mauremys mutica		94.2 (6)		64.7 (6)	
Rhinoclemys areolata		120.3 (3)	103 (1) at 26–27°C	66.3 (3)	
Testudinidae					
Testudo graeca			87.0, 85–90 at 26–28°C	62–71	Raynaud et al. (1970), Vivien-Roels and Petit (1973)
T. hermanni				63.3	Kirsche (1967)
T. marginata				74	Klingelhoffer, in Kirsche (1967)
Geochelone elegans			109 at 26–30°C 182.6 at 28.1–29.4°C		Tardent (1972) Shaw (1966),
G. elephantopus			96–106 at ca. 28.5°C		Throp (1969)

Table 4 (*Continued*)

Family and Species	Average Incubation Period (days) at Temperature Regime as Stated					Reference[a]
	Below 25°C	25–25.5°C	25–30°C	29.5–30°C	Mostly above 30°C	
G. pardalis	230 at 24°C			132		Schweizer, in Kirsche (1967), Rowe and Janulow (1973)
G. radiata	230 at 24°C		155 at ca. 27°C		145 at 29.5–33.7°C	Schweizer, in Kirsche (1967), Peters (1969), Zovickian (1973), Burchfield (1975)
Gopherus agassizi			176.3 at ca. 28°C 109.3 at ca. 26.7°C			Trotter (1973)

Malacochersus tornieri		178 at 28.1–29.4°C	Shaw (1970)
Trionychidae			
Trionyx ferox		82.7 (14) at 26–30°C	
T. muticus[b]			
Minnesota	101.7 (49)		
Texas		59.0 (165)	
		62.7 (12)	
T. spiniferus	95.4 (55)	69.0 (5) at 27.4°C	57.9 (34)	52.2 (22) at 30–32°C[c]
Cheloniidae			
Chelonia mydas		ca. 80 at 27°C	55.0	48.0 at 32°C	Bustard and Greenham (1968)

[a] For all personal data the sample size (number of eggs hatched) is given in parentheses following the incubation period.
[b] This species provides data substantiating an inverse relationship between latitude and length of incubation period.
[c] Eggs were incubated in situations conducive to retention of metabolic heat.
[d] May be in error because of metabolic heating.

Table 5 Incubation Periods Under Natural or Approximately Natural Conditions

Family and Species	Location	Incubation Period[a] (days)	Reference	Nesting Season
Pelomedusidae				
Podocnemis expansa	Venezuela	47 (N), 42–47 (N, P), 45 (N, P)	Mosqueira Manso (1945), Ramirez (1956), Roze (1964)	Tropical-dry
P. unifilis	Venezuela	75–90	Medem (1964)	Tropical-dry
Chelidae				
Chelodina expansa	Victoria, Australia	192–360 (N, E)	Goode and Russell (1968)	Autumn (few spring)
C. longicollis	Victoria, Australia	131–145 (N, E)	Goode and Russell (1968)	Spring
C. longicollis	Canberra, Australia	118–150 (N)	Vestjens (1969)	Spring-summer
C. oblonga	Perth, Australia	186 (N)	Burbidge (1967)	Spring
Emydura macquarri	Victoria, Australia	66–85 (N, E)	Goode and Russell (1968)	Spring
Pseudemydura umbrina	Perth, Australia	180 (N, E)	Ellenbogen (1971)	Spring
Kinosternidae				
Kinosternon leucostomum	Panama	126–148 (P?)	Moll and Legler (1971)	Tropical
Sternotherus minor depressus	Alabama	122 (P)	Estridge (1970)	Spring
Chelydridae				
Chelydra serpentina	Northwestern Minnesota	78–102 (P)	This account	Spring-summer
C. serpentina	Central Minnesota	83–105 (P, E), 101 (N, E)	Breckenridge (1944), this account	Spring

Species	Location		Reference	Season
C. serpentina	Southwestern South Dakota	91–125 (N, E)	Hammer (1969)	Spring
C. serpentina	Southeastern Wisconsin	67–73 (P)	This account	Spring
C. serpentina	Central Florida	48–118 (N)	Punzo (1975)	Spring-summer (also winter)
Macroclemys temmincki	Central Florida	100–108 (N, E)	Allen and Neill, in Carr (1952)	Spring
Emydidae				
Clemmys guttata	Southeastern Pennsylvania	76 (70–83) (N, E)	Ernst (1970c)	Spring-summer
C. insculpta	Pennsylvania	77 (N)	Allen (1955)	Spring-summer
C. insculpta	Wisconsin	60–65 (P)	This account	Spring
Emys orbicularis	Not stated	93–101	Mell (1938)	Spring
Terrapene carolina	Pennsylvania	74–99 (N, E)	Rosenberger (1972)	Spring-summer
T. carolina	Maryland	80–90 (69–136) (N, E)	Allard (1935, 1948)	Spring-summer
T. carolina	Washington, D.C.	99 (69–161) (N, P, E)	Ewing (1933)	Spring-summer
T. carolina	Southern Florida	ca. 60 (N)	Dickson (1953)	Winter-summer
T. ornata	Kansas	ca. 75 (estimated)	Legler (1960a)	Spring-summer
Graptemys geographica	Wisconsin	73–76 (P)	This account	Spring-summer
G. pseudo-geographica	Wisconsin	76–82 (P)	This account	Spring-summer
Malaclemys terrapin	Southern New Jersey	75 (67–104) (N, E)	Burger (1975)	Spring-summer
M. terrapin	North Carolina	60–66 (E), 75–90	Hildebrand (1929), Cunningham (1939)	Spring-summer
Chrysemys picta	Northwestern Minnesota	99 (N, P), 72–99 (P)	This account	Spring-summer
C. picta	Central Minnesota	98 (N, P), 79 (75–81) (N, P)	Breckenridge (1944), this account	Spring-summer

Table 5 (*Continued*)

Family and Species	Location	Incubation Period[a] (days)	Reference	Nesting Season
C. picta	Southeastern Wisconsin	60–65 (P)	This account	Spring-summer
C. picta	Southeastern Pennsylvania	76 (72–80) (N, P)	Ernst (1971b)	Spring-summer
Pseudemys floridana	Central Florida	122–150	Goff and Goff (1932)	Winter-summer
P. scripta	Mississippi	60–80 (45–130) (E)	Haga (1970)	Spring-summer
P. scripta	Tobasco, Mexico	60 (P)	Rosado (1967)	Tropical
P. scripta	Panama	73–94 (80 estimated) (P)	Moll and Legler (1971)	Tropical-dry
Emydoidea blandingi	Nova Scotia	88 (N, E)	Bleakney (1963)	Spring-summer
E. blandingi	Central Minnesota	77 (P)	This account	Spring-summer
Batagur baska	Perak, Malaysia	74–122 (E)	Balasingam and Khan (1969)	Tropical-dry
Mauremys caspica	Not stated	95–101	Mell (1938)	Spring-summer
Melanochelys trijuga	Ceylon	ca. 60	Deraniyagala (1939)	Tropical-dry
Rhinoclemys funerea	Panama	98–104 (P)	Moll and Legler (1971)	Tropical
Testudinidae				
Testudo graeca	Not stated	95–100	Mell (1938)	Spring-summer
T. hermanni	Southern France	73–104 (N, E)	Heron (1968)	Spring-summer
T. horsfieldi	Eastcentral Asia	80–100	Terent'ev (1965)	Spring
Chersina angulata	Southern South Africa	180, 360–420 (N, E)	Cairncross (1946), Rose (1962)	Winter
Gopherus agassizi	Mojave Desert, California	120 (P), 84–99 (N, P)	Grant (1936), Shade (1972)	Spring-summer

Species	Location	Values	Reference	Season
G. agassizi	San Diego, California	103 (N)	Nichols (1953)	Spring-summer
Homopus sp.	South Africa	210–240	Rose (1962)	Not stated
Geochelone elephantopus	Pizon, Galapagos Islands	120–150	De Vries (1968)	Tropical
G. elephantopus	Santa Cruz, Galapagos Islands	216	De Vries (1968)	Tropical
G. pardalis	Grahamstown, South Africa	440 (N)	Bowker (1926)	Summer
G. pardalis	South Africa	240–540 (N), 396	Archer (1948), Rose (1962)	Autumn-spring
G. pardalis	Eastern Zambia	189 (178–206) (N, E)	Wilson (1968)	Autumn-winter
G. pardalis	Lesotho	384–472 (N, P)	Jaques (1969)	Autumn-winter
G. pardalis	Pietermartzburg, South Africa	378–384 (N)	Rowe-Rowe (1970)	Autumn
G. sulcata	Khartoum, Egypt	212 (N, E)	Cloudsley-Thompson (1970)	Autumn
Trionychidae				
Trionyx muticus	Iowa	70–75 (N, E)	Muller (1921)	Spring-summer
T. muticus	Louisiana	65–77 (E)	Anderson (1958)	Spring-summer
T. sinensis	Tokyo	28–83 (N, E)	Mitsukuri (1895a)	Spring-summer
T. spiniferus	Wisconsin	82–84 (P)	This account	Spring-summer
Cheloniidae				
Chelonia mydas	Heron Island, Australia	65–71 (N, E)	Moorehouse (1933)	Spring-summer
C. mydas	Sarawak	55.0 (48–80) (E)	Hendrickson (1958)	Tropical
C. mydas	Costa Rica	55.6 (48–70) (N, E)	Carr and Hirth (1962)	Tropical
C. mydas	Ascension Island	59.5 (58–62) (N, E)	Carr and Hirth (1962)	Tropical
C. mydas	Malaysia	47–61 (E)	Balasingam (1965, 1967)	Tropical

Table 5 *(Continued)*

Family and Species	Location	Incubation Period[a] (days)	Reference	Nesting Season
C. mydas	Formosa	63	Fukada (1965)	Tropical
C. mydas	Yemen	48 (E)	Hirth, in Frazier (1971)	
C. mydas	Surinam	58.3 (47–64) (N, E)	Pritchard (1969a)	Tropical
C. mydas	Aldabra	47–69 (N, E)	Frazier (1971)	Tropical
Caretta caretta	South Carolina	64, 55 (49–62) (N, E) 48 (N, P)	Hildebrand and Hatsel (1927) Baldwin and Lofton (1959)	Spring-summer
C. caretta	Tongaland, South Africa	63.4 (61–67) (N, E)	Hughes and Brent (1973)	Spring-summer
Lepidochelys kempi	Tamaulipas, Mexico	53–56 (50–70) (E)	Chavez et al. (1969)	Spring-summer
L. olivacea	Surinam	55.7 (49–62) (N, E)	Pritchard (1969a)	Tropical
Eretmochelys imbricata	Micronesia	60–70	Fukada (1965)	Tropical
E. imbricata	Costa Rica	58.6 (52–74) (E)	Carr et al. (1966)	Tropical
Dermochelyidae *Dermochelys coriacea*	Surinam	62.4 (60–68) (N, E)	Pritchard (1969a)	Tropical
D. coriacea	Southern Florida	62	Friar (1971)	Spring-summer
D. coriacea	Tongaland, South Africa	67.5 (N, E)	Hughes and Brent (1973)	Spring-summer

[a]"N" indicates that incubation occurred in a nest made by a female turtle; "P," end of incubation was known to be the date of pipping or close to it; "E," end of incubation was date of emergence, in most cases within a week of pipping.

ally range from less than 100 to 110 days (Table 5). At controlled temperatures the periods are short but not excessively so relative to some warm temperate and tropical species. Natural incubation periods of south temperate species are generally greater than 100 days and may exceed a year. At controlled temperatures their durations also tend to be relatively long. South temperate species occur where winters are generally mild and the embryos may overwinter in the nest. Overwintering in nests of north temperate species is done by hatchlings (Ernst and Barbour, 1972; Ehrenfeld, Chapter 18). Among tropical species, incubation periods vary from 42 days in *Podocnemis expansa* to 212 days in *Geochelone elephantopus*.

Short developmental periods, usually less than 100 days in natural nests and less than 70 days at constant 30°C, are characteristic of Emydidae, Trionychidae, Cheloniidae, and Dermochelidae. Long periods, exceeding 100 days in nests and more than 70 days at 30°C, typify Kinosternidae and Testudinidae; in these two families *Sternotherus odoratus* and *Testudo hermanni* are the most northerly distributed and have the shortest periods. Among the Pleurodira, *Podocnemis expansa* (Roze, 1964) and *P. unifilis*, appear to have short periods. Most chelids appear to have long periods. In addition to data in Tables 4 and 5 there is a record of *Phrynops (Batrachemys) dahli* surviving in an egg, despite neglect, for a period of 297 days at approximately room temperature before it was opened to yield a near-term fetus (Medem, 1966). *Emydura macquarri* is a chelid with a short incubation period. Although some ordering of incubation periods according to family is possible, exceptions seem to be too numerous to permit the attribution of general phylogenetic significance to them. However chelonian incubation period can be related to eggshell form; only species with brittle-shelled eggs have long periods (>120 days in nests and >80 days at constant 30°C).

One relationship between incubation period and nesting habitat stands out. Species that nest on sandbars and beaches appear to include only those with short incubation periods. The selective advantage of a short incubation period is obvious; the substrate is seasonally ephemeral, and risk of death from flooding is high (e.g., Roze, 1964).

Romanoff (1967) referred to the early developmental period as that of differentiation and to late development as that of growth and explained how duration of the former has been shown to be more responsive to temperature in the chick. In *Chelydra serpentina* embryos Yntema (1968) observed that as development progressed, the differences over an incremental period (say, a week) at 30°C became less and less distinguishable from the same time span at 20°C until the last 3 weeks, one-third of the incubation period at 30°C, when in fact, temperature did not appear to affect rate.

One ramification of Yntema's observation is definition of an absolute minimum developmental period exclusive of temperature. The minimum I have observed in *C. serpentina* from Minnesota is 48 days. Speculation given by Yntema is that the developmental regime favors warm nest temperatures early in development and cooler ones later on. In *C. serpentina*, which is a winter nester in south Florida, a spring nester along the Gulf, and a spring-summer nester near northern limits of its range, the speculation does not seem to be justified, since increasingly warmer weather follows nesting throughout most of the range. Nest temperatures of *Chelonia mydas*, which have been monitored through incubation, tend to rise from 2 to 6°C during incubation as a result of metabolic heating (Hendrickson, 1958; Carr and Hirth, 1961; Bustard and Greenham, 1968). In autumn nesting species nest temperatures would decrease after laying, but inasmuch as hatching occurs in late spring or summer following winter, nest temperatures would have risen again.

An unexplained phenomenon is the extreme variation in lengths of incubation periods reported for the eggs of certain species of *Geochelone*, sometimes from the same clutch, maintained under similar conditions (Jayakar and Spurway, 1966; Burchfield, 1975; and Shaw, 1966 vs. Throp, 1969). Ranges in incubation periods are 50 to 100% of the median, which contrasts markedly with the 10 to 15% variation normally observed.

The Developmental Period Before Laying

Whereas abundant data are available on incubation periods after laying, there is very little information on lengths of prenesting developmental periods, namely, from ovulation to laying. Duration of prenesting development is of ecological interest for several reasons, including the strain of gravidity on the female, the necessity for her to remain near the nesting sites if she is to lay several clutches, and the differences in hazards faced by embryos in the oviducts as opposed to in the nests.

The developmental stage at laying has been described variously, but generally is agreed on as a late gastrula (Yntema, 1968; Mahmoud et al., 1973; Pasteels, 1957; Mitsukuri, 1893, 1894; Risley, 1933a, 1944; Goode and Russell, 1968). Statements and photographs provided by Roze (1964) suggest that if *Podocnemis expansa* is no more than a gastrula at laying, development proceeds extremely fast inasmuch as the embryo has the "unmistakable shape of a young turtle" at 10 days of 31 to 32°C. *C. serpentina* reaches this stage at 14^+ to 18 days at 30°C. Nevertheless, there are no substantial data to indicate that the normal freshly laid egg of any species is very far from a late gastrula, which is more advanced than the late blastula typical of the freshly laid avian egg.

There are currently three indirect approaches to estimation of duration of development in the oviduct.

1 Ratios of developmental time through gastrulation, as in the chelonian oviduct, to that from late gastrulation to birth have been computed for several vertebrates that hatch or are born precocially: the frog *Eleutherodactylus martinicensis* (which hatches as a froglet) at 30°C, time to complete gastrulation equals 5% of postgastrulation time (Hughes, 1962); chick 6% (Patten, 1958; Rol'nik, 1970); and pig, sheep, rhesus monkey (Witschi, 1972), and baboon (Hendrickx, 1971), 9 to 12%. In these diverse animals, development through gastrulation is usually a short but significant fraction of the total developmental period.
2 To use as evidence turtles with premature eggs in their oviducts, one must assume that the specimens had ovulated just before capture and that eggs had been maturing in the oviducts from dates of capture to dissection. It is also assumed that the eggs that have entered the oviducts will continue to mature at a normal rate. The extent to which stresses of captivity may alter maturation if at all, are not known. My observations on *Sternotherus odoratus, Terrapene ornata, Trionyx ferox*, and *T. muticus* suggest that embryonic development in the oviducts of some species probably lasts at least 2 weeks.
3 The third line of evidence, the minimum internesting interval, or at least an estimate of it, is available for some cryptodiran species (Table 6). Although we do not have direct evidence of when ovulation has occurred, some intervals are so short in contrast to the data on turtles captured with maturing oviducal eggs that these intervals may represent immediate ovulation following nesting in some species as well as optimal conditions for egg maturation. In other species with relatively long internesting intervals (e.g., *Geochelone* spp.), a period of vitellogenesis might precede ovulation. Assuming that the minimum internesting interval closely represents the developmental period of ovulation-fertilization to completed ripening, useful comparisons can be made with postlaying incubation periods (Table 6). It appears that no more than 11 to 26% (mean 15%) of the postlaying period of turtle embryos normally occurs in the oviducts. Comparisons made with constant 30°C may be less meaningful inasmuch as the temperature of the turtle before nesting is not known. The ratios based on the natural nests are somewhat large in contrast to those of the aforementioned nonchelonian vertebrates. Acceptance of a minimum period for prelaying development leads one to speculate that eggs following an atypically short internesting interval (say, 2–3 days)

Table 6 Internesting Intervals and Their Relationships to Incubation Periods[a]

Family and Species	Minimum Internesting Interval	Incubation Ratio Period at 30°C	Incubation Ratio %	Selected Natural Incubation Period	Ratio %	Authority on Internesting Interval
Emydidae						
Chrysemys picta	9 (W, L)	52	17	80	11	Ream (1967)
C. picta	14 (W)			80	18	Moll (1973)
Terrapene carolina	22 (C, E)	53	42	85	26	Ewing (1935)
Testudinidae						
Testudo hermanni	19 (W, E)	63	30	80	23	Heron (1968)
T. hermanni	15 (C, E)	63	24	80	19	Ehrengart (1971)
Geochelone elephantopus	37 (C, E)	175[b]	21	150	25	Shaw (1966)
G. pardalis	23 (C, E)	145	16	189	12	Rowe and Janulaw (1971)
Trionychidae						
Trionyx sinensis	10 (C, E)	60[c]	17	61	16	Mitsukuri (1895a)
Cheloniidae						
Chelonia mydas	9 (W, L)	55	16	50	18	Hendrickson (1958)
Caretta caretta	12 (W, L)			63	19	Caldwell (1962)
Dermochelidae						
Dermochelys coriacea	10 (W, L)			62	16	Pritchard (1969a)

[a] "W" indicates data from a wild or natural situation; "C," data from captivity; "L," laying of eggs during nestings is probable but not certain; "E," eggs were observed.
[b] Based on an incubation period of 182.6 days at 28.1 to 29.4°C.
[c] Based on incubation periods of 58 to 63 days for *T. muticus* and *T. spiniferus*.

may represent a portion of the oviducal complement left over from the initial nesting or they may be immature. The viability of such eggs might be an interesting topic for study.

MORPHOGENESIS

Major aspects of descriptive morphogenesis in Chelonia were completed during the nineteenth century. Rathke (1848) and Agassiz (1857) pioneered with extensive works; included in the latter are drawings by H. J. Clark and A. Sonrel of such high quality as to give the work significant reference value today. Later work providing more emphasis on the cellular level was completed by Mitsukuri (1891, 1893, 1894, 1895b), Mitsukuri and Ishikawa (1886), Mehnert (1892, 1895), and Will (1893). Mitsukuri included reviews of earlier work. Fisk and Tribe (1949) reviewed the dynamics between reptilian embryos and their extraembryonic membranes.

In conjunction with the increasingly experimental work of this century, exact knowledge of sequential stages of development has gained in importance. When work is conducted with large numbers of oviducal or freshly laid nest eggs, it is imperative in experimental planning to recognize quickly the morphogenic stages as well as to predict in advance when desired stages will be available. Pasteels (1957) outlines a series of stages in gastrulation of *Mauremys caspica, Pseudemys floridana*, and *Chrysemys picta*. Since embryonic development is related to temperature, prediction of occurrence of a desired stage involves controlling temperature. Such controlled series of stages are available for New York *Chelydra serpentina* at 20 to 30°C (Yntema, 1968) and Wisconsin *Chrysemys picta* at 21 to 23°C (Mahmoud et al., 1973).

Early Developmental Changes After Laying

Albumin surrounds the vitelline sac (i.e., the vitelline membrane and its enclosed contents) of infertile eggs, of those removed prematurely from oviducts, and of some freshly laid eggs. In symmetrically shaped eggs the vitelline sac occurs near the center; in tapered eggs it lies near the large end. The early vitelline sac occurs in two general shapes: elongate, with its width at 65 to 75% of its length, and approximately spherical, with its shortest diameter over 90% of its greatest. Elongate eggs with parchment or hard, expansible shells have elongate vitelline sacs, as in *Pelusios sinuatus, Clemmys insculpta, Terrapene carolina, Graptemys geographica*, and *Pseudemys scripta*. All other eggs have spherical vitelline sacs as found in any spherical eggs and in elongate brittle-shelled eggs of *Kinosternon scorpioides, K. subrubrum, Sternotherus odoratus, Melanochelys trijuga, Rhinoclemys areolata, R. pulcherrima*, and *Geochelone elegans*.

In fertile eggs between a few hours and 3 days old, the upper part of the vitelline sac, that is, the embryonic disk, which is now a gastrula, and adjacent vitelline membrane rise to the uppermost portion of the egg and adhere to the inner shell membrane. In elongate eggs with spherical vitelline sacs there can be exception to this statement, but union of vitelline and inner shell membranes occurs nevertheless. Adhesion as discussed by Mitsukuri (1894) is sufficiently strong in *Lepidochelys olivacea* to permit easy removal of the embryonic disk from the egg simply by cutting the egg shell around it, then lifting this piece up. The embryonic disk follows on its inner side. The eggshell at the point where adhesion has occurred becomes chalk white against the pale yellowness or pinkness of the freshly laid egg. It appears as a circular spot or occasionally as a minute ring corresponding in size to the area opaca of the embryonic disk.

Since the work of Agassiz (1857), little attention has been given to the process enabling the vitelline sac to rise. Some liquefaction of the albumin or, at least, movement of water to within the vitelline membrane, may cause it to expand. According to Agassiz, the entire vitelline sac first rises, but after a week in *Caretta caretta* and *Sternotherus odoratus* the vitelline membrane has expanded to such an extent as to embody most of the egg contents and to lie adjacent to the shell above and below. If such an egg is candled, yellow yolk material appears to occupy approximately the lower third of the egg, whereas clear fluid and the embryo occupy the upper two thirds. In elongate parchment-shelled eggs, as in *Terrapene carolina, Chrysemys picta*, and *Pseudemys scripta*, a thin layer of viscous albumin still remains beneath the vitelline sac when the eggs have advanced to the first appearance of blood islands forerunning vitelline circulation. Similarly, advanced elongate eggs of *Kinosternon subrubrum* and *S. odoratus* retain much albumin, but in these cases the albumin is at each end. The egg of *Rhinoclemys areolata* has albumin at its tapered end. Mitsukuri (1891) noted that albumin persists at least ventrally through to hatching in *Mauremys japonica* and *Trionyx sinensis*.

As noted by Agassiz, expansion of the vitelline sac begins before the egg is laid. This constitutes one of the better diagnostic features differentiating fresh oviducal eggs from those the turtle may have held for several weeks. In eggs that may be losing viability, expansion of the vitelline sac seems to have reached completion, the yellow yolk material moves freely within as one turns the egg, and the upper shell of chelydrid and parchment-shelled emydid eggs whitens in broad areas as opposed to a single point. The last aspect suggests that adhesion of the vitelline membrane as well as local action of the embryonic disk is involved in whitening of the shell. If the embryo dies as a blastula or gastrula, before or after laying, the vitelline sac often shrinks, and the egg assumes the superficial appearance of an unfertilized egg.

In normal eggs whitening of the shell follows patterns that seem best described in terms of egg shape and early vitelline sac shape. In spherical eggs and in the parchment to hard expansible-shelled elongate eggs of 16 species of Emydinae, whitening begins as a round spot and expands outward, more or less retaining its round shape. Brittle-shelled elongate eggs that have spherical vitelline sacs show whitening as a spot expanding into a band coincident with the least circumference of the egg. Band-type whitening occurs in *Melanochelys trijuga* (Deraniyagala, 1939), *Kinosternon bauri* (Einem, 1956), and several species of Australian chelids (Legler, personal communication) and has been confirmed personally in three genera of kinosternids and in the chelid *Elseya dentata*. In eggs of *S. odoratus* opened after the white band has formed, the vitelline sac appears to have expanded to touch the eggshell all along the white band. *R. areolata* has early whitening restricted to the large end of the tapered egg adjacent to the vitelline sac. In chelydrids, emydine emydids, and trionychids, expansion of the white area appears to slow when it reaches the lower part of the egg where yellow yolk lies close inside, but gradually the entire egg may become white. Presence of moist material touching the outside of the egg seems to retard whitening, especially at local contact points. Haga (1970) gave 24 hr for achievement of complete whitening by *Pseudemys scripta* eggs on his turtle farm and Vose (1964) 11 days for Minnesota species of *Trionyx* in natural nests. Both rates are rapid in my experience. The physical and possibly chemical changes involved in chelonian eggshell whitening have not been described.

Organogenesis

Descriptive organogenesis in turtles shows small variations from the basic vertebrate plan. Pasteels (1970) has summarized many aspects. More specialized papers, usually including reviews, are available for the nose (Parsons, 1970), eye (Underwood, 1970), optic tectum (Senn, 1971), pineal gland (Viviens-Roels and Petit, 1973), skull (Pehrson, 1945), limb bud development (Vasse and Pieau, 1970), alimentary canal (Shaner, 1925), urogenital system (Pieau, 1968; Raynaud and Pieau, 1970; Raynaud et al., 1970), thyroid gland (Lynn, 1970), and thymus (Bockman, 1970).

Yntema (1970a) removed portions of selected early embryonic somites and demonstrated experimentally that these extirpations were reflected in deletions from the carapace in a predictable manner. One notable conclusion was that the shell grows out around the limb girdles as opposed to the girdles retiring into the shell.

Extraembryonic Membranes

Their Morphogenesis. The extraembryonic membranes—the yolk sac, the chorion, the amnion, or amniochorion, and the allantois—constitute an important intermediary between the embryo and its environment. There are descriptive accounts by Mitsukuri (1891), Mehnert (1895), and Fisk and Tribe (1949), and several good illustrations in Agassiz (1857). Yntema (1968) and Mahmoud et al. (1973) staged early development of extraembryonic membranes according to time and temperature, but neither followed the membrane development beyond early expansion of the allantois.

According to the staging of *Chelydra serpentina* by Yntema and *Chrysemys picta* by Mahmoud et al., the yolk-sac membrane does not increase appreciably in size or give rise to blood islands forerunning vitelline circulation before the amniochorion has appeared as an enlarging fold or hood to cover the head region. Fisk and Tribe (1949) described spread of the amniochorion in *C. serpentina* and *C. picta* as relatively rapid in contrast to its progress in the chick (Patten, 1958) when, otherwise, embryos are at similar stages. The chelonian embryo touches the inner shell membrane at this stage and faces risk of excessively strong adhesion to it, especially if the egg dehydrates. Rapid establishment of the amniochorion decreases the duration of this risk. Progressively, the amniochorion forces the embryo away from the eggshell and eventually extends between the embryo and the shell when the amnionic cavity is complete.

In *C. serpentina* at 20°C and *C. picta* at 21 to 23°C the vitelline circulation is active in 2 weeks. The terminal sinus, the outer border of this area, gradually extends down over more of the yolk sac but does not necessarily spread to give vasculation to the entire yolk surface. In Chelydridae, Emydidae, and Trionychidae the expanding vascular area is approximately circular or slightly heart shaped with the anterior vitelline vein at the cleft. Most if not all of the vascularized area lies beneath the portion of the shell that has become chalk white. In Kinosternidae, where the white area is initially a band, the early vascular area also appears to be mostly restricted to this region. At first it is circular, but later in candled eggs it appears elongate, extending beneath the band down the sides of the egg from the embryonic region at the top.

The allantoic sac, an outgrowth of the hindgut, starts expanding at 35 days in *C. serpentina* at 20°C or about 9 days at 30°C and at 21 days in *C. picta* at 21 to 23°C. There is no published information to designate the stage at which it can function in a respiratory capacity. Eggs of most species I have candled during the last 2 to 3 weeks of development at 25°C suggest that the allantoic sac lines the chorion nearly completely,

MORPHOGENESIS

with the exception of the seroamniotic connection as described by Mitsukuri (1891).

Correlates with Oxygen Consumption. General accounts of respiration and respiratory membranes in eggs of reptiles and birds (Fisk and Tribe, 1949; Patten, 1958; Romanoff, 1967) suggest that initial oxygen uptake and carbon dioxide release occur through simple diffusion between the embryo and its surroundings. Vitelline circulation plays a dominant role in gas exchange for a brief while before the allantoic circulation assumes a major role.

Works on oxygen consumption by turtle eggs include those of Lynn and von Brand (1945), Ackerman and Prange (1972), and Prange and Ackerman (1974). The second listed work examined diffusion across empty shells of *Chelonia mydas* eggs that had just hatched. Permeability values were approximately twice the reported values for chicken eggs. Lynn and von Brand determined whole egg oxygen consumption for various developmental stages of *Kinosternon subrubrum, Chelydra serpentina, Terrapene carolina*, and *Chrysemys picta* maintained at continuous 25 to 25.5°C. Oxygen consumption initially amounted to 3 to 4% of final prehatching rates, with the period of most rapid increase in rates between the middle of incubation and the final third. Relatively low consumption rates probably persist well into the period of establishment of vitelline circulation. The most rapid increase in rates appears to come at approximately the time when the allantoic circulation has become fully established. Variations in nesting habitat and soil air circulation may require adjustments by eggs to deficiencies in air supply.

Membranes in Relation to Hatching. Prior to hatching the embryo is enveloped by several layers of extraembryonic membranes. From the embryo outward there is an amnion, then two layers of allantois, and a chorion. Agassiz (1857) and Mitsukuri (1891) described the fate of the membranes during hatching. The thin, nonvasculated amnion is torn to shreds, while the allantois most often is not torn. Rather, its folds are parted anteriorly at the seroamniotic connection and are worked posteriorly, presumably through movements of the forelimbs. In emydids the hind limbs at this stage are extended anteriorly such that the feet lie adjacent to the bridge. This position may aid in the backward movement of the allantois as it shrinks. When eggs are candled, parting of the allantois may be observed as early as 2 days before pipping in emydids and kinosternids and at least a day in chelydrids. In chelydrids and emydids beads of water appear on the shell of the egg nearest the turtle's head and forelimbs. Prior to pipping in kinosternids the allantois is frequently reduced to a small patch adjacent to the posterioventral inner surface of the egg shell. In other species a considerable amount of blood may be invested in the allantois at pipping. However hemorrhaging and loss of blood are only occasionally observed, and perhaps a little more often in *Trionyx* species than in emydids and chelydrids. I have washed freshly pipped turtles free of their eggshells and placed them in shallow water and noted no loss of blood. Apparently, blood is drawn into the turtle, and the allantois, which withers, soon detaches at the umbilicus.

Several days before pipping, sharp-tipped features on the embryo such as claws, shell margins in *Graptemys barbouri* and *G. oculifera*, and the caruncle are covered with gelatinlike sheaths. Shell margins of *Graptemys* species are folded ventrally such that the points lie tangential to the eggshell as opposed to pointing at it, as they would if in the fully unfolded state. Although never described, it is likely that the shell of *Heosemys spinosa*, which has the longest marginal spines known, is similarly folded and sheathed. Moll and Legler (1971) have reviewed the process of pipping in relation to the caruncle, which is used in some cases but not in others. The gelatinous sheath is lost from the caruncle and the claws at about the time of pipping. Some candled unpipped eggs of *Clemmys insculpta*, *Graptemys pseudogeographica*, and *Chrysemys picta* have revealed sets of scratch marks on the inner surface of the shell membranes at points adjacent to the claws. In other eggs similar scratches pierce the eggshell, thereby pipping it. In *Sternotherus odoratus* I have observed a single tiny breach of the shell adjacent to the caruncle in some cases and breaches next to each forefoot in others.

Once the eggshell is pipped and torn, the shell of the baby turtle begins to unfold. Photographs or drawings of baby turtles in such a condition are available for representative families as follows: Pelomedusidae (Medem, 1964:361), Chelidae (Goode, 1967b:108); Medem, 1966:488), Chelydridae (Yntema, 1968:plate 26), Emydidae (Agassiz, 1857:plate 18; Cagle, 1950:41–42; DeRooij, 1915:303; Mitsukuri, 1891:plate 8; Moll and Legler, 1971:33), Testudinidae (Auffenberg and Weaver, 1969:166; Trotter, 1973:27–28; Zovickian, 1973: cover), and Trionychidae (Koschmann, 1967:9). In Kinosternidae there is rarely any external yolk sac at pipping; in Pelomedusidae, Chelidae, Chelydridae, Emydidae, Testudinidae, and Trionychidae there is a large external yolk sac, generally 30 to 50% of the plastron length, and often with evidence of vitelline circulation. Marine turtles may have an external yolk sac at pipping (Hughes et al., 1967).

The unfolding of the carapace and plastron appears to aid in passage of the yolk into the body cavity or the converse may occur. Most freshly pipped turtles appear not at all ready to leave the eggshell, let alone the nest. Usually 1 to 4 days passes while the yolk is being withdrawn. Cagle (1942a) reported a rare observation of *Chelydra serpentina* hatchlings journeying from their nest to water while yolk sacs were still external. Another exception is Trionychidae, where crawling of the hatchling appears to aid in unfolding of its especially pliable shell and simultaneous withdrawal of yolk. Through manual assistance in unfolding the shell of *Trionyx muticus*, one can actually see the yolk disappear into the body cavity. Unfolding of the shell of kinosternids is relatively slight, though measurable (Adler, 1960). These turtles sometimes may be found crawling around incubation boxes within an hour of pipping.

CONVERSION OF MATTER INTO EMBRYO

Energetics

Bulk Conversions and Calorimetry. During the last 50 years many workers have given attention to the chemical and energetic conversion of egg into hatchling. Broad considerations of this approach as applied to amniote eggs are provided by Romanoff (1967) and Needham (1963), with the latter including turtles.

Table 7 provides fresh weight proportions of the basic morphological parts of unincubated eggs. Determinations involved infertile eggs whose yolks had solidified naturally and fertile eggs whose yolks had been solidified by heating. The matter of fresh weight proportions may appeal to casual interest, but it suffers in true value from the high variability caused by uptake of water by the vitelline sac, which can begin before laying. One interesting comparison involves a female *Trionyx muticus* that contained three ovulated ova in the coelomic cavity and 16 mature oviducal eggs. The average weight per ovum was 74.7% of that per matured egg.

The yolk is the primary nutritional source of developing embryos. In conversion of egg contents into hatchlings there is interest in the amount of yolk at outset in relation to hatchling at end. Table 8 provides complete information on three nonchelonians and available data for turtles. The wet weight of hatchling amniotes tends to be greater than the yolk of the freshly laid egg. Dry weight comparisons indicate that this difference is due to water in all cases except the chick, which derives much nonaqueous substance by imbibing albumin (Rol'nik, 1970). Turtles are not known to imbibe albumin. If they did, they would have available to them only about 4% more of the dry weight contents of the freshly laid egg (*C. caretta* data of Karashima, 1929a) than is contained in the yolk. The dry weight of yolk to fresh weight of whole egg ratios range from 0.139 to 0.230 (mean = 0.180) in 12 species of turtles, close to the ratio of 0.165 for the chick.

Energetic efficiencies for conversion of egg contents into hatchlings are available for three nonchelonian amniotes (Table 8). An average ashfree, dry weight caloric content of yolks of 12 diverse avian species is 8000 ± 100 cal/g (Slobodkin, 1962). For 10 species of small lizards an approximately corresponding determination is 6400 cal/g (data of Tinkle and Hadley, 1975). Determinations made from yolks of turtle eggs give 6600 cal/g for *Chelydra serpentina* and 6700 for *Pseudemys scripta* (Slobodkin, 1962), and by indirect calculation (West et al., 1966; Ricklefs and Cullen, 1973), 6800 for *C. caretta* (from data of Tomita, 1929).

An understanding of the chemical composition of yolk partially explains why unit weights of avian yolks have higher caloric contents than reptilian yolks and why the chick has a greater energy conversion efficiency than the lizards (Table 8) and probably some turtles. Total combustion of lipid through ignition in a calorimeter provides 9400 cal/g, whereas protein gives 5600 cal/g (West et al., 1966). Amounts of other combustible substances in amniote yolks are negligible. The avian yolk has the highest unit caloric content because it has the most lipid: 63.5% of dry weight, chick; 32%, *C. caretta* (Needham, 1963); 14.3–25.4%, *C. serpentina* (Lynn and von Brand, 1945) and calculated from Slobodkin (1962); 34%, *Chrysemys picta*, calculated from Chaikoff and Entenman (1946). During development the reptiles may use relatively more protein and relatively less lipid as an energy supply than the chick uses. However only 77% of the potential energy contained in proteins, as determined by combustion in a calorimeter, is utilizable by embryos, whereas more than 95% of the potential energy of lipids is. Hence correction for incomplete combustion of proteins could reduce the differences between energy conversion efficiencies of the organisms.

In offering an explanation for use of protein-rich as opposed to lipid-rich energy sources, Needham (1963) noted that metabolized protein gives less water than does metabolized lipid. Turtle eggs, such as those of *C. caretta*, can absorb water from their environment and also expel nitrogenous wastes into it; avian eggs do neither. What benefits, if any, the embryo or its mother derives from greater combustion of proteins are not evident.

Space Limitations. Developing embryos have space limitations imposed on them by the eggshell. Figures depicting term fetuses of birds and reptiles in the shell all show highly contorted fetuses that nearly fill the interior. Table 9 provides ratios in weight of neonate amniotes to their freshly laid eggs. My work exactly matches turtle hatchlings to their eggs and averages percentages computed in each instance to give a common mean percentage for each species or biotype. Other data tend

Table 7 Relative Proportions of Major Constituents of Eggs

Family and Species	Mean Weight of Eggs in Sample (g)	Percentage of Fresh Egg Weight		
		In Shell	In Albumin	In Yolk
Pelomedusidae				
Pelusios sinuatus	11.3	6.0	61.9	32.1
Kinosternidae				
Kinosternon bauri	4.9	17.2		
K. scorpioides	6.3	24.7	34.4	40.9
K. subrubrum	4.3	17.6		
Sternotherus minor	4.9	13.2		
S. odoratus	4.1	15.2	44.7	40.1
Chelydridae				
Chelydra serpentina	12.5	11.0	45.0	44.0
Macroclemys temmincki	27.1	9.4	52.7	37.9
Emydidae				
Clemmys insculpta	9.8	6.1		
Terrapene carolina	8.9	7.6	51.4	41.0
Chrysemys picta	6.3	8.1	44.2	47.7
Pseudemys concinna	15.8	6.6	56.3	37.1
Emydoidea blandingi	13.7	9.6		
Melanochelys trijuga	10.4	19.2	38.2	42.6
Rhinoclemys areolata	33.2	15.0	48.0	37.0
Testudinidae				
Geochelone elegans	27.4	22.9	41.1	36.0
Trionychidae				
Trionyx ferox	12.6	13.3	47.6	39.1
T. muticus	8.7	9.2	43.8	47.0
Cheloniidae				
Caretta caretta[a]	34.4	5.8	39.2	55.0
Dermochelyidae				
Dermochelys coriacea[b]	71.8	4.3	48.9	46.8
Precocial birds[c]				
Chicken	58	12.3	55.8	31.9
Duck	80	12.0	52.6	35.4

[a]*Source*: Tomita (1929).
[b]*Source*: Simkiss (1962).
[c]*Source*: Romanoff and Romanoff (1949).

Table 8 Data Pertaining to Energy Content of Amniote Eggs and Energy Conversion Efficiency During Development

Organism	Wet Weight (g)		Dry Weight (g)		Nonaqueous Content (%)		Caloric Content (kcal)		Energy Conversion Efficiency (%)	Reference
	Initial Yolk	Hatchling with Yolk Sac	Initial Yolk	Hatchling with Yolk Sac	Initial Yolk	Hatchling with Yolk Sac	Initial Yolk	Hatchling with Yolk Sac		
Chick (White Leghorn)	19.34	40.76	9.93	11.21	51.3	27.5 18.8[a]	70.2	64.0	60.4, 62.9, 67.0	Needham (1963), Romanoff (1967)
Lizards										
Iguana iguana	9	10	3.76	2.6	42	22–27	23	13	44.7–51.6	Ricklefs and Cullen (1973), Licht and Moberly (1965), Ballinger and Clark (1973)
Gerrhonotus coeruleus[b]	0.39	0.73	0.22	0.16	56.2	22.1	1.4	0.82	59.3[c]	Vitt (1974)

Turtles

Species								Reference
Caretta caretta	14.8	15.95	5.15	3.55	34.8	22.3	33.5	Karashima (1929a, 1929b), Sendju (1929), Tomita (1929), Needham (1942), Simkiss (1962)
Dermochelys coriacea	33.6	39.7						Personal determinations (3–12 replications)
Sternotherus odoratus	1.58	2.45	.70		44.1			
Chelydra serpentina	4.39	6.91	1.72		39.4			
Terrapene carolina	4.06	8.25	1.99		49.3			
Chrysemys picta	2.99	5.07	1.31		43.7			
Pseudemys concinna	4.89	10.81	2.26		46.1			
Rhinoclemys areolata	13.19	22.15	6.20		46.6			
Trionyx muticus	4.05	6.65	1.97		48.9			

[a] Embryo alone.
[b] Eggs have little substance in the shell and albumin.
[c] Not adjusted for unmetabolized yolk within hatchlings (a dry weight hatchling component of 35% yolk would reduce the figure to 50%).

Table 9 Weight Relationships Between Neonates and Freshly Laid Eggs

Family and Species	Weight of Neonate (g)	Percentage of Egg Weight	Reference[a]
Pelomedusidae			
Pelusios subniger	7.6	77.7	4 (1)
Podocnemis unifilis	16.2	70.3	Medem (1969)
P. vogli	11.2	72.6	Pardo (1969)
Chelidae			
Chelodina expansa	8.9	61.0	Legler (unpubl. data)
Elseya latisternum	4.7	47.9	Legler (unpubl. data)
Emydura sp.	3.2	53.4	Legler (unpubl. data)
Phrynops geoffroanus	11.6	60.5	Medem (1969)
Kinosternidae			
Staurotypus salvini	6.1	60.8	3 (1)
Kinosternon bauri	3.1	61.4	139 (70)
K. subrubrum	2.9	61.6	15 (5)
K. flavescens	3.1	70.1	39 (16)
K. leucostomum	4.6	64.8	2 (2)
K. scorpioides	3.7	60.9	5 (2)
Sternotherus carinatus	3.8	61.6	9 (3)
S. minor	3.2	65.9	64 (25)
S. odoratus			
Wisconsin	2.8	66.5	147 (33)
Southern Florida	1.7	60.3	16 (12)
Chelydridae			
Chelydra s. serpentina			
Minnesota	8.9	80.1	140 (7)
C. s. osceola			
Southern Florida	8.8	77.7	41 (4)
Macroclemys temmincki	17.9	69.3	97 (5)
Emydidae			
Clemmys guttata	4.7	75.2	9 (4)

Table 9 (*Continued*)

Family and Species	Weight of Neonate (g)	Percentage of Egg Weight	Reference[a]
C. insculpta	7.7	78.5	95 (10)
C. muhlenbergi	3.9	83.7	3 (1)
Terrapene c. carolina Indiana	8.4	82.7	74 (17)
T. c. major Western Florida	8.8	80.1	28 (9)
T. ornata	8.1	75.2	37 (11)
Graptemys barbouri	10.1	72.5	44 (5)
G. geographica	7.5	62.9	39 (6)
G. kohni	8.4	72.9	16 (3)
G. oculifera	9.5	80.1	4 (1)
G. pseudogeographica	7.0	68.5	102 (12)
G. pulchra	7.4	68.3	6 (1)
Chrysemys picta belli North Dakota	4.4	72.1	27 (3)
C. p. dorsalis Tennessee	4.6	79.4	30 (9)
Pseudemys concinna	12.3	70.2	11 (2)
P. floridana	9.0	77.3	49 (5)
P. nelsoni	8.9	76.7	4 (1)
P. rubriventris	7.7	64.7	Graham (1971a)
P. scripta	8.4	77.8	56 (9)
Emydoidea blandingi	9.0	66.2	4 (1)
Mauremys mutica	6.8	66.1	9 (4)
Rhinoclemys areolata	23.0	63.8	6 (5)
Testudinidae			
Testudo hermanni	7.8	61.2	Kirsche (1967)
Geochelone elephantopus	69.9	65.4	Shaw (1966)
G. sulcata	27.6	64.6	Cloudsley-Thompson (1970)
Malacochersus tornieri	16.0	68.2	Shaw (1970)
Trionychidae			
Trionyx ferox	10.0	71.3	91 (9)
T. muticus Minnesota	5.5	74.3	152 (8)
Texas	6.2	76.3	58 (16)

Table 9 *(Continued)*

Family and Species	Weight of Neonate (g)	Percentage of Egg Weight	Reference[a]
T. spiniferus			
Minnesota	8.2	75.6	17 (3)
Florida	9.6	74.1	41 (3)
Cheloniidae			
Chelonia depressa	37.3	57.1	Bustard and Limpus (1969)
C. mydas	25.1	48.6	Bustard and Limpus (1969)
Caretta caretta	16.0	46.4	Sendju (1929), Tomita (1929)
Eretmochelys imbricata	14.9	53.3	Deraniyagala (1939)
Lepidochelys kempi	16.1	49.7	Chavez et al. (1968)
L. olivacea	17.0	47.3	Deraniyagala (1939)
Dermochelyidae			
Dermochelys coriacea			
Ceylon	33.1 (51.0)	46.0 (70.9)	Deraniyagala (1939)
Malaysia	39.7	55.3	Simkiss (1962)
Other amniotes			
Crocodile			
Crocodylus porosus	79.6	71.0	Deraniyagala (1939)
Small lizards			
Amphibolurus barbatus	2.4	85	Badham (1971)
Cnemidophorus sexlineatus	0.78	107.7	Fitch (1958)
Large lizard,			
Varanus salvator	32	64.6	Kratzer (1973)
Snake,			
Spalerosophis cliffordi	14.1	81.5	Dmi'el (1967)
Precocial birds			
Chicken	35.0	60.7	Romanoff and Romanoff (1949)
Duck	41.6	60.9	Deth (1962)
Crane	109.8	60.3	Walkinshaw (1950)

[a]Personal data are represented by sample size: number of neonates (number of clutches).

to represent ratios of large sample means or of clutch means matched to neonate means.

Overall variation in ratios for turtles at 46.4 to 83.7% is somewhat broader than personally determined ratios of 60.3 to 83.7%. The latter hatchlings were held in shallow water for 5 to 10 days after pipping so that the yolk sac would be fully retracted and the hatchling partially unfolded into its functional baby form. Several determinations based on freshly pipped animals suggested that water imbibed posthatching was not significant.

Results verify an obvious expectation: larger neonates emerge from larger eggs; that is, ratios are rather similar in all species. Ratios for chelydrids and emydids (with lightweight, parchment shells) tend to be a bit higher than average, but also more variable; those for kinosternids and testudinids (with heavy, brittle shells) are relatively low (Table 7). Marine turtles are proportionately light yet come from parchment-shelled eggs. Hatchlings emerging from the nest may become dehydrated (Deraniyagala, 1939), however laboratory-hatched marine turtles show similar low ratios (Bustard and Limpus, 1969; Tomita, 1929; Senju, 1929; Simkiss, 1962).

The general similarity in proportions of hatchling to freshly laid egg in Table 9 favors the likelihood of the caloric content of these eggs also being the same on a per gram basis, given that such is true among hatchlings. Ecological studies on reproductive effort in turtles should be made easier by this fact inasmuch as the weight of a clutch becomes directly proportional to caloric input in interspecific comparisons, or at worst, requires minor adjustment.

Nitrogen Metabolism

Nitrogen is present in proteins occurring in the yolk, albumin and, presumably, the shell membranes. A single study (Nakamura, 1929) traced nitrogen content from outset to hatching and found that nitrogen in *Caretta caretta* egg contents decreased 15% during development. The internal accumulations and dissipations of nitrogen and their changes during development suggest that stresses imposed on the egg by its need to maintain its own reservoir for nitrogenous wastes may be reduced in turtles. Since the nitrogen loss phenomenon has not been studied beyond a single species laying parchment-shelled eggs in moist environments, one must be conservative about making broad speculations. Brittle-shelled eggs of tortoises laid in dry places may retain all their nitrogen, as do eggs of birds.

DEVELOPMENT OF THE IMMUNE SYSTEM

Recent work on the immune system is of general embryological interest because the research includes the first transplantation of chelonian embryonic tissues.

Cushing and Campbell (1957), Brent (1958), and Bellanti (1970) provide summaries of research with other vertebrates. Tissue antigens develop early in vertebrate embryos and tend to be unique to particular organs, first exhibiting their uniqueness in organ rudiments. Antibodies are initially supplied by the egg yolk in birds or by the mother in mammals. Both cases were investigated by immunizing mothers and testing for antibodies in fetuses or egg yolks (see Needham, 1963). Onset of antibody production varies with the species. During most of the embryonic period embryos accept implanted foreign tissue as if part of self, and such implants influence immune reactions toward favoring additional tissues from the same donor in postnatal life. A brief neutral period follows the embryonic acceptance period. The host rejects foreign tissue introduced during this period, but does not retain an immunogenic memory of having revieved this tissue. Foreign tissue received during the immunogenically competent period that follows not only is rejected but also is remembered immunogenically, and a second introduction of tissue from the host is rejected more rapidly than was the first.

Borysenko (1969b) studied the adult immunogenic capacity of *Chelydra serpentina*, which provides a baseline for comparison with embryos and juveniles. Hosts were from Wisconsin; donors included hatchlings and older snappers and also *Emydoidea blandingi* and *Chrysemys picta*. Borysenko made skin tissue allografts (= intraspecific grafts = homografts), xenografts (= interspecific grafts = heterografts), and autografts (between different parts of the same individual), the last serving as controls. Parameters sought included rejection time in days and rejection form (chronic or acute). At 25°C allograft destruction begins at 10 to 15 days following the transplant and reaches a point of gross cellular necrosis after an average of 40 days, with wide variation among individuals. Xenograft rejection begins in 8 to 10 days and is advanced at 24 days, hence is more rapid than for allografts. Immunogenic memory is demonstrated in that second allografts and xenografts from the same donor are rejected more rapidly than the first. All these rejections may be termed chronic, although that of second xenografts approaches acuteness, since they never completely heal in place. Rejection was more rapid at 33°C than at 25°C and was delayed for the duration of 100 day observation periods at 10°C. These findings indicate that the snapper immune system is similar to that of other poikilotherms (Borysenko, 1970; Cohen and Borysenko, 1970; Cohen 1971).

Borysenko (1969a) investigated maturation of the graft rejection capacity using seven age classes of snappers from hatchling to 6 months. Skin allografts were obtained from sibling and nonsibling donors of about the same ages as the hosts. Results obtained at 25°C indicated that rejection time decreased as age at transplantation increased from hatching to 4 months, when immunogenic maturity was reached. In turtles up to 1 month old, rejection times for grafts between siblings were greater than between nonsiblings. This difference, which was attributed to heredity similarities among sibling antigens, broke down between older turtles. Xenografts from *Emydoidea blandingi* were rejected more quickly than allografts at all ages. Spleen cell transplants (Borysenko and Tulipan, 1973) and *in vitro* spleen cell research (Sidkey and Auerback, 1968) confirm that immunogenic maturity in *C. serpentina* is reached at about 4 months.

Working with snapper embryos, Yntema (1970c) studied skin and carapace rudiment xenografts from embryonic *Chrysemys picta* and *Trionyx spiniferus*. Hosts were midway through somite formation when transplants were made. Hatchling chimeras with *T. spiniferus* and *C. picta* transplants were reared for 6 months or more. *T. spiniferus* tissue, which appeared as light-colored spotted skin on the carapace, began to undergo rejection within 6 months. *C. picta* tissue, which occurred as light-colored scutes on the carapace and as red pigmented blotches on the limbs, lasted longer, typically 1 year and up to 4 years. A few of the snappers that had rejected grafts from painted turtles were given second (skin) grafts from new painted turtle donors. These grafts lasted longer than painted turtle xenografts similarly planted on snappers without an embryonic graft history (Borysenko, 1969b). Yntema and Borysenko (1971) and Yntema (1974a) reported on allografted and xenografted (*C. picta* as donor) limb bud transplants made to limb bud stage snapper embryos. Several xenografts and allografts were rejected in 1 to 3 years. Two of six xenografts and one of five sibling allografts were completely healthy after 3 years and had grown in a manner typical of the donor species. Postnatal hosts having received limb bud allografts accepted an initial but not a second skin graft from the donor of the limb bud (Yntema, 1974b). These experiments indicate an embryonic acceptance period in turtles as in other vertebrates, when foreign tissue is designated as part of self. Yntema and Borysenko felt it unusual for grafts that had been accepted for very long periods not to have been retained indefinitely.

THE NATURAL ENVIRONMENT

Natural environmental relations of developing turtle embryos have received spotty attention. Incubation periods are so long that observers of nests must devote considerable time to making the repeated measurements that adequately describe nest microclimates. Bothersome precautions must also be taken to protect study nests against predation, which apparently is the fate of eggs in most nests in all but a few regions that have been studied.

Nests as Preliminary Indicators of Microclimate

Whereas a majority of species place their eggs in cavities they have dug in the soil, others do not, and the generalization itself includes some variation. *Sternotherus odoratus* occasionally buries its eggs partially (Risely, 1933b) or not at all; I have found some that appeared to have been developing for several days on the surface amid grass shaded by trees. Medem (1962) indicated that the emydids *Rhinoclemys annulata* and *R. punctularia* cover their eggs with leaves if at all. Moll and Legler (1971) offered similar evidence on *R. funerea*. Details on more complex nests are presented in Chapter 18.

The nest site may be deeply shaded (e.g., some *Macroclemys temmincki* nests) or under an open sky (many species, e.g., *Trionyx muticus*: Anderson, 1958; personal observations). The substrate forming the egg chamber may vary from sand (many species) to clay (*Chelodina* spp.: Goode and Russell, 1968; Vestjens, 1969; *Pseudemys scripta*: Cagle, 1944a; *Gopherus polyphemus*: Brode, 1959) or to rotting vegetation (*Sternotherus odoratus*: Risley, 1933b; *Dermatemys mawi*: Alvarez del Toro, 1960; *Chelydra serpentina*). It may have a high water-holding capacity because of an abundance of organic material, or it may present an exceedingly dry condition. In all, the varied locations of incubating eggs suggest diverse microclimates to be encountered by the developing embryos.

Temperature

Short-Term Variation. Eggs buried shallowly, especially in open situations, would be expected to face greater diel and short-term temperature variation, particularly as a result of solar heating or cold rain, than would eggs buried more deeply. In nests of Australian chelids at a depth of 18 cm Goode and Russell (1968) found a typical diel variation of 3°C and an extreme range of 8.5°C (18.3–26.8°C), while the soil surface varied from below freezing to 63°C. The data of Medem (1969) suggest diel variations of 2 to 6°C within the range of 28 to 34.5°C among eggs at a depth of 11 to 15 cm in *Podocnemis unifilis* nests. Surface temperatures varied from 27 to 39°C. In *P. vogli* nests at egg depths of 4.2 to 12 cm Pardo (1969) recorded 25 to 31°C at different times on different dates, with surface temperature of 28 to 34°C. In nests of Tennessee *Pseudemys scripta* (ca. 8–11 cm deep) Cagle (1937) found that temperatures ranged from 25.7 to 31.0°C. Moll and Legler (1971) reported variation from 22 to 30°C in nests of Panamanian *P. scripta* (11–18 cm deep). *Trionyx m. muticus* of the lower Sabine River of east Texas build shallow nests in broad sandbars at some of the most exposed situations available. Measurements on 19 nests in mid-June indicated that nest depths vary from 10 cm to the first egg to 17 cm to the bottom of the nest. The shallower eggs were generally at about 14 cm. On sunny days, early morning nest temperatures among the upper eggs varied in 10 nests from 28 to 30°C, the presumed diel low, to 34 to 36°C in 6 nests in late afternoon, the presumed diel high, giving a diel variation of 4 to 8°C.

Eggs of *Podocnemis expansa* lie deeply buried at 60 to 90 cm below a sandbar surface. Roze's (1964) data indicate a 1°C variation (31–32°C) among the eggs. Blohm and Fernandez-Yepez (1948) noted a 23°C variation (25–48°C) at the surface of *P. expansa* nests, and Roze stated that surface temperatures up to 60°C could occur. In sand adjacent to *Chelonia mydas* nests, also at a depth of 60 to 90 cm, Hendrickson (1958) recorded a 1 to 2°C diel variation about a mean of 29°C as opposed to a 5 to 23°C diel surface variation with temperatures between 26 and 50°C. Carr and Hirth (1961, 1962) gave 1°C or less as the diel variation at egg depths of *C. mydas* on Ascension Island.

Thermal Tolerances. Temperature extremes that affect survival of turtle eggs have been investigated in only a few species. Fresh eggs of *Chelydra serpentina* from Wisconsin and New York continue development at 20°C normally for about 3 months, but embryos die short of hatching. Some 90 day/20°C embryos have hatched into normal turtles after transfer to 30°C. Embryos started at 30°C can complete development at 20°C (Yntema, 1968). Development of fresh *Chrysemys picta* eggs from Minnesota and Wisconsin progresses only a few weeks at 20°C; then the embryos die (Ream, 1967), with vitelline circulation barely established. This observation is a little surprising in view of Ernst's (1972) estimation of an optimal temperature of 20.5°C for the species. Mahmoud et al. (1973) reported development through hatching at 21 to 23°C in Wisconsin *C. picta*; hence the threshold for continuous development must be near 21 to 22°C. Bustard and Greenham (1968) and Bustard (1971b) incubated 10 to 15 egg batches of *Chelonia mydas* eggs at each of several constant temperatures. Zero hatch occurred at 23°C, 20% at 25°C, and 60%, the maximum figure, hatched each time at 27 to 35°C.

In Minnesota the viability of *Trionyx spiniferus* eggs is approximately the same at 25 and 30°C, but at 25°C many hatchlings die within a few days of pipping. Data collected over a 3 year period indicate decreased viability for northern *T. muticus* at 25°C (e.g., 76.5% of 191 Minnesota eggs in 1973 died as opposed to 17% of 162 eggs at 30°C). Furthermore, 43 of the 45 *T. muticus* hatchlings from 25°C incubations died within a week; they typically survived 1 to 2 days. The remaining two behaved abnormally in more than 5 mm of water, each swimming franticly, flipping over after a few minutes, then lying nearly motionless. Freshly pipped specimens appear normally active but seem to lose rigor as the extraembryonic membranes regress. Possibly a respiratory defect is involved. Most of 28 species of *Trionyx* incubated at first at 22°C and later at 30°C developed at least slightly but failed to hatch. Hatchlings from eggs incubated entirely at 30°C have better than 90% survival during their first 2 weeks under the posthatching conditions provided.

The temperature of 25°C may represent a developmental threshold for *Trionyx* species; at this point serious defects have occurred, but they are not so serious as to prevent hatching in every case. This situation differs from that observed by Yntema (1960, 1968) inasmuch as his *Chelydra serpentina* embryos developed normally for long periods below that critical for hatching, yet could be hatched into normal animals following transfer to a warmer temperature. Yntema (1960) did induce gross abnormalities through early incubation at 15°C and observed general suspension of development at 10°C. Hatchlings resulted in both cases after transfer to room temperature; the 15°C treated were defective and the 10°C normal. Cagle (1950) reported that eggs of Louisiana *Pseudemys scripta* failed to survive a 2 week exposure to 10°C, but he was not convinced that temperature was the reason for death. Ripe eggs, in the oviducts of Tennessee *Sternotherus odoratus* and southern Indiana *Chrysemys picta marginata* have survived a 2 week exposure at 0 to 3°C. When these eggs were incubated at 30°C, 9 *S. odoratus* from 5 clutches and 31 *C. picta* eggs from 7 clutches hatched.

With regard to critical thermal maxima, Yntema (1960) reported that *C. serpentina* did not sustain development at constant 34°C, and Moll and Legler (1971) found that eggs of Panamanian *P. scripta* could not tolerate 35°C early in incubation. However Bustard (1971b) obtained 60% hatching of *C. mydas* eggs at 35°C after Bustard and Greenham (1968) found that continuous 38°C was lethal. Thus the range of constant temperatures known to yield healthy hatchlings in Cnelonia appears to lie between about 22 and 35°C. This range might be extended through study of *Platysternon megacephalum*, whose preference for cool situations has been observed by Mell (1938) and reviewed by DeKonigh (1968), and species of *Trionyx*, which experience relatively warm nest temperatures. Developmental rate in *T. sinensis* as reported by Mitsukuri (1895b) is remarkably coincident with warm conditions and can be accelerated to 28 days incubation, the most rapid reliably reported for turtles.

Reported temperature extremes survived by developing eggs in the soil, where exposure time may be brief, range from 4.9°C in *Chelodina expansa*, which nests in the autumn (Goode and Russell, 1968), to 46°C in artificial nest beds of *Malaclemys terrapin* eggs (Cunningham, 1939). The latter could not endure continuous 35 to 40.5°C. I have recorded temperatures of 39.2 to 40.0°C in three viable *C. p. belli* nests during an afternoon maximum relatively early in incubation.

Thermal tolerances of turtle eggs probably depend on a multitude of variables. Not only are some embryonic stages more temperature responsive than others (Yntema, 1968), but also there may be a diel periodicity in short-term tolerance as Kosh and Hutchison (1968) have found in adult *C. picta* as well as pre- and postlaying acclimation related to local environmental conditions. Autumn nesting species in south temperate climates with mild winters (e.g., *Chelodina expansa, Geochelone pardalis*) seem to have embryos capable of surviving extended periods of suboptimal temperatures at various developmental stages, as suggested by nesting data in relation to overwintering and spring-summer hatching. In contrast are north temperate species, which complete development during a single warm season (e.g., *Chelydra serpentina, Clemmys insculpta, Chrysemys picta*) with or without brief cool spells, as in northern Minnesota and Michigan, and species that bury their eggs deeply in warm soil beneath a body pit and face a narrower range in developmental temperatures.

Moisture

Depending mainly on the species, eggs have varying demands for soil moisture. Observations on artificially incubated eggs have indicated that parchment-shelled *Chrysemys picta* eggs partially in contact with an aqueous surface (wet cotton) will lose water if the remaining surface is exposed to normal laboratory air as opposed to a saturated humidity (Lynn and von Brand, 1945). In contrast, hobbyists generally place dishes of water in incubators with brittle-shelled eggs of *Gopherus agassizi*, but the eggs themselves rest exposed in circulating air (Lampkin, 1966; Anonymous, 1967).

From tolerances to humidity vaguely suggested by the foregoing information, one would expect wide variation in moisture levels of natural nests. Although quantitative data on normal nests are scarce, descriptive information exists. In initially preparing the egg chamber, the female characteristically voids cloacal fluid. Mucous liquid, possibly from the oviducts, accompanies laying in some species. Two of the more extreme cases appear to be *Lepidochelys olivacea*, whose eggs actually adhere together in a cluster (Deraniyagala, 1939), and *Geochelone pardalis*, where Wilson (1968) observed that eggs emerge within a mucous mass from the cloaca.

In some species the completed nest is obviously wetter than its surroundings. The plug of fresh *C. picta* nests in clay soil may have the consistency of soft mud. In nests of *Pseudemys scripta* there may be a drop in local substrate moisture as effects of cloacal fluid dissipate (Cagle, 1937). Where *Chelodina longicollis* and *C. expansa* nest in clay, Goode and Russell (1968) reported that eggs are "puddled" with mud such that some may become encased as the mud dries and hardens. This change may not affect embryonic development, but in prolonged absence of rain may delay emergence of hatchlings so long that they become emaciated from lack of food. Goode and Russell gave 324 to 360 days as the common natural incubation period of *C. expansa*. Russell (personal communication) found two nests of emaciated but living hatchlings 662 and 664 days after the eggs had been laid. Legler (1954) noted that egg chambers in clay soils had standing water in them after heavy rains, although he did not ascertain its duration or effect on the eggs. De Vries (1968) reported that heavy rains may cause mortality of newly hatched *Geochelone elephantopus* but did not specify if they were drowned in nests.

In nests of parchment-shelled eggs a stratification of moisture may occur. Cagle (1937) and Legler (1954) found that the uppermost eggs in nests of *P. scripta* and *C. p. belli*, respectively, were more often dented and dehydrated than the others. Allen and Neill (in Carr, 1952) observed that the bottommost eggs in a *Macroclemys temmincki* nest were larger and softer. In my experience with this species, uptake of water by the bottom eggs in contact with damper soil conditions is the most probable explanation for the stated differences in eggs.

Uptake of water by parchment-shelled reptilian eggs is well documented (Cunningham and Hurwitz, 1936; Cunningham and Heune, 1938; Dmi'el, 1967). As stated earlier a modest uptake of water may facilitate the combustion of proteins. Might one also consider a functional purpose of water uptake as a modifier of the eggs' external environments as opposed to the internal? In a natural nest of *C. p. belli*, where embryos had developed into extensively pigmented fetuses, soil within a few millimeters of each of the eggs was so dry that it fell off as dust. Soil more remote from the eggs in every direction was sufficiently damp for particles to adhere to each other. The egg shells had become completely chalk white, and water sprinkled on them formed beads, indicated lack of a wetting surface—a feature commonly observed during incubation at 30°C. Such a shield of low humidity might discourage molds and waterborne pathogens. By taking in water, eggs may increase in size, as some do markedly, hence add to their respiratory surface. At the same time they may remove water from clogged pores in surrounding soil.

Gas Exchange

One would expect that respiration of developing eggs would influence levels of oxygen and carbon dioxide in the nest chamber and neighboring soil. Ackerman and Prange (1972) and Prange and Ackerman (1974) took measurements in three nests of *Chelonia mydas*, which were within 2 weeks of hatching (Table 5), and found a decrease of 3 to 9% in oxygen over the 21% characteristic of normal air and a seventyfold increase over the usual ambient carbon dioxide level of 0.03%. Buckman and Bradey (1960) and Vilenskii (1960) indicated that oxygen percentages in soil air are typically lower than in air above ground and that carbon dioxide levels are severalfold higher. However it is probable that the *C. mydas* eggs had altered soil air locally.

Lynn and von Brand (1945) have shown that oxygen demands of the eggs of several turtle species increased as development advanced up to and through hatching. An interesting point is that a typical clutch of *C. mydas* eggs, by dint of its large size, mass, and relatively rapid developmental rate, probably demands more oxygen are exhausts more carbon dioxide than the average clutch of turtle eggs; thus such a clutch might be expected to exert more pressure for circulation of soil air. A rough estimation of required oxygen flow through egg chamber walls of *C. mydas* nests and those of two other species is possible if we assume that the egg chamber is a sphere and that ambient temperature is 25°C. Calculations involve egg chamber diameters (Hendrickson, 1958; Ernst and Barbour, 1972; Buso, unpublished data), clutch size (Prange and Ackerman, 1974; unpublished personal data), and prehatching rates of oxygen consumption at 25°C (Lynn and von Brand, 1945; Prange and Ackerman, 1974). A *C. mydas* clutch of 100 eggs in a 20 cm diameter egg chamber requires oxygen to flow through the egg chamber walls at a rate of 0.3 ml/cm² · hr. Corresponding data for *Chelydra serpentina* are 40 eggs, 14.5 cm diameter, and 0.05 ml/cm² · hr, and for *Chrysemys picta*, 10 eggs, 7 cm diameter, and 0.04 ml/cm² · hr.

C. mydas eggs are buried deeply in the ground and developmental temperatures are relatively high, even increasing as a result of metabolic heating toward the end (Bustard, 1972a). Offsetting these pressures are the coarseness of the sand, which enhances air circulation, and the probable low competition for oxygen with other soil flora and fauna, although tree roots are characteristic of many nest sites (Bustard and Greenham, 1968). Species laying especially massive clutches may be restricted to nest sites in sandy soils to ensure adequate porosity for air circulation which, Prange and Ackerman (1974) conclude, occurs primarily through simple diffusion.

Turning and Jarring of Eggs

It is said that chelonian eggs should not be turned, especially during the early part of incubation. The embryo that was once attached to the upper egg shell above the yolk mass will come to lie beneath it and be crushed or, at least, seriously distorted by it. The matter has not received systematic study, but should command some attention with regard to transportation and handling of eggs during chelonian research, husbandry, and conservation.

Personal observations are sketchy but allow qualitative statements on the matter. I have gently but completely inverted small numbers of eggs of *Sternotherus odoratus* (7 eggs), *Chelydra serpentina* (5), and *Pseudemys scripta* (6) on the second to fourth day of incubation at 25°C. Whitening of the egg shell indicated adhesion of the vitelline membrane, but blood islands of the vitelline circulation were not yet visible. A similar treatment at 30°C was given to four *Trionyx muticus* eggs that had just established vitelline circulation. The yolk settled over the embryonic region in all cases, yet none of the eggs died immediately. Disruption of development in one or more eggs of each species became evident during the next 2 weeks, and at least one egg from each species died, probably because of turning. Alternatively, the embryo gradually grew around and was able to rise above the yolk mass in two or more representatives, doing so in four of the five *C. serpentina* eggs. Development continued, although the extraembryonic membranes were not oriented normally. In each species one or more normal hatchlings resulted.

Turning of more advanced eggs involved eight *C. serpentina* on the twelfth day and eight on the forty-second day of incubation at 30°C (paddle stage limb bud, advanced skin pigmentation; Stages 14 and 24, respectively; Yntema, 1968), and 30 *C. p. belli* on the tenth and 30 on the forty-first days of incubation at 30°C. In the larger samples it is possible to say that hatching in turned eggs was as good as that of unturned controls (80–90%) in all cases except the *C. serpentina* turned on the twelfth day, where one in eight hatched. The embryos turned on the forty-first and forty-second day periods simply pipped upside down; none ruptured the yolk sac, an occasionally observed accident.

Fifteen eggs from four clutches of *C. p. marginata* were turned once daily from day one, while seven eggs from the same clutches served as unturned controls. Unexplained heavy mortality occurred in both groups, and the experiment was terminated on the thirtieth day. At this time seven embryos from the turned group and one control were still living, seemed to be normal, and had begun forming carapacial pigmentation, which is an advanced stage. In another experiment 27 eggs of *Macroclemys temmincki* selected from three clutches were inverted initially during the period of blood island formation, then once at 5 day intervals through day 50, and thereafter once every 7 to 10 days until hatching (90–100 days at ca. 27°C). Hatching survival was 63% for the turned eggs and 81% for 461 unturned control eggs from 13 clutches. One might expect damage to the embryo from pressures of the inverted yolk mass in these rather large (25–30 g) eggs.

It appears that the position effect of turning is not always lethal. Chelonian development has probably had to cope with turning in natural environments. Nests are not always tightly packed, and the upper eggs may be moved about as those beneath take up water, swell, and change in shape.

A different deleterious effect of turning may relate to torque applied to the egg contents during the act. Early in development, a shearing force might occur between the eggshell and the small attached embryonic region on the one hand and the remaining egg contents on the other. The result could be a tear in the vitelline membrane, allowing mixing of albumin with vitelline sac contents (yolk and fluid of the subgerminal space). These remarks stem from observations that eggs of *C. serpentina* and species of *Trionyx* 1 to 3 days old are often damaged merely by rolling them out of the egg chamber when excavating a nest. Immediately returning them to their original gravitational positions seems not to prevent damage or to correct any incurred. In contrast, carefully lifting the eggs out such that their original positions are retained usually does not harm them.

Turtle eggs may be subject to jarring while riding in a motor vehicle along a rough road, and this seems not to harm them. Natural turtle nests constructed along railroad tracks are subject to jarring by passing trains; railroad embankments, though not a natural environmental feature, are popular nesting areas. In one instance I found a nest of *C. serpentina* in sand and gravel about 50 cm from the ties. Five or more trains per day shook the neighboring earth noticeably. The eggs when discovered had reached an early limb bud stage and were successfully hatched under laboratory conditions. A *C. p. belli* nest between two ties and within 20 cm of one rail experienced a single freight each day, which jolted this track as a consequence of its poor state of repair. Ten eggs with fetuses well advanced in skin pigmentation were removed to the laboratory, where the nine unopened eggs hatched normally.

Although some turning and jarring apparently is tolerated, a careful study of how other turning and jarring harms eggs might help our understanding of eggs and how to work with them.

Microbial and Fungal Relationships

Several workers have remarked on fates of infertile eggs and inviable eggs failing early in incubation. As might be expected, there are numerous casual references to decay of dead eggs in natural situations (e.g., Archer, 1966), especially referring to molds (Legler, 1954; Goode and Russell, 1968; Moll and Legler, 1971). Regarding eggs under artificial incubation, Legler (1960a) mentioned that infertile eggs became slimy after several weeks, a suggestive indication of bacterial decay, and Lynn and von Brand (1945) and Bustard and Greenham (1968) refer to molds. Dead or moribund eggs can persist undecayed despite an environment richly endowed with bacteria and other decomposers. A most truly remarkable extent of this phenomenon is expressed by Goode and Russell (1968), who found that some eggs of *Chelodina expansa* in natural nests lacked any signs of embryonic development, yet were "clear and fresh" inside with no decomposition of yolk or albumin after more than 300 days. Deraniyagala (1939) noted that yolkless, albumin-filled, subnormal-sized eggs of *Dermochelys coriacea* remained fresh until normal eggs had hatched (after 60–70 days). In my experience numerous infertile eggs or others where, perhaps, the embryo has become arrested at a very early stage, have persisted undecayed beyond the time taken for neighboring eggs to hatch out. Examples include approximately 120 days for *Kinosternon bauri*, 100 days for *Chelydra serpentina osceola* and *Pseudemys scripta*, and 75 days for *Chrysemys picta*. It is logical to ask how eggs that have failed as functional units retain physical and chemical integrity for such long periods.

Although much is known about antimicrobial defenses of avian eggs (Board and Fuller, 1974), it appears that the only studies regarding bacteriocidal and fungicidal properties of turtle eggs have been reported on the steppe tortoise, *Testudo horsfieldi* (Movchan, 1964, 1966, 1967). Movchan (1964) found that the albumin had antibiotic properties, whereas the shell membranes and yolk did not. The albumin of unfertilized eggs and eggs at hatching destroyed the bacterial forms *Micrococcus lysodeikticus, Sarcina urea, S. lutea,* and *Mycobacterium album,* and the vegetative cells of *Bacillus mesentericus, B. megatherium,* and *B. subtilis.* The bacterialike organisms *Actinomyces albus, A. griseus,* and actinomyces of the tortoise's homeland soil of Tadjikistan were also destroyed. Albumin from fresh fertile eggs and eggs in early stages of postlaying development proved lethal to the aforementioned actinomyces as well as to *A. lavendulae, A. violaceus,* and *Streptomyces citreus.* The mold fungi *Aspergillus fumigatus, A. niger,* and *Penicillium granulatum* were also killed (Movchan, 1966). There were albumin-resistant bacteria: *Bacterium proteus, Pseudomonas fluorescens, P. pyocyanea,* and *Salmonella pullorum* (Movchan, 1964). Resistant fungi included the yeast forms *Tortula utilis* and species of *Rodoterula* (Movchan, 1964, 1966). Movchan (1967) isolated *Bacillus neapolitanum* from the egg contents of whole tortoise eggs. When fungicidal powers of different portions of the albumin were tested, the innermost from fresh eggs proved to be the most active, but after 8 days of incubation, more distal albumin became the most active portion and remained so (Movchan, 1966). These findings support Agassiz's (1857) observations of differential movement of various layers of the albumin as the vitelline sac absorbs substance from them.

It is common knowledge that turtles frequently harbor species of *Salmonella* and can pose a health hazard to ignorant or careless humans. Speculatively, a female turtle can pass salmonella from herself to her progeny as the eggs pass through her cloaca or become exposed to the fluid voided during nest building. Eggs and hatchlings have been contaminated by *Salmonella braenderup* and *S. typhinurium* (e.g., Feely and Treger, 1969).

Mortality from Abnormal Conditions in Nests

Although most reports attribute nest failure to predation, other causes have been indicated or proposed. Grasses and roots may pierce, erode, or encase the eggs (e.g., Baldwin and Lofton, 1959; Turkowski, 1972).

Several workers have related weather factors to fatalities, although the exact cause of death in each case was not proven. Dehydration is a suggested cause in *Chelodina longicollis* (Vestjens, 1969), *P. scripta* (Cagle, 1937), and *Chrysemys p. belli* (Gemmell, 1970). In each case dead eggs were found in a desiccated condition. However since these eggs tended to be nearer than the others to the soil surface, where the highest diel temperatures occur, or even in incompletely closed nests, the possibility that the critical thermal maximum was exceeded cannot be ruled out. Shells of dead eggs tend to lose water-holding capacity.

Excess water assisted in causing death by suffocation to embryos of *Caretta caretta* (Ragotskie, 1959) and *Podocnemis expansa* (Roze, 1964). In the former case heavy rains raised ground water levels, and some eggs were found in water in egg chambers under anaerobic conditions. The latter case, involving thousands of specimens, resulted from premature seasonal flooding of river sandbar nests.

Death due to low temperatures has been suggested for Michigan *Sternotherus odoratus* by Risley (1933b) and for Minnesota *Trionyx spiniferus* by Breckenridge (1960). In Minnesota the most probable cause of death was freezing, inasmuch as the frostline generally penetrates below *Trionyx* species' nest depths each winter. Dead term embryos of Minnesota *C. p. belli* unearthed in April (Jeff Lang, personal communication) and May (Breckenridge, 1944) were apparently overwintering mortalities. Dead *C. p. belli* hatchlings were found in nests, where in some cases I had measured subfreezing nest temperatures during the winter.

Effects of salinization on eggs of *Chelonia mydas* were studied by Bustard and Greenham (1968). Conceivably, a marine storm could greatly increase salt levels in beach sands that normally are not inundated. In laboratory tests, chlorinities reached 75 to 100% that of seawater before becoming lethal to eggs. Such levels were not observed in the nests they studied.

To determine the true significance of nonpredatory mortality in nests on turtle populations, we need quantitative studies that account for year-to-year climatic variation.

MORPHOLOGICAL VARIATION

Selected Terata and Other Morphological Variants

Developmental failures and atypical hatchlings are fairly common products of amniote eggs and may reflect the genetic constitution of the zygote in some cases and the maternal and incubational environments in others. So little work has been done on pathogenesis of chelonian embryos that simply to catalog observed deformities is a new undertaking. This section considers some obvious deformities and a small amount of variation that cannot justifiably be called deleterious. Criteria for selection included (1) probable identity with classical syndromes observed in other vertebrates (Dareste, 1891; Romanoff, 1972), (2) relevance to known occurrence in wild populations, and (3) special herpetological interest.

Longevities of terata, which may interest persons incubating eggs, vary; some specimens die before any sibling fetuses are ready to hatch. Another portion die during or just after the period when normal eggs hatch; such embryos may or may not exhibit features characteristic of term embryos. Occasionally, terata survive as fetuses long after the normal hatching period. Lynn and Ullrich (1950) documented a *Chrysemys p. picta* fetus they had removed from its egg 21 days after the average hatching time of 63 days for normal specimens. Two *Macroclemys temmincki* fetuses that survived 35 days beyond the normal hatching time of 120 days for their group were normal terminal embryos in some respects but seriously deformed in others. One *Clemmys insculpta* egg incubated at about 30°C remained viable for 20 days beyond the normal 43 day pipping period of 19 other eggs in the series, and for 16 days beyond the time the last normal hatchling had retracted its yolk sac. Upon opening the egg, I found a normal term fetus in all respects except for unusually heavy cornification of the shell and proximal yolk membrane. This specimen retracted its yolk after 15 days, ate, grew slightly, and was moderately active. It appears that terata will live as long as embryonic life support systems—most probably, respiratory, yolk supply, and excretory—continue to function. An embryo *Graptemys geographica* was about three-fourths as large as a normal term fetus when I removed it and its well-developed extraembryonic membranes from its egg at pipping time. Despite being so abnormal that it was difficult to discern anterior and posterior ends, it lived for a week at 22°C just resting on a wet paper towel.

Deformities of the Head. Under the combined various conditions in which I have incubated them, probably more than 2% of advanced embryos were afflicted with head deformities. In one sample involving 170 eggs, head deformities occurred in 7%. A single clutch of 10 *Pseudemys scripta elegans* eggs yielded eight term embryos with severe head deformities. Aside from deviant patterns in scute number and arrangement on the shell, anomalies of the head taken collectively are the most numerous terata evident externally among advanced embryos. Although the anteriodorsal region appears to be the most susceptible, overall extremes vary from slightly misshapen skulls (e.g., flattened face, microcephaly) to total absence of discernible head structures, a neck with a gap to the alimentary canal in one term-sized *P. s. elegans* fetus.

Classical terata of the head in turtles include encephalocoele (Figure 2a), true cyclopia (Figure 2b, c, d), false cyclopia or monophthalmia (Figure 2e, f), anophthalmia (eyelessness: Hildebrand, 1930; Sturn and Brattstrom, 1958; Mausolf and Wunder, 1974), cleft palate–harelip (Figure 3a, b, c, d) and a reduction of the mouth suggestive of synotia (Figure 4). Specimens with encephalocoele, true cyclopia, and severe reductions of the mouth seem to be unable to pip their eggs when they have survived as long as normal term fetuses. Specimens with false cyclopia, anophthalmia, and cleft palate–harelip have hatched vigorously and have attempted feeding. Anophthalmic specimens of *Chelydra serpentina* and *Clemmys insculpta* have grown slightly; another, *Pseudemys scripta*, kept with similarly aged, sighted *P. scripta* hatchlings by Mausolf and Wunder (1974) grew to 300 g in about 21 months. I have captured adult wild specimens of *Sternotherus odoratus*, *Chrysemys picta*, and *Trionyx muticus* with cleft palate–harelip.

Deformities of the Body. Celosomia, where the heart, liver, and other external organs are situated externally along the umbilical seam, has occurred in *Caretta caretta* (Coker, 1910) and *Chrysemys picta*. The overall body size has often been small relative to the head and organs. The few celosomic terata that pip soon die.

In term fetuses and hatchlings the body may be asymmetrically withered (Lynn and Ullrich, 1950). Parts of the body may be missing entirely, as exemplified by a *C. caretta* hatchling lacking its left front flipper (Baldwin and Lofton, 1959) and by *C. serpentina* and *T. muticus* missing parts of the shell.

Symmetrical reductions of the posterior end have occurred in *C. picta* (Figure 5a), *Graptemys geographica*, and *P. s. scripta*. Specimens surviving hatching experienced gangrenous deterioration of the affected region. Several species yielded hatchlings with kinky or spirally twisted tails. A single kink can render a tail that is normal in size functionally debilitating inasmuch as it may snag on objects as the hatchling moves about. Chelydrids and emydids often have bent tails; the tail is relatively larger at hatching in proportion to overall body size than it is in the adult. Deraniyagala (1939) reported a *Dermochelys coriacea* hatchling with a kinky tail.

Atypical Arrangements of Scutes of the Shell. Turtles with atypical numbers or peculiar arrangements of scutes or shields comprised 43% of 2220 specimens in the Chicago Natural History Museum (Zangerl and Johnson, 1957). Coker (1910) reported that 45% of 243 *Malaclemys terrapin* specimens had atypical scute arrangements, including 21% with

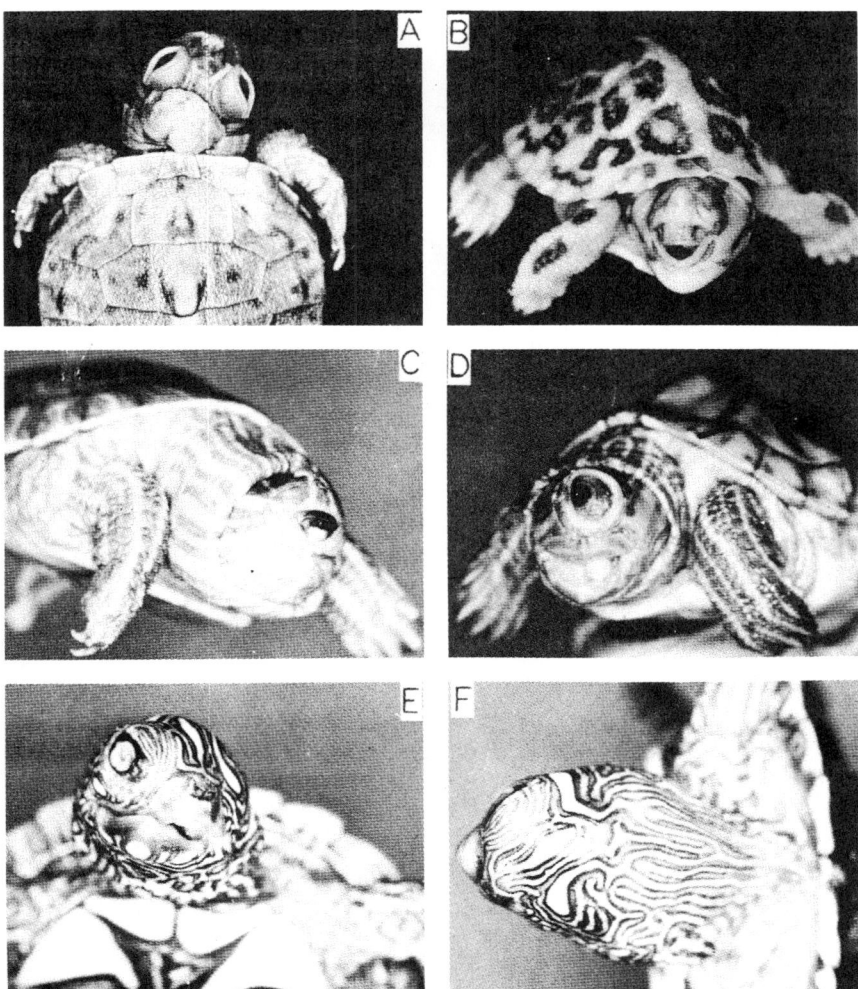

Figure 2 Deformities of the head. (*a*) Encephalocoele (hernia of the brain) in a term fetus of *Graptemys pseudogeographica*, one of three thus afflicted from a single clutch of eggs (also observed in *Sternotherus odoratus* and *Chrysemys picta*). The protruding nervous tissue may be covered with typical skin as opposed to being naked as shown. (*b*), (*c*), (*d*) True cyclopia (eyes fused anteriorly). (*b*) A late fetus of *C. picta* also showing a commonly associated deficiency in skin pigmentation (white areas). (*c*) A term fetus of *C. picta*. (*d*) A term fetus of *Pseudemys scripta elegans* with asymmetrical cyclopia, possibly involving monomicrophthalmia at an early developmental stage. The deformity also occurs in *Caretta caretta*, *Emydoidea blandingi*, *G. pseudogeographica*, and *Trionyx spiniferus*. (*e*), (*f*) False cyclopia or monophthalmia (lack of one eye with concomitant warping of skull) in a *G. pseudogeographica* hatchling. (*e*) Anterior view showing displaced upper jaw. (*f*) Dorsal view. The deformity also occurs in *S. odoratus*, *Chelydra serpentina*, *C. picta*, *P. scripta*, and *T. muticus*.

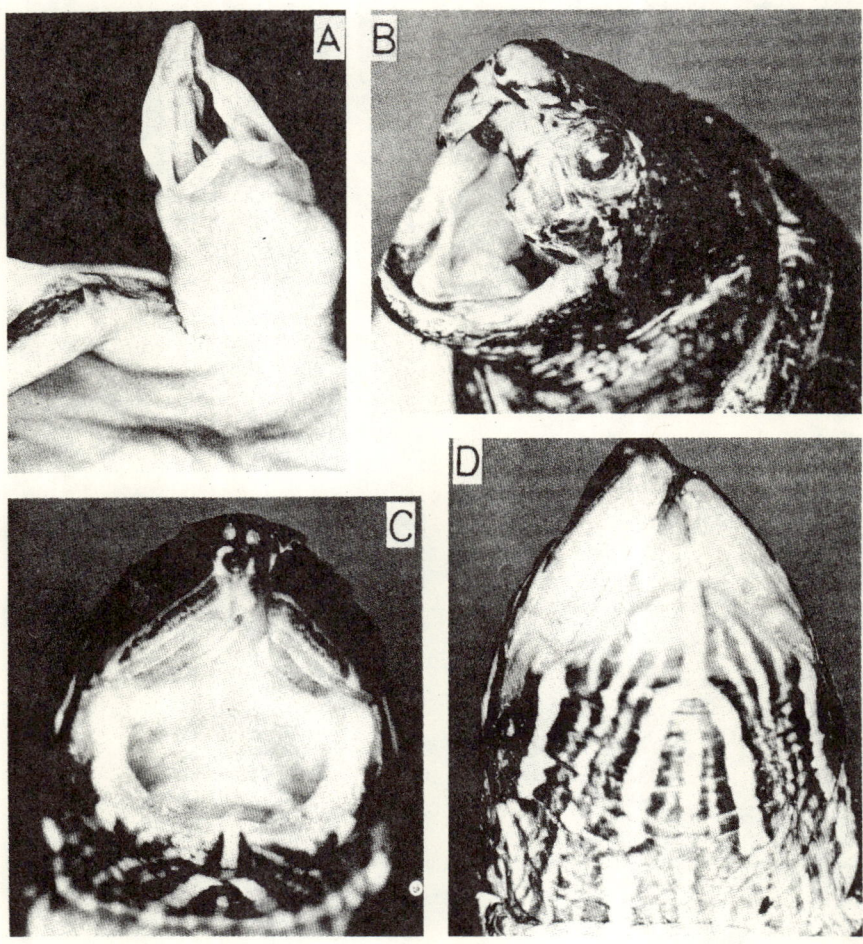

Figure 3 Cleft palate–harelip. (*a*) Hatchling *Trionyx muticus calvatus* with characteristic curvature of snout in direction of cleft. This specimen is one of 14 individuals (from 9 clutches) thus afflicted among 167 *T. muticus* hatchlings (42 clutches) from eggs incubated simultaneously. A possible although not proved cause is a 2 week period of cool temperature relatively early in incubation. (*b*) Wild-caught adult *Sternotherus odoratus* with harelip and degenerate nostril but without cleft palate. (*c*) Wild-caught adult *Chrysemys picta* with harelip and cleft palate. (*d*) Same specimen as in (*c*) but showing upper surface of oral cavity.

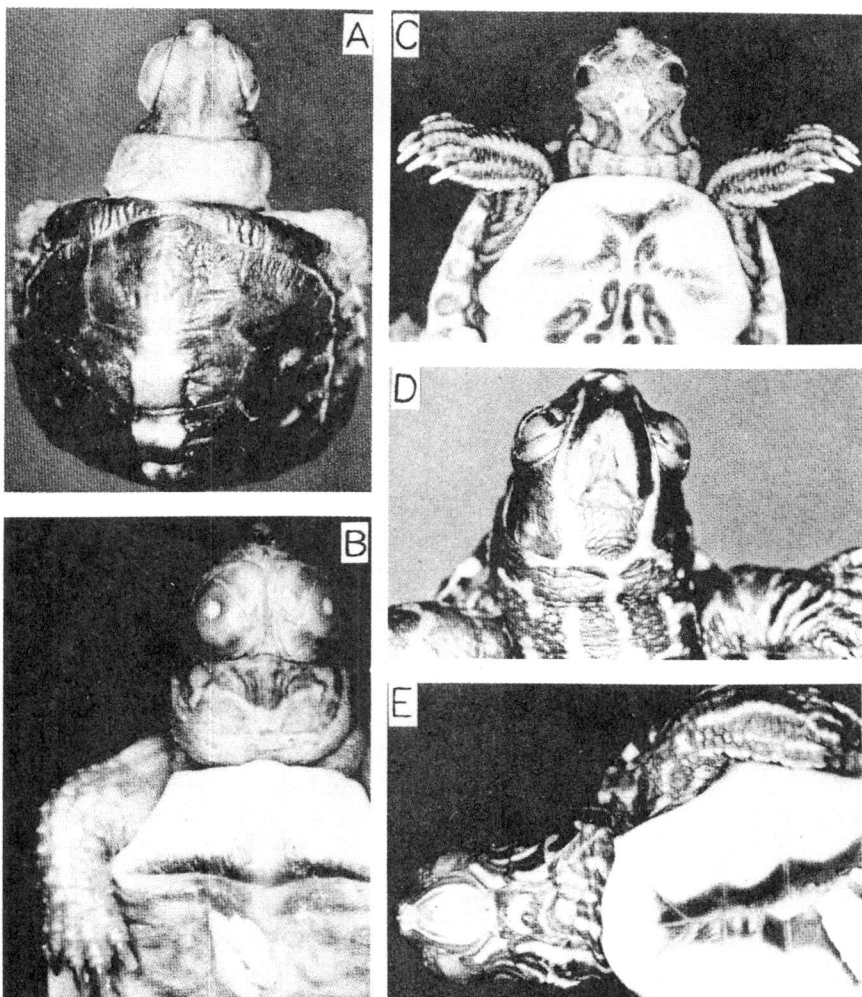

Figure 4 Absence or reduction of the mouth. (*a*) Near term fetus of *Terrapene carolina* showing characteristic narrowing of the head as viewed from above. (*b*) Ventral view of individual in (*a*); its eyes are displaced ventrally, and it lacks a mouth and jaws. (*c*) Term fetus of *Chrysemys picta* with mouth reduced to a tiny ventral slit continuously surrounded by a keratinized jaw structure, posterior to the interorbital area. (*d*) Term fetus of *Sternotherus odoratus*, which lacks a lower jaw but possesses a reduced and medially compressed upper jaw defining the mouth as a small vertical slit, which appeared to extend into a blind pocket. (*e*) Term fetus of *C. picta* with mouth only slightly more developed than in (*d*) but connected to an esophagus. The emydids were living when they were removed from their eggs, but seemed to be unable to breathe.

fewer or more than the standard number. Termed supernumerary scutes, those in excess of the normal number were present on shells of 10% of a sample of 476 *Graptemys geographica* (Newman, 1906b). Thus some kind of deviation from normal is relatively common, and as Newman and Zangerl and Johnson have concluded, this is not likely to be deleterious, since large and presumably old specimens are as likely to be deformed as hatchlings. In my experience of working closely with about 1500 adult specimens, the radical excesses in scute number commonly arising from artificially incubated eggs rarely occur among wild adults with one exception, the dovetail syndrome of carapacial anomalies (Figure 6). I have observed single supernumerary pleural scutes (cf., Zangerl, 1969; Ernst and Barbour, 1972) in all modern chelonian families bearing scutes. In *Lepidochelys olivacea* carapacial scute number is so variable that there is no single typical number (Deraniyagala, 1939; Pritchard, 1969a). Deraniyagala reported an aged *Chelonia mydas* individual bearing a single continuous horny plate on its carapace, possibly a senile as opposed to congenital type of variation.

At the turn of the century and shortly thereafter there were active discussions of whether scute anomalies reflected former stages in phylogenetic ancestry (atavisms) or simply terata (Gadow, 1899, 1905; Parker, 1901; Newman, 1906b, 1923; Coker, 1910; Grant, 1937). In updated considerations Zangerl and Johnson (1957) and Zangerl (1969) indicated that scute anomalies appear to reflect both genetic and teratogenic influences. Certain locally circumscribed patterns of variation were found to be much more common in some species than in others. Other patterns were not so defined and may even result in a high degree of asymmetry, as in regions where the shell appears withered. Certain symmetrical reductions in scute number such as the example given in Figure 5b (see also Willemsens, 1974) seem to be extremely rare. With regard to atavisms, the later works point out that the plastron and central regions of the carapace achieved their present scute arrangements during the Triassic era and have varied rather little since then. Only a few of the modern anomalies, such as appearance of inframarginals, reflect the fossil record.

The dovetail syndrome of carapacial anomalies (Figure 6), which appear to be benign terata, allows interesting speculation about embryogenesis of scute distribution. The syndrome is expressed in varying degrees of completeness in Chelidae, Kinosternidae (Figure 6a, b), Chelydridae (Figure 6c), Emydidae (Figure 6d, e; Parker, 1901; Coker, 1910; Lynn, 1937; Sturn and Brattstrom, 1958), Testudinidae (Baker, 1968), and Cheloniidae (Coker, 1910). In the minimum expression one vertebral scute is divided diagonally to give two in staggered formation;

Figure 5 Abnormalities of the body and epidermis. (*a*) *Chrysemys picta* individual lacking hind legs; it died while hatching. (*b*) Hatchling *Pseudemys scripta elegans* with unusual arrangement of scutes on its shell. For contrast, the specimen in (*d*) has the normal arrangement of scutes. (*c*) Pallid hatchling *Graptemys barbouri*. (*d*) Normal hatchling *Graptemys barbouri*.

in the maximum case there are five divided scutes to give a staggered vertebral row of 10 in place of the usual five. Sometimes the vertebral series exhibits incompletely divided elements as shown for *Chelonia mydas* by Deraniyagala (1939) and *Chelydra serpentina* in Figure 6c and *Chrysemys picta* in Figure 6e. Parker observed that the carapacial bones underlying the dovetailed scutes may not be anomalous in any way. My observations on shells of four kinosternids and four emydids have confirmed that the patternings of the keratin scute layer and the bony layer of the shell appear to rise independently. However the cellular pigment layer in between may reflect both symmetrical and asymmetrical features. The medial stripe of *C. p. dorsalis* in Figure 6d seems to be unaffected by asymmetry of the vertebral scutes; in Figure 6e it is. Anterior margins of the vertebral scutes are patterned according to the asymmetry.

Coker noted that pleurals and vertebrals met each other in a charac-

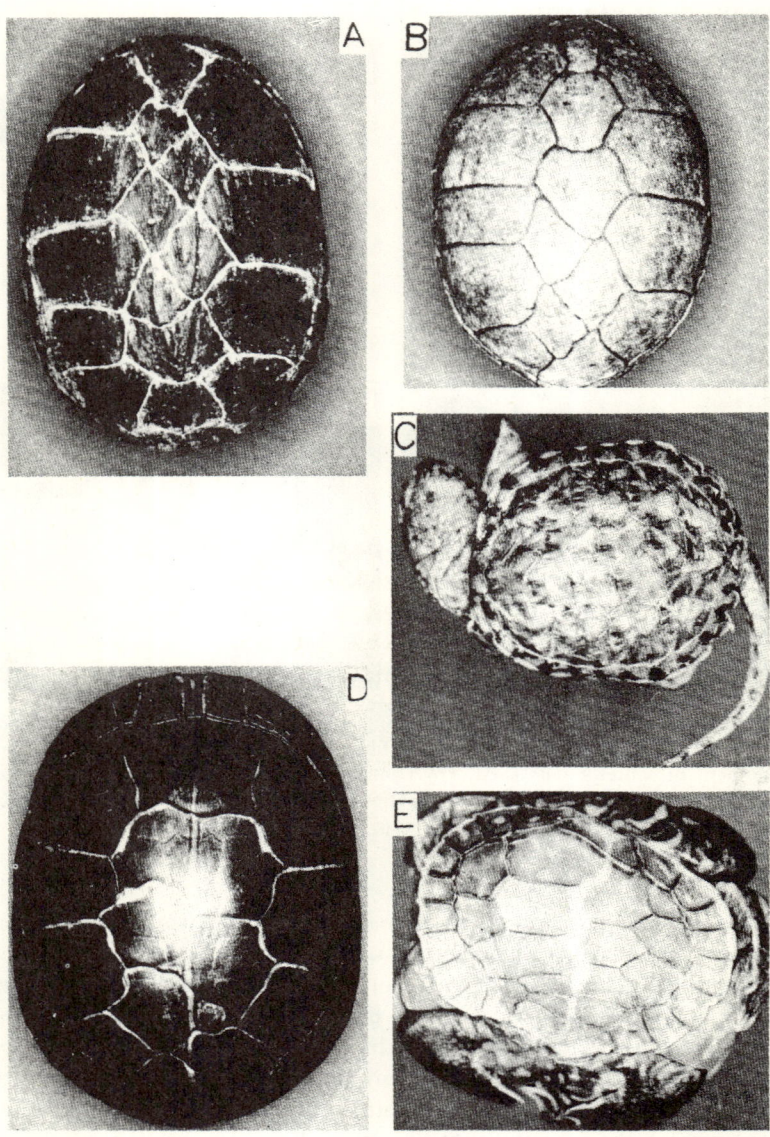

Figure 6 Dovetail syndrome in scutes of the carapace. (*a*) Adult *Sternotherus odoratus* with central region of carapace most affected. The shell has been dusted with white powder to bring out seams. (*b*) Adult *Kinosternon bauri* with posterior of carapace most affected. (*c*) Term fetus of *Chelydra serpentina* with incomplete division of vertebral scutes. (*d*) Adult *Chrysemys picta dorsalis* with posterior of carapace most affected. The medial stripe is straight as in normal specimens. (*e*) Term fetus of *C. p. dorsalis* with incomplete division of vertebral scutes and other gross deformities. The medial stripe zigzags.

teristic staggered fashion and that the dovetail syndrome arose when the pleurals on one side were not symmetrically arranged with regard to those on the other. Scute pattern may begin embryologically in two independent lateral series, which then extend medially. In normal cases the pleural scutes are symmetrical with regard to each other on opposing sides and carve off symmetrical vertebrals. In the dovetail syndrome, the pleural series typically become less and less symmetrical posteriorly, so much so that posterior vertebral scutes appear to have merged indistinguishably with pleurals from a given side, such as in Figure 6d. In quantitative terms, referring to the vertebral series, the posterior portion of the most posterior scute is asymmetrical in 45 of 58 dovetailed specimens examined and the anterior portion of the most anterior in 6 specimens. The dovetail syndrome tends not to affect the marginal series of scutes; 28 of 50 specimens were completely normal in this regard. Only 14 of the 50 had extra marginals on one or both sides, and 8 were deficient.

Variations in Pigmentation of the Skin and Shell. Totally albinistic turtles occur in at least four families, as indicated by the following documented species: *Chelydra serpentina* (Judd, 1971), *Malaclemys terrapin* (Hildebrand, 1930), *Testudo hermanni* (Wermuth, 1971), *Caretta caretta* (Hughes et al., 1967), and *Chelonia mydas* (Robinson, 1974). At least partially albinistic specimens have been reported in the soft-shell *Dogania subplana* and in *Heosemys grandis* (Anonymous, 1969), and *C. caretta* (Baldwin and Lofton, 1959; Hughes et al., 1967; Pond, 1972). Two of the reports on *C. caretta* indicated that some albinistic specimens exhibited deformed heads. In my observations deficiencies in the anteriodorsal region of the head are often accompanied with deficiencies in pigmentation, for example, tans in place of browns and blacks in chelydrids; xanthic *Pseudemys scripta*; *Emydoidea blandingi*, species of *Graptemys*, and *Chrysemys picta* with bizarre patterns including expanses of white (Figure 2b); and pallid *Trionyx spiniferus*. In these specimens a trend toward albinism could have resulted from environmentally induced defects in hormonal centers governing production of pigments. In contrast, a pigment deficiency may have become a genetically polymorphic trait in *G. barbouri* (Figure 5c, d), as shown by one clutch of eight eggs that hatched into five uniformly pallid and three typical specimens and by a pallid male captured as an adult.

In recent keys to *Graptemys* species (Carr, 1952; Ernst and Barbour, 1972; Conant, 1975) a single postocular crescent of light color extending behind and below the orbit is said to occur in *G. kohni*, but not in *G. pseudogeographica*, where the crescent is reduced or absent allowing addi-

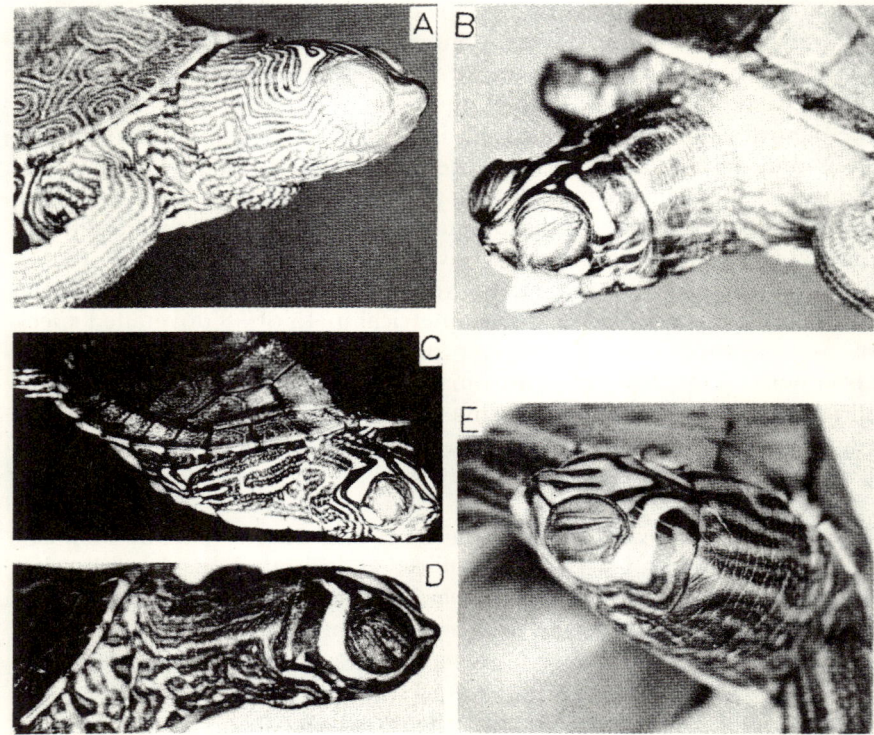

Figure 7 Variation in postorbital head stripes of *Graptemys pseudogeographica* from Winona, Minnesota. (*a*), (*b*) Pattern traditionally recognized as belonging to *G. pseudogeographica*. (*d*), (*e*) Pattern traditionally recognized as belonging to *G. kohni*. (*c*) A specimen with intermediate head striping. In the text (*a*) is also designated as the *pseudogeographica* morphotype and (*b*), (*c*), (*d*), (*f*) as the *ouachitensis* morphotype.

tional head stripes to reach the orbit from behind (Figure 7*a*, *b*). Personal collections of *Graptemys* adults from the Mississippi River at Winona, Minnesota, have contained a few specimens that key out as *G. kohni* because they exhibited complete postocular crescents similar to those in Figure 7*d*, *e*. Such a matter would invoke little embryological interest if incubation studies had not been conducted. Some clutches yielded both *pseudogeographica* and *kohni* types of head striping. The remarkable finding was that the *kohni* type of head striping appeared almost exclusively in eggs incubated at 25°C. Specifically, from 181 eggs representing 38 split clutches incubated at 25°C, 59 *kohni*-type specimens arose from eggs of 20 of the clutches, whereas at 30°C, 165 eggs representing the same clutches yielded only 5 *kohni*-type hatchlings from 3 of the clutches. A statistical test based on a 2 × 2 contingency table relating presence or absence of *kohni*-type head striping in the 38 clutches to temperature yielded a χ^2 value of 15.96 (P < .001).

In clutches yielding turtles with thin striping as in Figure 7a, *kohni*-type specimens have never appeared. The morphotype of *G. pseudogeographica* having a potential for *kohni*-type patterning seems to have thick individual stripes as in Figure 7b. Figure 7c represents an intermediate pattern developed from thick stripes where postocular and cheek stripes have almost united to give the *kohni*-type crescent. Vogt (1974, personal communication) has recognized the thin- and thick-striped morphotypes of *G. pseudogeographica*, has termed the former as *pseudogeographica* and the latter as *ouachitensis*, and has proposed that these distinctions be recognized at the specific level. Development of *kohni*-type crescents appears to be only one manifestation of expansion of light-colored areas at expense of dark areas in the *ouachitensis* morphotype. In some specimens incubated at 25°C the usually dark interorbital area can be as pale (Figure 7e) as those of *G. barbouri* and *G. pulchra*. The dark central plastral figure was also reduced at 25°C and contained an average of 1.17 complete breaks per turtle because of lack of invasions of dark color (18 hatchlings from four split clutches) in contrast to 0.41 break per turtle (19 hatchlings from the same split clutches) at 30°C.

Temperature-modified integumental pigmentation occurs in some mammals (Strickberger, 1968:174) and perhaps in snakes (Vinegar, 1974). Additional embryological work may reveal it in other chelonians, pinpoint the susceptible stage of development in the *ouachitensis* morphotype, and explain the process behind the observed differences.

Teratogenesis and the Natural Environment. Some terata are considered to arise spontaneously, either because the types of defects have been traced to abnormal genetic syndromes through inheritability or because the causes have not been identified. References to teratogenic influences of environmental extremes on poikilotherms are given by Needham (1963). Romanoff (1972) reviewed embryonic pathogenesis in birds. An environmental agent may first be identified qualitatively, such as temperature or humidity, then described quantitatively, and finally investigated as a function of embryonic time, for instance, the developmental stage of maximum sensitivity. Given conditions may be teratogenic at one stage but harmless at another. Wilson (1965) regarded predifferentiation, the period of cellular cleavages leading to gastrulation, as the least susceptible in mammalian embryogenesis. The period of differentiation and early organogenesis was the most susceptible, and that of advanced organogenesis leading to fetal growth was less so. Some of the classical deformities involve such fundamental distortions of structure that the stage at time of onset of disease can be guessed approximately, using the logic that organ rudiments will grow into a defective structure as opposed to a normal one to be replaced by a defective one later.

Each of the mortality factors discussed earlier might be expected to be teratogenic if occurring at sublethal levels at appropriate developmental stages. Perhaps the most widely recognized teratogenic agent, adverse temperature, has been applied to early developmental stages of *Chelydra serpentina* eggs. At exposure to 15°C followed by room temperature, numerous terata appeared including microphthalmia, otherwise deformed heads, and reduced carapace, hind legs, and tail (Yntema, 1960).

Lynn and Ullrich (1950) observed effects of dehydration on eggs of *C. serpentina*, *Terrapene carolina*, and *Chrysemys picta* incubated at room temperature. The conditions of the stress, which do not appear to have been rigorously controlled or defined, caused extensive mortality when applied to early developmental stages, but are claimed to have been largely teratogenic to midterm embryos. The report emphasized defects of the shell, but referred to defects of the head and gave examples of gross warping and withering of the body in photographs. Many portions of the body were irregularly withered and distorted as if portions of advanced embryos had been injured, then become misshapen through imperfect healing. I have observed a few cases of adhesion of the embryo to the inner shell membrane at expense of the normally intervening vascularized extraembryonic membranes. In advanced fetuses the attached area has appeared shrunken and withered. Localized drying relatively early in development may affect adhesion.

Postnatal turtles are amazingly tolerant of acute oxygen deprivation (Belkin, 1963), but the occurrence of an oxygen-deficiency edematus syndrome (Grabowski, 1970; Grabowski and Parr, 1958) should not be overlooked as a possible forerunner of deformities during embryonic development. It may be feasible to study the effects of graded oxygen supplies on parchment-shelled eggs partially submerged in water to determine rates of water absorption and subsequent effects on the embryo.

Diverse terata range from prenatally lethal to postnatally benign. Through identifying the main causes of terata observed in wild populations, we might gain additional insight into weaknesses in chelonian reproductive strategies.

ACKNOWLEDGEMENTS

Since 1962, several individuals and institutions have enabled me to collect the data included in this chapter. Craig E. Nelson, Arthur L. Koch, and Rollin C. Richmond kindly furnished facilities at Indiana University for the most recent research, and Craig E. Nelson provided space for the actual writing of this work. My earliest studies were conducted at the University of Minnesota, Minneapolis–St. Paul, in facilities managed by Huai C. Chiang and John R. Tester, and the University of Wisconsin–Washington County Campus (Delbert E. Meyer). Several field stations provided work space, lodging, or both: Archbold Biological Station, Richard Archbold and James N. Layne; Lake Itasca Forestry and Biological Station, William H. Marshall; Tall Timbers Research Station, Edwin V. Komarek and D. Bruce Means; University of Wisconsin–Milwaukee Field Station, Millicent S. Ficken. The private residence of Frank and Jean Chesley, Red Wing, Minnesota, became a field station and home away from home during several nesting seasons in the region. A portion of the work involved several museums, either through direct access to collections or through loans: American Museum of Natural History, Richard G. Zweifel; Carnegie Museum, C. J. McCoy; the Field Museum of Natural History, Hymen Marx; Harvard Museum of Comparative Zoology, Ernest E. Williams; Florida State Museum, Walter Auffenberg; University of Michigan Museum of Zoology, Charles F. Walker; United States National Museum, James A. Peters and George R. Zug; University of Utah, John M. Legler. The following individuals provided specimens in addition to those acknowledged in the text: W. Wilson Baker, James F. Berry, Larry A. Lantz, John M. Legler, Edward O. Moll, and P. Kelly Williams. Ronald W. Barrett, Harry Benson, Kathy Jansma, and Carolyn Wilhelm contributed substantially to the work of incubating eggs. Indiana University Department of Zoology contribution number 1030.

BEHAVIOR

CHAPTER 18
Behavior Associated with Nesting

DAVID W. EHRENFELD

The reader will quickly discover that this chapter is biased toward sea turtles. No doubt this reflects my own preferences, but it is also largely unavoidable. Most other details of their life cycles are hidden, but sea turtles become exhibitionists while nesting. Consequently, their nesting is better known than that of any other chelonian group, and the literature about it is, if not large, at least respectable.

Even the most thoroughly aquatic of turtles are tied to dry land, at least periodically, by the physiological and behavioral dictates of their reproductive habits. It is conceivable that there might be an evolutionary premium for aquatic turtles to be able to lay aquatic eggs, but none do so. This exclusively terrestrial phase of turtle reproduction is described next.

NESTING

The nesting behavior of turtles seems ready-made for the behaviorist interested in evolution. The constraints on the nesting process imposed by species-specific anatomy and by the typical physical characteristics of the nest site can often be identified and taken into account. One is then left with behavioral differences and similarities among species that may reflect, in part, phylogenetic events. Moreover, chelonian nest building occurs in a well-defined sequence of steps that lend themselves to precise behavioral characterization and interspecies comparisons, and the actual nest digging leaves a useful physical impression on the substrate. It is odd that turtle nesting behavior has been all but ignored as a subject of systematic comparative analysis; it could lend itself to the same sort of superb ethological analysis that was performed by Evans (1966) with sand wasps. Descriptions of turtle nesting fill the literature, but few offer precise information; there is also a high frequency of contradictions about basic information, so evidently at least some of the reports are not reliable. The only papers I know that provide careful, detailed ethological analyses of nesting in a series of species (sea turtles) with different degrees of relationship are those by Carr and his associates (Carr and Giovannoli, 1957; Carr and Ogren, 1959, 1960; Carr and Hirth, 1961, 1962; Carr, 1963a; Carr, Hirth, and Ogren, 1966; Carr and Carr, 1972).

The number of references has been reduced to a working minimum. Unless otherwise specified, statements about American turtles may be traced by consulting Carr (1952) or Ernst and Barbour (1972), and information about tortoises and side-necked turtles is documented by Pritchard (in press).

Nesting occurs in easily discernible steps. Carr and Ogren (1960) have identified 11 stages of nesting in the green turtle, *Chelonia mydas*:

1. Stranding, testing of stranding site, and emergence from wave wash.
2. Selecting of course and crawling from surf to nest site.
3. Selecting of nest site.
4. Clearing of nest premises.
5. Excavating of body pit.
6. Excavating of nest hole.
7. Oviposition.
8. Filling, covering, and packing of nest hole.
9. Filling of body pit and concealing of site of nesting.
10. Selecting of course and locomotion back to sea.
11. Reentering of wave wash and traversal of the surf.

I have condensed this useful sequence into fewer steps for the sake of generality, and with the exception of orientation to and from the nest site, the stages are discussed in turn.

Nest Site Selection and Conditions for Nesting

The only generalization that can be made about nest site selection is that, with a few exceptions, the criteria for suitable sites, both among and

NESTING

sometimes within species, are surprisingly broad. When populations or species are considered as a whole, this flexibility is especially marked in the case of selection of soil type, although it is possible that individuals have preferences that are consistent from year to year. Hirth and Carr (1970) analyzed sand from major green turtle nesting beaches in various parts of the world and found it highly variable in all characteristics analyzed. *Chrysemys picta picta* and *Malaclemys terrapin macrospilota* nest in nearly all kinds and textures of soils and sand, respectively. Where a preference is shown, it is usually for sandy and other friable, well-drained soils rather than clay and hard-packed soils (*Trionyx ferox, T. muticus, Pseudemys scripta, Graptemys geographica, Clemmys guttata, Gopherus polyphemus*).

Factors other than soil type thus appear to be the major determinants of site selection. However this leaves unexplained the "sand smelling" behavior that so dramatically characterizes the early phases of site selection in sea turtles of the genera *Eretmochelys, Lepidochelys, Caretta*, and *Chelonia* (Carr and Hirth, 1962; Carr et al., 1966; Hirth and Carr, 1970; and Bustard, 1973c). If this behavior is related to site selection, it may be olfactory identification of the general nesting area by turtles that have migrated a long distance; or it may have to do with the preference of individuals for a specific part of the beach or a certain sand type, either of which might have a distinctive odor or possibly texture (see Chapter 15). It is, of course, possible that the movements that have been characterized as "sand-smelling" are not associated with any special sensory discrimination.

Many species show preference for nest sites exposed to full sunlight. These include *Geochelone elephantopus porteri, Trionyx ferox, T. muticus, T. spiniferus, Graptemys pseudogeographica, G. geographica*, and the sea turtles. Among aquatic turtles, nest sites vary in distance from water. *Pseudemys concinna* and *T. muticus* tend to nest less than 100 ft from water, but *Malaclemys terrapin macrospilota* may travel more than half a mile from water, and *Platemys platycephala* also is known to travel a great distance. There does not seem to be any correlation between the extent of the aquatic habit and the distance traveled on land to nest. Ponderous sea turtles often drag themselves hundreds of feet over the beach, rejecting sites that appear to the casual observer to be quite suitable for nesting. The snapping turtle, *Chelydra serpentina*, more sprightly on land but still aquatic, does the same. One can only conclude that ecological factors related to reproductive success outweigh considerations of nesting efficiency or safety for the nesting female. Identification of these factors is usually difficult and speculative, yet there are exceptions. Ninety-four percent of the nests of the flatback turtles (*Chelonia depressa*) studied by

Limpus (1971) were on the tops of dunes or as high as possible on their seaward slopes. Similarly, the loggerhead turtles (*Caretta caretta*) studied by Baldwin and Lofton (in Caldwell et al., 1959) preferred to nest at the back of high, wide beaches, near the bases of low, rounded dunes, relatively far from the water. In both cases, the primary reason for this extra locomotor work seems to be to remove the eggs as far as possible from the destructive saltwater inundations caused by storm tides and heavy surf. Finally, egg and hatchling predation must also play a major, if obscure, role in determining nest site selection by all species.

A most comprehensive and precise investigation of factors affecting nest site selection is the report by Burger and Montevecchi (1975) about the northern diamondback terrapin (*Malaclemys t. terrapin*). They found strong correlations between nesting and such environmental variables as tidal phase, weather, percentage of vegetative cover, direction and degree of dune slope, and height of dunes. There was no mechanism for nest spacing or dispersion, and 2 of 40 females observed dug into nests of conspecifics. (For a discussion of this phenomenon as a possible regulator of population density, see Bustard and Tognetti, 1969.) Another extensive account of the relationship between nest site selection and environmental variables in the case of Australian sea turtles has been provided by Bustard (1973c).

A miscellany of observations about site selection and conditions for nesting in various species serves to underscore the variability in site preference and nesting conditions. *Gopherus berlandieri*, a desert grasslands tortoise, places its nests in the drip zone of bushes; sites may be used for years. *Geochelone denticulata*, a forest dweller, may deposit its eggs among dead leaves, without digging a nest at all. *Gopherus polyphemus* often nests in the mound of excavated sand in front of its own burrow. Several turtles, including those in the genera *Pseudemys* and *Kinosternon*, will deposit their eggs in the nesting mounds of alligators, presumably taking advantage of the protection and humidity regulation supplied by their host, and/or the elevated incubation temperature of the mound. Loggerhead turtles in Australia are thought by Limpus (1973) to nest near their best feeding grounds; perhaps this explains why they can lay abundant eggs with little weight loss, and, unlike most sea turtles, renest in successive years. A number of Australian freshwater species, including *Chelodina longicollis, C. expansa*, and *Emydura macquarri* tend to nest after heavy rains or during periods of high humidity (Goode, 1965; 1967b). Finally, most species are known to prefer to nest either during the day or at night—but not both. In general, the sea turtles nest by night, others mostly by day.

Preparation of the Nest Site

As we continue with the temporal analysis of nesting, we find that the interspecies variability becomes progressively reduced. Typically, in the species that prepare the nest site, the forefeet are used to scratch, scrape, or throw the dirt backward or to the side. This is best developed in some of the sea turtle genera, but not in *Eretmochelys*. In *Chelonia mydas* (Carr and Ogren, 1960), preparation of the nest site begins with first hesitant, then sweeping movements of the fore flippers, which snap toward the body from a lateral position, throwing sand backward 10 ft or more. Only after the "body pit" has taken form do the hind flippers join in the action; when the site is prepared the fore flippers gradually cease to move, and the sweeping strokes of the hind flippers are subtly converted to the digging movements that excavate the egg cavity. A nesting ground that has been extensively used looks as if it had been bombed; beach topography and vegetation are dramatically altered. I have seen green turtles leave craters with a volume of up to 80 ft^3 after the final covering of the eggs was completed. Other species that have been observed to use their front feet in the preparation of the nest site include *Chelydra serpentina osceola*, which may dig a body pit, *Sternotherus odoratus*, which may use all four feet and snout, and *Kinosternon subrubrum*, which begins by pushing dirt laterally with its forefeet, then turns around to use its hind feet.

Terrapene ornata is alleged to dig a body pit exclusively with the hind legs; others may do the same, however—this phase of nesting is often overlooked or badly described. It is my impression that a majority of turtles do very little preparing of the nest site before digging an egg cavity and consequently that most do not use their forefeet in this or perhaps any stage of nesting. But some of the strange and undoubtedly erroneous accounts of turtles digging egg cavities with their forefeet may be confused versions of nest site preparation, in which case the number of species that do this is larger than I have suggested. The oddest example of nest site preparation is probably that of *Gopherus berlandieri* that has been seen (in captivity) preparing the earth in its regular pallet by scraping with its epiplastral projection, moving its shell from side to side to shift dirt laterally, and pushing with its front feet to make a kind of body pit (Auffenberg and Weaver, 1969).

Urination over the proposed nest site (and in the egg cavity) is widespread among turtle families; a few of the species for which this behavior has been reported are *Chelodina longicollis, Geochelone elephantopus, G. pardalis, Trionyx ferox, Gopherus berlandieri, Pseudemys scripta, Chrysemys picta*, *Malaclemys terrapin*, and *Terrapene carolina*. Urination on the nest site is probably much more common than preliminary digging of body pits and the like. Patterson (1971), who studied urination as a mechanism of defense against egg predation in the desert tortoise, *Gopherus agassizi*, suggests six possible functions: (1) softening of the substrate, (2) providing moisture for the eggs, (3) bacteriostatic effect, (4) repulsion of egg predators through smell or taste, (5) compaction of dirt after nesting to ward off predators, and (6) camouflage. Patterson tested and confirmed effects 1, 4, and 5. Of these, I suspect that the softening of the dirt prior to and during digging may be the most important benefit. Many of the turtles that urinate on the nest site live in arid regions where the soil surface is baked into a hard crust. Indeed, in the Galapagos Islands the scarcity of soil suitable for nesting causes a reproductive migration of tortoises (*Geochelone elephantopus porteri*) from feeding to nesting ground. Even in the nesting areas, the soil is rendered suitable for digging only after it has been brought to a muddy consistency by urination. Pritchard (in press) feels that the rock-hard crust that forms afterward may protect the eggs in the nest. Of course when seasonal rains fail, it is probable that a sizable percentage of hatchlings are permanently entombed in the egg cavity.

Digging the Egg Cavity

The latter stages of nesting, beginning with the digging of the egg cavity, are the most stereotyped for chelonians as a group—exceptions to the general nesting pattern are very noticeable. All species dig the egg cavity exclusively with the hind feet. Even those species (*Geochelone denticulata*, *Sternotherus* spp., and *Kinosternon* spp.) that are known commonly to deposit eggs in litter or debris without digging any egg cavity will, on occasion, nest in the usual way. We do not know what evolutionary forces prompt a gopher tortoise, *Gopherus polyphemus*, which is capable of digging a 30 ft burrow with its forefeet, to dig an egg cavity with its elephantine hind legs, removing scarcely a few grains of dirt with each stroke; but these forces must be conservative indeed.

Not only are the hind legs used exclusively at this stage of nesting, but they are almost always used in the same sequence—alternately. This alternation may occur with every stroke in a very mechanical, rhythmic fashion, as is the case with sea turtles, or it may happen less precisely, after every one, two, or three strokes, as in the tortoises and freshwater turtles. An exception to this pattern is provided by *Geochelone radiata*, which reportedly uses each hind leg until it "tires" (Pritchard, in press). Unfortunately, this sort of painstaking information is usually ignored by

Table 1 Behavioral Differences in the Nesting of *Eretmochelys* and *Chelonia*[a]

Trait, Stage, or Feature	*Eretmochelys*	*Chelonia*
Gait	Diagonal legs move together	Paired legs move together
Oviposition	Edge of back flippers curl as eggs are extruded	Flippers remain at rest as eggs are extruded
Oviposition	Flippers spread beside nest during oviposition	Flippers spread together over nest during oviposition
Filling of nest	Sand dropped into egg cavity by separate flipperfuls	Sand scraped into nest by alternate hind flippers
"Sand smelling"	Muzzle repeatedly pressed against the sand throughout nesting emergence	Pressing of muzzle against sand mostly confined to initial stage of emergence
Visual appraisal of shore situation	More constant peering and craning of neck	More mechanical, dragging locomotion, with little peering or craning
Character of body pit	Body pit shallow, or desultory, or lacking	Body pit usually deep, often deeper then height of shell of turtle
Timing	Longer period of prospecting and filling	Longer period of digging pit and nest

[a]From Carr, Hirth, and Ogren (1966).

observers who write about the natural history of turtles. Table 1, which refers to all stages of nesting, is a rare example of interspecies comparisons that are sufficiently detailed to be of interest in an analysis of behavior.

The hole that results from this stereotyped digging behavior is usually flask shaped, wider at the bottom than at the top. At no time during the digging is the hole under visual inspection, yet one rarely sees any dirt brushed back into the nest by careless or imprecise movements. This holds true even during the final minutes of digging, when the entire body of the turtle is maneuvered by the front legs so as to give the hind legs greater reach. One is tempted to assume that tactile feedback from the hind legs and perhaps the tail must be highly developed to account for this remarkable performance, but I doubt that this is the case. From my observations of nesting turtles, I conclude that success in maintaining the integrity of the egg cavity is mostly related to the exceptional stereotyping and replication of the movements of the hind legs and the body. Casual experiments in rearranging the dirt behind nesting turtles, and observations of digging by turtles missing a hind foot or leg, where movements of the stump still alternate with movements of the intact leg, lend circumstantial support to this conclusion. On the other hand, some feedback must be involved during this phase of nesting, since trial holes, as many as seven or eight, are common in many species, and since obstructions often cause turtles to abandon an egg cavity before its completion. Goode (1967b) reports that Australian turtles may return to the water without nesting if the ground is very hard from the lack of rain, but he does not state whether trial digging is involved in the reconnaissance.

The only known major departure from the process described in this section occurs in *Pseudemys floridana peninsularis* (Carr, 1952). This species frequently digs (and deposits an egg in) one or more shallow "accessory side pockets" 2 to 5 in. from the main egg cavity, which contains about 20 eggs. Carr considered a decoy function for this bizarre habit but concluded that it was "more like a beacon" than a decoy. The only nests that escape predation are those dug before a heavy rain, which washes away the traces of nesting.

Egg Laying

One behavioral aspect of egg laying has drawn the attention of many observers. This is the oblivion to outside disturbance that seems to descend on many species as they lay their eggs, rendering the egg-laying females easy to observe. Deraniyagala (1939) struck an egg-laying *Dermochelys coriacea* on the head with a stick and even sat on her without deterring her. As anyone who has watched nesting sea turtles can verify, indifference to disturbance can be remarkably complete. Although both time of onset and duration are variable among individuals and among populations, it is always at a maximum during oviposition and the first stages of covering.

If one considers this phenomenon in terms of the "parental investment" of the nesting female in her eggs (Trivers, 1972), then the onset of the period of oblivion is well timed. In other words, there is no reason for the female to become indifferent to interference until she is thoroughly

NESTING

committed to egg laying (i.e. until she has laid a few eggs or until she has set in motion the physiological processes that immediately precede egg laying). Only at this point does it make genetic sense for her to take risks, because she has a reasonable chance of producing offspring, thus protecting her gene pool. If this hypothesis is correct, we might expect to find that populations nesting where they have no predators will have extended this period of oblivion to distraction, or "committedness" to egg laying. The following statement by Carr and Hirth (1962) comes as no surprise:

> The high alarm threshold of the emerging Ascension [Island] turtle is the most striking behavioral difference between it and the Tortuguero [Costa Rica] population. When a Costa Rican turtle strands and starts up the beach, she can be turned back into the surf by the slightest show of artificial light—a match struck 50 feet away, for example, or by the moving of a man or dog, even some distance up the shore, across the starlit sky. With the Ascension population the unswerving train of stereotypes, which in most turtles begins with the digging process and thereafter keeps the animal oblivious to outside interference, appears to take over at the time of stranding.

There are no significant predators of adult Ascension Island green turtles.

Another variable of the egg-laying stage is the use of the hind feet to position eggs in the egg cavity. I am not aware of any sea turtle that does this, but it is fairly common among other chelonians, including *Chelydra serpentina*, *Clemmys guttata*, *Chrysemys picta*, *Pseudemys floridana peninsularis*, *Gopherus agassizi*, *Trionyx muticus*, *T. spiniferus*, *Geochelone e. elephantopus*, *Testudo hermanni robertmertensi*, *Chelodina longicollis*, and *C. expansa*. In some species it is customary for the females to make two or three layers of eggs, with each layer separated by a thin partition of soft, friable dirt. (Caldwell et al., 1959, report this for *Caretta caretta*, but it does not seem to be typical behavior for the species.) Goode (1967b) states that in the case of *Chelodina expansa* the upper layers of eggs hatch first. The fact that the list above contains a number of tortoises indicates that flexibility of the hind foot is not a prerequisite for its use in rearranging eggs.

Covering the Nest

Continuing the same, widespread stereotyping as has been described for earlier stages of nesting, probably all turtles begin by filling the egg cavity with their hind legs. This behavior immediately follows oviposition, and the behavioral transition may appear quite abrupt. Although covering is sometimes described as a reversal of the hind leg movements seen in digging, this a misleading simplification. It is true, for example, that *Eretmochelys imbricata* picks up sand by the flipperful in filling the egg cavity, an apparent reversal of digging, but *Chelonia mydas* scrapes and pushes the sand over its eggs in a much coarser series of movements (Carr et al., 1966; see Table 1). Both species knead, poke, pat, and press the mound over the egg cavity in a manner quite unlike any behavior that comes before. Also, the tail is thrust deep into the dirt mound, perhaps in a tactile capacity or perhaps because that is the most convenient place for the turtle to keep it. In covering, individual turtles may use their hind feet both alternately and together.

After the initial filling of the egg cavity, each species seems to add its own embellishments to the covering and concealing process, but there are still a limited number of behaviors—perhaps, in part, because there is just so much that a turtle, with its particular anatomy, can do. These behaviors constitute a separate stage of nesting, but are included here for the sake of brevity. The sea turtle species that dig body pits fill them at this point, as the front flippers gradually join the hind flippers in throwing sand. Movements are very similar to those of initial pit digging, although experienced observers can tell the difference. As the turtle moves forward, the pit "moves" with it, leaving the eggs behind, hidden several feet under a pile of loose sand.

Turtles of many species use their plastrons in concealing the nest site. *Macroclemys temmincki, Clemmys guttata, Graptemys geographica,* and *Geochelone gigantea* have all been observed to smooth the nest site by dragging their plastrons over it. In the case of *G. gigantea*, this dragging continues for days. *Pseudemys rubriventris* raises itself up on all four legs and drops directly on the nest site. *Lepidochelys kempi* has a similar concealment technique, but there is more rocking of the plastron from side to side, thus packing the loose sand. A few species, such as *Trionyx spiniferus*, are said to scratch up the ground around the nest site, which possibly serves to conceal the nest site from predators. Turtles that dig little or no egg cavity, frequently leave the eggs after minimal or no concealment.

Summary

From the scanty data it is plain that the study of nesting offers great promise to ethologists. Some characteristics of modern turtles may have evolved rapidly, but this is not true of all aspects of nesting behavior. It is possible, of course, that the conservativism that one sees in certain nesting traits is the result of stringent environmental or anatomical limitation of alternative possibilities. On the other hand, it is conceivable that similarities among species are simply the consequence of some unexplained proclivity of turtles to keep ancestral habits and features, even when change might have been possible. The latter hypothesis, although perhaps not as intellectually satisfying, is better for ethologists because it minimizes the likelihood that phylogenetic patterns will be obscured by similarities that are the result of convergent or parallel evolution.

One often has the unusual opportunity of being able to identify many of the environmental and anatomical factors that bear on chelonian nesting, and to evaluate the possibility that they are responsible for the conservative (or nonconservative) nature of a particular trait. We can ask, what is the latitude for change that these factors might allow? The most obvious example of a clear-cut constraint on a behavioral component of nesting in any species is the length of the fully extended hind leg, which determines, after tilting of the shell is taken into account, the maximum depth of the egg cavity. Since most species seem to place their eggs at maximum depth, there is likely to be no ethological value in comparing egg cavity depths for various species (or in comparing partially dependent variables such as the number of strokes used in digging the cavity), because all one is really doing is comparing leg lengths.

When performing this kind of evaluation one occasionally finds a behavioral trait that does not seem to be *specifically* mandated by environment or anatomy. A possible candidate is the habit present in sea turtles, but not in other families, of kicking accumulated sand forward with one hind foot while the other removes sand from the egg cavity. The functional significance of the movement, which removes excavated sand from the lip of the cavity, is clear. The most reasonable conclusion is that this forward kick arose as a means of coping with a nesting substrate prone to minor avalanches. It is also clear that this task could be accomplished by turtles in many other ways, and its occurrence in two very dissimilar families is, in this case, good circumstantial evidence both for the antiquity of the trait and for the common origin of sea turtles. This is the kind of information that is required in analyses of behavioral evolution, and it is unfortunate that the nesting process among related species of freshwater and/or land turtles has rarely been described in such careful and comparative detail. The opportunity, however, is there.

EMERGENCE AND DISPERSAL OF HATCHLINGS

Social Facilitation of Emergence

Reproduction in turtles is fraught with hazards. Predators destroy a majority of nests in some species, occasionally while the eggs are still being laid. Excessively cold weather, storm tides, droughts, and other environmental disturbances take their toll. But if the turtles survive to emergence time, their problems have scarcely begun. First, there is the task of getting out of the nest.

Most hatchlings find themselves a few inches to several feet below the surface of the ground after hatching, and in many cases the struggle to reach the surface begins fairly soon. When there is long delay, it is usually related to the onset of cold weather or a prolonged drought. An unknown percentage of hatchlings die during the upward climb. It is not uncommon to find one or two dead hatchlings remaining in or slightly above an egg cavity from which other turtles have previously emerged.

In some cases, primarily among species with small egg clutches, the digging-out process is solitary. Among sea turtles and some others (Vanzolini, 1967) that have large clutches, digging out is socially facilitated, a process described in detail by Carr and Hirth (1961). These investigators found that when they reburied from one to ten green turtle eggs under natural conditions, percentage emergence was directly and significantly correlated with the size of the egg complement. From observations with glass-sided nests, Carr and Hirth concluded that the social facilitation of emergence is produced by mutual stimulation of locomotor activity, originating among the hatchlings crushed at the bottom of the pile, and spreading as a wave toward the animals on top. The wriggling of the latter turtles scratches sand from the roof of the cavity. The loose sand works its way down through the mass of hatchlings and is trampled into the floor of the cavity by the turtles on the bottom. Thus the whole nest cavity rises through the sand as a unit. Since only a few turtles are needed for this process, it is likely that the main advantage of having more than one hatchling in close contact is the mutual stimulation and reinforcement of the frenzied activity that is necessary to effect escape from the nest.

Effect of Temperature on Emergence

Hatchlings behave in various ways, according to species, when they reach the surface. It is my impression, unsubstantiated by any significant evidence, that hatchlings of the freshwater and land turtles that nest within a few feet of water or in well-vegetated areas may break through the surface of the ground at any time of day and disperse. I have seen newly emerged hatchlings of *Chelydra serpentina* and some emydid species heading for the water in full sunlight. It is nearly impossible, however, to make systematic observations of the time of emergence for species that nest singly or in small clusters; thus this information is missing for most turtles.

In the case of sea turtles, which often nest in large aggregates, this phase of the life cycle is much better known. With their nests located on open beaches, hatchling sea turtles are exposed to two greatly intensified hazards if they emerge during the day: first, sand temperatures are likely to be immediately lethal, and second, hatchlings moving on the bare sand are highly visible to predators. Accordingly, sea turtles generally disperse from the nest at night, and the phenomenon is temperature dependent, as first suggested by Hendrickson (1958). This remarkable mechanism has been experimentally elucidated by Bustard (1967), who worked with Australian green turtles. He described how the upward movement of the nest was inhibited when the topmost turtles reached a zone in which their temperatures rose to 30°C. At this point, generally still below the level of light penetration in the sand, the upper turtles became completely torpid, and this torpidity quieted the entire mass of hatchlings by interrupting the socially mediated waves of activity. At night, when sand and body temperatures fell below the critical threshold, activity would resume and the entire nest group would burst through the intervening inch or two of sand almost in unison. Bustard found that by removing the topmost, quiescent hatchlings from the nest he could induce the others to emerge in broad daylight before their body temperatures reached the inactivity threshold. This experiment, plus the frequent observations of sea turtles emerging on cool, rainy days, supports the temperature-dependence hypothesis and indicates that daylight itself plays no role in the usual inhibition of daytime emergence activity. Mrosovsky (1968), who studied both *Chelonia mydas* and *Eretmochelys imbricata*, has also confirmed the central role of temperature in hatchling emergence.

Predation on Hatchlings

After emergence from the nest, hatchling turtles are subject to intense predation pressures, especially those that must travel to an adjacent body of water before seeking concealment. The list of animals that regularly prey on dispersing freshwater and marine hatchlings is sobering, even to one who has great confidence in the survival abilities of turtles; a small sampling includes herons, raccoons, bears, king snakes, ghost, hermit, and robber crabs, dogs, frogs and toads, gulls, alligators and crocodiles, foxes, Australian tiger snakes, rats, kookaburras, buzzards, hogs, cats, flamingos, mongooses, frigate birds, cormorants, and goannas. Of course hatchlings that reach the water have no guarantee of safety—a whole new vertebrate class of predators awaits them there. Hirth (1971), who has summarized the data on predation of *Chelonia mydas*, states that the survival to maturity figure may be as low as 1%; some investigators feel that this is generous. Most freshwater and terrestrial species, with much smaller clutches, must do better than that.

ORIENTATION TO AND FROM THE NEST SITE

Early Studies

The orientation processes that enable adult and hatchling aquatic turtles to locate the water after nesting or emergence from the nest, have been a matter of great interest to scientists ever since Hooker (1908) first examined the mechanism in loggerhead sea turtles (*Caretta caretta*). Because of space limitations, I can only mention the more important findings from a number of papers. More elaborate reviews of the work prior to 1960 may be found in the following references: Carr and Ogren (1960); Caldwell and Caldwell (1962); Ortleb and Sexton (1964); Ehrenfeld and Carr (1967); Mrosovsky and Carr (1967); Ehrenfeld (1968); Mrosovsky and Shettleworth (1968); Mrosovsky (1972). In reading these works, three things become apparent: first, most of the studies have concerned sea turtles; second, because of experimental limitations it is generally assumed that the orientation away from water involves nothing more than a reversal of the normal water-finding mechanism; and third, no investigator has found evidence for a basic difference in orientation mechanism between hatchling and adult animals.

Out of the first studies of water finding came the general view that the process was primarily a visual one related to some color or intensity cue, or both (Hooker, 1911; Parker, 1922; Noble and Breslau, 1938; Daniel and Smith, 1947; Anderson, 1958). Some of the orientation cues proposed included blue light over water, the more open (brighter) horizon toward water, sharp outlines of dunes or foliage on the landward horizon, the more closed (darker) landward horizon, and the glitter of the surf. The only nonvisual behavior demonstrated to be of even secondary importance was geotaxis.

One of the reasons for confusion in the early studies was the poor definition of the orientation tasks that the turtles were actually performing, and the equally poor definition of the environmental cues available to the turtles and capable of being perceived by their sensory systems. Three examples must suffice. First, the statement by Daniel and Smith (1947) that young loggerheads were guided by reflections from the surf was based on their mistaken view that female loggerheads always nested in view of the sea. As noted, sea turtles often nest behind elevations in the terrain that completely block any view of the water at turtle eye level (Caldwell et al., 1959). Second, there is Hooker's notion that turtles were attracted to the blue light over water. Although it is a curiously widespread and persistent belief that the sky is more blue over water, Ehrenfeld and Carr (1967) were unable to find any spectral differences in the light intensities over the landward or seaward horizons, either by day or during moonlit nights. Third is the idea, advanced by Parker (1922), that turtles respond negatively to the sharp outlines of dunes and foliage. This may be true in the case of freshwater turtles, whose vision in air is good, but it is certainly not the case with sea turtles, which are extremely myopic when their eyes are out of water and are incapable of forming sharp images on their retinas (Walls, 1942; Ehrenfeld and Koch, 1967).

In considering the conflicting reports about geotaxis (Noble and Breslau, 1938; Parker, 1922), it should be remembered that all the investigators who have tested the effects of blindfolding on orientation under natural conditions (Daniel and Smith, 1947; Caldwell and Caldwell, 1962; Ehrenfeld and Carr, 1967) have observed that proper orientation ceases, and this is usually regardless of slope. In addition, Ehrenfeld (1968) and Mrosovsky and Shettleworth (in press) have been able to modify the direction of orientation *in progress* by using filters or partial blindfolds respectively, again regardless of slope. These blindfold and filter experiments demonstrate unequivocally that vision is far more important for water finding than any other sense. In recent years, *Chelonia mydas* has been demonstrated to have well-developed auditory (Ridgway et al., 1969) and olfactory (Manton et al., 1972a) sensitivities, but except for the "sand smelling" described previously, these senses probably play no role in the orientation of that species while on land.

Recent Studies

Since 1960, research on water finding has largely been directed at elucidating the precise visual cues used by turtles, and, to some extent, toward understanding how turtles process and analyze the environmental inputs. Carr and Ogren (1960) took hatchling green turtles from the Caribbean to the Pacific coast of Costa Rica and observed that they oriented toward the water without difficulty. These results eliminated the possibility of an inborn compass sense based on celestial (or magnetic) cues, at least for water finding. Burger (1976) similarly found an absence of compass orientation in hatchlings of the emydid species, *Malaclemys terrapin*.

By placing filter-holding "spectacles" over the eyes of adult, female green turtles Ehrenfeld and Carr (1967) and Ehrenfeld (1968) demonstrated that orientation could proceed when either the blue or the red component was removed from the naturally occurring incident light, either by day or by moonlit night. Depolarizing and diffusing filters had little effect, and turtles also oriented well with filters that transmitted light in the ultraviolet (300–400 nm) part of the spectrum. It was concluded that *Chelonia mydas* did not use color cues in this orientation, which was hardly surprising considering that none can be shown to be available. Discovery of the approximate spectral sensitivity of the green turtle, 350 to 650 nm, was a by-product of this work. Mrosovsky and Carr (1967) and Mrosovsky and Shettleworth (1968), using color and energy-controlled, projected lights in the natural orientation situation, also found that brightness discrimination was fundamental to water finding.

In tests with hatchlings placed in a walled arena, Ehrenfeld and Carr (1967) further characterized the orienting stimulus as the brightness pattern in the sky 4° or less above the horizon. This is essentially a measure of the "openness" of the horizon that was suggested previously (e.g. Parker, 1922; Anderson, 1957). Mrosovsky and Shettleworth (1968) and Mrosovsky (1972) have discussed at length the specific modulating effects on orientation of bright (rising or setting sun or moon) and dark (mountains, dark clouds over the water horizon) environmental features that might be expected to affect the orientation process under natural conditions.

In 1947 Daniel and Smith observed circus movements in two young loggerheads that had been unilaterally blindfolded. They interpreted this as evidence for a "phototropic and photokinetic organization of the animal." Ehrenfeld (1968) repeated these experiments with adult *C. mydas* with the same results. He was further able to test this phenomenon in a way that did not entirely impede the turtles' seaward progress, by placing neutral density filters instead of blindfolds over one eye. Deviations from a true seaward path occurred in the predicted directions, always toward the covered eye. It was concluded that the orientation mechanism was a classical tropotaxis (Fraenkel and Gunn, 1961), most often seen in invertebrates with simple, bilateral light receptors. The tropotactic mechanism was confirmed by Mrosovsky and Shettleworth (1968), using hatchling *C. mydas*. Subsequently, Mrosovsky and Shettleworth (in press) have begun to analyze the retinal and central nervous system mechanisms that comprise this tropotaxis, by blocking inputs to the temporal or nasal portions of the visual field with hemiblindfolds. Although they demonstrate that inputs from different directions contribute differently to sea-finding orientation, the results contain a number of inconsistencies and do not support any simple hypothesis.

Orientation Differences by Age and Species

In only one species, *Chelonia mydas*, have investigators looked extensively at both hatchlings and adults; and the vast size differential has necessitated the selection of rather different experiments with the two groups. A comparatively bright, open horizon is undoubtedly the main attractant for both hatchlings and adults moving toward water, but there are indications that the quality of the discrimination is cruder and more inflexible in the hatchlings. I find it much easier, for example, to lure newly emerged hatchlings away from the water by means of lights than to do so with adult females returning to the sea (including those returning by a route other than their emergence path). McFarlane (1963) has reported that hatchling loggerheads, but not adults, are attracted away from a seaward direction and on to shore roads by mercury vapor lighting. Whether this more sophisticated orientation capability of the adults is caused by greater experience, by a more highly developed nervous system, by an optically more competent eye, or by increased elevation of the eyes above ground level is not known. The only case I know of in which adult turtles (*C. caretta*) commonly become lost on their return to the water was reported by Baldwin and Lofton (in Caldwell et al., 1959), who were wardens at the wildlife refuge at Cape Romain, South Carolina. Here, in some areas, the beach is unusually deep, flat, and featureless, with only isolated dunes or none at all. Under these cueless conditions, orientation is impossible, even in adults. Most of these lost adults eventually wander into a portion of beach more favorable for orientation, but a few become exhausted far from water and are killed by the heat of the advancing day.

Our understanding of water-finding orientation in freshwater turtles is rudimentary, and there is no reason to assume that the mechanism is like that of sea turtles. Terrestrial orientation feats among some freshwater species are considerably more impressive than those of sea turtles because of the greater distances traveled on land, often in the absence of simple cues such as those provided by beach topography; moreover, their vision, as reported earlier, is more acute on land than that of their marine relatives. In addition to vision, perception of humidity gradients, magnetic fields, olfactory gradients, and temperature gradients has not been ruled out. Some form of proprioceptive memory or inertial guidance is also possible, but unlikely because it would be of no use to newly emerged hatchlings. Unpublished data of Robert Madden, who tracked terrestrial *Terrapene carolina* by radiotelemetry after displacement from their home ranges, and data of Carroll and Ehrenfeld (1978), who studied the movements of displaced *Clemmys insculpta* (semiterrestrial) by means of an extensive tagging-and-recovery program, indicate remarkable abilities of these species to orient in probably unfamiliar territory less than 2 km from home. This far exceeds the performance capabilities of any tropotactic orientation system, and it seems unreasonable to expect that the remaining emydids and other freshwater species (at least the adults) would be far behind with respect to water-finding ability.

SOCIAL BEHAVIOR

It has been known for centuries that migrating turtles may travel in groups composed of many individuals, but it is not certain that the members of the groups are in contact with one another. The possibility exists that they are simply following the same cues to the same place at the same time—independently.

Females of all observed species of turtles are capable of finding nest sites and of nesting solitarily; when aggregate nesting occurs there is no evidence of overt social contact among females. Nest construction is a similarly solitary process (although females may uncover another's eggs while digging their own nests). A captive Inaguan turtle (*Pseudemys malonei*), is supposed on several occasions to have released her hatchlings from their nests by scratching away the hard soil over them (Hodsdon and Pearson, 1943)—a behavior attributed to alligators (McIlhenny, 1935). I find this report intriguing and would like to see further study of *P. malonei* carried out in its natural habitat, where it is not especially difficult to find. Turtles have many praiseworthy and public-spirited characteristics, but they have not heretofore been known as good mothers.

In some species, as described earlier, hatchlings mutually facilitate their emergence from the nest. Finally, the travel of nesting females to the water is essentially a solitary process based on environmental cues perceived by each animal, although a number of individuals may be following parallel paths at the same time. Hatchlings of all species also seem to be capable of making the journey to water or other shelter by themselves, but Carr and Hirth (1961) claim that collisions between hatchling green turtles during the early phases of dispersion from the nest usually tend to increase the speed of locomotion and to improve the orientation direction of the members of the group.

In summary, in most cases sociality is neither a well-developed nor an indispensable feature of the various behaviors associated with nesting.

CHAPTER 19
Locomotion

WARREN F. WALKER, JR.

The habits and general pattern of movement of the various chelonian groups are known (Pritchard, 1967; Ernst and Barbour, 1972). All turtles can walk or crawl on the land, all can swim, and most can move slowly along the bottom of bodies of water.

Although the general patterns of movement are well known, a technology was not developed to record the precise movements of the limbs, which are usually quicker than the eye can follow, until the work of Muybridge (1887). In a period before cinéphotography, he devised a battery of still cameras that could be triggered in quick succession to produce a series of still pictures indicating the sequence of limb movements in mammals. Using cinéphotography, Howell (1944) extended Muybridge's work and recognized a wider variety of gaits. His studies included a few amphibians and reptiles. Gray (1968) has recently reviewed the extensive research on gaits and the mechanics of locomotion in many invertebrates and vertebrates completed by himself and his associates between 1933 and 1964. Hildebrand (1959, 1965, 1966) undertook a comprehensive study of quadrupedal gaits and developed convenient quantitative and graphic methods to facilitate their comparisons and analysis. Dagg and Vos (1968) and Zug (1972a) have added to Hildebrand's methods. Sukhanov (1968) and Alexander (1977) have published important reviews of terrestrial locomotion.

Howell (1944), Hildebrand (1966), and Sukhanov (1968) make brief mention of turtle locomotion; an unidentified tortoise walking was briefly described by de la Croix (1929), and other investigators have alluded to certain aspects of turtle limb movements in studies that were primarily morphological or physiological (Ruckes, 1929; Shaetfer, 1941; Zangerl, 1953; Friant, 1961; Gans and Hughes, 1967). I have made some preliminary reports on turtle walking (Walker, 1962, 1963), but Zug (1971) was the first to study locomotion in a wide range of species. As part of his investigations on the pelvic skeleton and musculature of cryptodiran turtles, he analyzed the terrestrial gaits of representatives of the Chelydridae (one species), Kinosternidae (four species), Emydidae (two terrestrial and nine semiaquatic species), Testudinidae (one species) and Trionychidae (two species). He also described the sequence of limb movements during swimming in 17 species representing all these families except for testudinids, and bottom walking in the species that use this method of progression. He included descriptions of the movement of the crus and pes (thigh movements are obscured by the shell). Zug (1972b) has reported further on the walking gaits of these turtles.

I have analyzed swimming in cheloniids (Walker, 1971b), made an X-ray study of limb movements during walking in *Chrysemys picta* (Walker, 1971a), related body form and locomotion in turtles (Walker, 1972), and reviewed the structure of the appendicular apparatus of chelonians (Walker, 1973). Other studies are in progress. John O. DeLancey, one of my research students, and I have studied variation in walking in 10 species, most of which are in the private collection of Miss Linda Polachek of Cleveland. We have photographed the following: Emydidae, *Clemmys insculpta, Cyclemys dentata, Emydoidea blandingi, Mauremys mutica, Terrapene c. carolina,* and *T. c. major*; Testudinidae, *Geochelone carbonaria, G. elegans, G. elongata, Gopherus polyphemus*; Platysternidae, *Platysternum megacephalum*.

ANALYSIS OF TERRESTRIAL LOCOMOTION

During terrestrial locomotion, the limbs must provide both support and propulsion. Many variables affect these seemingly simple considerations. A stationary or slow moving quadruped must be supported by at least a tripod of three feet within which its center of gravity falls (Figure 1a). As speed increases, an animal can tolerate brief periods of instability because, as it falls, it can catch and raise itself and move ahead to another moment of fall. There can be periods of two, one, or even no feet upon the ground. The degree to which a moving quadruped can tolerate periods of fall are affected by the distance the trunk is carried from the ground and the width of the stance (i.e., the distance between left and right feet). A cursorial mammal, for example, with long legs moving in the vertical plane and having a narrow stance, can tolerate more fall

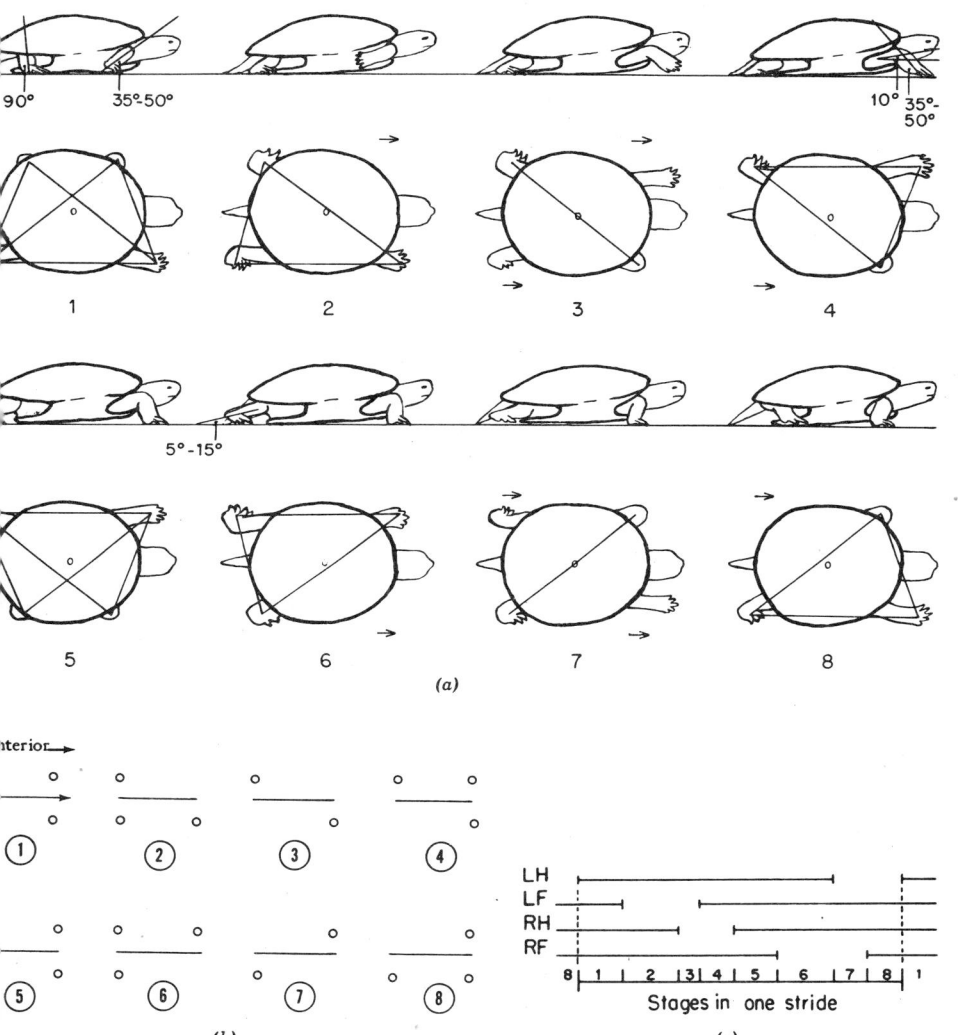

Figure 1 The eight stages in a single stride of a lateral sequence, diagonal couplet walk of *Chrysemys picta marginata* portrayed in three different ways. (*a*) Drawings from cinéphotographs of lateral and dorsal views. Circles indicate the animal's center of gravity; corners of support triangles are in the center of the feet on the ground; arrows indicate protracting feet; numbers represent the eight stages of a stride that differ in the number and combination of supporting feet. (*b*) Support pattern of the type that would have been used by Muybridge or Howell. The anterior end of the animal is toward the right, circles represent supporting feet in the eight successive stages, the line separates left from right feet. (*c*) Gait diagram of a stride with the formula 80-38. The solid lines represent duration of foot contact; LH = left hind foot, LF = left front foot, RH = right hind foot, RF = right front foot (Parts *a* and *c* from Walker, 1971b.) With permission of the publisher.

during its movement than a turtle, in which the trunk is carried close to the ground, the humerus and the femur move in the horizontal plane, and the stance is broad.

Propulsion requires that the limbs develop a force on the ground needed to move the center of gravity in the appropriate direction and at the appropriate speed. In accordance with Newton's third law of motion, a quadruped exerts a backward force against the ground with one or more legs and moves forward because the ground resists this with an equal but opposite force. When moving at a steady speed, the resultant of all of the longitudinal, horizontal forces acting on the body equals zero (Gray, 1944; Barclay, 1946). Since friction from wind resistance is negligible at the speeds most quadrupeds travel, the propulsive force of legs moving posteriorly must be opposed by the retarding effects of legs just placed on the ground. When an animal accelerates or decelerates, the resultant of all the longitudinal, horizontal forces will be respectively greater or less than zero. Similar considerations apply to vertical forces (the rise and fall of an animal will cancel and the resultant vertical force will be zero), and if the animal is moving in a straight line, they apply also to turning forces acting in the horizontal plane.

Given the appropriate forces, speed of travel is affected by the length and rate of the stride. The length of the stride is the distance traveled between the placement of one foot on the ground to the next placement of that same foot. The left hind foot is commonly used as the reference one. Variables that affect stride length are limb length and the distance a foot is carried forward when off the ground. Cursorial mammals, with long limbs moving in the vertical plane, clearly have longer strides than turtles. Their greater stability also enables cursorial mammals to use gaits with fewer feet on the ground at any moment; thus feet off the ground can be carried farther forward during a stride than in a turtle, where at least three feet need to be on the ground during most of a stride. Many vertebrates can also increase stride length by trunk undulations, but a turtle cannot. Stride rate is the frequency with which limbs oscillate back and forth in a unit of time. As it increases, so does speed.

Many of the variables affecting locomotion can be integrated in an analysis of gaits. The type of gaits commonly used by terrestrial vertebrates are described as symmetrical because the two hind (or front) feet tend to be evenly spaced in time; for example, a right hind foot is placed midway between the first and second placement of a left hind foot. Examples are walks, trots, and paces. (Asymmetric gaits, such as gallops and canters, have more irregular foot placements.) Muybridge (1887) and Howell (1944) described gaits in terms of "support" or "footfall patterns," which are pictorial representations of the number and combination of feet upon the ground during the stages of one stride (Figure 1b). Sequence of limb movement and support are shown, but not the length of time feet are on and off the ground. Hildebrand (1965, 1966) includes these variables in a gait diagram, which indicates the duration of foot contact with the ground by the length of horizontal lines (Figure 1c).

Hildebrand expresses the information in a gait diagram by two numbers in a gait formula. The first number is the percentage of the stride interval that the left hind foot is on the ground. Other feet would have a similar duration of contact in the more common symmetrical gaits. Foot contact correlates with speed, but as noted, it is not the only variable in speed. The second number relates the movement of the left front foot to the left hind foot; it represents the percentage of the stride interval that the placement of the left front foot follows, or lags behind, the placement of the left hind foot. The relationship of the placement of the right hind foot to the left hind foot is known in a symmetrical gait, and always comes about the middle of the stride.

Gaits can be compared by graphing gait formulas; duration of contact on the abscissa and lag of front foot placement behind ipsilateral hind foot placement on the ordinate (Figure 2). By applying descriptive terms to speed and relationship between hind and front foot placement, Hildebrand has refined our terminology of symmetrical gaits and developed a scheme that can be applied easily to all terrestrial vertebrates.

Gait formulas and diagrams for all possible gaits can be reconstructed from the two parameters of the graph. Not all the theoretically possible gaits are used. In a study of 400 gait formulas representing 15 orders of mammals and "a less complete representation of amphibians and reptiles," Hildebrand (1966) has found that natural symmetrical gaits fall within the clear area of the graph. There are no natural gaits with foot contact greater than 90% of the stride interval; rate of movement would be too slow when foot contact exceeds this. At the other end of the abscissa, the fastest symmetrical gaits appear to need a foot contact of at least 20 to 25% of the stride interval.

Factors limiting sequence of foot placement are more complex. When a front foot is placed on the ground at the same time or very shortly after the ipsilateral hind foot, the animal is pacing, supported much of the time only by two ipsilateral feet. This is a relatively unstable gait and is used only during fast walks and runs, chiefly by cursorial animals. As lag in front foot placement increases, the left front foot moves increasingly out of phase with the left hind foot, but it is still the next foot to be placed. Such gaits are described as lateral sequence gaits. When the lag is about 6% of the stride interval, ipsilateral front and hind feet, although

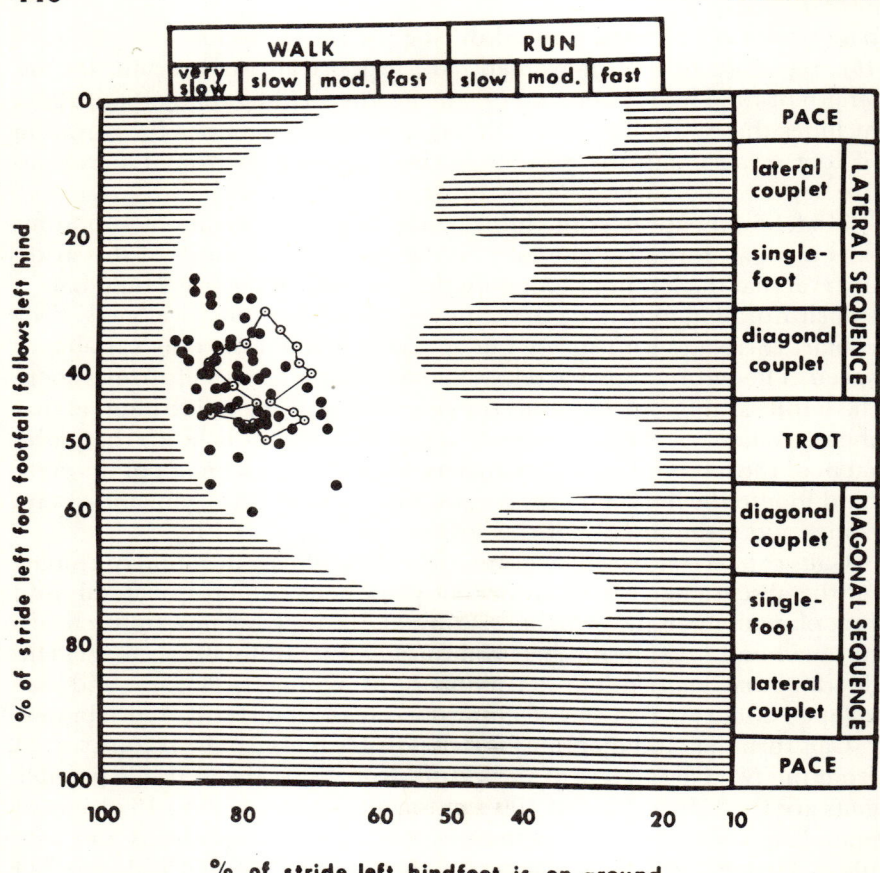

Figure 2 Distribution of gait formulas. The clear area shows the region in which Hildebrand (1965) found natural gaits of terrestrial vertebrates to fall. Black circles indicate Zug's findings for 19 turtle species. The polygon outlined by open circles encompasses 58 additional observations for 10 species observed in my laboratory. (Partly after Zug, 1971.) With permission of the University of Michigan Museum of Zoology.

slightly out of phase, are moving together most of the time. This is a lateral couplet gait. As lag increases to about 18%, the feet move independently (single-footed gaits), and when lag is about 31% a front foot begins to move most of the time with a contralateral hind foot (diagonal couplet gait). The lateral sequence gaits have greater stability than a pace for the animal is supported much of the time by three feet or by two diagonal feet (e.g., left hind and right front) so that a line of support passes under the animal's center of gravity. Most walks fall into this category. As lag increases from 44 to 56%, contralateral front and hind feet move together and the animal is trotting. This can be a fast gait; two contralateral feet provide greater stability than two ipsilateral feet, so the trot can also be used at walking as well as running speeds. As lag increases beyond 56%, the right front foot is placed after the left hind, and gaits become diagonal sequence. Diagonal sequence gaits are relatively rare.

Hildebrand reduces a gait diagram into a gait formula—two variables that describe the duration of foot contact and the sequence of limb movements. Support patterns are not immediately evident in the formulas, but can be seen by reconstructing the gait diagrams. Dagg and Vos (1968) emphasize the different combinations of supporting feet by deriving a walk pattern from the gait diagram. This is a statement of the percentage of the stride interval in which the animal is supported by different combinations of feet. In symmetrical walks the combinations are two ipsilateral feet, two diagonal contralateral feet, one hind and both front feet, one front and both hind feet, all four feet. The walk pattern for the gait diagrammed in Figure 1 is 0, 17, 21, 34, 28. The degree of stability is shown clearly in a walk pattern but, as Zug (1972a) points out, not the sequence of limb movements. Zug proposes that both a gait formula and a walk pattern be used to quantify the information in a gait diagram for purposes of analysis.

LIMB MOVEMENTS IN WALKING

The morphology of turtles restricts their gaits and their speed of walking. Turtles have a very short and broad trunk, which gives them an exceptionally wide stance, and the trunk is encased in a rigid shell. Their humerus and femur move back and forth close to the horizontal plane, but the shell and morphology of the bones impose certain limits to their excursions, as described by Walker (1971b) for the semiaquatic emydid, *Chrysemys picta marginata*.

Figure 3 gives the positions assumed by the skeletal elements when the feet are placed on the ground at the start of retraction in (*a*) and (*c*) and those they assume just before the feet are removed at the beginning of retraction in (*b*) and (*d*). The glenoid cavity of the pectoral girdle is located rather far laterally relative to the acetabulum of the pelvic girdle. This, coupled with modifications of the humerus, permit the brachium to be turned in front of the head, with the elbow at the midline of the body, when the turtle withdraws into its shell. When the front foot is

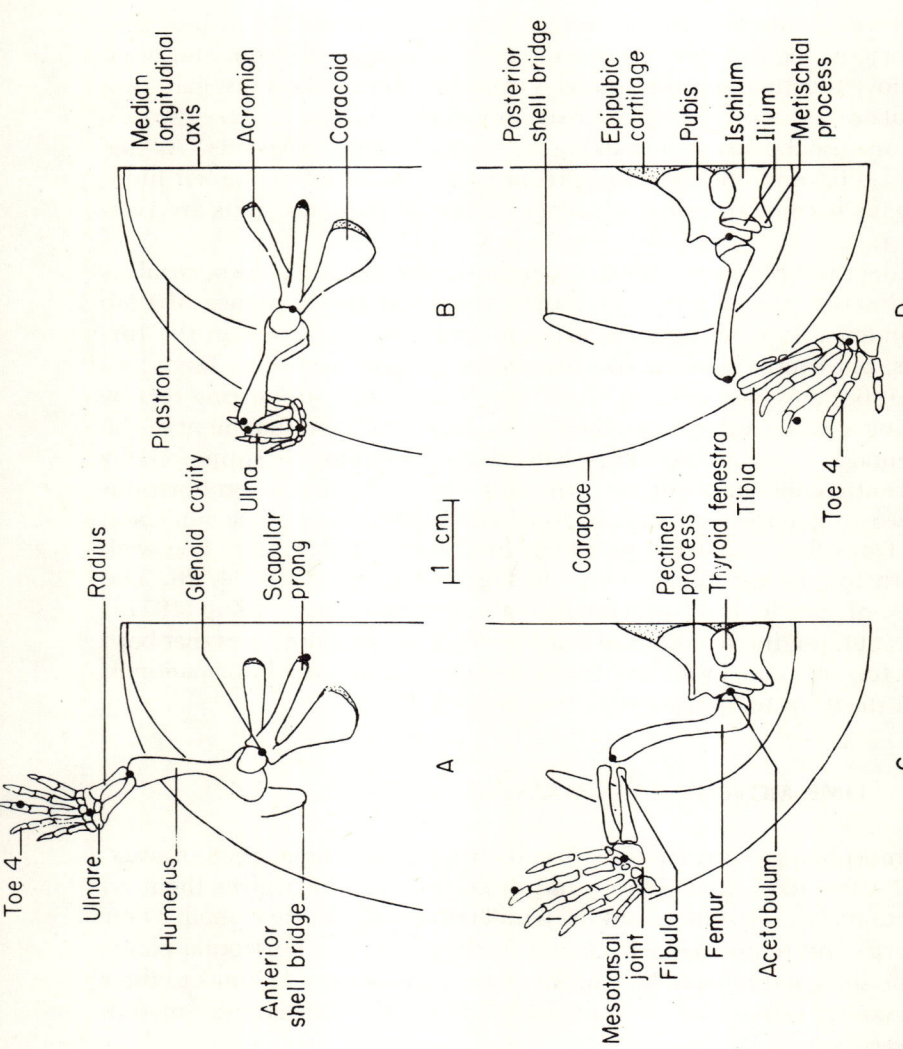

Figure 3 Dorsal view drawings made from X-rays of *Chrysemys picta* of the positions of the pectoral (*a*), (*b*) and pelvic (*c*), (*d*) girdles and limbs just after foot placement on the

placed on the ground, the humerus also is rotated rather far forward so that its longitudinal axis nearly parallels that of the body. The humerus is retracted through an arc of about 60°, by which time it is close to the bridge of the shell. Total excursion of the humerus is assisted by a modest rotation of the pectoral girdle, which is articulated to the shell by ligaments extending to the acromion and scapular prongs. The femur is also modified so that the hind limb can be withdrawn into the inguinal pocket. When the hind foot is placed on the ground, the femur too is directed forward, although not quite to the degree of the humerus. The morphology of the proximal end of the femur permits it to be retracted through an arc of slightly over 60°, by which point its longitudinal axis is nearly perpendicular to the longitudinal axis of the body. The arcs in the turtle are also directed more anteriorly; that is, a turtle's limb is rotated farther forward at foot placement and not retracted so far as in other amphibians and reptiles, which have approximately 90° arcs.

As humerus and femur are retracted, they also rotate about their longitudinal axes. Barclay (1946) referred to it as the crank effect because the turning of these elements resembles the rotation of the shaft of a crank in that it helps to swing the lower limbs, or crank handle, back and forth.

Changes in the positions of the antebrachium (forearm), crus (shin), and feet can be seen in Figures 1 and 3. A front foot is placed on the ground well anterior and slightly lateral to the position of the elbow so that, as seen in a side view, the antebrachium makes a posterior angle with the ground of 35° to 50° (Figure 1, stage 4). A hind foot is placed on the ground lateral to, and nearly in the same transverse plane as, the knee so that, as seen in a lateral view, the crus is nearly perpendicular to the ground (posterior angle with the ground of 85–90°, Figure 1, stage 1). In both cases the toes are directed forward.

As the body pivots forward during limb retraction, the feet maintain their general position on the ground. The antebrachium flexes and the elbow swings over the carpus. As retraction ends, the antebrachium subtends an anterior angle with the ground of 35° to 50°, and the foot extends slightly (Figure 1, stage 1). The crus also flexes at the knee, but since the knee lies so far medial to the tarsus, it does not come into the same longitudinal, vertical plane as the tarsus until the end of retraction, when the heel swings medially and the toes point anterolaterally. At the end of retraction, the crus makes a very low angle with the ground (anterior angle of 5–15°, Figure 1, stage 6). During protraction, both antebrachium and crus are flexed. They extend again shortly before the feet are placed on the ground.

Comparable movements of the crus have been reported by Zug

(1971). In genera with small plastrons (*Chelydra* and *Trionyx*) the hind foot is placed on the ground at a level anterior to the knee. It also lies well lateral to the knees in all genera except for *Gopherus*, where the foot is placed under the knee as seen both in a lateral and posterior view.

Chrysemys picta carries its trunk very close to the ground, indeed the posterior corners of the plastron may drag during part of a stride. Zug reports that other semiaquatic or aquatic emydids also carry the trunk very low, but he reports no drag. The terrestrial genera, together with the chelydrids, kinosternids, and sometimes the trionychids, hold their trunk farther above the substratum.

TERRESTRIAL GAITS

The closeness with which turtles carry their trunk to the ground and the width of their stance, which places the line of footfall well to the side of the center of gravity, precludes them from utilizing paces and lateral couplet gaits. The unsupported side of the body would topple and drag heavily on the ground. Turtles utilize gaits that maximize stability (Figure 2). With the exception of one diagonal sequence gait, observed in a single stride of *Clemmys insculpta* (Zug, 1971), turtles utilize lateral sequence, single-footed or diagonal couplet walks, and walking trots.

Zug (1971) found little correlation between type of gait and species. Later (Zug, 1972b) he analyzed the walk patterns of the turtles observed in 1971. He found five statistically defined similarity groups, but since again they did not correlate with species, with gait, or with such behavioral groups as terrestrial, bottom walking, or swimming, their significance is not clear. All the species observed in my laboratory, and most of Zug's, use the lateral sequence, diagonal couplet walk. This is by far the most common gait and is probably ubiquitous except for the highly aquatic sea turtles and *Carettochelys*, in which the pectoral limbs are modified as flippers. Many of the species that we have observed can shift from a walk to a walking trot, but the trot is used less frequently than the walk. All the trots we observed were in the more terrestrial species: *Geochelone carbonaria, G. elongata, Clemmys insculpta, Cyclemys dentata, Terrapene carolina*. Zug has observed trots in a wider spectrum, including representatives of all the families he studied, except for the Chelydridae. Although used less frequently than the diagonal couplet walk, the walking trot is not an uncommon gait among turtles. The lateral sequence, single-footed gait is rarely used; Zug observed it and diagonal couplet walks in *Chelydra serpentina* and *T. carolina*.

A close analysis of the gaits used by turtles show the high degree of stability they afford. In a pictorial representation of the lateral sequence, diagonal couplet walk (Figure 1a), one can see that there are eight stages that represent different numbers and combinations of feet on the ground. In four of them a turtle is securely supported by triangles of three feet within which its center of gravity falls. The other four stages are transitions between the triangles. In two of these, which represent transitions between triangles whose base is on the same side of the body, the turtle is supported by four feet; a new triangle is established before the previous one is given up. In the other two transitions the base of the supporting triangle shifts from one side of the body to the other, and the turtle is supported only by a pair of diagonal, contralateral feet. These are the only periods when the turtle is not statically stable, and then the animal is dynamically stable because its center of gravity is crossing the line between the two supporting feet as it moves from one triangle to the next. There are two support cycles in the gait, since the last four stages are mirror images of the first four. The number of feet on the ground in successive stages is 4-3-2-3-4-3-2-3. Each foot is on the ground for six stages and off of the ground for two as it is moved forward.

Analysis of a gait diagram (Figure 1c) and walk pattern (Figure 5) emphasizes the high degree of stability of the lateral sequence, diagonal couplet walk. Stages are not of equal length. Zug (1972b) reports observations similar to those in Figure 1 for lateral sequence, diagonal couplet walks in 14 taxonomic samples. Averaging his data I compute values for the respective walk patterns of 0, 15, 23, 32 and 30. The two stages in which the turtle is supported only by contralateral feet are the shortest support patterns used, totaling an average of 15% of the stride. The longest stages are the tripods involving two hind feet and one front foot. On average, the turtle is securely supported by three or four feet for 85% of the stride.

In very slow lateral sequence, single-footed walks, the turtle is supported all the time by three or four feet (Figure 4a). The two-footed phase of the typical diagonal couplet walk is replaced by a four-footed phase. Average values for the walk pattern based on two taxonomic samples given by Zug (1972b) are: 0, 0, 32, 32, and 35. Support is divided nearly equally between the two types of tripodal and the quadrupodal stages (Figure 5).

In an occasional very slow lateral sequence, diagonal couplet walk in which the lag in front foot placement is also relatively short—for example, gait formula of 88-36 (Zug, 1972b), quadrupodal support stages also replace the stages with two diagonal feet. Although support patterns of this type provide maximum stability, turtles seldom use them.

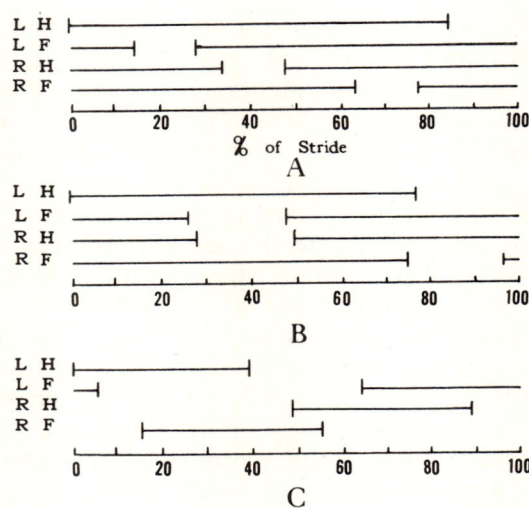

Figure 4 Representative gait diagrams. (*a*) A slow, lateral sequence, single-footed walk of *Terrapene carolina* (gait formula 86-28). (*b*) A walking trot of *Cyclemys dentata* (gait formula 78-48). (*c*) Pattern of limb retraction in swimming (formula 40-65). [(*a*) Based on data in Zug, 1972b; (*c*) based on data in Zug, 1971.]

The trot is the least stable gait for periods of support provided only by two contralateral feet increases (Figure 4*b*). In the running trot of a cursorial mammal, the transitions between stages of support on contralateral feet can be made with no feet on the ground, but in the walking trot of the turtle, there are four feet on the ground during all or much of the transitions. If front foot lag is exactly 50% there are only quadrupodal transitions, but if the lag deviates from 50%, as is usually the case, there are also tripodal stages. Average values for the walk patterns based on seven taxonomic samples of walking trots given by Zug (1972b) are 0, 28, 15, 28, and 29 (Figure 5).

Although there is a great deal of variation among individual walks and trots, certain trends emerge. The shift from single-footed through diagonal couplet gaits to walking trots involves a decrease in stability, for the percentage of the stride interval occupied by a pair of diagonal, contralateral feet increases (Figure 5). There is necessarily a concomitant decrease in certain other types of support. In turtles the tripods involving two front feet and one hind foot decrease considerably more than other support patterns.

The unequal duration of support patterns is also reflected in velocity fluctuations within a stride (Figure 6, *Terrapene carolina*). There are two

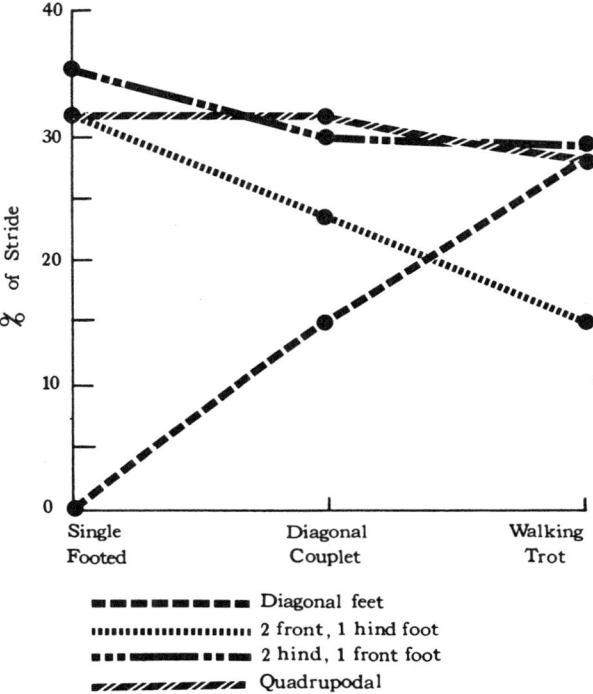

Figure 5 Average percentage of the stride interval occupied by the four support patterns in the gaits of turtles. (Based on data from Zug, 1972b.)

velocity cycles that correlate with the two cycles of support. Velocity is at a minimum at the start (stage 1) or middle of a stride (stage 5), when four feet are on the ground. Acceleration occurs in stages 2 and 3 as the turtle shifts from four to three to two supporting feet. Velocity reaches a maximum as the turtle lunges forward on two contralateral feet. Marked deceleration occurs in stage 4 as the turtle catches its fall and reestablishes a triangle of support by placing a front foot on the ground well anterior to the trunk. This cycle is repeated in the second half of the stride. We have observed similar fluctuations in the walks and walking trots of six additional species. Although no data are available, the frequent alternation between periods of acceleration and deceleration must be expensive in terms of energy consumption.

Velocity analysis of a stride suggests that the tripods with two front feet on the ground are the least desirable of the stable support stages, for they also involve considerable deceleration. Possibly this is one reason

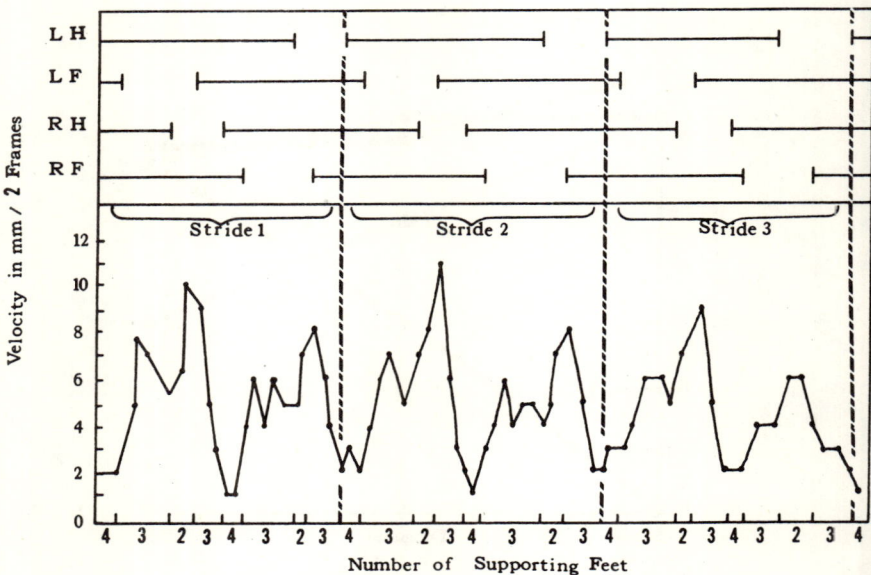

Figure 6 Gait diagram and velocity changes within three successive strides of a lateral sequence, diagonal couplet walk of *Terrapene c. carolina*. Photographs were taken at 32 frames per second. (Observations made by John O. DeLancey.)

that this support pattern is nearly always of shorter duration than the tripod with two hind feet and one front foot, and that this is the pattern that decreases the most as turtles shift to gaits that involve longer periods of diagonal support. A shorter duration for stages with two front feet and one hind foot than for stages with two hind feet and one front foot is only possible if front foot contact with the substratum is less than hind foot contact (Zug, 1971, 1972b).

FACTORS AFFECTING SPEED

Gaits used by turtles provide good stability at overall speeds ranging from a very slow to a moderate walk. The range of speeds is quite limited, since turtles have fewer options for increasing speed than many other tetrapods. Stride length is limited by trunk rigidity, short limbs, and the relatively short arcs through which the limbs are retracted. In many tetrapods stride length and speed can be increased by decreasing the duration of foot contact, which increases the distance the foot is carried forward by the movement of the body while it is off the ground,

FACTORS AFFECTING SPEED

but stability considerations preclude turtles from using gaits in which each foot is in contact with the ground for much less than 70% of the stride interval (Figure 2). Yet within narrow limits some species can increase speed by decreasing foot contact. DeLancey found that average speed increases in *Cyclemys dentata* from 1.2 carapace lengths per second to 1.8 lengths as foot contact decreases from 79% of the stride interval to 73%. A correlation between duration of foot contact and speed was less clear in three other species he examined. An increasing lag in the placement of the front foot, which would shift the gait from a walk toward a trot, also had little effect on increasing speed.

The most important variable in increasing speed in turtles is by increasing stride frequency. In all species DeLancey studied, stride frequency increased as speed increased (Figure 7). But increasing the number of limb oscillations clearly requires more energy expenditure than would be necessary if speed were increased by other means.

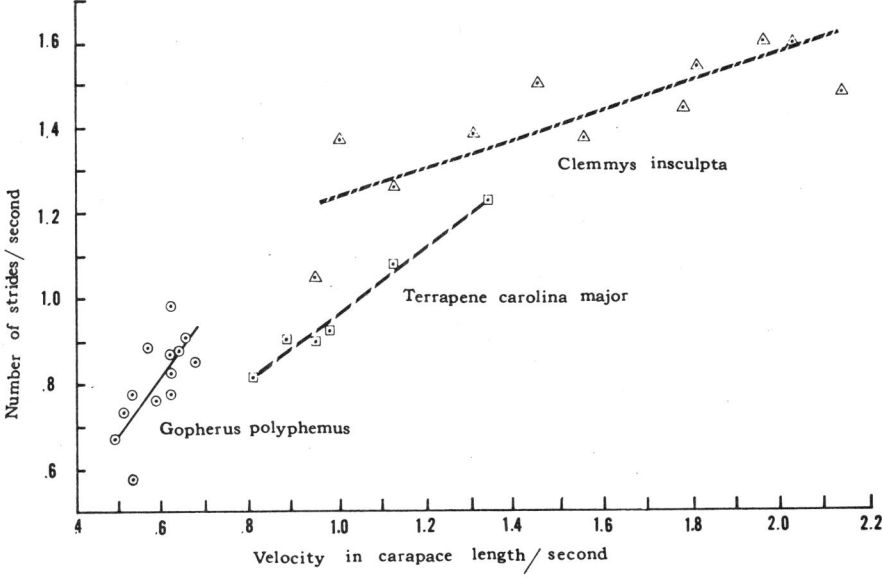

Figure 7 Linear regression lines showing change in frequency of stride (the dependent variable) relative to velocity in three turtle species. The lines were computer fitted to the data. Absolute velocities range from 0.06 meters per second for the slowest stride of *Gopherus polyphemus* to 0.35 meters per second for the fastest stride of *Clemmys insculpta*. (Observations by John O. DeLancey.)

AQUATIC LOCOMOTION

In water, turtles move by bottom walking (Zug, 1971), or they swim. Most species swim by paddling with all limbs, but sea turtles, and *Carettochelys insculpta* of New Guinea rivers, swim by simultaneous movements of enlarged pectoral flippers.

Zug found that bottom walking often supplements swimming in emydids and is the primary method of aquatic locomotion among chelydrids and kinosternids. Propulsion derives primarily from thrusts of the feet or sometimes only the toes against the bottom, but the toes are usually abducted and the web spread so that additional thrusts derive from the reaction of the water to limb movements. The body, largely supported by the water, appears to be propelled with great ease. During the retractive phase the crus moves posteriorly in the vertical plane. At the end of retraction there is a medial sweep of the limb, which brings the toes posteriorly. During protraction the foot is held well clear of the substratum and sometimes is feathered.

In swimming by paddling, the body is supported by the water, and propulsive forces derive only from the limbs pushing against the water (Zug, 1971). Speed and sequence of limb movements are not affected by considerations of support and stability either in swimming or bottom walking. The limbs are moved faster than in walking; sometimes the retractive phase, which is the equivalent of foot contact with the ground in a terrestrial gait, is as low as 30% of the stride interval (Figure 8). There is a shift from the lateral sequence, diagonal couplet pattern of limb movement through the trot to diagonal sequence patterns, the diagonal couplet being a common one.

A gait diagram and swimming pattern (analogous to a walk pattern) of a representative diagonal sequence, diagonal couplet swim at a rate equivalent to a moderate run (formula 40-65) reveals the need for support by water, and certain advantages of the pattern for swimming (Figure 4c). The percentage of the stride interval occupied by different numbers and combinations of feet in this case is one foot = 40, two ipsilateral diagonal feet = 10, two diagonal contralateral feet = 50. At no point in the stride are more than two feet being retracted simultaneously, and in this example there are four periods, which add up to 40% of the stride, when only one foot is being retracted. On land, a turtle could not support itself in this manner. During half the stride, contralateral front and hind feet are retracting concurrently, thereby providing a good thrust while counteracting any tendency to yaw (turn right or left) that would derive from feet acting on only one side of the body. The short periods when only one foot is retracting are separated from each

AQUATIC LOCOMOTION 451

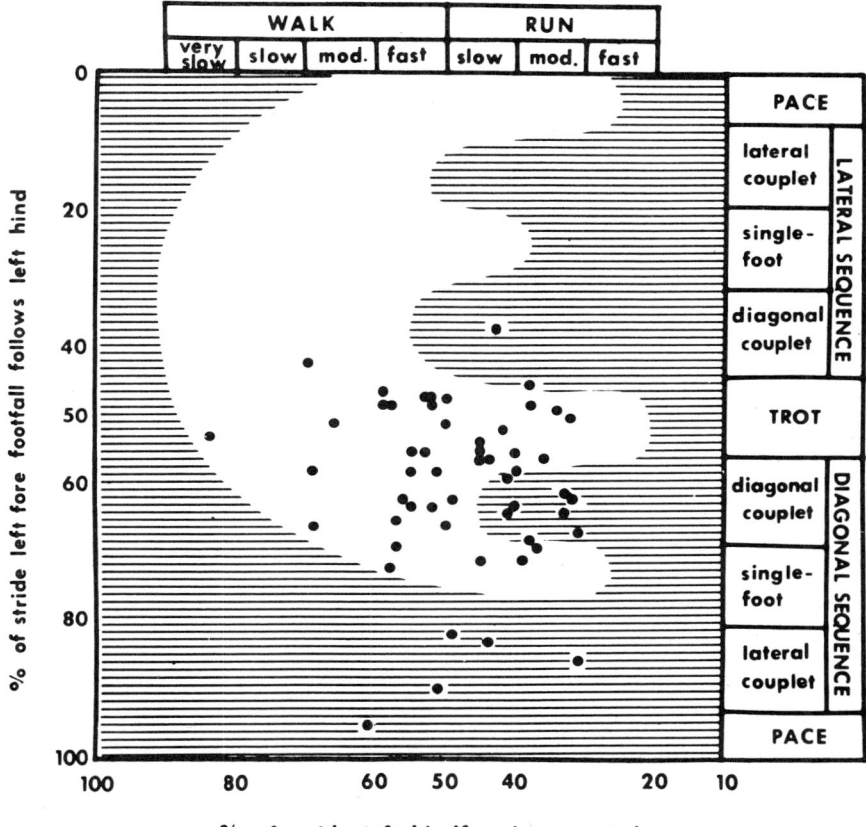

Figure 8 Distribution of the gait formula for 17 species of swimming turtles. The clear area is the region occupied by terrestrial gaits of vertebrates. (From Zug, 1971.) With permission of the University of Michigan Museum of Zoology.

other and alternate between the sides of the body. Periods with two ipsilateral feet acting are very brief.

The hind limb is fully extended early in retraction in a swimming limb cycle, and the thigh and crus move posteriorly in the horizontal plane (*Trionyx spiniferus* Figure 9), or slightly ventrolateral to this plane (other genera studied by Zug). In no case is the crus held as vertically as in walking. *T. spiniferus* can retract its thigh to the point that the entire limb is nearly parallel to the longitudinal body axis. In other genera the crus is parallel to the body axis at the end of retraction, but the thigh extends posterolaterally. During retraction the foot is held in the vertical plane,

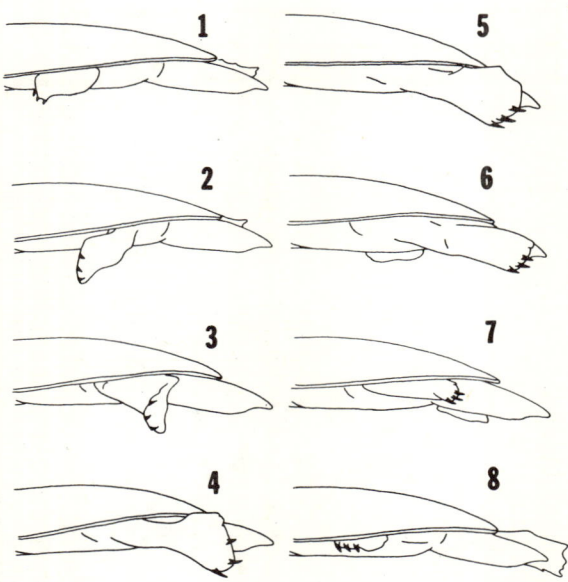

Figure 9 Lateral view of the left hind limb of *Trionyx spiniferus* during a single stride of swimming; 1–4, retraction; 5–8, protraction. (From Zug, 1971.) With permission of University of Michigan Museum of Zoology.

the toes are abducted, and the web is spread. The limb flexes at the knee during protraction and moves quickly forward into the inguinal pocket. The toes are adducted and the foot is feathered.

The antebrachium and hand of sea turtles and *Carettochelys insculpta* is modified as a flipper-shaped blade, whose use in cheloniids has been commented on by Parrish (1958) and analyzed by Walker (1971a). When moving slowly across the bottom, the pectoral flippers may be used alternately, but when the animal is swimming in open water they move up and down together. During a down-upstroke cycle the blade follows an overall course that is inclined from 40° to 70° from the horizontal plane of the body (Figure 10), so that on a downstroke it also moves posteriorly and on an upstroke it moves anteriorly. The angle of inclination from the horizontal plane is greater on the downstroke than on the upstroke so that, if the excursions of the blade are plotted with the animal in a fixed position, the distal end of the blade transcribes an elongated and inclined figure 8. This accentuates the posterior thrust on the downstroke and reduces the anterior thrust on the upstroke.

Thrusts are made more effective by changes in the angle of attack of the blade. On the downstroke its leading edge is ventral and its distal

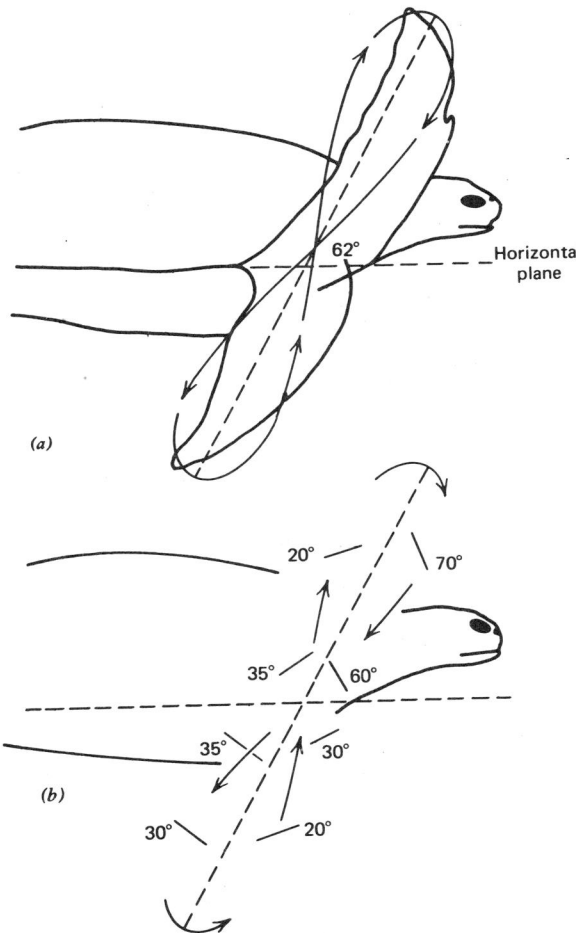

Figure 10 Composite drawings from cinéphotographs of the pectoral limbs of *Chelonia m. mydas* during a representative cycle. (*a*) Line of excursion of blade. (*b*) Movements of the distal end of the blade and its angles of attack with turtle in a fixed position. The heavy dashed lines show the overall axis of blade movement. (From Walker, 1971a.) With permission of the publisher.

surface inclined 70° to 80° from the horizontal. The inclination decreases near the end of the downstroke and reverses on the upstroke so that the leading edge of the blade is dorsal and the inclination from the horizontal is about 35° in the opposite direction. The upstroke thus appears to have a propulsive component, but it would not be as great as in the downstroke.

A typical cycle has a duration of 2 to 4 sec. Most of this time is evenly divided between down and upstrokes, since turns at the top and bottom of the cycle are made very quickly.

The blade flexes slightly at the elbow near the end of a downstroke, but extends again during the upstroke and remains extended through most of the downstroke. Humeral movements are the main driving force of the blade. Although the humerus has a greater vertical component to its movement than in freshwater species, the proximal articular surface of the humerus is modified in such a way that simple retraction causes some of the brachial adduction and rotation that is observed.

The extensively webbed and flipperlike hind limbs are not used for propulsion in normal swimming, but drag behind as elevators and rudders.

SOME UNANSWERED QUESTIONS

In recent years we have learned a great deal about the locomotion of turtles and the constraints on their movements; however we have found little correlation between the different gaits and either speed or taxon. The circumstances that lead to shifts between lateral sequence, diagonal couplet gaits, and walking trots, are not clear. Sea turtles drag themselves along the ground in a rather clumsy fashion, but the pattern of limb movements has not been described in detail. In general we know more about terrestrial locomotion than about swimming. Except for sea turtles, the details of the movements of individual front limbs in swimming have not been described.

The marked fluctuations in velocity within a stride suggest that the gaits of turtles are rather expensive in terms of energy consumption, but no data are available on this subject. Studies are in progress in my laboratory of the forces acting on the feet of turtles using Barclay's (1946) methods; this provides an indirect measure of muscular effort, but there may be more direct methods for acquiring the same information.

Differences in limb skeletal and muscular anatomy of chelonian groups have been reviewed (Walker, 1973), but the adaptive nature of many of the differences can only be surmised. We need more information on the movements of individual limbs in different taxa, and precise data from electromyography on the muscles that are active under different circumstances, before many of the adaptations can be interpreted.

CHAPTER 20
Learning

HENRY MORLOCK

In recent years a number of investigations have been conducted in which turtles were trained to respond to new stimuli or to make new responses. Investigators were primarily interested in establishing behavioral indicators of visual function, or in testing the generality of principles of learning established for other animals. Few researchers have been specifically interested in how behaviors are learned by turtles per se or in the roles of learning in natural environments. This chapter brings together the diverse studies in order to begin to develop an understanding of learning processes in the lives of turtles.[1]

Three broad categories of learning research have been carried out using turtles as subjects. Most of the investigations have involved one of these, namely, instrumental conditioning and issues relevant to it. The small amount of research on habituation and classical conditioning is presented first.

HABITUATION

Habituation is an inhibitory form of learning that consists of a decline in responding to a repeated stimulus; it is not the result of peripheral factors, such as fatigue or sensory adaptation. Several general analyses and literature reviews are available: Thompson and Spencer (1966),

---

[1]Although this review is not exhaustive, it is representative, particularly of the literature published in English.

Groves and Thompson (1970), and Razran (1971); Goodman and Weinberger (1973) deal mainly with habituation in reptiles and amphibians.

In one of the first demonstrations of habituation in turtles, Honigman (1921) presented food in bags and bottles and noted a decrease in snapping, which was ineffective in obtaining food from the containers. Humphrey (1933) reported a decline in the frequency of leg retraction of a musk turtle to repeated tapping of its shell. More recently Hayes, Hertzler, and Hogberg (1968) and Hertzler and Hayes (1969) obtained habituation of head and eye movements in *Pseudemys scripta*. The movements were initially elicited by a revolving field of vertical black stripes on the inside of a white drum within which the subject was suspended. Hayes and Saiff (1967) and Ireland, Hayes, and Laddin (1969) also found habituation of head withdrawal, which was originally elicited by a "looming" circular shadow. Crampton and Schwam (1962) did not find habituation of a head-turning response to angular acceleration, nor did Hayes and Ireland (1972) show habituation of an optomotor response of swimming after or along with rotating vertical stripes. Habituation, then, is found in turtles, but its generality, dynamics, and relationships with stimuli in natural environments require further study.

CLASSICAL CONDITIONING

Classical conditioning is the familiar Pavlovian form of learning in which an original eliciting stimulus and a "neutral" stimulus are associated. This learning with the unconditioned (US) and conditioned (CS) stimuli is also known as respondent conditioning. There has been little interest in the topic among turtle researchers beyond a few examples in the literature that appear to show that turtles can be so conditioned. Parschin (1929), Poliakoff (1930), and Hasratjan and Alexanjan (1933) reported successful conditioning, and Razran (1961) published a short description of a method to condition limb flexion. Peretti and Zrout (1975) describe conditioning in spinal subjects. Other references and short descriptions of results obtained by Russian investigators of classical conditioning and related topics can be found in Razran (1961) and Voronin (1962). As with habituation, it appears safe to say that this form of learning is found in turtles, but little is known beyond that.

INSTRUMENTAL CONDITIONING

Instrumental conditioning includes those situations in which the organism's learned behavior is instrumental in bringing about a stimulus change.[2] In this type of behavior modification the responses come under the control of, or are at least influenced by, their consequents.

Maze Learning

In maze learning, the response required of the turtle is traversing an apparatus, usually simply by walking or swimming. Learning is said to have occurred when the turtle reaches a designated criterion in terms of reduced latency to leave a start area, increased speed in traversing the maze, and/or a high proportion of entries into a part of the maze marked by a distinctive position or stimulus.

One of the first demonstrations of learning in a turtle was reported by Yerkes (1901), who trained a spotted turtle (*Clemmys gutatta*) to take a route around several obstructions to get to a home box. Since then, investigators have reported successful maze learning in another *Clemmys* species and also in several other genera: *Chrysemys* (Kirk and Bitterman, 1963; Spigel, 1965, 1966; Heidt and Burbridge, 1966; Morlock, Brothers, and Shaffer, 1968; Hart, Cogan, and Williamson, 1969; Ellis and Barcik, 1972), *Clemmys* (Tinklepaugh, 1932; Fink, 1959; Harless, 1970a), *Gopherus* (Fink, 1959), *Pseudemys* (Seidman, 1949; Fink, 1959), and *Terrapene* (Fink, 1959). Mazes are not very efficient apparatuses for collecting data on learning and, perhaps because of this, the list of turtles studied is short and the number of investigations of the determinants of maze behavior few.

Operant Conditioning

With operant conditioning, the behavior required of the animal is typically a low probability response that is selectively reinforced until it becomes a high probability response. The more information the investigator has at hand on the usual behavior of the species, the optimal environmental requirements, and any possible reinforcers, the more readily apparatuses can be constructed and animals trained. In free operant conditioning the turtle is "free" to respond at any time once it is in the apparatus, as opposed to the discrete trials of maze situations where the investigator traditionally places the animal in a start area before each traversal. The free operant setting permits high numbers of responses to be made in each session.

[2]Informative discussions and descriptions of these techniques can be found in Sidowski (1966) and D'amato (1970). An excellent demonstration of shaping new responses by successive approximations is found in E. Reese's film *Behavior theory in practice* (available from Prentice-Hall, Englewood Cliffs, N.J.).

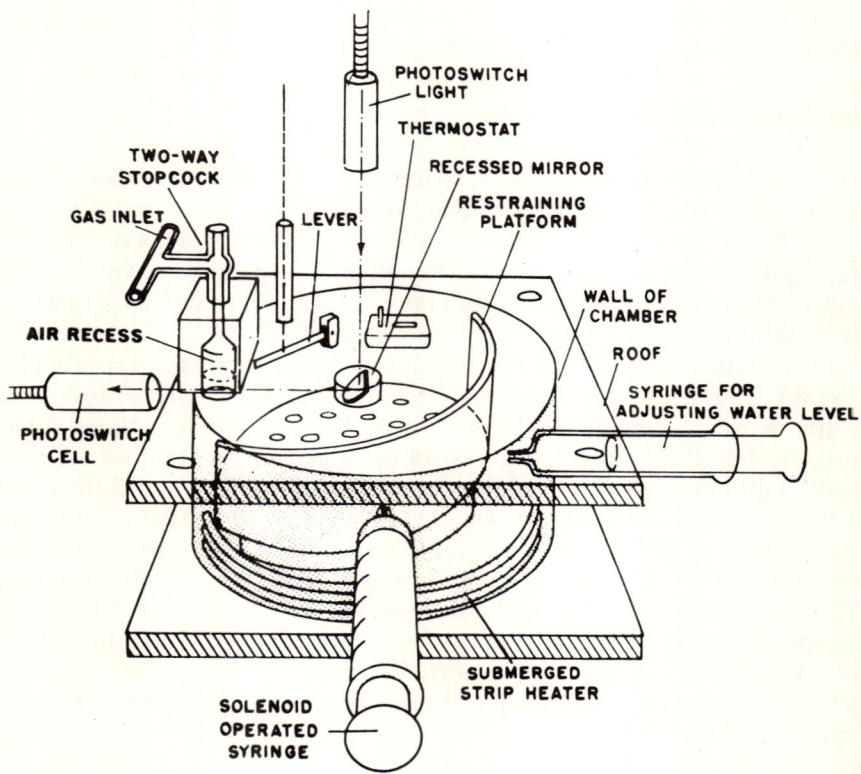

Figure 1 Diagram of operant conditioning apparatus used to train turtles to press a lever for access to air or other gases. (From van Sommers, 1963.) Copyright 1963 by the American Psychological Association. Reprinted by permission of APA and P. van Sommers.

Van Sommers (1963) constructed an elegant device within which young, submerged *Pseudemys scripta elegans* were trained to depress a lever for access to air. Stable rates of response required 4 to 16 hr of training. Rate declined when the duration of access to air was increased from 5 to 10 sec, when the gas mixture used for reinforcement contained 10% carbon dioxide, and when responding was no longer rewarded. Rate increased when the diet of the subjects was augmented with uncooked beef, and when the temperature of the water was raised from 10 to 22°C.

Bitterman (1964) described a two-choice, key pressing device in which correct responses were followed by access to small pieces of hamburger.

Well-tamed *Chrysemys p. picta* were readily trained to press the keys, and one subject worked steadily for more than 6 months with 6 training sessions per week. Its typical response rate was 10 responses per minute on a ratio schedule of about 20 responses per reinforcement.

Crawford and Seibert (1964) utilized shrimp pellets to train *Pseudemys floridana suwanniensis* to press a lever. Response rate increased with water temperature in the apparatus. In further studies with this technique Crawford, Adams, and Whitt (1966) observed an increase in response rate when the schedule of reinforcement was shifted from continuous (reward for each response) to fixed ratio 2 (reward for every other response). Further increases in the ratio did not affect response rate until the ratio requirement reached 10, when all subjects showed a decline. Crawford and Adams (1968) obtained higher response rates with variable ratio and fixed and variable interval schedules and the most consistent responding with a variable ratio schedule.

Pert and Bitterman (1969) described a pump used to inject controlled amounts of beef paste through a nipple into the testing chamber. *Chrysemys p. picta* were easily trained to eat from the nipple, and the rate of nipple touching responses was rather constant under a variable interval schedule. Rate declined progressively as the average interval between reinforcements was increased.

In an impressive demonstration of the way the behavior of the green turtle, *Chelonia mydas*, can be brought under control of discriminative stimuli and reinforcement schedules, Manton, Karr, and Ehrenfeld (1972a, 1972b) developed an operant conditioning technique for determining the detectability of waterborne olfactory stimuli. (Also see Chapter 15.) The subjects were first trained to press a panel or key on the right to obtain a reward consisting of a cube of meat. Following this, they were trained to press a panel on the left before pressing the panel on the right to obtain the reward. At first the cue for shifting from the left panel to the right panel was onset of a light above the left panel. When the subjects were responding consistently and correctly, the light signal was faded as an olfactory stimulus was introduced simultaneously in its place. In this way the authors achieved olfactory stimulus control. The requirement of responding to the left key before responding to the right key assured that the subjects were in the proper position to receive the olfactory stimulus when it was presented and minimized false reports at the right key. Several reinforcement schedules were used in the training of the key pressing responses. Stable rates were obtained with a variable interval schedule, and rate was increased by raising the response demand from three to five responses per reinforcement on the average with a variable ratio schedule.

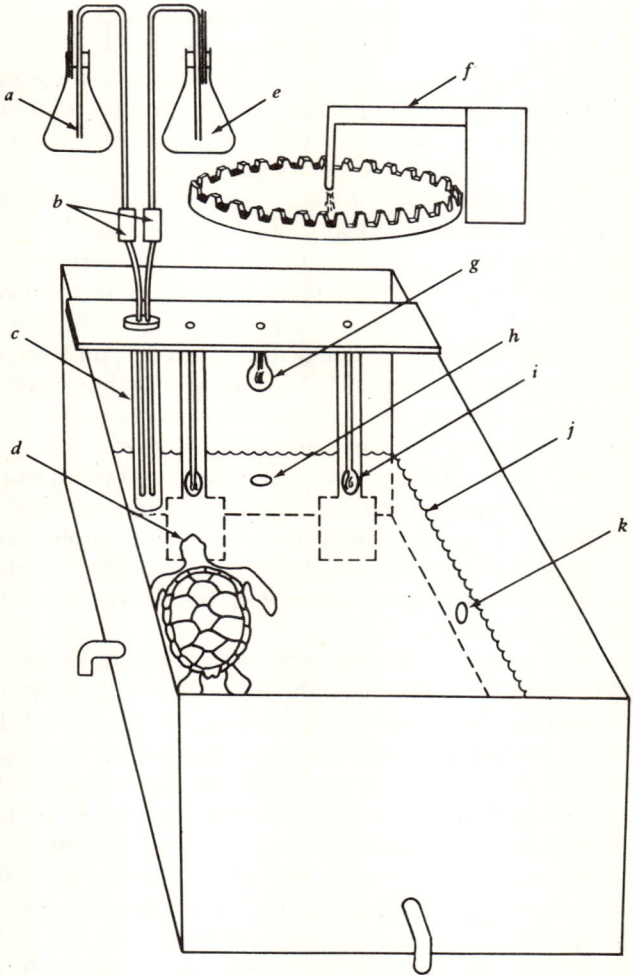

Figure 2 Panel pressing apparatus used by Manton et al. (1972): *a*, chemical or water reservoir; *b*, release valves; *c*, delivery tubes; *d*, turtle pressing left key; *e*, second reservoir; *f*, feeder; *g*, overhead light; *h*, water inlet; *i*, key light; *j*, water level; *k*, water outlet. Reprinted by permission of S. Karger AG, Basel, and M. Manton.

Pert and Gonzales (1974) first trained *C. p. picta* to make five responses to a response key within 1 min before making a final response to a magazine key, which provided the subjects with a small amount of beef paste. Then the responses to the response key were brought under the control of the color of the response key by differentially associating the two colors of the light that illuminated the key with two different amounts of beef paste, obtained when the magazine key was pressed after the appropriate number of responses to the response key. The turtles responded with lower latency when the color of the response key was the one that was associated with the larger reward. Reversing the relationship between key color and magnitude of reward reversed the difference in response latencies.

ISSUES IN INSTRUMENTAL CONDITIONING

Following are several special issues that have traditionally interested psychologists studying learning.

Discrimination Learning

Casteel (1911) was the first to carry out studies of discrimination learning in turtles. *Chrysemys picta marginata* learned to discriminate between black and white panels, vertical and horizontal lines, and vertical lines of varying width, but not between complex visual patterns. His method involved simultaneous presentation of the discriminanda in a modified discrimination box. The subjects received a small amount of meat for approaching the correct, positive stimulus, and on some occasions an electric shock for approaching the incorrect, negative stimulus.

Since this initial study, many investigations have been carried out to determine what stimuli turtles can discriminate. Table 1 summarizes the studies in which, as in Casteel's investigation, the discriminanda were presented simultaneously in different spatial positions.[3] All the studies of this type involved visual stimuli and the usual reinforcer was food. These studies were successful in that the turtles did learn differential responses to the stimuli.

Investigators who were primarily interested in the *detectability* of stimuli have often used a different technique: one in which the stimulus in question is presented or not and positive or negative reinforcers are contingent on whether the subject makes an appropriate response. These studies asked whether certain stimuli could serve as discriminanda for responding. In terms of the distinction between simul-

---

[3]Only studies whose purpose was the investigation of sensory capabilities are included. Other studies have been done in which turtles learned to discriminate between stimuli, but the aim of the investigator focused on issues such as habit reversal. Such studies are reported elsewhere in this chapter. In some cases questions may be raised about stimuli confounded with the cue of principal interest to the researcher in the tabled studies. Space does not permit detailed discussion of the individual studies in this regard.

Table 1 Discrimination Learning with Simultaneous Presentation of Cues

Species	Method or Task	Cues	Incentive	Investigators
Chrysemys picta marginata	Discrimination box	Light intensity, line width, and orientation pattern	Meat; shock for error	Casteel (1911)
Emys orbicularis, Mauremys caspica	Approach to disks held on forks	Light wavelength and intensity	Meat	Wojtusiak (1933)
Geochelone elephantopus	Choice between colored panels	Light wavelength	Banana skin	Quaranta (1952)[a]
Chinemys reevesi, Cyclemys amboinensis, M. caspica	Approach to disks held on forks	Light pattern	Meat	Mylnarski (1952)
E. orbicularis	Approach to disks in and out of water	Horizontal vs. vertical stripes	Not determined	Dudziak (1955)
C. p. marginata	Discrimination box	Light intensity	Access to water; turned over for error	Spigel (1963)
Caretta c. caretta	T-maze	Light wavelength	Fish	Fehring (1972)
Podocnemis unifilis	Discrimination box	Light pattern	Fish	Pritz, Bass, and Northcutt (1973)
Terrapene carolina	Discrimination box	Light area	Reduced temperature	Brosgole (1976)

[a] See also Quaranta and Evans (1949).

taneous and successive presentation of discriminanda, only the early study of Andrews (1915) is a classic example of the successive method. Turtles were first taught to grasp food from a forceps. Then, the investigator shocked the turtles for grasping the food when a whistle was sounding, but not when a bell was sounding. Since her subjects came, for the most part, to grasp the food only when the bell was sounding, Andrews concluded that they could discriminate between some sounds. In a follow-up study Kuroda (1925) was unable to train *Mauremys japonica* to discriminate between the presence and absence of a bell sound, so the basis of the discrimination in Andrew's study is left somewhat in doubt. Her turtles, for example, could have solved the problem on the basis of the whistle alone.

Table 2 summarizes representative studies that involved the presentation of only one stimulus at a time. These studies, in comparison with those of Table 1, more often involved shock, using avoidance learning paradigms, and more simple, reflexive or respondent behaviors.

Each method (i.e., simultaneous vs. successive presentation of stimuli) has its advantages and disadvantages. A problem with the simultaneous method resides in dealing with response biases. Turtles, like other animals, sometimes come into training, or subsequently develop through training, habits of going to one side of the apparatus, and these are extremely hard to break. Many investigators, beginning with Casteel (1911), have commented on the stubborn nature of such biases once they appear. A problem with the successive method of presentation is the occurrence of false positive responses on trials when the positive stimulus is not presented. Manton et al. (1972a) and Pert and Gonzalez (1974), reduced responding during times when the discriminative stimulus was absent by having false responses postpone the occurrence of the discriminative stimulus. The extent to which such measures must be programmed into the procedure and time must be given to the turtle to learn the contingencies lessens some of the advantage associated with this method. Further study of discrimination learning in the successive presentation paradigm might be quite worthwhile, however, since it allows one to observe independently the development of approach responses to the positive stimulus and the inhibition of responses to the negative stimulus.

Not all methods work equally well in all applications. Sokol and Muntz (1966), for example, had success with a key pressing technique (Bitterman, 1964) when they used it to study spectral sensitivity under scotopic conditions, but not under photopic conditions. In that case they used a variant of Wojtusiak's (1933) method which involves locomotion toward one or the other of two simultaneously presented visual stimuli. A bit of

Table 2 Studies of Stimulus Detectability

Species	Method or Task	Cue	Incentive	Investigators
Pseudemys scripta elegans	Discrimination box	Light wavelength	Meat	Armington (1954)
P. s. elegans	Approach to food	Olfactory	Shock for eating in presence of olfactory cue	Boycott and Guillery (1962)
P. s. elegans	Head retraction	Light wavelength	Shock for failing to retract head in presence of cue	Granda, Matsumiya, and Stirling (1965), Zwick and Granda (1968), Maxwell and Granda (1972), Granda, Maxwell, and Zwick (1972)
P. s. elegans	Head retraction	Sound frequency	Shock for failing to retract head in presence of cue	Patterson (1966a, b)
Chrysemys picta picta	Key press	Light wavelength	Fish	Sokol and Muntz (1966)
C. p. belli, P. s. elegans	Discrimination box	Light wavelength	Meat	Muntz and Sokol (1967)
C. p. picta	Key press	Light flicker	Meat	Graf (1967, 1972)
C. p. belli	Four choice panel approach	Light intensity	Meat	Muntz and Northmore (1968)
Chelonia mydas	Key press	Light and olfactory	Meat	Manton, Karr, and Ehrenfeld (1972a, b)

food is hung in front of each display, but if the subject approaches the "incorrect" stimulus, the food is withdrawn.

Investigators of visual discrimination in turtles should also be warned of the possible presence of brightness and color biases in their subjects. Several studies have indicated the presence of a positive phototaxis in hatchling and mature turtles (Noble and Breslau, 1938; Anderson, 1958; Ortleb and Sexton, 1964; Mrosovsky and Boycott, 1966; Mrosovsky and Carr, 1967; Ehrenfeld and Carr, 1967; Ehrenfeld, 1968). It should not be assumed that all species are positively phototactic, nor among those that are, that the behavior occurs under all conditions. For example, Anderson (1958) observed hatchling *Trionyx muticus* burrow into sand and hatchling *Graptemys pulchra* and *G. oculifera* move into shade when released in bright sun light, but when they were released in the evening they moved toward water, presumably a relatively bright area of the environment. Such observations indicate that life history data may be useful in revealing the conditions under which phototaxes may appear. Of course most investigators of visual discrimination give preliminary tests for light biases before beginning discrimination training. Mrosovsky and Boycott (1966), however, suggested that past tests may not have been extensive enough to reveal existing biases. It is possible, for example, that several trials are necessary before turtles begin to attend to the visual stimuli. Furthermore, initial preferences may change as the subjects become more familiar with the procedure and apparatus. A wary animal may initially prefer black, then with increased experience, it may switch to white. Thus an investigator who took the change in behavior to indicate that his experimental manipulation of stimuli was responsible for the response shift to white would be mistaken.

In regard to color biases, a bias to approach short wavelength light has been reported for hatchling green turtles (Mrosovsky and Carr, 1967, and Mrosovsky and Shettleworth, 1968), hatchling *Chrysemys picta*, *Chelydra serpentina*, and *Sternotherus odoratus* (Noble and Breslau, 1938), and hatchling loggerhead turtles (Fehring, 1972). Quaranta and Evans (1949) found an orange preference in *Geochelone elephantopus*. Grant (1960), Eglis (1962), and Auffenberg and Weaver (1969) also report preferences for reds by tortoises, with some seasonal changes.

Complicating the issue, changes in wavelength bias with changes in brightness have also been found. Graf (1972) found that *C. picta* preferred long wavelength light when the stimuli were bright, but not when the stimuli were dim. In one test with *C. serpentina*, Noble and Breslau (1938) found that at high intensities red was preferred, but at low intensities blue was preferred. A possibly related finding is that of Sokol and Muntz (1966), who discovered that under high levels of key (target) illumination, rate of response to the key was greater when longer wavelengths were used for the illumination of the key, but under low levels of key illumination, rate of response to the key was greater when shorter wavelengths were used. It is possible that these color preferences are simply manifestations of a basic positive phototaxis coupled with changes in spectral sensitivity of retinal elements with changes in light levels (Granda, Maxwell, and Zwick, 1972; Chapter 13). Anyone contemplating learning research with colors as discriminanda would be well advised to read Hailman and Jaeger (1971) in addition to readily available basic information on color discrimination.

Motivation

About two-thirds of the investigations have used food as a positive reinforcer or reward. Ground beef has been employed most often, but other foods have been tried, depending on the judgment of the experimenter in regard to the species used. Food is useful for painted turtles and young sea turtles, for they are reliable eaters. Furthermore, access to food can be readily controlled with devices described by Bitterman (1964) and Pert and Bitterman (1969), which also standardize such delivery variables as location, magnitude, and times. On the other hand, we know little about measuring or manipulating hunger in turtles of any species. Thus we do not know what the maximum or the optimum level of hunger might be, nor how to maintain it through an experiment. Most investigators have established hunger by depriving their subjects for 24 hr, but there are no data indicating the species with which 24 hr deprivation is sufficient, how much food an animal on such a deprivation schedule should be given, how often the deprivation can be repeated, and so on. A good deal of parametric research and observation needs to be done.

A number of investigators have experimented with other positive reinforcers. Spigel (1963) and Gonzalez and Bitterman (1962) found immersion in water following water deprivation to be an effective reinforcer for *Chrysemys picta marginata* and *C. p. picta*. Matsumiya and Granda (1964), Van Sommers (1963), and Morlock et al. (1968) found that access to air worked well as a reinforcer for submerged *Pseudemys scripta elegans* and *C. p. picta*. Figure 3 shows the subject's chamber in Matsumiya's apparatus. When the subject pressed the overhead panel, the water level was lowered, permitting the subject to breathe or to get its head above the water, which in itself may be sufficient reinforcement for an animal confined below water. Harless (1970) observed that opportu-

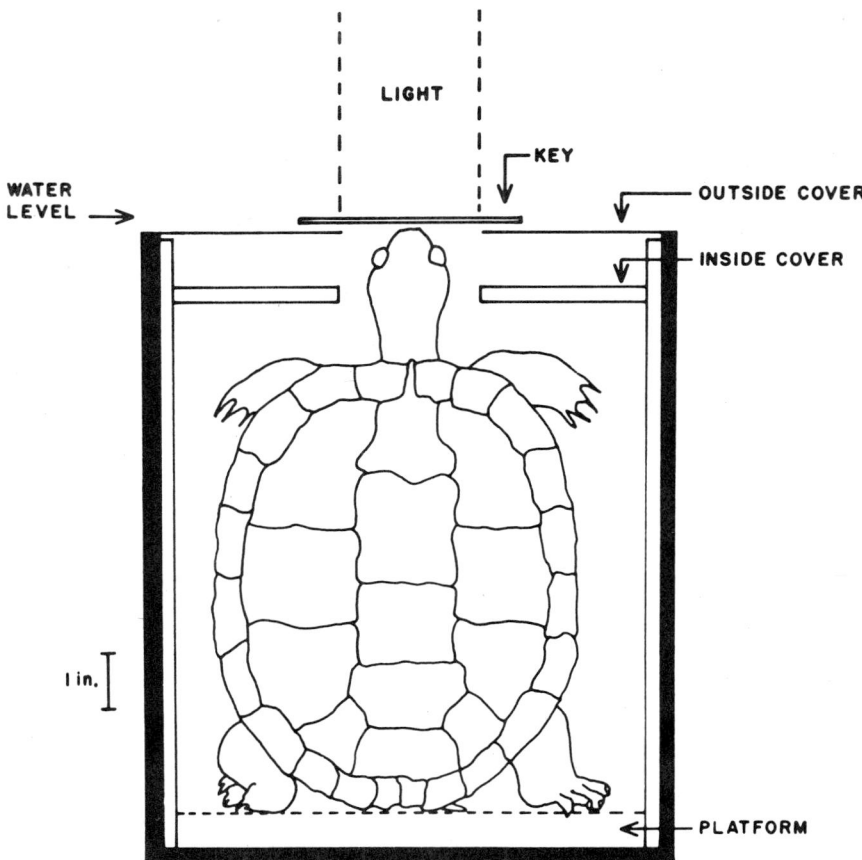

Figure 3 Subject's chamber in apparatus used by Matsumiya and Granda (1964) to train turtles to press an overhead panel for access to air. Courtesy of Y. Matsumiya.

nity to leave a maze and explore the surrounding environment was reinforcing to *Clemmys insculpta*.

Electric shock has been used as a negative reinforcer in a few studies of avoidance and escape training. Mrosovsky (1964) shocked *Pseudemys ornata callirostris* that dove into a white instead of a black tank, and this measure increased the frequency of diving into the black tank. Spigel and Ellis (1965b, 1966) suppressed climbing by shocking *C. p. marginata* when they reared against the side of the small box in which they were confined. Boycott and Guillery (1962) suppressed snapping at bits of food when an olfactory stimulus was presented by shocking subjects for

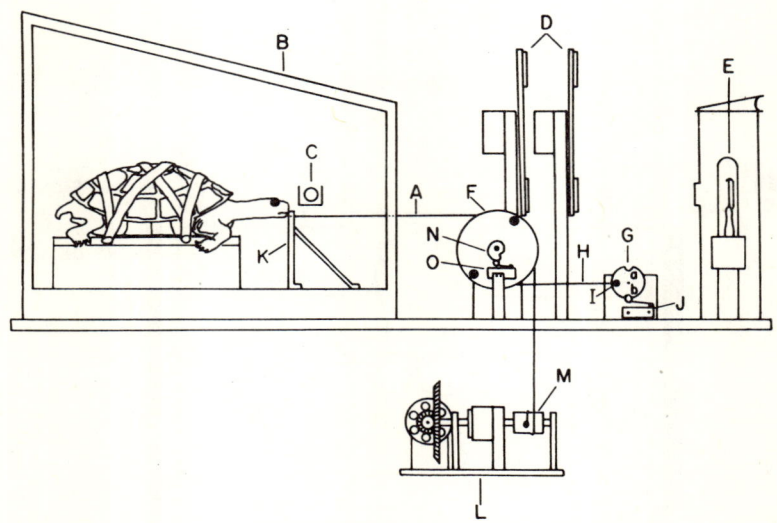

Figure 4 Apparatus used by Maxwell and Granda (1975) in which turtles learned to retract their heads to avoid electric shock to the chin. The apparatus here is patterned after that used by Granda et al. (1965) and is applied in this case to the study of visual thresholds. *A*, leader line; *B*, chamber; *C*, milk-glass target; *D*, filter wheels; *E*, lamp; *F*, pulley; *G*, motorized disk; *H*, line; *I*, swivel; *J*, microswitch; *K*, stop; *L*, microwinch; *M*, takeup spool; *N*, cam; *O*, microswitch. Reprinted by permission of *Physiology and Behavior*.

eating in the presence of the odor. Ellis and Spigel (1966) had some *C. picta* learn a shuttling avoidance response to avoid shock, but apparently (see Chapter 22) these turtles do not acquire such active avoidance responses readily. Granda, Matsumiya, and Sterling (1965) shocked their subjects, *P. scripta elegans*, on the lower jaw whenever they failed to retract their head within 5 sec after a light stimulus was presented; these investigators obtained consistent head retraction responses to the presentation of lights. The apparatus is illustrated in Figure 4.

Aversive stimuli have occasionally been used in combination with positive reinforcers in discrimination learning situations. Casteel (1911) shocked his subjects for incorrect choices and fed them for correct ones. Spigel (1963) turned his *C. picta* over when they committed an error and gave them a 15 sec swim when they made a correct response. There are no data, however, to indicate that either of these combinations is any more effective than any other type of reinforcement; moreover, there is no evidence that such attempted punishment does not interfere with the production of the desired behaviors. For example, Ellis and Barcik (1972) found that inverting subjects produced increases in wall climbing and immobility, and decreases in traversing the maze. Immature *Pseudemys ornata callirostris* that were given a small number of moderately strong electric shocks showed less than normal weight gain, which suggests that shock may have injurious effects (Edwards and Mrosovsky, 1964).

A possibly safer type of punishment is the time-out procedure used by Graf (1967), Manton et al. (1972), and Pert and Gonzales (1972). With this procedure incorrect responses result in a forced delay (time-out) before a correct response can be made.

Another type of mild punishment that has been used in some maze research with turtles is confinement within the incorrect goal box when the wrong alley is chosen. Graf and Tighe (1971), for example, used a 10 sec confinement following incorrect choices. Such periods of confinement should be short (but how short remains to be determined), to avoid the introduction of interfering, climbing-escape responses which, once begun, may appear on subsequent trials and disrupt the desired behavior (see Spigel, 1964).

Research has not yet indicated how motivation for any species may be best controlled and manipulated. To facilitate progress it might be useful if investigators would publish measures of the effectiveness of their motivational conditions. Although some measures may be inconvenient or impossible to use in some studies, there are simple measures, such as latency to eat or drink (Bolles, 1962), which can be easily obtained even during the course of a study, without producing much, if any, interference.

Amount of Reward

Three studies have been reported in which amount of reward has been manipulated: Pert and Bitterman (1970), Pert and Gonzalez (1974), and Wolach and Latta (1974). In general, responses are faster for larger amounts of reward, and response times change accordingly when reward magnitudes are changed.

Reversal and Probability Learning

In the standard reversal study, the initial response-reinforcement contingency is reversed after the subject has attained a criterion performance or after a predetermined number of trials. This is repeated many times, and the subjects may come to eliminate errors faster and faster over the reversals. Many animals are capable of doing this (Warren, 1973) with various discriminative stimuli.

Seidman (1949) showed progressive improvement in reversal learning of *Pseudemys scripta elegans* in a T-maze. But in Siedman's apparatus it was possible for the subjects to see which response would be reinforced from the choice point, and some investigators (Holmes and Bitterman, 1966) have accordingly questioned the extent to which learning was actually demonstrated.

Kirk and Bitterman (1963) reinvestigated the problem with *C. p. picta*, using a maze free of the problem just mentioned. The subjects were trained with a correction procedure: if the subject made an incorrect choice, it was permitted to retrace its path and to enter the correct arm. In terms of first or initial choice on a trial, the subjects showed evidence of progressive improvement when the reversals were carried out daily, regardless of their performance on the previous day; but there was no evidence of improvement when the subjects were required to learn each problem to a certain criterion. Why this difference in arranging the reversals should make a difference is not known, but it indicates the need to take the specific conditions of training into account when the behavior is interpreted.

Holmes and Bitterman (1966) used the two-choice key pressing apparatus described by Bitterman (1964) to study reversal learning in visual and spatial situations. In the visual tasks the keys were illuminated differentially by red and green bulbs. In the spatial tasks both keys were the same color. Clear progressive improvement was seen in all the spatial and visual tasks except one: a visual task in which the contingency was reversed every 4 days. This peculiar difference was replicated, but the reasons for it were not clear. Thus turtles seem to be capable of progressive improvement in habit reversal, but task-related variables enter into the determination of the performance in as yet mysterious ways.

Graf and Tighe (1971) compared performance following a reversal shift (RS) with performance following an extradimensional shift (EDS) in the cue-reward contingency. The already described reversal shift involves changing the cue-reward contingency so that the negative stimulus now becomes the positive stimulus, and vice versa. The stimulus dimension that was relevant before the shift remains relevant. An EDS also involves changing the cue reward relationship, but now a previously irrelevant dimension becomes relevant, with one value being associated with reward and the other with nonreward (e.g., a shift from red + and green − to bright + and dim −). The subjects required fewer trials to learn the EDS than the RS, and comparing the dimensions, brightness discriminations were learned more readily than hue discriminations. But as Graf and Tighe point out, it should be understood that in switching from one dimension to another, one of the combinations of hue and brightness was previously reinforced so that the subject had only to learn one new subproblem in the EDS, whereas in the RS he had to learn two new subproblems. Subproblem analysis indicated that the subjects required as many trials to master the new problem in the EDS as they did in mastering each of the new subproblems in the RS. Graf and Tighe interpreted the overall difference in ease of learning between EDS and RS as the result of the subjects' having learned each of the two pairs separately; that is, in the case of hue relevant, the subjects learned to approach red-bright when it was presented in combination with green-dim and separately learned to approach red-dim when it was presented in combination with green-bright. Thus the subjects did not learn to approach the more red stimulus (i.e., they were not responding during the preshift phase on a relational or dimensional basis but on an absolute basis), which is consistent with the previous results on transposition tests conducted by Kuroda (1933) and Spigel (1963).

In the only study of probability learning reported for turtles Kirk and Bitterman (1965) found a difference in performance among *C. p. picta* depending on the cues. When visual cues were available the subjects seemed to employ a "matching" strategy in which the proportionate distribution of their choices matched the probability of payoff for those choices. With spatial cues, the subjects used a maximizing or other nonrandom strategy, such as reward following.

Extinction

Extinction, the reduction in response probability that follows the termination of reinforcement for responding, has been of considerable interest, but apparently not to students of turtle learning, who have, understandably, been more concerned with getting their subjects to do something than with stopping them from doing it. Most of the interest in extinction of turtle behavior has centered on the effect of partial reinforcement during acquisition; however there is a little work on other issues and some scattered observations.

Pert and Bitterman (1970) trained two groups of *C. p. picta* to traverse a runway for a large or a small reward. When both groups were subsequently given 30 extinction trials, both showed an expected large, progressive decrement in running speed. Contrary to responding in rats, responses of the large reward group did not extinguish faster than those of the low reward group. Pert and Gonzalez (1974) also failed to find a large reward–extinction effect; a large reward group responded more rapidly in extinction than a small reward group, but the rate of extinction was about the same for both.

Ellis and Barcik (1972) reported that following the termination of reward for entering one of the arms of a T-maze, the turtles came to increase their frequency of entering the previously incorrect arm over the extinction trials. These results for appetitive tasks may differ, however, from those of aversive conditioning situations, in which extinction has often been found to be rather slow. Granda et al. (1965) and Patterson and Gulick (1966), who used basically the same avoidance task as Granda et al., found the responses of their *P. s. elegans* to be extremely resistant to extinction.

Responses of laboratory rats that are reinforced on a random 50% of the acquisition trials show a greater resistance to extinction than rats reinforced on every one of the acquisition trials. This greater resistance to extinction of responses of the partially reinforced subjects is referred to as the partial reinforcement effect (PRE) and has been studied in turtles.

Eskin and Bitterman (1961) trained young *P. s. elegans* to traverse a runway. During extinction the continuous and partial reinforcement groups did not differ in running speed nor in the number of times the subjects of the two groups failed to reach the goal box at all. Subsequent studies with *C. p. picta* in runways by Gonzalez and Bitterman (1962) and Pert and Bitterman (1970), also failed to find the usual PRE.

In contrast to these studies are two investigations using *P. s. troostii* in which rather clear PREs appeared. Murillo, Dierks, and Capaldi (1961) again used a runway situation and obtained a clear PRE, although it did not appear at the outset of the extinction trials. Unlike the other studies, Wise and Gallagher (1964) used a choice technique and obtained a clear PRE in terms of the number of choices in the direction of the previously rewarded alternative. Their discussion of the literature enumerated several differences among the studies reported up to that time.

DISCUSSION

Our knowledge of learning in turtles is still fragmentary, but it is encouraging to note the increasing sophistication of the recent techniques and the refinement of the questions current researchers are asking. We are now well past the time of asking whether turtles can learn anything and are at the point of asking about the specific conditions under which learning takes place and the specific ways behaviors change.

We urgently need work on constraints on learning. What do turtles learn especially rapidly and what do they learn painfully slowly? A

DISCUSSION

related question concerns the relative facility with which associations can be formed to different sensory dimensions. Several informative reviews and discussions are available (Rozin and Kalat, 1971; Shettleworth, 1972; Bolles, 1973; Garcia, Clarke, and Hankins, 1973).

Additional work on the importance of particular conditions of training is also needed, along with more intensive analysis of the specific changes in behavior brought about by them. More attention to individuals should be given with within-subject changes given as much importance as between-subject changes. It may be time to stop discarding the stubborn subjects and to pay more attention to the conditions that have brought about their behaviors.

We also need naturalistic observations directed toward learning. We need to know the problems the different species face, such as homing, learning what to eat and what to avoid, and the nature of the sensorimotor coordinations that must be learned. Students of learning should find existing accounts such as those summarized by Carr (1952), Pritchard (1967), and Ernst and Barbour (1972) valuable. It goes without saying that our learning investigations would be more fruitful in regard to application to the field if we knew what demands are placed on our subjects in their native habitat.

It is hoped that laboratory investigators of learning in turtles will feel a kinship to workers in other areas and will find some usefulness in the findings of their partners. In particular, studies of visual processes should be of value to those using visual discrimination methods, and seasonal and maturational changes in feeding habits along with food preferences should be of interest, since food is so often used as a reward.

Many areas of inquiry have scarcely been touched, but should be fruitful. There is little work on memory and imprinting, which may be important in regard to long-term dietary preferences and the selection of nesting sites (Chapters 12 and 15). Very little attention has been paid to processes involved in inhibitory control of behavior. It might be especially interesting to combine neurophysiological research with behavioral research as Spigel and Ellis (1966) have reported (see Chapter 22). Given the retiring nature of many turtles, it is surprising that more studies of passive avoidance have not been reported. Certainly many species, such as the box turtles, have exploited this alternative to active avoidance in response to danger. Comparative work might indicate differences in avoidance behaviors, which are related to body structures and habitats of various taxonomic groups.

In general the many adaptations seen in turtles provide fruitful opportunities for analysis of the processes by which behavior changes to promote survival. There is much to do, and for many species the time may be short.

ACKNOWLEDGMENTS

I am grateful to F. W. Irwin, M. E. Bitterman, N. Mrosovsky, I. M. Spigel, and D. Ehrenfeld for their helpful comments on an earlier draft of this chapter.

CHAPTER 21
Social Behavior

MARION HARLESS

Generally speaking, social responses have been described in terms of interactions during emergence from the nest, feeding, basking, moving (including migrations, emigrations, and foraging), shelter-associated activities (including estivation and brumation or "hibernation"), and mating—essentially in all aspects of turtles' lives. As in reports of social facilitation at hatching (Chapter 18), a narrow segment of the population may be involved. More frequently a range of ages and both sexes are represented in a given social activity. Agonistic behavior occurs with most of these activities, and there are reports of aggregations for all the activities above (plus nesting; see Chapter 18), though not all of these necessarily occur for a given species.

Some responses frequently included as social behavior have topographies such that if a stimulus other than a turtle were involved, the behavior would not be labeled social (e.g., climbing on one another when basking). Other responses, such as "titillation" by males of some species, appear to be restricted to clearly social situations.

By 1970 more than a hundred observations of social interactions had been recorded in scientific literature; primarily they were, and continue to be, descriptions of responses for small numbers of adult turtles. Little attention has been paid to specific stimuli that set the occasion for, elicit, or reinforce responses; the same is true of physiological correlates, ontogenetic development, and speculations or hypotheses regarding evolutionary development and comparative relationships. At this point in turtle behavior analysis, it does not seem to matter whether social behaviors are discussed in terms of fixed action patterns, respondents and operants, and so on, as long as the responses and their antecedents and consequences are accurately described. Videotapes and still and motion photographs aid immeasurably in description and analysis (Chapter 5).

STIMULI INVOLVED IN SOCIAL INTERACTION

Stimuli Involved in Setting the Occasion for Social Behavior

Here we are primarily concerned with aspects of the physical setting. Under what conditions are turtles likely to encounter other turtles? Are there conditions that increase or decrease the probability of social responding? To answer these questions, field-collected life history data of daily and seasonal movements and correlated physical stimuli are needed. There are probably more field data for the box turtles (e.g., Nichols, 1939; Minton, 1944; Stickel, 1950; Carpenter, 1957; Legler, 1960a; Williams, 1961; Webb, Minckley, and Craddock, 1963; Dolbeer, 1969; Metcalf and Metcalf, 1970; Reagan, 1974; Brown, 1974; Schwartz and Schwartz, 1974; Yahner, 1974) than for any other group, but no one has yet suggested that we have sufficient information to understand their behavior. The physical variables discussed next are virtually inseparable in the natural environment and remain little studied in laboratory settings where the stimuli can be independently manipulated.

Light, Temperature, and Moisture. For some species, daily activity begins at dawn (see Chapters 23–26 for species information). Although the onset of light is perhaps sufficient to elicit movement, it does not necessarily bring turtles together. Foraging turtles are, of course, likely to encounter other turtles. Later in the day, patches of sunlight or sunlight in combination with features such as rocks or partially submerged logs may provide basking areas. For tortoises especially, shade may be a feature that brings the animals together (e.g., *Geochelone gigantea*, Bourn, 1976). Decreasing daylight is accompanied by another bout of activity for many species as they forage and move to sleeping places. Seasonal changes in light may determine whether animals move into or out of sunshine or shade. Seasonal changes in the amount of light also probably play a role in determining reproductive readiness and subsequent mating behavior, as do seasonal changes in moisture and temperature (see Chapter 16).

When there is an adequate amount of something, there is no advantage in fighting for first access. Inadequate amounts should lead to behavioral adaptations that ensure that sufficient numbers of the animals gain access to the resource and stay healthy enough to maintain the population in good condition. Bury and Wolfheim (1973) described open mouth threats of Pacific pond turtles in gaining and maintaining well-spaced basking sites. This agonistic behavior of *Clemmys marmorata* is in marked contrast to intra- and interspecies conviviality of most basking species. *C. marmorata* is a west coast North American species found even at high elevations, and the threat behavior may have evolved as a response to inadequate amounts of sunshine or warmth. At this time, however, there are insufficient data to state that the behavior is more common at basking time, that it occurs throughout the species' range, or that it occurs throughout the year. *C. marmorata* may travel some distance to lay eggs in sunny spots (Storer, 1930), but social behavior has not been recorded at these sites.

Decreasing temperatures are important in bringing about brumation, and aggregations may be found in choice spots (see Chapter 11). High temperatures and lack of moisture may bring about even larger aggregations during summer months. Numerous kinds of information have been collected at these sites, but there are no data on behavioral interactions per se.

In addition to aggregations in moist places during dry weather (e.g., *Terrapene carolina*, Minton, 1944), moisture in the form of rainfall may play a role in increased activity and movement (e.g., Gibbons, 1970). Ornate box turtles in field and captive settings are highly active just before, during, and after rainfall; in captivity the time after a rainfall in warm weather is marked by heightened social activity, probably because all the turtles are active at once.

To observe interactions in laboratories, investigators frequently choose times convenient to their own schedules; the ambient temperature and humidity are generally the same in turtle housing and test areas, but light is frequently brighter in the latter. Social responding might occur more reliably if investigators would take advantage of the physical stimuli that set the occasion for increased social activity in the wild. Weaver (1970) noted that *Gopherus polyphemus* are more active in the field after a rain and sprinkled captives before pairing them for observations. Boice (1970a) similarly sprinkled the entire living area of captive box turtles before each observation period.

Topography and Terrain. Topographic features may bring turtles together; in some cases, increased humidity, shade, or light is probably responsible. Allard (1948) and Knowlton (1943), for example, described captive turtles concentrated in areas of decaying vegetation. Stickel

(1950), Carpenter (1957), Metcalf and Metcalf (1970), and Schwartz and Schwartz (1974) identified areas with abundant leaf cover and humus as determinants in locating box turtles in the wild; Stickel commonly found six to ten turtles in a particular thicket. Auffenberg (1969b) found that brush piles and a log served as aggregation points for captive *Geochelone denticulata*.

Man-made features may alter natural environments in such a way that turtles meet more often than they would otherwise. Metcalf and Metcalf (1970) commented on box turtles traveling along fence lines, and Boice (personal communication) noted encounters in such areas. Ravines and rock outcrops similarly channel animals along more confined routes where chances of meeting are enhanced.

Turtles follow animal trails, including trails made by the turtles themselves (Stickel, 1950). In the case of *Geochelone elephantopus*, the tortoises drastically altered the environment in the Galapagos Islands. Worn single-file passageways forced a literal straight line hierarchy in which no tortoise passed the animal in front of it during seasonal movements (Van Denburgh, 1914; Townsend, 1925). Zoo research with large tortoises yielded rather consistent data on order of entering and leaving a shelter (Evans and Quaranta, 1951; Quaranta and Evans, 1949a). Orderly behavior would appear to be an energy-conserving adaptation in these large herd-type herbivores. It seems that such behavior would be learned quickly if the outcome were access to abundant sunshine, shade, plant food, quiet lagoon waters or other nonmoving reinforcing stimuli.

Food. Turtles pass much of their active time in foraging. Vegetarian species of sea turtles are commonly found in feeding aggregations, but the stimuli that guide the turtles to the feeding grounds are no more evident than the stimuli associated with nesting beaches (see Chapters 15, 18, 24; Skinner, 1975).

Turtles often described as solitary may also aggregate where food is abundant; for example, Dolbeer (1969) described a number of *Terrapene carolina* feeding on grapes. Freshwater species may gather to feed on garbage (Carr, 1952) or carrion (Hatt, 1932). Whether the animals merely encounter the food while foraging or whether the approach responses are elicited by olfactory stimuli, turtles learn quickly to return to a site with abundant food (Metcalf and Metcalf, 1970).

There is usually no social conflict in feeding aggregations. Legler (1960a) reported two *T. ornata* feeding on insects in the same pile of cow dung. Perhaps there is absolutely no attention paid to conspecifics, or perhaps the animals are all already familiar with each other. Species, amount of food present, and amount of food deprivation probably determine whether agonistic behaviors occur.

There are comments in the literature of turtles chasing other turtles, especially among captive specimens, to procure food the first turtle has. Some turtles will definitely take food and move away from other turtles, but it is probably more accurate to describe the following by other turtles as "food chasing" and not "turtle chasing." Whether the movement itself or enhanced odor is responsible, moving food is chosen over food items of the same kind that remain stationary. Captive individuals of *Clemmys insculpta* and *Terrapene* species frequently bite at each other's head or forefeet in a back-and-forth fashion to procure "crumbs" until all the food particles are gone. There may be much head and limb retraction and many biting responses, but these feeding episodes rarely change into agonistic behavior once the food is gone.

Hormonal Influence

External variables play roles in reproductive physiology (Chapter 16), but specific relationships between the variables, hormones, and social behaviors have been little studied. Evans (1940a, 1940b, 1946, 1951a, 1951b, 1952a, 1952b, 1960, 1963) has completed virtually all the published research and has determined that hormonal injections enhance secondary sex characteristics, general activity, and agonistic and sexual behaviors in immature as well as adult turtles. Weaver (1970) found external glandular changes and a generalized pushing response in hormonally injected adult male *Gopherus berlandieri*. In field settings Stickel (1950) and Schwartz and Schwartz (1974) found more box turtles in close proximity to each other during October than during other months when they were active. Although three different species of *Terrapene* are involved, the reproductive physiology data of Legler (1960a), Hansen (1938), and Hansen and Riley (1941) support the idea of hormonal involvement in the increased box turtle interactions in the natural environment during the early fall.

Stimuli Involved in Eliciting Social Behavior

In laboratory situations, shock may be used to elicit escape and agonistic behaviors (Fraser and Spigel, 1971; Chapter 22), but typically the stimuli that elicit social responses originate with a conspecific turtle.

Visual Stimuli. At least some of the morphological differences that humans use in distinguishing species must be visual stimuli used by turtles themselves in species discrimination. The various size, shape, and color differences found in secondary sex characteristics (Chapter 4) may be important in sex identification within a species. Postural and other behavioral differences are also likely to be perceived by turtles. These aspects can be manipulated and investigated in field and laboratory settings, but the studies are few in number and we remain essentially ignorant of the role of visual stimuli.

Auffenberg stated that head movements are important visual stimuli in *Geochelone carbonaria* and *G. denticulata* (1965), but not in *G. travancorica* (1964). Both sexes of *G. travancorica* undergo color changes of the head during the breeding season, but the head color apparently does not serve as a visual stimulus.

Coloration of nonshell body areas varies among individual *Clemmys insculpta*; the color also varies in individuals of both sexes over time, but the roles of light, hormonal influence, diet, and so on have not yet been established. Dominant males and females are frequently more brightly colored than their subordinates. In this species the male pulsates his salmon to bright orange throat while mounted on the female, but it is not certain how distinctly the female can see the pulsating gular area because her head is retracted during mating (Harless, 1970b).

Evans (1960, 1968) considered the pulsating orange throat of *Terrapene carolina triunguis* to be an eliciting stimulus for approach behavior by the female. The other subspecies of *T. carolina* have yellow throats, and although pulsating occurs, biting and pushing play larger roles in premounting behavior (Evans, 1968).

Bury and Wolfheim (1973) described the open-mouth responses of *Clemmys marmorata*, which has yellow edging of the jaw and bright pink buccal cavity and tongue. In many cases the open-mouth response was followed by an escape response of a nearby turtle. Evans (1961a) described the white jaw edging in *C. insculpta* as a releaser with effects on withdrawal or escape in conspecifics. The open-mouth response necessarily precedes biting, but the duration of the open-mouth response may be very brief and viewed as part of biting at or biting, or it may be quite long and viewed as an "intention movement" or a "threat." Similarly, the magnitude of the response may vary from a slight dropping of the lower jaw to the fully open mouth. Although the lighter edgings of the jaw may be important in this and other genera, it is not clear that they are eliciting stimuli (or "releasers") because the responses that follow are not completely uniform. Captive animals of both species show both escape and avoidance behavior without the open-mouth response occurring in another individual, indicating that there are additional controlling stimuli.

Accounts in the popular press tell of green turtles mounting or attempting to mount scuba divers, simple rounded decoys used by turtle fishermen, and other animate and inanimate portions of the environment.

Although there are now many observations on behavior in this species (see Table 1), it is not clear what specific visual stimuli elicit approach behavior in male *Chelonia mydas*. Surely at least the upper and lower limits of "large object" could easily be determined with decoys or models.

Evans (1956b) introduced a large preserved conspecific into an established group of seven *Terrapene carolina* and noted that the dominant male retreated and a smaller male touched snouts with the model. In research on visual stimuli and approach behavior, Weaver (1970) tested individual *Gopherus berlandieri* males and females with male and female preserved *G. berlandieri* displayed in "lifelike" stances in a sealed aquarium and also with a live tortoise in the aquarium. Males tended to approach all stimuli, including the empty aquarium used as a control, with about the same frequency. Females tended to approach the moving tortoises more than the control.

Placing a mirror in a turtle's environment can provide data on approach (Davis and Jackson, 1973), head extending, bobbing, biting, and similar behaviors (reviewed for other vertebrates by Gallup, 1968). Evans (1956b) found that male *T. carolina* attacked their images, whereas females looked at their images and walked off. Unpublished observations on *C. insculpta*, which had a large mirrored surface placed in their extensive living area, include the following. The dominant male fought with his image almost constantly from dawn to dusk for 3 days and at the end of 2 weeks still made occasional responses to his image. Other adult males engaged in agonistic behavior to a lesser degree, and the two female and two immature turtles exhibited initial head extending and nothing more. Mirror image work, free of confounding olfactory cues, is especially useful when determining response categories.

Olfactory Stimuli. Auffenberg (1965) considered olfactory discrimination to be more primitive than visual discrimination in tortoise social interactions, and although comparative and evolutionary data are absent, this conclusion seems to be applicable for all turtle groups.

Olfactory stimuli may remain in the environment long after the turtle itself has moved, and they probably have more immediate widespread effects on entire populations, whereas visual stimuli may play a greater role in interactions between individuals and subsequent long-term population effects.

Nichols (1957) noted that captive *Gopherus agassizi* found cloacal excretions aversive and avoided areas with fresh deposits. Patterson (1971a) altered aggregation and sleeping behavior by adding large amounts of collected fresh feces to the usual sleeping areas of captive *G. agassizi*. When the odor diminished, the tortoises returned. Woodbury and

Hardy (1948) had observed that in free-ranging *G. agassizi*, the larger specimens entered the communal winter dens later than the smaller individuals. Patterson (1971a) suggested that the cloacal odors of excreta of dominant males are especially aversive. When the larger and presumably more dominant animals enter the den last, their cloacal odors do not prevent smaller tortoises from entering the brumation area.

Evans and Quaranta (1951) noted that most of the tortoises in their zoo study eliminated in the morning before leaving the indoor shelter and that excreta may have played a role in emergence from the shelter each morning. If the odors are aversive, in a natural environment dispersal would result. In a zoo, of course, the tortoises could not disperse and returned each evening to the attendant-cleaned area. There are numerous field data of box turtles returning or not returning to the "forms" used for sleep and rest. There are apparently no data on fresh excreta as determinants of form use on subsequent nights or by other box turtles.

Evans suggested that social "imprinting" occurs and cited olfactory stimuli as being important. He originally (1958) found that young *Chelydra serpentina* remained in closer proximity to the tank partner of 15 months than turtles they had been with for an intervening period. Thirty sibling pairs of young *Chrysemys picta* also spent more time near each other than did nonfamiliar animals observed as pairs (Evans, Preston, and Hardy, 1973). Pairs of *C. serpentina* housed together for 9 months and pairs of *Pseudemys scripta* that were maintained for 5 months in separate containers with water exchanged between pair members daily also spent more time in close proximity during extended observations than did nonpartners (Evans, Preston, and McGeary, 1974).

Hatchling turtles are normally not in the presence of their siblings in the limited space or for the extended times of Evans's laboratory work, but early learning about olfactory stimuli may occur. The strong odors of young turtles commented on by Neill (1948a, 1948b), for example, may play some role in the development of perception of more subtle odors by adults. It is doubtful whether any olfactory presence of cloacal fluids released by females during deposition of eggs remains at hatching, but the possibility of this cue for species familiarization is certainly testable.

It seems likely that at least some odors serve as ever-present species or even population markers. In sea turtle populations that come together yet maintain their integrity (Pritchard, 1973a), permanent mingling may be prevented by the unfamiliar odors of turtles from the respective populations. (See Marler, 1976, for a recent review of familiarity.)

Odor production and discrimination undoubtedly play a major role in sex and species identification. The type of research carried out by

Weaver (1970) in determining the role of olfactory stimuli in adult social interaction is needed for other species; see Manton (Chapter 15) for some details and for additional information on olfactory behavior.

Tactile Stimuli. While visual and olfactory stimuli are important in initial approach and discrimination, tactile stimuli probably have increasing significance in final stages of agonistic and mating behavior. In keeping with general analyses of behavior (e.g., Breland and Breland, 1966), it is likely that as more species are studied, the eliciting stimuli for tail relaxation (and eversion of the cloaca where it occurs), intromission, and the final stages of copulation will be found to be similar for all species, whereas eliciting and discriminative stimuli successively earlier in mating and agonistic behavior will be increasingly dissimilar across species.

Evans (1953) described virtually all the "releasers" in matings of *T. carolina* as tactile. Other descriptions of courtship and mating (see Table 1) point to increased tactile stimulation in later stages of mating.

Tactile stimuli include the actual contact or vibration of touching, stroking, titillating or drumming feet or tail of another animal, bumping or ramming by head or shell (from any direction), and bites. Species that have sexually dimorphic gular plates, claws, or leg or tail scales seem to be clear candidates for research on tactile stimuli. Males of some species that mate in the water "nibble" at the water (e.g., *Pseudemys scripta elegans*, Jackson and Davis, 1972b) and may "pump" the water at the female (e.g., *Clemmys insculpta*, Evans, 1967). Agitation of the water caused by these responses may provide additional tactile stimulation to the female.

Auditory Stimuli. Turtles are able to hear (see, e.g., Wever and Vernon, 1956a, 1956b; Patterson, 1966; Patterson and Gulick, 1966), and notes about sound production exist for freshwater (e.g., Allen, 1950), sea (e.g., Carr, 1952), and land (e.g., DeSola, 1930) species. Tortoises seem to be most vocal, and auditory stimuli may play some role in their mating behavior (see Weaver, 1970; Campbell and Evans, 1967, 1972, for literature reviews). It seems that most vocalizations (from "teakettle whistles" to "chirps" to "groans") simply occur in stressful situations and may or may not be eliciting or discriminative stimuli for other turtles.

SOCIAL RESPONDING

Traditional concepts in social behavior include territoriality, dominance hierarchies, aggregations, and mating. Mating (Table 1) and aggrega-

Table 1 Representative Reports of Courtship and/or Mating

Family and Species	Reference
Chelydridae	
Chelydra serpentina	Conant (1938), Hamilton (1940), Hammer (1969), Legler (1955), Pell (1941), Taylor (1933)
Kinosternidae	
Kinosternon flavescens	Mahmoud (1967), Taylor (1933)
K. subrubrum	Carr (1952), Mahmoud (1967)
Sternotherus carinatus	Mahmoud (1967)
S. odoratus	Lagler (1941)
Testudinidae	
Homopus areolatus	Eglis (1962)
Chersina angulata	Rose (1950)
Malacochersus tornieri	Loveridge and Williams (1957)
Geochelone carbonaria	Auffenberg (1965)
G. denticulata	Auffenberg (1965), Beltz (1954)
G. pardalis	Leakey (1948)
G. gigantea	Gaymer (1968)
G. sulcata	Cloudsley-Thompson (1970), Grubb (1971)
G. travancorica	Auffenberg (1964b)
Gopherus agassizi	Nichols (1953, 1957), Stuart (1954), Tomko (1972)
G. berlandieri	Hamilton (1944), Weaver (1970)
G. flavomarginatus	Legler and Webb (1961)
G. polyphemus	Auffenberg (1966b)
Testudo graeca	Noel-Hume and Noel-Hume (1954), Watson (1962)
Emydidae	
Pseudemys concinna	Jackson and Davis (1972a), Marchand (1944)
P. scripta	Cagle (1950, 1955), Conant (1938), Davis and Jackson (1970), Ernst and Barbour (1972), Grant (1936), Jackson and Davis (1972b), Taylor (1933)
Chrysemys picta	Ernst (1971), Gibbons (1968), Taylor (1933)
Graptemys barbouri	Wahlquist (1970)
G. flavimaculata	Wahlquist (1970)
G. kohni	Ernst and Barbour (1972)
G. pseudogeographica	Ernst and Barbour (1972)
Emydoidea blandingi	Carr (1952), Conant (1951a), Richmond (1970)
Clemmys guttata	Carr (1952), Ernst (1967, 1970c), Finneran (1948), Mertens (1960)
C. insculpta	Carr (1952), Evans (1961a, 1967), Fisher (1945), Harless (1970b), Knowlton (1943), Wright (1918)

Table 1 (*Continued*)

Family and Species	Reference
C. muhlenbergi	Barton and Price (1955)
Terrapene carolina	Allard (1949), Cahn and Conder (1932), Dolbeer (1969), Evans (1953, 1956a, 1960, 1968), Ewing (1935), Knowlton (1943), Penn and Potthurst (1940), Rosenberger (1936), Schwartz and Schwartz (1974)
T. ornata	Brumwell (1940), Legler (1960a), Norris and Zweifel (1950)
Rhinoclemys annulata	Mittermeier (1971)
Mauremys caspica	Eglis (1962)
Trionychidae	
Trionyx muticus	Legler (1955), Plummer (1976a)
T. sinensis	Mitsukuru (1905)
Cheloniidae	
Chelonia mydas	Booth and Peters (1972), Bustard (1973a), Carr (1952), Carr and Giovannoli (1957), Frazier (1971), Harrison (1954), Hendrickson (1958), Whitham (1970)
Eretmochelys imbricata	Hornell (1927)
Caretta caretta	Wood (1953)
Lepidochelys kempi	Carr (1967), Chavez, Contreras, and Hernandez (1968)

tions certainly occur in natural settings, but, once again, we have few data to determine the stimuli that bring or keep turtles together.

Cagle (1944b), Lardie (1964), and Pell (1941) described what might be territories for some freshwater turtles, but field researchers by and large conclude that although definite home ranges exist for many species, territories do not (Stickel, 1950; Legler, 1960a). Reports of agonistic responding in the natural environment do not indicate whether the behavior occurs at the edges of home ranges. It is probable that turtles do not "defend" or "patrol" boundaries of home ranges, but for most species there may be dominance relationships among turtles that have overlapping home ranges. In some species the dominance relationships may hold only for a segment of the year, such as the breeding season. To date there are no published studies of year-around dominance relationships under relatively natural conditions, although there are reports of persistent hierarchies in captive animals.

Dominance Hierarchies

In addition to dominance recorded in earlier mentioned studies by Evans (e.g., 1940b), Quaranta and Evans (1949a) found dominance in eating to be related to shelter-associated behaviors in zoo-maintained tortoises. Weaver (1970) tested captive adult male tortoises in pairs following field observations of agonistic behavior in tortoises; he described dominant and subordinate postures and stated that a dominant tortoise "inhibits feeding" in a subordinate, but presented no data on a hierarchy.

A total of 2280 1 min food acquisition trials were completed for snapping turtles by Froese and Burghardt (1974). Of 36 possible pairs, 17 were observed until they had significant "win-loss" ratios; of these, 15 were in the order of "dominance" in total "win-losses."

Ornate box turtles with the shortest latencies to eat meat and banana also were the fastest and most active on five other measures (Harless and Lambiotte, 1971). Not designed as a test of dominance, the study indicated stability of rankings of activity in each of eight trios.

Boice (1970a) found that food acquisition data could not be arranged in any reliable hierarchy in two groups of *Terrapene c. carolina* studied over a 15 week period. Agonistic behaviors during food competition were variable for the first 5 weeks, but beyond that time a hierarchy was evident for each group.

A subsequent study (Boice, Quanty, and Williams, 1974) with three five-turtle groups of *T. c. triunguis* over 14 months showed similar results. Boice (1970b) and Froese and Burghardt (1974) stated that similar unpublished findings hold, respectively, for *T. ornata* and *T. c. carolina*.

Two groups of wood turtles (*Clemmys insculpta*) were studied for 6 months to determine whether dominance hierarchies exist in captivity for this species (Harless, 1970b). One group of 12 was housed in individual cages. All pair combinations were observed for 15 minute periods in water, in a small enclosure, and in the same enclosure with a single piece of banana. The turtles were then placed in a large enclosure for further observations.

The second group was initially placed in a large enclosure; 270 quarter-hour formal observations were carried out over 90 days, at which time pair comparisons in the three settings were begun, with each pair returned to the group enclosure following each test. Observations continued on group behavior. For these turtles, most of the agonistic responding was over by the end of 3 hr and virtually none occurred after the fifth day. Their social responses were fewer and less agonistic when they met in pair comparison tests and reflected the dominance hierarchy

SOCIAL RESPONDING

established in the enclosure, with the exception that the dominant animal, a female, was bitten by two males and mounted by a third in the water tests. Only rarely did a female bite at or bite another turtle; it is not clear how females establish dominance.

When the individually housed turtles eventually met in the group setting, there was much more overall agonistic behavior. The dominant turtle continued biting with decreasing frequency for the first 90 minutes, by which time all turtles except the dominant female had made subordinate responses to him. On subsequent days the "new" group looked no different from the first set of animals in the adjacent enclosure.

Acquisition of food in the pair comparison tests was worthless as an indicator of social behavior. In group settings however, after sometimes attempting to move off with food, subordinates usually relinquished it to more dominant animals.

For an established group of *C. insculpta* at most times of the year, there is little social interaction beyond approach, some appendage retraction, and occasional head extending. As a collection of wood turtles becomes a group, however, there are many social responses that lead to extremely stable dominance hierarchies (Harless, 1970b).

Although dominant males in captive groups may be more likely to mate (e.g., Boice, 1970a), it is not clear whether dominance plays any part in mating in the wild. In some breeding aggregations a 1:1 sex ratio has been noted (*C. guttata*, Ernst, 1967), but in others (*Chelonia mydas*, Booth and Peters, 1972) there may be 10 to 40 females to one male; it is not known how male *C. mydas* that have access to breeding females differ from other males in the population.

Subtle Responses

Some sequences of activity have been successfully analyzed in captivity, but we are probably missing more subtle social responses, especially those of initial approach, choice among a number of available animals for social interaction, avoidance, and leaving the situation. Boice (1970b) described looking away from the gaze of a dominant turtle as a subtle response.

Female wood turtles in captive settings frequently approach and stay near a given male, moving when he moves. A female *Clemmys insculpta* may also approach a male, extend her head to the base of his tail, then follow directly behind him, move rapidly to a position directly in front of or partially under him each time he stops, and carry out the sequence repeatedly when he goes over or around her. In my observations this behavior has occurred only outdoors in the fall. Typically the sequence continues until they enter the water; then the male mounts the female. Knowlton (1943) reported similar behavior of a pair of captive wood turtles that moved to an indoor pool each fall, and Fisher (1945) observed a female *C. insculpta* crawling under a male in a field setting. In the narrow confines of most laboratory testing situations, such behaviors simply do not have the opportunity to occur. Evans (1961) noted that a captive female *C. insculpta* circled the male (in water) with her carapace tilted toward him while she turned her head to look at the male and that the male then mounted; the tilted carapace or the head-carapace configuration seems to be an eliciting stimulus for male agonistic behavior in any setting (Harless, 1970b).

Subordinate male wood turtles in captivity will occasionally bite a dominant male who is otherwise occupied. The subordinate, for example, moves his head and appears to see a certain dominant male engaged in mutual head extending with another turtle. The subordinate circles around, perhaps passing many other turtles, and slowly approaches the dominant male from behind. A very subordinate male makes one quick bite to a rear leg or the tail, then moves off quite rapidly. He is some distance away by the time the dominant male turns his head and there is virtually never any return attack by the dominant male. When a hierarchy is still in the process of being established, a less subordinate male may approach in the same manner and bite the other male, but maintain his grip until dislodged, when fighting between the two may continue. Males mounted on females are particularly vulnerable to this type of agonistic behavior.

Booth and Peters (1972) described attacks on the male of mating green turtle pairs by "escort" or "attendant" males. In this species, the female may bite back at an intruding male.

Ultimately the female turtle of a pair determines whether mating occurs because she must relax her tail and, in some cases, relax her rear plastron or elevate the rear carapace before intromission can take place. At least some female turtles have definite preferences in mates and will ward off all males but one.

A common defense against a male is escape by walking, running, or swimming. Because a female is moving rapidly in front of a male does not necessarily mean that the female is escaping, however. This is easy to determine in animals whose fear responses to humans have habituated, at least with small species, by removing the female some distance from the male. A female ready to mate will walk, swim or run just as quickly back to the male and resume her "escape." Otherwise the female will continue to move away from the male, frequently into an area the male cannot enter, or under something. Booth and Peters (1972) found a portion of a lagoon that was not entered by pursuing male *Chelonia mydas*; physical characteristics of the area that set the occasion for cessation of pursuit behavior were not identified. Female *C. mydas* may also beach themselves to escape males (Bustard, 1972a). Pursued females of some species, including *C. mydas* (Booth and Peters, 1972), turn to confront a male when escape responses to not deter him. In ornate box turtles, an escaping female may stop abruptly and turn sideways to the oncoming male. If it is a large female and a small male, this position is usually sufficient to deter the male, who stops or turns away. Sometimes there is considerable vacillation on the part of the male. He shifts back and forth from his head-high, elevated-shell pursuit position to a lowered head and body position. If the female begins walking before he has turned away, he will once again pursue. Females are less likely to present this lateral position to large males and they, in turn, are less likely to be deterred by the female's posture. Occasionally female box turtles simply close their shells while males attempt to mount, sometimes for more than 2 hr.

Brumwell (1940) observed that a free Kansas *Terrapene ornata* turned and snapped at three of her four pursuers, but then copulated with number one in line. Biting at has not been recorded for female *T. ornata* from Texas populations until a male is mounted, and the response is then ineffective in repelling the male (Harless and Lambiotte, 1971). Biting at and biting have been noted in females of numerous other species, from feeble snaps of freshwater *Pseudemys scripta elegans* (Jackson and Davis, 1972b) to scar-producing bites of sea turtles (Booth and Peters, 1972). Booth and Peters (1972) also showed the position of females who had just completed either copulation or egg laying; the planar surfaces of the hind flippers are brought together, effectively covering the tail area, while the female rests on the bottom.

The "presentation posture" described by Weaver (1970) for a female *Gopherus berlandieri* harassed by an aggressive courting male is not dissimilar in appearance from the elevated rear carapace "defensive posture" pictured by Dodd and Brodie (1975) for *Chelydra serpentina*. This is a common behavior, and species that do not have much protection from the shell are especially likely to assume this stance when provoked from the front, and may walk off backward or sideways. If attack or harassment continues, the cloaca may be voided; some animals evert the cloaca, and males may also extend the penis.

Penis extension by males not engaged in copulation has evidently not been reported for tortoises, although Boice (1970a) has noted this behavior for *Terrapene carolina* in the presence of a female on land. Evans

(1953) and Harless and Lambiotte (1971) reported extrusion and tumescence for *T. carolina* and *T. ornata*, respectively, when the turtles were in or near water. Captive *C. insculpta* also make this response in water. It is most frequently seen in males who seem "newly" adult—those who have not previously been observed in agonistic or sexual interactions.

Female *C. insculpta* sometimes approach a strange male or a familiar male with whom they do not spend time, extend the head, then lift it, make a flipping motion with one or both front feet, and then move off. The entire response topography is the same as that made by individuals of both sexes when they encounter food of a type the particular individual does not eat; the flipping portion is also used when foraging in shallow streams.

COMPARATIVE AND EVOLUTIONARY ASPECTS

Few comparative studies have been completed. Mahmoud's (1967) description of courtship and mating in turtles of two genera of Kinosternidae from Oklahoma showed no large differences among the four species. He did not observe the plastron-to-plastron copulation recorded by Finneran (1948) for *Sternotherus odoratus* in a Connecticut observation. There is no reason to suspect any geographical differences in sexual behavior when morphological and ecological considerations remain the same. For *S. odoratus*, however, Tinkle (1961) suggested greater sexual dimorphism in individuals from southern populations; Gibbons (1970a) confirmed this, but there has been no investigation of behavioral differences in various populations. On the basis of hybridization data (Folkerts, 1967), penial analyses (Zug, 1966), and notation of similarities between behavior of *S. odoratus* and *S. carinatus* (Mahmoud, 1967), it is unlikely that the sexual behavior of the unstudied *S. minor* differs greatly from that of *S. odoratus*. An analysis of the social behavior of the three *S. minor* subspecies might show some interesting subtle differences reflecting disparities in morphology and activity patterns.

For many other genera (see Tables 1 and 2), there are scattered data to serve as bases for more complete comparative research. In addition, records exist for interactions between individuals of different species and even different genera: *Terrapene ornata* and *T. carolina* hybrids (Blaney, 1968); *T. ornata* and *T. c. triunguis* (Rodeck, 1949); *Trionyx spiniferus* and *T. muticus* (Legler, 1955); *Geochelone elegans* and *Testudo hermanni* attending to *Malacochersus tornieri* (Eglis, 1962); *Gopherus berlandieri* and *G. agassizi* (Householder, 1950; cf. Weaver, 1970); *Graptemys*

Table 2 Representative Reports of Dominance and/or Agonistic Behavior

Family and Species	Reference
Chelydridae	
Chelydra serpentina	Conant (1938), Dodd and Brodie (1975), Evans (1952), Froese and Burghardt (1974), Hammer (1969), Linsdale (1927), Raney and Josephson (1954)
Macrochelys temmincki	Allen and Neill (1950)
Kinosternidae	
Kinosternon bauri	Carr (1952)
K. flavescens	Mahmoud (1967)
K. subrubrum	Pope (1939), Rigley (1974)
Sternotherus minor	Jackson (1969)
Testudinidae	
Chersine angulata	Rose (1950)
Geochelone carbonaria	Auffenberg (1965)
G. elephantopus	Evans and Quaranta (1951), Quaranta and Evans (1949a)
Gopherus agassizi	Grant (1936), Miller (1932), Woodbury and Hardy (1948)
G. berlandieri	Weaver (1970)
G. polyphemus	Carr (1952), Ernst and Barbour (1972), Weaver (1970)
Emydidae	
Chrysemys picta	Evans (1940b), Fraser and Spigel (1971)
Graptemys flavomarginatus	Legler and Webb (1961)
Clemmys guttata	Babbitt (1932), Ernst (1970)
C. insculpta	Dinkins (1954), Evans (1961a, 1967), Harless (1970b), Knowlton (1943)
C. marmorata	Bury and Wolfheim (1973)
Terrapene carolina	Allard (1935), Boice (1970a), Boice, Quanty, and Williams (1974), Evans (1956a, 1956b, 1960), Latham (1917), Stickel (1950)
T. ornata	Brumwell (1940), Harless and Lambiotte (1971), Legler (1960a)
Trionychidae	
Trionyx muticus	Plummer (1976a)
T. spiniferus	Lardie (1964)
Cheloniidae	
Chelonia mydas	Booth and Peters (1972), Bustard (1973a), Loveridge and Williams (1957)
Eretmochelys imbricata	Parrish (1958)

flavimaculata and *G. pseudogeographica* (Ernst, 1974); *Gopherus agassizi, G. polyphemus, G. berlandieri* (Ernst and Barbour, 1972).

In natural environments, sexual and agonistic responses seem to be confined primarily to members of the same species or even subspecies. For example, as Taylor (1933), Jackson and Davis (1972), and Ernst and Barbour (1972) pointed out, courtship is remarkably similar for *Chrysemys picta* and *Pseudemys scripta*, yet even in the many sympatric populations there is no interbreeding. Perhaps in this instance the stimuli involved in initial approach for the two species are sufficiently dissimilar that the turtles never reach courtship stages of social behavior with other than conspecifics.

General comments on comparative and evolutionary aspects have been made by Evans (1961b, 1966, 1968) and Brattstrom (1974). Weaver (1970) has summarized data on social behavior in tortoises and speculated, along with Eglis (1962) and Auffenberg (1965), on the evolution of tortoise behaviors.

Some conclusions in the literature (e.g., that dominance is related to size) may hold for some, but certainly not all species—even among the few studied to date. Other conclusions (e.g., that biting is a primitive form of courtship) seem to be more substantive in terms of both ontogenetic and phylogenetic development. Conclusions about "all turtles" must remain tentative until we have more widespread field and laboratory, observational and experimental studies of a greater number of species. At this stage, it is obvious that turtles do engage in social behavior.

ACKNOWLEDGMENTS

Dorothy Adalis, George Breiding, Maurice Brooks, Peter A. Smith, John Orth, James N. Shafer, Thomas B. Collins, F. D. Klopfer, and my parents, knowingly or not, encouraged my work with turtles somewhere along the way. Mary Ann Benson, Marilyn Munchmeyer, and JoAnne Peale, who carried out more than 200 hours of systematic observations of wood turtle social behavior with me in 1961 and 1962, have never been publicly thanked. Their dedicated and thorough work is hereby finally and gratefully acknowledged.

CHAPTER 22
Emotional Reactivity

IRWIN M. SPIGEL

It has been said that a turtle exemplifies metabolism in slow motion. To physiologists that notion must constitute as much a half-truth as the idea that turtles exemplify behavior in slow motion. But like all half-truths, certain basic and obvious consistencies with reality are evident—though more than locomotion is of interest to the behavioral scientist. While a turtle is more than a curio for the scientifically voyeuristic, who find the characteristic attenuation of much of its activity advantageous, these qualities often become agonizingly frustrating in the examination of its emotional reactivity.

Although the most unambiguous definitions of "stress" are generally those given in terms of physiological response, for turtles, at least, such descriptions are undramatic at best, and often totally unsatisfactory to those genuinely interested in animal behavior. Use of terms such as "emotional responsiveness" does not improve the situation, since these are more definitive of the stimulus complex than the behavior to be observed. The physiological events that are consequent upon the application of operationally stressful stimuli are of broad interest, but so also are the behavioral reactions. Of stimuli there are no dearth: electric shock, disorientation, cold, and confinement, to name but a few. But how to relate the quality and magnitude of these to a behavioral repertoire that is characterized by little more than decrement or attenuation of ongoing activity? Simple: measure the decrement! The obvious alternative, of course, is to apply established procedures and protracted measurement scales, inadequate or inappropriate as these may be.

All the experiments performed in our laboratory employed two subspecies of the freshwater turtle *Chrysemys picta* (*C. p. picta* or *C. p. mar-*

ginata). Observation of the subjects (*S*s) during their first month of captivity yielded evidence of exceptionally wide individual variation in both activity level and habituation to handling. In attempting to habituate the head withdrawal of *C. picta* to mechanical stimulation, scores of 2 to more than 200 trials were obtained, with some *S*s never ceasing to retract at all. Many turtles withdrew the head immediately upon being picked up; such animals eventually proved useless in experiments, since they would not move, regardless of the motivational contingency. In all our later experiments *S*s that continued to withdraw their heads upon being handled following laboratory habituation were discarded. This screening procedure resulted in the loss of between 5 and 10% of those obtained through commercial channels.

Hayes and Saiff (1967) reported a visual alarm reaction (in the form of head withdrawal) in *Pseudemys scripta, Terrapene carolina,* and *Chrysemys picta* to a suddenly enlarged circular shadow ("looming stimulus") and its subsequent habituation; however, we were unable to obtain any such responsiveness in studying the latter species. It must be noted, however, that our *S*s faced an illuminated disk controlled by a diaphragm, whereas Hayes and Saiff used a cast shadow.

DETOUR LEARNING

Not surprisingly, initial experimental insight into contingent emotional reactivity was incidental to the main thrust of early research. In an effort to determine a satisfactory motivational complex within which to study learning, appropriate positive and negative reinforcers were examined. One of the most striking behavioral observations during early study of *C. picta* was the tenacity of *S*s attempting to escape from enforced confinement in dry, high-walled enclosures. A second aversive situation was disorientation by turning the turtle on its back. In one early study (Spigel, 1963) it was shown that the speed with which turtles ran down a 36 in. runway to a tank of water following 2 hr and 6 hr confinement in small, high-walled cells was more than double that of those given the opportunity to run after immediate withdrawal from their home tank. A second experiment utilizing antecedent confinement and contingent natation or disorientation, respectively, for correct and incorrect choices successfully demonstrated brightness discrimination in these animals. During the latter study, strong position fixations were shown by many *S*s given the more difficult of two discriminations. Despite repeated seemingly aversive stimulation on successive trials, turtles with such position habits showed virtually no tendency to change direction.

These studies seemed to demonstrate that walled confinement might be successfully employed as both motivational antecedent and stressful stimulus in an examination of the emotional reactivity of *C. picta*. This may well have been apparent to Yerkes (1901), who used escape from confinement as an aversive condition for a demonstration of maze learning in *Clemmys guttata*.

Behavioral manifestations of emotional reactivity must be determined for each species studied, along with the particular events that give rise to them. Ireland and Gans (1972), for example, noted in a study of the adaptive significance of the flexible shell of the tortoise *Malacochersus tornieri* that disturbed individuals dig in the foreclaws and rotate the forelimbs outward; in the natural habitat, these movements would facilitate wedging the body into narrow rock crevices. The awareness of such species-specific response patterns in advance of complex experimental stimulus manipulation can aid considerably in more accurate assessment of laboratory observation.

C. picta's consistent effort to escape from walled confinement, and its equally apparent approach preference for an expanse of daylight (Spigel, 1964a), suggested that a detour situation might be useful in exploring a wide variety of behaviors. Both learning and retention of detour performance by *C. picta* were easily established (Spigel, 1964c). By the sixth trial, turtles were performing with maximum efficiency in turning from a three-sided, barred enclosure facing a window in order to reorient to the light, and making a subsequent return to their home tank by going around one of the side walls. Ss in a second experiment, confined for 2 hr in a walled cell prior to trial 8, showed a severe decrement in their performance (Figure 1); the turtles reverted on the

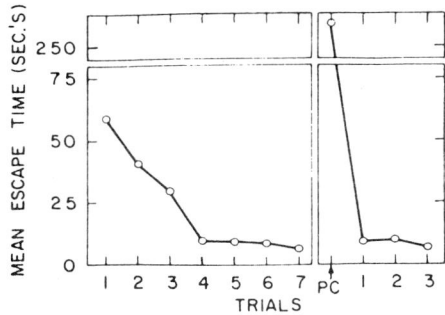

Figure 1 Detour learning by turtles, showing the severe decrement in performance following a period of walled confinement (PC indicates first post-confinement trial). Copyright 1964 by the American Psychological Association. Reprinted by permission.

first postconfinement trial to more direct but inadequate modes of escape, in some cases more primitive than the responses observed in early learning trials.

A third study in this series examined the hypothesis that the walled confinement produced an emotional state that generated in the animal escape response patterns that interfered with the previously acquired detour-learning pattern. This induced decrement was minimized when Ss were confined under chlorpromazine or ammobarbital administration and potentiated following amphetamine injection.

As a further test of the interference hypothesis (Spigel, 1964a), the detour learning of turtles confined prior to their trials was compared with that of animals confined for the same length of time following their daily trials. The considerably attenuated performance of Ss confined prior to their detour trials once again tended to support the position that the deficit was attributable to the persistence of response patterns established during the confinement period. Posttrial confinement failed to adversely affect the progress of the animals' detour performance.

Since general activity in the form of sustained climbing, scaling, and exploratory effort is also potentiated by walled confinement, another study (Spigel, 1964b) sought to assess more objectively the influence of drugs on this observed output. An apparatus that recorded both locomotor and climbing movement was employed with turtles injected intraperitoneally with chlorpromazine, amphetamine, or saline solution. This experiment also examined the effect of stimulus change—a sequencing of light and darkness—on activity level. Marked reduction of locomotor and climbing activity followed administration of the tranquilizer under all illumination sequences. However, the possibility of complex interaction between the effect of the drug and the test situation remained, inasmuch as the turtles, virtually motionless in the test apparatus, were normally active immediately upon return to their home tank. For turtles given the placebo, activity following a change from either light to dark, or dark to light, was greater than for continued illumination or darkness. Turtles administered amphetamine, however, remained highly active regardless of illumination sequence.

It might be well to note that Daniel and Smith (1947) reported that neonate loggerhead turtles, after remaining in darkness for some time, became more active at light onset but returned to quiescence with darkness. The particular activity in progress, however, could alter this response pattern.

A demonstration of light-contingent operant behavior (Spigel, 1965) in which turtles pressed a panel to simply change the surroundings from light to dark, or the reverse, raised the possibility that for *C. picta*, at least, such illumination facilitates habituation to the test surround.

ESCAPE AND AVOIDANCE

It was at this point that we began direct examination of avoidance behavior in *C. picta*. A traditional one-way shuttle-box situation, modified for turtles, was constructed, and electric shock was selected as the aversive stimulus. Fewer than 10% of the Ss failed to learn rather quickly to escape from the shock side, over the barrier to the nonshock half, yet none of our Ss showed any tendency to actively *avoid* during a 5 sec interval following placement, even after an extended series of trials. Turtles often scampered directly up to the barrier immediately upon placement in the apparatus, then remained there until the onset of shock before shuttling over to the other side. However, no further signal, other than placement in the box, was employed.

Passive avoidance, on the other hand, presented no such problem (Spigel, 1965). An apparatus was designed so that when a mercury switch taped to the forward part of the carapace reached a lie of 20°, the animal would receive electric shock until it returned to a less vertical position. The avoidance of shock by refraining from attempting to climb from the enclosure was examined under varying stimulus intensities. The expected negatively accelerated gradients as a function of shock intensity were observed for two measures of acquisition: number of trials to criterion and mean seconds of climbing per trial. The five-trial extinction procedure, however, failed to differentiate the intensity groups.

ENDOCRINE INVOLVEMENT

Investigations of the role of the adrenal-pituitary axis in avoidance learning had suggested that exogenously administered ACTH reduced the effectiveness of anxiety-producing stimuli (Mirsky, Miller, and Stein, 1953). ACTH had also been implicated in the acquisition of passive avoidance in the rat (Levine and Jones, 1965) but not in the acquisition of shuttle-box avoidance (Murphy and Miller, 1955; Miller and Ogawa, 1962). It has also been shown that under stressful stimulation the adrenal glands of rattlesnakes (*Crotalus atrox*) increased in weight (Allen, Russell, Gumbreck, and Shetlar, 1959). Such information suggested exploration of endocrine function in the acquisition of active and passive avoidance behavior by *C. picta*.

Neither ACTH nor cortisone administration influenced passive avoidance performance, but exogenously administered cortisone facilitated the acquisition of active avoidance (Spigel and Ellis, 1965b). This was both surprising and puzzling. Hydrocortisone is produced naturally in reptiles (Phillips, Chester-Jones, and Bellamy, 1962; Sandor, Lamoureaux, and Lanthier, 1964), but a pilot study failed to yield a standard dosage of the acetate of this hormone amenable for use with *C. picta*. Most *S*s sickened after the third or fourth injection, and many failed to recover. Cortisone, which has not been found in any quantifiable amount in reptiles, produced no such observable debility. We had no satisfactory explanation for this, nor did a great deal of unpublished follow-up study clarify the situation.

Although the results generally extend the implication of the adrenal role in the vertebrate response to stress, the data differ from those obtained with laboratory rats in that the facilitation of active rather than passive avoidance was observed. This could conceivably reflect a marked difference in the organization of species-specific response systems. It is also possible that the particular mode of passive avoidance, which is actually a climbing suppression, is not analogous to that employed for mammals. From these experiments, however, it appeared that the corticoid facilitated only systems involved in active and directed movements by *S*s under stress. Exogenous corticoid administration may stimulate mechanisms that act to reduce the disruptive consequences of stress. Some support for this is found in the 11 day retention performance of the cortisone animals. In some as yet unspecifiable manner, these *S*s were apparently rendered amenable to efficient acquisition of the appropriate motor patterns leading to shock avoidance. It is also possible that the effects of cortisone on the behavior may have been mediated by its anesthetic properties.

The failure of ACTH to yield facilitation may possibly be attributable to the differences in molecular structure between the mammalian ACTH used and the analogous hormone in the reptile. It is also possible that the pathways by which this material is metabolized and the speed with which it is utilized are quite different in reptile and mammal.

F. T. Crawford (personal communication) suggested that because of our screening procedure, we may have been using a select, highly active sample. This may indeed be the case, but the alternative was no data, and indeed, no experiment at all. Even with the sample as chosen, the results seem to require explanation at a more molecular level. In any case, passive avoidance appeared to offer a more productive dimension for further examination of emotional reactivity and the focus of the research shifted to central neural mechanisms.

CEREBRAL INVOLVEMENT

Several experiments confirmed the loss of spontaneous movement with massive cerebral damage in the turtle, but less extensive lesions failed to affect either locomotor activity levels or the acquisition of escape in a one-way shuttle box.

Spigel and Ellis (1966) examined the effects of forebrain lesions on passive avoidance under varying shock intensities. The results suggested that damage to cerebral tissue impaired the climbing response independent of shock level, rather than facilitating the acquisition of avoidance behavior. Figure 2 gives schematic representations of the lesions.

Climbing under the conditions imposed represented an attempt to escape, and as such apparently involved emotionally integrative mechanisms. An interpretation of the observed climbing deficit in terms of interference with, or inhibition of, a response more ubiquitously generated or aroused by confinement seems to be more appropriate. The lesion-induced climbing decrement in a situation that has been shown to be relatively aversive to members of this species is consistent with the observation of the loss of appropriate motor responses in cerebrally lesioned snakes (Goldby and Gamble, 1957) under emotional stimulation. To this extent, the participation of reptile hemispheric tissue in emotionally integrative behavior patterns is suggested. Some support for this conclusion may be found in Morlock's (1972) observations that

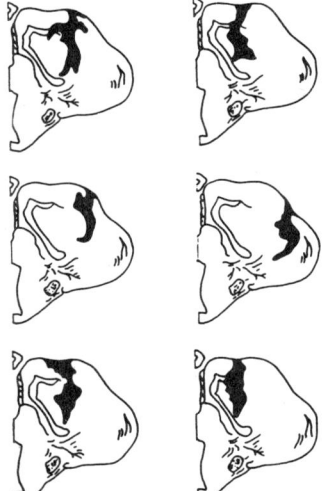

Figure 2 Typical hemispheric lesions in operated Ss.

dorsal cortex ablation in turtles had little effect on open-field activity, latency-to-eat, and two-choice spatial discrimination learning.

Results almost identical to those obtained from cerebrally lesioned turtles with respect to climbing suppression were observed by Spigel and Ellis (1967a) with amphetamine administration. Climbing, with or without shock contingency, was considerably reduced, with half the stimulant-injected Ss failing even to attempt to climb.

If it is assumed that the sympathomimetic amphetamine acts on subcortical structures, the combined observations of the earlier lesion study and the current experiment suggest that in a reptile, cerebrum and lower centers may be somewhat antagonistic. That is, lesioning the former and stimulating the latter may produce similar results. It is further possible that the escape climbing in confinement is a response to some limited range of arousal stimulation. When exceeded, arousal results in withdrawal and immobility. There is certainly evidence suggestive of such biphasic reactivity in gross observation of Ss in the laboratory and from unpublished observations of this species in active avoidance situations. It may be that the amphetamine dosage employed resulted in an arousal level in excess of the upper limit for the climbing response in the present situation.

Ellis and Barcik (1972) examined the consequences of disorientation on the suppression of food-reinforced maze arm selection. Although complete elimination of escape activity and immobility was observed in all Ss during the course of learning, suppression training (turning the animals in this group on their backs for selection of the previously rewarded goal choice), resulted in a restoration of the typical escape pattern. Ellis and Barcik reported reduced traverse time, reversals, stereotyped efforts to climb from the apparatus, increased breathing rate, and periods of inactivity—virtually the same repertoire observed in previous experiments in turtles under stress. In an effort to explore the neural nature of these events, bilateral anteromedial lesions were made in several Ss to portions of the brain analogous to mammalian hippocampal and septal tissue. Lesions were made following acquisition but prior to suppression training. Lesioned turtles did not differ from normal animals in percentage correct or latency, but they failed to show the increased climbing that was characteristic of normal suppression behavior. Such observations remained consistent with the findings of Spigel and Ellis (1966) with more laterally placed cerebral lesions. It was suggested that systematic destruction of these cerebral areas may provide information with respect to the neural organization of reptilian defensive behavior. The necessity for careful control and specification of the previous experience of the animals in the interpretation of turtle performance is clear. The demonstration of learning requires elimination of prepotent avoidance reactions.

EXCRETORY ELECTROLYTES

A somewhat different approach to the study of emotional reactivity grew out of a parallel series of experiments devoted to the relation of electrolyte balance and drinking (Spigel, Ellis, and Kaiser, 1967; Spigel and Ellis, 1968). Flame photometric analysis for sodium and potassium, useful in the determination of serum cations, was employed in the study of the turtle's habitat water to obtain excretory electrolyte levels.

A preliminary experiment (Spigel and Ellis, 1967b) relating the Na^+ excretion during the course of the turtle's habituation to a new surround, and/or a change in feeding schedule, revealed the presence of higher cation levels only when both conditions were simultaneously imposed. In addition, a gradual decline in the electrolyte levels was observed over a 20 day period, providing what appeared to be a physiological index of habituation. The excretory pattern appears in Figure 3.

Whereas the adaptive roles of the adrenal glands and other internal systems have been extensively studied in connection with the response to stress (Selyé, 1946, 1950), systematic efforts to relate secretory patterns to experience are of more recent vintage, and the findings are less

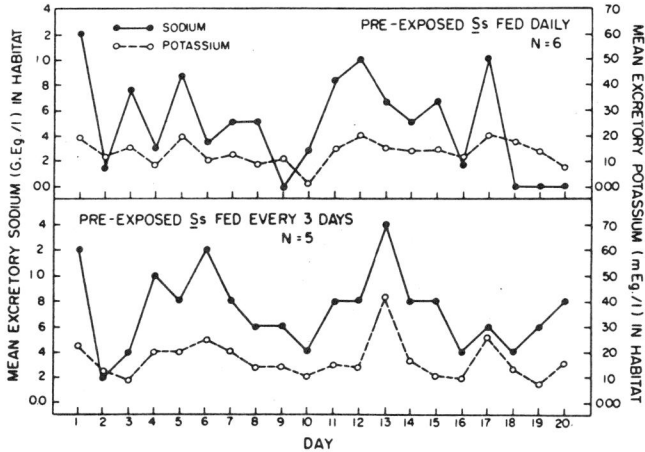

Figure 3 The 20-day record of excretory Na^+ and K^+ for Ss not previously habituated to test dishes. Reprinted with permission.

amenable to clear interpretation. The data with reference to mineralocorticoid activity and the response to stress are particularly equivocal.

Plasma sodium retention and potassium loss have been reported in humans under stress (Hollander, Chobian, and Williams, 1960), presumably reflecting hypersecretion of adrenal steroids. However, Hoagland (1961) found increases in *both* sodium and potassium content of the urine of men placed under competitive stress as compared with controls, but only in those over 20 years of age. To complicate matters further, Paré (1964), using rats in a conditioned fear situation, failed to find significant changes in either excretory Na^+ or K^+, but did observe consequent depletion of blood eosinophils, as well as adrenal hypertrophy and elevated adrenal ascorbic acid levels in experimental animals. He concluded that mineralocorticoids did not participate in the stress response. The implication of aldosterone had been minimized at the outset owing to the independence of its secretory activity from ACTH levels, thus from the hypothalamic-hypophyseal-adrenal axis. Consideration of Paré's results in conjunction with investigations of other species summarized by Phillips and Bellamy (1963) suggests that there may be little phylogenetic consistency in the relation of adrenal corticoid activity to excretory electrolytes. Among vertebrates, the relationship is most obscure among reptiles (Phillips, Jones, and Bellamy, 1962), although they produce both corticosterone and aldosterone.

In a comprehensive series of preliminary experiments, potentiated levels of sodium and potassium were consistently observed in both daily and bidaily analyses of the water in the tanks of shocked Ss as compared with that of nonshocked Ss for periods of 1 to 6 days. Increments in cation concentrations of periodically shocked Ss were even more significant when values were corrected for individual urinary volume. Although Na^+ and K^+ levels were also higher in the residual bladder material of shocked turtles, collected and analyzed immediately following sacrifice, no such differences in cation concentration were evident in the serum of experimental and control Ss.

Elevated urinary alkali metal (combined Na^+ and K^+) was observed in the habitat water of electrically shocked turtles as compared with nonshocked controls, the cation excretion continuing even in the absence of shock when Ss were retained in the same surround on subsequent days (Spigel and Ramsay, 1969a). Excretory Na^+ and K^+, however, fell to the level of controls if Ss were removed to a new environment (Figure 4). It appeared that elevated urinary alkali metal was the consequence of a conditioned emotional response to the antecedent shock situation.

Reptiles appear to show at least two sources of species-specific differences with respect to adrenal secretory patterns that require further

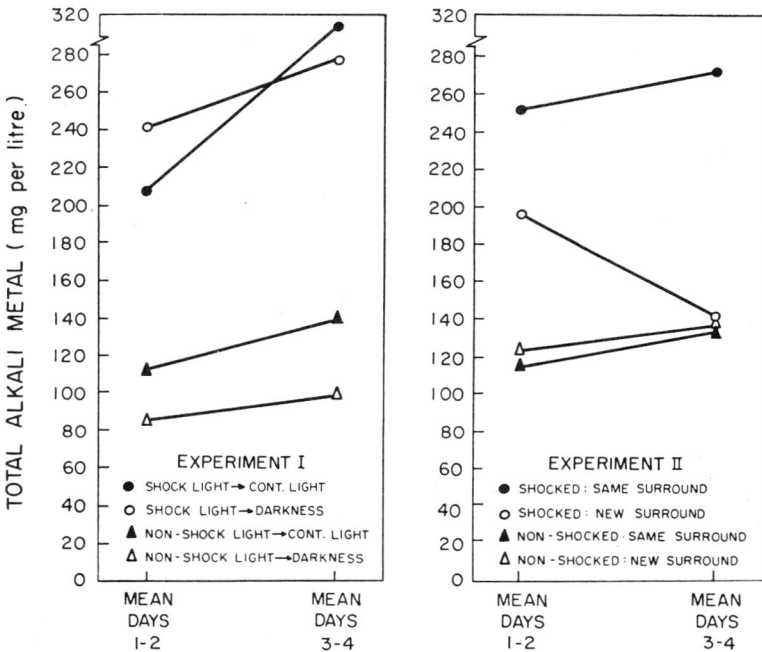

Figure 4 *Left:* Comparison of the urinary alkali metal level of Ss shocked on days 1 and 2 with that of nonshocked controls. (The sustained increase in excretory cations for shocked Ss is also observed on days 3 and 4 when no shock is administered, and the only other change for two of the groups is in the illumination.) *Right:* Results of removing previously shocked Ss to an entirely new environment for the 2 nonshock days. Urinary alkali metal levels for these Ss drop to those of controls as compared with the higher concentrations for Ss retained in the shock-contingent surround on no-shock days. Copyright 1969 by the American Psychological Association. Reprinted by permission.

investigation. On one hand, if one were to discount aldosterone because of its independence of ACTH output, the *in vitro* determination (Sandor, Lamoureaux, and Lanthier, 1964) of the presence of 11-desoxycorticosterone in *C. picta* could conceivably focus attention on active hypothalamic-hypophyseal involvement in potentiated mineralocorticoid levels, and the consequent rise in excretory electrolyte. However this mineralocorticoid has not yet been isolated *in vivo*. On the other hand, aldosterone may be critically involved in the electrolyte response to stress by way of a mechanism similar to the renin-angiotensin system of mammals. It is conceivable that stress-induced changes in blood pressure lead to alterations in glomerulofiltration rate and aldosterone levels, causing reduced tubular reabsorption of sodium and a consequently higher excretory electrolyte level.

Such a mechanism operating independently of the hypophyseal-adrenal axis may be necessary in accounting for the determination of only partial consistency of climbing decrement and excretory alkali metal concentrations in cerebrally lesioned turtles. Unlike the climbing deficit apparent in the lesioned animals, which was independent of contingent shock, Spigel, Ramsay, and Seggie (1970) found that both lesioned and sham-operated Ss alike produced elevated urinary Na^+ and K^+ levels when under periodic shock. Some consistency with the earlier behavioral observations was evident in the significantly lower cation concentrations of shocked, lesioned Ss for the first 2 days of the 6 day study. For the remaining 4 days, the excretory alkali metal of shocked sham operates fell to the level of shocked, lesioned Ss. It is possible, of course, that periodic shock constituted far more stressful stimulation for *C. picta* than did confinement, exceeding any limits of control by cerebral tissue alone. Still demanding further study is the *initial* reduction of excretory sodium and potassium observed in shocked, lesioned Ss, since this condition continues to implicate the cerebral hemispheres in some overall emotionally integrative response.

Extinction of the excretory alkali metal response (EAMR) to stress as a function of differentially applied antecedent shock was explored in further studies (Spigel and Ramsay, 1970). Not only did the urinary Na^+ and K^+ levels prove to be sensitive to varied antecedent shock experience, but they also produced extinction curves typical of those observed for conditioned response decrements in a variety of species. Antecedent shock for periods of 1, 3, 6, and 12 hr were administered in one study (Figure 5), and varied periods of either consistently or randomly applied shock constituted the experimental treatment in subsequent work. The time of return to normal excretory cation levels differed for turtles given as little as 3 or 6 hr of periodic antecedent shock.

The question of the discrepancy in stress-induced alkali metal excretion between reptiles and rats remains unresolved. It is possible, of course, that the conditioned emotional response (CER) procedure employed with rats is not sufficiently stressful, or that the EAMR is itself insufficiently sensitive to either the qualitative or quantitative stress implicit in the CER experience. There is some evidence for the latter view in Paré's determinations that other physiological indices of CER-produced stress that required sacrifice of the animal were positive despite the failure to find increasing excretory concentrations in experimental Ss.

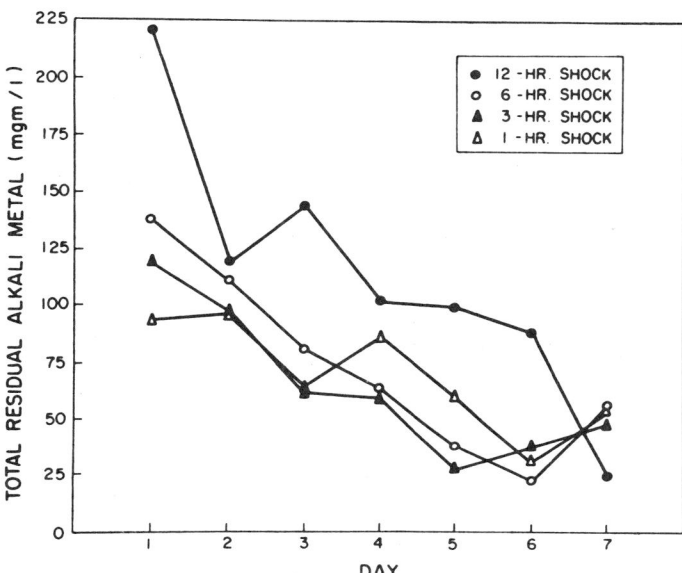

Figure 5 Comparison of the extinction of the excretory alkali metal response in Ss given varying amounts of electric shock.

An even greater species or procedural difference appears to loom as a result of other observations. Turtles shocked in a random fashion over a period of 6 days showed rapid and clearly evident reduction of the EAMR during extinction. Animals administered the same amount of shock in a consistent daily pattern over the same period of time, however, revealed almost no extinction whatsoever of the EAMR during the subsequent nonshock days (Figure 6). The absence of any appreciable difference between levels of urinary Na^+ or K^+ between shock and nonshock for these Ss suggests that any extinction of the excretory cation response in nonrandomly shocked turtles would at the very least have been considerably prolonged. On none of the 10 subsequent nonshock days was there any overlap of mean urinary alkali metal levels for Ss systematically or randomly shocked during the preliminary 6 day treatment phase.

Earlier study (Spigel, Ellis, and Kaiser, 1967; Spigel and Ellis, 1968) had demonstrated remarkably precise water regulatory mechanisms in *C. picta*. Initial water intake of Ss following 72 hr dehydration was equal to the total consumption by nondeprived Ss as compared with the 2 day

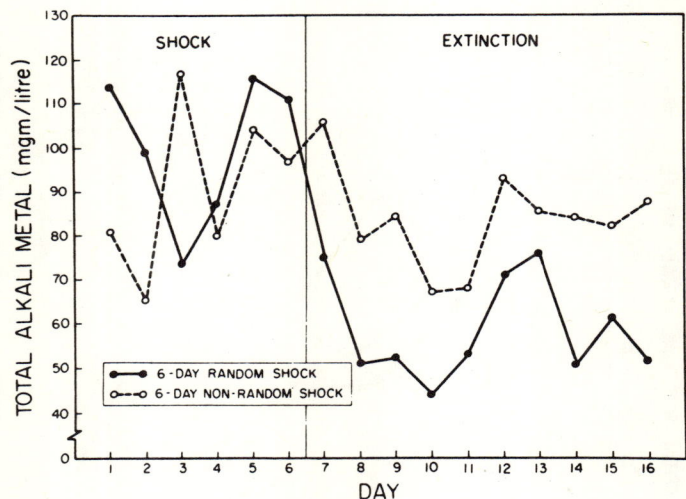

Figure 6 Excretory alkali metal levels of Ss given a total of 48 hr of periodic shock either systematically or randomly over a 6 day period, and the subsequent urinary cation output for the 10 following nonshock days.

drinking of a nondeprived group. In an effort to study the relationship between stress and water consumption in *C. picta*, Spigel and Ramsay (1969b) determined that electric shock administered outside the drinking situation failed in any general way to affect water intake in differentially hydrated Ss; nor did such treatment disturb the precise aquastatic mechanism previously observed in this species. Deprived turtles did, however, tend to consume more water than controls when some degree of familiarity with the drinking surround had been obtained. Although shock administration within the water-available situation likewise failed to produce a general drinking deficit, initial intake by nondeprived Ss was reduced, providing the first suggestion of a disturbance in the water-regulatory mechanism.

SOCIAL–EMOTIONAL REACTIVITY

Emotional reactivity in social situations has received only minimal study in the case of turtles (Chapter 21). Evans (1952) observed greater aggressiveness in male *Terrapene c. carolina* than in female. *T. c. carolina*, *Pseudemys scripta troosti*, *Chrysemys picta*, and *Chelydra serpentina* all showed increased aggressiveness when administered male hormone.

Boice (1970) observed that in captive box turtles (*T. c. carolina*) the establishment of dominance hierarchies was accompanied by a range of behaviors from biting to pushing. Harless and Lambiotte (1971) observed in autumn open field testing that female *Terrapene ornata* were more active and variable than males with and without the presence of a novel stimulus. With only a few exceptions by females during mating, no biting was observed over the course of laboratory maintenance. A number of recent underwater observations of the green turtle (*Chelonia mydas*) by Booth and Peters (1972) revealed a great deal of aggressive reactivity, especially with respect to mating and reproductive interaction.

Shock-induced threat and biting by *C. picta* has been studied by Fraser and Spigel (1971). When restrained facing each other, the turtles threatened and bit others in response to electric shock. Shock alone caused turtles to threaten an unshocked S, while the movements of a shocked animal were sufficient to cause an unshocked but restrained turtle to threaten. When the turtles were free to move, they avoided an encounter when shocked, even reversing a strong position preference to do so. In short, *C. picta* tends to show little aggressiveness in the laboratory; biting is more common than threat when both Ss are shocked, but the reverse is true when only one is shocked.

Because of the relatively small sample of research given over to the emotional responsiveness of turtles, cross-species comparison would be premature at best. Nevertheless, differences are sufficiently evident to warrant extensive study along a number of dimensions. The role in emotional reactivity of both physiological and social variables is especially demanding of increased attention. Greater coordination between field and laboratory examinations of an increased number of species should prove to be productive.

ACKNOWLEDGMENT

Research conducted by the author and his associates since 1966, and cited in this chapter, was supported by grants from the National Research Council of Canada.

CHAPTER 23
Rhythms

EUGENE V. GOURLEY

One factor underlying the behavior of an animal in its natural environment is the rhythmicity of that environment. Each species adapts to the diel and seasonal changes of light and temperature in a specific fashion, which is reflected in the activity pattern of the species. This rhythmicity has been studied in a variety of organisms, but relatively little work has been done with turtles. The subject of rhythmicity can be considered from three aspects: determination of the activity patterns of turtles in their natural environment; determination of the factors, either internal or external, that control these rhythms; and determination of the adaptive significance of these patterns.

ACTIVITY PATTERNS IN THE NATURAL ENVIRONMENT

The naturally occurring rhythms of particular importance to turtles are diurnal rhythms and seasonal rhythms, although others such as lunar rhythms may affect certain species. Our information regarding these rhythms is general: there are few precise data concerning individual or population variability. Most species of turtles are diurnally active, but some are nocturnal—for example, *Chelydra serpentina osceola*, *Terrapene ornata* (Carr, 1952), and *Chelonia mydas* hatchlings (Moorhouse, 1933). Others engage in nocturnal activity under extreme conditions (*e.g.*, *Sternotherus odoratus*) or during a specific part of their life cycle (e.g., *Chelonia mydas* adults, for nesting: Carr, 1952). Some—such as *Gopherus agassizi*—are nocturnal and crepuscular (Woodbury and Hardy, 1948), whereas others, such as *Gopherus polyphemus* (Oliver, 1955) and *Geochelone sulcata* (Cloudsley-Thompson, 1970), may be crepuscular only during extreme weather conditions.

Seasonal activity patterns are also generally described for turtles, particularly egg-laying, hibernation, and migration periods. Oliver (1955), for example, summarized the time of egg laying for 10 species of turtles. Carpenter (1957) described the time of hibernation for *Terrapene carolina triunguis* in eastern Oklahoma as beginning in October and ending in mid-March. Woodbury and Hardy (1948) reported that *Gopherus agassizi* migrated from the summer ranges to winter dens in November, where they remained until the first of March. Gibbons (1970d) reported terrestrial movements in *Kinosternon subrubrum, Sternotherus odoratus, Pseudemys floridana, P. scripta, Chelydra serpentina*, and *Deirochelys reticularia* from March to October in South Carolina.

Biweekly patterns of activity during the nesting season are reported for nesting *Chelonia mydas* (Carr and Ogren, 1960), as are 2 year or 3 year nesting cycles for individual females of the same species (Carr, 1967a). (See Chapters 16 and 17.)

DETERMINATION OF CONTROLLING FACTORS

Many organisms demonstrate a persistent rhythm of locomotor activity, metabolic rate, or some other parameter in the absence of the time-giving stimuli, *zeitgebers*, in the natural environment. The most prominent of the natural zeitgebers are light and temperature fluctuations. The rhythms that persist in the absence of zeitgebers apparently originate within the organism and are considered to be endogenous rhythms. Endogenous rhythms may be of long or short period lengths, but the most widely studied are the circadian rhythms with period length of approximately 24 hr. Based on evidence that hatchling *Chelonia mydas* could find directions by the position of the sun, Fischer (1964) suggested that this species must possess an internal clock to compensate for the movement of the sun across the sky. Brett (1971) demonstrated a persistent locomotor activity rhythm in *Pseudemys scripta* under relatively constant conditions (22–25°C and 0.5 lux light intensity). Graham (1972) found such a rhythm in *Chrysemys picta* (15°C and 5 ft-candles illuminance). Cloudsley-Thompson (1970) reported a similar locomotor activity rhythm in *Geochelone sulcata*, but noted that there was no advance of the cycle in constant light (500 lux at 35°C) nor retardation in constant darkness (at 25°C) (Figure 1). Without these characteristic shifts, the criteria for an endogenous, circadian rhythm are not met (Aschoff, 1960). Cloudsley-Thompson (1970) points out that this may be attrib-

DETERMINATION OF CONTROLLING FACTORS

uted to the small annual variation in the length of daylight in the tropics where this species is found.

How widespread are circadian rhythms (or any other endogenous rhythms) in turtles? Different results have been obtained under different sets of environmental conditions. Graham (1972) found that *C. picta* activity became arrhythmic under constant light and temperature conditions. Gourley (1972) demonstrated a persistent locomotor rhythm in *Gopherus polyphemus* under constant conditions (26°C and 350 lux illumination) that was more pronounced with a refuge into which the animals could escape to avoid the constant light conditions (Figures 2 and 3). Harker (1960) had suggested that a refuge would have this effect. In most of the studies above only two or three individuals were tested, and the need for larger samples is obvious.

Of considerable interest is the determination of external factors that control these rhythms. Since a particular species may be considered to have a diel activity pattern—diurnal, nocturnal, or crepuscular—it is strongly suggested that light may exert a significant effect. Light onset or termination, photoperiod, and light intensity seem to be most important in this respect.

Brett (1971) and Gourley (1972) have reported that activity rhythms of *Pseudemys scripta* and *Gopherus polyphemus* can be kept phase synchronized under alternating light and dark conditions. Both these studies utilized a sudden light-dark transition. Graham (1972) reported similar findings with *Chrysemys picta*. Utilizing a gradual twilight transition eliminated an abnormal activity peak at the transition time and suggested that this more closely simulates natural conditions.

Light intensity may have an effect on the activity pattern of an animal under constant conditions (Hoffmann, 1965), but apparently this parameter has not been systematically examined with turtles.

The photoperiod length also has a marked effect on many organisms. Burger (1937) found that in male *Pseudemys scripta elegans* artificially lengthened photoperiods arrested the spermatogenic cycle and initiated a new cycle. Graham (1972) reported differences in activity patterns of *Sternotherus odoratus*, *Chrysemys picta*, and *Clemmys guttata* under different photoperiod conditions. *Testudo hermanni* subjected to a 4 hr light period stopped their locomotor activity even at temperatures at which they were normally active (Thinès, 1966).

Much of the seasonal activity of turtles is attributed to the direct effect of temperature, but few studies have examined this premise critically. Hoffmann (1968) demonstrated that lizards could be kept in phase under otherwise constant conditions with a diurnal fluctuation of only 0.9°C, suggesting that very subtle environmental changes may have an

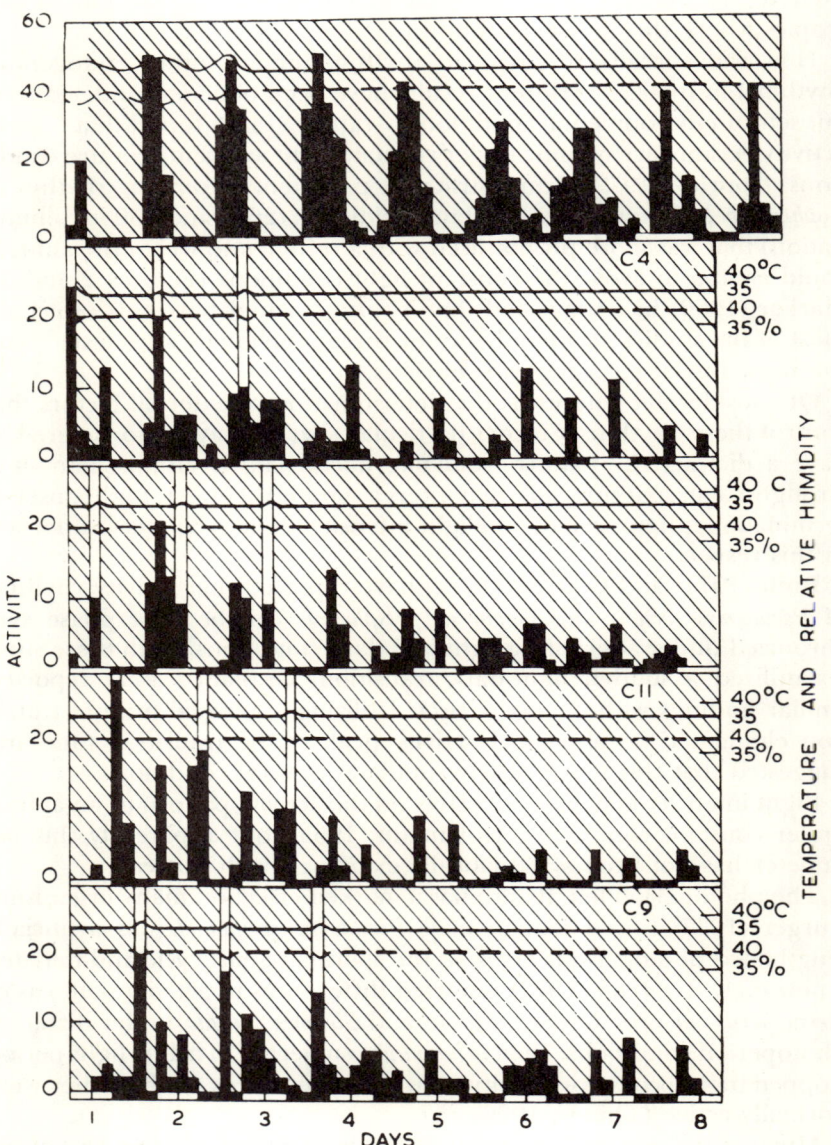

Figure 1 *Top row:* Circadian rhythm of locomotory activity in *Geochelone sulcata* (no. B3), at first in room conditions, then in darkness for several days, apart from one 3 hr period of light; finally in constant light. *Lower rows:* Effect of 3 hr periods of light on the activity rhythm of *G. sulcata* kept otherwise in darkness (*left*) and of 3 hr periods of darkness (*right*) on the rhythm of tortoises kept otherwise in constant light (*right*). *Ordinates:* Activity in the

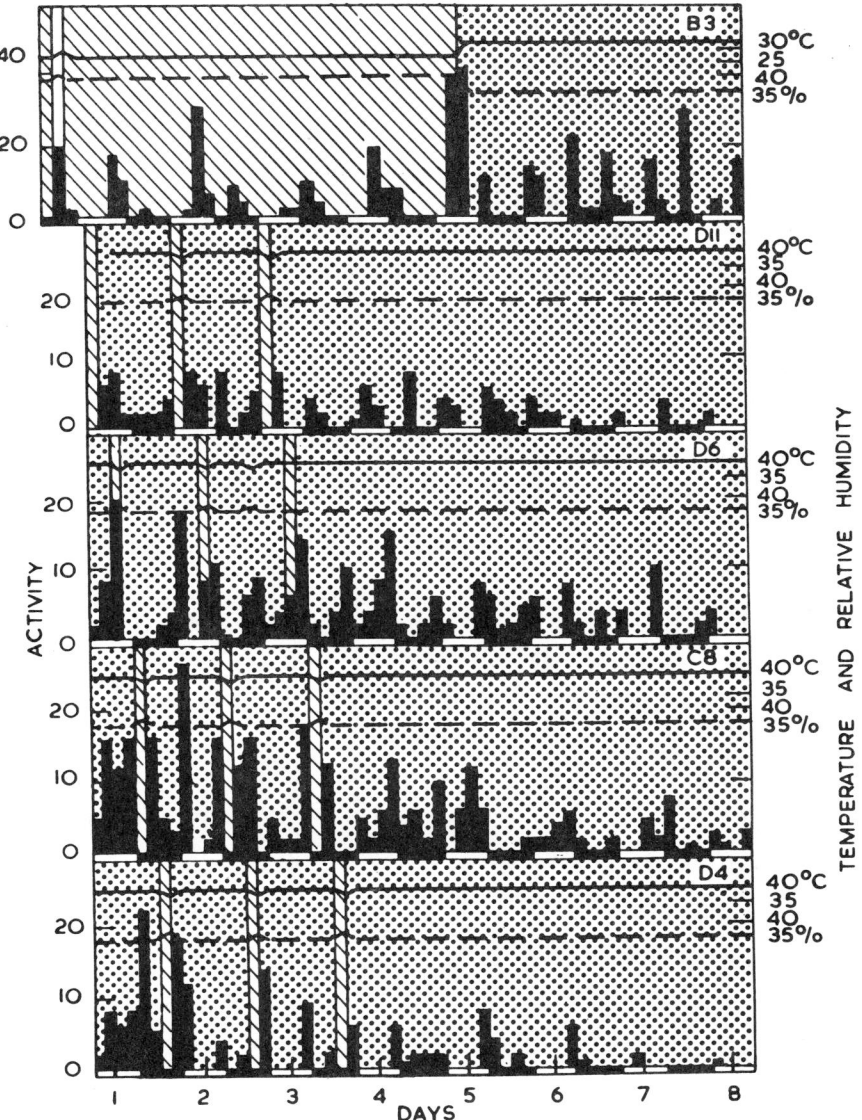

left, temperature (solid lines) and relative humidity (broken lines) on the right. *Abscissa:* Time (days). The black strips represent 12 hr periods from 1800 to 0600 hours. Although light stimulated activity and darkness inhibited it, neither reset the rhythm. (From Cloudsley-Thompson, 1970. Reprinted with the permission of the Zoological Society of London and the author.)

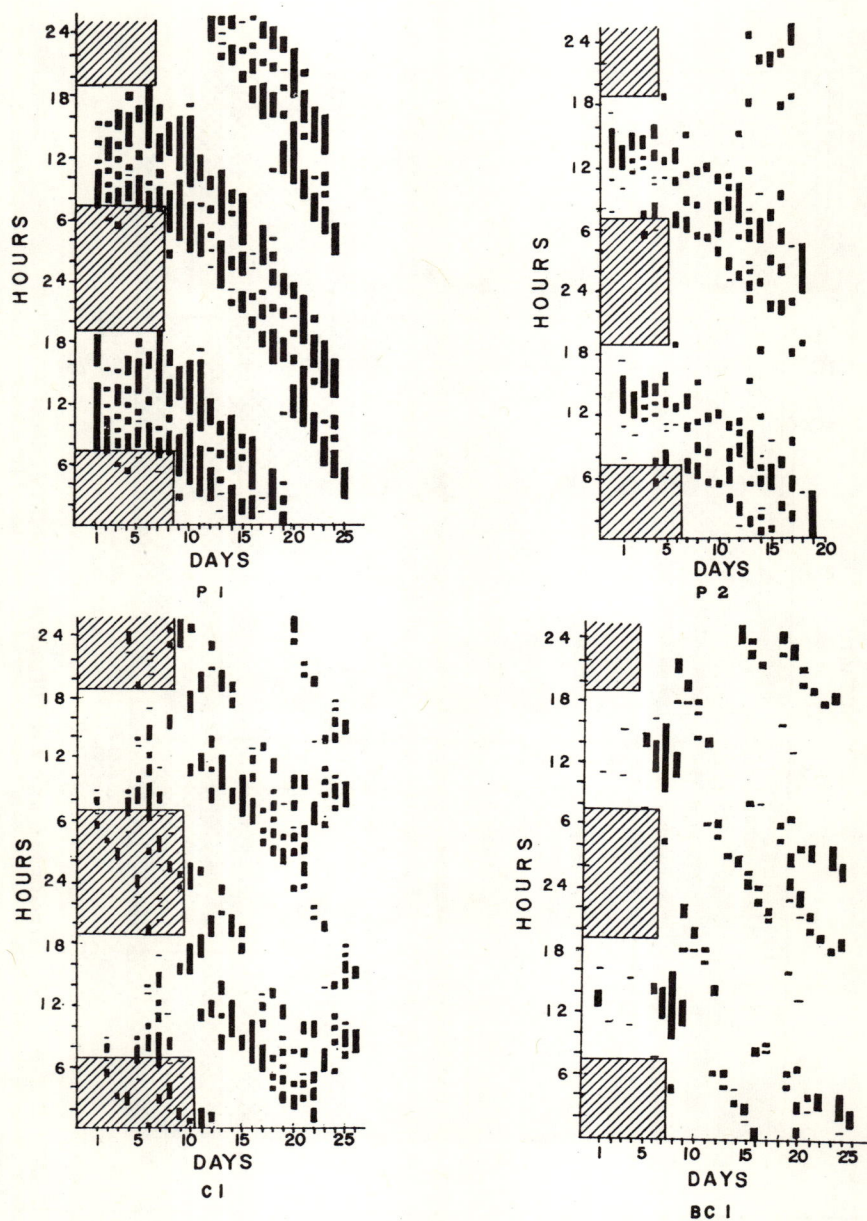

Figure 2 Activity patterns of gopher tortoises, *Gopherus polyphemus*, under alternating light-dark conditions and under constant light with light refuge present. Vertical bars represent periods during which the tortoises were active; shaded area, the periods of imposed darkness. To illustrate more clearly rhythmicity, the succeeding day is shown above the numbered day. (From Gourley, 1972.)

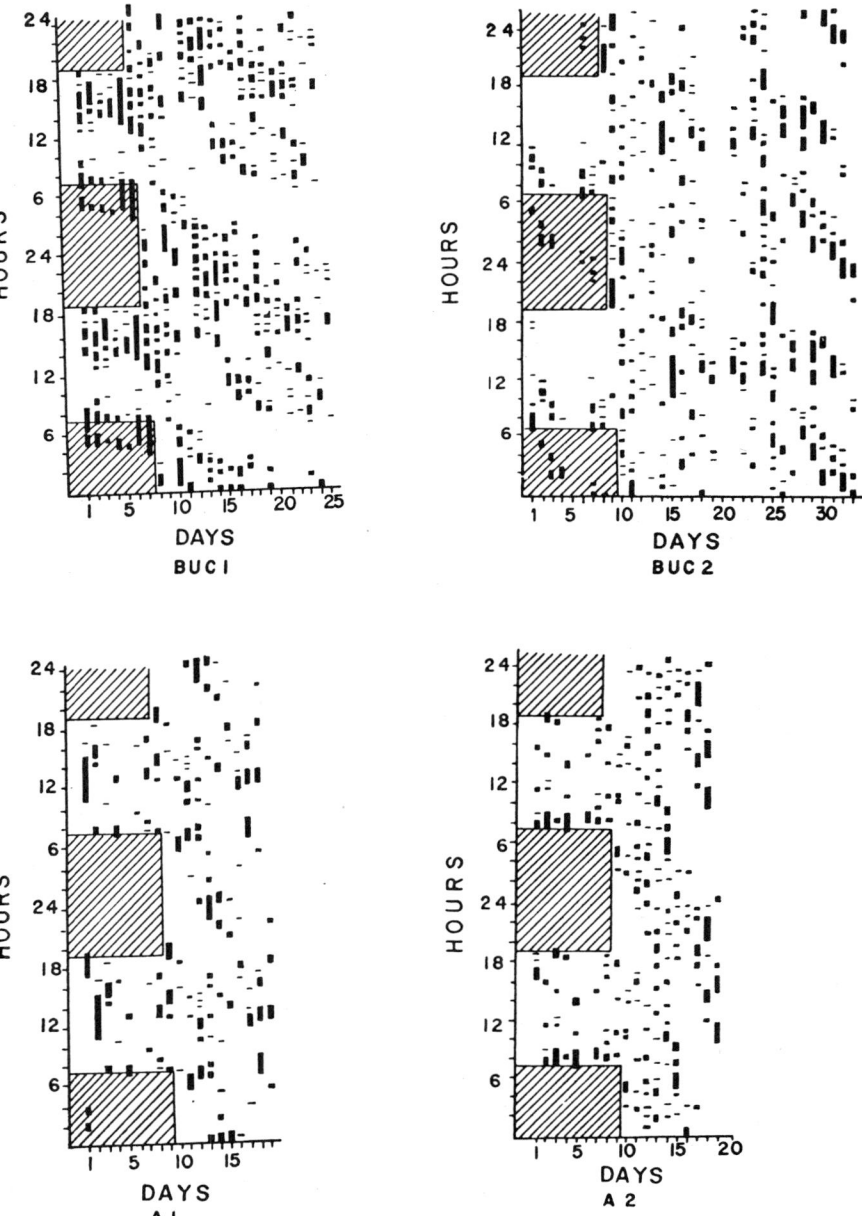

Figure 3 Activity patterns of *G. polyphemus* under alternating light-dark conditions and under constant light with no refuge from the light. Data presented as explained for Figure 2. (From Gourley, 1972.)

effect on the activity and behavior of reptiles. Gourley (1972) reported that an environmental temperature fluctuation of 5°C had a marked effect on the onset of activity in *Gopherus polyphemus* under constant light conditions. Quay (1971), however, found that in *Sternotherus odoratus* a natural thermoperiod (daily rise and fall of temperature within the photoperiod) and an inverted thermoperiod both resulted in similar activity patterns. By gradually shifting the photoperiod and temperature cycle out of phase with each other, Graham (1972) found a sudden alteration in the activity pattern, indicating that the locomotor activity pattern in *C. picta* is closely coupled to the biological clock.

Many other variables have been shown to influence the rhythmicity of organisms—geomagnetism, background radiation variations, and cosmic radiation (Brown, 1973); but these apparently have not been examined in turtles.

Lunar-related rhythms have been reported by Brett (1971) in *Pseudemys scripta*. Sea turtles demonstrate a biweekly nesting periodicity that might prove to be a lunar-month rhythm. Another form that might demonstrate a tidal rhythm is the estuarine *Malaclemmys terrapin*.

Annual rhythms have been recently demonstrated in birds but have not been investigated in turtles. The previously mentioned annual stability in hibernation and egg laying for several species may show persistence when studied in individuals.

ADAPTIVE SIGNIFICANCE

The rhythmic activity patterns of organisms have important adaptive functions, including synchronization with physical and biological factors in the environment (Cloudsley-Thompson, 1961; Bünning, 1964).

The physical factors keep the organisms in phase with the environment, even with a continuously running biological clock. The adaptiveness of this phase synchronization obviously allows animals to encounter the appropriate temperature and humidity for effective locomotion and physiological function and to prevent undue dessication. The separation of activity patterns in nocturnally and diurnally active species may help to reduce competition between species with similar food requirements.

The few studies on turtles suggest a decided correlation between rhythmic activity and the environment. Bustard (1967) points out that hatchling *Chelonia mydas* emerge at night. This allows them to avoid diurnal predators and the high temperatures characteristic of the nesting beaches, although numerous nocturnal predators attack the hatchlings. Kosh and Hutchison (1968) found that the critical thermal

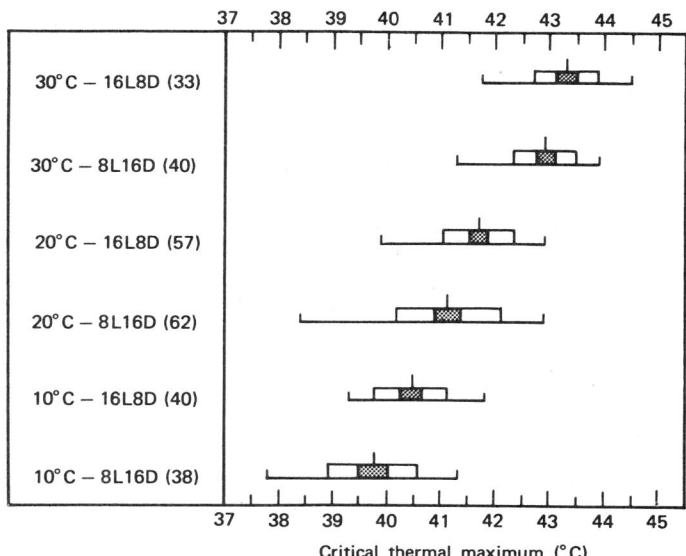

Figure 4 The critical thermal maximum of *Chrysemys picta* at different acclimation temperatures and photoperiods. The numbers in parentheses are sample sizes. The letter and number combinations represent the photoperiod regimes used in acclimation; for example, 16L8D = 16 hr of light alternating with 8 hr of darkness. The horizontal line represents the range; the thin vertical line, the mean; one black and one white rectangle combined on each side of the mean, one standard deviation; one black rectangle on each side of the mean, two standard errors. If the black rectangles of any two sets of data do not overlap, the difference between the two means may be considered to be statistically significant. (From Hutchison and Kosh, 1964.)

maximum showed a diel pattern in *Chrysemys picta*, with higher temperatures tolerated during the light phase in alternating light-dark conditions. Hutchison and Kosh (1964) also showed a higher critical thermal maximum under a 16 hr photoperiod than under an 8 hr photoperiod in *C. picta* (Figure 4). Graham (1972) found that the mean preferred temperatures of *C. picta* and *Clemmys guttata* were higher when acclimated under 16 hr photoperiod rather than 8 hr photoperiod at the same temperatures. Such preferences and shifts in critical thermal maxima may be adaptations that facilitate acclimation of tolerance to seasonal changes. Graham (1972) also reported that at 15°C the pattern was bimodal. This change in pattern may be adaptive in obtaining maximum thermal input at low temperatures and reducing thermal stress at higher temperatures. Quay (1967) has reported a 24 hr rhythm in the neural transmitter 5-hydroxytryptamine in *Pseudemys scripta*. This chemical rhythmicity probably affects the activity and behavior of this species, but the effect is not yet known.

One behavior pattern that is intimately associated with endogenous rhythms is that of long-range celestial orientation: compass orientation or bicoordinate navigation. Compass orientation is manifested by an individual or a population showing a preferred direction of travel; when displaced, they continue to maintain the original preferred direction. Bicoordinate navigation also involves a preferred direction; but after being displaced, the animals prefer a direction that would take them to the same point as the original preferred direction before displacement. Many investigators have found evidence that the environmental cue that provides the animals with the necessary directional information is often the position of a celestial body: the sun, the moon, or star positions. For long-range celestial orientation to be effective, it must be time compensated to allow for the movement of these celestial bodies. This time compensation apparently involves the animal's internal clock. Considerable evidence supports such a relationship. The most rigorous indication involves artificially phase shifting the animal's internal clock 6 hr out of phase with natural conditions by shifting the phase of the artificial photoperiod (a process that takes about a week) and noting a 90° shift in direction (Hoffman, 1954).

Several reports attest to the presence of such a time-compensated solar orientation in turtles. Gould (1957, 1959) reported navigation in *Terrapene carolina* and *Chrysemys picta*. He displaced individuals of these two species and recorded a direction preference in a homeward direction. Fischer (1964) suggested similar capabilities in hatchling *Chelonia mydas*. Emlen (1969) reported landmark orientation but no evidence of celestial orientation in *C. picta*. Murphy (1970) demonstrated celestial orientation in *Pseudemys scripta* when landmarks were excluded by the walls of an arena and noted that sun-compass orientation could be demonstrated only in the absence of landmarks. Gourley (1974) reported time-compensated solar orientation in *Gopherus polyphemus* in open fields and in an arena that excluded landmarks; he noted only slight differences in preferred directions under these two sets of conditions. When the activity cycle was phase-shifted 6 hr, however, the characteristic 90° shift in preferred direction was obtained (Figure 5). This clearly indicates that *G. polyphemus* is capable of time-compensated solar orientation.

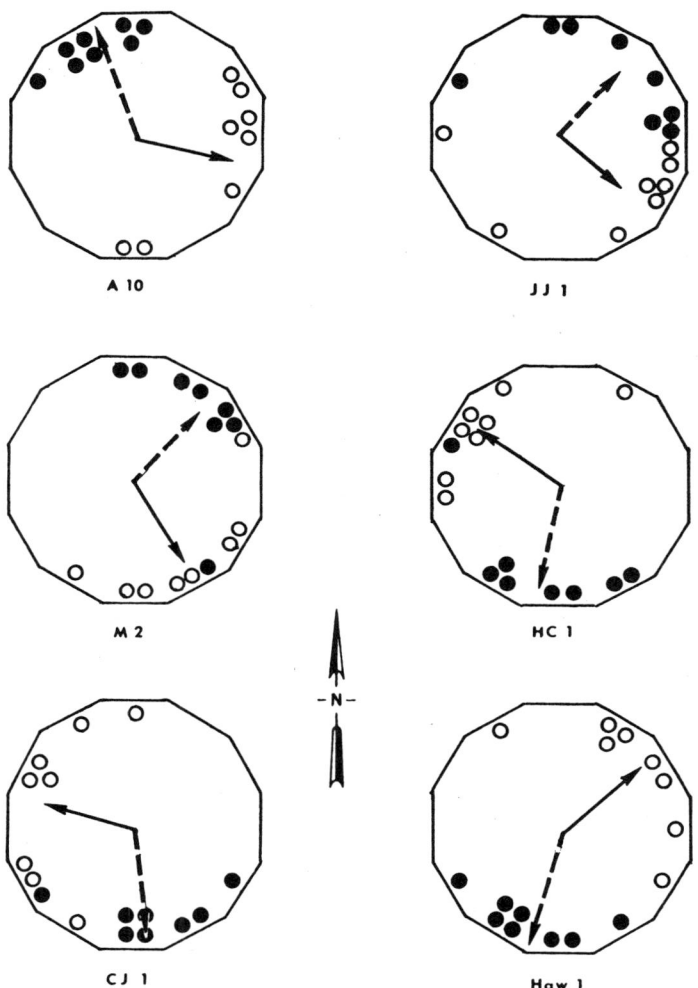

Figure 5 Comparison of original preferred directions of G. *polyphemus* in an arena during the morning testing period and the preferred directions after the tortoises had been subjected to a phase shift 6 hr advanced. Open circles, original directions; solid circles, directions after phase shifting. Solid arrows indicate original mean preferred directions; dashed arrows indicate shifted mean preferred directions; arrow lengths indicate r_c values. (From Gourley, 1974.)

CONCLUSION

The few papers pertaining to rhythms in turtles present conflicting results, but not enough information is available to determine whether these conflicts are due to procedural differences or to species differences. Rhythms have a marked effect on the behavior of turtles, and this area must be accorded considerably more attention, if the behavior patterns of turtles are to be understood.

POPULATION DYNAMICS

CHAPTER 24
Population Dynamics of Sea Turtles

H. ROBERT BUSTARD

The marine turtles are placed in two families, the Dermochelyidae with the single species *Dermochelys coriacea* (luth or leathery turtle) and the Cheloniidae, to which the following six species belong: *Chelonia mydas* (green turtle), *C. depressa* (flatback), *Caretta caretta* (loggerhead), *Lepidochelys kempi* (Kemp's ridley), *L. olivacea* (Pacific or olive ridley), and *Eretmochelys imbricata* (hawksbill or tortoiseshell turtle). Included in the group are a vegetarian browser on algae and angiosperms (*C. mydas*), a specialized jellyfish feeder (*D. coriacea*: Brongersma, 1969), and carnivorous or predominantly carnivorous species (*C. depressa, C. caretta, Lepidochelys* spp., and *E. imbricata*).

Zoogeographic information is presented in Chapter 1, and additional studies would be rewarding. Not only might they help to explain the "local" distribution of *L. kempi* and *C. depressa*, but they also could be expected to shed light on the major rookery areas. Preliminary results of our research indicate that microzoogeographical studies could prove to be especially rewarding. In the Torres Strait region are hawksbill populations, readily distinguishable on carapace morphology, coloration, and shell thickness, yet they nest on coral cays only a few miles apart. The biological implication of virtually "sessile" *E. imbricata* populations each living on its own reef and nesting on the associated cay with extremely little contact with neighboring populations offers completely new insights into the known behavior of sea turtles.

Despite the commercial importance of sea turtles, their population dynamics are not well known, largely because of the difficulty of adequate sampling at sea. For this reason research has been largely concentrated on the tiny but vital period spent ashore during nesting.

Major impetus for postwar research on sea turtles resulted from Carr's predominantly Caribbean work (Carr, 1952; Carr and Giovannoli, 1957; Caldwell, Carr, and Ogren, 1959; Carr and Ogren, 1960; Carr and Hirth, 1962; Carr, Hirth, and Ogren, 1966; Carr, 1967; Hirth and Carr, 1970; Carr and Carr, 1970; Carr, Ross, and Carr, 1974). Also important was the work of Harrisson, who "invented" the use of the monel metal cow eartag to mark turtles in Sarawak, now Eastern Malaysia (Harrisson, 1951, 1952, 1954, 1956a, 1956b, 1962), and Hendrickson (1958), also in Malaysia. Hendrickson's exhaustive paper, the first attempt to consider the overall population ecology of a marine turtle species (*C. mydas*), has been a major contribution to marine turtle research. Most valuable work on the South African populations of loggerhead and leathery turtles has been carried out by Hughes and co-workers (Bass, McAllister, and van Schoor, 1965; Hughes, Bass, and Mentis, 1967; Hughes and Mentis, 1967a; Hughes, 1970). The early work of Carr and Harrisson stimulated me, and in the extremely favorable environment of Australia's 1250 mile long Great Barrier Reef, which is free of human exploitation, I commenced a long-term population study in 1964 (Bustard, 1966; Bustard and Greenham, 1969; Bustard, 1972a).

This account sets out to highlight specific, profitable areas of endeavor in marine turtle population biology. A number of these, though pure research projects, are doubly valuable because the work would have immediate applied use. In view of the current state of sea turtle conservation and management, some priority toward these aspects is warranted.

ACTIVITIES

Data on activities of sea turtles are piecemeal, frequently being the result of cursory observation. We examine the daily and seasonal cycles separately.

Daily Activity Cycle

A synthesis of available information for the green turtle (*Chelonia mydas*) indicates that this species is a diurnal browser that retires to safe retreats in the late afternoon. Mating also takes place by day; however Hendrickson (1958) refers to nocturnal mating in Malaysian *C. mydas*. Behavior in the green turtle is so variable geographically that experts have for some years referred to the "green turtle species group," indicating that extensive studies may delineate several distinct species.

In remote areas of the Pacific such as islands off northern Australia (Bustard, 1972a) and in the outer Hawaiian Islands (Kenyon and Rice, 1959), adult green turtles haul out by day to lie in the sun. Records go back several hundred years (Dampier, 1697). Basking turtles are invariably females that haul out in the morning and return to the water in the late afternoon (Bustard, 1972a). This period coincides with the time during which male turtles actively seek mates. The hauling out of the females may be avoidance behavior to elude promiscuous male attentions. Groups of adult *C. mydas* strand themselves in small tidal pools presumably for the same purpose.

There are fairly precise daily nesting cycles in many parts of the world. Although these are frequently associated with tidal cycles—particularly where these are well marked—time of day may also be a key factor. In *Chelonia mydas* of the Great Barrier Reef, nesting takes place during darkness. When a high tide falls in the early evening, turtles await dusk before emerging. On the early morning high tide turtles emerge before daybreak, though those with a successful egg chamber will complete the nesting process during daylight. Since green turtles' body temperatures increase by several celsius degrees during the nesting process as a result of the arduous digging activity, there appears to be selective advantage for this timing. It is puzzling, therefore, that it is not exhibited by all species. Barrier Reef *Caretta caretta* likewise avoid the heat of the day, but nest in the late afternoon if the tide is high and again in the early morning as well as by night. *Chelonia depressa*, in the heat of the Australian tropics (about 10°S) will nest at any time of the day or night in the absence of a strong tidal cycle, and Australian *Eretmochelys imbricata* populations at similar latitudes also nest throughout the 24 hr period, with only slight preferences for nocturnal nesting indicated. The huge arribidas of ridleys in Mexico and Central America may also occur by day. *Dermochelys coriacea*, in contrast, behaves like *C. mydas* and usually nests under cover of darkness.

Seasonal Cycle

The more constant marine environment of sea turtles is a moderating influence on seasonal cycles. No hibernation or estivation, for instance, takes place in any marine turtle species, although there is now evidence of winter dormancy in some Mexican populations of *C. mydas* (Felger, Clifton, and Regal, 1976).

The major seasonal cycles are related to nesting. Pritchard (1973) reported that some Surinam green turtles have been observed to renest after only 1 year; usually, however, there is no annual reproductive cycle in sea turtles. Nesting takes place repeatedly in a single nesting season after an interval of 2, 3, or 4 years, with the modal interval being 3 years, depending on the population. The reproductive effort expended in a nesting season, both in terms of egg production and lengthy migration undertaken, is the major cyclical factor in the lives of adult female turtles.

Migrations from feeding grounds to rookeries and back again may be extremely long. Carr (1967) has shown on the basis of tag recoveries that the Costa Rica green turtle population moves as far afield as the Yucatan peninsula, into the Bay of Campeche, the Florida Keys, and Isla de Margarita, Venezuela. Turtles tagged by Carr and his associates on Ascension Island have been recaptured in feeding grounds off the Brazilian coast and have also been recaptured nesting again at Ascension. Green turtles nesting on the beaches of Surinam also use the feeding grounds off the coast of Brazil. Thus turtles from at least two populations come together for feeding and many months later depart to two widely separated nesting areas (Pritchard, 1973).

The speed with which migrations occur indicates that they are active movements, not the result of random wanderings. For instance, Hughes, Bass, and Mentis (1967) recorded the recapture of a Natal (South Africa) tagged loggerhead 1650 miles to the north on the coast of East Africa after 91 days. Bustard and Limpus (1970, 1971) reported on recoveries after only 63 days of Queensland tagged turtles, including a loggerhead, recaptured in the Trobriand Islands off eastern Papua, New Guinea, a distance of 1000 miles in a straight line from the tagging site, but approximately 2000 miles by the more likely turtle route following the coastline. Green and loggerhead turtles tagged at Heron Island at the extreme south of the Great Barrier Reef and loggerheads tagged at Bundaberg further south on the Queensland mainland are captured northward along a 1000 mile stretch of the East Queensland coastline. A number reach Papua, and a disproportionately large percentage round the "heel" of eastern Papua to reach the more northerly Trobriand Islands. Several turtles have been captured in the Gulf of Carpentaria, a migration of at least 1500 miles, and a number of *C. mydas* have been recaptured in New Caledonia up to 1000 miles east of Heron Island.

How sea turtles accomplish their long-distance migrations between feeding grounds and nesting grounds is not known. Koch, Carr, and Ehrenfeld (1969) proposed that the nestward migration of the green turtle population that nests on Ascension Island might be guided by a combination of olfactory cues from waterborne chemicals and a compass orientation sense based either on the position of the sun or on the direction of the prevailing waves (see also Carr, 1967; Chapter 15). Carr and Coleman (1974) speculated that this arduous long-range migration resulted from sea floor spreading, the nesting area on Ascension Island moving gradually away from the feeding grounds off the Brazilian coast.

While at the nesting beach, the female green turtle nests two or more times; the usual range of the internesting interval is 12 to 14 days. These females show what Carr (1975) termed "site tenacity" in that successive emergences occur in approximately the same location on the beach where they were originally tagged years earlier. *Caretta caretta* also shows nest site tenacity, but LeBuff (1974) reported that nest site tenacity may be less well developed in the Florida Gulf Coast population of loggerheads than in other sea turtle populations.

The intervals between nesting emergences in a given season are taken up with courtship and mating. Bustard (1972a, 1973a) and Booth and Peters (1972) describe and include photographs of such internesting behavior for Australian populations. Carr, Ross, and Carr (1974) describe the internesting behavior of female green turtles off the shore of Ascension Island (see also Carr, 1967).

ECOLOGICAL RESEARCH

Nesting

The gross physical aspects of beach features affecting nesting as well as beach microenvironments all need further investigation to determine their significance in turtle nesting and egg ecology (Hendrickson, 1958; Bustard and Greenham, 1968; Chapters 17, 18).

The sheer density of nesting turtles at a rookery may result in massive egg destruction. Bustard and Tognetti (1969) carried out detailed computer simulations of turtle populations of various sizes on nesting beaches. Following this, Bustard and Matters (in prep.) used the computer to study beach configurations and the effects of intra- and interspecific turtle competition on the incubating eggs. The interaction between nesting populations of green and loggerhead turtles is shown in Bustard (1972a). On a coral cay with 40 nesting loggerheads, the probability of nest destruction by intraspecific disturbance is .15. However if the nesting green turtle population on the same cay is taken into account, the same small loggerhead population has a combined probability of nest destruction (resulting from intra- and interspecific effects) approaching 40% of nests laid.

Such studies are important in demonstrating underlying causative factors affecting population parameters. If management was required to potentiate the loggerhead in the example above, control of green turtles would be most helpful. Furthermore, a redistribution of features on the high beach platform would help to spread out the nesting, thereby reducing intraspecific nest destruction.

Human interference in managing nesting may also have a deleterious effect. Workers operating sea turtle hatcheries have reported a regular substantial reduction in percentage hatch when eggs are moved. It was at first assumed that this resulted from rough treatment by unskilled labor (Hendrickson, 1958). However it is now known that such reduction occurs under the most rigorous conditions, as when the eggs are collected as laid, carried a few hundred yards in sealed plastic bags, and quickly reburied in the sand. Percentage hatch, which under natural conditions in undisturbed nests may average around 90%, drops to 60 to 65% (Bustard, 1972a). Such a marked drop has important conservation and economic overtones and requires careful investigation (see Chapter 17).

Hatchlings

The mechanism of synchronous emergence from the nest is described by Carr and Hirth (1961) and Bustard (1967). From a predation standpoint, the importance of this sudden bursting forth of a complete batch of young is obvious (see Chapter 18).

Whatever the level of land predation on the nocturnally emerging hatchlings, the level is very considerably higher once the hatchlings reach the water, especially if they have to cross a shallow reef platform (Hendrickson, 1958). Percentage survival during this hazardous period may depend on the overall numbers making the journey (Bustard, 1972a, 1976). Hence enormous rookeries may be important for the survival of hatchlings, even if the densities achieved result in substantial levels of destruction for the incubating eggs by the turtles themselves. This loss may be small compared to that sustained when the newly hatched turtles enter the water. It may be necessary to flood the sea surrounding the breeding beaches with baby turtles because few may get away to deeper water until the predators have taken their fill.

If this hypothesis is supported—and it should be tested by computer studies following lengthy field input collation—the faster the turtles can get their young into the water, the better. Thus short nesting seasons, in which the turtles arrive synchronously and produce their egg lay as quickly as possible, should be favored. It is interesting to see how closely field observations fit the hypothesis. In the case of the green turtle, clutch size is enormous. About 110 eggs are laid in clutches deposited at fortnightly intervals. This would appear to be the minimum time required to mature the next clutch of eggs, and each clutch certainly seems to be the maximum the turtle can produce at a time in regard to sheer physical accommodation, quite apart from production requirements in terms of resources. Incidentally, the arribada phenomenon of the ridleys (*Lepidochelys*) must have marked survival value for the nests and hatchlings: in the case of *L. kempi*, most of the breeding females emerge on a single beach at one time to nest, taking these limits to their ultimate extreme. Hildebrand (1963) recorded an estimated 40,000 turtles on 1 mile of beach at one time.

High-density nesting must also reduce overall percentage destruction of incubating eggs by nest predators, though the destruction of nests by other nesting turtles will increase. The egg predation is directly analogous to that on the reef platform. The predators become satiated after devouring only a percentage of the production, the percentage decreasing as production increases.

This phenomenon, if valid, has important lessons for conservation attempts at recreating rookeries that have been destroyed. Though clearly new rookeries can be created by the turtles, a new rookery may develop only under most exceptional sets of circumstances, analogous to tree regeneration in a desert resulting from a series of above-average rainfall years. This may explain why Carr's worthy attempts in this direction in the Caribbean have failed to produce detectable results—the numbers released at any one site have been insufficient to allow for postpredation survivors. It may be more important to maintain the current levels of some of our present rookeries rather than making efforts to recreate extinct ones.

Hatchling coloration also requires investigation. Dark carapaces may function as important radiant heat absorbers in varying degrees in all species, but most notably in hatchling green turtles (Bustard, 1970). The lighter dorsal coloration of hatchling *Chelonia depressa* may reflect the high levels of sunshine and tropical environment in the shallow seas inhabited by this species. This is, however, sheer speculation at present.

Young Turtles

After the hatchling turtles enter the sea and disperse, they are seldom seen during their first year. This has been referred to as the "lost year" [International Union for the Conservation of Nature and Natural Resources (IUCN), 1969, 1971]. This is not a helpful concept, making a mystery where none exists, and channeling research into a literal search for a needle in a haystack. It is obvious that hatchling turtles are so small that they will seldom be seen in the world's oceans. Furthermore, while still small enough to fall victim to many predacious fish species, there is presumably survival value in remaining in the planktonic phase in deep water. Their presence in considerable numbers in deep water offshore is attested to by the large numbers washed up on beaches in certain areas by onshore storms (Hughes, 1970, 1971). I have similar data for Perth beaches in Western Australia, where very large numbers of loggerheads about 3 months old were stranded during onshore storms. In view of their age and the location of the nesting grounds far to the north, it is assumed that the hatchlings drift southward on the prevailing current.

Virtually nothing is known about survival or growth of wild sea turtles. The difficulties of such work are immense; for instance, how is one to tag or otherwise mark an animal, weighing 20 to 30 g, which will not return to the beach (if it does survive to return and home successfully) until it weighs about 100 kg (Chapter 2)? Furthermore, enormous numbers must be marked. In a preliminary study I marked more than 65,000 green turtle hatchlings by a marginal removal method allowing year class identification, during the years 1965–1968. One individual that seems unquestionably to belong to the 1966 group was recaptured by Limpus. At 8½ years, the turtle, still subadult and tentatively identified as a male, weighed 51.8 kg and had a shell length of 77 cm. Minimum breeding size of females of this population is 88 cm. The figure for males is unknown. The weight was less than a captive-reared individual of the same population, which attained that weight when between 6 and 7 years old.

If one postulates a survival to first breeding size of around 1 in 1000, then the 65,000 marked hatchlings would result in 65 adults and, if the sex ratio were equal, slightly more than 30 breeding females. This gives an indication of the scale on which tagging operations on hatchlings should be carried out. To ensure an adequate sample at maturity, taking adult nesting dispersal into account, a million or more should be marked. Though possible, this task would be formidably expensive. (In the above-quoted study, government controls limited the number of hatchlings that could be marked each year).

Growth and survival remain unexplored fields, and data on these key aspects of life history are needed before accurate life tables can be prepared.

Adult Turtles

Considerable information results from tagging adult females on nesting beaches. A sound tagging program must be extremely long term, on the order of 20 years, and the area chosen should be one allowing maximum scrutiny of the population with minimal disturbance. A coral cay is much easier to use than a stretch of mainland beach for the simple reason that having completed a traverse of the beach on a cay, one is back at the starting point, ready to commence another traverse.

Any disturbance introduces a possible interference factor that may affect return nestings, thus the apparent population parameters of the population being studied. Our staff are issued dim three-cell head torches for use only behind cover or in special circumstances. Lights are never shone on the sea because this may frighten turtles homing in on the beach. (Behavioral responses to disturbances vary greatly among individuals of a single population and among different populations of the same species, quite apart from interspecific differences.) Tagging and measurements, the latter kept to a minimum, are done as soon as the turtle has started covering the eggs. Monel, self-piercing tags of National Band and Tag Company are used, and the operation takes less than a second.

Extremely little is known about the feeding habits of most species at anything other than an anecdotal level (see also Ehrenfeld, 1974). This is a field to which major attention should be directed in several parts of the world.

An illustrated book on tameness in wild green turtles, both on land and in the sea, appeared recently (Bustard, 1973). Turtles under such close observation offer the opportunity to gather information on such previously unstudied behaviors as courtship and mating (Booth and Peters, 1972; Bustard, 1973).

Failure to investigate some of the obvious research problems undoubtedly reflects in part the expense of mounting marine turtle studies, often at a considerable distance from the base institution, and the need in many instances for the studies to be pursued over many years.

POPULATION STRUCTURE

An attempt to describe the population structure of even the green turtle at once illustrates the paucity of information on key data basic to the construction of a life table. With our present knowledge, it is not possible to make species comparisons.

Growth and Age

Except for occasional fragmentary evidence from individual turtles, present data are based entirely on captive-reared turtles and have questionable relevance to the natural situation. *Chelonia mydas* apparently reach sexual maturity at an age of 4 to 6 years (Hendrickson, 1958). This, however, may be optimistic unless the top of the estimate is taken and only strictly tropical populations are considered. There are likely to be considerable climatically linked differences in growth patterns.

Under favorable conditions in captivity, the smaller *Eretmochelys imbricata* may attain sexual maturity at 3 to 4 years. No data are available for the largest species, *Dermochelys coriacea*.

Data on age and age-related growth are absent because of lack of hatchling tagging, extreme paucity of subsequent recaptures, and absence of a technique for calculating the age of adult turtles. Ideally a technique that can be applied to living turtles on the beach is required, but even a skeletal aging technique of demonstrated reliability would be a major advance.

Data are required on the relative abundance of each age (size) class in order to calculate recruitment figures.

Sex Ratio

No reliable information is available on adult sex ratio. Comments are widespread in the literature on the presence of several attendant males around a copulating pair. Our own observations indicate that this does not imply an excess of males, as has been suggested, but merely a combination of the males' promiscuity during the breeding season and the fact that most females do not permit repeated copulations, with the result that males are continually actively searching out receptive females. Male activity at this time also may bias direct counts.

Data from turtle fisheries are likely to be biased. Turtles are most readily taken during the breeding season when the females are ashore laying their eggs, and if turtling is carried out only on the beaches, the sample will be entirely female. If taken on the reef, the sample is likely to be biased in favor of males. If lures are used, the sample will consist almost entirely of males. To date no scientific study has been carried out on adult sex ratio.

Sex ratio at hatching is not known for any marine turtle species.

Density

No density calculations have been made for any sea turtle species, although many authors have referred to the huge numbers nesting on small beaches. Carr (1967) suggested a figure of 50 million *C. mydas* in the Caribbean before the advent of European turtle fisheries, and an estimated 10,000 now. There can be no doubt that in areas of shallow warm seas where turtle grass beds are abundant, *C. mydas* formerly were extremely numerous. In favorable areas of the world today where there has been little organized fishery, it is still possible to count large numbers browsing on a small area of reef.

POPULATION DYNAMICS

Population Studies

Population studies should, of course, be carried out in areas where human predation is minimal. If this is not possible, it is worthwhile to operate only where the likelihood of tags being returned from butchered turtles is high throughout the entire area from which the turtles are likely to be taken during their major migrations. Failure to appreciate this will cast doubt on survival values. Tag recoveries, a regular feature of fisheries operations, do aid the accumulation of growth and migration data.

The production data are well known for the green turtle and in varying degrees for other species. Main emphasis should be concentrated on both juvenile and adult loss factors, on growth, and on longevity. Adult tagging can give considerable information on the latter two points.

Reproductive Effort

Extreme fecundity offsets the substantial, but geographically highly variable, level of egg predation and the enormous toll taken of newly hatched turtles during very early life in the water.

Turtles are disadvantaged as archaic reptiles because all are oviparous. In the case of sea turtles this necessitates not only long migrations to the nesting beaches, but also the process of hauling out and completing the laborious nesting ritual.

The number of clutches laid shows species differences and also highly significant intraspecific geographical variations, and even marked interseason differences at the same rookery (Bustard, 1972a). Clutch size, also showing interspecific differences, is remarkably constant geographically within a species, at least in the widely studied green turtle with an average of around 110 eggs per clutch. However this constancy appears to occur at the expense of considerable variations in mean egg weight (Hendrickson, 1958; Bustard and Greenham, 1968, 1969).

A number of factors affect clutch size in the green turtle (Bustard, 1972a). One of these is turtle size, with larger turtles producing larger clutches of eggs. There are also unexplained but substantial differences in the number of clutches laid by females using the same rookery in different years. In three successive years at Heron Island, south Queensland, the mean number of clutches laid was between 3 and 4 in the first year, between 5 and 6 in the second (a marked increase resulting in an overall increase in reproductive effort of 73%), and intermediate in the third year. Even greater interyear differences have been recorded for *C. caretta* (Bustard, 1972a). The observation period was the same in each year, and there is no reason to doubt the efficiency of the tag recapture operation on this small cay in any year. The variation in reproductive effort could reflect population differences (i.e., between the "populations" that may be quasi-isolated genetically, nesting on the cay in different years), or perhaps it is a reflection of variation in the quantity or quality of food supply in the preceding season(s).

Reproductive physiology would appear to be an extremely profitable field that could throw light on the foregoing question, as well as being of major interest in its own right. Even behavioral accounts of mating are sparse, and such questions as sperm retention in sea turtles have not been adequately tackled.

Recruitment

Data are not available on recruitment, a key aspect of the population dynamics for any species of sea turtle, and the available data from which inferences can be drawn are subject to conflicting interpretation because of the absence of firm data on other factors, notably mortality/longevity and growth.

Mortality

Quantitative data on mortality, essential to a proper understanding of population ecology, are lacking; however relevant data are scattered through the literature.

Adult turtles, by virtue of their large size, are safe from most predation, but shark predation is a continuing threat to even the largest turtles. Bustard (1972a) cited lethal attack on a large loggerhead by a hammerhead shark and a tiger shark; a tiger shark is also recorded by Travis (1959), and Caldwell and Caldwell (1969) recorded the remains of leathery turtle in the stomach of three killer whales taken in the West Indies.

Duncan (1943), Travis (1959), and Bustard (1972a) refer to behaviors of adult turtles that minimize risk of attack. These include seeking out safe retreats, such as under ledges of coral or shelving rock, in the late afternoon. The turtle moves as far under the protection as possible and always faces inward. This may in part explain why major shark damage invariably occurs to the rear of the turtle. Another possibility is that the turtle was fleeing when attacked. Hendrickson (1958) probably comes closest to the truth when he says that only the lucky ones survive to be recorded as victims of shark attack. Except in the most exceptional circumstances, successful anterior attacks must result in the turtle's death; thus our sample of injured turtles is biased. We have observed one regularly nesting female that had lost the entire left front flipper, but was in visibly good condition.

Since damage such as results from shark attack (viz., loss of limbs or parts thereof) is not cumulative throughout life, at least in the green turtle (Bustard, 1976), such damage must in some way debilitate the turtle sufficiently to make it more prone to future attack. Furthermore, the size-frequency distributions (Bustard, 1972a, 1976) peak in the middle region instead of showing a large cohort at the top end of the observed growth curve, as might be expected. This provides further evidence of loss from the population after the middle of the nesting female size distribution is attained. These comments are based on the assumption that green turtles either grow throughout life like other reptiles studied, albeit very slowly in adult life, or that growth virtually ceases after attainment of a certain size. The limited data available (Bustard 1972a, 1976) suggest that postmaturation growth, even in females toward the lower end of the nesting size distribution, is extremely slow. An alternative explanation is that females mature at a certain age and that the size at this age reflects feeding opportunities during immature life. Available evidence does not permit a decision on whether the large size discrepancy between nesting females is the result of a long period of postmaturational growth, probably at a slow rate, therefore a great age and consequent low mortality, or whether the discrepancy results from different prematurational growth rates and low or nil growth rates thereafter, consequently the possibility of low age reflecting continuing high mortality levels. It should be noted, however, that if the latter hypothesis is correct, "growth" data collected on breeding females are grossly misleading. Furthermore, the latter hypothesis would explain why the incidence of shark attack is not found to be cumulative throughout life, looked at from a size-distribution viewpoint.

Mortality is known to be high for hatchling turtles, but once again quantitative data are lacking. The smaller the hatchlings, the greater the range and, therefore, abundance of potential predators. After only a matter of weeks at sea, growth in *Chelonia mydas* must make the young turtles immune from attack from many of the important predators on the newly hatched turtles. *C. depressa* is the only species producing a markedly smaller number of significantly larger eggs. Clutch size in *C. depressa* is about half that of *C. mydas*, and the individual eggs are about 50% heavier than local *C. mydas* eggs, themselves large by world standards (Bustard and Limpus, 1969). Silver gulls (*Larus novaehollandiae*), important predators on any young *C. mydas* that atypically emerge by day, cannot swallow hatchling *C. depressa*, and ghost crabs (*Ocypoda ceratophthalma*), which take a heavy toll of hatchlings of all other species, cannot hold back the larger and much more powerful *C. depressa* hatchlings.

Longevity

No accurate data are available on longevity in the wild of any species, though inferences of doubtful significance may be drawn from zoo data (e.g., Flower, 1938). Eventually longevity data will accumulate from the recapture of tagged adults if the tags prove sufficiently long-lived.

One field requiring detailed research vital to longevity studies is the percentage of the adult population actually returning to nest in subsequent nesting years. It has been assumed, perhaps with insufficient data, that turtles possess a strong sense of return to the rookery of their hatching. The best data for subsequent return of turtles marked *as adults* to the same breeding beach are those of Carr (1967). For the Costa Rican population, Carr recaptured 447 individuals back at the nesting beach in subsequent years from a tagged population of 5758 green turtles. It should be noted that 55 of these returns did not occur until between 5 and 9 years had elapsed (Carr and Carr, 1970), an observation that indicates the importance of long-term studies. Of 635 green turtles tagged on Ascension Island, only 8 were recaptured at Ascension after periods of 2 to 4 years (Carr, 1967). Similarly, Harrisson in Sarawak recorded a low level of recaptures in subsequent years. Low return figures of adults between nesting seasons (not intraseasonally, where the return is extremely strong) pose problems. Certainly the turnover in populations cannot be sufficient to account for returns at the 1 to 2% level (Bustard, 1972a). But what happens to the remainder? Nothing is known except that return may sometimes be long delayed (Carr and Carr, 1970). We do not know if these turtles have failed to breed or if they have nested elsewhere in the interim, although there have been a few recaptures at nest sites other than the one at which tagging occurred

MANAGEMENT

(Carr, 1975; LeBuff, 1974; Bustard, 1976). Until further data are available, it seems unlikely that loss figures, therefore conclusions about longevity, can be accurately developed.

Species Accounts

Species accounts have been published for *Dermochelys coriacea* (Pritchard, 1971b) and *Lepidochelys kempi* (Pritchard and Marquez, 1973). A recent account of the conservation status of the green turtle is given by Bustard (1974).

MANAGEMENT

Management, in its narrow sense, has the dual aim of producing turtles commercially and thereby, also independently, conserving world turtle stocks. In the wider sense it means the application of knowledge to manipulate the population for any purpose. Some of the best work in management, though at a less rigorous level than would be expected today, is quite old. Outstanding accounts are provided by Hornell (1927), Banks (1937), and Ingle and Smith (1949).

Population ecologists can play a key role in conservation and management defined in the broadest sense. A number of recent population surveys have been reported including those of Pritchard (1969a) for the Guianas, Carr (1969) in the Caribbean, Hirth and Carr (1970) in the Gulf of Aden and the Seychelles, Bustard (1970) in Fiji, and Frazier (1971) at Aldabra. Although mostly tentative in their conclusions, they do indicate areas for future work. The proceedings of the IUCN Turtle Specialist Group meetings (IUCN, 1969, 1971) contain much valuable geographical information not referred to under specific authors. There is a need for continued applied investigations along the lines of Bustard (1971, 1973).

Though a precise life table is not yet available for any species, it is possible to arrive at a very simple rule-of-thumb calculation providing extensive insight into population parameters. Let us consider the simplest calculation: a green turtle nesting four times in a breeding season and having a 3 year nesting cycle. This turtle, laying an average of 110 eggs per clutch, produces 440 eggs each season, that is, during each 3 year cycle. If one arbitrarily assumes a breeding lifespan of three cycles, the production is approximately 1320 eggs. These were produced by one pair of turtles, and if the population is to remain stable they must, on the average, replace themselves; that is, 2 of these 1320 eggs (1 in

660) must survive to replace the parents in the breeding population. On the other hand, green turtles probably survive more than three cycles, and a more accurate estimate might be between three and six cycles. So we arrive at an approximate figure of about 1 in 1000. As pointed out by Hendrickson (1958), marine turtle eggs are therefore ecologically cheap and adult turtles ecologically expensive. This is another way of saying that marine turtle populations are geared to a large wastage or loss of eggs and hatchlings, but cannot withstand a large sustained loss of the adult cohort of the population. All this could change, however, if the egg/hatchling loss was to be greatly curtailed by management.

Marquez and Doi (1973) have used modern methods of population estimation to determine the level of exploitation in a wild population (unmanaged) of *C. mydas* in the Gulf of California (Mexico). They predict the likeliest situation as follows: the present population is 1600 tons, the initial population was 8500 tons, and the maximum sustainable yield is 164 tons. The optimum population level on which the maximum sustainable yield can be provided would be 4100 tons. The present sustainable yield is 115 tons.

Marquez and Doi point out that the present status is much below the optimum level because of overfishing for 15 years, and that large egg loss by predators and human extraction is also important. Their work is important because the best management of sea turtles will harness wild populations, as Ehrenfeld (1974) stressed. Management in the example above would immediately reduce fishing for adults combined with better egg and hatchling survival as a result of legislation and predator control. This would be aimed at building up the population to the level offering maximum sustainable annual harvest. Management, by promoting much greater hatchling survival through hatcheries, could then increase the annual harvest substantially.

The form of the life table would suggest that the way to build up populations would be to protect eggs and captive-reared groups of hatchlings through the most vulnerable stage, when they are newly hatched and extremely small. However some researchers take the view that imprinting may be responsible for the return to the nesting beach in adult life, and any tampering with release of newly emerged turtles is likely, if not certain, to destroy this imprinting.

Clearly an increased harvest would be permissible only when it had been demonstrated conclusively that management had succeeded in producing a larger adult population. Management techniques cannot be used to argue the case for an increased harvest now or a further egg take now, which would have similar effects.

MANAGEMENT

There has been considerable controversy about the wild survival of hatchlings that have been reared in captivity for even a few months. Some have claimed that it should be possible to increase survival by the order of up to 100 times normal wild survival figures, but others expect the captives to do poorly when released. We have shown that crocodiles, members of the other archaic reptile group, do very well when captive reared, then released.

The experiment would not be easy in sea turtles because it would have to be done on a large scale to produce convincing results. Interpretation of the results would be difficult, since no natural survival figures are presently available for young sea turtles. Included in this research, therefore, would be the need to work on natural survival of hatchlings over periods of, say, 1 to 2 weeks, 1 month, and 3, 6, and 9 month periods. This would be extremely difficult. If a more or less closed area were used for experimental work, survival data on captive-reared hatchlings would not be as difficult to obtain. Location of such a natural area might, however, present problems. Large samples would be essential, but this would simply require captive-rearing facilities on a sufficiently large scale. In a restricted area the liberation of 30,000 to 60,000 hatchlings in each of the age groups of 3, 6, 9, and 12 months, would provide most valuable data.

During the course of field studies in Queensland we have witnessed countless attacks on hatchling green and loggerhead turtles by the black-tipped reef shark (*Carcharinus spallanzani*). This small shark, seldom exceeding 6 ft, is a voracious feeder on hatchlings. In limited trials releasing both green and hawksbill turtles in Torres Strait these sharks quickly appeared when releases were in progress. Turtles of at least 6 in. carapace length (attainable in less than 6 months) were selected and released at the beach. The sharks inspected but never attacked the turtles; presumably now outside the prey size range for this small species. In small series releases, done during the day at half-tide on an extensive reef platform, it was possible to watch the movements toward the reef edge of every member of several dozen such turtles released individually. Had they been newly hatched young, it is most unlikely that any of the group would have survived the sharks to reach the reef edge.

"Turtle Farming"

The application of the word "farming" (i.e., management of a wild population) to marine turtles is an unfortunate choice, but it is now widely accepted. The term "free range management" is greatly to be preferred. The ecological potential for free range management has been aptly stated by Ehrenfeld (1974):

> The most obvious significance of the green turtle in our time is that it occupies the second trophic level in an increasingly hungry world, utilizing an otherwise nearly unexploited habitat in the biosphere and thus not competing directly or indirectly with man for food. Furthermore, it is known to be capable of existing in large numbers near what are now the most densely populated and protein-hungry portions of the inhabited earth.

The other sea turtle species, as noted earlier, also feed on items not currently utilized by man.

We can be sure that strength of economic demand will result in considerable effort being directed towards sea turtle farming. It will go on whether we support it or not; but by lending our support, we can prevent a great many mishaps. We also have the opportunity to channel and educate the farming effort so that it achieves the maximum good for conservation of wild turtles. Unless this approach is adopted as a matter of urgency, we are likely to end with a series of "closed" farms in which the turtles spend their entire lives in concrete tanks in order to provide a high-priced gourmet steak, completely failing to utilize the enormous beds of naturally occurring turtle grass, utilizing considerable energy for the turtle production, providing no protein for the tropic poor and, incidentally, having a minimal effect on conservation (because high production costs make the captive turtles noncompetitive with wild caught turtles).

Considerable developmental effort should be directed at cottage-industry farming for people who would ordinarily be the hunters and turtle egg collectors (Bustard, 1972a, 1972b). Properly organized, this is likely to bring about an important grass roots interest in turtles as an asset, thereby making conservation practicable in poor countries. Commercial interests are already involved in farming using closed concrete pens built on land. The conservation lobby is actively involved but far from united. Free range management needs broad government backing to succeed because regulations must be drawn up and enforced. Welding these conflicting forces into a useful team to take a positive stand on fast-disappearing sea turtle resources is a formidable task.

It is crucial to realize that enterprises of turtle farming or turtle population management or ranching are unlikely to succeed without the inputs of well-trained turtle biologists. Repeated failures could prove lethal to the turtle populations.

CHAPTER 25
Demography of Terrestrial Turtles

WALTER AUFFENBERG and JOHN B. IVERSON

The materials presented here include data drawn from two closely related turtle groups: the pond turtles, Emydidae, and the true tortoises, Testudinidae. Of the former, only the most terrestrial forms are included (*Terrapene* spp., *Clemmys insculpta*, etc.). Both groups of terrestrial turtles are faced with the same environmental problems and limitations, as well as the same opportunities. The tortoises are believed to have evolved from the family Emydidae, and the ancestral forms were probably very similar to the extant terrestrial emydid species. In general, major emphasis has been placed on members of the family Testudinidae because their habitat requirements are more restricted, and they are so obviously adapted for life on land.

Zoogeography and taxonomic relationships are discussed in Chapter 1; additional information is included in Bienz (1895), Auffenberg (1966), Khozatsky and Mlynarski (1966), and Vuillemin (1972). Although tortoise species were never native to Australia or Melanesia, waif dispersal across marine barriers is common. During the Tertiary the range of tortoises extended throughout what are now temperate latitudes. This is believed due to a high degree of climatic equability, enabling "tropical" and "temperate" biotas to intermingle. As the climatic trend toward lowered temperature progressed, the distribution of tortoises, for the most part, tended to move toward the present equatorial belt, accompanied in many cases by maladjustments leading to extinction. All extant temperate species of higher latitudes excavate extensive burrows, which they inhabit for much of the year (*Gopherus agassizi, G. flavomarginatus, G. polyphemus,* and apparently some populations of *Testudo horsfieldi*).

Most testudinid species are found in xeric habitats, ranging from tropical deciduous forests, thorn bush, beach scrub, savanna of several types, and steppe, to near desert conditions. These species (predominantly grazers) are found in both tropical and subtropical areas. Mesic-forest inhabiting species are found only in tropical primary and secondary forests, where fallen fruit is sufficiently common during the entire year in the absence of grasses. Table 1 gives the generalized environmental distribution of tortoise species.

ACTIVITIES

Periodicity

Seasonal Cycle. The yearly activity pattern is affected by the necessity (or lack) of a period of hibernation (or brumation) and/or estivation, and if present, the duration. High in the mountains or in areas with long dry seasons, the activity period may be only a few months long, whereas in areas of annual uniformity the activity period may last the entire year.

Regardless of its seasonal duration, activity is never uniformly distributed temporally, and marked variations occur. In temperate areas most terrestrial turtles have the highest activity peak in spring, regardless of whether they undergo hibernation. From this point, activity gradually dwindles to a summer low, generally followed by a minor activity peak in fall. This pattern is clearly shown in the annual cycle of *Gopherus polyphemus* in northern Florida and is quite different from that of *G. agassizi* in southwestern Utah. In the latter there is a long hibernation period and individuals are active only from April to early October, with a peak period in August (apparently correlated with rainfall). In this part of the range individuals breed immediately as they come out of hibernation and before they disperse to their summer activity ranges where there is no breeding.

In Florida *G. polyphemus* do not hibernate in communal groups, and reproduction takes place in May and June. This coincides with peak annual activity, when adults (particularly males) wander for greater distances than at any other time of the year. These seasonally enlarged activity ranges are evidently due to both mate search and territorial factors.

Activity cycles of terrestrial turtles in xeric habitats are usually corre-

Table 1 Generalized Biotopes of Land Tortoises

Species	Tropical Evergreen	Tropical Deciduous	Dry Woodland	Seaside Scrub Thorn Brush	Savanna	Steppe	Desert
Gopherus							
polyphemus			XXXX	XXXX	XXXX		
G. berlandieri				XXXX			
G. agassizi				XXXX			XXXX
G. flavomarginatus				XXXX			
Testudo							
hermanni			XXXX	XXXX			
T. graeca			XXXX	XXXX			
T. kleinmanni			XXXX	XXXX			XXXX
T. marginata			XXXX				
T. horsfieldi				XXXX		XXXX	XXXX
Malacochersus							
tornieri					XXXX		
Kinixys							
belliana		XXXX	XXXX	XXXX	XXXX		
K. erosa		XXXX	XXXX	XXXX	XXXX	XXXX	
K. homeana		XXXX	XXXX	XXXX	XXXX		
Pyxis							
arachnoides		XXXX	XXXX				
Geochelone (*Indotestudo*)							
travancorica	XXXX						
G. (I.)							
elongata	XXXX	XXXX					
G. (I.)							
forsteni	XXXX						
G. (Geochelone)							
platynota		XXXX	XXXX	XXXX			
G. (G.) elegans		XXXX	XXXX	XXXX	XXXX		
G. (G.)							
pardalis		XXXX	XXXX	XXXX			
G. (G.)							
sulcata			XXXX	XXXX	XXXX	XXXX	
G. (Asterochelys)							
radiata			XXXX	XXXX			
G. (A.) yniphora			XXXX	XXXX			
G. (Aldabrachelys)							
gigantea			XXXX	XXXX			

Table 1 (*Continued*)

Species	Tropical Evergreen	Tropical Deciduous	Dry Woodland	Seaside Scrub Thorn Brush	Savanna	Steppe	Desert
G. (*Manouria*) emys	XXXX						
G. (*M.*) impressa	XXXX						
G. (*Chelonoides*) elephantopus		XXXX	XXXX	XXXX			
G. (*C.*) denticulata	XXXX	XXXX					
G. (*C.*) carbonaria	XXXX	XXXX	XXXX	XXXX	XXXX		
G. (*C.*) chilensis				XXXX	XXXX	XXXX	XXXX
Psammobates tentorius			XXXX	XXXX	XXXX		XXXX
P. geometricus			XXXX	XXXX			
P. oculifer				XXXX			
Homopos areolatus		XXXX	XXXX	XXXX			
H. boulengeri			XXXX	XXXX			
H. femoralis			XXXX	XXXX			

lated with rainfall patterns, with the highest peaks occurring during the monsoon or similar rainy periods.

For a total of 13 months the activities of several adult *Gopherus polyphemus* were carefully monitored in Alachua County, Florida (Auffenberg, unpubl. ms). Part of the automatic recording system was comprised of microswitches placed within and outside the burrow to record tortoise movements for in-depth analysis of activity cycles. Table 2 shows the results of such a tabulation on one adult male. He was active (left his burrow and walked at least 1 m from the burrow mouth) slightly more than half the days of the year, with an activity peak in April. Most movement was restricted to the burrow vicinity, for activity was considerably reduced beyond 1 m from the burrow entrance.

Daily Activity Cycle. Seasonal activity reflects the sum total of daily activities throughout the season. The same factors that influence seasonal activity also pertain to daily activity. Sleeping must be added to this list, for it marks the primary period of daily inactivity.

Table 2 Monitored Movements of Adult Male *Gopherus polyphemus*, Alachua County, Florida: August 1963–September 1964

	January	February	March	April	May	June	July	August	September	October	November	December
Number of days active/month	12	7	18	29	29	21	12	11	18	21	11	0.05 = 194 days/year
Average number of hours active/month for active days (>1 m from burrow)	1.6	1.3	4.2	8.3	6.3	3.2	1.6	2.5	3.0	2.8	2.4	2.0
Total hours available (12 hr day, 0700–1900)	372	336	372	341	372	341	372	341	341	372	341	372
Total hours of daylight used	19.2	9.1	75.6	240.7	182.7	67.2	19.2	27.5	54.0	58.8	26.4	10.0
Percent used (>1 m from burrow)	5.2	2.7	20.3	70.6	49.1	19.7	5.2	8.1	15.8	15.9	7.7	2.7

The daily activity cycle is in large part a response to temperature and moisture conditions rather than to light (although light is important insofar as it must be present). Because of the temperature and moisture factors, the daily activity cycle changes from one season to another and from one area to another. For example, in many species the daily cycle shows a unimodal distribution during the cooler months, when only the midday is sufficiently warm for normal activity. However the same species may show a marked bimodal activity cycle in the hottest and driest parts of the year, when midday conditions are simply too hot and the thermal load too great (e.g., Legler, 1960a; Auffenberg and Weaver, 1969). *Gopherus agassizi* is even active at night, in rare instances, to escape hot daytime temperatures (Woodbury and Hardy, 1948).

Although many individuals feed regularly every day during the part of the year when they are active, some become inactive for no apparent reason and may go for fairly long periods of time without food, despite good climatic conditions and plentiful food. Individuals of *Gopherus polyphemus* have remained deep in their burrows for as long as 1 month in summer, though other tortoises in the same colony were continually active during the same period (see Legler, 1960a; Stickel, 1950, for analogous rest periods in *Terrapene ornata* and *T. carolina*). During the coolest parts of the year the daily activity of most land tortoises may consist almost entirely of basking.

Hibernation and Estivation. Even the commonly inhabited tropical and subtropical areas have their characteristic seasonal changes. They may be extreme, as in annual dry and wet seasons, or very slight, as in many lowland tropical areas. Regardless of the degree or type of seasonal changes, terrestrial turtles are more susceptible to them than are aquatic species, and a greater level of adjustment is apparently necessary.

Terrestrial turtles are usually inactive during the winter months in the temperate and outer portions of the subtropical zones; they are abroad only during the warmer parts of spring, summer, and early fall. Hibernation in temperate areas is obviously a more regular and more distinct phenomenon than short-term cool weather inactivity in subtropical forms. Thus gopher tortoises of all species are often active on warm days during the winter, the length of inactivity depending on the extent and severity of the cold spell. However, *Terrapene* species, *Clemmys insculpta*, as well as populations of *Testudo horsfieldi* in southern Russia have a clearly defined, long period of inactivity that is not related to temperature alone but to temporal factors as well.

Estivation during the hot, often dry periods of the year has been reported in one form or another in all species for which activity information is available. As might be expected, estivation is most prevalent in areas of considerable seasonal variation in rainfall.

In either hibernation or estivation, the animal must locate a site providing not only adequate protection from temperature extremes, but also sufficient moisture. These specific conditions are usually obtained by burrowing underground. Most often the shell is barely but completely covered with earth. This has even been reported in tortoises with a carapace length of 1400 mm and is the general pattern in temperate emydid species. Temperate tortoises often dig a burrow, either open to the surface, or plugged [*Gopherus berlandieri*, especially young specimens (Auffenberg and Weaver, 1969)]. The plugged burrow is apparently utilized by *Testudo horsfieldi* in Russia. *Geochelone chilensis*, in the southern parts of its range in extreme northern Patagonia, spends the winter in open burrows (Auffenberg, field notes). In more equable zones, most tortoises simply pass the generally short and intermittent poor weather conditions in grass clumps, under brush piles and other debris.

Some tortoise species (*Gopherus agassizi*, etc.) hibernate regularly in the colder parts of their range (Woodbury and Hardy, 1948), but not in warmer parts (Auffenberg and Weaver, 1969). The same pattern is found from higher to lower latitude populations of *Testudo horsfieldi* and *Geochelone chilensis*. In certain widely distributed emydids, such as *Terrapene carolina*, populations in southern Florida and Mexico remain active all year.

Species with well-developed inherent rhythms that are not dependent on a temperature stimulus to enter hibernation or seek shelter are in less danger of being caught and killed by a sudden weather change than are those stimulated only by a change in temperature. Tortoises living near the distributional limits of the familial range often exist with a dangerously slight margin between them and freezing.

Under these conditions two alternatives seem to be available from the standpoint of adaptation to temperate conditions: (1) a small body allows some tortoise species to burrow or find adequate shelter under debris, leaf litter, and so on, so that small size is often selected for, and (2) large size. In the latter case the great mass of a giant tortoise, heated by the sun during the day, loses heat much more slowly than a similar smaller mass. Even small bushes afford protection from heat loss to the open sky. Thus nocturnal heat loss in an adult giant tortoise sleeping under a dense shrub will be so slow that its deep core temperatures remain sufficiently high to avoid chilling during the night. Repeated warming during the following day will replace that heat lost the night before. It is this physiological mechanism that explains why giant tortoises were found at high latitudes in continental areas during even glacial periods of the Pleistocene.

Populations of *Gopherus agassizi* near northern distributional limits in Utah move into the hibernation dens well before the arrival of dangerously low temperatures. Whether these populations are responding to slight changes in temperature, photoperiod, or other cues, remains unknown.

Most terrestrial turtles hibernate singly, but in some parts of its range individuals of *G. agassizi* hibernate communally (see Auffenberg and Weaver, 1969, for discussion). Woodbury and Hardy (1948) have shown that in Utah these communal dens are large tunnels extending into the gravelly soils of dry creek beds for distances as great as 20 ft, sometimes accommodating a large number of tortoises. Patterson (1971a) reports that social position may determine the sequence in which tortoises enter the den. *Terrapene carolina* occasionally hibernate in small congregations (Carpenter, 1957). Similar congregations also occur in tropical species during certain times of the year (Williams, 1960).

Activity Range

None of the principal published studies on home range (Stickel, 1950; Woodbury and Hardy, 1948; Legler, 1960a; Auffenberg and Weaver, 1969), with the exception of Schwartz and Schwartz (1974) and Gould's (1957) work on *Terrapene*, demonstrated precisely defined home ranges or well-established homing ability. *Gopherus polyphemus* has a rather well-defined activity range within which all feeding and reproduction take place. Populations of *G. agassizi* that retire to burrows instead of forms or pallets show a well-defined home range (Auffenberg and Weaver, 1969).

In *G. polyphemus* the method found most useful in determining home range is that known as the probability density function (Dice and Clark, 1953). In this particular application the burrow entrance was used as the geographic center of all activity, and recaptures were plotted as a series of radii from this point. A mean capture radius was then computed for each individual. This measurement, plus 2 standard deviations, may be used as the radius of a circle that will inscribe an area within which $95^+\%$ of the movements of an individual will most likely fall. The computed area of the circle is the estimated average home range size.

Table 3 shows that in *G. polyphemus* the mean recapture radius is largely independent of the number of recaptures. This is because the

Table 3 Mean Movement Radii (m) for *Gopherus polyphemus*

	Males		Females	
Recapture Number	N	Recapture Radius (O.R.[a])	N	Recapture Radius (O.R.[a])
1	31	42.8 (0–813)	31	27.1 (0–610)
2–10	53	48.3 (0–972)	63	32.6 (0–670)
11–25	59	43.9 (0–1060)	59	29.3 (0–711)
26–50	72	44.6 (0–1201)	83	23.8 (0–761)
51–100	66	51.3 (0–1832)	66	30.2 (0–931)
All captures	281	46.8	302	28.3

[a]Observed range.

burrow tends to dominate the movement pattern. There is a significant difference in mean recapture radii of adult males and females, with males tending to have larger home ranges (brought about mainly by wider ranging during April and May). Table 4 indicates that home range size is directly correlated with available food resources, and Table 5 makes it clear that activity range is correlated with both size and season. During May through June adult male *G. polyphemus* frequently enlarge their home range to take in the activity range of several additional females (Table 5).

Table 4 Mean Recapture Radii (m) of *Gopherus polyphemus* Compared to Amount of Herbaceous Ground Cover

Community	Site	Ground Cover Basal Density (%) (All Plant Species Combined)	Recapture Radius (Sexes Combined, N = 312)
Long leaf pine–turkey oak	A	92	17.8
	B	43	23.6
Xeric live oak hammock	A	68	18.2
	B	31	36.3
Sand pine scrub	A	13	53.5
	B	03	65.4

Table 5 Average Monthly Movements (m) of *Gopherus polyphemus* in Alachua County, Florida: 1963–1969 ($N = 383$)

Tortoise Carapace Length (mm)	January–February	March–April	May–June	July–August	September–October	November–December
Hatchling–100						
Male	3.5	5.0	7.7	5.8	5.0	5.4
Female	4.3	4.7	6.5	5.9	4.7	5.6
101–200						
Male	6.3	7.9	13.3	21.6	26.6	15.3
Female	9.2	8.8	15.2	18.9	27.1	17.8
201–300						
Male	11.2	31.4	78.6	48.8	43.7	21.6
Female	13.1	22.5	61.5	44.2	51.8	38.1
Over 301						
Male	10.6	41.3	69.3	46.7	45.5	32.3
Female	12.3	38.2	56.1	53.3	56.2	21.8

Studies on this and other *Gopherus* species show a positive correlation between individual use of a burrow and a home or activity range. For example, throughout most of its range, *G. berlandieri* is decidedly nomadic and does not dig burrows. Where it does dig burrows, well-defined trails leading from their entrances show a regular traffic to and from them, suggesting in turn that these individuals have a more restricted activity range than those that do not dig burrows. Beck (1903) reports very well-worn trails between tree cacti made by *Geochelone elephantopus*, suggesting similar well-specified feeding trails.

In the nonburrowing populations of *G. berlandieri* in southeastern Texas, individuals move randomly throughout the entire activity season, remaining for a few days at each of many substations along the entire route. Presumably, each movement phase is initiated by extrinsic factors such as a specialized food resource (*Opuntia* fruit) or specific social interaction.

The mean daily movement of individual tortoises, regardless of species, seems to be greatest in populations where the shelter is apart from the feeding ground, or where food plants are scarce or widely scattered. Food is so plentiful in most parts of the range of *G. polyphemus* that most individuals usually move less than 50 m from their burrows to feed. On the other hand, within the thorn brush communities of the Tamaulipan Biotic Province, food plants are widely scattered and individuals of *G. berlandieri* may move as much as 400 m in a day. In the same way, populations of *G. agassizi* differ greatly in the extent of their daily activity, which is determined largely by the degree of their dependence on a particular shelter, as well as food availability (Auffenberg and Weaver, 1969).

Movement and Orientation

There are very few studies on the long-range movements of terrestrial turtles marked in the wild. Work done with species of the genus *Gopherus* suggests that they rarely move more than 2 miles from their hatching spot during their lives, though there are a few exceptions. Some populations of the Galapagos tortoises (*Geochelone elephantopus*) move seasonally from near the base of the mountains to several thousand feet higher, sometimes along miles of paths worn into the rock by thousands of previous migrants. These migratory movements are associated with climatic changes whereby the tortoises are found at lower elevations during the rainy season, when they lay their eggs. During the dry season the vegetation near sea level becomes dry and unpalatable, whereas plants are still turgid with water at higher elevations, particularly in interior-drained caldra of extinct volcanos (all after Van Denburgh, 1914).

Nichols (1939) on Long Island, New York, showed that the normal daily movement of individual *Terrapene c. carolina* has a diameter of about 750 ft. The actual boundaries of these daily ranges shift with time. Approximately 90% of the turtles that were moved from one point to another returned to the area from which they were moved up to a distance of three-quarters of a mile. Adult turtles showed a more marked homing ability than did young individuals. Stickel (1950) studied populations of the same turtle in Maryland, where she found the average home range diameter of males and females to be about 330 and 370 ft, respectively; some individuals had two home ranges and traveled between them at frequent intervals. Medsger (1919) reported an easily recognized individual that was recaptured 18 years later within 100 yards of the original capture point, and 17 years later it was found within 150 yards of the original capture. Schneck (1886) reports taking a marked turtle within a half-mile of the place where it was captured 62 years earlier.

In *Terrapene coahuila*, Brown (1974) found that males tended to move farther than females. Woodbury and Hardy (1948) found that in *Gopherus agassizi* populations in Utah there was a regular seasonal migration between the winter den area and the summer foraging range. The migration involved relatively short distances, and these workers estimated that the total activity range of individual turtles was within 10 to 100 acres.

The orienting mechanisms that permit terrestrial turtles to direct their movements are poorly understood. *Terrapene ornata* can effectively home after displacement of 1 to 2 miles, 15 to 30 times the radius of the normal activity range (Legler, 1960a; Metcalf and Metcalf, 1970). Homeward headings were observed in *T. carolina* removed up to 5.8 miles from their homes by Gould (1957) and in *Clemmys insculpta* removed more than a kilometer (Carroll and Ehrenfeld, 1978).

Metcalf and Metcalf (1970) suggested three mechanisms by which box turtles find their way: visual recognition of landmarks, scent utilization, and celestial navigation. The ability to use the first two cues is of obvious value within the normal range of activities of a terrestrial turtle, but it is the role of the last that must be most important in orienting outside familiar territory. While investigating the utilization of celestial navigation by *Terrapene carolina*, Gould (1957, 1959) found that homing ability broke down when skies were overcast and suggested that box turtles employ the sun for orientation much like certain birds (Matthews, 1968). Box turtles are furthermore able to move unidirectionally for considerable distances and periods of time (Lemkau, 1970). As in some birds, this ability permits the maintenance of a directional course when normal navigational cues are obscured.

Evidence presented by Auffenberg and Weaver (1969) suggests that *Gopherus berlandieri* uses visual and, possibly, olfactory cues to orient within its normal activity range. Experimentation with *G. polyphemus* by Gourley (1965) has yielded the most complete analysis of orientation in terrestrial turtles. His data suggest the employment of several basic mechanisms: (1) landmarks (with shape, color, and/or size characteristics) may be important cues for both long- and short-range orientation; (2) a sun compass may also function in long-range orientation; (3) a single light source may possibly provide some directional information; (4) olfactory cues appear to be utilized by the tortoises in directing movements within their activity ranges; (5) the presence of a population-specific directional preference is indicated; and (6) a true sense of a bicoordinate navigational system appears not to be used by gopher tortoises over long distances.

POPULATION STRUCTURE

Species Comparison

There are few areas in the world in which several species of terrestrial turtles are found in the same area and habitat. When this is the case, the species often belong to different genera. At present such patterns are found only in Africa and Asia; indeed, as many as four sympatric testudinids exist in the southern part of Africa. This pattern was much more widespread in the Pleistocene, and sympatry was a more common phenomenon.

There is no evidence suggesting ecological incompatibility among sympatric tortoises. Among grazing tortoises there may be more grass than needed, in view of low tortoise density and estivation patterns during critical periods. There is, however, no overlap among frugivorous species, suggesting the possibility that interspecific competition may be an important factor in this instance.

Growth and Age

Growth in reptiles is usually intermittent; for example, growth is greatly reduced, or stops, when turtles are hibernating or estivating (see Chapter 4 for techniques).

Repeated measurements of wild individuals that can be identified with certainty over a period of time have been attempted for short periods in *Gopherus agassizi* (Bogert, 1937). Data on captive tortoises have been reported by Townsend (1931), Miller (1932), Goin and Goff (1941), and Patterson and Brattstrom (1972). The only long-term information is available for *G. polyphemus* and *G. agassizi* (Figure 1).

Among terrestrial emydid turtles, the most complete growth study is by Legler (1960a). In this case major emphasis was placed on study of growth rings on the abdominal scutes. Nichols (1939) found that the number of growth rings formed in marked individuals of *Terrapene carolina* did not correspond to the number of growing seasons elapsed; he concluded that growth rings were unreliable indicators of age and that box turtles frequently skipped seasons of growth. Woodbury and Hardy (1948) and Miller (1955) came to approximately the same conclusions regarding *Gopherus agassizi*. However Legler (1960a) points out that these workers were studying turtles of all sizes and ages, some of which were past the age of regular annular growth. Recent studies on testudinid growth (Medica, Bury, and Turner, ms; Patterson and

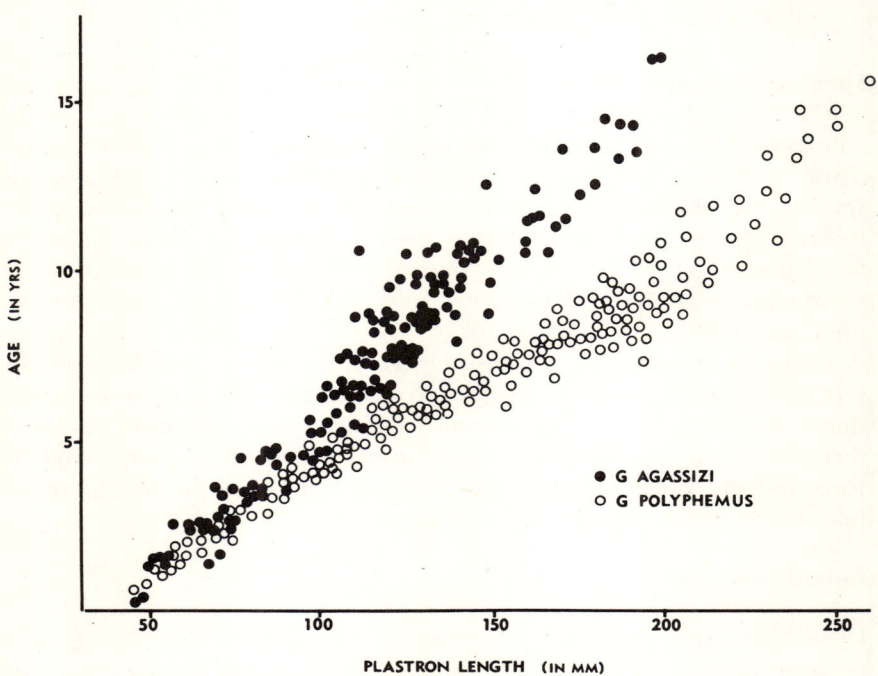

Figure 1 Growth in wild *Gopherus polyphemus* (data from present study) and *G. agassizi* (after Medica et al., in prep.)

Brattstrom, 1972) indicate that a single annulus is formed each year during early life.

From the population point of view, it is not growth per se that is important, but the proportion of time between the reproductive and prereproductive periods. Therefore age at sexual maturity is one of the most important population characteristics. It determines how much energy is required, and how long an individual is exposed to predation and disease before it becomes a reproductive member of the population. Unfortunately, very few data are available. In general, males seem to mature earlier than females in emydids, but this tendency may be reversed in at least some testudinids (Woodbury and Hardy, 1948).

There is a relationship between maximum size and size at hatching, size at sexual maturity, and rate of growth and longevity (Table 6). However there is only a poor correlation between number of offspring and size.

Medica, Bury, and Turner (ms) have clearly shown a strong correlation between growth in *G. agassizi* during April to July and rainfall in the desert areas in which the species is found and rainfall of the preceding winter, which resulted in increased vegetation in spring. It is not surprising that the growth pattern in the southeastern representative of the genus *Gopherus* living in much more mesic habitats is more rapid (Figure 1). This difference in growth rate is undoubtedly due to the more general seasonal distribution of food plants during much of the year, resulting in a longer annual growth period (April through September). There are no other data for long-term growth in other tortoises, though they are desperately needed, particularly in monsoon and rain forest habitats.

Many studies have shown that growth rate declines with age. Young individuals in the first few years of life increase in length 10 to 25% per year, whereas those past the onset of maturity tend to grow much more slowly, and sometimes not at all. Therefore reliability of age estimates based on size alone is usually reduced in proportion to increasing age.

Sexual Difference in Growth and Size

Many species of tortoises show pronounced differences between the sexes, largely associated with differences in growth rate. In most species the female is the larger. In several species the female may possess a much thinner shell, with fontanelles common at wear points, such as over the pelvis (*Geochelone chilensis*, among others). Males often have longer tails than females, enabling the cloaca of the male to more easily reach that of the female during copulation (Auffenberg, 1964).

Sex Ratio

The sex ratio of most reptile populations is approximately 1:1. Because variation in sex ratio may have important consequences for population regulation, it is important to analyze this ratio in natural populations of land turtles, but few wild populations have been investigated from this standpoint.

Part of the problem is that it is often difficult to determine the sex of live tortoises. Individuals of most species can be sexed only at or after the onset of sexual maturity. Even then, secondary sexual characters are not always reliable, often showing considerable individual variation. Morphologic characters used to determine sex in living turtles include the tail, the plastron, and sometimes the head (see Chapter 4).

Table 6 Size,[a] Growth, and Longevity[b] Statistics of Terrestrial Turtles

Family and Species	Average Hatchling Size/Maximum Adult Size (mm)	Ratio of Hatchling Size to Maximum Size	Size at Sexual Maturity[c] (mm)	Longevity, Captive (years)	Major Reference
Emydidae					
Terrapene ornata	30/146	1:4.9	100–110 M 120–130 F	50	Legler (1960a)
T. carolina	28/198	1:7.1	100–130 M,F	138	Oliver (1955)
Clemmys insculpta	36/260	1:7.4	140–190 M,F	58+	Oliver (1955)
Testudinidae					
Chersine angulata	32/267	1:8.3	100–185 M,F	11+	Rose (1950)
Gopherus agassizi	42/370	1:8.8	250–316 M,F 230–365 F	20+	Woodbury and Hardy (1948)
G. berlandieri	45/220	1:4.9	105–128 M,F	?	Auffenberg and Weaver (1969)

G. polyphemus	45/370	1:8.2	230–341 M 238–368 F	25+ 3+	Auffenberg notes Auffenberg notes
Homopus areolatus	31.5/115	1:3.7	95–115 M,F	3+	Auffenberg notes
Geochelone radiata	34/458	1:13.5	300–360 M,F	?85	Zovikian (1973)
G. chilensis	59/283	1:4.8	—	—	Auffenberg (1969a)
G. pardalis	55/502	1:4.6	—	?42+	Loveridge and Williams (1957)
G. elephantopus	56/1120	1:20.0	—	?150	Rothschild (1915)
G. elegans	37/280	1:7.6	—		Jayakar and Spurway (1966)
Testudo graeca	35/300	1:8.6	—	102+	Lambert (1970)
T. hermanni	34/165	1:4.9	—		Heron (1968)
T. kleinmanni	32/135	1:4.2	—	21+	Loveridge and Williams (1957)
Malacochersus tornieri	42/175	1:4.2	135–175 M,F	8+	Rose (1950)
Psammobates tentorius	25.5/141+	1:5.5	95–130 M,F	7+	Rose (1950)

[a]Straight line carapace length.
[b]Largely after Flower (1925).
[c]M = male; F = female.

Tail characters useful in determining sex include the usually longer tail of the male (most genera), sometimes with a conelike enlarged scale at the tail tip (emydids and some tortoises), a terminal hook (*Geochelone*, subgenus *Indotestudo*), a flattened tip (*Geochelone*, subgenus *Asterochelys*), and so on. Males of the high-domed species in particular have a deep concavity on the plastron to fit the dome of the female when the male is mounted in the breeding position. In the more flattened species (*Malacochersus tornieri*, etc.) the sexes are difficult and often impossible to separate by this means. A change in head color of males is known during the breeding season of *Geochelone travancorica* (Auffenberg, 1964).

The available sex ratio data suggest that most turtles have a 1:1 ratio. However asymmetric ratios have been reported (Auffenberg and Weaver, 1969), showing that females are more common than males. We found no information on sex ratios in land tortoises other than *Gopherus*, so we took the opportunity to check the extensive skeletal collections of the Florida State Museum and drew from this resource as well as from the associated data. This information is largely based on samples of adult specimens and is therefore not totally satisfactory, but nevertheless represents most of the information presently available. Table 7 suggests that in several species the ratios are considerably out of line and should be investigated.

Density

It is abundantly clear that for all land tortoises, environmental limitation, so obvious in the xeric habitats in which these reptiles live, is more related to density than to total tortoise population. This is clearly shown in transects through *Gopherus polyphemus* populations (Table 8) in several Florida and Georgia localities. Differences in density are related to two major factors: removal rate by humans, and habitat quality (of which food resource is most important). In this species the limiting food resource is largely grass, whereas in other tortoise species herbs and fruits may be important. Table 8 shows that in *G. polyphemus* habitat grass cover is clearly correlated with tortoise density and is probably the major factor regulating density in undisturbed populations. In low-lying areas, particularly near the coast, groundwater level may be a determining factor in both spatial distribution and population density, for this species tends to be absent from areas where deeper soils are permanently flooded.

Figure 2 shows that even within the same community, density of resident tortoises may vary considerably throughout the population, dependent on several ecological factors, of which grass biomass is the most important. The distribution of grass type and density varies on the

Table 7 Sex Ratios in Terrestrial Turtles

Family and Species	Sex Ratio, Males to Females	(N)	Major Reference
Emydidae			
Terrapene ornata	1.00:1.69	(164)	Legler (1960a)
T. carolina	1.00:1.09	(245)	Stickel (1950)
T. nelsoni	1.00:1.18	(37)	Milstead and Tinkle (1967)
T. coahuila	1.00:1.34	(164)	Brown (1971, 1974)
Clemmys insculpta	1.33:1.00	(28)	Ernst and Barbour (1972)
Testudinidae			
Gopherus agassizi	1.00:1.50	(252)	Woodbury and Hardy (1948)
G. berlandieri	1.62:1.00	(141)	Auffenberg and Weaver (1969)
G. flavomarginatus	1.00:1.10	(21)	This study[a]
G. polyphemus	1.06:1.00	(101)	This study[a]
Geochelone radiata	1.00:1.55	(38)	This study
G. pardalis	1.00:1.35	(20)	This study
G. elegans	1.08:1.00	(25)	This study
G. carbonaria	1.90:1.00	(29)	This study
G. denticulata	1.05:1.00	(41)	This study
G. chilensis	1.00:4.40	(27)	This study
Homopus areolatus	1.00:1.00	(22)	This study
Kinixys belliana	1.00:1.00	(13)	This study
K. homeana	2.50:1.00	(7)	This study
Testudo graeca	1.30:1.00	(30)	This study
T. hermanni	1.00:1.50	(10)	This study
T. horsfieldi	1.00:8.50	(19)	This study

[a]Skeletal material in the Florida State Museum Collection, often associated with eggs and other collections.

hillsides, based on edaphic factors, such as soil chemistry, moisture, and shade. In areas with less relief, edaphic features often vary less markedly, so that density zonation is a less common phenomenon. However near the boundary of differing soil types, zonation of this type is sometimes readily obvious.

Density change in time that is not due to mortality has not previously been demonstrated in land tortoise populations. Long-term studies on G. polyphemus in Florida now make it clear that population change due to

Table 8 Density of *Gopherus polyphemus* Burrows and Grass Cover

Locality	Habitat (See Carr, 1940, for definitions)	Density (1 tortoise/m^2)	Grass Cover (% Basal Cover)
Pitts, Georgia	Long leaf pine–turkey oak	1/485	92
Abbeville, Georgia	Xeric hammock (red oak)	1/1066	86
Alachua, Florida	Xeric hammock (red oak)	1/1696	83
Archer, Florida	Long leaf pine–turkey oak	1/1833	67
Astor Park, Florida (B)	Sand pine scrub	1/1440	63
Astor Park, Florida (A)	Sand pine scrub	1/1838	51
Dania, Florida	Sand pine scrub	1/3214	49
Salerno, Florida	Sand pine scrub	1/2812	38
Trenton, Florida	Long leaf pine–turkey oak	1/11,303	17
Manatee Springs, Florida	Xeric hammock (live oak)	1/8321	0.05
Howard, Georgia	Long leaf pine–turkey oak	1/26,645	0.03

natural processes are common in this species. For the most part such changes are due to primary community succession. The best data are from a study site in Lake County, Florida, where a single population has been under observation since 1960. During the intervening years the study area changed from an old field community to mature sand pine scrub. During this time, gopher tortoise density in the area changed from one tortoise per 1330 m^2 in 1960 to one tortoise per 26,230 m^2 in 1971; a reduction of 1800% in 11 years. There was no evidence of increased mortality, but only of emigration, increasing in intensity as open habitat grasses were replaced with subcanopy herbs and finally a thick pine needle duff.

Very few data are available on density of other species of tortoises. Auffenberg and Weaver (1969) report *Gopherus berlandieri* densities in southeastern Texas as a maximum of 1 tortoise per 82 m^2 (or 55/acre) and minimum density of 1 per 430 m^2 (11/acre). On the basis of studies

POPULATION DYNAMICS

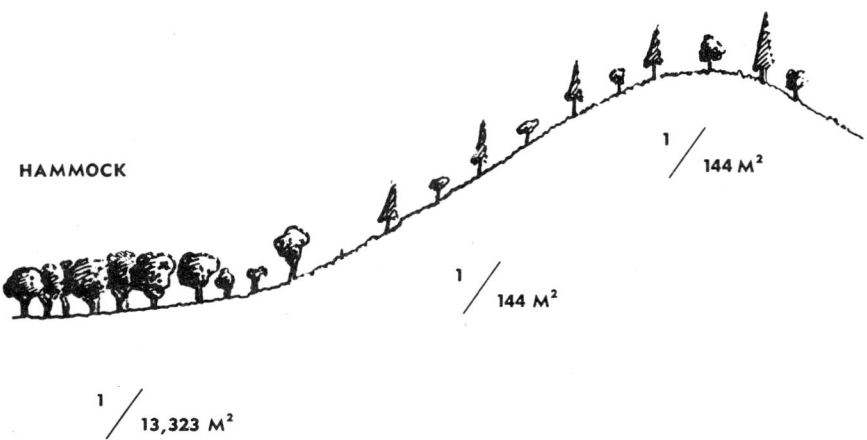

Figure 2 Density of *Gopherus polyphemus* as related in habitat near Jacksonville, Telfair County, Georgia.

conducted by Grubb (1968) and others on *Geochelone gigantea* on the Aldabra Islands, Grubb concludes a density of approximately 42,000 tortoises for 11 km² or about 3800/km². With a mean weight per tortoise of about 80 kg, the biomass per square kilometer is approximately 300,000 kg (extrapolated from Grubb, 1968, and others). Of course this huge amount occurs largely because on these and similar oceanic islands there are few, and usually no, native competitors or significant predators of the adult tortoises. These factors combine to maintain a biomass at much lower values in mainland tortoises. For example, biomass of the most dense *G. polyphemus* colonies in northern Florida is only about 22,000 kg/km².

POPULATION DYNAMICS

Few data are available on the long-range population dynamics of any turtle species, except for *Gopherus* (though data are rapidly accumulating for the Galapagos tortoise, *Geochelone elephantopus*). It is not known, for instance, at what level recruitment rate operates in the different species of tortoises. A state of adjustment must exist between members lost and those gained. But the nature and the extent of fluctuation in population density and the nature of the factors responsible for density alteration are not known at all. Is recruitment through hatching and immigration somehow adjusted to compensate for natality and emigration? Probably this is true only in the tortoise species that inhabit mature, stable environments. In species inhabiting ecotonal or successional stages, such as *Gopherus polyphemus*, population numbers are not stable, but may fluctuate greatly depending on local conditions, of which recurrent fire is one of the most important. Without fire in what would ordinarily be savanna-type communities, tortoise populations tend to be short-lived in terms of normal colony life expectancy, population reduction being correlated directly with degree and rate of successional habitat modification. Much of this change in population density is brought about by emigration of adults.

Emigration and Immigration

In gopher tortoises, population loss in a colony through emigration is due to movement of adults, juveniles, or both, to new locations, though these occur as individual, noncoordinated movements. These two age classes must be emigrating for somewhat different reasons, though there may be some overlap under specific conditions, such as habitat deterioration. This is clearly shown by emigration patterns of both age classes during normal periods of population growth.

Adults of *Gopherus polyphemus* emigrate at a proportionally much higher rate than hatchlings, since the latter are largely nomadic and movements are often random, carrying some into the depths of the colony as well as away from it. However the movements of adults are more directed, often involving proportionally greater distances. In periods of habitat modification, the rate of emigration is not necessarily greater, but it is sustained for a much longer time.

Juveniles also slowly move away from the area of hatching. If the movement is sufficiently far, it might be considered emigration. Presumably this movement is reasonably rapid and probably random; the rate is largely determined by available local food resources (most hatchling terrestrial turtles are omnivorous for the first few years of life). However for more or less colonial forms, particularly *G. polyphemus*, this outward movement of hatchlings may be slow indeed, and may not take place for years, if at all. Figure 3 shows the movement through the colony of a small group of hatchling *G. polyphemus* in northern Florida. The entire process leading to emigration of at least two individuals took several years. To date, some individuals appear to be completely incorporated into the adult population of the colony, though emigration remains a probability during extensive movements prior to courtship and breeding.

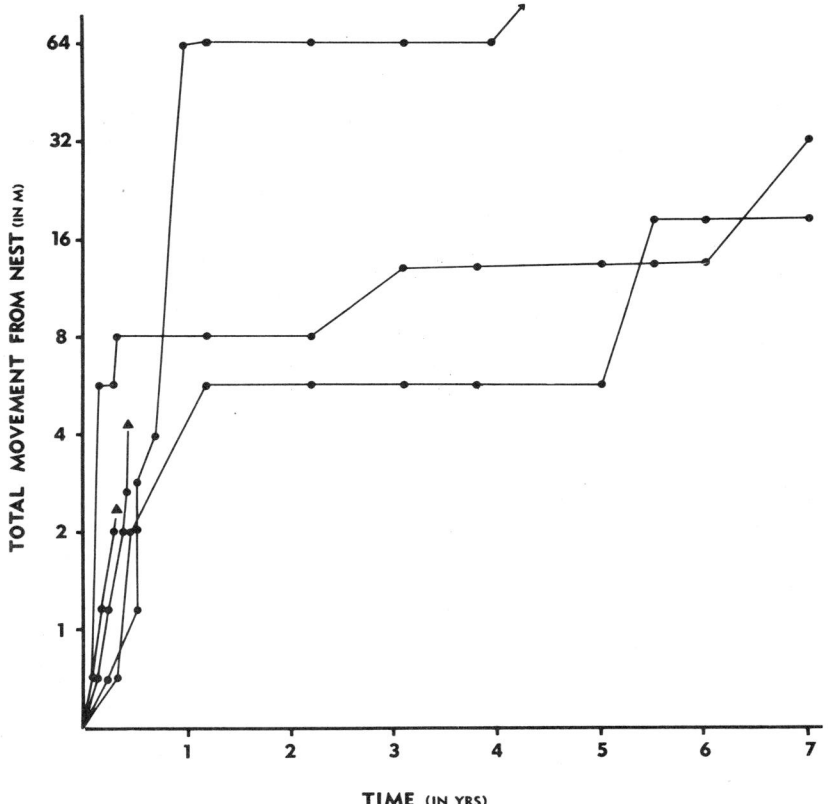

Figure 3 Movement of *Gopherus polyphemus* siblings from nest site in Alachua County, Florida. Triangle indicates death of individual.

Immigration takes place by means of individuals entering an established colony or group, or by the establishment of new groups or colonies. The latter seems to be characteristic of the movements and population dynamics of at least some mesic tropical forest forms. Thus Auffenberg noted that *Geochelone travancorica* in India and *G. carbonaria* in Panama form temporary aggregations near important food resources, such as the fruiting of particular trees. At this time individuals may gather in small groups close to the resource, often sharing a single appropriate shelter, such as a natural brush pile, a fallen tree, or thicket. The same type of immigration takes place near the hibernacula utilized by some populations of *Gopherus agassizi* in southern Utah (Woodbury and Hardy, 1948). In both places immigration into particular areas brings about the formation of temporary aggregations of feeding, breeding, estivating, or hibernating individuals. These aggregations may last from several days to several months, depending on the reasons for immigration in the first place.

Immigration in colonial species such as *Gopherus polyphemus* is different in that it may result in permanent residency and long-term interaction with other members of the colony. In this species, immigration is most likely during the beginning of the reproductive period, when individuals (usually males) may leave their own colony and enter another located some distance away. The changes may or may not be permanent: they rarely account for more than 5% of the local population at any one time and are usually less. Unlike the formation of aggregations around fruiting trees or a hibernaculum, this type of movement and recruitment to a particular population cannot be predicted.

Habitat modification by either destroying or opening a formerly dense habitat by fire or secondary succession is predictable, since emigration and immigration are inevitable. Unfortunately, we have no information on the immigrative aspects of such movement for any tortoise species.

Mortality

Here we are concerned with the kinds of things that result in an early death, so that the individual fails to attain its reproductive potential. Few data are available on the extent of disease and predation in wild populations, and their real importance remains largely unknown.

The most obvious, and perhaps most extensive predation occurs in the early life stages, particularly in the nest. A large number of small predators, including snakes, rodents, canids, felids, viverids, mustelids, and even edentates (Auffenberg, 1969a) have been reported as preying on turtle eggs, and there is extensive, largely repetitive literature available. However there are almost no data available on the percentage of successful layings (i.e., the nests that remain undisturbed through hatching). Data from populations of both *Gopherus berlandieri* and *G. polyphemus* make it clear that at least in these two species egg predation is apparently highest for about 7 days after deposition (Auffenberg and Weaver, 1969). Scents associated with nest construction, such as soil that has been dampened with cloacal fluids and the presence of the female in a small area for a reasonably long time, are probably more important in leading predators to a nest than the scent of the eggs themselves, in spite of the demonstration by Patterson (1971b) that the urine of *G. agassizi* may mildly and temporarily discourage potential egg predators. Another apparently very short period of high mortality occurs just at hatching, when the egg shells have been breached and the young are moving to the surface (Auffenberg, field notes).

Of the posthatchling stages in land tortoises, mortality is undoubtedly highest during the first few days and weeks of life on the surface. Desiccation (and predation?) is apparently partially avoided by the hatchlings burrowing into the soil, often enlarging the burrows of large insects or mice. Sometimes these small burrows are plugged between the tortoise and the outside world (Auffenberg and Weaver, 1969). Rodents, canids, and mustelids have been reported as feeding on newly hatched tortoises. Some snakes also do so, such as *Drymarchon corais*, which is known to feed on juvenile *G. polyphemus* (R. Mount, personal communication) and *Geochelone carbonaria* (Auffenberg, field notes). Birds of various groups feed on the young of several species of tortoises in Africa (Loveridge and Williams, 1957). Insects, particularly certain tabanid flies, sometimes infest living land turtles and may cause their death (King and Griffo, 1958). Fires are certainly a serious threat to nonburrowing forms, and may be responsible for severe injury and even death in several species (e.g. Legler, 1960a). The nine-banded armadillo (*Dasypus novemcinctus*) accidentally seals adult gopher tortoises into the recesses of their burrows by packing vegetable debris into the hole during the winter when the armadillo may use it as a hibernation den, and particularly when it excavates side passages from the tortoise burrow, entombing the turtle in the deeper portions. Entombment is particularly common when the burrow is next to a fence; the armadillo digs away the back of the burrow to pass under the fence and throws the earth into the tortoise burrow. The damage caused in the southeast by this nondeliberate entombment is difficult to ascertain, but it is sufficiently common in areas of high armadillo density that the total number of tortoises killed each year must be very high. In one instance, activity of a fox squirrel (*Sciurus niger*) brought about entombment (Auffenberg, field notes).

Because of their shell, habits, and protective behavioral patterns, adult turtles are apparently relatively free of major predation. This is particularly true of the larger species. Suids, canids, felids, and rodents have been known to kill adult tortoises, though mortality from this cause is undoubtedly very low.

Introduced domestic animals in areas that had no large predators or competitors previously are extremely important in the maintenance of insular populations of tortoises. This has been abundantly shown in islands such as Galapagos, where feral dogs and introduced rats feed on young *Geochelone elephantopus*. Goats compete with the tortoise for food and apparently trample the nesting sites. Pigs not only destroy the nests, but also may kill specimens up 9 kg in weight (e.g., MacFarland, 1972).

The detrimental influence of these introduced man-associates, as well as the activities of man himself in both habitat modification and primary predation, are the most critical factors in present management problems (MacFarland, Villa, and Toro, 1974).

An extremely common form of death for tortoises of all ages in areas of karst topography is tumbling into open fissures, caverns, and sinkholes. The thousands of fossil tortoises now known from extensive karst areas attest to the importance of this factor in population dynamics in areas of appropriate surface geology. In Florida these openings in the underlying limestone trap many box turtles and tortoises, and their remains are often found scattered in the recent surface of caves and sinkholes.

Longevity

At present most of our knowledge of the maximum age reached by different animals has been gained from captive individuals. Data on wild populations are incomplete and cannot be adequately compared with life spans of captive individuals. Where information is available, captive animals appear to live longer than those in the wild. Growth rings are usually useless for establishing the life span of old individuals; usually because the laminae are so badly worn that not all the rings can be counted. As a result, all longevity records available so far have been obtained from captive animals. Some of these data are assembled in Table 6. Additional information is available in Fowler (1925, and other years) and in Conant and Hudson (1949). The large elephantine tortoises (*Geochelone elephantopus* and *G. gigantea*) are often presumed to be very old, largely on the basis of great size, but this may not be the case, for it is known that they grow very rapidly.

The oldest tortoise for which we have reliable data is "Marion's tortoise," an example of the now extinct *Geochelone sumeirei* from the Seychelles Islands. This famous tortoise was brought as a gift to Mauritius in 1766 and remained there until its accidental death in 1918, for a known age of 152 years. However even some small species of terrestrial turtles, including emydids, may exceed 100 years of age.

Reproductive Effort

It is evident that in every tortoise population remaining at constant density, the rate of recruitment must equal mortality. Since all terrestrial turtles are seasonal breeders, the number of eggs produced during only

POPULATION DYNAMICS 567

a part of the year must equal total losses during the entire year. There is no present evidence that recruitment is dependent on mortality, as is apparently true of other vertebrate classes. Acceptable terrestrial turtle recruits can obviously reach the population through immigration and through hatchlings that grow up there.

Like many other reptile groups, terrestrial turtles often lay several clutches of eggs each year. In tropical regions of uniform climate, some species, such as *Geochelone pardalis*, may lay as many as five clutches in a year—providing a potentially large number of young annually. However most temperate species have only one clutch most of the time (see Chapters 16 and 17 for data summaries). In either case, no terrestrial turtles are acyclic in laying eggs; reproduction is climatically correlated. There are also large differences in clutch size, with a general correlation between average adult size and number of eggs laid per clutch. Smaller species of *Homopus*, *Psammobates*, and *Chersine* appear to mature early, are small, and tend to lay one to two eggs per clutch and to have relatively few clutches per year. The large species tend to mature later and to lay multiple clutches of many eggs. Several of the latter (*Geochelone pardalis*, *G. radiata*, and others inhabiting xeric environments) have a remarkably long developmental stage in the egg (up to 1.5 years in some populations of *G. pardalis*). Eggs of these species are thus exposed to predation for a long period, and having large numbers of eggs per clutch in multiple annual clutches is apparently an adaptation to the high probability of predation during the long developmental period. Although there is apparently some genetically controlled variation of egg developmental period (Jayakar and Spurway, 1964), it is also clear that environmental factors, including both temperature and moisture, can hasten or retard development. Tortoises have several adaptations to counter the productive loss imposed by delayed maturity. One of these is the production of larger clutches, achieved partly by a larger size at maturity, and the other is by a long reproductive life expectancy.

Since most terrestrial turtles are long-lived, the actual total reproductive effort of the individual on the basis of long-range studies is very important. There are no published accounts of such studies, but some work on *Gopherus polyphemus* by Auffenberg sheds at least some light on this question.

Continued study of several colonies of this species in northern Florida shows quite clearly that in only a few is actual egg laying even close to the potential. This is because many of the colonies are relatively small (average number per colony in a sample of 27 colonies is 22), and not all individuals are necessarily active at the same time. A high percentage of the random meetings between males and females may bring together individuals not physiologically ready to breed at that moment. Furthermore, relatively few of the individuals are mature. Last, given the small numbers found in many colonies, the adult sex ratio may be quite uneven. The females of some species of turtles are known to store sperm for several years, but very little is known regarding this ability in tortoises. In any event, evidence suggests that in many years there is no egg laying at all in some *G. polyphemus* colonies. In addition, the high predation of the eggs that are laid means that in most colonies very few young are produced annually. There are several colonies in Alachua County where no young have successfully hatched during periods of as long as 7 years. The distribution of hole sizes in other localities often shows that one or more size classes are completely missing, suggesting long periods of nonrecruitment. It is not known whether other tortoise species show similarly long unproductive periods, but it is highly likely that this is the case in at least some of the species.

Fertility and hatching success in the wild are known only for *Geochelone elephantopus*. In this species, fertility is apparently near 80% (McFarland, Villa, and Toro, 1974).

CONCLUSION

The true tortoises are especially interesting because of their habits and physiology. Unfortunately, because they are so obviously harmless, and quaint in many ways, tortoises often fall prey to the pet trade. We hope that new import laws at least in the United States will reduce the continual drain on many populations. A number of forms have been so assiduously hunted for their delicate flesh that their populations have been reduced to the point of extinction. This is particularly true of insular forms. When the species that became extinct in prehistoric times are added to those of historic times, the number of extirpated forms becomes very significant. Many populations are being destroyed through extensive land modification—primarily because of conversion of their habitat to agricultural and housing uses. Irrigation projects, encroaching fields, and hungry people will surely account for the demise of many additional tortoise species in the next several decades. A number of species are already on internationally recognized endangered lists, and many others should clearly be added. Protective legislation is being passed in a number of countries, but it is, unfortunately, far from being sufficiently effective.

CONCLUSION

One of the most pressing problems from both the scientific and conservation standpoints is that none of these interesting and threatened species is well known—particularly in view of the danger that present tortoise populations over the entire world will be drastically reduced in the future. It is indeed ironic that in many instances we may need the very information we are just beginning to accumulate to save at least some of the species from extirpation. It is reasonably clear that information of the appropriate type is accumulating only for a few species, mainly in the genera *Terrapene* and *Gopherus*. We sincerely hope that efforts can be mounted in the near future to study other tortoises in the same detail.

CHAPTER 26
Population Ecology of Freshwater Turtles

R. BRUCE BURY

Turtles are adapted for life in aquatic habitats through many ecological and morphological specializations (Walker, 1973; Zangerl, 1969). Because turtles are large and often locally abundant, they may be the major vertebrate in total biomass. Freshwater species may be omnivorous, herbivorous, or carnivorous. Some prey heavily on specific foods such as snails or fishes, while others are scavengers and opportunistic feeders. Turtles, especially their eggs and hatchlings, may be important food for other wildlife.

Until recently, most research on the population biology of these animals consisted of fragmentary, short-term, or descriptive studies, and most of our knowledge is based on a few species of North American emydid turtles. The bulk of the available data concerns distribution, systematics, and natural history (Table 1), which provides the foundation for ecological research.

POPULATION STRUCTURE

Species Composition

Impressions of relative abundance of species in turtle communities are obfuscated by collecting biases, species' habitat differences or preferences, seasonal changes, geographic variability, and local conditions.

Table 1 A Selected Synopsis of Literature on Freshwater Turtles

Topic or Region	Reference
General	Bellairs and Carrington (1966), Carr (1963b), Ditmars (1928), Goin and Goin (1971), Porter (1972)
Freshwater turtles, world	
General	Gans and Parsons (1970), Pritchard (1967), Schmidt and Inger (1957), Terent'ev (1961), Wermuth and Mertens (1961), Zangerl (1969)
Regional	
Mexico	Alvarez del Toro (1960), Smith and Smith (1973), Smith and Taylor (1950)
Central America	Breder (1946), Stuart (1948, 1963)
South America	Medem (1962), Vaz-Ferreira and de Soriano (1960)
Asia	Bourret (1941), Deraniyagala (1939), DeRooij (1915), Mao (1971), Pope (1935), Smith (1930), Taylor (1970)
Australia	Goode (1967b), Worrell (1963)
Africa	Loveridge (1941), Loveridge and Williams (1957), Rose (1950), Villiers (1958)
Russia	Nikol'skii (1915)
Europe	Dottrens and Aellen (1963), Mertens and Wermuth (1960)
North America	Bleakney (1963), Carr (1952), Cochran and Goin (1970), Conant (1975), Ernst and Barbour (1972), Logier and Tonier (1961), Oliver (1955), Pope (1939), Stebbins (1954, 1966)

However some qualitative and quantitative features of the species composition of freshwater turtle populations are becoming evident (Table 2).

Throughout its range, the slider, *Pseudemys scripta*, apparently is the predominant species in ponds and lakes (Table 2). There may be competition between the painted turtle, *Chrysemys picta*, and *P. scripta* in large, deep bodies of water, since *C. picta* is always in the minority when *P. scripta* is also present (Cagle and Chaney, 1950; Ernst, 1971c). Moll

Table 2 Percentage Species Composition of Five United States Turtle Populations

Reference:	Cagle (1942b)	Cagle (1950)	Gibbons (1970d)	Tinkle (1961)	Ernst (1971c)
Locality:	Illinois	Louisiana	South Carolina	Gulf Coast	Pennsylvania
N:	1902	835	568	1106	1190
Species					
Pseudemys scripta	72	74	74	11	
P. floridana		1	5	16	
Chrysemys picta	13				78
Deirochelys reticularia		<1	16		
Clemmys spp.					13
Graptemys spp.		<1		48	
Sternotherus spp.	12	2	3	23	5
Trionyx spp.	2	15		1	
Chelydra serpentina	1	4	1	1	4
Other		3	1		

(1973) called for detailed comparisons of behavior and food habits between these suspected competitors to indicate their level of coexistence. The interactions of *C. picta* with other species should be examined; *C. picta* seems to predominate in certain situations.

In habitats other than ponds and lakes, other species may predominate. For example, species of *Graptemys* are more abundant in flowing rivers. Also, species of aquatic genera such as *Chelydra*, *Trionyx*, and *Sternotherus* may constitute a higher proportion of turtle communities than has been reported because investigators seldom spend much time sampling these turtles.

Age Composition

According to Odum (1971), age distribution is an important population characteristic that influences both natality and mortality. Differences in sampling methods, seasonal factors (particularly the influx of hatchlings), and other variables affect the estimated proportion of age classes in freshwater turtle populations and make it difficult to obtain a random sample from a natural population. Varied sampling techniques are used for different species, sex, and size classes of turtles living in different habitats, from temporary pools to large lakes and from trickles to rivers (see Chapter 2). Reported proportions of juveniles in different populations vary from zero to 70% (Moll and Legler, 1971; Ream and Ream, 1966). In a composite of samples (N = 6637 individuals), there were 31% juveniles (Table 3).

It is generally accepted that adult turtles may live for decades, but the upper limit of life and reproductive span are virtually unknown. The reproductive potential of freshwater turtles is high because the reproductive age for females is relatively long (see Wilbur, 1975a), clutch size in most species is large (Ernst and Barbour, 1972), and some species have multiple clutches each year (Moll, 1975). But an apparent high mortality of eggs and hatchlings greatly reduces recruitment into the juvenile age class. Once a turtle reaches the juvenile stage, its chances to reach maturity are high, since increases in size and shell hardness make a juvenile less vulnerable to predation, dessication, and other decimating factors affecting hatchlings. Juveniles may take from 2 to 18 years to reach sexual maturity and adult size.

In general, juveniles appear to comprise a variable but low percentage of most populations, whereas there is a high proportion of long-lived mature individuals. These attributes represent the evolutionary pattern of "k strategists" (MacArthur, 1972) which usually optimize adaptations in stable environments over long periods of time.

Sexual Dimorphism

Male and female freshwater turtles usually differ in coloration, size, shell proportions, and other features (Carr, 1952; Ernst and Barbour, 1972; McCoy, 1968; Viosca, 1933; Graham, Chapter 4). The reasons and energetics for the size differences of sexes are poorly understood. Presumably the dimorphism in some species reflects factors important in social interactions (e.g., fighting or courtship), survival, or reproduction. For example, in a species with larger females the diets of the sexes may differ, reducing intraspecific competition for food. Greater reproductive fitness is obtained because the large female body size permits more eggs to develop in each gravid female. Species with little or no size dimorphism may lack the evolutionary or environmental constraints to develop marked size differences. Known examples of larger male size include snapping turtles, in which individuals occasionally engage in aggressive

Table 3 Comparison of the Age Composition in Selected Freshwater Turtles

Species	Locality	Juveniles Total	%	Adults Total	%	Ratio (Juv = 1)	Reference
Clemmys guttata	Pennsylvania	67	32	140	68	2.09	Ernst (1976)
C. marmorata	California	248	35	456	65	1.83	Bury (1972)
Chrysemys picta	Illinois	98	41	139	59	1.41	Cagle (1942b)
C. picta	Illinois	124	51	120	49	0.97	Cagle (1954)
C. picta	Michigan	141	58	102	42	0.72	Cagle (1954)
C. picta	Wisconsin	51	10	479	90	9.39	Ream and Ream (1966)
C. picta	Pennsylvania	180	19	374	81	2.08	Ernst (1971c)
C. picta	Michigan	305	39	480	61	1.57	Gibbons (1968b)
Pseudemys scripta	Illinois	263	19	1113	81	4.23	Cagle (1942b)
P. scripta	Illinois	377	31	825	69	2.18	Cagle (1950)
P. scripta	Panama	121	37	208	63	1.71	Moll and Legler (1971)
Sternotherus odoratus	Illinois	78	35	148	65	1.89	Cagle (1942b)
		2053	31	4584	69	2.23	

encounters (Froese and Burghardt, 1974; Hammer, 1971), but there is no evidence indicating that larger male size leads to any mating advantages.

Sexual Maturity

Demographic studies require a definition of size classes and accurate determination of the age and size at sexual maturity. Typically, the age classes of freshwater turtles consist of adults (sexually mature), juveniles (sexually immature), and hatchlings (first year), which are often determined by size measurement. (See Chapters 16 and 4 for additional information.)

The present evidence on the attainment of sexual maturity in freshwater turtles suggests the following:

1. Males often mature earlier and at a smaller size than females.
2. Growth is most rapid before maturity is reached.
3. In temperate regions individuals in southern populations mature earlier than northern conspecifics.
4. Individuals of smaller bodied species mature at a younger age and at a smaller size than large bodied forms.
5. Sexual maturity usually is related to obtaining a certain size rather than a certain age.

Research on nonemydid turtles and tropical species may significantly alter these tentative conclusions about attainment of sexual maturity in freshwater turtles (Table 4).

Sex Ratio

There are several reports of unequal sex ratios, especially of females outnumbering males, in adult freshwater turtles (Forbes, 1940; Carr, 1952; Timken, 1968; Tinkle, 1961); however the observed differences may be due to improper methodology and selective sampling (Gibbons, 1970c; Ream and Ream, 1966).

Hildebrand (1929) reported that only 14% of diamondback terrapins raised to maturity in captivity were males. However both Cagle (1952) and Gibbons (1970c) suggested that many of those presumed to be females were in fact juvenile males that did not display male secondary sex characters because of the conditions.

Table 5 and Figure 1 compare 39 studies. For the 23 samples with more than 100 individuals, 15 (65%) are not significantly different from

Table 4 Comparison of the Size and Age of Freshwater Turtles at the Attainment of Sexual Maturity

Species	Locality	♂♂ PL (mm)	♂♂ Age (years)	♀♀ PL (mm)	♀♀ Age (years)	Reference
Chrysemys picta	Southern Minnesota	85+	—	125+	—	Legler (1954)
C. picta	Wisconsin	95–100	4–5	135–140	7–8	Christiansen and Moll (1973)
C. picta	New Mexico	80–90	3	130	5	Christiansen and Moll (1973)
C. picta	Northern Michigan	90	—	120–130	—	Cagle (1954)
C. picta	Southern Michigan	80	5	110–120	7–10	Gibbons (1968b)
C. picta	Southern Michigan	100–105	4–5	ca. 120	7	Wilbur (1975a)
C. picta	Central Illinois	80–85	3–4	130	4–6	Moll (1973)
C. picta	Southern Illinois	70	—	120–125	—	Cagle (1954)
C. picta	Pennsylvania	80–90	4	100	4–6	Ernst (1971c, d)
C. picta	Tennessee	65	2–3	105	4–5	Moll (1973)
C. picta	Louisiana, Arkansas	60–65	2–3	100	4	Moll (1973)
Pseudemys scripta	Illinois, Tennessee, Louisiana	90–100	2–5	150–195	—	Cagle (1944b, 1950)
P. scripta	Oklahoma	90–100	3	175	4	Webb (1961)
P. scripta	Panama	125–135	2–3	250–260	5–8	Moll and Legler (1971)
P. rubriventris	Massachusetts	ca. 220	9+	ca. 220	9+	Graham (1971b)
P. concinna	Florida	145	—	170	—	Jackson (1970)

Table 4 (*Continued*)

Species	Locality	♂♂ PL (mm)	Age (years)	♀♀ PL (mm)	Age (years)	Reference
Clemmys guttata	Pennsylvania	80	10	80	10	Ernst (1970c, 1975)
C. marmorata	Northern California	100–120	5–8	100–120	6–8	Bury (1975)
Emys orbicularis	Southern Russia	—	5–6	—	5–6	Lukina (1971)
E. orbicularis	Ukraine area	—	7–9	—	7–9	Lukina (1971)
Deirochelys reticularia	Southern Carolina	75–85	—	150–160	—	Gibbons (1969)
Graptemys pulchra	Alabama	80	3–4	220	ca. 14	Shealy (1975)
G. pseudogeographica	Oklahoma	70	3	150	5–6	Webb (1961)
Malaclemys terrapin	Eastern United States	—	—	135	4–8	Hildebrand (1932)
M. terrapin	Louisiana	160+	—	160+	—	Cagle (1952)
Terrapene coahuila	Northern Mexico	95–110	—	90–100	—	Brown (1974)
Chelydra serpentina	Quebec	200	—	200	—	Mosimann and Bider (1960)
C. serpentina	Tennessee	145	—	145	—	White and Murphy (1973)
Macroclemys temmincki	Southern United States	370+	11–13	330+	11–13	Dobie (1971)
Trionyx sinensis	Japan	—	—	125–145	6	Mitsukari (1905)

Species	Location					Reference
Trionyx spiniferus	Minnesota	—	—	90–105	3–4	Breckenridge (1955)
T. s. emoryi	Southern Texas, Northern Mexico	80–90	4+	200	8–9	Webb (1962)
T. s. pallidus	Southern United States	90–100	4+	160–185	8–9	Webb (1962)
T. muticus	Central United States	80–90	4+	140–160	6–7	Webb (1962)
Sternotherus odoratus	Michigan	—	3–4	—	9–11[a]	Risley (1933b)
S. odoratus	Oklahoma	ca. 55–75	4–7	55–75	5–8	Mahmoud (1967)
S. carinatus	Southern United States	ca. 90–115	5–6	ca. 90	4–5	Tinkle (1958, 1961)
S. carinatus	Oklahoma	ca. 60–80	4–7	65–80	5–8	Mahmoud (1967)
Kinosternon subrubrum	Florida	60+	3–4	60+	3–4	Ernst et al. (1973)
K. subrubrum	Oklahoma	ca. 60–80	4–7	ca. 65–80	5–8	Mahmoud (1967)
K. flavescens	Oklahoma	ca. 60–80	4–7	ca. 65–80	5–8	Mahmoud (1967)

[a]Tinkle (1961) considers this to be an overestimate.

Table 5. A Comparison of Sex Ratios

Species	Locality	N	♂♂	♀♀ (%)	Ratio (♀♀ = 1)[a]	Significance[b]	Reference
Phrynops dahli	Colombia	32	13	19(60)	0.68	ns	Medem (1966)
Clemmys marmorata	California	456	246	210(47)	1.17	ns	Bury (1972)
C. guttata	Pennsylvania	140	61	79(56)	0.77	ns	Ernst (1976)
Pseudemys scripta	Panama	208	137	71(34)	1.92	*	Moll and Legler (1971)
P. scripta	Illinois	825	401	424(52)	0.94	ns	Cagle (1950)
P. scripta	Oklahoma	80	38	42(52)	0.90	ns	Webb (1961)
P. scripta	Louisiana	225	123	102(45)	1.20	ns	Viosca (1933)
P. concinna	Florida	123	66	57(47)	1.15	ns	Jackson (1970)
Chrysemys picta	Southern and SouthEastern United States	249	125	124(50)	1.00	ns	Moll (1973)
C. picta	New Mexico	107	53	54(51)	0.98	ns	Christiansen and Moll (1973)
C. picta	Pennsylvania	749	374	375(50)	1.00	ns	Ernst (1971c)
C. picta	New York	90	62	28(31)	2.21	*	Raney and Lachner (1942)
C. picta	New York	71	42	29(41)	1.45	ns	Bayless (1975)
C. picta	Minnesota	54	32	22(41)	1.45	ns	Ernst and Ernst (1972)
C. picta	Michigan	480	265	215(45)	1.23	*	Gibbons (1968b)
C. picta	Michigan	604	242	362(60)	0.66	*	Sexton (1959b)
C. picta	Wisconsin	479	270	209(44)	1.29	*	Ream and Ream (1966)

Species	Location						Reference
Deirochelys reticularia	South Carolina	74	48	26(36)	1.84	*	Gibbons (1969)
Graptemys pseudogeographica	Missouri River	72	36	36(50)	1.00	ns	Timken (1968)
Malaclemys terrapin	Louisiana	70	57	13(19)	4.38	*	Cagle (1952)
Terrapene coahuila	Coahuila, Mexico	164	70	94(57)	0.74	ns	Brown (1974)
Chelydra serpentina	Quebec	55	27	19(51)	0.96	ns	Mosimann and Bider (1960)
C. serpentina	Michigan	151	74	77(51)	0.96	ns	Lagler and Applegate (1943)
C. serpentina	Tennessee	22	14	8(36)	1.75	ns	Froese and Burghardt (1975)
Sternotherus odoratus	Michigan	255	77	178(70)	0.43	*	Risley (1933b)
S. odoratus	Illinois	148	61	87(59)	0.70	*	Cagle (1942b)
S. odoratus	Eastern and Southern United States	647	308	339(52)	0.91	ns	Tinkle (1961)
S. carinatus	Oklahoma	58	22	36(62)	0.61	*	Mahmoud (1969)
S. carinatus	Eastern and Southern United States	123	50	73(59)	0.68	*	Tinkle (1958)
S. m. minor	Eastern and Southern United States	347	169	178(51)	0.94	ns	Tinkle (1958)
S. m. depressus	Southern United States	42	10	32(76)	0.31	*	Tinkle (1958)
S. m. peltifer	Southern United States	62	32	30(48)	1.06	ns	Tinkle (1958)

Table 5. (*Continued*)

Species	Locality	N	♂♂	♀♀ (%)	Ratio (♀♀ = 1)[a]	Significance[b]	Reference
S. m. peltifer	Alabama	193	88	105(54)	0.83	ns	Folkerts (1968)
Kinosternon subrubrum	Oklahoma	202	79	123(61)	0.64	*	Mahmoud (1969)
K. flavescens	Oklahoma	243	107	136(56)	0.78	ns	Mahmoud (1969)
K. flavescens	New Mexico	84	44	40(48)	1.10	ns	Christiansen and Dunham (1972)
Trionyx spiniferus	Illinois	41	24	17(42)	1.41	ns	Cagle (1942b)
T. spiniferus	Texas	39	19	20(52)	0.95	ns	Webb (1962)
T. ferox	Minnesota	163	71	92(56)	0.77	ns	Breckenridge (1955)
Grand total		8227	4037	4190(51)	0.96		

[a]Chi square test ($p < .05$).
[b]* = significant; ns = not significant.

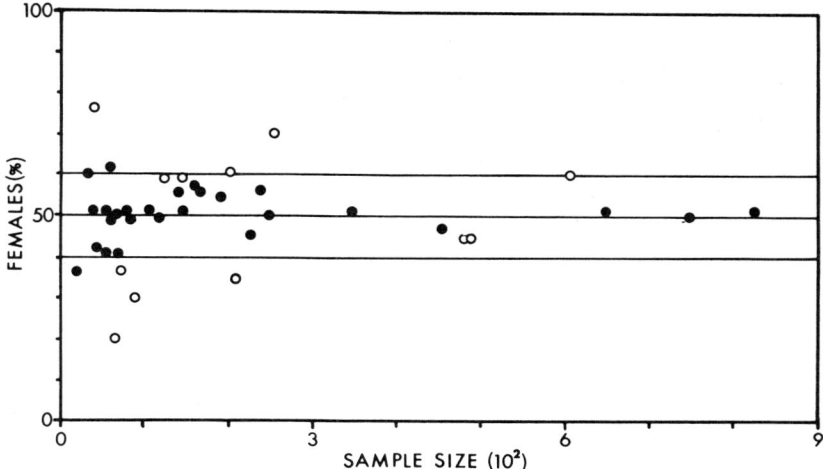

Figure 1 Proportions of females in 39 samples of freshwater turtles. Open circles represent significant differences from the 1:1 sex ratio ($p < .05$).

1:1. In one of these samples, Risley (1933b) reported a high proportion of females (70%) in a sample of 255 *Sternotherus odoratus*; this marked difference is attributed to a sampling bias for more females than males (Gibbons, 1970c). In another case, there was a small proportion of females (34%) in a sample of 208 sliders, *P. scripta*, examined by Moll and Legler (1971). They state that this was probably due to a differential mortality of females (nesting females are captured for food by local people). Excluding these two cases, 71% of the large samples have a 1:1 ratio, which appears to be the norm; caution should be exercised in accepting unequal sex ratios, especially for small samples.

Growth

Changes of form influence the habits of animals, and rates of growth may indicate adaptive strategies of different species or of the same species in varying environmental conditions. Mosimann (1958) and Jolicoeur and Mosimann (1960) reviewed differential growth in the shell of freshwater turtles and described the ontogenetic changes and divergences in the shape of males and females during growth. In *Chrysemys picta*, *Graptemys geographica*, and *Sternotherus odoratus*, the relative rates of posthatchling growth tend to produce a more elongate and streamlined body form that would increase the speed and ease propulsion of a swimming turtle. (Chapter 4 discusses techniques for determining age.)

Several growth patterns are demonstrated by studies of *C. picta* (Cagle, 1954; Ernst, 1971a, 1971c; Ernst and Barbour, 1972; Ernst and Ernst, 1972; Gibbons, 1968b, 1968d; Pearse, 1923b; Wilbur, 1975b). Eggs laid in early summer produce late summer hatchlings, which have a short first growing season. Eggs deposited in the latter part of the summer produce hatchlings that overwinter in the nest and, in the next spring, have a long first growing season, attaining larger size. Individuals may double in length and weight during the first season, but the rate of growth decreases as size increases, with a marked decline in growth once individuals reach maturity. Some old turtles show no growth at all. An apparent switch from a carnivorous to a herbivorous diet when turtles mature might account for the slower rate of growth (Clark and Gibbons, 1969), and food quality in different habitats may be an important determinant of rate of growth (Gibbons, 1967). Northern subspecies of *C. picta* have accelerated early growth rates as compared to southern populations, but the reasons for such rapid growth in northern latitudes are not known.

Similar growth patterns occur in *Pseudemys scripta* (Cagle, 1946, 1948, 1950; Moll and Legler, 1971; Webb, 1961). Males mature earlier than females and, on the average, remain smaller. Growth is responsive to and limited by ecological factors. Sliders taken in a ditch grew faster than ones from a lake, presumably as a result of warm shallow water and abundant food (Cagle, 1946). Gibbons (1970d) found that sliders from a thermally polluted reservoir in South Carolina grew faster than those from other nearby populations. Presumably, the hot water increased productivity at lower trophic levels, ultimately resulting in a large and more varied food supply. The length of the growing season appears to affect the amount of growth in north temperate populations. In the tropics, the most important environmental variables are amount of rainfall (causing inactivity and slowing growth) and sunshine (increasing activity and growth). Individuals of about the same size in similar ecological conditions may exhibit wide differences in the rate of growth.

Jackson (1970), who followed the growth of the Suwannee terrapin (*P. concinna*) in Florida, and Graham (1971b), who studied the red-bellied turtle (*P. rubriventris*) in Massachusetts, found progressive declines in the growth rate associated with increase in size, as well as noticeable variation in the growth rates of turtles of similar size. For *Deirochelys reticularia* Gibbons (1969) reported a growth rate of 25 to 30 mm per year up to sexual maturity, when growth drastically decreased. Webb (1961) found that the growth of *Graptemys pseudogeographica* was about 20 mm per year for the first 2 years. Some males matured in their second year, after which their rate of growth markedly decreased.

Diamondback terrapins, *Malaclemys terrapin*, were the subject of several studies because of their commercial value earlier in the century (Barney, 1922; Hildebrand, 1929, 1932). Alteration of diet and prolonged activity produced more growth per year. Some of the turtles took 12 to 15 years to reach a marketable size (about 125 mm long), but others obtained this size in 5 to 6 years. Cagle (1952) reported that individuals in a natural population of Louisiana *M. terrapin* slowed their growth rate following the attainment of sexual maturity (males, third year; females, sixth). Louisiana turtles grew faster than captive terrapins raised in outdoor pens in North Carolina.

Rhode Island *Clemmys guttata* increase about 7.5 mm in plastron length (PL) per year up to the fifth year of life (Graham, 1970). Ernst (1975) found that the growth of Pennsylvania *C. guttata* in the first season depended on whether hatchlings emerged in late summer or overwintered in the nest. The growth rate decreased as size increased and leveled off at about 10 years of age (80 mm PL); thereafter, mature adults appeared to grow at a rate of 2 to 3% per year. Growth in *C. guttata* was slower and more uniform than that of *Chrysemys picta* from the same pond. The major period of growth for *C. guttata* extended from April to June, whereas *C. picta* in the same area increased in size from June to August. Temperature appeared to be an important limiting factor affecting growth in *C. guttata*, and rainfall seemed the critical factor in *C. picta*.

Risley (1933b) reported that the growth rate in the musk turtle, *Sternotherus odoratus*, decreased with age. Tinkle (1958) demonstrated that the growth curves of *S. carinatus*, *S. depressus*, and *S. minor* were similar up to 3 to 4 years of age, when the growth rate in *S. minor* declined while that of the others continued at about the same rate. Mahmoud (1969) compared the growth rate in four species of kinosternids in Oklahoma and found that juveniles grew rapidly up to 60 mm carapace length (CL) when the growth rate markedly decreased. He attributed intraspecific variation in growth rate to variation in the size of hatchlings, food availability, genetic factors, and the degree of suitability of habitat. Ernst et al. (1973) found that females of Florida *Kinosternon subrubrum* were slightly larger than males throughout life, but the rate of growth for both sexes was similar. Individual variation and the rate of growth decreased with larger size. Growth rates apparently do not slow as rapidly in kinosternids as in emydid turtles following sexual maturity.

Hammer (1969) reported rapid growth in *Chelydra serpentina* up to 7 years of age, followed by continued slow growth to 20 years. Dobie

Table 6 Comparison of Estimated Population Densities[a]

Species	Locality	Habitat	Density per Acre	Reference
Clemmys guttata	Pennsylvania	Pond	16–32	Ernst (1976)
C. marmorata	California	Stream	170	Bury (1972)
Pseudemys scripta	Panama	Pond	77	Moll and Legler (1971)
P. scripta	South Carolina	Freshwater bay	35.6	Gibbons (1970d)
Chrysemys picta	Pennsylvania	Pond	239	Ernst (1971c)
C. picta	New York	Pond	9–11	Bayless (1975)
C. picta	Michigan	Pond	40–160	Sexton (1959b)
C. picta	Michigan	Pond	233	Gibbons (1968b)
C. picta	Wisconsin	Lake	5–20	Pearse (1923a)
Deirochelys reticularia	South Carolina	Freshwater bay	16.2	Gibbons (1970d)
Chelydra serpentina	South Dakota	Marsh	0.5	Hammer (1969)
C. serpentina	Wisconsin	Lake	0.7	Pearse (1923a)
C. serpentina	Illinois	Lake	2	Lagler (1943a)
C. serpentina	Tennessee	Pond	48	Froese and Burghardt (1975)
Terrapene coahuila	Coahuila, Mexico	Marsh	60	Brown (1974)
Sternotherus odoratus	Oklahoma	Creek	60.7	Mahmoud (1969)
S. carinatus	Oklahoma	River	92.6	Mahmoud (1969)
Kinosternon subrubrum	Oklahoma	Creek	64.5	Mahmoud (1969)
K. subrubrum	Oklahoma	Creek	104.6	Mahmoud (1969)
K. subrubrum	South Carolina	Pond	6.9	Gibbons (1970d)
K. flavescens	Oklahoma	Pond	11.3	Mahmoud (1969)

[a]Most are based on Lincoln Index estimates (see Chapter 4).

(1971) found that in the alligator snapping turtle, *Macroclemys temmincki*, growth rate was rapid for the first 11 to 13 years of age, then after maturity decreased significantly to a slow, constant rate.

Breckenridge (1955) found that younger soft-shell turtles, *Trionyx spiniferus*, in Minnesota grew the fastest and it appeared that the growth rate decreased at the time of sexual maturity. Webb (1962) also suggested that there is an apparent decrease in the rate of growth of *Trionyx* species when they become sexually mature.

Density and Biomass

There is marked variation in the estimated densities of freshwater turtles from different aquatic ecosystems and geographic areas (Table 6). Painted turtles in ponds may number from 230 to 240 individuals per acre, the greatest density of any known turtle population. Freshwater turtles may aggregate in large numbers near favorable sites for feeding and basking, thereby reaching high local densities (Bury, 1972; Cagle, 1950; Ernst, 1971c; Sexton, 1959b).

Based on the weights of 100 *Clemmys marmorata* (density of 170 turtles per acre), I calculated a biomass of 137 kg/hectare (121 lb/acre) of stream. This estimate suggests that freshwater turtles have a significant role in energy flow and nutrient cycling in aquatic environments.

The few studies on the subject indicate that body weight in freshwater turtles may be correlated with carapace length, which in turn is related to many other linear measurements of the shell (Lagler and Applegate, 1943; Mosimann, 1958; Dunson, 1967a; Yntema, 1970; Graham, 1971b). Such relationships can provide convenient and accurate determinations of biomass in a population of turtles.

ACTIVITIES

Daily Activities

A day in the active life of a freshwater turtle consists of basking, feeding, resting, moving (within a habitat or migrating), and reproductive effort (courtship, mating, egg laying). The times and duration of each of these behaviors differ among various kinds of freshwater turtles, and, of course, with time of the year.

The activity of the musk turtle, *Sternotherus odoratus*, is mostly nocturnal (Lagler, 1943a; Tinkle, 1958). Mahmoud (1969) reported that Oklahoma species of *Kinosternon* and *Sternotherus* were crepuscular and photophobic during the summer months but basically diurnal in cooler parts of the year. Female snapping turtles display a crepuscular activity pattern during the nesting season (Hammer, 1969). Both *Chelydra serpentina* and *Macroclemys temmincki* seem to be nocturnal in habits (Ernst and Barbour, 1972). North American *Trionyx* species may be nocturnal but do occasionally bask during the day (Lagler, 1943a; Webb, 1962; Ernst and Barbour, 1972). Sliders have been observed emerged on logs and banks at night, and these individuals may be basking (if the air is warmer than the water) or resting (Neill and Allen, 1954; Cagle, 1950; Boyer, 1965).

Emydid turtles have well-developed daytime basking habits, by which they raise and maintain body temperatures at an optimal operating level (Boyer, 1965; Bury, 1972; Ernst, 1972; Waters, 1974; Chapter 11). Cagle (1950) reported that feeding in *Pseudemys scripta* in the United States may occur at any time, but is usually restricted to early morning and late afternoon and that basking occurs mainly during midmorning and midafternoon. At other times sliders rest on the bottom or float on the surface of the water. Moll and Legler (1971) found that, except for nesting females, the slider in Panama is diurnal. Basking generally occurred between 0900 and 1500 with a peak at 1300.

Chrysemys picta are also diurnal, even when migrating from ponds (Sexton, 1959b; Ernst, 1971c). They do not feed when water temperatures are below 15°C and generally become inactive at less than 10°C. Ernst (1971c) found that once the sun rose, turtles basked several hours before foraging. After feeding, they basked from 1100 to 1400, then remained inactive until late afternoon or early evening foraging. Environmental conditions affected the time and duration of each behavior; for example, in cool weather they bask almost all day, feeding sparingly, if at all (Ernst, 1972).

Pritchard and Greenhood (1968) found that *P. nelsoni*, *P. floridana*, and *P. concinna* in Florida basked from 0730 to 1645, mostly at midday. Also in Florida, Auth (1975) reported that basking in *P. scripta* and *P. floridana* started about 0800 and reached a daily maximum between 1000 and 1100 each clear day in August and September. Basking decreased during overcast periods. He also reported that water temperature is probably more important than air temperature in determining whether a turtle does or does not bask.

Spotted turtles, *Clemmys guttata*, in southeastern Pennsylvania are diurnal except for females, which nest after dusk (Ernst, 1976). When the sun rises, spotted turtles either bask until warm or forage for food. Alternate periods of feeding and basking take place during the day with no apparent pattern. In cool weather spotted turtles either bask for most of the day and feed only sparingly or become inactive. Bury (1972) reported that Pacific pond turtles, *C. marmorata*, in a northern California stream began foraging at sunrise (0600) and when the sun first fell on basking sites, most turtles left the water to bask. Some turtles basked periodically between 0800 and dusk, but the peak period was from 0900 to 1000. At other times these turtles foraged or were inactive. Similarly, Waters (1974) reported that map turtles, *Graptemys nigrinoda*, actively swam and foraged between 0500 and 0800, basked from 0800 to 1600 with peaks around 0930 and 1300, and foraged when not basking.

Seasonal Changes

Most temperate species of freshwater turtles show marked changes in the intensity and periodicity of reproductive, feeding, and other daily behavior related to seasonal shifts in environmental conditions. Feeding activity apparently decreases with falling temperatures (Cagle, 1946). Boyer (1965) reported that in the southern United States when conditions in December were favorable, large numbers of freshwater turtles were observed basking, while during a cold period in April, few turtles appeared.

The annual cycle of the painted turtle is well documented throughout most of its range (Bayless, 1975; Conant, 1951; Christiansen and Moll, 1973; Ernst, 1971c, 1972; Ernst and Ernst, 1972; Gibbons, 1968b, 1968d; Lagler, 1943a; Moll, 1973; Powell, 1967; Sexton, 1959b, 1965; Wilbur, 1975a). Inactivity usually occurs in winter, but during warm spells turtles may become temporarily active and engage in basking. Feeding and basking become more frequent when ambient temperatures rise to 20 to 30°C. Southern populations seem to be active most of the year. In late spring there may be mass migrations from main water bodies to outlying smaller waterways. Courtship and mating occur in late spring; egg laying occurs primarily from June 1 to mid-July. Numbers and activity levels are both highest in midsummer.

In Alabaman *Graptemys nigrinoda*, Waters (1974) found that basking occurred only when water temperatures were above 10°C. Basking in the winter occurred mostly on sunny days. In summer, sawback turtles basked or foraged from sunrise to late afternoon. Terrestrial activity in the chicken turtle, *Deirochelys reticularia*, in South Carolina was related to periods of seasonal rainfall (Gibbons, 1969).

Seasonal activity of *Clemmys guttata* apparently is controlled by water temperature and the reproductive cycle (Conant, 1951; Ernst, 1976; Nemuras, 1966). *C. guttata* may be active during most of the year (except in winter), but most are to be found from March to mid-June. Spotted turtles apparently prefer cool weather, being active at water temperatures ranging from 8.5 to 32°C. Most *C. guttata* bask following cold weather dormancy; however feeding does not start until water temperatures exceed 15°C. Courtship is in late spring. Egg-laying occurs in June. Bog turtles, *C. muhlenbergi*, are found from April to October and seem to be most active from April to June (Barton and Price, 1955; Nemuras, 1967). According to Storer (1930), the Pacific pond turtle, *C. marmorata*, usually is active for about 6 months during the warmer portion of the year, with egg laying mostly occurring in June or July. In northern California, these turtles are active from about late May to October (Bury, 1972).

In northern Mexico, Brown (1974) found that the aquatic box turtle, *Terrapene coahuila*, was active throughout the year except during short periods of cold weather in the winter.

Among Oklahoma kinosternid turtles, Mahmoud (1969) reported considerable variation in the seasonal periodicity of activity; the broader the thermoactivity range of a species, the longer the annual period of activity. *K. flavescens* estivated during the hot days of the summer. Periods of inactivity apparently were responses to environmental conditions; turtles kept warm in captivity were active all year. Gibbons (1970d) found that *K. subrubrum* in a South Carolina freshwater bay was commonly captured on land during the summer and fall. At the same site, Bennett (1972) found that this species also spent the winter on land, and generally became dormant below 21°C.

In *Chelydra serpentina*, individuals are mostly active from April to November (Froese and Burghardt, 1975; Lagler, 1943a; White and Murphy, 1973). Although individuals have been noted moving under ice in winter, the time of activity in the southern United States is apparently longer than at northern latitudes.

Length of the season of activity for North American *Trionyx* species also seems to increase with corresponding decreases in latitude (Webb, 1962). The normal period of activity of *T. spiniferus* between about 40° and 43° latitudes is from April to September. Southern populations of *Trionyx* species may be active all year.

SPATIAL DISTRIBUTION

Brown and Orians (1970) stated that spacing in mobile animals has important effects on the population dynamics, population genetics, and

SPATIAL DISTRIBUTION

evolution of species, and is brought about to a considerable degree by the manner in which different individuals of a species react to each other. Environmental factors such as temperature, nutrient supply, and cover are also important determinants of patterns of spatial distribution.

Regular (even, uniform), clumped (aggregated), and random distributions may be found within a species during different seasons or activities. For example, a population may be regularly distributed in an area when territoriality is operative, clumped during periods of resting or basking, and random when the animals are foraging or moving. Most populations of turtles seem to display a clumped distribution during daily and seasonal activities; however this may be attributed largely to habitat preferences.

In the past, conclusions regarding habitat preferences were based on chance discoveries and field observation. Animals with sensory abilities and a perceptual world akin to our own are those most easily located (Klopfer, 1969), which biases our understanding of habitat preferences. The occurrence of a species in a particular habitat is not a priori evidence that such a set of conditions is optimal or actively selected.

Aquatic and semiaquatic turtles are stressed in the absence of water because their rate of evaporative water loss is high as compared to terrestrial forms (Ernst, 1968a; also see Chapters 12 and 22). Freshwater turtles apparently prefer certain conditions in aquatic ecosystems. Many authors have described habitat conditions (see Table 1), but the following studies indicate habitat selection.

Cagle (1950) set traps in 15 lakes and found that only the traps placed in areas of shallow water (less than 5 ft deep) and abundant vegetation yielded many *Pseudemys scripta*; the major needs of this turtle were sunlight, aquatic vegetation, and shallow, quiet water. Similarly, painted turtles frequent static waters and have preferences for areas with floating aquatic vegetation (Sexton, 1959b). Ernst (1976) reported that *Clemmys guttata* in Pennsylvania forage in shallow water and principally live in marsh situations. In a stream community, deep, large pools with logs, branches, or boulders were favored sites of *C. marmorata* (Bury, 1972). Waters (1974) found that adult sawbacks, *Graptemys nigrinoda*, selected stationary snags separated from the shore by open water as basking sites; such sites determined what areas along a river were occupied by adults. Juveniles were not as selective as adults. Many other authors report emydid turtles aggregating at specific basking sites.

Kinosternids also appear to select specific habitat conditions. Tinkle (1961) stated that *Sternotherus m. minor*, *S. m. peltifer*, and *S. carinatus* are most frequently associated with running water or with permanent bodies of water connected with running water, such as oxbow lakes, whereas *S. odoratus* is most abundant in ponds, lakes, and other lentic situations. Skorepa and Ozment (1968) reported that in southern Illinois the eastern mud turtle, *Kinosternon subrubrum*, mainly lives in temporary, shallow ponds in woodlands, but the musk turtle, *S. odoratus*, prefers deeper, more permanent bodies of water and the two species are rarely found together. In Oklahoma, *K. flavescens* preferred temporary bodies of water such as mudholes and farm ponds, whereas *S. odoratus* and *S. carinatus* preferred running water and *K. subrubrum* displayed a broad preference for almost any type of water (Mahmoud, 1969); all were most abundant in areas of heavy aquatic vegetation. In the laboratory, he reported that *Kinosternon* species selected a sandy bottom and shallow water, but *Sternotherus* species preferred a gravel substrate and deeper water. In Florida, *K. bauri* usually inhabits flowing water more than 2 ft deep but also frequents deeper, quieter water, whereas *K. subrubrum* more frequently is found in standing water of depths of less than 30 in. (Ernst et al., 1972).

Adult alligator snappers prefer deep, dark rivers (Allen and Neill, 1950). The common snapper occurs in a wide range of conditions but is most abundant where there is vegetation or cover (Sexton, 1958). Australian and South American pleurodine turtles also have differing habitat requirements (Goode, 1967; Medem, 1960, 1962). There are also seasonal and ontogenetic shifts in selectivity. For example, Moll and Legler (1971) discovered that the slider, *P. scripta*, in Panama undergoes ontogenetic changes in habitat preference.

Shifts in spatial distributions may be related to the occurrence of two or more species in the same locality. A single body of water may be occupied by an array of different turtles representing several genera (Figure 2). In the southern United States up to 17 species may occur in one area. Ecological equivalents occur on other continents but have not been studied. Description of multispecies communities is poorly developed, and we know little about interspecific relations. This is important because the distribution and success of some species may be due to the absence or low abundance of other forms.

MacArthur (1972) reported that in temperate zones more species seem to have ranges limited by habitat than by other factors; however in the tropics competition or predation may impose greater restrictions on the range of a species. Temperate species of turtles seem to fit this idea, but not enough is known about the lives of tropical species to allow meaningful discussion.

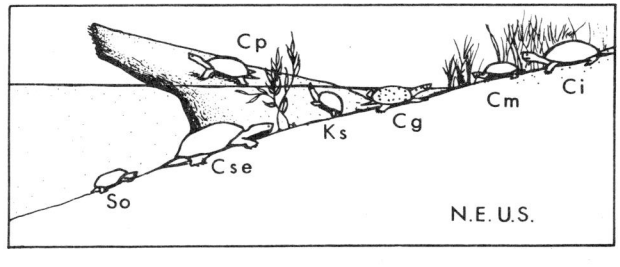
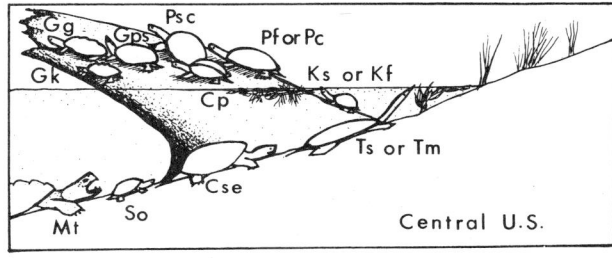
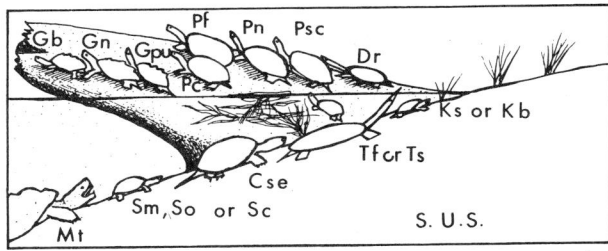

Figure 2 Generalized niche separation of freshwater turtles in three regions of the United States. Note increasing complexity of the community southward. The drawings may represent the edge of a river and of a pond. Not all of the *Graptemys* and *Pseudemys* species occur at any one locality.

Abbreviations:

Cg	*Clemmys guttata*	Kf	*K. flavescens*
Ci	*C. insculpta*	Ks	*K. subrubrum*
Cm	*C. muhlenbergi*	Mt	*Macroclemys temmincki*
Cp	*Chrysemys picta*	Pc	*Pseudemys concinna*
Cs	*Chelydra serpentina*	Pf	*P. floridana*
Dr	*Deirochelys reticularia*	Pn	*P. nelsoni*
Gb	*Graptemys barbouri*	Ps	*P. scripta*
Gg	*G. geographica*	Sc	*Sternotherus carinatus*
Gk	*G. kohni*	Sm	*S. minor*
Gn	*G. nigrinoda*	So	*S. odoratus*
Gps	*G. pseudogeographica*	Tf	*Trionyx ferox*
Gpu	*G. pulchra*	Tm	*T. muticus*
Kb	*Kinosternon bauri*	Ts	*T. spiniferus*

MOVEMENTS

Home Range

There are numerous definitions of "home range." The prominent ones applied to reptiles include the following:

1   The distance between the two farthest points of capture (Nichols, 1939; Stickel, 1950; Sexton, 1959b).
2   The area, usually around a home site, within which the individual ordinarily moves about in the course of its daily activities (Burt, 1943; Stickel, 1950). Mahmoud (1969) preferred the term "activity range" (Carpenter, 1952) to "home range" because the former describes the type of limited range undertaken in day-to-day existence by most reptiles, which may lack a home site or nesting site.
3   The area occupied by an animal and utilized for feeding, reproduction, and other purposes (Tinkle, 1967); the area in which an animal normally lives, exclusive of migrations, emigrations, or unusual erratic wanderings (Brown and Orians, 1970); the total area traversed during seasonal migrations (Moll and Legler, 1971); last, the area of the smallest subregion that accounts for a specified proportion (e.g., 95%) of its total utilization (Jennrich and Turner, 1969).

One of the simplest methods of estimating home range consists of drawing the smallest convex polygon that contains all the capture points for an individual and using the area of this polygon as an index of the home range (Jennrich and Turner, 1969). This method has been used to calculate the home or activity ranges of *Clemmys guttata* (Ernst, 1970a) and kinosternids (Mahmoud, 1969). A minimum polygon index for home range (Mohr, 1947; Mohr and Stumpf, 1966) has been used to estimate the home ranges of *Pseudemys scripta* (Moll and Legler, 1971), and *Clemmys guttata* (Ernst, 1970a); this method probably includes areas that are rarely used by an individual. A modified minimum area method (Harvey and Barbour, 1965; Ernst, 1970a) excludes points that possibly represent sallies outside the main home area and eliminates large areas of no captures.

Linear representation is useful for defined habitats, such as streams or rivers, where the animals are known to traverse certain areas. The area of occupation is calculated by multiplying the width of the used habitat by the distance between capture points. This method has been used to

MOVEMENTS

measure the home range of *Clemmys marmorata* in a stream (Bury, 1972) and of *P. scripta* adults in a river (Moll and Legler, 1971).

Several procedures are based on recapture radii (Jennrich and Turner, 1969) for animals that presumably have a circular home range. Brown (1974) averaged the distances between successive points of capture as an estimate of the radius of the home range of *Terrapene coahuila*. Ernst (1970a) found this method useful for analyzing limited amounts of recapture data on *C. guttata*, but it gave consistent overestimates of home range.

Jennrich and Turner (1969) discussed the inherent problems with these methods and suggested that the convex polygon method be used for quick graphical estimates of home ranges. They introduced a new probabilistic model that adequately characterizes the home range of animals in homogeneous habitats. Van Winkle, Martin, and Sebetich (1973) provided a probability model that may be important in analyzing home ranges of turtles that utilize ecotonal situations between land and water.

Sizes and shapes of home ranges are influenced by physical and ecological barriers (Brown, 1974; Moll and Legler, 1971). Ernst (1976) reports much shifting in the home ranges of *Clemmys guttata* as a result of normal ecological succession that caused turtles to move to more favorable conditions. Reported estimates of home range in freshwater turtles are few and vary considerably (Table 7), owing partly to the diverse procedures and methods used to estimate areas and the small sample sizes, limited diversity of species examined, variable environments, and lack of information on differences between sexes and sizes.

Cagle (1944b, 1946) reported that *P. scripta* utilized restricted home ranges closely associated with feeding and basking areas. Marchand (1945a) reported that the activity range of *P. floridana*, a large Florida slider, probably does not exceed 300 yards.

Pearse (1923a) suggested that *Chrysemys picta* is rather sedentary and, if environmental factors remain favorable, will remain in one locality for years. The limited home range of *C. picta* has also been documented by Sexton (1959b), Gibbons (1968d), and Ernst (1971c).

Mahmoud (1969) found that in kinosternids, habitat types and conditions influenced the size and shape of the activity range. Breckenridge (1955) reported that a Minnesota population of *Trionyx ferox* had limited activity ranges. Webb (1962) confirmed that soft-shelled species wander little.

In some species, juveniles apparently wander farther than adults, presumably during a dispersal phase. In others, there is little movement

Table 7 Estimated Home Range Sizes[a]

				Home Range		
Species	Locality	Habitat	Size (N)	Acres	Length (m)	Reference
Pseudemys scripta	Panama	River	Ad (9)	8.84 (4.34–17.2)	287 (137–503)	Moll and Legler (1971)
		Pond	Juv (24)	1.03	61 (27–99)	Moll and Legler (1971)
		Pond	Baby (10)	0.01 (0–0.02)	34 (10–62)	Moll and Legler (1971)
Clemmys marmorata	California	Stream	Ad ♂ (19)	2.41 (0.55–5.99)	976 (275–2425)	Bury (1972)
			Ad ♀ (23)	0.61 (0–1.85)	248 (0–750)	Bury (1972)
			Juv (18)	0.90 (0–2.84)	363 (0–1150)	Bury (1972)
C. guttata	Pennsylvania	Pond	Ad ♂ (6)	1.30 (1.24–1.38)	—	Ernst (1970a)
			Ad ♀ (5)	1.31 (1.23–1.39)	—	Ernst (1970b)
Terrapene coahuila	Mexico	Marsh	Ad (54)	0.13	12.8 (3–51)	Brown (1974)

Species	Location	Habitat	Sex/Age (n)	Length	Area	Reference
Chelydra serpentina	Pennsylvania	Pond	Ad (9)	4.55	74.5	Ernst (1971c)
	South Dakota	Marsh	Ad (107)	—	1104 (0–6000)	Hammer (1969)
Macroclemys temmincki	Oklahoma	River	Ad (1)	—	9600	Wickham (1922)
Sternotherus carinatus	Oklahoma	River	Ad ♂ (4)	—	39	Mahmoud (1969)
			Ad ♀ (1)	—	17	Mahmoud (1969)
			Juv (7)	—	35	Mahmoud (1969)
S. odoratus	Oklahoma	Creek	Ad ♂ (39)	0.06	67	Mahmoud (1969)
			Ad ♀ (37)	0.12	44	Mahmoud (1969)
	Pennsylvania	Pond	Ad ♂ (15)	—	166	Ernst and Barbour (1972)
			Ad ♀ (6)	—	113	Ernst and Barbour (1972)
			Juv (2)	—	185	Ernst and Barbour (1972)
Kinosternon subrubrum	Oklahoma	Creek	Ad ♂ (79)	0.12	52	Mahmoud (1969)
			Ad ♀ (115)	0.13	62	Mahmoud (1969)
			Juv (6)	—	16	Mahmoud (1969)
K. flavescens	Oklahoma	Pond	Ad ♂ (4)	0.26	198	Mahmoud (1969)
			Ad ♀ (11)	0.31	214	Mahmoud (1969)
			Juv (17)	—	180	Mahmoud (1969)

[a]Length is the distance between recaptures. The range in hectares equals acres × 0.4047.

in juveniles (Table 7). Sizes of home ranges in males and females may be equal or different; however there are insufficient data to generalize on this aspect. Overlap in home ranges is known for *P. scripta* (Moll and Legler, 1971), *C. picta* (Ernst, 1971c), *Clemmys guttata* (Ernst, 1976), and *C. marmorata* (Bury, 1972). Territoriality has not been adequately demonstrated for any freshwater turtles.

Homing Behavior

Cagle (1944b) reported that many *P. scripta* and *C. picta* released some distance from the point of capture would return to the original area of collection, presumably indicating homing behavior. Turtles released at strange bodies of water appeared to wander at random. Williams (1952) found that about half of the *C. picta* he released showed homing behavior; one individual returned 15 times within 1 month. The *Sternotherus odoratus* he studied also tended to return to the original capture site.

Ortleb and Sexton (1964) found that *C. picta* can discriminate between different environmental cues. Gibbons and Smith (1968) reported that orientation of freshwater turtles on land is necessary because aquatic turtles travel on land to lay eggs, or move to different water bodies during certain times of the year. These authors found that several species show a directional orientation when on land.

When Emlen (1969) released *C. picta* away from their home pond to test orientation and homing behavior, the animals homed successfully when displaced 100 m in various directions. Topographic cues may be important in the homing process, since homeward orientation was absent at a distance of 1 mile. Gould (1959) provided evidence that *C. picta* can orient from new and different homeward directions when taken into unfamiliar territory, presumably by using the sun for directional bearings. Although Emlen considered 100 m release sites to be outside the normal home range, Ernst (1970b) thought that these release sites would be familiar to *C. picta*. Data presented in Table 7 clearly indicate that a distance 100 m from a capture point is likely to be within the home range of many freshwater turtles. Twenty-five of 50 *C. picta* that were moved 1 mile upstream, and 22 of 50 that were displaced 1 mile downstream, were recaptured in the pond within 2 years, most within 1 year (Ernst 1970b). Only 12 of 60 turtles that were moved 2 miles downstream returned. Topographic and celestial cues were suggested as the chief means of orientation. Ernst (1968b) found similar homing behavior in *Clemmys guttata* in the same pond.

Mahmoud (1969) noted that the kinosternids he displaced in Oklahoma moved toward the point of original capture, which suggests an ability for local orientation ability. Recognition of the familiar area could result from topographic or environmental cues. Turtles readily learn visual and spatial discriminations in laboratory situations (see Chapter 20).

Storer (1930) reported that female *Clemmys marmorata* may move up to a quarter-mile from water for egg laying, and Moll and Legler (1971) found female *P. scripta* up to 400 m from water. Most turtles may move less than the maximum (Table 7), yet the total familiar area of such long-lived animals may encompass more terrain then we now assume. Some individual turtles may traverse relatively long distances in a few days (Williams, 1952; Ernst, 1970b; Bury, 1972), and these movements seem to be directional. This suggests homing behavior or long familiarity with the areas covered.

NEW DIRECTIONS FOR RESEARCH

What are the adaptive roles and strategies of freshwater turtles? What are the morphological, physiological, and evolutionary constraints of these animals, and how do these relate to population ecology? How can studies on freshwater turtles help us understand the function and roles of organisms in aquatic and semiaquatic ecosystems? What are their problems of competition, diversity, and stability of populations in both natural and man altered environments? Can freshwater turtles provide meaningful indicators of environmental quality and be used as biological monitors of changing aquatic ecosystems?

Do certain types of turtles (e.g., large baskers, small scavengers, specialized feeders) have advantages in different aquatic ecosystems? Do freshwater turtles select habitat conditions, or are their local distributions partly or wholly influenced by competitive interactions with similar and dissimilar species? What is the "carrying capacity" in numbers and biomass of turtles for specific environments? Do turtles compete with other animals for resources? Is there an optimal number or combination of species in defined ecosystems? Are there seasonal and yearly changes in the proportion of species? Do congeneric forms displace one another and, if so, through what mechanism (reproductive fitness, interspecific competition) is this accomplished? What are the limiting factors in multispecies turtle communities? Do exotic species replace or compete with native species?

Schoener (1974) points out that studies of resource partitioning analyze the limits that interspecific competition places on the stable coexistence of multispecies communities. Such work on turtle communities is virtually unknown. We lack comparative studies on freshwater turtles that might reveal the survival strategy of different species in the same habitat and indicate the adaptations of single species within different ecosystems.

Work ought to center on growth as a dynamic, evolutionary process and compare intra- and interspecific differences in growth rates. Both local and geographic factors affecting growth await further investigation, for here lie the means for evaluating the determining factors such as food, genetics, or temperature. Comparisons of growth rates and patterns in multispecies communities may reveal important strategies that function to minimize competition or predation. Experimental work is needed to determine what factors in many species cause marked decrease in growth rate at the attainment of sexual maturity, what causes such wide differences in the rates of growth in individuals of the same population of turtles, and whether crowding reduces the rate.

Besides intensive short-term studies, there are opportunities to follow natural populations of freshwater turtles marked in earlier studies; these animals can be checked at intervals to determine losses, recruitment, age, and other parameters over several decades.

Little is known about the densities of multispecies communities and whether there are optimal numbers of turtles per unit area or volume of water. Suspected limiting factors (food, predators, parasites, and cover) can be tested in natural situations by experimental manipulations. We are learning about varying activity peaks of freshwater turtles but have yet to relate temporal shifts in activity and density in one species to the activities of other species.

Researchers can profitably use rare but information-rich events, such as the draining of ponds and lakes or diversions of channels and streams, to measure large numbers of turtles and compute biomass in different ecosystems. Such population data are increasingly important in determining environmental changes resulting from man's activities. There are virtually no studies on the population ecology of a turtle community before and after (1–10 years) events such as stream channelization, dredging, dam construction, eutrophication, oil pollution, or changes in use of the water body (recreation, industrial, or "recovered").

Better biomass estimates will help define the energetics in both single and multispecies communities of freshwater turtles. Presently, the organization and composition of turtle populations is almost entirely expressed in terms of the number of individuals, which is misleading. For example, one adult snapping turtle is equivalent in weight to many of its own, younger conspecifics or to several other adult turtles of smaller species. Do the large turtles interfere with the activities of smaller species

NEW DIRECTIONS FOR RESEARCH

in the same ecosystem? Do the smaller turtles remove food from the lower parts of the trophic level, thereby reducing the available resources for the larger forms?

We ought to assess the time-energy budgets of various species in the same situation or of a single species over a wider area. Shifts in the daily schedule of marked turtles can be followed over several months and related to environmental or seasonal changes. Differences in the daily activities of different species may reduce competition for food, basking sites, or resting areas. Tests could be designed to demonstrate interference or avoidance of resources in the presence of other species of turtles. Nocturnal habits in freshwater turtles should be related to resource utilizations and interactions with diurnal forms.

The periods of inactivity are poorly known, and location and study of north temperate turtles in the cooler parts of the year should be undertaken.

Much remains to be learned about the spatial relationships within turtle populations. We need to know whether turtles select particular conditions when they are available, and which alternatives are preferred in situations where suboptimal conditions prevail. Precise definition of the selective abilities and advantages to such habitat partitioning is a challenging field of research that may explain the success of many species in complex, competitive environments.

Turner (1971) urges that home range estimates contribute to the solution of specific problems or to the understanding of particular ecological processes. For example, Tinkle (1967) stated that studies of home range are important in indicating the size of an area necessary to sustain an individual and that the degree of overlap of home ranges indicates the social structure of the population. Improvements in home range estimates may be obtained by selecting a population in a defined habitat and using radiotelemetry or some other tracking device (Bury, 1972; Kroll et al., 1973; McGinnis, 1967a, 1967b; Moll and Legler, 1971; Templeton, 1970; Werker, 1970; Chapter 3) to ensure relocation of every marked individual at every sampling. More turtles should be tested in orientation arenas (Emlen, 1969; Gourley, 1974).

Further information is needed regarding the behavioral ecology of freshwater turtles. Evidence of aggression and social interaction (see Chapter 21) suggests that social behavior is more complex than previously assumed and may be important in competition for food, basking sites, or other resources.

Freshwater turtles are excellent subjects for scientific studies of animal biology. In addition, some species are commercially valuable and others are endangered or declining in numbers. Although the commercial farming of turtles once seemed attractive, it has not proved to be economically feasible. Sustained yield and management of wild stocks may prove to be more productive and less expensive than raising animals in captivity. However freshwater turtles are currently harvested with little knowledge of the demography of the species taken, and given the relatively low recruitment rates in most populations, there is danger of depleting the resource. Present information on the population ecology of freshwater turtles is wholly inadequate for conservation and management of these resources. Future research should be both basic and applied, so that freshwater turtle populations will remain as subjects for biological research and as viable components of our aquatic and semiaquatic environments.

ADDENDUM

Since completion of this chapter, several pertinent references were published, as follows: Berry (1975), Burger (1976a, 1976b, 1977), Ewert (1976), Fitch and Plummer (1975), Gibbons and Coker (1977), Hulse (1976), Major (1975), Moll (1976), Plummer (1976, 1976b), Plummer and Shirer (1975), Punzo (1975), and Ward et al. (1976).

ACKNOWLEDGMENTS

Dr. Carl H. Ernst kindly offered comments on the paper and provided stimulating discussions of turtle biology that greatly aided this chapter. I also thank Drs. Alfred L. Gardner, Jaclyn H. Wolfheim, and George R. Zug for their reviews of the manuscript.

Bibliography

Abel, J. H. and R. A. Ellis. 1966. Histochemical and electron microscopic observations on the salt secreting lachrymal glands of marine turtles. *Am. J. Anat.* **18**:337–358.

Ackerman, R. A. and H. D. Prange. 1972. Oxygen diffusion across a sea turtle (*Chelonia mydas*) egg shell. *Comp. Biochem. Physiol.* **43A**:905–909.

Ackman, R. G., S. N. Hooper, and W. Frair. 1971. Comparison of the fatty acid compositions of depot fats from fresh-water and marine turtles. *Comp. Biochem. Physiol.* **40B**:931–944.

Adler, K. K. 1960. Notes on lateral expansion of the periphery in juveniles of *Sternotherus odoratus*. *Copeia.* **1960**:156.

Adrian, E. G. 1950. The electrical activity of the mammalian olfactory bulb. *EEG Clin. Neurophysiol.* **2**:377–388.

Adriani, J. 1955. *Selection of anesthesia*. Springfield, Ill.: Thomas.

Adriani, J. 1962. *The chemistry and physics of anesthesia*. Springfield, Ill.: Thomas.

Adriani, J. 1970. *The pharmacology of anesthetic drugs*. Springfield, Ill.: Thomas.

Agassiz, L. 1857. *Contributions to the natural history of the United States of America*. Vols. 1, 2. Boston: Little, Brown.

Alberts, J. R. and B. G. Galef, Jr. 1971. Acute anosmia in the rat: A behavioral test of peripherally induced olfactory deficit. *Physiol. Behav.* **62**:619–621.

Alexander, F. 1969. *An introduction to veterinary pharmacology*. Edinburgh: Livingstone.

Alexander, R. McN. 1977. *Mechanics and scaling of terrestrial locomotion* in T. J. Pedley (Ed.), *Scale effects in animal locomotion*, pp. 93–110. London: Academic Press.

Ali, M. A. 1974. *Vision in fishes*. New York: Plenum Press.

Allard, H. A. 1935. The natural history of the box turtle. *Sci. Mon.* **41**:325–338.

Allard, H. A. 1948. The eastern box turtle and its behavior. *J. Tenn. Acad. Sci.* **23**:307–321.

Allard, H. A. 1949. The eastern box turtle and its behavior. *J. Tenn. Acad. Sci.* **24**:146–152.

Allee, W., A. Emerson, O. Park, T. Park, and K. Schmidt. 1949. *Principles of animal ecology*. Philadelphia: Saunders.

Allen, B. M. 1906. The origin of sex cells of *Chrysemys*. *Anat. Anz.* **29**:217–236.

Allen, E. R. 1950. Sounds produced by the Suwannee terrapin. *Copeia.* **1950**:62.

Allen, E. R. and W. T. Neill. 1950. The alligator snapping turtle, *Macrochelys temminckii*, in Florida. *Ross Allen's Reptile Institute, Spec. Pub.* 4.

Allen, J. F. and R. A. Littleford. 1955. Observations on the feeding habits and growth of immature diamondback terrapins. *Herpetologica*. **11**:77–80.

Allen, R., L. G. Gumbreck, and M. R. Stellar. 1959. Some preliminary studies of "stress" in the western diamond-backed rattlesnake (*Crotalus atrox*). *Proc. Okla. Acad. Sci.* **40**:19–21.

Allen, W. B. 1955. Some notes on reptiles. *Herpetologica*. **11**:228.

Allison, A. C. 1953. The morphology of the olfactory system in the vertebrates. *Biol. Rev.* **28**:195–244.

Altland, P. D. 1951. Observations on the structure of the reproductive organs of the box turtle. *J. Morphol.* **89**:599–616.

Altland, P. D., B. Highman, and B. Wood. 1951. Some effects of X-radiation in turtles. *J. Exp. Zool.* **118**:1–17.

Altland, P. D. and M. Parker. 1955. Effects of hypoxia upon the box turtle. *Am. J. Physiol.* **180**:421–427.

Alvarez del Toro, M. 1960. *Los reptiles de Chiapas*. Chiapas, Mexico: Instituto Zoológico del Estado, Tuxtla Gutiérrez.

Andersen, H. T. 1966. Physiological adaptations in diving vertebrates. *Physiol. Rev.* **46**:212–243.

Anderson, N. G. and K. M. Wilbur. 1948. Gastric carbonic anhydrase and acid secretion in turtles. *J. Cell. Comp. Physiol.* **31**:293–302.

Anderson, P. K. 1958. The photic responses and water-approach behavior of hatchling turtles. *Copeia.* **1958**:211–215.

Anderson, P. K. 1965. *The reptiles of Missouri*. Columbia: University of Missouri Press.

Andreev, S. F. 1948. Adaptations of reptiles to high temperatures of deserts. *Uch. Zap. Biol. Fak. Univ. Chernovits.* **1**:109–118.

Andrews, O. 1915. The ability of turtles to discriminate between sounds. *Wisc. Nat. Hist. Soc. Bull.* **13**:189–195.

Anonymous. 1937. *A guide to the vertebrate fauna of the eastern Cape Province, South Africa*, Part 2. Grahamstown: Albany Museum.

Anonymous. 1967. Spring is here. *Int. Turtle Tortoise Soc. J.* **1**(4):10–12.

Anonymous. 1969. 1969 California turtle and tortoise club exhibit. *Int. Turtle Tortoise Soc. J.* **3**(4):20–25.

Anonymous. 1972. A good food. *Int. Turtle Tortoise Soc. J.* **6**(3):29.

Anonymous. 1973. It takes more than lettuce. . . . *Int. Turtle Tortoise Soc. J.* **7**(2):34.

Archer, W. H. 1948. The mountain tortoise. *Afr. Wildl.* **2**(2):75–78, (3):74–77.

Archer, W. H. 1967a. The angulated tortoise. *Afr. Wildl.* **21**(2):137–143.

Archer, W. H. 1967b. The geometric tortoise. *Afr. Wildl.* **21**(4):321–329.

Archer, W. H. 1968a. More notes on the angulated tortoise. *Afr. Wildl.* **22**(2):141–146.

Archer, W. H. 1968b. The Padlopers. *Afr. Wildl.* **22**(1):29–35.

Ariens-Kappers, C. U., G. C. Huber, and E. C. Crosby. 1936. *The comparative anatomy of the nervous system of vertebrates*. New York: Hafner.

Armington, J. C. 1954. Spectral sensitivity of the turtle, *Pseudemys*. *J. Comp. Physiol. Psychol.* **47**:1–6.

BIBLIOGRAPHY

Aschoff, J. 1960. Exogenous and endogenous components in circadian rhythms. *Cold Spring Harbor Symp. Quant. Biol.* **1960**:25-28.

Ashley, L. 1955. *Laboratory anatomy of the turtle.* Dubuque: Wm. C. Brown.

Auffenberg, W. 1963. A note on the drinking habits of some land tortoises. *Anim. Behav.* **11**:72-73.

Auffenberg, W. 1964a. A first record of breeding colour changes in a tortoise. *J. Bombay Nat. Hist. Soc.* **61**:191-192.

Auffenberg, W. 1964b. Notes on the courtship of the land tortoise *Geochelone travancorica* (Boulenger). *J. Bombay Nat. Hist. Soc.* **61**:247-253.

Auffenberg, W. 1965. Sex and species discrimination in two sympatric South American tortoises. *Copeia.* **1965**:335-342.

Auffenberg, W. 1966a. The carpus of land tortoises (Testudinae). *Bull. Fla. State Mus., Biol. Sci.* **10**(5):159-191.

Auffenberg, W. 1966b. On the courtship of *Gopherus polyphemus*. *Herpetologica.* **22**:113-117.

Auffenberg, W. 1969a. Land of the Chaco tortoise, *Geochelone chilensis*. *Int. Turtle Tortoise Soc. J.* **3**(3):16-19, 36-37.

Auffenberg, W. 1969b. Social behavior of *Geochelone denticulata*. *Quart. J. Fla. Acad. Sci.* **32**:50-58.

Auffenberg, W. 1969c. *Tortoise behavior and survival.* Skokie, Ill.: Rand-McNally.

Auffenberg, W. 1971. A new fossil tortoise, with remarks on the origin of South American testudinines. *Copeia.* **1971**:106-117.

Auffenberg, W. 1974. Checklist of fossil land tortoises (Testudinidae). *Bull. Fla. State Mus., Biol. Sci.* **18**:121-251.

Auffenberg, W. and W. G. Weaver, Jr. 1969. *Gopherus berlandieri* in southeastern Texas. *Bull. Fla. State Mus., Biol. Sci.* **13**:141-203.

Auth, D. L. 1975. Behavioral ecology of basking in the yellow-bellied turtle, *Chrysemys scripta scripta* (Schoepff). *Bull. Fla. State Mus., Biol. Sci.* **20**:1-45.

Babcock, H. L. 1937. A new species of the red-bellied terrapin, *Pseudemys rubriventris* (LeConte). *Occas. Pap. Boston Soc. Nat. Hist.* **8**:293-294.

Babcock, H. L. 1971. *Turtles of the northeastern United States.* New York: Dover.

Backhaus, D. 1963. Zur Behandlung und Pflege von Schlangen. In W. Ludwig, H. v. Behring, H. E. Schultze, and W. Freiherr v. Polnitz (Eds.), *Die Giftschlangen der Erde*, pp. 223-253. Marburg: Behringwerk-Mitt. Spec. Suppl.

Backhaus, D. 1965. Gegen *Strongyloides* infektionen bei Reptilien. *Salamandra.* **1**:27-28.

Badham, J. A. 1971. Albumin formation in eggs of the agamid *Amphibolurus barbatus barbatus*. *Copeia.* **1971**:543-545.

Baker, E. 1968. It can happen. *Int. Turtle Tortoise Soc. J.* **2**(6):4-5.

Baker, J. R. 1938. The evolution of breeding seasons. In G. deBeer (Ed.), *Evolution*, pp. 161-177. Oxford: Clarendon.

Baker, L. A. and F. L. White. 1970. Redistribution of cardiac output in response to heating in *Iguana iguana*. *Comp. Biochem. Physiol.* **35**:253-262.

Baker-Cohen, K. F. 1968a. Comparative enzyme histochemical observations on submammalian brains. I. Striatal structures in reptiles and birds. *Ergebn. Anat. Entwickl.-Gesch.* **40**(6):1-41.

Balasingam, E. and Mohamed Khan Bin Momin Khan. 1969. Conservation of the Perak River terrapins (*Batagur baska*). *Malay. Nature J.* **23**:27–29.

Balazs, G. H. 1974. Observations on the basking habits in the captive juvenile Pacific green turtle. *Copeia.* **1974**:542–544.

Baldwin, F. M. 1925a. Body temperature changes in turtles and their physiological interpretations. *Am. J. Physiol.* **72**:210–211.

Baldwin, F. M. 1925b. The relation of body to environmental temperatures in turtles, *Chrysemys marginata belli* (Gray) and *Chelydra serpentina* (Linn.). *Biol. Bull.* **48**:432–445.

Baldwin, W. P. and J. P. Lofton. 1959. The loggerhead turtles of Cape Romain, South Carolina. In D. K. Caldwell (Ed.), The Atlantic loggerhead sea turtle, *Caretta caretta caretta* (L.), in America. III. *Bull. Fla. State Mus., Biol Sci.* **4**:319–348.

Ballinger, R. E. and D. R. Clark. 1973. Energy content of lizard eggs and the measurement of reproductive effort. *J. Herpetol.* **7**:129–132.

Banks, E. 1937. The breeding of the edible turtle (*Chelonia mydas*). *Sarawak Mus. J.* **4**:523–532.

Bantli, H. 1974. Analysis of differences between potentials evoked by climbing fibers in cerebellum of cat and turtle. *J. Neurophysiol.* **37**:573–593.

Barclay, O. R. 1946. The mechanics of amphibian locomotion. *J. Exp. Biol.* **23**:177–203.

Barney, R. L. 1922. Further notes on the natural history and artificial propagation of the diamond-back terrapin. *Bull. U.S. Bur. Fish.* **38**:91–111.

Bartlett, R. D. 1971. The effect of Gro-Lux lighting. *Int. Turtle Tortoise Soc. J.* **5**(2):32.

Barton, A. J. and J. W. Price. 1955. Our knowledge of the bogturtle, *Clemmys muhlenbergi*, surveyed and augmented. *Copeia.* **1955**:159–165.

Barwick, R. E. and P. J. Fullager. 1967. A bibliography of radio telemetry in biological studies. *Proc. Ecol. Soc. Australia.* **2**:27–49.

Bass, A. H. 1976. Retinal projections in the painted turtle. *Neurosci. Abstr.* **1976**:177.

Baumann, T. W. 1966. A study of brain and cervical spinal cord in *Chrysemys picta*. Ph.D. dissertation, St. Louis University.

Baur, G. 1893. Notes on the classification and taxonomy of the Testudinata. *Proc. Am. Phil. Soc.* **31**:210–225.

Bayless, L. E. 1975. Population parameters for *Chrysemys picta* in a New York pond. *Am. Midl. Nat.* **93**:168–176.

Baylor, D. A. and A. L. Hodgkin. 1973. Detection and resolution of visual stimuli by turtle photoreceptors. *J. Physiol.* **234**:163–198.

Baylor, D. A. and R. Fettiplace. 1975. Light path and photon capture in turtle photoreceptors. *J. Physiol.* **248**:433–464.

Baze, W. B. and F. R. Horne. 1970. Ureogenesis in Chelonia. *Comp. Biochem. Physiol.* **34**:91–100.

Beck, R. H. 1903. In the home of the giant tortoise. *Ann. Rept. N.Y. Zool. Soc.* **7**:1–17.

Belekhova, M. G. and A. A. Kosareva. 1971. Organization of the turtle thalamus: Visual, somatic and tectal zones. *Brain Behav. Evol.* **4**:337–375.

Belkin, D. A. 1962. Anaerobiosis in diving turtles. *Physiologist.* **5**:105.

Belkin, D. A. 1963. Anoxia: Tolerance in reptiles. *Science.* **139**:492–493.

Belkin, D. A. 1964. Variations in heart rate during voluntary diving in the turtle, *Pseudemys concinna*. *Copeia.* **1964**:321–330.

Belkin, D. A. 1965. Reduction of metabolic rate in response to starvation in the turtle *Sternothaerus minor. Copeia.* **1965**:367-368.

Belkin, D. A. 1968. Aquatic respiration and underwater survival of two freshwater turtle species. *Respir. Physiol.* **4**:1-14.

Belkin, D. A. and C. Gans. 1968. An unusual Chelonian feeding niche. *Ecology.* **49**:768-769.

Bellairs, A. 1970. *The life of reptiles.* Vol. 1. New York: Universe.

Bellairs, A. and R. Carrington. 1966. *The world of reptiles.* New York: Elsevier.

Bellanti, J. A. 1971. *Immunology.* Philadelphia: Saunders.

Benbrook, E. A. and M. W. Sloss. 1961. *Veterinary clinical parasitology.* Ames: Iowa State University Press.

Bennett, A. F. 1973. Ventilation in two species of lizards during rest and activity. *Comp. Biochem. Physiol.* **46A**:653-671.

Bennett, D. H. 1972. Notes on the terrestrial wintering of mud turtles (*Kinosternon subrubrum*). *Herpetologica.* **28**:245-247.

Bennett, D. H., J. W. Gibbons, and J. C. Franson. 1970. Terrestrial activity in aquatic turtles. *Ecology.* **51**:738-740.

Bentley, P. J. 1962. Studies on the permeability of the large intestine and urinary bladder of the tortoise *Testudo graeca*, with special reference to the effects of neurohypophysial and adrenocortical hormones. *Gen. Comp. Endocrinol.* **2**:323-328.

Bentley, P. J., W. L. Bretz, and K. Schmidt-Nielsen. 1967. Osmoregulation in the diamondback terrapin, *Malaclemys terrapin centrata. J. Exp. Biol.* **46**:161-167.

Bentley, P. H. and K. Schmidt-Nielsen. 1966. Cutaneous water loss in reptiles. *Science.* **151**:1547-1549.

Betz, T. W. 1962. Surgical anesthesia in reptiles, with special reference to the water snake, *Natrix rhombifera. Copeia.* **1962**:284-287.

Bider, J. R. and W. Hoek. 1971. An efficient and apparently unbiased sampling technique for population studies of painted turtles. *Herpetologica.* **27**:481-484.

Bienz, A. 1895. *Dermatemys mawii* Gray; eine osteologische Studia mit Beitragen aus Kenntniss von Baue der Schildkroten. *Rev. Suisse Zool., Ann. Mus. Hist. Nat. Geneve* **3**:61-135.

Bishop, S. C. and W. J. Schoonmacher. 1921. Turtle hunting in midwinter. *Copeia.* **1921**:37-38.

Bitterman, M. E. 1964. An instrumental technique for the turtle. *J. Exp. Anal. Behav.* **7**:189-190.

Bitterman, M. E. 1965. Phyletic differences in learning. *Am. Psychol.* **20**:396-410.

Blair, W., A. Blair, P. Brodkorb, F. Cagle, and G. Moore. 1957. *Vertebrates of the United States.* New York: McGraw-Hill.

Blanc, C. P. and C. C. Carpenter. 1969. Studies of the Iguanidae of Madagascar. III. Social and reproductive behavior of *Chalaradon madagascariensis. J. Herpetol.* **3**:125-134.

Blaney, R. M. 1968. Hybridization of the box turtles *Terrapene carolina* and *T. ornata* in western Louisiana. *Proc. La. Acad. Sci.* **31**:54-57.

Bleakney, J. S. 1963. Notes on the distribution and life histories of turtles in Nova Scotia. *Can. Field Nat.* **77**:67-76.

Blohm, T. and A. Fernandez-Yépez. 1948. La Sociedad de Ciencias Naturales de La Salle en Pararuma. *Mem. Soc. Cienc. Nat. La Salle* (Caracas) **8**(21):35-69.

Board, R. G. and R. Fuller. 1974. Non-specific antimicrobial defences of the avian egg, embryo and neonate. *Biol. Rev. Camb. Phil. Soc.* **49**:15–49.

Bockman, D. E. 1970. The thymus. In C. Gans and T. S. Parsons (Eds.), *Biology of the reptilia*, Vol. 3, pp. 111–133. New York: Academic Press.

Boeckh, J. 1969. Electrical activity in olfactory receptor cells. In C. Pfaffman (Ed.), *Olfaction and taste*, Vol. III, pp. 34–51. New York: Rockefeller University Press.

Bogert, C. M. 1937. Notes on the growth rate of the desert tortoise, *Gopherus agassizi. Copeia.* **1937**:191–192.

Bogert, C. M. and R. B. Cowles. 1947. Results of the Archbold Expeditions. No. 58. Moisture loss in relation to habitat selection in some Floridian reptiles. *Am. Mus. Novit.* **1368**:1–34.

Boice, R. 1970a. Competitive feeding behaviors in captive *Terrapene c. carolina. Anim. Behav.* **18**:703–710.

Boice, R. 1970b. Social behaviors and hierarchies in box turtles (*Terrapene*). *Proc. 78th Ann. Conv., APA.* **1970**:229–230.

Boice, R., C. B. Quanty, and R. C. Williams, 1974. Competition and possible dominance in turtles, toads, and frogs. *J. Comp. Physiol. Psychol.* **86**:1116–1131.

Bojanus, L. H. 1819. *Anatome Testudinis Europaeae.* Vilnae. [Reprinted in 1970. Society for the Study of Amphibians and Reptiles. Facsimile Reprints in Herpetology, No. 26.]

Bolles, R. C. 1962. The readiness to eat and drink: The effect of deprivation conditioning. *J. Comp. Physiol. Psychol.* **55**:230–234.

Bolles, R. C. 1973. The comparative psychology of learning: The selective association principle and some problems with "general" laws of learning. In G. Bermant (Ed.), *Perspectives on animal behavior*, pp. 280–306. Glenview, Ill.: Scott, Foresman.

Booth, J. and J. A. Peters. 1972. Behavioral studies on the green turtle (*Chelonia mydas*) in the sea. *Anim. Behav.* **20**:808–812.

Borysenko, M. 1969a. The maturation of the capacity to reject skin allografts and xenografts in the snapping turtle, *Chelydra serpentina. J. Exp. Zool.* **170**:359–364.

Borysenko, M. 1969b. Skin allograft and xenograft rejection in the snapping turtle, *Chelydra serpentina. J. Exp. Zool.* **170**:341–358.

Borysenko, M. 1970. Transplantation immunity in Reptilia. *Transplant. Proc.* **2**:229–306.

Borysenko, M. and P. Tulipan. 1973. The graft-*versus*-host reaction in the snapping turtle, *Chelydra serpentina. Transplantation.* **16**:496–504.

Boulenger, G. A. 1966. *Catalogue of the chelonians, rhynchocephalians, and crocodiles in the British Museum.* Codicote, Herts.: Wheldon & Wesley.

Bourn, D. 1976. Giant tortoises do almost too well on island reserve. *Smithsonian* **7**(2):82–90.

Bourret, R. 1941. Les tortues de l'Indochine. *Inst. Océanogr. Indochine.* **38**:1–235.

Bowker, F. 1926. Tortoise eggs and nest. *S. Afr. J. Nat. Hist.* **6**:37.

Boycott, B. B. and R. W. Guillery. 1962. Olfactory and visual learning in the red-eared terrapin, *Pseudemys scripta elegans* (Weid). *J. Exp. Biol.* **39**:567–577.

Boyer, D. R. 1965. Ecology of the basking habit in turtles. *Ecology* **46**:99–118.

Bradley, R. M. 1971. Tongue topography. In L. M. Biedler (Ed.), *Handbook of sensory physiology*, Vol. IV, Part 2, pp. 1–30. New York: Springer-Verlag,

Braid, M. R. 1974. A bal-chatri trap for basking turtles. *Copeia.* **1974**:539–540.

BIBLIOGRAPHY

Bramble, D. M. 1974. Emydid shell kinesis: Biomechanics and evolution. *Copeia.* **1974**:707–727.

Brander, R. and W. Cochran. 1971. Radio-location telemetry. In R. Giles, Jr. (Ed.), *Wildlife management techniques*, pp. 95–103. Washington, D.C.: Wildlife Society.

Brattstrom, B. H. 1965. Body temperatures of reptiles. *Am. Midl. Nat.* **73**:376–422.

Brattstrom, B. H. 1970. Amphibia. In B. C. Whittow (Ed.), *Comparative physiology of thermoregulation*, Vol. I. *Invertebrates and nonmammalian vertebrates*, pp. 135–166. New York: Academic Press.

Brattstrom, B. H. 1974. The evolution of reptilian social behavior. *Am. Zool.* **14**:35–49.

Brazenor, C. W. and G. Kaye. 1953. Anesthesia for reptiles. *Copeia.* **1953**:165–170.

Breazile, J. E. and R. L. Kitchell. 1969. Euthanasia for laboratory animals. *Fed. Proc.* **28**:1577–1579.

Breckenridge, W. J. 1944. *Reptiles and amphibians of Minnesota*. Minneapolis: University of Minnesota Press.

Breckenridge, W. J. 1955. Observations on the life-history of the soft-shelled turtle *Trionyx ferox*, with especial reference to growth. *Copeia.* **1955**:5–9.

Breckenridge, W. J. 1960. A spiny soft-shelled turtle nest study. *Herpetologica.* **16**:284–285.

Breder, C. M. Jr. 1946. Amphibians and reptiles of the Rió Chucunaque drainage, Darién, Panamá, with notes on their life histories and habits. *Bull. Am. Mus. Nat. Hist.* **86**:375–436.

Breder, R. B. 1927. Turtle trailing: A new technique for studying the life habits of certain Testudinata. *Zoologica.* **9**:231–243.

Breland, K. and M. Breland. 1966. *Animal behavior.* New York: Macmillan.

Brenner, F. J. 1970. The influence of light and temperature on fat utilization in female *Clemmys insculpta. Ohio J. Sci.* **70**:233–237.

Brent, J. 1958. Tissue transplant immunity. *Progr. Allergy* **5**:271–348.

Brett, W. 1971. Persistent rhythms of locomotion activity in the turtle *Pseudemys scripta. Comp. Biochem. Physiol.* **40A**:925–934.

Bricker, N. S. and S. Klahr. 1966. Effects of dinitrophenol and oligomycin on the coupling between anaerobic metabolism and anaerobic sodium transport by the isolated turtle bladder. *J. Gen. Physiol.* **49**:483–499.

Bridges, C. D. B. 1965. Absorption properties, interconversions, and environmental adaptation of pigments from fish photoreceptors. *Cold Spring Harbor Symp. Quant. Biol.* **30**:317–334.

Bridges, C. D. B. 1972. The rhodopsin-porphyropsin visual system. In H. J. A. Dartnall (Ed.), *Photochemistry of vision.* Vol. VII/1. *Handbook of sensory physiology*, pp. 417–480. New York: Springer-Verlag.

Brode, W. E. 1959. Notes on behavior of *Gopherus polyphemus. Herpetologica.* **15**:101–102.

Brodsky, W. A. and T. P. Schilb. 1963. Anion transport in isolated turtle bladder. *Fed. Proc.* **22**:322.

Brodsky, W. A. and T. P. Schilb. 1966. Ionic mechanisms for sodium and chloride transport across turtle bladders. *Am. J. Physiol.* **210**:987–996.

Brongersma, L. D. 1969. Miscellaneous notes on turtles. II. *Proc. K. Ned. Akad. Wetensch.*, C. **72**:76–102.

Brongersma, L. D. 1972. European Atlantic turtles. *Zool. Verhand.*, Leiden. No. 121:1–318.

Brown, F. 1973. Biological rhythms. In C. Prosser (Ed.), *Comparative animal physiology*, pp. 429–456. Philadelphia: Saunders.

Brown, G. W., Jr. and P. P. Cohen. 1960. Comparative biochemistry of urea synthesis. *Biochem. J.* 75:82–91.

Brown, J. L. and G. H. Orians. 1970. Spacing patterns in mobile animals. *Ann. Rev. Ecol. Syst.* 1:239–262.

Brown, K. T. 1969. A linear area centralis extending across the turtle retina and stabilized to the horizon by non-visual cues. *Vis. Res.* 9:1053–1062.

Brown, W. S. 1971. Morphometrics of *Terrapene coahuila* (Chelonia, Emydidae), with comments on its evolutionary status. *Southwest. Nat.* 16(2):171–184.

Brown, W. S. 1974. Ecology of the aquatic box turtle, *Terrapene coahuila* (Chelonia, Emydidae), in northern Mexico. *Bull. Fla. St. Mus., Biol. Sci.* 19:1–67.

Brumwell, M. J. 1940. Notes on the courtship of the turtle, *Terrapene ornata*. *Trans. Kans. Acad. Sci.* 43:391–392.

Buckman, H. O. and N. C. Brady. 1960. *The nature and properties of soils*. New York: Macmillan.

Bünning, E. 1964. *The physiological clock*. New York: Springer-Verlag.

Burbidge, AV. A. 1967. The biology of southwestern Australian tortoises. Ph.D. dissertation. University of Western Australia, Nedlands.

Burchfield, P. M. 1975. Hatching the radiated tortoise, *Testudo radiata*, at Brownsville Zoo. *Int. Zoo Yearb.* 15:90–92.

Burda, D. J. 1965. Development of intracranial arterial patterns in turtles. *J. Morphol.* 116:117–187.

Burger, J. W. 1937. Experimental sexual photoperiodicity in the male turtle, *Pseudemys elegans* (Wied). *Am. Natur.* 71:481–487.

Burger, J. 1975. Factors affecting survival and hatching of the eggs of the diamondback terrapin. *Am. Soc. Ichthyol. Herpetol. Meet. Program.* 55:31–32 (Abstr.).

Burger, J. 1976a. Behavior of hatchling diamondback terrapins (*Malaclemys terrapin*) in the field. *Copeia.* 1976:742–748.

Burger, J. 1976b. Temperature relationships in nests of the northern diamondback terrapin, *Malaclemys terrapin terrapin. Herpetologica.* 32:412–418.

Burger, J. 1977. Determinants of hatching success in diamondback terrapin, *Malaclemys terrapin. Am. Midl. Nat.* 97:444–464.

Burger, J. and W. A. Montevecchia. 1975. Tidal synchronization and nest site selection in the northern diamondback terrapin *Malaclemys terrapin terrapin* Schoepff. *Copeia.* 1975:113–119.

Burghardt, G. M. 1967. The primacy effect of the first feeding experience in the snapping turtle. *Psychon. Sci.* 7:383–384.

Burghardt, G. M. 1970. Chemical perception in reptiles. In J. W. Johnston, D. G. Mouton, and A. Turk (Eds.), *Communication by chemical signals*, pp. 241–308. New York: Appleton-Century-Crofts.

Burghardt, G. M. and E. H. Hess. 1966. Food imprinting in the snapping turtle. *Science.* 151:108–109.

Burke, T. J. 1970. Hypovitaminosis A. *Int. Turtle Tortoise Soc. J.* 4(3):8–9.

Burne, R. H. 1905. Notes on the muscular and visceral anatomy of the leathery turtle (*Dermochelys coriacea*). *Proc. Sci. Meet. Zool. Soc. London.* **1**:291–324.

Burt, W. H. 1943. Territoriality and home range concepts as applied to mammals. *J. Mammal.* **24**:346–352.

Bury, R. B. 1972. Habits and home range of the Pacific pond turtle, *Clemmys marmorata*, in a stream community. Ph.D. dissertation, University of California, Berkeley.

Bury, R. B. 1975. Ecology and behavior of *Clemmys marmorata* in a stream community. *Am. Soc. Ichthyol. Herpetol. Meet. Program.* **55**:32 (Abstr.).

Bury, R. B. and J. Wolfheim. 1973. Aggression in free-living pond turtles, *Clemmys marmorata*. *BioScience.* **23**:659–662.

Butler, C. G. 1970. Chemical communication in insects: Behavioral and ecological aspects. In J. Johnston, Jr., D. G. Moulton, and A. Turk (Eds.), *Advances in chemoreception*, Vol. I, pp. 35–78. New York: Appleton-Century-Crofts.

Butler, D. G. 1972a. Antidiuretic effect of arginine vasotocin in the western painted turtle, *Chrysemys picta belli*. *Gen. Comp. Endocrinol.* **18**:121–125.

Butler, D. G. 1972b. Effect of hypophysectomy on renal function in the western painted turtle, *Chrysemys picta belli*. *Gen. Comp. Endocrinol.* **19**:377–383.

Butler, D. G. and W. H. Knox. 1970. Adrenalectomy of the painted turtle (*Chrysemys picta belli*): Effect on ionoregulation and tissue glycogen. *Gen. Comp. Endocrinol.* **14**:551–566.

Bustard, H. R. 1966. Turtle biology at Heron Island. *Austl. Nat. Hist.* **15**:262–264.

Bustard, H. R. 1967. Mechanism of nocturnal emergence from the nest in green turtle hatchlings. *Nature.* **214**:317.

Bustard, H. R. 1969. Defensive behavior and locomotion of the Pacific boa, *Candoia aspera*, with a brief review of head concealment in snakes. *Herpetologica.* **16**:164–174.

Bustard, H. R. 1970a. The adaptive significance of coloration in hatchling green sea turtles. *Herpetologica.* **26**:224–227.

Bustard, H. R. 1970b. Turtles and an iguana in Fiji. *Oryx.* **10**:316–322.

Bustard, H. R. 1971a. Conservation and rational exploitation of the leathery turtle (*Dermochelys coriacea*). *Oryx.* **11**:233–239.

Bustard, H. R. 1971b. Temperature and water tolerances of incubating sea turtle eggs. *Brit. J. Herpetol.* **4**:196–198.

Bustard, H. R. 1972a. *Sea turtles: Their natural history and conservation.* London: Collins.

Bustard, H. R. 1972b. Farming turtles and crocodiles in Torres Strait. *Austl. Fish.* **31**:6–8.

Bustard, H. R. 1973a. *Kay's turtles.* London: Collins.

Bustard, H. R. 1973b. Saving the hawksbill turtle. *Oryx.* **13**:93–98.

Bustard, H. R. 1973c. *Sea turtles: Natural history and conservation.* New York: Taplinger.

Bustard, H. R. 1974. The green turtle. In N. Sitwell (Ed.), *Wildlife '74. The world conservation yearbook*, pp. 80–87. London: London Editions Ltd.

Bustard, H. R. 1976. Turtles of coral reefs and coral islands. In O. A. Jones and R. Endean (Eds.), *Biology and geology of coral reefs*, Vol. 3, pp. 343–368. New York: Academic Press.

Bustard, H. R. and P. M. Greenham. 1968. Physical and chemical factors affecting hatching in the green sea turtle, *Chelonia mydas* (L.). *Ecology.* **49**:269–276.

Bustard, H. R. and P. M. Greenham. 1969. The nesting behavior of the green sea turtle *Chelonia mydas* (L.) on a Great Barrier Reef island. *Herpetologica.* **25**:93–102.

Bustard, H. R. and C. Limpus. 1969. Observations on the flatback turtle *Chelonia depressa* Garman. *Herpetologica.* **25**:27–34.

Bustard, H. R. and C. Limpus. 1970. First international recapture of an Australian tagged loggerhead turtle. *Herpetologica.* **26**:358–359.

Bustard, H. R. and C. Limpus. 1971. Loggerhead turtle movements. *Brit. J. Herpetol.* **4**:228–230.

Bustard, H. R., K. Simkiss, and N. K. Jenkins. 1969. Calcium, magnesium and phosphorus in eggs and hatchlings of green and loggerhead sea turtles. *J. Zool. London.* **158**:311–315.

Bustard, H. R. and K. P. Tognetti. 1969. Green sea turtles: A discrete simulation of density dependent population regulation. *Science.* **163**:939–941.

Cagle, F. R. 1937. Egg laying habits of the slider turtle (*Pseudemys troosti*), the painted turtle (*Chrysemys picta*), and the musk turtle (*Sternotherus odoratus*). *J. Tenn. Acad. Sci.* **12**:87–95.

Cagle, F. R. 1939. A system for marking turtles for future identification. *Copeia.* **1939**:170–173.

Cagle, F. R. 1942a. Herpetological fauna of Jackson and Union Counties, Illinois. *Am. Midl. Nat.* **28**:164–200.

Cagle, F. R. 1942b. Turtle populations in southern Illinois. *Copeia.* **1942**:155–162.

Cagle, F. R. 1944a. Activity and winter changes of hatchling *Pseudemys. Copeia.* **1944**:105–109.

Cagle, F. R. 1944b. Home range, homing behavior, and migration in turtles. *Misc. Publ. Mus. Zool. Univ. Mich.* **61**:1–34.

Cagle, F. R. 1944c. A technique for obtaining turtle eggs for study. *Copeia.* **1944**:60.

Cagle, F. R. 1944d. Sexual maturity in the female of the turtle, *Pseudemys scripta elegans. Copeia.* **1944**:149–152.

Cagle, F. R. 1946. The growth of the slider turtle, *Pseudemys scripta elegans. Am. Midl. Nat.* **36**:685–729.

Cagle, F. R. 1948. The growth of turtles in Lake Glendale, Illinois. *Copeia.* **1948**:197–203.

Cagle, F. R. 1950. The life history of the slider turtle, *Pseudemys scripta troostii* (Holbrook). *Ecol. Monogr.* **20**:31–54.

Cagle, F. R. 1952. The status of the turtles, *Graptemys pulchra* Baur and *Graptemys barbouri* Carr and Marchand with notes on their natural history. *Copeia.* **1952**:223–234.

Cagle, F. R. 1953. An outline for the study of a reptile life history. *Tulane Stud. Zool.* **1**:31–52.

Cagle, F. R. 1954. Observations on the life cycles of painted turtles (genus *Chrysemys*). *Am. Midl. Nat.* **52**:225–235.

Cagle, F. R. 1955. Courtship behavior in juvenile turtles. *Copeia.* **1955**:307.

Cagle, F. R. and A. H. Chaney. 1950. Turtle populations in Louisiana. *Am. Midl. Nat.* **43**:383–388.

Cagle, F. R. and J. Tihen. 1948. Retention of eggs by the turtle *Deirochelys reticularia. Copeia.* **1948**:66.

Cahn, A. R. 1937. The turtles of Illinois. *Illinois Biol. Monogr.* **35**:1–128.

Cahn, A. R. and E. Condor. 1932. Mating of box turtles. *Copeia.* **1932**:86–88.

Cain, W. S. and T. Engen. 1969. Olfactory adaptation and the scaling of odor intensity. In

C. Pfaffman (Ed.), *Olfaction and taste*, Vol. III, pp. 127-141. New York: Rockefeller University Press.

Cairncross, B. L. 1946. Notes on South African tortoises. *Ann. Transvaal Mus.* **20**:395-397.

Cajal, S. R. 1955. *Studies on the cerebral cortex.* Chicago: Year Book Publishers.

Calderwood, H. W. 1971. Anesthesia for reptiles. *J. Am. Vet. Med. Assoc.* **159**:1618-1625.

Caldwell, D. 1959. The loggerhead turtles of Cape Romain, South Carolina. *Bull. Fla. State Mus., Biol. Sci.* **4**:319-348.

Caldwell, D. K. 1962. Comments on the nesting behavior of Atlantic loggerhead sea turtles, based primarily on tagging returns. *Quart. J. Fla. Acad. Sci.* **25**:287-302.

Caldwell, D. K., A. Carr, L. Ogren, F. H. Berry, and R. A. Ragotzkie. 1959. The Atlantic loggerhead sea turtle, *Caretta caretta caretta* (L.), in America. *Bull. Fla. State Mus., Biol. Sci.* **4**:293-348.

Caldwell, M. C. and D. K. Caldwell. 1962. Factors in the ability of the northeastern Pacific green turtle to orient toward the sea from the land, a possible coordinate in long-range navigation. *Contrib. Sci.* **60**:1-27.

Camin, J. H. 1948. Mite transmission of a hemorrhagic septicemia in snakes. *J. Parasitol.* **34**:345-354.

Campbell, H. W. 1967. A word of caution. *Int. Turtle Tortoise Soc. J.* **1**(3):38-39.

Campbell, H. W. and K. Campbell. 1972. Subdivision turtle style. *Int. Turtle Tortoise Soc. J.* **6**(2):16-19, 35.

Campbell, H. W. and W. E. Evans. 1967. Sound production in two species of tortoises. *Herpetologica.* **23**:204-209.

Campbell, H. W. and W. E. Evans. 1972. Observations on the vocal behavior of chelonians. *Herpetologica.* **28**:277-280.

Carpenter, C. C. 1952. Comparative ecology of the common garter snake (*Thamnophis s. sirtalis*), the ribbon snake (*Thamnophis s. sauritus*) and Butler's garter snake (*Thamnophis butleri*) in mixed populations. *Ecol. Monogr.* **22**:236-258.

Carpenter, C. C. 1955. Sounding turtles: A field locating technique. *Herpetologica.* **11**:120.

Carpenter, C. C. 1957. Hibernation, hibernacula and associated behavior in the three-toed box turtle (*Terrapene carolina triunguis*). *Copeia.* **1957**:278-282.

Carpenter, C. C. 1961. Patterns of social behavior in the desert iguana, *Dipsosaurus dorsalis. Copeia.* **1961**:396-405.

Carpenter, C. C. 1962. A comparison of the patterns of display of *Urosaurus*, *Uta*, and *Streptosaurus. Herpetologica.* **18**:145-152.

Carpenter, C. C. 1963. Patterns of behavior in three forms of the fringe-toed lizards (*Uma*-Iguanidae). *Copeia.* **1963**:406-412.

Carpenter, C. C. 1965. The display of the Cocos Island anole. *Herpetologica.* **21**:256-260.

Carpenter, C. C. 1966a. Comparative behavior of the Galapagos lava lizards (*Tropidurus*). In R. I. Bowman (Ed.), *The Galapagos: Proceedings of the Galapagos International Project*, pp. 269-273. Berkeley: University of California Press.

Carpenter, C. C. 1966b. The marine iguana of the Galapagos Islands, its behavior and ecology. *Proc. Calif. Acad. Sci.* **34**:329-376.

Carpenter, C. C. 1967. Display patterns of the Mexican iguanid lizards of the genus *Uma. Herpetologica.* **23**:285-293.

Carpenter, C. C. 1969. Behavioral and ecological notes on the Galapagos land iguana. *Herpetologica.* **25**:155-164.

Carpenter, C. C., J. A. Badham, and B. Kimble. 1970. Behavior patterns of three species of *Amphibolurus* (Agamidae). *Copeia*. **1970**:497–505.

Carpenter, C. C. and G. Grubitz, III. 1961. Time-motion study of a lizard. *Ecology*. **42**:199–200.

Carpenter, C. C. and K. T. Mosley, Jr. 1967. Display patterns of iguanid lizards. 16 mm. motion picture. *Psychological Cinema Register*, Pennsylvania State University, 24 minutes, silent. Code PCR-128K.

Carpenter, C. C. and D. T. Stalling. 1968. Motion picture recording of rodent behavior under red light. *Am. Zool.* **6**:739 (Abstr.).

Carr, A. F. 1935. The identity and status of two turtles of the genus *Pseudemys*. *Copeia*. **1935**:147–148.

Carr, A. F. 1937a. A new turtle from Florida, with notes on *Pseudemys floridana mobiliensis* (Holbrook). *Occas. Pap. Mus. Zool. Univ. Mich.* **348**:1–7.

Carr, A. F. 1937b. The status of *Pseudemys scripta* (Schoepff) and *Pseudemys troostii* (Holbrook). *Herpetologica*. **1**:75–77.

Carr, A. F. 1938a. A new subspecies of *Pseudemys floridana*, with notes on the *floridana* complex. *Copeia*. **1938**:105–109.

Carr, A. F. 1938b. Notes on the *Pseudemys scripta* complex. *Herpetologica*. **1**:131–135.

Carr, A. F. 1938c. *Pseudemys nelsoni*, a new turtle from Florida. *Occas. Pap. Boston Soc. Nat. Hist.* **8**:305–310.

Carr, A. F. 1940. A contribution to the herpetology of Florida. *Univ. Fla. Press Pub., Biol. Sci. Ser.* **3**(1):1–118.

Carr, A. F. 1942a. A new *Pseudemys* from Sonora, Mexico. *Am. Mus. Nov.* No. 1181:1–4.

Carr, A. F. 1942b. Notes on sea turtles. *Proc. N. Engl. Zool. Club* **21**:1–16.

Carr, A. F. 1942c. The status of *Pseudemys floridana texana*, with notes on parallelism in *Pseudemys*. *Proc. N. Engl. Zool. Club* **21**:69–76.

Carr, A. F. 1952. *Handbook of turtles*. Ithaca, N.Y.: Comstock.

Carr, A. F. 1962. Orientation problems in high seas travel and terrestrial movements of marine turtles. *Am. Sci.* **50**:358–374.

Carr, A. F. 1963a. Panspecific reproductive convergence in *Lepidochelys kempi*. *Ergebn. Biol.* **26**:297–303.

Carr, A. F. 1963b. *The reptiles*. New York: Time.

Carr, A. F. 1965. The navigation of the green turtle. *Sci. Am.* **212**(5):78–86.

Carr, A. 1967a. Adaptive aspects of the scheduled travel of *Chelonia*. In R. Storm (Ed.), *Animal orientation and navigation*, pp. 35–53. Corvallis: Oregon State University Press.

Carr, A. F. 1967b. *So excellente a fishe*. New York: Natural History Press.

Carr, A. 1969. Sea turtle resources of the Caribbean and Gulf of Mexico. *IUCN Bull.* **2**:74–75.

Carr, A. 1975. The Ascension Island green turtle colony. *Copeia*. **1975**:547–555.

Carr, A. and M. H. Carr. 1970. Modulated reproductive periodicity in *Chelonia*. *Ecology*. **51**:335–337.

Carr, A. and M. H. Carr. 1972. Site fixity in the Caribbean green turtle. *Ecology*. **53**:425–429.

Carr, A. and P. J. Coleman. 1974. Sea floor spreading theory and the odyssey of the green turtle from Brazil to Ascension Island, Central Atlantic. *Nature*. **249**:128–130.

Carr, A. F. and J. W. Crenshaw. 1957. A taxonomic reappraisal of the turtle *Pseudemys alabamensis* Baur. *Bull. Fla. State Mus., Biol. Sci.* **2**:25–42.

Carr, A. F. and L. Giovannoli. 1957. The ecology and migrations of sea turtles. 2. Results of field work in Costa Rica, 1955. *Am. Mus. Nov.* No. 1835:1–32.

Carr, A. and H. Hirth. 1961. Social facilitation in green turtle siblings. *Anim. Behav.* **9**:68–70.

Carr, A. and H. Hirth. 1962. The ecology and migrations of sea turtles. 5. Comparative features of isolated green turtle colonies. *Am. Mus. Nov.* No. 2091:1–42.

Carr, A., H. Hirth, and L. Ogren. 1966. The ecology and migrations of sea turtles. 6. The hawksbill turtle in the Caribbean Sea. *Am. Mus. Nov.* No. 2248:1–29.

Carr, A. and L. Ogren. 1959. The ecology and migrations of sea turtles. 3. *Dermochelys* in Costa Rica. *Am. Mus. Nov.* No. 1958:1–29.

Carr, A. and L. Ogren. 1960. The ecology and migrations of sea turtles. 4. The green turtle in the Caribbean Sea. *Bull. Am. Mus. Nat. Hist.* **121**:1–48.

Carr, A., P. Ross, and S. Carr. 1974. Internesting behavior of the green turtle, *Chelonia mydas*, at a mid-ocean breeding ground. *Copeia.* **1974**:703–706.

Carroll, R. L. 1969. Origin of reptiles. In C. Gans, A. d'A. Bellairs, and T. S. Parsons (Eds.), *Biology of the Reptilia*, pp. 1–44. New York: Academic Press.

Carroll, T. E. and D. W. Ehrenfeld. 1978. Intermediate-range homing in the wood turtle, *Clemmys insculpta*. *Copeia.* **1978**:117–126.

Casteel, D. B. 1911. The discriminative ability of the painted turtle. *J. Anim. Behav.* **1**:1–28.

Chaikoff, I. L. and C. Entenman. 1946. The lipids of blood, liver, and egg yolk of the turtle. *J. Biol. Chem.* **166**:683–689.

Chaney, A. and C. L. Smith. 1950. Methods for collecting map turtles. *Copeia.* **1950**:323–324.

Chavez, H. 1968. Marcado y recaptura de individuos de tortuga lora. *Inst. Nac. Inv. Biol. Pes., Mexico* No. 19:1–28.

Chavez, H. 1969. Tagging and recapture of the lora turtle, *Lepidochelys kempi*. *Int. Turtle Tortoise J.* **3**(4):14–19, 32–36.

Chavez, H., M. Contreras G., and T. P. E. Hernandez D. 1968. On the coast of Tamaulipas. Parts 1 and 2. *Int. Turtle Tortoise Soc. J.* **2**(4):20–29, 37; (5):16–19, 27–34.

Cherchi, M. A. 1956. Termoregolazione in *Testudo hermanni* Gmelin. *Boll. Mus. Inst. Biol. Genova* **26**:5–46.

Cherchi, M. A. 1960. Ulteriori recherche sulla termoregolazione in *Testudo hermanni* Gmelin. *Boll. Mus. Inst. Biol. Genova* **30**:35–60.

Chesley, L. C. 1934. The influence of temperature upon the analysis of cold and warm blooded animals. *Biol. Bull.* **66**:330–338.

Christiansen, J. L. and A. E. Dunham. 1972. Reproduction of the yellow mud turtle (*Kinosternon flavescens flavescens*) in New Mexico. *Herpetologica.* **28**:130–137.

Christiansen, J. L. and E. O. Moll. 1973. Latitudinal reproductive variation within a single subspecies of painted turtle, *Chrysemys picta belli*. *Herpetologica.* **29**:152–163.

Ckhikvadze, V. M. 1970. On the origin of the modern Palearctic land tortoises. *Bull. Acad. Sci. Georgian S.S.R.* **57**(1):245–247.

Ckhikvadze, V. M. 1973. Tretichnie cherepaki Zaisanskoi Kotlovini. *Acad. Sci. Georgian S.S.R.* **1973**:1–100.

Clark, D. B. and J. W. Gibbons. 1969. Dietary shift in the turtle *Pseudemys scripta* (Schoepff) from youth to maturity. *Copeia.* **1969**:704-706.

Clark, D. R., Jr. 1971. Branding as a marking technique for amphibians and reptiles. *Copeia.* **1971**:148-151.

Clarke, G. L. and H. R. James. 1939. Laboratory analysis of the selective absorption of light by sea water. *J. Opt. Soc. Am.* **29**:43-55.

Clementi, O. 1929. Carattere ureotelico del metabolismo dei Chelonia. *Arch. Soc. Biol.* **14**:171.

Cloudsley-Thompson, J. L. 1961. *Rhythmic activity in animal physiology and behaviour.* New York: Academic Press.

Cloudsley-Thompson, J. L. 1968. Thermoregulation in tortoises. *Nature.* **217**:575.

Cloudsley-Thompson, J. L. 1969. *The zoology of tropical Africa.* London: Weidenfeld and Nicolson.

Cloudsley-Thompson, J. L. 1970. On the biology of the desert tortoise *Testudo sulcata* in Sudan. *J. Zool., London.* **160**:17-33.

Cloudsley-Thompson, J. L. 1971. *The temperature and water relations of reptiles.* London: Merrow.

Cloudsley-Thompson, J. L. 1974. Physiological thermoregulation in the spurred tortoise (*Testudo graeca* L.). *J. Nat. Hist.* **8**:577-587.

Cochran, D. M. and C. J. Goin. 1970. *The new field book of reptiles and amphibians.* New York: Putnam.

Cochran, W. W. and R. D. Lord, Jr. 1963. A radio-tracking system for wild animals. *J. Wildl. Manag.* **27**:9-24.

Cochran, W. W. and E. M. Nelson. 1963. The Model D-11 direction-finding receiver. *Mus. Nat. Hist., Univ. Minn.*, Tech. Rep. No. 2:1-14.

Cochran, W. W., D. W. Warner, J. R. Tester, and V. B. Kuechle. 1965. Automatic radio-tracking system for monitoring animal movements. *BioScience.* **15**:98-100.

Cogger, H. G. 1970. First record of the pitted-shell turtle, *Carettochelys insculpta* from Australia. *Search.* **1**:41.

Cohen, N. 1971. Reptiles as models for the study of immunity and its phylogenesis. *J. Am. Vet. Med. Assoc.* **159**:1662-1671.

Cohen, N. and M. Borysenko. 1970. Acute and chronic graft rejection: Possible phylogeny of transplantation antigens. *Transplant. Proc.* **2**:333-336.

Coker, R. E. 1910. Diversity of scutes of Chelonia. *J. Morphol.* **21**:1-75.

Colbert, E. H., R. B. Cowles, and C. M. Bogert. 1946. Temperature tolerances in the American alligator and their bearing on the habits, evolution and extinction of the dinosaurs. *Bull. Am. Mus. Nat. Hist.* **87**:327-374.

Collins, J. I. 1970. The chelonian *Rhinochelys* Seeley from the Upper Cretaceous of England and France. *Paleontology.* **13**(3):355-378.

Combescot, C. 1954a. Sur le cycle mâle et not amment la spermatogénèse, chez la tortue d'eau algérienne (*Emys leprosa* Schw.). *C. R. Soc. Biol.* **148**:2021-2023.

Combescot, C. 1954b. Sexualité et cycle génital de la tortue d'eau algérienne, *Emys leprosa* Schw. *Bull. Soc. Hist. Nat. Afr. Nord.* **45**:366-377.

Combescot, C. 1955. Données histophysiologiques sur l'oviducte de la tortue d'eau algérienne (*Emys leprosa* Schw.). *C. R. Soc. Biol.* **149**:93-95.

Conant, R. 1938. The reptiles of Ohio. *Am. Midl. Nat.* **20**:1-200.
Conant, R. 1951a. The red-bellied terrapin, *Pseudemys rubriventris* (LeConte), in Pennsylvania. *Ann. Carnegie Mus.* **32**:281-289.
Conant, R. 1951b. *The reptiles of Ohio.* Notre Dame, Ind.: University of Notre Dame Press.
Conant, R. 1975. *A field guide to reptiles and amphibians of eastern and central North America.* Boston: Houghton Mifflin.
Conant, R. and K. G. Hudson. 1949. Longevity records for reptiles and amphibians in the Philadelphia Zoological Garden. *Herpetologica.* **5**:1-8.
Conant, S. 1973. Visual and acoustic communication in the bluejay (*Cyanocitta cristata*—Aves, Corvidae). Ph.D. dissertation, University of Oklahoma, Norman.
Cordier, R. 1928. Études histophysiologiques sur la tube urinaire des reptiles. *Arch. Biol.* **38**:111-171.
Cosgrove, G. E. and M. L. Davis. 1965. Radiosensitivity of snakes and box turtles. *Radiat. Res.* **25**:706-712.
Cowan, D. F. 1968. Diseases of captive reptiles. *J. Am. Vet. Assoc.* **153**:848-859.
Cowan, F. B. M. 1969. Gross and microscopic anatomy of the orbital glands of *Malaclemys* and other emydine turtles. *Can. J. Zool.* **47**:723-729.
Cowan, F. B. M. 1971. The ultrastructure of the lachrymal "salt" glands and the Harderian gland in the euryhaline *Malaclemys* and some closely related stenohaline emydines. *Can. J. Zool.* **49**:691-697.
Cowles, R. B. and C. M. Bogert. 1944. A preliminary study of the thermal requirements of desert reptiles. *Bull. Am. Mus. Nat. Hist.* **83**:261-296.
Crampton, G. H. and W. J. Schwam. 1962. Turtle vestibular responses to angular acceleration with comparative data from cat and man. *J. Comp. Physiol. Psychol.* **55**:315-321.
Crawford, F. T. and P. M. Adams. 1968. The effect of schedules of reinforcement upon the response rates of turtles. *Psychon. Sci.* **11**:153-154.
Crawford, F. T., P. M. Adams, and J. M. Whitt. 1966. Response rate of turtles to fixed ratio reinforcement. *Psychon. Sci.* **6**:19-20.
Crawford, F. T. and L. E. Siebert. 1964. Operant rate in the turtle as a function of temperature. *Psychon. Sci.* **1**:215-216.
Crenshaw, J. W. 1962. Variation in serum albumins and other blood proteins of turtles of the Kinosternidae. *Physiol. Zool.* **35**:157-165.
Crescitelli, F. 1972. The visual cells and visual pigments of the vertebrate eye. In H. J. A. Dartnall (Ed.), *Photochemistry of vision*, pp. 245-363. New York: Springer-Verlag.
Crespi, L. P. 1942. Quantitative variation in incentive and performance in the white rat. *Am. J. Psychol.* **55**:467-517.
Croix, D. M. de la. 1929. Filogenia de las locomóciones cuadrupedal y bipedal en los vertebrados y evolución de la forma consecutiva de la evolución de la locomoción. *An. Soc. Cient. Argent.* **108**:383-406.
Crosby, E. C. 1969. Comparative anatomy of the preoptic and hypothalamic areas. In W. Haymaker, E. Anderson, and J. H. Nauta (Eds.), *The hypothalamus*, pp. 52-169. Springfield, Ill.: Thomas.
Crosfil, M. L. and J. G. Widdicombe. 1961. Physical characteristics of the chest and lungs and the work of breathing in different mammalian species. *J. Physiol., London.* **158**:1-14.

Cruce, W. L. R. and R. Nieuwenhuys. 1974. The cell masses in the brain stem of the turtle *Testudo hermanni*; a topographical and topological analysis. *J. Comp. Neurol.* **156**:277–306.

Cullum, L. and J. T. Justus. 1973. Housing for aquatic animals. *Lab. Anim. Sci.* **23**:126–129.

Cunningham, B. 1939. Effect of temperature upon the development rate of the embryo of the diamondback terrapin (*Malaclemys concentrata* Latreille). *Am. Nat.* **73**:381–384.

Cunningham, B. and E. Huene. 1938. Further studies in water absorption by reptile eggs. *Am. Nat.* **72**:380–385.

Cunningham, B. and A. P. Hurwitz. 1936. Water absorption by reptile eggs during incubation. *Am. Nat.* **70**:590–595.

Cunningham, B., M. W. Woodward, and J. Pridgen. 1939. Further studies on incubation of turtle (*Malaclemys concentrata* Lat.) eggs. *Am. Nat.* **73**:285–288.

Curwen, A. O. and R. N. Miller. 1939. The pretectal region of the turtle. *J. Comp. Neurol.* **71**:99–120.

Czajka, A. F. and M. A. Nickerson. 1974. State regulations for collecting reptiles and amphibians in the fifty United States. *Milwaukee Pub. Mus. Spec. Publ.* **1**:1–79.

Dagg, A. I. and A. D. Vos. 1968. The walking gaits of some species of *Pecora*. *J. Zool.* **155**:103–110.

D'amato, M. R. 1970. *Experimental psychology: Methodology, psychophysics, and learning*. New York: McGraw-Hill.

Dampier, C. W. 1697. *A new voyage around the world*. London: James Knapton.

Daniel, R. S. and K. U. Smith. 1947. The sea-approach behavior of the neonate loggerhead turtle (*Caretta caretta*). *J. Comp. Physiol. Psychol.* **40**:413–420.

Dantzler, W. H. and B. Schmidt-Nielsen. 1966. Excretion in fresh-water turtle (*Pseudemys scripta*) and desert tortoise (*Gopherus agassizii*). *Am. J. Physiol.* **210**:198–210.

Dareste, C. 1891. *Recherches sur la production artificielle des monstruosités; ou, Essais de tératogénie expérimentale*. Paris: Reinwald.

Davis, D. 1963. Estimating the numbers of game populations. In H. Mosby (Ed.), *Wildlife investigational techniques*, pp. 89–118. Washington, D.C.: Wildlife Society.

Davis, J. D. and C. G. Jackson, Jr. 1970. Copulatory behavior in the red-eared turtle, *Pseudemys scripta elegans* (Wied). *Herpetologica.* **26**:238–239.

Davis, W., Jr. and G. Sartor. 1975. A method of observing movements of aquatic turtles. *Herpetol. Rev.* **6**:13–14.

Dawson, W. R. and G. A. Bartholomew. 1958. Metabolic and cardiac responses to temperature in the lizard *Dipsosaurus dorsalis*. *Physiol. Zool.* **31**:100–111.

Deane, H. W., C. Enroth-Cugell, M. A. Gongaware, M. Neyland, and A. Forbes. 1958. Electroretinogram of freshwater turtle. *J. Neurophysiol.* **21**:45–61.

Deck, R. 1927. Longevity of *Terrapene carolina* (Linné). *Copeia*. **1927**:179.

Deevey, E. 1947. Life tables for natural populations of animals. *Quart. Rev. Biol.* **22**:283–314.

De Koningh, H. L. 1968. An oriental big-head. *Int. Turtle Tortoise Soc. J.* **2**(4):14–17.

Deraniyagala, P. E. P. 1930. Testudinate evolution. *Proc. Zool. Soc., London.* **1930**:1057–1070.

Deraniyagala, P. E. P. 1939. *The tetrapod reptiles of Ceylon*. Vol. I. *Testudinates and crocodilians*. London: Dulau.

BIBLIOGRAPHY

De Rooij, N. 1915. *The reptiles of the Indo-Australian Archipelago.* Vol. I. *Lacerta, Chelonia, Emydosauria.* London: Brill.

De Sola, C. R. 1930. The Liebespiel of *Testudo vandenburghi*, a new name for the mid-Albemarle Island Galapagos tortoise. *Copeia.* **1930**:79–80.

Dessauer, H. C. 1970. Blood chemistry of reptiles: Physiological and evolutionary aspects. In C. Gans and T. S. Parsons (Eds.), *Biology of the reptilia.* Vol. III. *Morphology C*, pp. 1–72. New York: Academic Press.

Deth, J. H. M. G. van. 1962. Changes in water, sodium and potassium distribution in the duck's egg during incubation. *Aust. J. Exp. Biol. Med. Sci.* **40**:173–189.

De Vries, T. 1968. On the spot. *Int. Turtle Tortoise Soc. J.* **2**(2):16, 18–19, 33.

Dice, L. R. and P. J. Clark. 1953. The statistical concept of home range as applied to the recapture radius of the deermouse (*Peromyscus*). *Contrib. Lab. Vert. Biol. Univ. Mich.* **62**:1–15.

Dickson, J. D., III. 1953. The private life of the box turtle. *Everglades Nat. Hist.* **1**(2):58–62.

Dieterich, R. A. 1967. What's wrong with your turtle? *Int. Turtle Tortoise Soc. J.* **1**(3):20–21, 42.

Dill, C. D. 1972. Reptilian core temperatures: Variation within individuals. *Copeia.* **1972**:577–579.

Dinkins, A. 1954. A brief observation of male combat in *Clemmys insculpta. Herpetologica.* **10**:20.

Ditmars, R. L. 1928. *Reptiles of the world.* New York: Macmillan.

Dix, M. W. and J. I. Richardson. 1972. Reproductive periodicity of the loggerhead sea turtle, *Caretta caretta* (Reptilia: Chelonia). *ASB Bull.* **19**(2):65.

Dmi'el, R. 1967. Studies on reproduction, growth and feeding in the snake *Spalerosophis cliffordi* (Colubridae). *Copeia.* **1967**:332–346.

Dobbs, J. S. 1973. *Reptilian disease: Recognition and treatment.* Hollywood, Fla.: Ralph Curtis.

Dobie, J. L. 1971. Reproduction and growth in the alligator snapping turtle, *Macroclemys temmincki* (Troost). *Copeia.* **1971**:645–658.

Dobzhansky, T. 1950. Evolution in the tropics. *Am. Sci.* **38**:209–221.

Dodd, C. K., Jr. and E. D. Brodie, Jr. 1975. Notes on the defensive behavior of the snapping turtle, *Chelydra serpentina. Herpetologica.* **31**:286–288.

Dolbeer, R. A. 1969. A study of population density, seasonal movements and weight changes, and winter behavior of the eastern box turtle, *Terrapene c. carolina* L., in eastern Tennessee. Master's thesis. University of Tennessee.

Dottrens, E. and V. Aellen. 1963. *Batraciens et reptiles d'Europe.* Neuchatel, Switzerland: Delachaux et Niestlé.

Doyle, R. E. and A. F. Moreland. 1968. Diseases in turtles. *Lab. Anim. Dig.* **4**:3–6.

Doyle, R. E. and A. F. Moreland. 1969. Diseases of turtles. *Int. Turtle Tortoise Soc. J.* **3**(2):29–31.

Dudziak, J. 1955. Visual acuity in the turtle. *Folia Biol.* **3**:205–228.

Duke-Elder, S. 1958. *System of ophthalmology.* Vol. I. *The eye in evolution.* St. Louis: Mosby.

Duncan, D. D. 1943. Capturing giant turtles in the Caribbean. *Nat. Geogr. Mag.* **84**(2):177–190.

Dunlop, C. E. 1955. Notes on the visceral anatomy of the giant leatherback turtle *Dermochelys coriacea* (Linnaeus). *Bull. Tulane Med. Fac.* **14**:55–69.

Dunson, W. A. 1960. Aquatic respiration in *Trionyx spinifer asper*. *Herpetologica*. **16**:277–283.

Dunson, W. A. 1967a. Relationship between length and weight in the spiny soft-shell turtle. *Copeia*. **1967**:483–485.

Dunson, W. A. 1967b. Sodium fluxes in freshwater turtles. *J. Exp. Zool*. **165**:171–182.

Dunson, W. A. 1970. Some aspects of electrolyte and water balance in three estuarine reptiles, the diamondback terrapin, American and "saltwater" crocodiles. *Comp. Biochem. Physiol*. **32**:161–174.

Dunson, W. A. and R. D. Weymouth. 1965. Active uptake of sodium by softshell turtles, *Trionyx spinifer*. *Science*. **149**:67–69.

Ebbesson, S. O. E. 1969. Brain stem afferents from the spinal cord in a sample of reptilian and amphibian species. *Ann. N.Y. Acad. Sci*. **167**:80–101.

Edgren, R. A. 1960. Ovulation time in the musk turtle, *Sternothaerus odoratus*. *Copeia*. **1960**:60–61.

Edgren, R. A. and M. K. Edgren. 1955. Thermoregulation in the musk turtle, *Sternothaerus odoratus* Latreille. *Herpetologica*. **11**:213–217.

Edgren, R. A. and L. H. Tiffany. 1953. Some North American turtles and their epizoophytic algae. *Ecology*. **34**:733–740.

Edwards, A. A. and N. Mrosovsky. 1964. Electric shock and reduced weight gain in the terrapin. *Brit. J. Herpetol*. **3**:182–183.

Eglis, A. 1962. Tortoise behavior: A taxonomic adjunct. *Herpetologica*. **18**:1–8.

Eglis, A. 1963. Nesting of a parrot-beaked tortoise. *Herpetologica*. **19**:66–68.

Ehrenfeld, D. W. 1968. The role of vision in the sea-finding orientation of the green turtle (*Chelonia mydas*). 2. Orientation mechanism and range of spectral sensitivity. *Anim. Behav*. **16**:281–287.

Ehrenfeld, D. W. 1974. Conserving the edible sea turtle: Can mariculture help? *Am. Sci*. **62**(1):23–31.

Ehrenfeld, D. W. and A. Carr. 1967. The role of vision in the sea-finding orientation of the green turtle (*Chelonia mydas*). *Anim. Behav*. **15**:25–36.

Ehrenfeld, D. W. and A. L. Koch. 1967. Visual accommodation in the green turtle. *Science*. **155**:827–828.

Ehrenfeld, J. and D. W. Ehrenfeld. 1973. Externally secreting glands of freshwater and sea turtles. *Copeia*. **1973**:305–314.

Ehrengart, W. 1971. Zur Pflege and Zucht der griechischen Landschildkröte (*Testudo h. hermanni*). *Salamandra*. **7**:71–80.

Einem, G. E. 1956. Certain aspects of the natural history of the mud turtle, *Kinosternon bauri*. *Copeia*. **1956**:186–188.

Ellenbogen, M. 1971. The western swamp tortoise of Australia. *Int. Turtle Tortoise Soc. J*. **5**(4):16–17, 39.

Ellis, K. R. and J. D. Barcik. 1972. Acquisition and suppression of an appetitive task in the freshwater turtle *Chrysemys picta marginata*. *Am. Zool*. **12**:646 (Abstr.).

Ellis, K. R. and I. Spigel. 1966. Effects of cortisone on active and passive avoidance acquisition in the turtle. *Proc. 74th Ann. Conv. APA*. **1966**:115–116.

Emlen, S. T. 1969. Homing ability and orientation in the painted turtle, *Chrysemys picta marginata*. *Behaviour*. **33**:58–76.

Ernst, C. H. 1965. Bait preferences of some freshwater turtles. *J. Ohio Herpetol. Soc*. **5**:53.

BIBLIOGRAPHY

Ernst, C. H. 1967. A mating aggregation of the turtle *Clemmys guttata*. *Copeia*. **1967**:473–474.

Ernst, C. H. 1968a. Evaporative water-loss relationships of turtles. *J. Herpetol.* **2**:159–161.

Ernst, C. H. 1968b. Homing ability of the spotted turtle, *Clemmys guttata* (Schneider). *Herpetologica*. **24**:77–78.

Ernst, C. H. 1970a. Home range of the spotted turtle, *Clemmys guttata* (Schneider). *Copeia*. **1970**:391–393.

Ernst, C. H. 1970b. Homing ability in the painted turtle *Chrysemys picta* (Schneider). *Herpetologica*. **26**:399–403.

Ernst, C. H. 1970c. Reproduction in *Clemmys guttata*. *Herpetologica*. **26**:228–232.

Ernst, C. H. 1971a. Growth of the painted turtle, *Chrysemys picta* in southeastern Pennsylvania. *Herpetologica* **27**:135–141.

Ernst, C. H. 1971b. Observations on the egg and hatchling of the American turtle, *Chrysemys picta*. *Brit. J. Herpetol.* **4**:224–228.

Ernst, C. H. 1971c. Population dynamics and activity cycles of *Chrysemys picta* in southeastern Pennsylvania. *J. Herpetol.* **5**:151–160.

Ernst, C. H. 1971d. Sexual cycles and maturity of the turtle, *Chrysemys picta*. *Biol. Bull.* **140**:191–200.

Ernst, C. H. 1972. Temperature-activity relationships in the painted turtle, *Chrysemys picta*. *Copeia*. **1972**:217–222.

Ernst, C. H. 1974. Observations on the courtship of male *Graptemys pseudogeographica*. *J. Herpetol.* **8**:377–378.

Ernst, C. H. 1975. Growth of the spotted turtle, *Clemmys guttata*. *J. Herpetol.* **9**:313–319.

Ernst, C. H. In press. Ecology of the spotted turtle, *Clemmys guttata*, in southeastern Pennsylvania. *J. Herpetol.*

Ernst, C. H. and R. W. Barbour. 1972. *Turtles of the United States*. Lexington: University Press of Kentucky.

Ernst, C. H., R. W. Barbour, and J. R. Butler. 1972. Habitat preferences of two Florida turtles, genus *Kinosternon*. *Trans. Ky. Acad. Sci.* **33**:41–42.

Ernst, C. H., R. W. Barbour, B. M. Ernst, and J. B. Butler. 1973. Growth of the mud turtle, *Kinosternon subrubrum*, in Florida. *Herpetologica*. **29**:247–250.

Ernst, C. H. and B. M. Ernst. 1972. Biology of *Chrysemys picta bellii* in southwestern Minnesota. *Proc. Minn. Acad. Sci.* **38**:77–80.

Eskin, R. M. and M. E. Bitterman. 1961. Partial reinforcement in the turtle. *Quart. J. Exp. Psychol.* **13**:112–116.

Estridge, R. E. 1970. The taxonomic status of *Sternotherus depressus* (Testudinata, Kinosternidae) with observations on its ecology. Master's thesis. Auburn University.

Evans, H. E. 1966. *The comparative ethology and evolution of the sand wasps*. Cambridge: Harvard University Press.

Evans, L. T. 1940a. Effects of light and hormones upon the activity of young turtles, *Chrysemys picta*. *Biol. Bull.* **79**:370–371.

Evans, L. T. 1940b. Effects of testosterone propionate upon social dominance in young turtles, *Chrysemys picta*. *Biol Bull.* **79**:371.

Evans, L. T. 1946. Endocrine effects upon the claws of immature turtles, *Pseudemys elegans*. *Anat. Rec.* **94**:64.

Evans, L. T. 1949. The reproductive behavior of the giant tortoises, *T. vicina* and *T. vandenburghii*. *Anat. Rec.* **105**:579–580.

Evans, L. T. 1951a. Effects of male hormone upon the tail of the slider turtle *Pseudemys scripta troostii. Science.* **114**:277–279.

Evans, L. T. 1951b. Endocrines and secondary sex characters of the box tortoise, *Terrapene carolina. Anat. Rec.* **111**:139.

Evans, L. T. 1952a. Endocrine relationships in turtles, II. Claw growth in the slider *Pseudemys scripta troostii. Anat. Rec.* **112**:251–264.

Evans, L. T. 1952b. Endocrine relationships in turtles, III. Some effects of male hormones in turtles. *Herpetologica.* **8**:11–14.

Evans, L. T. 1953. The courtship pattern of the box turtle *Terrapene c. carolina. Herpetologica.* **9**:189–192.

Evans, L. T. 1956a. Study of the social and diurnal behavior of the turtle, *Terrapene c. carolina. Anat. Rec.* **125**:609–610.

Evans, L. T. 1956b. The use of models and mirrors in a study of *Terrapene c. carolina. Anat. Rec.* **125**:610.

Evans, L. T. 1958. Social conditioning in young snapping turtles. *Chelydra serpentina. Anat. Rec.* **131**:549.

Evans, L. T. 1960. Neuroendocrine mechanisms in courtship of a turtle. *Anat. Rec.* **138**:347.

Evans, L. T. 1961a. Aquatic courtship of the wood turtle, *Clemmys insculpta. Am. Zool.* **1**:353.

Evans, L. T. 1961b. Structure as related to behavior in the organization of populations of reptiles. In F. Blair (Ed.), *Vertebrate speciation*, pp. 148–178. Austin: University of Texas Press.

Evans, L. T. 1963. Innate behavior (courtship) in relation to endocrines in turtles. *Am. Zool.* **3**:541.

Evans, L. T. 1966. Morphology and kinetics of courtship in certain Testudinidae. *Am. Zool.* **6**:573.

Evans, L. T. 1967. How are age and size related to mating in the wood turtle, *Clemmys insculpta? Am. Zool.* **7**:799.

Evans, L. T. 1968. The evolution of courtship in the turtle species, *Terrapene carolina. Am. Zool.* **8**:695–696.

Evans, L. T., W. Preston, and E. Hardy. 1973. Imprinting in *Chrysemys* turtles. *Am. Zool.* **13**:1269.

Evans, L. T., W. C. Preston, and M. E. McGeary. 1974. Imprinting in turtles. *Am. Zool.* **14**:1278.

Evans, L. T. and J. Quaranta. 1949. Patterns of cooperative behavior in a herd of 14 giant tortoises of the Bronx Zool. *Anat. Rec.* **105**:106.

Evans, L. T. and J. Quaranta. 1951. A study of the social behavior of a captive herd of tortoises. *Zoologica.* **36**:171–181.

Evans, W. E. and J. Bastian. 1969. Marine mammal communication: Social and ecological factors. In H. T. Anderson (Ed.), *Biology of marine mammals*, pp. 462–470. New York: Academic Press.

Ewert, M A. 1976. Nests, nesting and aerial basking of *Macroclemys* under natural conditions and comparisons with *Chelydra* (Testudines: Chelydridae). *Herpetologica.* **32**:150–156.

Ewing, H. E. 1933. Reproduction in the eastern box-turtle *Terrapene carolina carolina* (Linné). *Copeia.* **1933**:95–96.

Ewing, H. E. 1935. Further notes on the reproduction of the eastern box-turtle, Terrapene carolina (Linné). *Copeia.* **1935**:102.

Ewing, H. E. 1943. Continued fertility in female box turtle following mating. *Copeia.* **1943**:112–114.

Feeley, J. C. and M. D. Treger. 1969. Penetration of turtle eggs by *Salmonella braenderup. Pub. Health Rep.* **84**(2): 156–158.

Fehring, W. K. 1972. Hue discrimination in hatchling loggerhead turtles (*Caretta caretta caretta*). *Anim. Behav.* **20**:632–636.

Felger, R. S. K. Clifton, and P. J. Regal. 1976. Winter dormancy in sea turtles: Independent discovery and exploitation in Gulf of California by two local cultures. *Science.* **191**:283–285.

Ferguson, G. W. 1970. Mating behavior of the side-blotched lizards of the genus *Uta* (Sauria-Iguanidae). *Anim. Behav.* **18**:65–72.

Ferguson, G. W. Variation and evolution of the push-up displays of the side-blotched lizard genus *Uta* (Iguanidae). *Syst. Zool.* **20**:79–101.

Feuer, R. C. 1966. Variation in snapping turtles, *Chelydra serpentina* Linnaeus: A study in quantitative systematics. Ph.D. dissertation. University of Utah, Salt Lake City.

Fink, H. K. 1959. *Mind and performance.* New York: Vantage.

Finneran, L. C. 1948. Reptiles at Branford, Connecticut. *Herpetologica.* **4**:123–126.

Fischer, K. 1964. Spontanes Richtungsfinden nach dem Sonnenstand bei *Chelonia mydas* L. (Suppenschildkröte). *Naturwissenschaften.* **51**:203.

Fisher, C. 1945. Early spring mating of the wood turtle. *Copeia.* **1945**:175–176.

Fisk, A. and M. Tribe. 1949. The development of the amnion and chorion of reptiles. *Proc. Zool. Soc. London.* **119**:83–114.

Fitch, H. 1949. Outline for ecological life history studies of reptiles. *Ecology.* **30**:520–532.

Fitch, H. S. 1956. Temperature responses in free-living amphibians and reptiles of northeastern Kansas. *Univ. Kans. Publ. Mus. Nat. Hist.* **8**:417–476.

Fitch, H. S. 1963. Natural history of the racer, *Coluber constrictor. Univ. Kans. Publ., Mus. Nat. Hist.* **15**:351–468.

Fitch, H. S. 1970. Reproductive cycles of lizards and snakes. *Univ. Kans. Publ., Mus. Nat. Hist.* **52**:1–247.

Fitch, H. S. and M. V. Plummer. 1975. A preliminary ecological study of the soft-shelled turtle *Trionyx muticus* in the Kansas River. *Isr. J. Zool.* **24**:28–42.

Fitch, H. S. and H. W. Shirer. 1971. A radiotelemetric study of spatial relationships in some common snakes. *Copeia.* **1971**:118–128.

Flanigan, W. F., Jr. 1974. Sleep and wakefulness in chelonian reptiles. II. The red-footed tortoise, *Geochelone carbonaria. Arch. Ital. Biol.* **112**:253–277.

Flanigan, W. F., Jr., C. P. Knight, K. M. Hartse, and A. Rechtschaffen. 1974. Sleep and wakefulness in chelonian reptiles. I. The box turtle, *Terrapene carolina. Arch. Ital. Biol.* **112**:227–252.

Florkin, M. 1949. *Biochemical evolution.* New York: Academic Press.

Flower, S. S. 1925. Contributions to our knowledge of the duration of life in vertebrate animals. III. Reptiles. *Proc. Zool. Soc. London.* **1925**:911–981.

Flower, S. S. 1933. Notes on the recent reptiles and amphibians of Egypt with a list of the species recorded from that kingdom. *Proc. Zool. Soc. London.* **103**:735–851.

Flower, S. S. 1938. Further notes on the duration of life in animals. III. Reptiles. *Proc. Zool. Soc. London.* **107**:1–39.

Folkerts, G. W. 1967. Notes on a hybrid musk turtle. *Copeia.* **1967**:479–480.

Folkerts, G. 1968. Food habits of the stripe-necked musk turtle, *Sternotherus minor peltifer* Smith and Glass. *J. Herpetol.* **2**:171–173.

Forbes, A. M. 1972. Distribution of injected europium in hatchling *Chelonia mydas* L. and its considerations as a tag. Master's thesis. University of Rhode Island.

Forbes, T. R. 1940. A note on reptilian sex ratios. *Copeia.* **1940**:132.

Forbes, T. R. 1961. Endocrinology of reproduction in cold-blooded vertebrates. In W. C. Young and G. W. Corner (Eds.), *Sex and internal secretions*, pp. 1035–1087. Baltimore: Williams and Wilkins.

Foster, R. E. and W. C. Hall. 1975. The connections and laminar organization of the optic tectum in a reptile (*Iguana iguana*). *J. Comp. Neurol.* **163**:397–426.

Foster, R. E. and T. L. Peele. 1975. Thalamotelencephalic auditory pathways in the lizard (*Iguana iguana*). *Anat. Rec.* **181**:530.

Fraenkel, G. S. and D. L. Gunn. 1961. *The orientation of animals.* New York: Dover.

Frair, W. 1963. Blood group studies with turtles. *Science.* **140**:1412–1414.

Frair, W. 1964. Turtle family relationships as determined by serological tests. In C. A. Leone (Ed.), *Taxonomic biochemistry and serology*, pp. 535–544. New York: Ronald Press.

Frair, W. 1971. The world's largest living turtle. *Int. Turtle Tortoise Soc. J.* **5**(2):22–25, 31.

Frankel, H. M., A. Spitzer, J. Blaine, and E. P. Schoener. 1969. Respiratory responses of turtles (*Pseudemys scripta*) to changes in arterial blood gas composition. *Comp. Biochem. Physiol.* **31**:535–546.

Fraser, D. and I. M. Spigel. 1971. Shock-induced threat and biting by the turtle. *J. Exp. Anal. Behav.* **16**:349–353.

Frazier, J. 1971. Observations on sea turtles at Aldabra Atoll. *Phil. Trans. Roy. Soc. London*, Ser. B. **260**:373–410.

Frazzetta, T. H. 1962. A functional consideration of cranial kinesis in lizards. *J. Morphol.* **111**:287–320.

Freiberg, M. A. 1967. Tortugas de la Argentina. *Cienc. Invest.* **23**:351–363.

Freiberg, M. A. 1971. *El mundo de las tortugas.* Buenos Aires: Ediciones Albatross.

Freiberg, M. A. 1972. Taxonomic relations among chelydrid and kinosternid turtles elucidated by serological tests. *Copeia.* **1972**:97–108.

Freiberg, M. A. 1973. Dos nuevas tortugas terrestres de Argentina. *Bol. Soc. Biol. Concep.* **46**:81–93.

Friant, M. 1961. Recherches sur la ceinture scapulaire des chelonians. *Acta Anat.* **45**:143–154.

Friar, W., R. G. Ackman, and N. Mrosovsky. 1972. Body temperature of *Dermochelys coriacea*: Warm turtle from cold water. *Science.* **177**:791–793.

Froese, A. D. and G. M. Burghardt. 1974. Food competition in captive juvenile snapping turtles, *Chelydra serpentina*. *Anim. Behav.* **22**:735–740.

Froese, A. D. and G. M. Burghardt. 1975. A dense natural population of the common snapping turtle (*Chelydra s. serpentina*). *Herpetologica.* **31**:204–208.

BIBLIOGRAPHY

Fry, F. E. J. 1967. Responses of vertebrate poikilotherms to temperature. In A. H. Rose (Ed.), *Thermobiology*, pp. 375-409. New York: Academic Press.

Frye, F. L. 1973. *Husbandry, medicine and surgery in captive reptiles*. Bonner Springs, Kans.: VM Publications.

Fukada, H. 1965. Breeding habits of some Japanese reptiles (critical review). *Bull. Kyoto Gakugei Univ.*, Ser. B. **27**:65-82.

Gadow, H. 1899. Orthogenetic variation in the shells of Chelonia. *Willey Zool. Results*. **3**:207-222.

Gadow, H. 1901. *The Cambridge natural history*. Vol. 8. *Amphibia and reptiles*. London: Macmillan.

Gadow, H. 1905. Orthogenetic variation. *Science*. **22**:637-640.

Gaffney, E. S. 1972a. An illustrated glossary of turtle skull nomenclature. *Am. Mus. Nat. Hist.* No. 2486:1-33.

Gaffney, E. S. 1972b. The systematics of the North American family Baenidae (Reptilia, Cryptodira). *Bull. Am. Mus. Nat. Hist.* **147**(5):241-320.

Gage, S. H. and S. P. Gage. 1886. Aquatic respiration in soft-shelled turtles: A contribution to the physiology of respiration in vertebrates. *Am. Nat.* **20**:233-236.

Gallup, G. G., Jr. 1968. Mirror-image stimulation. *Psychol. Bull.* **70**:782-793.

Gamble, H. J. 1956. An experimental study of the secondary olfactory connections in *Testudo graeca*. *J. Anat.* **90**:15-29.

Gandal, C. P. 1958. Cardiac puncture in anesthetized turtles. *Zoologica*. **43**:93-94.

Gans, C. 1962. Terrestrial locomotion without limbs. *Am. Zool.* **2**:167-182.

Gans, C. 1969. Comments on inertial feeding. *Copeia*. **1969**:855-857.

Gans, C. and G. M. Hughes. 1967. The mechanism of lung ventilation in the tortoise *Testudo graeca* Linné. *J. Exp. Biol.* **47**:1-20.

Gans, C. and T. M. Parsons. 1970. Taxonomic literature on reptiles. In C. Gans and T. S. Parsons (Eds.), *Biology of the reptilia*, Vol. 2, pp. 315-333. New York: Academic Press.

Garcia, J., C. Clarke, and W. Hankins. 1973. Natural responses to scheduled rewards. In P. P. G. Bateson and P. H. Klopfer (Eds.), *Perspectives in ethology*, pp. 1-41. New York: Plenum Press.

Gaunt, A. S. and C. Gans. 1969. Mechanics of respiration in the snapping turtle, *Chelydra serpentina* (Linné). *J. Morphol.* **128**:195-228.

Gaymer, R. 1968. The Indian Ocean giant tortoise *Testudo gigantea* on Aldabra. *J. Zool., London*. **154**:341-363.

Gaymer, R. 1973. A marking method of giant tortoises, and field trials in Aldabra. *J. Zool., London*. **169**:393-401.

Gegenbaur, C. 1901. *Vergleichenden Anatomie der Wirbelthiere*. Vol. 2. Leipzig: W. Engelmann.

Gemmell, D. J. 1970. Some observations on the nesting of the western painted turtle, *Chrysemys picta belli*, in northern Minnesota. *Can. Field-Nat.* **84**:308-309.

Gibbons, J. W. 1967. Variations in growth rates in three populations of the painted turtle, *Chrysemys picta*. *Herpetologica*. **23**:296-303.

Gibbons, J. W. 1968a. Growth rates of the common snapping turtle, *Chelydra serpentina*, in a polluted river. *Herpetologica*. **24**:266-267.

Gibbons, J. W. 1968b. Observations on the ecology and population dynamics of the Blanding's turtle, *Emydoidea blandingi*. *Can. J. Zool.* **46**:288-290.

Gibbons, J. W. 1968c. Population structure and survivorship in the painted turtle, *Chrysemys picta. Copeia.* **1968**:260–268.

Gibbons, J. W. 1968d. Reproductive potential, activity, and cycles in the painted turtle, *Chrysemys picta. Ecology.* **49**:399–409.

Gibbons, J. W. 1969. Ecology and population dynamics of the chicken turtle, *Deirochelys reticularia. Copeia.* **1969**:669–676.

Gibbons, J. W. 1970a. Reproductive characteristics of a Florida population of musk turtles (*Sternothaerus odoratus*). *Herpetologica.* **26**:268–270.

Gibbons, J. W. 1970b. Reproductive dynamics of a turtle (*Pseudemys scripta*) population in a reservoir receiving heated effluent from a nuclear reactor. *Can. J. Zool.* **48**:881–885.

Gibbons, J. W. 1970c. Sex ratio in turtles. *Res. Pop. Ecol.* **12**:252–254.

Gibbons, J. W. 1970d. Terrestrial activity and the population dynamics of aquatic turtles. *Am. Midl. Nat.* **83**:404–414.

Gibbons, J. W. and J. W. Coker. 1977. Ecological and life history aspects of the cooter, *Chrysemys floridana* (Le Conte). *Herpetologica.* **33**:29–33.

Gibbons, J. W. and M. H. Smith. 1968. Evidence of orientation by turtles. *Herpetologica.* **24**:331–333.

Gibbons, J. W. and D. W. Tinkle. 1969. Reproductive variation between turtle populations in a single geographic area. *Ecology.* **50**:340–341.

Gilles-Baillien, M. 1970. Urea and osmoregulation in the diamondback terrapin *Malaclemys centrata centrata* (Latreille). *J. Exp. Biol.* **52**:691–697.

Gillett, W. G. 1923. The histologic structure of the eye of the soft-shelled turtle. *Am. J. Ophthalmol.* **6**:955–973.

Giordano, R. V. and D. C. Jackson. 1973. The effect of temperature on ventilation in the green iguana (*Iguana iguana*). *Comp. Biochem. Physiol.* **45A**:235–238.

Girgis, S. 1961a. Aquatic respiration in the common Nile turtle, *Trionyx triunguis* (Forskål). *Comp. Biochem. Physiol.* **3**:206–217.

Girgis, S. 1961b. Observations on the heart in the family Trionychidae. *Bull. Brit. Mus. (Nat. Hist.) Zool.* **8**:73–107.

Girgis, S. 1962a. Anatomical and functional adaptations in the venous system of a diving reptile, *Trionyx triunguis* (Forskål). *Proc. Zool. Soc. London.* **138**:355–377.

Girgis, S. 1962b. Studies on the arteries in the common Nile turtle, *Trionyx triunguis. Proc. Zool. Soc. London.* **142**:191–216.

Goff, C. C. and D. S. Goff. 1932. Egg laying and incubation of *Pseudemys floridana. Copeia.* **1932**:92–94.

Goin, C. J. 1942. A method for collecting vertebrates associated with water hyacinths. *Copeia.* **1942**:183–184.

Goin, C. J. and C. C. Goff. 1941. Notes on the growth rate of the gopher turtle, *Gopherus polyphemus. Herpetologica.* **2**:66–68.

Goin, C. J. and O. B. Goin. 1971. *Introduction to herpetology*, 2nd ed. San Francisco: Freeman.

Goldby, F. and H. J. Gamble. 1957. The reptilian cerebral hemispheres. *Biol. Rev.* **32**:384–420.

Gonzalez, R. C. and M. E. Bitterman. 1962. A further study of partial reinforcement in the turtle. *Quart. J. Exp. Psychol.* **14**:109–112.

Goode, J. 1965. Nesting behavior of freshwater tortoises in Victoria. *Vict. Nat.* **82**:218–222.

Goode, J. 1967a. Artificial incubation of eggs of Victorian chelid tortoises. *Int. Turtle Tortoise Soc. J.* **1**(3):6–11, 41.

Goode, J. 1967b. *Freshwater tortoises of Australia and New Guinea.* Melbourne: Landsowne.

Goode, J. and J. Russell. 1968. Incubation of eggs of three species of chelid tortoises and notes on their embryological development. *Aust. J. Zool.* **16**:749–761.

Goodman, D. A. and N. M. Weinberger, 1973. Habituation in "lower" tetrapod vertebrates: Amphibia as vertebrate model systems. In H. V. S. Peeke and M. J. Herz (Eds.), *Habituation.* Vol. 1. New York: Academic Press.

Goodrich, E. S. 1919. Notes on the reptilian heart. *J. Anat. London.* **53**:298–304.

Gorman, G. C. 1968. The relationships of *Anolis* of the Roquet species group (Sauria: Iguanidae). III. Comparative study of display behavior. *Brevoria.* No. 284:1–31.

Gould, E. 1957. Orientation in box turtles, *Terrapene c. carolina* (Linnaeus). *Biol. Bull.* **112**:336–348.

Gould, E. 1959. Studies on the orientation of turtles. *Copeia.* **1959**:174–176.

Gourley, E. 1965. A preliminary study of orientation in gopher tortoises. Master's thesis. University of Florida.

Gourley, E. 1972. Circadian activity rhythm of the gopher tortoise (*Gopherus polyphemus*). *Anim. Behav.* **20**:13–20.

Gourley, E. 1974. Orientation of the gopher tortoise, *Gopherus polyphemus. Anim. Behav.* **22**:158–169.

Grabowski, C. T. 1970. Embryonic oxygen deficiency—A physiological approach to the analysis of teratological mechanisms. *Adv. Teratol.* **4**:125–167.

Grabowski, C. T. and J. Paar. 1958. The teratogenic effects of graded doses on hypoxia on the chick embryo. *Am. J. Anat.* **103**:313–347.

Graf, V. 1967. A spectral sensitivity curve and wavelength discrimination for the turtle, *Chrysemys picta picta. Vis. Res.* **7**:915–928.

Graf, V. 1972. Behavioral visual functions for *Chrysemys picta picta*. Preferences and frequency responses. *Brain Behav. Evol.* **5**:155–175.

Graf, V. and T. Tighe. 1971. Subproblem analysis of discrimination shift learning in the turtle (*Chrysemys picta picta*). *Psychon. Sci.* **25**:257–259.

Graham, T. E. 1969. Pursuit of the Plymouth turtle. *Int. Turtle Tortoise Soc. J.* **3**(1):10–13.

Graham, T. E. 1970. Growth rate of the spotted turtle, *Clemmys guttata*, in southern Rhode Island. *J. Herpetol.* **4**:87–88.

Graham, T. E. 1971a. Eggs and hatchlings of the red-bellied turtle, *Chrysemys rubriventris*, from Plymouth, Massachusetts. *J. Herpetol.* **5**:59–60.

Graham, T. E. 1971b. Growth rate of the red-bellied turtle, *Chrysemys rubriventris*, at Plymouth, Massachusetts. *Copeia.* **1971**:353–356.

Graham, T. E. 1972. Temperature-photoperiod on diel locomotor activity and thermal selection in the turtles, *Chrysemys picta* (Schneider), *Clemmys guttata* (Schneider), and *Sternotherus odoratus* (Latreille). Ph.D. dissertation. University of Rhode Island, Kingston.

Graham, T. E. and V. Hutchison. 1969. Centenarian box turtles. *Int. Turtle Tortoise Soc. J.* **3**(3):25–29.

Granda, A. M. 1962. Electrical responses of the light- and dark-adapted turtle eye. *Vis. Res.* **2**:343–356.

Granda, A. M. and K. W. Haden. 1970. Retinal oil globule counts and distributions in two species of turtles: *Pseudemys scripta elegans* (Wied) and *Chelonia mydas mydas* (Linnaeus). *Vis. Res.* **10**:79–84.

Granda, A. M., Y. Matsumiya, and C. E. Stirling. 1965. A method for producing avoidance behavior in the turtle. *Psychon. Sci.* **2**:187–188.

Granda, A. M., J. H. Maxwell, and H. Zwick. 1972. The temporal course of dark adaptation in the turtle, *Pseudemys*, using a behavioral paradigm. *Vis. Res.* **12**:653–672.

Granda, A. M. and P. J. O'Shea. 1972. Spectral sensitivity of the green turtle (*Chelonia mydas mydas*) determined by electrical responses to heterochromatic light. *Brain Behav. Evol.* **5**:143–154.

Granda, A. M. and C. E. Stirling. 1965. Differential spectral sensitivity in the optic tectum and eye of the turtle. *J. Gen. Physiol.* **48**:901–917.

Granda, A. M. and C. E. Stirling. 1966. The spectral sensitivity of the turtle's eye to very dim lights. *Vis. Res.* **6**:143–152.

Grant, C. 1936. The southwestern desert tortoise, *Gopherus agassizii. Zoologica.* **21**:225–229.

Grant, C. 1937. Orthogenetic variation. *Proc. Indiana Acad. Sci.* **46**:240–245.

Grant, C. 1960. Differentiation of the southwestern tortoises (genus *Gopherus*) with notes on their habits. *Trans. San Diego Soc. Nat. Hist.* **12**:441–448.

Gray, C. W., J. Davis, and W. G. McGarten. 1966. Treatment of *Pseudomonas* infections in the snake and lizard collection at Washington Zoo. *Int. Zoo Yearb.* **6**:278.

Gray, J. 1944. Studies in the mechanics of the tetrapod skeleton. *J. Exp. Biol.* **20**:88–116.

Gray, J. 1968. *Animal locomotion*. London: Weidenfeld and Nicolson.

Graziadei, P. P. C. 1971. The olfactory mucosa of vertebrates. In L. M. Biedler (Ed.), *Handbook of sensory physiology*, Vol. IV, pp. 27–58. New York: Springer-Verlag.

Graziadei, P. P. C. and R. L. Pierantoni. 1969. Interreceptor contacts in olfactory mucosa of frog and turtle. In C. J. Arcenaux (Ed.), *Proceedings: Electron Microscopy Society of America*, pp. 302–303. Baton Rouge, La.: Claitor's.

Graziadei, P. P. C. and D. Tucker. 1970. Vomeronasal receptors in turtles. *Z. Zellforsch. Mikro. Anat.* **105**:498–514.

Green, H. H., P. R. Steinmetz, and H. S. Frazier. 1970. Evidence for proton transport by turtle bladder in presence of ambient bicarbonate. *Am. J. Physiol.* **218**:845–850.

Greer, A. E., J. D. Lazell, and R. M. Wright. 1973. Anatomical evidence for a countercurrent heat exchanger in the leatherback turtle (*Dermochelys coriacea*). *Nature.* **244**:181.

Gregory, W. K. 1946. Pareiasaurs versus placodonts as near ancestors to the turtles. *Bull. Am. Mus. Nat. Hist.* **86**:279–326.

Grignon, G. and M. Grignon. 1962. La vascularisation de l'hypophyse chez la tortue terrestre (*Testudo mauritanica*). *Anat. Ann.* **109**:492–506.

Groves, P. M. and R. F. Thompson. 1970. Habituation: A dual-process theory. *Psychol. Rev.* **77**:419–450.

Grubb, P. 1968. Interim report on tortoises. January 1968. Royal Society. Appendix A to Aldabra Island Project, pp. 1–7.

Grubb, P. 1971. Comparative notes on the behavior of *Geochelone sulcata*. *Herpetologica.* **27**:328–333.

Gulick, W. L. and H. Zwick. 1966. Auditory sensitivity of the turtle. *Psychol. Rec.* **16**:47–53.

Gunning, G. E. and W. M. Lewis. 1957. An electrical shocker for the collection of amphibians and reptiles in the aquatic environment. *Copeia.* **1957**:52.

BIBLIOGRAPHY

Haga, J. 1970. Turtle farming. *Int. Turtle Tortoise Soc. J.* **4**(2):6–9.

Hailman, J. P. and R. G. Jaeger. 1971. On criteria for color preferences in turtles. *J. Herpetol.* **5**:83–85.

Hall, F. G. 1924. The respiratory exchange in turtles. *J. Metabol. Res.* **6**:393–401.

Hall, L. W. 1971. *Veterinary anesthesia and analgesia.* London: Bailliere Tindall.

Hall, W. C. and F. F. Ebner. 1970a. Parallels in the visual afferent projections of the thalamus in the hedgehog (*Paraechinus hypomelas*) and the turtle (*Pseudemys scripta*). *Brain Behav. Evol.* **3**:135–154.

Hall, W. C. and F. F. Ebner. 1970b. Thalamotelencephalic projections in the turtle (*Pseudemys scripta*). *J. Comp. Neurol.* **140**:101–122.

Hamilton, W. J. 1940. Observations on the reproductive behavior of the snapping turtle. *Copeia.* **1940**:124–126.

Hammel, H. T., F. T. Caldwell, Jr., and R. M. Abrams. 1967. Regulation of body temperature in the blue-tongued lizard. *Science.* **156**:1260–1262.

Hammer, D. A. 1969. Parameters of a marsh snapping turtle population. La Creek Refuge, South Dakota. *J. Wildl. Manag.* **33**:995–1005.

Hammer, D. A. 1971. The durable snapping turtle. *Nat. Hist.* **80**(6):59–65.

Haning, Q. C. and A. M. Thompson. 1965. A comparative study of tissue carbon dioxide in vertebrates. *Comp. Biochem. Physiol.* **15**:17–26.

Hansen, I. B. 1938. Studies on the reproductive tract of the male box turtle. *Anat. Rec.* **72**:121.

Hansen, I. B. 1941. The breathing mechanism of turtles. *Science.* **96**:64.

Hansen, I. B. and A. Riley. 1941. Seasonal changes in the oviduct of the turtle, *Terrapene carolina. Anat. Rec. Suppl.* **81**:116.

Hardy, J. D. 1961. Physiology of temperature regulation. *Physiol. Rev.* **41**:521–606.

Harker, J. 1960. In discussion. *Cold Spring Harbor Symp. Quant. Biol.* **25**:354.

Harless, M. 1970a. Position discrimination by *Clemmys insculpta.* Paper presented at the meeting of the American Society of Ichthyologists and Herpetologists, New Orleans.

Harless, M. 1970b. Social behavior in wood turtles. *Am. Zool.* **10**:289 (Abstr.).

Harless, M. D. and C. W. Lambiotte. 1971. Behavior of captive ornate box turtles. *J. Biol. Psychol.* **13**(2):17–23.

Harris, D. R. 1965. A technique for collecting aquatic reptiles and amphibians. *J. Ohio Herpetol. Soc.* **5**:35–36.

Harrisson, T. 1951. The edible turtle in Borneo. 1. Breeding seasons. *Sarawak Mus. J.* **5**:593–596.

Harrisson, T. 1952. Breeding of the edible turtle. *Nature.* **169**:198.

Harrisson, T. 1954. The edible turtle (*Chelonia mydas*) in Borneo. 2. Copulation. *Sarawak Mus. J.* **6**:126–128.

Harrisson, T. 1956a. The edible turtle (*Chelonia mydas*) in Borneo. 4. Growing turtles and growing problems. *Sarawak Mus. J.* **7**:233–239.

Harrisson, T. 1956b. Tagging green turtles. 1951–56. *Nature.* **178**:1479.

Harrisson, T. 1962. Notes on the green turtle. II. West Borneo numbers, the downward trend. *Sarawak Mus. J.* **10**:614–623.

Hart, R. R., D. C. Cogan, and L. L. Williamson. 1969. Maze path selection in the turtle (*Chrysemys*): A quasi-comparative study. *Psychol. Rec.* **19**:301–304.

Hartse, K. M. and A. Rechtschaffen. 1974. Effect of atropine sulfate on the sleep-related EEG spike activity of the tortoise, *Geochelone carbonaria*. *Brain Behav. Evol.* **9**:81–94.

Harvey, M. J. and R. W. Barbour. 1965. Home range of *Microtus ochrogaster* as determined by a modified minimum area method. *J. Mammal.* **46**:398–402.

Hasler, A. D. 1966. *Underwater guideposts: Homing of salmon.* Madison: University of Wisconsin Press.

Hasratjan, E. and A. Alexanjan. 1933. Beitrage zür frag über die bedingten Reflexe bei Schildkroten. *Fiziol. Zh. U.S.S.R.* **16**:451–456.

Hatt, R. T. 1932. Turtles eat heron. *Copeia.* **1932**:37.

Hausmann, P. 1968. *Claudius angustatus.* *Int. Turtle Tortoise Soc. J.* **2**(3):14–15.

Hausmann, P. 1968. Mata mata. *Int. Turtle Tortoise Soc. J.* **2**(4):18–19.

Hayes, W. N. and D. R. Herzler. 1967. Role of the optic tectum and general cortex in reptilian vision. *Psychon. Sci.* **9**:521–522.

Hayes, W. N., D. R. Hertzler, and D. K. Hogberg. 1968. Visual responsiveness and habituation in the turtle. *J. Comp. Physiol. Psychol.* **65**:331–335.

Hayes, W. N. and L. C. Ireland. 1972. A study of visual orientation mechanisms in turtles and guinea pigs. *Brain Behav. Evol.* **5**:226–239.

Hayes, W. N. and E. J. Saiff. 1967. Visual alarm reactions in turtles. *Anim. Behav.* **15**:102–106.

Hayne, D. 1949. Two methods for estimating population from trapping records. *J. Mammal.* **30**:399–411.

Heath, J. E. 1964. Reptilian thermoregulation: Evaluation of field studies. *Science.* **146**:784–785.

Heath, J. E. 1965. Temperature regulation and diurnal activity in horned lizards. *Univ. Calif. Publ. Zool.* **64**:97–129.

Heath, J. E. 1970. Behavioral regulation of body temperature in poikilotherms. *Physiologist.* **13**:399–410.

Heath, J. E., E. Gasdorf, and R. G. Northcutt. 1968. The effect of thermal stimulation of anterior hypothalamus on blood pressure in the turtle. *Comp. Biochem. Physiol.* **26**:509–518.

Heidt, G. A. and R. G. Burbridge. 1966. Some aspects of color preference, substrate preference, and learning in hatchling *Chrysemys picta. Herpetologica.* **22**:288–292.

Heisey, S. R. 1970. Cerebrospinal and extracellular fluid spaces in turtle brain. *Am. J. Physiol.* **219**:1564–1567.

Hendrickson, J. R. 1958. The green sea turtle, *Chelonia mydas* (Linn.) in Malaya and Sarawak. *Proc. Zool. Soc. London.* **130**:455–535.

Hendrickx, A. G. 1971. *Embryology of the baboon.* Chicago: University of Chicago Press.

Heron, K. 1968. Tortoises in a French garden. *Int. Turtle Tortoise Soc. J.* **2**(1):18–19, 30–33, 39–40.

Hertzler, D. R. and W. N. Hayes. 1967. Cortical and tectal function in visually guided behavior of turtles. *J. Comp. Physiol. Psychol.* **63**:444–447.

Hertzler, D. R. and W. N. Hayes. 1969. Effects of monocular vision and midbrain transection on movement detection in the turtle. *J. Comp. Physiol. Psychol.* **67**:473–478.

Hess, E. H. 1964. Imprinting in birds. *Science.* **146**:1128–1139.

Hewitt, J. 1933. On the Cape species and subspecies of the genus *Chersinella* Gray. Part I. *Ann. Natal Mus.* **7**:255–293.

Hewitt, J. 1934. On the Cape species and subspecies of the genus *Chersinella* Gray. Part II. *Ann. Natal Mus.* **7**:303-349.

Hildebrand, H. H. 1963. Hallazgo del area de anidación de la tortuga marina "lora," *Lepidochelys kempi* (Garman) en la costa occidental del Golfo de Mexico. *Ciencia, Mexico.* **22**:105-112.

Hildebrand, M. 1959. Motions of the running cheetah and horse. *J. Mammal.* **40**:481-495.

Hildebrand, M. 1965. Symmetrical gaits of horses. *Science.* **150**:701-708.

Hildebrand, M. 1966. Analysis of the symmetrical gaits of tetrapods. *Folio Biotheor.* **6**:9-22.

Hildebrand, S. F. 1929. Review of experiments on artificial culture of diamond-back terrapin. *Bull. U.S. Bur. Fish.* **45**:25-70.

Hildebrand, S. F. 1930. Duplicity and other abnormalities in diamond-back terrapins. *J. Elisha Mitchell Sci. Soc.* **46**:41-53.

Hildebrand, S. F. 1932. Growth of diamond-back terrapins. Size attained, sex ratio, and longevity. *Zoologica.* **9**:551-563.

Hildebrand, S. F. and C. Hatsel. 1927. On the growth, care, and behavior of loggerhead turtles in captivity. *Proc. Nat. Acad. Sci., U.S.A.* **13**:374-377.

Hirth, H. F. 1962. Cloacal temperatures of the green and hawksbill sea turtles. *Copeia.* **1962**:647-648.

Hirth, H. F. 1971. Synopsis of biological data on the green turtle *Chelonia mydas* (Linnaeus) 1758. *FAO Fisheries Synopsis No. 85.* Rome: UN Food & Agricultural Organization.

Hirth, H. F. and A. Carr. 1970. The green turtle in the Gulf of Aden and Seychelles Islands. *Verh. K. Ned. Akad. Wet.* **58**:1-44.

Hirth, H. F. and W. Schaffer. 1974. Survival rate of the green turtle, *Chelonia mydas*, necessary to maintain stable populations. *Copeia.* **1974**:544-546.

Hoagland, H. 1961. Some endocrine responses in man. In A. Simon (Ed.), *The physiology of emotions.* Springfield, Ill.: Thomas.

Hodgson, E. S. and R. F. Mathewson. 1971. Chemosensory orientation in sharks. In H. E. Adler (Ed.), *Orientation: Sensory basis*, pp. 175-182. New York: New York Academy of Science.

Hodgson, E. S. and K. D. Roeder. 1956. Electrophysiological studies of arthropod chemoreception. I. General properties of the labellar chemoreceptors of Diptera. *J. Cell. Comp. Physiol.* **48**:51-75.

Hodson, L. A. and J. F. W. Pearson. 1943. Notes on the discovery and biology of two Bahaman fresh-water turtles of the genus *Pseudemys. Proc. Fla. Acad. Sci.* **6**(2):17-23.

Hoessle, C. 1969. Simple incubators for reptile eggs at St. Louis Zoological Park. *Int. Zoo Yearb.* **9**:13-14.

Hoffman, K. 1954. Versuche zu der im Richtungsfinden der Vogel enthaltenen Zeitschatzung. *Z. Tierpsychol.* **11**:453-475.

Hoffman, K. 1965. Overt circadian frequencies and circadian rule. In J. Aschoff (Ed.), *Circadian clocks*, pp. 87-94. Amsterdam: North Holland.

Hoffman, K. 1968. Synchronisation der circadianen Aktivitatsperiodik von Eideschen durch Temperaturcyclen verschiedener Amplitude. *Z. Vergl. Physiol.* **58**:225-228.

Hollander, W., A. V. Chobian, and R. W. Williams. 1960. Potential of the aldosterone antagonists in hypertension. In F. C. Bartter (Ed.), *The clinical use of aldosterone antagonists*, pp. 169-177. Springfield, Ill.: Thomas.

Holman, J. A. 1964. Observations on dermatemyid and staurotypine turtles from Veracruz, Mexico. *Herpetologica.* **19**:277-279.

Holmes, P. H. and M. E. Bitterman. 1966. Spatial and visual habit reversal in the turtle. *J. Comp. Physiol. Psychol.* **62**:328–331.

Holmes, W. N. and R. L. McBean. 1964. Some aspects of electrolyte excretion in the green turtle, *Chelonia mydas mydas*. *J. Exp. Biol.* **41**:81–90.

Holmes, W. N., J. G. Phillips, and I. Chester Jones. 1963. Adrenocortical factors associated with adaptation of vertebrates to marine environments. *Recent Prog. Hormone Res.* **19**:619–672.

Honegger, R. E. 1967. Beobachtungen an den Riesenschildkroten (*Testudo gigantea* Scheweigger) der Inseln im indischen Ozean. *Salamandra.* **3**:101–121.

Honigman, H. 1921. Zur Biologie der Schildkroten. *Biol. Zentralbl.* **41**:241–250.

Hooker, D. 1908. Preliminary observations on the behavior of some newly hatched loggerhead turtles (*Thalassochelys caretta*). *Yearb. Carnegie Inst. Wash.* **6**:111–112.

Hooker, D. 1911. Certain reactions to color in the young loggerhead turtle. *Pap. Tortugas Lab.* **3**:69–76.

Hornell, J. 1927. *The turtle fisheries of the Seychelles Islands.* London: H. M. Stationery Office.

Householder, V. H. 1950. Courtship and coition of the desert tortoise. *Herpetologica.* **6**:11.

Howell, A. B. 1944. *Speed in animals.* Chicago: University of Chicago Press.

Howell, B. J., F. W. Baumgardner, K. Bondi, and H. Rahn. 1970. Acid-base balance in cold-blooded vertebrates as a function of body temperature. *Am. J. Physiol.* **218**:600–606.

Hubbard, R. and G. Wald. 1952a. *Cis-trans* isomers of vitamin A and retinene in the rhodopsin system. *J. Gen. Physiol.* **36**:269–315.

Hubbard, R. and G. Wald. 1952b. *Cis-trans* isomers of vitamin A and retinene in vision. *Science.* **115**:60–63.

Hughes, A. F. W. 1962. An experimental study of the relationship between limb and spinal cord in the embryo of *Eleutherodactylus martinicensis*. *J. Embryol. Exp. Morphol.* **10**:575–601.

Hughes, D. A. and J. D. Richard. 1974. The nesting of the Pacific ridley *Lepidochelys olivacea* on Playa Nancite, Costa Rica. *Mar. Biol.* **24**:97–107.

Hughes, G. R. 1970. Further studies on marine turtles in Tongaland. III. *Lammergeyer.* **12**:7–25.

Hughes, G. R. 1971. Sea turtle research and conservation in south east Africa. In *Marine turtles: Proc. 2nd Meet. Mar. Turtle Specialists,* IUCN Pub. (New Ser.) No. 31: 57–67.

Hughes, G. R. 1974. The sea turtles of south east Africa. II. The biology of the Tongaland loggerhead turtle *Caretta caretta* L. with comments on the leatherback turtle *Dermochelys coriacea* L. and the green turtle *Chelonia mydas* L. in the study region. *South Afr. Assoc. Mar. Biol. Res. Inv. Rep.* No. 36:1–96.

Hughes, G. R., A. J. Bass, and M. T. Mentis. 1967. Further studies on marine turtles in Tongaland. I. *Lammergeyer.* **3**:5–54.

Hughes, G. R. and B. Brent. 1973. The marine turtles of Tongaland. 7. *Lammergeyer.* **17**:40–62.

Hughes, G. R. and M. T. Mentis. 1967a. Further studies on marine turtles in Tongaland. II. *Lammergeyer.* **3**:55–72.

Hughes, G. R. and M. T. Mentis. 1967b. The marine turtles of Tongaland. *Afr. Wildl.* **21**:286–297.

Hulse, A. C. 1976. Growth and morphometrics of *Kinosternon sonoriense* (Reptilia, Testudines, Kinosternidae). *J. Herpetol.* **10**:341-348.

Humphrey, G. 1933. *The nature of learning.* New York: Harcourt Brace.

Hunsaker, D. II. 1966. Parasitism in turtles. *Int. Turtle Tortoise Soc. J.* **1**(1): 6-7, 23, 37, 46.

Hunt, T. J. 1957. Notes on diseases and mortality in testudines. *Herpetologica.* **13**:19-23.

Hunt, T. J. 1964. Anesthesia of the tortoise. In O. Graham-Jones (Ed.), *Small animal anesthesia*, pp. 71-77. New York: Macmillan.

Hutchison, V. H. 1961. Critical thermal maxima in salamanders. *Physiol. Zool.* **34**:92-125.

Hutchison, V. H. and R. J. Kosh. 1965. The effect of photoperiod on the critical thermal maxima of painted turtles (*Chrysemys picta*). *Herpetologica.* **20**:233-238.

Hutchison, V. H. and J. L. Larimer. 1960. Reflectivity of the integuments of some lizards from different habitats. *Ecology.* **41**:199-209.

Hutchison, V. H. and A. Vinegar. 1963. The feeding of the box turtle, *Terrapene carolina*, on a live snake. *Herpetologica.* **19**:284.

Hutchison, V. H., A. Vinegar, and R. J. Kosh. 1966. Critical thermal maxima in turtles. *Herpetologica.* **22**:32-41.

Hutton, K. E. 1960. Seasonal physiological changes in the red-eared turtle, *Pseudemys scripta elegans. Copeia.* **1960**:360-362.

Hutton, K. E., D. R. Boyer, J. C. Williams, and P. M. Campbell. 1960. Effects of temperature and body size upon heart rate and oxygen consumption in turtles. *J. Cell. Comp. Physiol.* **55**:87-93.

Hutton, K. E. and C. J. Goodnight. 1957. Variations in the blood chemistry of turtles under active and hibernating conditions. *Physiol. Zool.* **30**:198-207.

Inchaustegui Miranda, S. J. 1973. Las tortugas Dominicanas de agua dulce *Chrysemys decussata vicina* y *Chrysemys decorata* (Testudinidae, Emydidae). Tesis sustentada. Universidad Autonoma de Santo Domingo.

Ingle, R. M. and F. G. Walton Smith. 1949. *Sea turtles and the turtle industry of the West Indies, Florida and the Gulf of Mexico with annotated bibliography.* Miami, Fla.: University of Miami Press.

IUCN. 1969. *Marine turtles.* IUCN Publ. NS. Suppl. Pap. No. 20.

IUCN. 1971. *Marine turtles.* IUCN Publ. NS. Suppl. Pap. No. 31.

Ireland, L. C. and C. Gans. 1972. The adaptive significance of the flexible shell of the tortoise *Malacochersus tornieri. Anim. Behav.* **20**:778-781.

Ireland, L. C., W. N. Hayes, and L. H. Laddin. 1969. The relation between frequency and amplitude of visual alarm reactions in *Pseudemys. Anim. Behav.* **17**:386-388.

Iverson, J. B. 1977. Reproduction in freshwater and terrestrial turtles of North Florida. *Herpetologica.* **33**:205-212.

Iwase, Y. and D. Lisenby. 1965. Olfactory bulb responses in the turtle, with special reference to the deep negative spike. *Jap. J. Physiol.* **15**:331-341.

Jacobs, W. 1939. Die Lunge der Seeschildkröte *Caretta caretta* (L.) als Schwebeorgan. *Z. Vergl. Physiol.* **27**:1-28.

Jackson, C. G. 1969. Agonistic behavior in *Sternotherus minor minor* Agassiz. *Herpetologica.* **25**:53-54.

Jackson, C. G., Jr. 1970. A biometrical study of growth in *Pseudemys concinna suwanniensis*. I. *Copeia.* **1970**:528-534.

Jackson, C. G., Jr. and J. D. Davis. 1972a. Courtship display behavior of *Chrysemys concinna suwanniensis*. *Copeia*. **1972**:385–387.

Jackson, C. G., Jr. and J. D. Davis. 1972b. A quantitative study of the courtship display of the red-eared turtle, *Chrysemys scripta elegans* (Wied). *Herpetologica*. **28**:58–63.

Jackson, C. G. and M. Fulton. 1970. A turtle colony epizootic apparently of microbial origin. *J. Wildl. Dis.* **6**:466–468.

Jackson, C. G. and M. M. Jackson. 1971. The frequency of *Salmonella* and *Arizona* microorganisms in zoo turtles. *J. Wildl. Dis.* **7**:130–132.

Jackson, D. C. 1968. Metabolic depression and oxygen depletion in the diving turtle. *J. Appl. Physiol.* **24**:503–509.

Jackson, D. C. 1969. Buoyancy control in the freshwater turtle, *Pseudemys scripta elegans*. *Science*. **166**:1649–1651.

Jackson, D. C. 1971a. The effect of temperature on ventilation in the turtle *Pseudemys scripta elegans*. *Respir. Physiol.* **12**:131–140.

Jackson, D. C. 1971b. Mechanical basis for lung volume variability in the turtle *Pseudemys scripta elegans*. *Am. J. Physiol.* **220**:754–758.

Jackson, D. C. 1973. Ventilatory response to hypoxia in turtles at various temperatures. *Respir. Physiol.* **18**:178–187.

Jackson, D. C., S. E. Palmer, and W. L. Meadow. 1974. The effects of temperature and CO_2 breathing on ventilation and acid-base status of turtles. *Respir. Physiol.* **20**:131–146.

Jackson, D. C. and K. Schmidt-Nielson. 1966. Heat production during diving in the fresh water turtle, *Pseudemys scripta*. *J. Cell. Physiol.* **67**:225–232.

Jackson, D. C. and H. Silverblatt. 1974. Respiration and acid-base status of turtles following experimental dives. *Am. J. Physiol.* **226**:903–909.

Jackson, M. M., C. G. Jackson, and M. Fulton. 1969. Investigation of the enteric bacteria of the Testudinata: I. Occurrence of the genera *Arizona*, *Citrobacter*, *Edwardsiella* and *Salmonella*. *Bull. Wildl. Dis.* **5**:328–329.

Jaques, J. 1969. Hatching and early life of the mountain tortoise. *Afr. Wildl.* **23**(2):95–104.

Jayakar, S. D. and H. Spurway. 1966. Contribution to the biology of the Indian starred tortoise *Testudo elegans* Schoepff. *J. Bombay Nat. Hist. Soc.* **63**:83–113.

Jennrich, R. J. and F. B. Turner. 1969. Measurement of non-circular home range. *J. Theoret. Biol.* **22**:227–237.

Jenssen, T. A. 1970. Female response to filmed displays of *Anolis nebulosus* (Sauria, Iguanidae). *Anim. Behav.* **18**:640–647.

Jenssen, T. A. 1971. Display analysis of *Anolis nebulosus* (Sauria, Iguanidae). *Copeia*. **1971**:197–209.

Jerlov, N. G. 1968. *Optical oceanography*. Amsterdam: Elsevier.

Johansen, K., C. Lenfant, and D. Hanson. 1970. Phylogenetic development of pulmonary circulation. *Fed. Proc.* **29**:1135–1140.

Johlin, J. M. and F. B. Moreland. 1933. Studies of the blood picture of the turtle after complete anoxia. *J. Biol. Chem.* **103**:107–114.

Johnston, J. B. 1915. The cell masses in the forebrain of the turtle, *Cistudo carolina*. *J. Comp. Neurol.* **25**:393–468.

Johnston, J. B. 1916. Evidence of a motor pallium in the forebrain of reptiles. *J. Comp. Neurol.* **26**:475–479.

Johnston, J. B. 1923. Further contributions to the study of the evolution of the forebrain. *J. Comp. Neurol.* **35**:337–481.

Jolicoeur, P. and J. E. Mosimann. 1960. Size and shape variation in the painted turtle. A principal component analysis. *Growth.* **24**:339–354.

Jordan, D. S. 1899. *Manual of the vertebrate animals of the northeastern United States.* 8th ed. New York: World Book.

Judd, W. W. 1971. A young albino snapping turtle, *Chelydra serpentina* L., in southern Ontario, Canada. *Can. Field-Nat.* **85**:254–255.

Juhasz-Nagy, A., M. Szentivanyi, M. Szabo, and B. Vamosi. 1963. Coronary circulation of the tortoise heart. *Acta Physiol. Acad. Sci. Hung.* **23**:137A.

Kaplan, H. M. 1957a. The care and diseases of laboratory turtles. *Proc. Anim. Care Panel.* **7**:259–272.

Kaplan, H. M. 1957b. Septicemic, cutaneous ulcerative disease of turtles. *Proc. Anim. Care Panel.* **7**:273–277.

Kaplan, H. M. 1958a. Disease in laboratory frogs and turtles. *Am. Biol. Teacher* **20**:160–162.

Kaplan, H. M. 1958b. Treatment of escherichiosis in turtles, frogs, and rabbits. *Proc. Anim. Care Panel.* **8**:101–106.

Kaplan, H. M. 1969. Anesthesia in amphibians and reptiles. *Fed. Proc.* **28**:1541–1546.

Kaplan, H. M. and R. Taylor. 1957. Anesthesia in turtles. *Herpetologica.* **13**:43–45.

Karamian, A. I., N. P. Vesselkin, M. G. Belekhova, and T. M. Zagorulko. 1966. Electrophysiological characteristics of tectal and thalamocortical divisions of the visual system in lower vertebrates. *J. Comp. Neurol.* **127**:559–576.

Karashima, J. 1929a. Beiträge zur Embryochemie der Reptilien. V. Über das Verhalten der anorganischen Bestanteile bei der Bebrütung des Meerschildkröteneies. *J. Biochem.* (Tokyo). **10**:369–374.

Karashima, J. 1929b. Beiträge zur Embryochemie ser Reptilien. VI. Über das Verhalten der Fette bei der Bebrütung von Meerschildkröteneiern. *J. Biochem.* (Tokyo). **10**:375–377.

Karrer, P. and E. Jucker. 1950. *Carotenoids.* Amsterdam: Elsevier.

Kastle, W. 1963. Zur Ethologie des Grasanolis (*Norops auratus*) (Daudin). *Z. Tierpsychol.* **20**:16–33.

Kastle, W. 1965. Zur Ethologie des Anden-Anolis *Phenacosaurus richteri. Z. Tierpsychol.* **22**:751–769.

Kaushiva, B. S. 1940. The arterial system of the pond turtle *Lissemys punctata* (Bonaterre). *Proc. Indian Acad. Sci.* **12B**:84–94.

Kellog, F. E., D. E. Frizel, and R. H. Wright. 1962. The olfactory guidance of flying insects. IV. Drosophila. *Can. Entomol.* **94**:884–888.

Kenyon, K. W. and D. W. Rice. 1959. Life history of the Hawaiian monk seal. *Pac. Sci.* **23**:215–252.

Kenyon, W. A. 1925. Digestive enzymes in poikilothermal vertebrates, with comparative studies of these in amphibians, reptiles and mammals. *Bull. U.S. Bur. Fish.* **41**:179–200.

Kerr, D. I. B. and K. E. Hagbarth. 1955. An investigation of olfactory centrifugal fiber system. *J. Neurophysiol.* **18**:363–374.

Khalil, F. 1947. Excretion in reptiles. I. Non-protein nitrogen constituent of the urine of the sea turtle *Chelonia mydas*. *J. Biol. Chem.* **171**:611–616.

Khalil, F. and G. Haggag. 1955. Ureotelism and uricotelism in tortoises. *J. Exp. Zool.* **130**:423–432.

Khan Bin Momin Khan, M. 1964. A note on *Batagur baska* (the river terrapin or tuntong). *Malay. Nat. J.* **18**:184–186.

Khozatsky, L. I. and M. Mlynarski. 1966. *Agrionemys*—Nouveau genre de tortues terrestris (Testudinidae). *Bull. Acad. Pol. Sci.* **14**(2):123–125.

Kimball, P. 1923. A contribution to the anatomy and the development of the arterial and venous system in turtles. *Anat. Rec.* **25**:201–223.

King, F. W. 1971. Housing, sanitation and nutrition of reptiles. *J. Am. Vet. Med. Assoc.* **159**:1612–1615.

King, W. and J. V. Griffo, Jr. 1958. A box turtle fatality apparently caused by *Sarcophaga cistudinis* larvae. *Fla. Entomol.* **4**(1):97.

Kirk, K. L. and M. E. Bitterman. 1963. Habit reversal in the turtle. *Quart. J. Exp. Psychol.* **15**:52–57.

Kirk, K. L. and M. E. Bitterman. 1965. Probability learning by the turtle. *Science.* **148**:1484–1485.

Kirsche, W. 1967. Zur Haltung. Zucht und Ethologie der griechischen Landschildkröte (*Testudo hermanni hermanni*). *Salamandra.* **3**:36–66.

Klahr, S. and N. S. Bricker. 1964. Na^+ transport by isolated turtle bladder during anaerobiosis and exposure to KCN. *Am. J. Physiol.* **206**:1333–1339.

Kleerekoper, H. 1967. Some aspects of olfaction in fishes with special reference to orientation. *Am. Zool.* **7**:385–395.

Kleerekoper, H. 1972. Orientation through chemoreception in fishes. In S. R. Galler, K. Schmidt-Koenig, G. J. Jacobs, and R. E. Belleville (Eds.), *Animal orientation and navigation*, pp. 459–468. Washington, D.C.: National Aeronautics and Space Administration.

Klinghammer, E. and E. H. Hess. 1964. Imprinting in an altricial bird: The ring dove (*Streptopelia risoria*). *Science.* **146**:265–266.

Klopfer, P. H. 1969. *Habitats and territories. A study of the use of space by animals.* New York: Basic Books.

Knapp, H. and D. S. Kang. 1968a. The retinal projections of the side-necked turtle (*Podocnemis unifilis*) with some notes on the possible origin of the pars dorsalis of the lateral geniculate body. *Brain Behav. Evol.* **1**:369–404.

Knapp, H. and D. S. Kang. 1968b. The visual pathways of the snapping turtle (*Chelydra serpentina*). *Brain Behav. Evol.* **1**:19–42.

Knowlton, J. G. 1943. *My turtles.* Washington, D.C.: Author.

Koch, A. L., A. Carr, and D. W. Ehrenfeld. 1969. The problem of open-sea navigation: The migration of the green turtle to Ascension Island. *J. Theor. Biol.* **22**:163–179.

Kohler, A. J. 1972a. The common books on reptile care: A comparative overview. Part I. *Bull. Md. Herpetol. Soc.* **8**:14–26.

Kohler, A. J. 1972b. The common books on reptile care: A comparative overview. Part II. *Bull. Md. Herpetol. Soc.* **8**:41–51.

Kohler, A. J. 1972c. The common books on reptile care: A comparative overview. Part III. *Bull. Md. Herpetol. Soc.* **8**:66–79.

Korschgen, L. 1971. Procedures for food-baits analysis. In R. Giles (Ed.), *Wildlife management techniques*, pp. 233-250. Washington, D.C.: Wildlife Society.

Kosareva, A. A. 1967. Projection of optic tract fibers to visual centers in a turtle. *J. Comp. Neurol.* **130**:263-276.

Koschmann, G. 1967. A softshell comes to life. *Int. Turtle Tortoise Soc. J.* **1**(6):9.

Kosh, R. J. and V. H. Hutchison. 1968. Daily rhythmicity of temperature tolerance in eastern painted turtles, *Chrysemys picta. Copeia.* **1968**:244-246.

Kratzer, H. 1973. Beobachtungen über die Zeitigungsdauer eines Eigeleges von *Varanus salvator* (Sauria, Varanidae). *Salamandra.* **9**:27-33.

Krause, W. 1863. Über die Endigung der Muskelnerven. *Z. Rat. Med.* **20**:1-18.

Kroll, J. C., D. R. Clark, Jr., and J. W. Albert. 1973. Radiotelemetry for studying thermoregulation in free-ranging snakes. *Ecology.* **54**:454-456.

Kuroda, R. 1925. A contribution to the subject of the hearing of tortoises. *J. Comp. Psychol.* **5**:285-291.

Kuroda, R. 1933. Studies on visual discrimination in the tortoise *Clemmys japonica. Acta Psychol. Keijo.* **2**:31-59.

Lack, D. 1954. *The natural regulation of animal numbers.* Oxford: Clarendon.

Lack, D. 1968. *Ecological adaptations for breeding in birds.* London: Methuen.

Lagler, K. F. 1941. Fall mating and courtship of the musk turtle. *Copeia.* **1941**:268.

Lagler, K. F. 1943a. Food habits and economic relations of the turtles of Michigan with special reference to fish management. *Am. Midl. Nat.* **29**:257-312.

Lagler, K. F. 1943b. Methods of collecting freshwater turtles. *Copeia.* **1943**:21-25.

Lagler, K. F. 1956. *Freshwater fishery biology.* Dubuque, Iowa: Wm. C. Brown.

Lagler, K. F. and V. C. Applegate. 1943. Relationship between the length and the weight in the snapping turtle *Chelydra serpentina* Linnaeus. *Am. Nat.* **77**:476-478.

Lambert, M. R. K. 1970. Tortoise drain in Morocco. *Int. Turtle Tortoise Soc. J.* **4**(5):16-19, 35-37.

Lampkin, W. H. 1966. Hatching tortoise eggs. *Int. Turtle Tortoise Soc. J.* **1**(1):4-5.

Lampkin, W. and A. Turner. 1967. Giant vs. time. *Int. Turtle Tortoise Soc. J.* **1**(2):24-28, 46-48.

Landreth, H. F. 1972. Physiological responses of *Elaphe obsoleta* and *Pituophis melanoleucus* to lowered ambient temperatures. *Herpetologica.* **28**:376-380.

Lardie, R. L. 1964. Pugnacious behavior in the soft-shell *Trionyx spinifer pallidus* and implications of territoriality. *Herpetologica.* **20**:281-284.

Lardie, R. L. 1975. Observations on reproduction in *Kinosternon. J. Herpetol.* **9**:260-264.

Larsell, O. 1918. Studies on the nervus terminalis: Mammals. *J. Comp. Neurol.* **30**:3-68.

Larsell, O. 1932. The cerebellum of reptiles: Chelonians and alligators. *J. Comp. Neurol.* **56**:299-345.

Larsell, O. 1967. *The comparative anatomy and histology of the cerebellum from myxinoids through birds.* Minneapolis: University of Minnesota Press.

Latham, R. 1917. Studying the box turtle. *Copeia.* **1917**:15-16.

Latham, R. and F. Schlauch. 1969. Inscribed eastern box turtles. *Int. Turtle Tortoise Soc. J.* **3**(4):13.

Laurent, R. F. 1965. A contribution to the knowledge of the genus *Pelusios. Mus. Roy. l'Afr. Cent., Tervuren*, Ser. 8 (Sci. Zool.) No. 135:1-33.

Lawler, A. R. and J. A. Musick. 1972. Sand beach hibernation by a northern diamondback terrapin, *Malaclemys terrapin terrapin* (Schoepff). *Copeia*. **1972**:380–390.

Layne, J. N. 1952. Behavior of captive loggerhead turtles *Caretta c. caretta* (Linnaeus). *Copeia*. **1952**:115.

LeBuff, C. R. 1974. Unusual nesting relocation in the loggerhead turtle, *Caretta caretta*. *Herpetologica*. **30**:29–31.

LeBuff and R. W. Beatty. 1971. Some aspects of nesting of the loggerhead turtle, *Caretta caretta caretta* (Linne) on the Gulf coast of Florida. *Herpetologica*. **27**:153–156.

Legler, J. M. 1954. Nesting habits of the western painted turtle, *Chrysemys picta belli* (Gray). *Herpetologica*. **10**:137–144.

Legler, J. M. 1955. Observations on the sexual behavior of captive turtles. *Lloydia*. **18**:95–99.

Legler, J. M. 1956. A simple and practical method of artificially incubating reptile eggs. *Herpetologica*. **12**:290.

Legler, J. M. 1958. Extra-uterine migration of ova in turtles. *Herpetologica*. **14**:49–52.

Legler, J. M. 1960a. Natural history of the ornate box turtle, *Terrapene ornata ornata* Agassiz. *Publ. Mus. Nat. Hist. Univ. Kans.* **11**:527–669.

Legler, J. M. 1960b. A simple and inexpensive device for trapping aquatic turtles. *Utah Acad. Sci. Proc.* **37**:63–66.

Legler, J. M. 1977. Stomach flushing: A technique for chelonian dietary studies. *Herpetologica*. **33**:281–284.

Legler, J. M. and R. G. Webb. 1961. Remarks on a collection of Bolson tortoises, *Gopherus flavomarginatus*. *Herpetologica*. **17**:26–37.

Legler, W. K. 1969. Analysis of radiotelemetry system for tracking small vertebrates. Ph.D. dissertation, University of Kansas, Lawrence.

Legler, W. K. 1971. Radiotelemetric observations of cardiac rates in the ornate box turtle. *Copeia*. **1971**:760–761.

Lehmann, H. D. 1972. Zur Behandlung der Coccidiose bei Reptilien. *Salamandra*. **8**:48–49.

Lemkau, P. J. 1970. Movement of the box turtle, *Terrapene c. carolina* (Linnaeus) in unfamiliar territory. *Copeia*. **1970**:781–783.

Lenfant, C., K. Johansen, J. A. Petersen, and K. Schmidt-Nielsen. 1970. Respiration in the freshwater turtle, *Chelys fimbriata*. *Respir. Physiol.* **8**:261–275.

Levine, S. and I. E. Jones. 1965. Adrenocorticotropic hormone (ACTH) and passive avoidance learning. *J. Comp. Physiol. Psychol.* **59**:357–360.

Lewis, H. B. 1918. Some analyses of the urine of reptiles. *Science*. **48**:376.

Licht, P. 1972a. Environmental physiology of reptilian breeding cycles: Role of temperature. *Gen. Comp. Endocrinol. Suppl.* No. 3:477–488.

Licht, P. 1972b. Problems in experimentation on timing mechanisms for annual physiological cycles in reptiles. In F. E. South (Ed.), *Hibernation and hypothermia, perspectives and challenges*, pp. 681–711. Amsterdam: Elsevier.

Licht, P., W. R. Dawson, V. H. Shoemaker, and A. R. Main. 1966. Observations on the thermal relations of Western Australian lizards. *Copeia*. **1966**:97–110.

Licht, P. and W. R. Moberly. 1965. Thermal requirements for embryonic development in the tropical lizard *Iguana iguana*. *Copeia*. **1965**:515–517.

Liebman, P. A. 1972. Microspectrophotometry of photoreceptors. In H. J. A. Dartnall (Ed.), *Photochemistry of vision*. Vol. VII/1. *Handbook of sensory physiology*, pp. 481-528. New York: Springer-Verlag.

Liebman, P. A. and A. M. Granda. 1971. Microspectrophotometric measurements of visual pigments in two species of turtle, *Pseudemys scripta* and *Chelonia mydas*. *Vis. Res.* **11**:105-114.

Liebman, P. A. and A. M. Granda. 1975. Superdense carotenoid spectra resolved in single cone oil droplets. *Nature*. **253**:370-372.

Limpus, C. 1971. The flatback turtle, *Chelonia depressa* Garman, in southeast Queensland, Australia. *Herpetologica*. **27**:431-446.

Limpus, C. J. Loggerhead turtles (*Caretta caretta*) in Australia: Food sources while nesting. *Herpetologica*. **29**:42-45.

Lin, W. C. and W. H. Ko. 1968. A study of microwatt-power pulsed carrier transmitter circuits. *Med. Biol. Eng.* **6**:309-317.

Linsdale, J. M. 1927. Amphibians and reptiles of Doniphan County, Kansas. *Copeia*. **1927**:75-81.

Loch, J. H. 1950. Notes on the Perak River turtle. *Malay. Nat. J.* **5**:157-160.

Lofts, B. 1968. Patterns of testicular activity. In E. J. Barrington and C. B. Jorgenson (Eds.), *Perspectives in endocrinology*, pp. 239-304. London: Academic Press.

Lofts, B. 1969. Seasonal cycles in the reptilian testes. *Gen. Comp. Endocrinol. Suppl.* **2**:147-155.

Lofts, B. and C. Boswell. 1961. Seasonal changes in the distribution of the testis lipids of the Caspian terrapin, *Clemmys caspica*. *Proc. Zool. Soc. London.* **136**:581-592.

Lofts, B. and H. Tsui. 1977. Histological and histochemical changes in the gonads and epididymides of the male soft-shelled turtle *Trionyx sinensis*. *J. Zool. London.* **181**:57-68.

Logier, E. B. S. and G. C. Toner. 1961. Check-list of the amphibians and reptiles of Canada and Alaska. *Royal Ont. Mus. Contrib.* No. **53**:1-92.

Lopes, N. 1955. The action of alloxan in the turtle *Pseudemys d'orbignyi* D and B. *Acta Physiol. Latinoam.* **4**:190-199.

Loveridge, A. 1941. Revision of the African terrapin of the family Pelomedusidae. *Bull. Mus. Comp. Zool. Harvard.* **88**:467-524.

Loveridge, A. and E. E. Williams. 1957. Revision of the African tortoises and turtles of the suborder Cryptodira. *Bull. Mus. Comp. Zool. Harvard.* **115**:161-557.

Lovern, J. A. 1940. Fat metabolism in fishes. 14. The utilization of ethyl esters and fatty acids by the eel and their effect on depot fat consumption. *Biochem. J.* **23**:704-708.

Lovibond, S. H. 1968. The aversiveness of uncertainty: An analysis in terms of activation and information theory. *Austl. J. Psychol.* **20**:85-91.

Lowe, C. H. and H. J. Vance. 1955. Acclimation of the critical thermal maximum of the reptile *Urosaurus ornatus*. *Science*. **122**:73-75.

Lowenstein, W. R., C. A. Terzuolo, and Y. Washizu. 1963. Separation of transducer and impulse-generating processes in sensory receptors. *Science*. **142**:1180.

Lucey, E. C. 1974. Heart rate and physiological thermoregulation in a basking turtle, *Pseudemys scripta elegans*. *Comp. Biochem. Physiol.* **48A**:471-482.

Lüdicke, M. 1936. Über die Atmung von *Emys orbicularis* L. *Zool. Jahrb. Abt. Allgem. Zool. Physiol.* **56**:83–106.

Luederwaldt, H. 1926. Os chelonios Brasileiros. *Rev. Mus. Paulista* **14**:1–66.

Lukina, G. P. 1971. Reproductive physiology of the pond tortoise in the eastern Azov region. *Sov. J. Ecol.* **2**(3):99–100.

Lumb, W. V. 1963. *Small animal anesthesia.* Philadelphia: Lea and Febiger.

Lumsden, T. 1924. Chelonian respiration (tortoise). *J. Physiol. London.* **57**:354–367.

Lyman, R. A., Jr. 1945. The anti-haemolytic function of calcium in the blood of the snapping turtle, *Chelydra serpentina*. *J. Cell. Comp. Physiol.* **25**:65–73.

Lynn, W. G. 1937. Variation in scutes and plates in the box-turtle, *Terrapene carolina*. *Am. Nat.* **71**:421–426.

Lynn, W. G. 1970. The thyroid. In C. Gans and T. S. Parsons (Eds.), *Biology of the reptilia*. Vol. 3, pp. 201–234. New York: Academic Press.

Lynn, W. G. and T. von Brand. 1945. Studies on the oxygen consumption and water metabolism of turtle eggs. *Biol. Bull.* **88**:112–125.

Lynn, W. G. and M. C. Ullrich. 1950. Experimental production of shell abnormalities in turtles. *Copeia.* **1950**:253–262.

MacArthur, R. H. 1972. *Geographical ecology.* New York: Harper & Row.

MacArthur, R. and E. O. Wilson. 1967. *The theory of island biogeography.* Princeton, N.J.: Princeton University Press.

MacFarland, C. 1972. Goliaths of the Galapagos. *Nat. Geogr.* **142**(5):632–649.

MacFarland, C., J. Villa, and B. Toro. 1974. The Galapagos giant tortoises (*Geochelone elephantopus*). Part 1. Status of the surviving populations. *Biol. Conserv.* **6**(2):117–133.

MacKay, R. S. 1964. Galapagos tortoise and marine iguana deep core body temperature measured by radiotelemetry. *Nature.* **204**:355–358.

MacKay, R. S. 1968. Observations on peristaltic activity versus temperature and circadian rhythms in undisturbed *Varanus flavescens* and *Ctenosaura pectinata*. *Copeia.* **1968**:252–259.

MacKay, R. S. 1970. *Bio-medical telemetry: Sensing and transmitting biological information from animals and man*, 2nd ed. New York: Wiley.

Madison, D. M., A. Scholz, and A. D. Hasler. 1972. Behavioral evidence of "imprinting" to chemical cues in salmon. *Am. Zool.* **12**:643 (Abstr.).

Magnus, R. and J. Müller. 1835. Untersuchung eines Schildkrotenharns. *Arch. Anat. Physiol. Wissensch. Med.* **1835**:214–219.

Mahmoud, I. Y. 1967. Courtship behavior and sexual maturity in four species of kinosternid turtles. *Copeia.* **1967**:314–319.

Mahmoud, I. Y. 1968a. Feeding behavior in kinosternid turtles. *Herpetologica.* **24**:300–305.

Mahmoud, I. Y. 1968b. Nesting behavior in the western painted turtle, *Chrysemys picta belli*. *Herpetologica.* **24**:158–162.

Mahmoud, I. Y. 1969. Comparative ecology of the kinosternid turtles of Oklahoma. *Southwest. Nat.* **14**:31–66.

Mahmoud, I. Y., G. L. Hess, and J. Klicka. 1973. Normal embryonic stages of the western painted turtle, *Chrysemys picta belli*. *J. Morphol.* **141**:269–279.

Mahmoud, I. Y. and J. Klicka. 1971. Post-hatching changes in glycogen concentration in tissues of fed and unfed snapping turtles. *Am. Midl. Nat.* **86**:248–252.

BIBLIOGRAPHY

Mahmoud, I. Y. and J. Klicka. 1972. Seasonal gonadal changes in kinosternid turtles. *J. Herpetol.* **6**:183-189.

Mahmoud, I. Y. and N. Lavenda. 1969. Establishment and eradication of food preferences in red-eared turtles. *Copeia.* **1969**:298-300.

Mahoney, J. J. and V. H. Hutchison. 1969. Photoperiod acclimation and 24-hour variations in the critical thermal maxima of a tropical and a temperate frog. *Oecologia.* **2**:143-161.

Major, P. D. 1975. Density of snapping turtles, *Chelydra serpentina* in western West Virginia. *Herpetologica.* **31**:332-335.

Manton, M. L., A. Karr, and D. W. Ehrenfeld. 1972a. Chemoreception in the migratory sea turtle, *Chelonia mydas. Biol. Bull.* **143**:184-195.

Manton, M. L., A. Karr, and D. W. Ehrenfeld. 1972b. An operant method for the study of chemoreception in the green turtle, *Chelonia mydas. Brain Behav. Evol.* **5**:188-201.

Mao, S. H. 1971. *Turtles of Taiwan. A natural history of the turtles.* Taiwan: Commercial Press.

Marchand, L. J. 1942. A contribution to a knowledge of the natural history of certain freshwater turtles. Master's thesis. University of Florida.

Marchand, L. J. 1944. Notes on the courtship of a Florida terrapin. *Copeia.* **1944**:191-192.

Marchand, L. J. 1945a. The individual range of some Florida turtles. *Copeia.* **1945**:75-77.

Marchand, L. J. 1945b. Water goggling: A new method for the study of turtles. *Copeia.* **1945**:37-40.

Marchand, R. F. 1845. Über die Zusammensetzung des Harns der Schildkrote. *J. Prakt. Chem.* **34**:244.

Marchiafava, P. C. 1976. Centrifugal actions on amacrine and ganglion cells in the retina of the turtle. *J. Physiol.* **255**:137-155.

Marcus, L. C. 1971. Infectious diseases of reptiles. *J. Am. Vet. Med. Assoc.* **159**:1626-1631.

Marler, P. 1976. On animal aggression: The roles of strangeness and familiarity. *Am. Psychol.* **36**:239-246.

Marquez, R. and T. Doi. 1973. Ensayo téorico sobre el analisis de la población de tortuga prieta, *Chelonia mydas carrinegra* Caldwell, en aguas de Golfo de California, México. *Bull. Tokai Reg. Res. Lab.* No. **73**:1-22.

Marshall, W. K., Jr. and H. W. Smith. 1930. The glomerular development of the vertebrate kidney in relation to habitat. *Biol. Bull.* **59**:135-153.

Mathews, D. F. 1972. Response patterns of single neurons in the tortoise olfactory epithelium and olfactory bulb. *J. Gen. Physiol.* **60**:166-180.

Mathews, D. F. and D. Tucker. 1966. Single unit activity in the tortoise olfactory mucosa. *Fed. Proc.* **25**:329.

Mathur, P. N. 1944. The anatomy of the reptilian heart. Part 1. *Varanus monitor* (Linné). *Proc. Indian Acad. Sci.* **20B**:1-29.

Mathur, P. N. 1946. The anatomy of the reptilian heart. Part 2. Serpentes, Testudinata and Loricata. *Proc. Indian Acad. Sci.* **23B**:129-152.

Matsumiya, Y. M. and A. M. Granda. 1964. Free operant conditioning in the turtle. Paper presented at the 55th Annual Meeting of the Eastern Psychological Association, Philadelphia.

Matthews, G. V. T. 1968. *Bird navigation.* London: Cambridge University Press.

Mattox, N. 1935. Annular rings in the long bones of turtles and their correlation with size. *Trans. Ill. Acad. Sci.* **28**:255-256.

Mausolf, F. A. and C. C. Wunder. 1974. Growth of an anophthalamic turtle. *Copeia.* **1974**:548-550.

Maxwell, F. D. 1911. *Reports on inland and sea fisheries in the Thongwa, Myaungmya, and Bassein Districts and turtle banks of the Irrawaddy Division.* Rangoon.

Maxwell, J. H. and A. M. Granda. 1972. Dark adaptation in the turtle, *Pseudemys. Brain Behav. Evol.* **5**:176-187.

Maxwell, J. H. and A. M. Granda. 1975. An automated apparatus for the determination of visual thresholds in turtles. *Physiol. Behav.* **15**:131-132.

Mayhew, W. 1965. Hibernation in the horned lizard, *Phrynosoma m'Calli. Comp. Biochem. Physiol.* **16**:103-119.

Mayhew, W. 1968. Biology of desert amphibians and reptiles. In G. W. Brown, Jr. (Ed.), *Desert biology*, pp. 195-356. New York: Academic Press.

McCoy, C. J. 1968. The development of melanism in an Oklahoma population of *Chrysemys scripta elegans* (Reptilia: Testudinidae). *Proc. Okla. Acad. Sci.* **47**:84-87.

McCoy, R. H. and R. J. Seidler. 1973. Potential pathogens in the environment: Isolation, enumeration, and identification of seven genera of intestinal bacteria associated with small green pet turtles. *Appl. Microbiol.* **25**:534-538.

McCutcheon, F. H. 1943. The respiratory mechanism in turtles. *Physiol. Zool.* **16**:255-269.

McCutcheon, F. H. 1947. Specific O_2 affinity of hemoglobin in elasmobranchs and turtles. *J. Cell. Comp. Physiol.* **29**:333-344.

McDowell, S. B. 1961. On the major arterial canals in the ear-region of testudinoid turtles and the classification of Testudinoidea. *Bull. Mus. Comp. Zool. Harvard.* **125**:23-39.

McDowell, S. B. 1964. Partition of the genus *Clemmys* and related problems in the taxonomy of the aquatic Testudinidae. *Proc. Zool. Soc. London.* **143**:239-279.

McGinnis, S. M. 1967a. The adaptation of biotelemetry technique to small reptiles. *Copeia.* **1967**:472-473.

McGinnis, S. M. 1967b. Telemetry applied to studies of thermoregulation in reptiles. *Proc. 1967 Nat. Telem. Conf.* **1967**:252-254.

McGinnis, S. M. 1968. Respiration rate and body temperature of the Pacific green turtle, *Chelonia mydas agassizii. Am. Zool.* **8**:766.

McGinnis, S. M. and W. G. Voigt. 1971. Thermoregulation in the desert tortoise, *Gopherus agassizi. Comp. Biochem. Physiol.* **40A**:119-126.

McFarlane, R. W. 1963. Disorientation of loggerhead hatchlings by artificial road lighting. *Copeia.* **1963**:153.

McIlhenny, E. A. 1935. *The alligator's life history.* Boston: Christopher.

McNab, B. 1971. On the ecological significance of Bergman's rule. *Ecology.* **52**:845-854.

McNeil, E. and W. R. Hinshaw. 1946. *Salmonella* from Galapagos turtles, a gila monster, and an iguana. *Am. J. Vet. Res.* **7**:62-63.

Mebs, D. 1963. Beobachtungen an *Platysternon megacephalum. Aquar. Terrar. Z.* **16**:20-23.

Medem, F. 1960. Datos zoogeográficos y ecológicos sobre los Crocodylia y Testudinata de los Ríos Amazonas, Putumayo y Caquetá. *Caldasia.* **8**:341-351.

Medem, F. 1962. La distribución geográfica y ecología de los Crocodylia y Testudinata en el Departmento del Chocó. *Rev. Acad. Colomb. Cienc. Exactas Fis. Nat.* **11**:279-303.

Medem, F. 1964. Morphologie, Öcologie und Verbreitung der Schildkröte. *Podocnemis unifilis* in Kolumbien (Testudinata, Pelomedusidae). *Senckenb. Biol.* **45**:353-368.

Medem, F. 1966. Contribuciones al conocimiento sobre la ecologia y distribución geográfica de *Phrynops* (*Batrachemys*) *dahli*; (Testudinata, Pleurodira, Chelidae). *Caldasia.* **9**:467–489.

Medem, F. 1969. Estudios adicionales sobre Crocodylia y Testudinata del Alto Caquetá y Rio Caguán. *Caldasia.* **10**:329–353.

Medem, F. 1973. Beiträge zur Kenntnis über die Fortpflanzung der Buckel-Schildkröte, *Phrynops* (*Mesoclemmys*) *gibbus*. *Salamandra.* **9**:91–98.

Medica, P. A., R. B. Bury, and F. B. Turner. (ms) Age and growth of the desert tortoise (*Gopherus agassizii*) in southern Nevada.

Medsger, O. P. 1919. Notes on the first turtle I ever saw. *Copeia.* **1919**(69):29.

Mehnert, E. 1892. Gastrulation und Kleimblätterbildung der *Emys lutaria taurica*. *Morphol. Arb.* **1**:365–495.

Mehnert, E. 1895. Über Entwickelung, Bau und Funktion des Amnion und Amnionsganges nach Untersuchungen an *Emys lutaria taurica* (Marsilii). *Morphol. Arb.* **4**:207–224.

Mell, R. 1929. *Betrage sur fauna sinica. IV. Grundzüge einer Oekologie der chinesischen Reptilian und einer herpetologischen Reptilien tiergeographie Chinas.* Berlin.

Mell, R. 1938. Aus der Biologie chinesischer Schildkröten. Beiträge zur Fauna sinica. VI. *Z. Syst. Zool.* (Leipzig). **7**:390–476.

Mertens, R. 1960. *The world of amphibians and reptiles.* London: G. G. Harrap.

Mertens, R. 1967. Bemerkenswerte Sübwasserschildkröten aus Brasilien. *Senckenb. Biol.* **48**:71–82.

Mertens, R. and H. Wermuth. 1960. *Die Amphibien und Reptilien Europas.* Frankfurt: Verlag M. Kramer.

Metcalf, E. and A. L. Metcalf. 1970. Observations on ornate box turtles (*Terrapene ornata ornata* Agassiz). *Trans. Kans. Acad. Sci.* **73**:95–117.

Millen, J. E., H. V. Murdaugh, Jr., C. B. Baver, and E. D. Robin. 1964. Circulatory adaptations to diving in the freshwater turtle. *Science.* **145**:591–593.

Miller, L. 1932. Notes on the desert tortoise (*Testudo agassizii*). *Trans. San Diego Soc. Nat. Hist.* **7**:187–208.

Miller, L. 1955. Further observations on the desert tortoise, *Gopherus agassizii* of California. *Copeia.* **1955**:113–118.

Miller, M. R. 1959. The endocrine basis for reproductive adaptations in reptiles. In A. Gorbman (Ed.), *Comparative endocrinology*, pp. 499–516. New York: Wiley.

Miller, R. E. and N. Ogawa. 1962. The effect of adrenocorticotropic hormone (ACTH) on avoidance conditioning in the adrenalectomized rat. *J. Comp. Physiol. Psychol.* **55**:211–213.

Mills, T. W. 1886. Notes on the urine of the tortoise with special reference to uric acid and urea. *J. Physiol.* **7**:453–457.

Milstead, W. M. and D. W. Tinkle. 1967. *Terrapene* of western Mexico with comments on the species groups in the genus. *Copeia.* **1967**:180–187.

Minton, S. A. 1966. A contribution to the herpetology of West Pakistan. *Bull. Am. Mus. Nat. Hist.* **134**:29–184.

Minton, S., Jr. 1944. Introduction to the study of the reptiles of Indiana. *Am. Midl. Nat.* **32**:438–477.

Mirsky, I. A., R. E. Miller, and M. Stein. 1953. Relation of adrenocortical activity and adaptive behavior. *Psychosom. Med.* **15**:574–584.
Mitsukuri, K. 1891. On the foetal membranes of Chelonia. *J. Coll. Sci. Imp. Univ. Tokyo.* **4**:1–53.
Mitsukuri, K. 1893. Further studies on the formation of the germinal layers in Chelonia. *J. Coll. Sci. Imp. Univ. Tokyo.* **5**:35–52.
Mitsukuri, K. 1894. On the process of gastrulation in Chelonia. *J. Coll. Sci. Imp. Univ. Tokyo.* **6**:227–277.
Mitsukuri, K. 1895a. How many times does the snapping turtle lay eggs in one season? *Zool. Mag. Tokyo.* **6**:143–147.
Mitsukuri, K. 1895b. On the fate of the blastopore, the relations of the primitive streak, and the formation of the posterior end of the embryo in Chelonia, together with remarks on the nature of the mesoblastic ova in vertebrates. *J. Coll. Sci. Imp. Univ. Tokyo.* **10**:1–118.
Mitsukuri, K. 1905. The cultivation of marine and fresh-water animals in Japan. The snapping turtle, or soft-shell tortoise, "suppon." *Bull. U.S. Bur. Fish.* **24**:260–266.
Mitsukuri, K. and C. Ishikawa. 1886. On the formation of the germinal layers in Chelonia. *J. Coll. Sci. Imp. Univ. Tokyo.* **1**:211–246.
Mittermeier, R. A. 1971. Notes on the behavior and ecology of *Rhinoclemys annulata* Gray. *Herpetologica.* **27**:485–488.
Mlynarski, M. M. 1952. The ability to recognize complete forms from their fragments in the water tortoises: Emydinae. *Bull. Int. Acad. Pol. Sci.* **1952**:253–270.
Mlynarski, M. 1956. Studies on the morphology of the shell of recent and fossil tortoises. I–II. *Acta Zool. Cracov.* **1**:1–18.
Mlynarski, M. 1966. Morphology of the shell of *Agrionemys horsfieldi* (Gray, 1944) (Testudines, Reptilia). *Acta Biol. Cracov., Ser. Zool.* **9**:219–223.
Mlynarski, M. 1969. *Fossile Schildkröten.* Wittenberg: A. Ziemsen Verlag.
Mlynarski, M. 1972. *Zangerlia testudinimorpha* N. Gen., N. Sp. A primitive land tortoise from the Upper Cretaceous of Mongolia. *Paleontol. Pol.* No. **27**:85–92.
Mohr, C. O. 1947. Table of equivalent populations of North American small mammals. *Am. Midl. Nat.* **37**:223–249.
Mohr, C. O. and W. A. Stumpf. 1966. Comparison of methods for calculating areas of animal activity. *J. Wildl. Manag.* **30**:293–304.
Moll, E. O. 1973. Latitudinal and intersubspecific variation in reproduction of the painted turtle, *Chrysemys picta. Herpetologica.* **29**:307–318.
Moll, E. O. 1975. Patterns of chelonian reproductivity. *Am. Soc. Ichthyol. Herpetol. Meet. Program* 54–55 (Abstr.).
Moll, D. 1976. Environmental influence on growth rate in the Ouachita map turtle, *Graptemys pseudogeographica ouachitensis. Herpetologica.* **32**:439–443.
Moll, E. O. and J. M. Legler. 1971. The life history of a neotropical slider turtle, *Pseudemys scripta* (Schoepff) in Panama. *Bull. Los Angeles Co. Mus. Nat. Hist. (Sci.)* **11**:1–102.
Mondolfi, E. 1955. Anotaciones sobre la biología de tres quelónios de los llanos de Venezuela. *Mem. Soc. Cienc. Nat. La Salle* (Caracas). **15**:177–183.
Moorehouse, F. W. 1933. Notes on the green turtle (*Chelonia mydas*). *Rep. Gt. Barrier Reef Comm.* **4**:1–22.

Mora, J., R. Martuscelli, J. Ortiz-Pineda, and G. Soberon. 1965. The regulation of urea-biosynthesis enzymes in vertebrates. *Biochem. J.* **96**:28-35.

Morlock, H. C. 1972. Behavior following ablation of the dorsal cortex of turtles. *Brain Behav. Evol.* **5**:256-263.

Morlock, H. C., N. Brothers, and L. Shaffer. 1968. Access to air as a reinforcer for turtles. *Psychol. Rep.* **23**:1222.

Mosby, H. S. and D. E. Cantner. 1955. The use of Avertin in capturing wild turkeys and as an oral-based anesthetic for other wild animals. *Southwest Vet.* **9**:132-136.

Mosimann, J. E. 1956. Variation and relative growth in the plastral scutes of the turtle, *Kinosternon integrum* Leconte. *Misc. Publ. Mus. Zool., Univ. Mich.* **97**:1-43.

Mosimann, J. E. 1958. An analysis of allometry in the chelonian shell. *Rev. Can. Biol.* **17**:137-228.

Mosimann, J. E. and J. R. Bider. 1960. Variation, sexual dimorphism, and maturity in a Quebec population of the common snapping turtle, *Chelydra serpentina*. *Can. J. Zool.* **38**:19-38.

Mosquiera Manso, J. M. 1945. The tortoises of the Orinoco. Biological notes of *Podocnemis expansa*. *3rd Interam. Conf. Agr. Caracas* **29**:5-43.

Moulton, D. G. and L. M. Biedler. 1967. Structure and function of the peripheral olfactory system. *Physiol. Rev.* **47**:1-52.

Movchan, N. A. 1964. Antibiotic properties of the egg albumin of the steppe tortoise (*Testudo horsfieldi* Grav.) *Vestn. Leningr. Univ., Biol.* **15(3)**:18-25.

Movchan, N. A. 1966. Fungicidal properties of the albumin of the eggs of the steppe tortoise *Testudo horsfieldi*. *Vestn. Leningr. Univ. Biol.* **3(1)**:59-69.

Movchan, N. A. 1967. Bacteria in the egg of the steppe tortoise. *Vestn. Leningr. Univ., Biol.* **9(2)**:155-157.

Moyle, V. 1949. Nitrogenous excretion in chelonian reptiles. *Biochem. J.* **44**:581-584.

Mrosovsky, N. 1964. Modification of the diving-in response of the red-eared terrapin, *Pseudemys ornata callirostris*. *Quart. J. Exp. Psychol.* **16**:166-171.

Mrosovsky, N. 1968. Nocturnal emergence of hatchling sea turtles: Control by thermal inhibition of activity. *Nature.* **220**:1338-1339.

Mrosovsky, N. 1972. The water-finding ability of sea turtles. Behavioral studies and physiological speculations. *Brain Behav. Evol.* **5**:202-225.

Mrosovsky, N. and B. Boycott. 1966. Intra- and interspecific differences in phototactic behavior of fresh water turtles. *Behaviour.* **26**:215-227.

Mrosovsky, N. and A. Carr. 1967. Preference for light of short wavelengths in hatchling green sea turtles, *Chelonia mydas*, tested on their natural nesting beaches. *Behaviour.* **28**:217-231.

Mrosovsky, N. and P. C. H. Pritchard. 1971. Body temperature of *Dermochelys coriacea* and other sea turtles. *Copeia.* **1971**:631.

Mrosovsky, N. and S. J. Shettleworth. 1968. Wavelength preferences and brightness clues in the water finding behaviour of sea turtles. *Behaviour.* **32**:211-257.

Mrosovsky, N. and S. J. Shettleworth. In press. Further studies of the sea-finding mechanism in green turtle hatchlings. *Behaviour.*

Muller, J. F. 1921. Notes on the habits of the soft-shell turtle, *Amyda mutica*. *Am. Midl. Nat.* **7**:180-183.

Mulvaney, B. D. and H. E. Heist. 1971. Regeneration of rabbit olfactory epithelium. *Am. J. Anat.* **131**:241–252.

Munson, J. P. 1904. Researches on the oogenesis of the tortoise, *Clemmys marmorata*. *Am. J. Anat.* **3**:311–347.

Muntz, W. R. A. and D. P. M. Northmore. 1968. Background light, temperature, and visual noise in the turtle. *Vis. Res.* **8**:787–800.

Muntz, W. R. A. and S. Sokol. 1967. Psychophysical thresholds to different wavelengths in light adapted turtles. *Vis. Res.* **7**:729–741.

Munzel, P. 1938. Untersuchungen über die Harnstoffbidlung bei Wirbeltieren. *Zool. J.* **59**:113.

Murillo, N. R., J. K. Diercks, and E. J. Capaldi. 1961. Performance of the turtle, *Pseudemys scripta troostii*, in a partial reinforcement situation. *J. Comp. Physiol. Psychol.* **54**:204–206.

Murphy, G. 1970. Orientation of adult and hatchling red-eared turtles, *Pseudemys scripta elegans*. Ph.D. dissertation. Mississippi State University.

Murphy, J. B. 1969. Notes on iguanids and varanids in a mixed exhibit at Dallas Zoo. *Int. Zoo Yearb.* **9**:39–41.

Murphy, J. B. 1972. A reptile haven in Texas. *Int. Turtle Tortoise Soc. J.* **6**(1):6–9, 29.

Murphy, J. B. 1973a. A review of diseases and treatment of captive Chelonians. Part I. *HISS News-J.* **1**:5–8.

Murphy, J. B. 1973b. A review of diseases and treatment of captive Chelonians. Part II. Bacterial and viral infections. *HISS News-J.* **1**:77–81.

Murphy, J. B. 1973c. A review of diseases and treatment of captive Chelonians. Part III. Protozoal infections, vectors. *HISS News-J.* **1**:123–128.

Murphy, J. B. 1973d. A review of diseases and treatment of captive Chelonians. Part IV. Turbellarians, trematodes, cestodes, acanthocephalan worms, nematodes, insects, myiasis. *HISS News-J.* **1**:139–150.

Murphy, J. B. 1975. A brief outline of suggested treatments for diseases of captive reptiles. *Misc. Publ., Soc. Study Amphib. Reptiles.* **4**:ii, 1–13.

Murphy, J. V. and R. E. Miller. 1955. The effect of adrenocorticotropic hormone (ACTH) on avoidance conditioning in the rat. *J. Comp. Physiol. Psychol.* **48**:47–49.

Musacchia, X. J. 1959. The viability of *Chrysemys picta* submerged at various temperatures. *Physiol. Zool.* **32**:47–50.

Musacchia, X. J. and M. L. Sievers. 1956. Effects of induced cold torpor on blood of *Chrysemys picta*. *Am. J. Physiol.* **187**:99–102.

Muybridge, E. 1887. *Animal locomotion*. London: Chapman and Hall. [Selection of plates republished by L. S. Brown (Ed.). 1957. *Animals in motion*. New York: Dover.]

Nakajima, J. 1920. *Eretmochelys* of Palau. *Suisan Kenkysûshi* **15**:145–152.

Nakajima, S. 1964. Adaptation in stretch receptor neurons of crayfish. *Science.* **146**:1168.

Nakamura, Y. 1929. [Chemical embryology of reptiles.] *J. Biochem.* (Tokyo). **10**:357–360.

Nauta, W. J. H. and H. J. Karten. 1970. A general profile of the vertebrate brain, with sidelights on the ancestry of the cerebral cortex. In F. O. Schmitt (Ed.), *The neurosciences: Second study program*, pp. 7–26. New York: Rockefeller University Press.

Needham, J. 1942. *Biochemistry and morphogenesis*. London: Cambridge University Press.

Needham, J. 1963. *Chemical embryology.* New York: Hafner.

Neill, W. T. 1948a. Odor of young box turtles. *Copeia.* **1948**:130.

Neill, W. T. 1948b. Use of scent glands by prenatal *Sternotherus minor*. *Herpetologica*. **4**:148.

Neill, W. T. and E. R. Allen. 1954. Algae on turtles: Some additional considerations. *Ecology*. **35**:581–584.

Nelson, D. J. 1950. A treatment for helminthiasis in ophidia. *Herpetologica*. **6**:57–59.

Nemuras, K. 1966. Genus *Clemmys*. *Int. Turtle Tortoise Soc. J.* **1**(1):26–27, 39, 44.

Nemuras, K. T. 1967. Notes on the natural history of *Clemmys muhlenbergi*. *Bull. Md. Herpetol. Soc.* **3**:90–96.

Netting, M. G. 1936. Hibernation and migration of the spotted turtle, *Clemmys gutatta* (Schneider). *Copeia*. **1936**:112.

New, D. A. T. 1966. *The culture of vertebrate embryos*. London: Logos.

Newman, H. H. 1906a. The habits of certain tortoises. *J. Comp. Neurol*. **16**:126–152.

Newman, H. H. 1906b. The significance of scute and plate "abnormalities" in Chelonia. *Biol. Bull.* **10**:68–114.

Nichols, J. T. 1939. Range and homing of individual box turtles. *Copeia*. **1939**:125–127.

Nichols, U. G. 1953. Habits of the desert tortoise, *Gopherus agassizii*. *Herpetologica*. **9**:65–69.

Nichols, U. G. 1957. The desert tortoise in captivity. *Herpetologica*. **13**:141–144.

Nielsen, B. 1961. On the regulation of the respiration in the reptiles. I. The effect of temperature and CO_2 on the respiration of lizards (*Lacerta*). *J. Exp. Biol.* **38**:301–314.

Nikol'skii, A. M. 1915. *Fauna of Russia and adjacent countries. Reptiles*. Vol. I. Chelonia and Sauria. Translation (1963), Smithsonian Institution, Washington, D.C.

Nikolskii, A. M. 1963. *Chelonia and Sauria*. Translated by L. and E. Lochva, Jerusalem. U. S. Department of Commerce.

Noble, G. K. 1937. The sense organs involved in the courtship of *Storeria*, *Thamnophis*, and other snakes. *Bull. Am. Mus. Nat. Hist.* **73**:673–725.

Noble, G. K. and A. M. Breslau. 1938. The senses involved in the migration of young fresh water turtles after hatching. *J. Comp. Psychol.* **25**:175–193.

Noble, G. K. and Clausen, H. J. 1936. The aggregation behavior of *Storeria dekayi* and other snakes with special reference to the sense organs involved. *Ecol. Monogr.* **6**:271–316.

Nöel-Hume, I. and A. Nöel-Hume. 1954. *Tortoises, terrapins and turtles*. London: Foyle.

Norris, K. S. 1967. Color adaptation in desert reptiles and in thermal relationships. In W. W. Milstead (Ed.), *Lizard ecology: A symposium*, pp. 162–299. Columbia: University of Missouri Press.

Norris, K. S. and R. G. Zweifel. 1950. Observations on the habits of the ornate box turtle, *Terrapene ornata* (Agassiz). *Chicago Acad. Sci. Nat. Hist. Misc.* No. 58:1–4.

Northcutt, R. G. 1970. The telencephalon of the western painted turtle. *Ill. Biol. Monogr.* **43**:1–113.

O'Donoghue, C. H. 1917. A note on the ductus caroticus and ductus arteriosus and their distribution in the Reptilia. *J. Anat. London.* **51**:137–149.

O'Donoghue, C. H. 1918. The heart of the leathery turtle, with a note on the septum ventriculorum in the Reptilia. *J. Anat. London.* **52**:467–480.

Odum, E. P. 1971. *Fundamentals of ecology*, 3rd ed. Philadelphia: Saunders.

Ohr, E. A. 1976. Tricaine methanesulfonate. I. pH and its effects on anesthetic potency. *Comp. Biochem. Physiol.* **54C**:13–17.

Oliver, J. D. 1905. "Old 1844" the Hope Valley turtle. *Nat. Mag.* **47**:71–74, 108–109.

Oliver, J. A. 1955. *The natural history of North American amphibians and reptiles.* Princeton, N.J.: Van Nostrand.

Olson, E. C. 1947. The family Diadectidae. *Fieldiana, Geol.* **11**:3–53.

Organ, J. 1961. Studies of the population dynamics of the salamander genus *Desmognathus* in Virginia. *Ecol. Monogr.* **31**:189–220.

Orrego, F. 1961a. The reptilian forebrain. I. The olfactory pathways and cortical areas in the turtle. *Arch. Ital. Biol.* **99**:425–445.

Orrego, F. 1961b. The reptilian forebrain. II. Electrical activity in the olfactory bulb. *Arch. Ital. Biol.* **99**:446–465.

Orrego, F. 1962. The reptilian forebrain. III. Cross connections between the olfactory bulbs and the cortical areas in the turtle. *Arch. Ital. Biol.* **100**:1–16.

Ortleb, E. P. and O. J. Sexton. 1964. Orientation of the painted turtle, *Chrysemys picta*. *Am. Midl. Nat.* **71**:320–333.

Ottoson, D. 1956. Analysis of electrical activity of the olfactory epithelium. *Acta Physiol. Scand.* **35**(Suppl. 122):1–83.

Ottoson, D. and G. M. Shepard. 1967. Experiments and concepts in olfactory physiology. In Y. Zotterman (Ed.), *Progress in brain research*. Vol. 23, pp. 83–138. New York: Elsevier.

Overton, W. 1971. Estimating the number of animals in wildlife populations. In R. Giles, Jr. (Ed.), *Wildlife management techniques*, pp. 403–456. Washington, D.C.: Wildlife Society.

Page, L. A. 1961. Experimental ulcerative stomatitis in king snakes. *Cornell Vet.* **51**:259–266.

Page, L. A. 1966. Diseases and infections of snakes: A review. *Bull. Wildl. Dis. Assoc.* **2**:111–116.

Papez, J. W. 1935. The thalamus of turtles and thalamic evolution. *J. Comp. Neurol.* **64**:433–475.

Papez, J. W. 1936. Evolution of the medial geniculate body. *J. Comp. Neurol.* **64**:41–61.

Parcher, S. R. 1974. Observations on the natural histories of six Malagasy Chamaeloeontidae. *Z. Tierpsychol.* **34**:500–523.

Pardo, H. A. Contribución al conocimiento de la morfología, ecología, comportamiento y distribución geográfica de *Podocnemis vogli*, Testudinata (Pelomedusidae). *Rev. Acad. Colomb. Cienc. Exactas Fis. Nat.* **13**:303–326.

Paré, W. 1964. Electrolyte balance and chronic environmental stress. Paper presented at the meeting of the Eastern Psychological Association, Philadelphia.

Parent, A. 1973. Distribution of monoamine-containing nerve terminals in the brain of the painted turtle, *Chrysemys picta*. *J. Comp. Neurol.* **148**:153–166.

Parent, A. 1976. Striatal afferent connections in the turtle (*Chrysemys picta*) as revealed by retrograde axonal transport of horseradish peroxidase. *Brain Res.* **108**:25–36.

Parent, A. and D. Poitras. 1974. Morphological organization of monoamine-containing neurons in the hypothalamus of the painted turtle (*Chrysemys picta*). *J. Comp. Neurol.* **154**:379–394.

Parker, G. H. 1901. Correlated abnormalities in the scutes and bony plates of the carapace of the sculptured tortoise. *Am. Nat.* **35**:17–24.

Parker, G. H. 1922. The crawling of young loggerhead turtles toward the sea. *J. Exp. Zool.* **36**:323–331.

BIBLIOGRAPHY

Parrish, F. K. 1958. Miscellaneous observations on the behavior of captive sea turtles. *Bull. Mar. Sci. Gulf Caribb.* **8**:348–355.

Parschin, A. N. 1929. Bedingte reflexe bei Schildkroten. *Pflug. Arch. Ges. Physiol.* **222**:328–333.

Parsons, J. J. 1962. *The green turtle and man.* Gainesville: University of Florida Press.

Parsons, T. S. 1959a. Nasal anatomy and phylogeny of reptiles. *Evolution.* **13**:175–187.

Parsons, T. S. 1959b. Studies in the comparative embryology of the reptilian nose. *Bull. Mus. Comp. Zool., Harvard.* **120**:101–277.

Parsons, T. S. 1968. Variations in the choanal structure of recent turtles. *Can. J. Zool.* **46**:1235–1263.

Parsons, T. S. 1970. The nose and Jacobson's organ. In C. Gans and T. S. Parsons (Eds.), *Biology of the reptilia*, Vol. 2, pp. 99–191. New York: Academic Press.

Parsons, T. S. and E. E. Williams. 1961. Two Jurassic turtle skulls: A morphological study. *Bull. Mus. Comp. Zool., Harvard.* **125**:43–107.

Pasteels, J. J. 1957. Une table analytique du développement des reptiles. I. Stades de gastrulation chez les Chéloniens et les Lacertiliens. *Ann. Soc. Roy. Zool. Belg.* **87**:217–241.

Pasteels, J. J. 1970. Développement embryonnaire. In P.-P. Grassé (Ed.), *Traité de zoologie.* 14. *Reptiles. Glandes endocrines—Embryologie, systematique–paleontologie*, pp. 893–971. Paris: Masson.

Patten, B. M. 1958. *Foundation of embryology.* New York: McGraw-Hill.

Patterson, R. 1970. A request for information on hinge flexibility and calcium metabolism in female tortoises. *Int. Turtle Tortoise Soc. J.* **4**(5):39.

Patterson, R. 1971a. Aggregation and dispersal behavior in captive *Gopherus agassizi. J. Herpetol.* **5**:214–216.

Patterson, R. 1971b. The role of urination in egg predator defense in the desert tortoise (*Gopherus agassizi*). *Herpetologica.* **27**:197–199.

Patterson, R. and B. Brattstrom. 1972. Growth in captive *Gopherus agassizi. Herpetologica.* **18**:169–171.

Patterson, W. C. 1966. Hearing in the turtle. *J. Aud. Res.* **6**:453–464.

Patterson, W. C. and W. L. Gulick. 1966. A method for measuring auditory thresholds in the turtle. *J. Aud. Res.* **6**:219–227.

Patterson, W. C., F. C. Evering, and C. L. McNall. 1968. The relationship of temperature to the cochlear response in a poikilotherm (*Pseudemys scripta elegans*). *J. Aud. Res.* **8**:439–448.

Pauley, J. D. 1971. A pulse modulated biotelemetry system for monitoring heart rate and body temperature in free-ranging vertebrates. Ph.D. dissertation. University of Kansas, Lawrence.

Paxson, D. W. 1962. An observation of eggs in a tortoise shell. *Herpetologica.* **17**:278–279.

Pearse, A. S. 1923a. The abundance and migration of turtles. *Ecology.* **4**:24–28.

Pearse, A. S. 1923b. The growth of the painted turtle. *Biol. Bull.* **45**:145–148.

Pehrson, T. 1945. Some problems concerning the development of the skull of turtles. *Acta Zool., Stockholm.* **26**:157–184.

Peiponen, V. A. 1964. Zur Bedeutung der Ölkugeln in Farbensehen der Sauropsiden. *Ann. Zool. Fenn.* **1**:281–302.

Pell, S. M. 1941. Notes on the habits of the common snapping turtle *Chelydra serpentina* (Linné) in central New York. Master's thesis. Cornell University, Ithaca, N. Y.

Pellegrini, G. 1925. Sulle modificazione degli elementi interstiziali del testicolo negli animali ad attivita sessuale periodica. *Arc. Ital. Anal. Embriol.* **22**:550–585.

Penn, G. H. and K. E. Pottharst. 1940. The reproduction and dormancy of *Terrapene major* in New Orleans. *Herpetologica.* **2**:25–29.

Perry, S. F. 1972. The lungs of the red eared turtle, *Chrysemys* (*Pseudemys*) *scripta elegans*, as a gas exchange organ: A histological and quantitative morphological study. Ph.D. dissertation. University of Michigan, Ann Arbor.

Pert, A. and M. E. Bitterman. 1969. A technique for the study of consummatory behavior and instrumental learning in the turtle. *Am. Psychol.* **24**:258–261.

Pert, A. and M. E. Bitterman. 1970. Reward and learning in the turtle. *Learn. Motiv.* **1**:121–128.

Pert, A. and R. C. Gonzales. 1974. The behavior of the turtle (*Chrysemys picta picta*) in simultaneous, successive, and behavioral contrast situations. *J. Comp. Physiol. Psychol.* **87**:526–538.

Peters, U. 1969. Some observations on the captive breeding of the Madagascar tortoise *Testudo radiata* at Sydney Zoo. *Int. Zoo Yearb.* **9**:29.

Peters, U. 1970. Die Papua-Schildkröte (*Carettochelys insculpta*) in Australien. *Aquar. Terrar. Z.* **23**(6):182–183.

Phillips, J. G. and D. Bellamy. 1963. Adrenocortical hormones. In U.S. van Euler and H. Heller (Eds.), *Comparative endocrinology*, Vol. 1, pp. 208–257. New York: Academic Press.

Phillips, J. G., I. Chester-Jones, and D. Bellamy. 1962. Biosynthesis of adrenocortical hormones by adrenal glands of lizards and snakes. *J. Endocrinol.* **25**:233–237.

Pieau, C. 1968. Les principales étapes de la morphogénèse du cloaque chez l'embryon de tortue grecque (*Testudo graeca* L.). *C. R. Acad. Sci., Ser. D.* **266**:2452–2455.

Pieau, C. 1971. Sur la proportion sexuelle chez les embryons de deux Chéloniens (*Testudo graeca* L. et *Emys orbicularis* L.) issus d'oeufs incubés artificiellement. *C. R. Acad. Sci., Ser. D.* **272**:3071–3074.

Pieau, C. 1974a. Différenciation du sexe en fonction de la température chez les embryons d'*Emys orbicularis* L. (Chelonien): Effets des hormones sexuelles. *Ann. Embryol. Morphol.* **7**:365–394.

Plummer, M. V. 1976a. Notes on the courtship and mating behavior of the softshell turtle, *Trionyx muticus* (Reptilia, Testudines, Trionychidae). *J. Herpetol.* **11**:90–92.

Plummer, M. V. 1976b. Some aspects of nesting success in the turtle, *Trionyx muticus*. *Herpetologica.* **32**:353–359.

Plummer, M. V. and H. W. Shirer. 1975. Movement patterns in a river population of the softshell turtle, *Trionyx muticus*. *Occas. Pap., Univ. Kans., Mus. Nat. Hist.* No. 43:1–26.

Poliakoff, K. L. 1930. Zur Physiologie des Reich- und Horanalysators bei der Schildkrote, *Emys orbicularis*. *Russ. J. Physiol.* **13**:161–178.

Pope, C. H. 1935. *Natural history of Central Asia*. Vol. X. *The reptiles of China*. New York: American Museum of Natural History.

Pope, C. H. 1939. *Turtles of the United States and Canada*. New York: Knopf.

Porter, K. R. 1972. *Herpetology*. Philadelphia: Saunders.

Pough, F. H. 1970. A quick method for permanently marking snakes and turtles. *Herpetologica.* **26**:428–430.

Powell, C. 1967. Female sexual cycles of *Chrysemys picta* and *Clemmys insculpta* in Nova Scotia. *Can. Field-Nat.* **81**:134–140.

Prange, H. D. and R. A. Ackerman. 1974. Oxygen consumption and mechanisms of gas exchange of green turtle (*Chelonia mydas*) eggs and hatchlings. *Copeia.* **1974**:758–763.

Price, J. 1951. A half-century old box turtle, *Terrapene carolina carolina* (Linnaeus) from northern Ohio. *Copeia.* **1951**:312.

Pritchard, P. 1967. *Living turtles of the world.* Jersey City: TFH Publications.

Pritchard, P. C. H. 1969a. Sea turtles of the Guianas. *Bull. Fla. State Mus. Biol. Sci.* **13**:85–140.

Pritchard, P. C. H. 1969b. Studies of the systematics and reproductive cycles of the genus *Lepidochelys.* Ph.D. dissertation. University of Florida, Gainesville.

Pritchard, P. C. H. 1971a. Galapagos sea turtles—Preliminary findings. *J. Herpetol.* **5**:1–9.

Pritchard, P. C. H. 1971b. The leatherback or leathery turtle *Dermochelys coriacea. IUCN Monogr.* No. 1.

Pritchard, P. C. H. 1973. International migrations of South American sea turtles (Cheloniidae and Dermochelidae). *Anim. Behav.* **21**:18–27.

Pritchard, P. C. H. In press. *Encyclopedia of turtles.* Jersey City: TFH Publications.

Pritchard, P. C. H. and W. F. Greenwood. 1968. The sun and the turtle. *Int. Turtle Tortoise Soc. J.* **2**(1):20–25, 34.

Pritchard, P. C. H. and R. Marquez M. 1973. Kemp's ridley turtle or Atlantic ridley, *Lepidochelys kempi. IUCN Monogr.* No. 2:1–30.

Pritz, M. B. 1974a. Ascending connections of a midbrain auditory area in a crocodile, *Caiman crocodilus. J. Comp. Neurol.* **153**:179–197.

Pritz, M. B. 1974b. Ascending connections of a thalamic auditory area in a crocodile, *Caiman crocodilus. J. Comp. Neurol.* **153**:199–213.

Pritz, M. B., A. H. Bass, and R. G. Northcutt. 1973. A simple apparatus and training techniques for teaching turtles to perform a visual discrimination task. *Copeia.* **1973**:181–183.

Punzo, F. 1975. Studies on the feeding behavior, diet, nesting habits and temperature relationships of *Chelydra serpentina osceola* (Chelonia: Chelydridae). *J. Herpetol.* **9**:207–210.

Purdue, J. R. and C. C. Carpenter. 1972a. A comparative study of the body movements of displaying males of the lizard genus *Sceloporus* (Iguanidae). *Behaviour.* **41**:68–91.

Purdue, J. R. and C. C. Carpenter. 1972b. A comparative study of the display motion in the iguanid genera *Sceloporus, Uta,* and *Urosaurus. Herpetologica.* **28**:137–140.

Quaranta, J. V. 1952. An experimental study of the color vision of the tortoise. *Zoologica.* **37**:295–311.

Quaranta, J. V. and L. T. Evans. 1949a. A study of the dominance order in a herd of captive giant tortoises (Testudinidae) resident at the Bronx Zoo. *Anat. Rec.* **105**:511.

Quaranta, J. V. and L. T. Evans. 1949b. The visual learning of *Testudo vicina. Anat. Rec.* **105**:580.

Quay, W. 1967. Twenty-four hour rhythms in cerebral and brainstem contents of 5-hydroxytryptamine in a turtle, *Pseudemys scripta elegans*. *Comp. Biochem. Physiol.* **20**:217–221.

Quay, W. 1971. Relative effects of daily photoperiod and thermoperiod on timing of the 24-hour activity rhythm of the musk turtle (*Sternothaerus odoratus*). *Am. Zool.* **11**:670 (Abstr.).

Ragotskie, R. A. 1959. Mortality of loggerhead turtle eggs from excessive rainfall. *Ecology*. **40**:303–305.

Rahn, H. 1967. Gas transport from the external environment to the cells. In A. V. S. de Reuck and R. Porter (Eds.), *Development of the lung*, pp. 3–23. Boston: Little, Brown.

Rainey, D. G. 1953. Death of an ornate box turtle parasitized by dipterous larvae. *Herpetologica*. **9**:109–110.

Ramirez, E. M. V. 1956. Estudio biologico de la tortuga "Arrau" del Orinoco. *Agr. Venez.* **190**:44–63.

Randall, W. C., D. E. Stullken, and W. A. Hiestand. 1944. Respiration of reptiles as influenced by the composition of the inspired air. *Copeia.* **1944**:136–144.

Raney, E. C. and R. A. Josephson. 1954. Record of combat in the snapping turtle, *Chelydra serpentina*. *Copeia.* **1954**:228.

Raney, E. C. and E. A. Lachner. 1942. Summer food of *Chrysemys picta marginata* in Chautauqua Lake, New York. *Copeia.* **1942**:83–85.

Rapatz, G. L. and X. J. Musacchia. 1957. Metabolism of *Chrysemys picta* during fasting and cold torpor. *Am. J. Physiol.* **188**:456–460.

Ratcliffe, H. L. 1966. *Annual report Penrose Research Laboratory*. Philadelphia: Zoological Society.

Rathke, H. 1848. *Über die Entwickelung der Schildkroten.* Braunschweig: F. Veiweg.

Rawson, K. S. and P. H. Hartline. 1964. Telemetry of homing behavior by the deermouse, *Peromyscus*. *Science*. **146**:1596–1597.

Raynaud, A. and C. Pieau. 1970. Contribution a l'étude des premiers stades de la formation des organes copulateurs chez les reptiles. *Mem. Mus. Nat. Hist. Paris, Ser. A.* **58**:144–187.

Raynaud, A., C. Pieau, and J. Raynaud. 1970. Étude histologique comparative de l'allongement des canaux de Müller, de l'arrêt de leur progression en direction caudale et de leur destruction, chez les embryons males de diverses espèces de reptiles. *Ann. Embryol. Morphol.* **3**:21–47.

Razran, G. 1961. Recent Soviet phyletic comparisons of classical and of operant conditioning. *J. Comp. Physiol. Psychol.* **54**:357–365.

Razran, G. 1971. *Mind in evolution.* Boston: Houghton Mifflin.

Reagan, D. P. 1974. Habitat selection in the three-toed box turtle, *Terrapene carolina triunguis*. *Copeia.* **1974**:512–527.

Ream, C. H. 1967. Some aspects of the ecology of painted turtles, Lake Mendota, Wisconsin. Ph.D. dissertation. University of Wisconsin, Madison.

Ream, C. and R. Ream. 1966. The influence of sampling methods on the estimation of population structure in painted turtles. *Am. Midl. Nat.* **75**:325–338.

Regal, P. J. 1971. Long-term studies with operant conditioning techniques of temperature regulation patterns in reptiles. *J. Physiol.* **63**:403–406.

Reichenbach-Klinke, H. and E. Elkan. 1965. *The principal diseases of lower vertebrates.* New York: Academic Press.

Reiner, A. and Karten, H. J. 1978. A bisynaptic retinocerebellar pathway in the turtle. *Brain Res.* **150**:163–169.

Richards, P. W. *The tropical rain forest: An ecological study.* Cambridge: Cambridge University Press.

Richmond, G. 1970. The marsh awakens. *Int. Turtle Tortoise Soc. J.* **4**(1):28–29.

Ricklefs, R. E. and J. Cullen. 1973. Embryonic growth of the green iguana, *Iguana iguana*. *Copeia.* **1973**:296–305.

Richter, A. and E. J. Simon. 1974. Electrical responses of double cones in the turtle retina. *J. Physiol.* **242**:673–683.

Ricker, W. 1958. *Handbook of computations for biological statistics of fish populations.* Fisheries Research Board of Canada. Bull. 119.

Ridgway, S. H., E. G. Wever, J. G. McCormick, J. Palin, and J. H. Anderson. 1969. Hearing in the giant sea turtle, *Chelonia mydas. Proc. Nat. Acad. Sci., U.S.A.* **64**:884–890.

Riedesel, M. L. 1973. Hibernation, spring 1973. *Physiologist.* **16**:565–579.

Riedesel, M. L., J. L. Cloudsley-Thompson, and J. A. Cloudsley-Thompson. 1971. Evaporative thermoregulation in turtles. *Physiol. Zool.* **44**:28–32.

Rigley, L. 1974. Agonistic behavior of the eastern mud turtle *Kinosternon subrubrum subrubrum. Bull. Md. Herpetol. Soc.* **10**:22–23.

Riley, V. 1960. Adaptation of orbital bleeding technique to rapid serial blood studies. *Proc. Soc. Exp. Biol. Med.* **104**:751–754.

Risley, P. L. 1933a. Contributions on the development of the reproductive system in *Sternotherus odoratus* (Latreille). I. Embryonic origin and migration of the germ cells. II. Gonadogenesis and sex differentiation. *Z. Zellforsch. Mikrosk. Anat.* (Berlin). **18**:459–543.

Risley, P. L. 1933b. Observations on the natural history of the common musk turtle, *Sternotherus odoratus* (Latreille). *Pap. Mich. Acad. Sci. Arts Lett.* **17**:685–711.

Risley, P. L. 1933c. The spermatogenic and seasonal cycle of the testis of the musk turtle, *Sternotherus odoratus* (Latreille). *Anat. Rec.* **57** Suppl.:78.

Risley, P. L. 1936. Centrioles in germ cells of turtles, including observations on the "manchette" in spermiogenesis. *Z. Wiss. Zool.* **148**:133–158.

Risley, P. L. 1938. Seasonal changes in the testes of the musk turtle, *Sternotherus odoratus* L. *J. Morphol.* **63**:307–317.

Risley, P. L. 1944. Arrested development of turtle embryos. *Anat. Rec.* **88**:454–455.

Robbins, D. O. 1972. Coding of intensity and wavelength in optic tectal cells of the turtle. *Brain Behav. Evol.* **5**:124–142.

Robin, E. D. 1962. Relationship between temperature and plasma pH and carbon dioxide tension in the turtle. *Nature.* **195**:249–251.

Robin, E. D., J. W. Vester, H. V. Murdaugh, Jr., and J. E. Millen. 1964. Prolonged anaerobiosis in a vertebrate: Anaerobic metabolism in the freshwater turtle. *J. Cell. Comp. Physiol.* **63**:287–297.

Robinson, K. M. and G. G. Murphy. 1975. A new method for trapping softshell turtles. *Herpetol. Rev.* **6**:111.

Robinson, P. 1974. Albino green turtle is thriving nicely. *Christian Sci. Monitor.* April 15:F6, fig.

Robson, D. and H. Regier. 1964. Sample size in Petersen mark-recapture experiments. *Trans. Am. Fish. Soc.* **93**:215–226.

Rodbard, S., F. Samson, and D. Ferguson. 1950. Thermosensitivity of the turtle brain as manifested by blood pressure changes. *Am. J. Physiol.* **160**:402–408.

Rodeck, H. G. 1949. Notes on box turtles in Colorado. *Copeia.* **1949**:32–34.

Rogers, C. G. 1938. *Textbook of comparative physiology.* New York: McGraw-Hill.

Rogers, L. J. 1966. The nitrogen excretion of *Chelodina longicollis* under conditions of hydration and dehydration. *Comp. Biochem. Physiol.* **18**:249–260.

Rol'nik, V. V. 1970. *Bird embryology.* Jerualsem: Israel Progress in Science Translations.

Romanoff, A. L. 1967. *Biochemistry of the avian embryo.* New York: Wiley-Interscience.

Romanoff, A. L. 1972. *Pathogenesis of the avian embryo.* New York: Wiley-Interscience.

Romanoff, A. L. and A. J. Romanoff. 1949. *The avian egg.* New York: Wiley.

Romer, A. S. 1956. *Osteology of the reptiles.* Chicago: University of Chicago Press.

Rooij, N. de. 1915. *The reptiles of the Indo-Australian Archipelago.* I. *Lacertilia, Chelonia, Emydosauria.* Leiden: Brill.

Root, R. W. 1949. Aquatic respiration in the musk turtle. *Physiol. Zool.* **22**:172–178.

Rosado, R. D. 1967. La "Jicotea." *Int. Turtle Tortoise Soc. J.* **1**(3):16–19.

Rose, F. L. 1969. Dessication rates and temperature relationships of *Terrapene ornata* following scute removal. *Southwest Nat.* **14**:67–72.

Rose, F. L. 1970. Tortoise chin gland fatty acid composition: Behavioral significance. *Comp. Biochem. Physiol.* **32**:577–580.

Rose, F. L., R. B. Drotman, and W. G. Weaver. 1969. Electrophoresis of chin gland extracts of *Gopherus* (tortoises). *Comp. Biochem. Physiol.* **29**:847–851.

Rose, W. 1962. *Reptiles and amphibians of South Africa.* Capetown: Maskew Miller.

Rosenberg, M. E. 1970. Excitation and inhibition of motoneurons in the tortoise. *J. Physiol.* **221**:715–730.

Rosenberger, R. C. 1936. Notes on some habits of *Terrapene carolina* (Linné). *Copeia.* **1936**:177.

Rosenberger, R. C. 1972. Interesting facts about turtles. *Int. Turtle Tortoise Soc. J.* **6**(4):4–7.

Rosskopf, W. J. 1975. Botfly larvae infestation in a western box turtle. *Tortuga Gaz.* **11**(11):8–9.

Rothschild, F. R. S. 1915. The giant tortoises of the Galapagos Islands in the Tring Museum. *Nov. Zool.* **22**:403–417.

Rowe, J. and J. Janulaw. 1971. A noteworthy conservation achievement. *Int. Turtle Tortoise Soc. J.* **5**(4):20–30.

Rowe-Rowe, D. T. 1970. Tortoise incubation: Berg tortoise *Geochelone pardalis. Lammergeyer.* **11**:86.

Roze, J. A. 1964. Pilgrim of the river. *Nat. Hist.* **73**(7):34–41.

Rozin, P. and J. W. Kalat. 1971. Specific hungers and poison avoidance as adaptive specializations of learning. *Psychol. Rev.* **78**:459–486.

Ruckes, H. 1929. Studies in chelonian osteology. I. Truss and arch analogies in chelonian pelves. *Ann. N.Y. Acad. Sci.* **31**:31–80.

Russell, W. S., A. P. Mead, and J. S. Hayes. 1954. A basis for the quantitative study of the structure of behavior. *Behaviour.* **6**:153–205.

Sachsse, W. 1974. Zum Fortpflanzungsverhalten von *Clemmys muhlenbergii* bei weitgehender Nachahmung der natürlich Lebensbedingungen in Terrarium (Testudines, Emydidae). *Salamandra*. **10**:1-14.

Sadleir, R. 1973. *The reproduction of vertebrates.* New York: Academic Press.

Salthe, S. N. 1969. Reproductive modes and the number and sizes of ova in the urodeles. *Am. Midl. Nat.* **81**:467-490.

Sandor, T., J. Lamoureaux and A. Lanthier. 1964. Adrenocortical function in reptiles: The *in vitro* biosynthesis of adrenal cortical steroids by adrenal slices of two common North American turtles, the slider turtle (*Pseudemys scripta elegans*) and the painted turtle (*Chrysemys picta picta*). *Steroids.* **4**:213-227.

Santos, E. 1955. *Anfibios e repteis do Brazil.* Rio de Janeiro: Ed. Briquiet & Cia.

Schaeffer, B. 1941. The morphological and functional evolution of the tarsus in amphibians and reptiles. *Bull. Am. Mus. Nat. Hist.* **78**:395-472.

Schepers, G. W. H. 1939. The blood vascular system of the brain of *Testudo geometrica*. *J. Anat.* **73**:451-495.

Schiff, H. 1835. Zur Kenntnis des Schildkrotenharns. *Ann. Chem. Pharm.* **3**:368.

Schilb, T. P. and W. A. Brodsky. 1972. CO_2 gradients and acidification by transport of HCO_3 in turtle bladders. *Am. J. Physiol.* **222**:272-281.

Schladweiler, J. L. and I. J. Ball, Jr. 1968. Telemetry bibliography emphasizing studies of wild animals under natural conditions. *Bell Mus. Nat. Hist.*, Tech. Rep. 15.

Schmidt, A. A. 1970. Zur Fortpflanzung der Kreuzbrustschildkröte (*Staurotypus salvinii*) in Gefangenschaft. *Salamandra*. **6**:3-10.

Schmidt, K. P. and R. F. Inger. 1957. *Living reptiles of the world.* Garden City, N.Y.: Hanover House.

Schmidt-Nielsen, K. 1964. Organ systems in adaptation: The excretory system. In *Handbook of physiology*, pp. 215-243. Washington, D.C.: American Physiological Society.

Schmidt-Nielsen, K. and P. J. Bentley. 1966. Desert tortoise, *Gopherus agassizii*: Cutaneous water loss. *Science.* **154**:911.

Schmidt-Nielsen, K. and W. R. Dawson. 1964. Terrestrial animals in dry heat. In *Handbook of physiology*, pp. 467-480. Washington, D.C.: American Physiological Society.

Schmidt-Nielsen, K. and R. Fange. 1958. Salt glands in marine reptiles. *Nature.* **182**:783-785.

Schnabel, Z. 1938. Estimation of the total fish population of a lake. *Am. Math. Mon.* **45**:348-352.

Schneck, J. 1886. Longevity of turtles. *Am. Nat.* **20**:897.

Schoener, T. W. 1974. Resource partitioning in ecological communities. *Science.* **135**:27-39.

Schultze, M. 1866. Zur Anatomie und Physiologie der Retina. *Arch. Mikrosk. Anat. Entw. Mech.* **2**:175-286.

Schwartz, C. W. and E. R. Schwartz. 1974. The three-toed box turtle in Central Missouri: Its population, home range, and movements. Missouri Department of Conservation, Terrestrial Series (5):1-28.

Seidman, E. 1949. Relative ability of the newt and terrapin to reverse a direction habit. *J. Comp. Physiol. Psychol.* **42**:320-327.

Seligman, M. E. P. 1968. Chronic fear produced by unpredictable electric shock. *J. Comp. Physiol. Psychol.* **66**:402-411.

Selyé, H. 1946. The general adaptation syndrome and the diseases of adaptation. *J. Clin. Endocrin. Metabol.* **6**:117–230.

Selyé, H. 1950. *Stress.* Montreal: Acta.

Sendju, Y. 1929. Über die Bildung von die Mulchsäure der Bebrütung von Meerschildkröteneiern. *J. Biochem. (Tokyo).* **10**:361–363.

Senn, D. G. 1971. Structure and development of the optic tectum of the snapping turtle (*Chelydra serpentina* L.). *Acta Anat.* **80**:46–57.

Seo, H. 1926. Turtles. In *Dobutsu Kyôzai no Kompon feki Kenkyu*, pp. 201–232. Tokyo: Bunyosha.

Sergeev, A. 1937. Some materials to the problem of reptilian post-embryonic growth. *Zool. Zh.* **16**:723–735.

Sergeev, A. M. 1941. On the biology and reproduction of the steppe tortoise (*Testudo horsfieldi* Gray). *Zool. Zh.* **20**:118.

Sergeyev, A. 1939. The body temperature of reptiles. *Doklady.* **22**:49–52.

Sexton, O. J. 1958. The relationship between habitat preferences of hatchling *Chelydra serpentina* and the physical structure of the vegetation. *Ecology.* **39**:751–754.

Sexton, O. J. 1959a. A method of estimating the age of painted turtles for use in demographic studies. *Ecology.* **40**:716–718.

Sexton, O. J. 1959b. Spatial and temporal movements of a population of the painted turtle, *Chrysemys picta marginata* (Agassiz). *Ecol. Monogr.* **29**:113–140.

Sexton, O. J. 1960. Notas sobre la reproducción de una tortuga Venezulana, la *Kinosternon scorpiodes*. *Mem. Soc. Cienc. Nat.* **20**:189–197.

Sexton, O. J. 1965. The annual cycle of growth and shedding in the midland painted turtle, *Chrysemys picta marginata*. *Copeia.* **1965**:314–318.

Shade, D. M. 1972. A hatching is recorded. *Int. Turtle Tortoise Soc. J.* **6**(3):20–22.

Shah, R. V. 1962. A comparative study of the respiratory muscles in Chelonia. *Breviora.* **161**:1–16.

Shaner, R. F. 1925. The development of the digestive tract and its arteries in reptiles. *Anat. Rec.* **30**:259–276.

Shaw, C. E. 1966. Breeding the Galapagos tortoise—Success story. *Oryx.* **9**:119–126.

Shaw, C. E. 1970. The hardy (and prolific) soft shelled tortoise. *Int. Turtle Tortoise Soc. J.* **4**(1):6–9, 30–31.

Shealy, R. M. 1975. Growth and age determination in *Graptemys pulchra*. *Am. Soc. Ichthyol. Herpetol. Meet. Program.* **55**:60.

Shealy, R. M. 1976. The natural history of the Alabama map turtle, *Graptemys pulchra*, in Alabama. *Bull. Fla. State Mus., Biol. Sci.* **21**:47–111.

Shettleworth, S. J. 1972. Constraints on learning. In D. Lehrman, R. Hinde, and E. Shaw (Eds.), *Advances in the study of behavior.* Vol. 4, pp. 1–68. New York: Academic Press.

Shibuya, T. 1964. Dissociation of olfactory neural response and mucosal potential. *Science.* **143**:1338–1340.

Shibuya, T. 1969. Activities of single olfactory receptor cells. In C. Pfaffman (Ed.), *Olfaction and taste.* Vol. III, pp. 109–116. New York: Rockefeller University Press.

Shibuya, T. and S. Shibuya. 1963. Olfactory epithelium: Unitary responses in the tortoise. *Science.* **140**:495–496.

Shirer, H. W. and J. F. Downhower. 1968. Radio tracking of dispersing yellow bellied marmots. *Trans. Kans. Acad. Sci.* **71**:463–479.

BIBLIOGRAPHY 657

Sidkey, Y. A. and R. Auerback. 1968. Tissue analysis of immunological capacity of snapping turtles. *J. Exp. Zool.* **167**:187–196.

Sidowski, J. (Ed.). 1966. *Experimental methods and instrumentation in psychology.* New York: McGraw-Hill.

Simkiss, K. 1962. The source of calcium for the ossification of the embryos of the giant leathery turtle. *Comp. Biochem. Physiol.* **7**:71–79.

Singh, D. P. 1977. Annual sexual rhythm in relation to environmental factors in a tropical pond turtle, *Lissemys punctata granosa. Herpetologica.* **33**:190–194.

Skinner, B. F. 1975. The shaping of phylogenic behavior. *J. Exp. Anal. Behav.* **24**:117–120.

Skoloda, T. E. 1973. Ipsilateral visual evoked potentials in the turtle, *Pseudemys scripta elegans.* Ph.D. dissertation. University of Delaware, Newark.

Skorepa, A. C. and J. F. Ozmont. 1968. Habitat, habits, and variation of *Kinosternon subrubrum* in southern Illinois. *Trans. Ill. State Acad. Sci.* **61**:247–251.

Slater, L. E. 1963. *Bio-telemetry: The use of telemetry in animal behavior and physiology in relation to ecological problems.* New York: Pergamon Press.

Slobodkin, L. B. 1961. *Growth and regulation of animal populations.* New York: Holt, Rinehart, and Winston.

Slobodkin, L. B. 1962. Energy in animal ecology. *Adv. Ecol. Res.* **1**:69–101.

Smith, D. V. 1975. Time course of the rat chorda tympani response to linearly rising current. In D. A. Denton and J. P. Coughlan (Eds.), *Olfaction and taste*, Vol. 5, pp. 191–194. New York: Academic Press.

Smith, G. M. A. and C. Coates. 1938. Fibroepithelial growths of the skin in large marine turtles, *Chelonia mydas* (Linnaeus). *Zoologica* (N.Y.) **23**:93–98.

Smith, H. M. 1904. Notes on the breeding of the yellow-bellied terrapin. *Smithsonian Misc. Collect.* **45**:252–253.

Smith, H. M. 1956. Handbook of amphibians and reptiles of Kansas. *Univ. Kans. Mus. Nat. Hist. Misc. Publ.* **9**:1–356.

Smith, H. M. 1961. Function of the choanal rakers of the green sea turtle. *Herpetologica.* **17**:214.

Smith, H. M. and L. F. James. 1958. The taxonomic significance of cloacal bursae in turtles. *Trans. Kans. Acad. Sci.* **61**:86–96.

Smith, H. and R. B. Smith. 1973. *Synopsis of the herpetofauna of Mexico.* Vol. II. *Analysis of the literature exclusive of the Mexican axolotl.* Augusta, W. Va.: F. Lunberg.

Smith, H. M. and E. H. Taylor. 1950. An annotated checklist and key to the reptiles of Mexico exclusive of the snakes. *Bull. U.S. Nat. Mus.* **199**:1–253.

Smith, H. W. 1929. The inorganic composition of the body fluids of the Chelonia. *J. Biol. Chem.* **82**:651–661.

Smith, M. A. 1930. The Reptilia and Amphibia of the Malay Peninsula. *Bull. Raffles Mus.* **3**:1–149.

Smith, M. A. 1931. *The fauna of British India, including Ceylon and Burma. Reptilia and Amphibia.* Vol. I. *Loricata, Testudines.* London: Taylor, Francis.

Smith, P. W. 1947. The reptiles and amphibians of eastern central Illinois. *Bull. Chicago Acad. Sci.* **8**:21–40.

Snedigar, R. and E. J. Rokosky. 1950. Courtship and egg laying of captive *Testudo denticulata. Copeia.* **1950**:46–49.

Sokol, S. and W. R. A. Muntz. 1966. The spectral sensitivity of the turtle, *Chrysemys picta picta*. *Vis. Res.* **6**:285-292.

Soma, L. R. 1971. Intubation of the trachea. In L. R. Soma (Ed.), *Textbook of veterinary anesthesia*, pp. 229-246. Baltimore: Williams and Wilkins.

Southwort, F. D., Jr. and A. C. Redfield. 1926. The transport of gas by the blood of the turtle. *J. Gen. Physiol.* **9**:387-403.

Spaet, R. H. 1973. Seasonal variation in the tubular and interstitial areas of the testes in *Sternotherus odoratus* (Latreille). Master's thesis. Eastern Illinois University.

Spigel, I. M. 1963. Running speed and intermediate brightness discrimination in the fresh water turtle (*Chrysemys*). *J. Comp. Physiol. Psychol.* **56**:924-928.

Spigel, I. M. 1964a. Antecedent confinement and detour learning in the turtle. *Psychol. Rep.* **14**:915-918.

Spigel, I. M. 1964b. Effect of drugs and illumination change on activity in the turtle. *Psychol. Rec.* **14**:305-310.

Spigel, I. M. 1964c. Learning, retention and disruption of detour behavior in the turtle. *J. Comp. Physiol. Psychol.* **57**:108-112.

Spigel, I. M. 1965. Light-contingent operant behavior in the turtle. *Psychon. Sci.* **3**:133-134.

Spigel, I. M. 1966. Variability in maze-path selection by the turtle. *J. Gen. Psychol.* **75**:21-27.

Spigel, I. M. and K. R. Ellis. 1965a. Climbing suppression: Passive avoidance in the turtle. *Psychon. Sci.* **3**:215-216.

Spigel, I. M. and K. R. Ellis. 1965b. Effects of cortisone on active and passive avoidance acquisition in the turtle. *Proc. 74th Ann. Meet. Am. Psychol. Assoc.* **1965**:115-116.

Spigel, I. M. and K. R. Ellis. 1965c. An emergent albedo preference in the turtle. *Percept. Mot. Kills* **21**:113-114.

Spigel, I. M. and K. R. Ellis. 1966. Cerebral lesions and climbing suppression in the turtle. *Psychon. Sci.* **5**:211-213.

Spigel, I. M. and K. R. Ellis. 1967a. D-Amphetamine and climbing suppression in the turtle. *Psychol. Rep.* **20**:1257-1258.

Spigel, I. M. and K. R. Ellis. 1967b. Excretory electrolytes and habituation in the turtle. *Psychon. Sci.* **8**:381-382.

Spigel, I. M. and K. R. Ellis. 1968. Electrolyte balance and deficit drinking in a reptile. *J. Comp. Physiol. Psychol.* **65**:384-387.

Spigel, I. M., K. R. Ellis, and Y. E. Kaiser. 1967. Electrolyte balance and drinking in the freshwater turtle. *J. Comp. Physiol. Psychol.* **64**:313-317.

Spigel, I. M. and A. Ramsay. 1969a. Excretory electrolytes and response to stress in a reptile. *J. Comp. Physiol. Psychol.* **68**:18-21.

Spigel, I. M. and A. Ramsay. 1969b. Stress and water consumption in a reptile. *Psychon. Sci.* **15**:29-30.

Spigel, I. M. and A. Ramsay. 1970. Extinction of the excretory alkali metal response (EAMR) to stress in a reptile. *Psychon. Sci.* **19**:261-263.

Spigel, I. M., A. Ramsay, and J. Seggie. 1970. Cerebral lesions and the excretory alkali metal response (EAMR) in the reptile. *Psychon. Sci.* **18**:59-60.

Spoczynska, J. O. 1970. *Dermochelys coriacea*, a challenge. *Int. Turtle Tortoise Soc. J.* **4**(1):24-26.

Spoczynska, J. O. 1971. A green turtle study. *Int. Turtle Tortoise Soc. J.* **5**(3):12–14, 36.
Spray, D. C. 1972. Weight shifts in the intact turtle during heating and cooling. *Comp. Biochem. Physiol.* **43A**:491–494.
Spray, D. C. and M. L. May. 1972. Heating and cooling rates in four species of turtles. *Comp. Biochem. Physiol.* **41A**:507–522.
Stebbins, R. C. 1954. *Amphibians and reptiles of western North America.* New York: McGraw-Hill.
Stebbins, R. C. 1966. *A field guide to western reptiles and amphibians.* Boston: Houghton Mifflin.
Steggerda, F. R. and H. E. Essex. 1957. Circulation and blood pressure in the great vessels and heart of the turtle (*Chelydra serpentina*). *Am. J. Physiol.* **190**:320–326.
Stickel, L. F. 1950. Populations and home range relationships of the box turtle, *Terrapene c. carolina* (Linnaeus). *Ecol. Monogr.* **20**:351–378.
Stimpson, J. H. 1965. Comparative aspects of the control of glycogen utilization in vertebrate livers. *Comp. Biochem. Physiol.* **15**:187–197.
Storer, T. I. 1930. Notes on the range and life-history of the Pacific fresh-water turtle, *Clemmys marmorata.* *Univ. Calif. Publ. Zool.* **32**:429–441.
Strejckova, A. and Z. Servit. 1973. Isolated head of the turtle: A useful model in the physiology and pathophysiology of the brain. *Physiol. Bohemoslov.* **22**:37–41.
Strickberger, M. W. 1968. *Genetics.* New York: Macmillan.
Strobel, G. E. and H. Wollman. 1969. Pharmacology of anesthetic agents. *Fed. Proc.* **28**:1386–1403.
Straughan, P. 1970. *The salt water aquarium in the home.* South Brunswick, N.J.: A. S. Barnes.
Stuart, G. S. 1954. Observations on reproduction in the tortoise *Gopherus agassizi* in captivity. *Copeia.* **1954**:61–62.
Stuart, L. C. 1948. The amphibians and reptiles of Alta Verapaz, Guatemala. *Misc. Publ. Mus. Zool., Univ. Mich.* **69**:1–109.
Stuart, L. C. 1963. A check-list of the herpetofauna of Guatemala. *Misc. Publ. Mus. Zool., Univ. Mich.* **122**:1–150.
Sturbaum, B. A. and M. L. Riedesel. 1974. Temperature regulation responses of ornate box turtles, *Terrapene ornata*, to heat. *Comp. Biochem. Physiol.* **48A**:527–538.
Sturn, A. and B. H. Brattstrom. 1958. A serial abnormality in the painted turtle. *Herpetologica.* **13**:277–278.
Sud, B. N. 1956. Studies on reptilian spermatogenesis. II. Spermatogenesis of the fresh-water turtle, *Lissemys punctata* Bonnaterre, with observations on the living material under the phase-contrast microscope. *Res. Bull. Punjab Univ.* (86):31–47.
Sukhanov, V. B. 1968. *General system of symmetrical locomotion of terrestrial vertebrates and some features of movement of lower tetrapods.* Leningrad: Science.
Sullivan, B. and A. Riggs. 1967. Structure, function and evolution of turtle hemoglobins. III. Oxygenation properties. *Comp. Biochem. Physiol.* **23**:459–474.
Sutton, O. G. 1947. The problem of diffusion in the lower atmosphere. *Quart. J. Roy. Metereol. Soc. London.* **73**:257–280.
Swindells, R. J. and F. C. Brown. 1964. Ability of *Testudo elongata* Blyth to withstand excessive heat. *Brit. J. Herpetol.* **3**:166.
Swingland, I. R. 1977. Reproductive effort and life history strategy of Aldabran giant tortoise. *Nature* **269**:402–404.

Snyder, R. C. 1952. Quadrapedal and bipedal locomotion of lizards. *Copeia.* **1952**:64–70.

Takagi, S. F. and T. Yajima. 1964. Electrical responses to odours of degenerating olfactory epithelium. *Nature.* **202**:1220.

Takagi, S. F. and T. Yajima. 1965. Electrical activity and histological change in the degenerating olfactory epithelium. *J. Gen. Physiol.* **48**:559–569.

Tamm, I. 1952. Agglutination of fish and turtle erythrocytes by viruses. *Biol. Bull.* **102**:149–156.

Tardent, P. 1972. Haltung und Zucht der Sternschildkröte *Testudo elegans. Salamandra.* **8**:165–175.

Tavernor, W. D. 1971. Muscle relaxants. In L. R. Soma (Ed.), *Textbook of veterinary anesthesia*, pp. 111–120. Baltimore: Williams and Wilkins.

Taylor, E. H. 1933. Observations on the courtship of turtles. *Univ. Kans. Sci. Bull.* **21**:269–271.

Taylor, E. H. 1970. The turtles and crocodiles of Thailand and adjacent waters, with a synoptic herpetological bibliography. *Univ. Kans. Sci. Bull.* **49**:87–179.

Taylor, J. 1969. *Salmonella* in wild animals. *Symp. Zool. Soc. London.* **24**:53–73.

Taylor, S. J. 1952. Vascularity of the hypophis of lower vertebrates. The painted turtle, *Chrysemys picta marginata* Agassiz. *Can. J. Zool.* **30**:134–143.

Telford, S. R., Jr. 1967. What do we know about the parasites in reptiles? *Int. Turtle Tortoise Soc. J.* **1**(6):10–13, 37–38.

Telford, S. R., Jr. 1971. Parasitic diseases of reptiles. *J. Am. Vet. Med. Assoc.* **159**:1644–1652.

Templeton, J. R. 1970. Reptiles. In G. C. Whittow (Ed.), *The comparative physiology of thermoregulation*, pp. 167–221. New York: Academic Press.

Templeton, J. R. and W. R. Dawson. 1963. Respiration in the lizard *Crotaphytus collaris. Physiol. Zool.* **36**:104–121.

Ten Donkelaar, H. J. 1976a. Descending pathways from the brain stem to the spinal cord in some reptiles. I. Origin. *J. Comp. Neurol.* **167**:421–442.

Ten Donkelaar, H. J. 1976b. Descending pathways from the brain stem to the spinal cord in some reptiles. II. Course and site of termination. *J. Comp. Neurol.* **167**:443–464.

Tenney, S. M. and J. B. Tenney. 1970. Quantitative morphology of cold-blooded lungs: Amphibia and Reptilia. *Respir. Physiol.* **9**:197–215.

Terent'ev, P. V. 1961. *Herpetology: A manual on amphibians and reptiles.* Moscow: Gosudarstvennoe Izdatel'stvo "Vyshava Sbkola." (Translation: 1965, U.S. Department of Commerce.)

Terent'ev, P. V. 1965. *Herpetology: A manual on amphibians and reptiles.* Jerusalem: Israel Progress in Science Translations.

Thinès, G. 1968. Activity regulation in the tortoise *Testudo hermanni* Gmelin. *Psychol. Belg.* **8**:131–138.

Thompson, R. F. and W. A. Spencer. 1966. Habituation: A model phenomenon for the study of neuronal substrates of behavior. *Psychol. Rev.* **73**:16–43.

Thomson, J. S. 1932. The anatomy of the tortoise. *Sci. Proc. Royal Dublin Soc.* **20**:359–461.

Thorp, J. L. 1969. Notes on breeding the Galapagos tortoise, *Testudo elephantopus*, at Honolulu Zoo. *Int. Zoo Yearb.* **9**:30–31.

Timken, R. L. 1968. *Graptemys pseudogeographica* in the upper Missouri River of the north central United States. *J. Herpetol.* **1**:76–82.

BIBLIOGRAPHY

Tinkle, D. W. 1958a. Experiments with censusing of southern turtle populations. *Herpetologica.* **14**:172-175.

Tinkle, D. W. 1958b. The systematics and ecology of the *Sternotherus carinatus* complex (Testudinata, Chelydridae). *Tulane Stud. Zool.* **6**:1-56.

Tinkle, D. W. 1959a. Additional remarks on extra-uterine migration of ova in turtles. *Herpetologica.* **15**:161-162.

Tinkle, D. W. 1959b. The relation of the fall line to the distribution and abundance of turtles. *Copeia.* **1959**:167-170.

Tinkle, D. W. 1961. Geographic variation in reproduction, size, sex ratio and maturity of *Sternotherus odoratus* (Testudinata: Chelydridae). *Ecology.* **42**:68-76.

Tinkle, D. W. 1967. The life and demography of the side-blotched lizard, *Uta stansburiana*. *Misc. Publ. Mus. Zool., Univ. Mich.* **132**:1-182.

Tinkle, D. W. 1969. The concept of reproductive effort and its relation to the evolution of life histories of lizards. *Am. Nat.* **103**:501-513.

Tinkle, D. W. and N. F. Hadley. 1975. Lizard reproductive effort; caloric estimates and comments on its evolution. *Ecology.* **56**:427-434.

Tinkle, D. W., H. M. Wilbur, and S. G. Tilley. 1970. Evolutionary strategies in lizard reproduction. *Evolution.* **24**:55-74.

Tinklepaugh, O. L. 1932. Maze learning of a turtle. *J. Comp. Psychol.* **13**:201-206.

Tomita, M. 1929. Beitrage zur Embryochemie der Reptilien. *J. Biochem. Tokyo.* **10**:351-356.

Tomko, D. S. 1972. Autumn breeding of the desert tortoise. *Copeia.* **1972**:895.

Townsend, C. H. 1925. The Galapagos tortoises in their relation to the whaling industry. *Zoologica.* **4**(3):55-135.

Townsend, C. H. 1931. Growth and age in the giant tortoise of the Galapagos. *Zoologica.* **9**:459-466.

Travis, W. 1959. *Beyond the reefs.* London: George Allen & Unwin Ltd.

Trefethen, P. 1956. Sonic equipment for tracking individual fish. U.S. Fish and Wildlife Service, Special Scientific Report on Fisheries No. 179.

Trivers, R. L. 1972. Parental investment and sexual selection. In B. Campbell (Ed.), *Sexual selection and the descent of man, 1871-1971*, pp. 136-179. Chicago: Aldine.

Trobec, T. N. and J. G. Stanley. 1971. Ionoregulatory response of the hypophysectomized turtle *Chrysemys picta* to osmotic stresses. *Gen. Comp. Endocrinol.* **17**:479-482.

Trotter, J. 1973. Incubation made easy. *Int. Turtle Tortoise Soc. J.* **7**(1):26-31.

Tucker, D. 1963a. Olfactory, vomeronasal, and trigeminal receptor responses to odorants. In Y. Zotterman (Ed.), *Olfaction and taste*, pp. 45-69. New York: Pergamon Press.

Tucker, 1963b. Physical variables in the olfactory stimulation process. *J. Gen. Physiol.* **46**:453-489.

Tucker, D. 1971. Nonolfactory responses from the nasal cavity: Jacobson's organ and the trigeminal system. In L. M. Beidler (Ed.), *Handbook of sensory physiology*, Vol. IV, pp. 27-58. New York: Springer-Verlag.

Tucker, D. and T. Shibuya. 1965. A physiologic and pharmacologic study of olfactory receptors. *Cold Spring Harbor Symp. Quant. Biol.* **30**:207-215.

Tuckerman, F. 1892. On the terminations of the nerves in the lingual papillae of the Chelonia. *Int. Monatsschr. Anat. Physiol.* **9**:1-5.

Tuge, H. 1932. Somatic motor mechanisms in the midbrain and medulla oblongata of *Chrysemys elegans* (Wied). *J. Comp. Neurol.* **55**:185–271.

Turkowski, F. J. 1972. Grass sprout grows through embryo of yellow-bellied turtle *Chrysemys scripta. Herpetol. Rev.* **4**:165.

Turner, F. B. 1971. Estimating lizard home ranges. *Herpetol. Rev.* **3**:77.

Ulinski, P. S. 1972. Tongue movements in the common boa (*Constrictor constrictor*). *Anim. Behav.* **20**:373–382.

Underwood, G. L. 1970. The eye. In C. Gans and T. S. Parsons (Eds.), *Biology of the Reptilia*, Vol. 2, pp. 1–97. New York: Academic Press.

Urban, E. K. 1965. Quantitative study of locomotion in Teiid lizards. *Anim. Behav.* **13**:513–529.

Vandeford, A. D. 1968. The utilization of an anthelmintic for reptiles. *Am. Assoc. Zookeepers Newsl.* **1968**:9–11.

Van Denburgh, J. 1914. The gigantic land tortoises of the Galapagos Islands. *Proc. Calif. Acad. Sci.* 4th Ser. **2**:203–374.

Van Sommers, P. 1963. Air motivated behavior in the turtle. *J. Comp. Physiol. Psychol.* **56**:590–596.

Van Winkle, W., Jr., D. C. Martin, and M. J. Seketick. 1973. A home range model for animals inhabiting an ecotone. *Ecology.* **54**:205–209.

Vanzolini, P. E. 1967. Notes on the nesting behavior of *Podocnemis expansa* in the Amazon Valley (Testudines, Pelomedusidae). *Pap. Avuls. Zool. São Paulo.* **20**:191–215.

Vasse, J. and C. Pieau. 1970. Les premiers stades de la formation de l'ébauche du membre antérieur chez l'embryon de tortue mauresque (*Testudo graeca* L). *Ann. Embryol. Morphog.* **3**:399–409.

Vaz-Ferreira, P. and B. S. de Soriano. 1960. Notas de reptiles de Uruguay. *Rev. Fac. Human. Cienc. Univ. Repub. Montevideo.* **18**:133–206.

Verts, B. J. 1963. Equipment and technique for radio-tracking striped skunks. *J. Wildl. Manag.* **27**:325–339.

Vestjens, W. J. M. 1969. Nesting, egg-laying and hatching of the snake-necked tortoise at Canberra, A. C. T. *Austl. Zool.* **15**:141–149.

Vilenskii, D. G. 1960. *Soil Science.* Jerusalem: Israel Progress in Science Translations.

Villiers, A. 1958. Tortues et crocodiles de l'Afrique Noire Française. *Init. Afr.* **15**:1–354.

Vinegar, A. 1974. Evolutionary implications of temperature induced anomalies of development in snake embryos. *Herpetologica.* **30**:72–74.

Vinegar, A., V. H. Hutchison, and H. G. Dowling. 1970. Metabolism, energetics and thermoregulation during brooding of snakes of the genus *Python* (Reptilia, Boidae). *Zoologica.* **55**:19–48.

Viosca, P. 1933. The *Pseudemys troosti-elegans* complex, a case of sexual dimorphism. *Copeia.* **1933**:208–210.

Vitt, L. J. 1974. Reproductive effort and energy comparisons of adults, eggs and neonates of *Gerrhonotus coeruleus principis. J. Herpetol.* **8**:165–168.

Vivien-Roels, B. and A. Petit. 1973. Embryogénèse du complexe épiphysaire chez la tortue maureque (*Testudo graeca* L.). *Ann. Embryol. Morphog.* **6**:151–168.

Vogt, R. C. 1974. Systematics of the *Graptemys pseudogeographica* complex. *Herpetol. Rev.* **5**:79.

Voight, W. G. 1975. Heating and cooling rates and their effects upon heart rate and subcutaneous temperatures in the desert tortoise, *Gopherus agassizii*. *Comp. Biochem. Physiol.* **52A**:527-531.

Voight, W. G. and C. R. Johnson. 1976. Aestivation and thermoregulation in the Texas tortoise, *Gopherus berlandieri*. *Comp. Biochem. Physiol.* **53A**:41-44.

Voight, W. G. and C. R. Johnson. 1977. Physiological control of heat exchange rates in the Texas tortoise, *Gopherus berlandieri*. *Comp. Biochem. Physiol.* **56A**:495-498.

Volanschi, D. and Z. Servit. 1969. Epileptic focus in the forebrain of the turtle. *Exp. Neurol.* **24**:137-146.

Voronin, L. G. 1962. Some results of comparative-physiological investigations of higher nervous activity. *Psychol. Bull.* **59**:161-195.

Vose, R. N. 1964. Nesting habits of the soft-shelled turtles (*Trionyx* sp.). *J. Minn. Acad. Sci.* **31**:122-124.

Vuillemin, S. 1972. Note sur *Madakinixys domerguei* n. gen., n. sp. (Testudinidae). *Ann. Univ. Madagascar, Ser. Sci. Nat. Math.* (9):169-192.

Wahlquist, H. 1970. Sawbacks of the Gulf Coast. *Int. Turtle Tortoise Soc. J.* **4**(4):10-13, 28.

Wahlquist, H. and G. W. Folkerts. 1973. Eggs and hatchlings of Barbour's map turtle, *Graptemys barbouri* Carr and Marchand. *Herpetologica*. **29**:236-237.

Wald, G. 1939a. On the distribution of vitamin A_1 and A_2. *J. Gen. Physiol.* **22**:391-415.

Wald, G. 1939b. The porphyropsin visual system. *J. Gen. Physiol.* **22**:775-794.

Wald, G. 1941. The visual systems of euryhaline fishes. *J. Gen. Physiol.* **25**:235-245.

Wald, G. 1958. Retinal chemistry and the physiology of vision. In *Visual problems of colour.* National Physical Laboratory Symposium, No. 8, pp. 7-61. London: H. M. Stationery Office.

Wald, G. 1959. The photoreceptor process in vision. In J. Field (Ed.), *Handbook of physiology. I. Neurophysiology*, Vol. 1, pp. 671-692. Washington, D.C.: American Physiology Society.

Wald, G., P. K. Brown, and P. H. Smith. 1953. Cyanopsin, a new pigment of cone vision. *Science.* **118**:505-508.

Wald, G. and H. Zussman. 1938. Carotenoids of the chicken retina. *J. Biol. Chem.* **122**:449-460.

Walker, J. M. and R. J. Berger. 1973. A polygraphic study of the tortoise (*Testudo denticulata*). *Brain Behav. Evol.* **8**:453-467.

Walker, W. F. 1959. Closure of the nostrils in the Atlantic loggerhead and other sea turtles. *Copeia.* **1959**:257-259.

Walker, W. F., Jr. 1962. Aspects of the functional anatomy of the chelonian pectoral girdle and limb. *Am. Zool.* **2**:566 (Abstr.).

Walker, W. F., Jr. 1963. An analysis of forces developed at the feet of turtles during walking. *Am. Zool.* **3**:488 (Abstr.).

Walker, W. F., Jr. 1971a. A structural and functional analysis of walking in the turtle, *Chrysemys picta marginata*. *J. Morphol.* **134**:195-214.

Walker, W. F., Jr. 1971b. Swimming in sea turtles of the family Cheloniidae. *Copeia.* **1971**:229-233.

Walker, W. F., Jr. 1972. Body form and gait in terrestrial vertebrates. *Ohio J. Sci.* **72**:177-183.

Walker, W. F., Jr. 1973. The locomotor apparatus of testudines. In C. Gans and T. S. Parsons (Eds.), *Biology of the Reptilia*, Vol. 4, pp. 1-100. New York: Academic Press.

Walkinshaw, L. H. 1950. The sandhill crane in the Bernard W. Baker Sanctuary, Michigan. *Auk.* **67**:38–51.

Wallach, J. D. 1969. Medical care of reptiles. *J. Am. Vet. Med. Assoc.* **155**:1017–1034.

Wallach, J. D. 1970. Nutritional diseases of exotic animals. *J. Am. Vet. Med. Assoc.* **157**:583–599.

Walsh, J. V., J. C. Houk, R. L. Atluri, and E. Mugnaini. 1972. Synaptic transmission at single glomeruli in the turtle cerebellum. *Science.* **178**:881–883.

Walls, G. L. 1942. *The vertebrate eye and its adaptive radiation.* Bloomfield Hills, Mich.: Cranbrook Institute of Science.

Walls, G. L. and Judd, H. D. 1933. The intra-ocular coulour filters of vertebrates. *Brit. J. Ophthalmol.* **17**:641–675.

Warren, J. M. 1973. Learning in vertebrates. In D. A. Dewsbury and D. A. Rethlingshafer (Eds.), *Comparative psychology: A modern survey*, pp. 471–509. New York: McGraw-Hill.

Waters, J. C. 1974. The biological significance of the basking habit in the black-knobbed sawback, *Graptemys nigrinoda* Cagle. Master's thesis. Auburn University, Auburn, Ala.

Watrous, F. T. 1971. Care of turtles in captivity. I. The use of Trout Chow as food. *Bull. N. Va. Herpetol. Assoc.* **1**(1):18–19.

Watson, D. M. S. 1914. *Eunotosaurus africanus* Seeley and the ancestry of the Chelonia. *Proc. Zool. Soc. London.* **1914**:1011–1020.

Watson, G. E. 1962. Notes on the copulation and distribution of Aegean land tortoises. *Mus., Biol. Sci.* **15**:1–43.

Weathers, W. W. and F. N. White. 1971. Physiological thermoregulation in turtles. *Am. J. Physiol.* **221**:704–710.

Weary, C. C. 1969. An improved method of marking snakes. *Copeia.* **1969**:854–855.

Weaver, W. G. 1970. Courtship and combat behavior in *Gopherus berlandieri*. *Bull. Fla. State Mus., Biol. Sci.* **15**:1–43.

Weaver, W. G. and F. L. Rose. 1967. Systematics, fossil history, and evolution of the genus *Chrysemys*. *Tulane Stud. Zool.* **14**:63–73.

Webb, G. J. W. and C. R. Johnson. 1972. Head-body temperature differences in turtles. *Comp. Biochem. Physiol.* **43A**:593–611.

Webb, G. J. W. and G. J. Witten. 1973. Critical thermal maxima of turtles, validity of body temperature. *Comp. Biochem. Physiol.* **45A**:829–832.

Webb, R. G. 1961. Observations on the life histories of turtles (genus *Pseudemys* and *Graptemys*) in Lake Texoma, Oklahoma. *Am. Midl. Nat.* **65**:193–214.

Webb, R. G. 1962. North American Recent soft-shelled turtles (family Trionychidae). *Publ. Mus. Nat. Hist. Univ. Kans.* **13**:429–611.

Webb, R. G., W. L. Minckley and J. E. Craddock. 1963. Remarks on the Coahuilan box turtle, *Terrapene coahuila* (Testudines, Emydidae). *Southwest Nat.* **8**(2):89–99.

Werber, M. 1970. *A bibliography of wildlife movements and tracking systems.* Washington, D.C.: George Washington Medical Center, Biological Science, Communications Project.

Wermuth, H. 1971. Eine totalabinotische Landschildkroete *(Testudo hermanni)*. *Aquar. Terrar. Z.* **24**:276.

Wermuth, H. and R. Mertens. 1961. *Schildkröten, Krokodile, Brückeneschen.* Jena: G. Fischer.

West, E. S., W. R. Todd, H. S. Mason, and J. T. Van Bruggen. 1966. *Textbook of biochemistry.* New York: Macmillan.

Wever, E. G. and J. Vernon. 1956a. Auditory responses in the common box turtle. *Proc. Nat. Acad. Sci. U.S.A.* **42**:962-965.

Wever, E. G. and J. Vernon. 1956b. Sound transmission in the turtle's ear. *Proc. Nat. Acad. Sci. U.S.A.* **42**:292-299.

White, F. N. 1959. Circulation in the reptilian heart (Squamata). *Anat. Rec.* **135**:129-134.

White, F. N. 1968. Functional anatomy of the heart of reptiles. *Am. Zool.* **8**:211-219.

White, F. N. 1973. Temperature and the Galapagos marine iguana—Insights into reptilian thermoregulation. *Comp. Biochem. Physiol.* **45A**:503-513.

White, F. N. and G. Ross. 1965. Blood flow in turtles. *Nature.* **208**:759-760.

White, F. N. and G. Ross. 1966. Circulatory changes during experimental diving in the turtle. *Am. J. Physiol.* **211**:15-18.

White, J. B. and G. G. Murphy. 1973. The reproductive cycle and sexual dimorphism of the common snapping turtle, *Chelydra serpentina serpentina. Herpetologica.* **29**:240-246.

Whitford, W. G. and V. H. Hutchison. 1963. Cutaneous and pulmonary gas exchange in the spotted salamander, *Ambystoma maculatum. Biol. Bull.* **124**:344-354.

Whitham, R. 1970. Breeding of a pair of pen-reared green turtles. *Quart. J. Fla. Acad. Sci.* **33**:288-290.

Whitson, M. A. 1971. Field and laboratory investigations of the ethology of courtship and copulation in the greater roadrunner *(Geococcyx californianus—*Aves, Cuculidae). Ph.D. dissertation. University of Oklahoma, Norman.

Whittow, G. C. 1973. Evolution of thermoregulation. In G. C. Whittow (Ed.), *Comparative physiology of thermoregulation*, Vol. III. Special aspects of thermoregulation, pp. 201-258, New York: Academic Press.

Wickham, M. M. 1922. Notes on the migration of *Macrochelys lacertina. Proc. Okla. Acad. Sci.* **2**:20-22.

Wilbur, H. M. 1975a. The evolutionary and mathematical demography of the turtle *Chrysemys picta. Ecology.* **56**:64-77.

Wilbur, H. M. 1975b. A growth model for the turtle *Chrysemys picta. Copeia.* **1975**:337-343.

Wilbur, H., D. Tinkle and J. Collins. 1974. Environmental certainty, trophic level and resource availability in life history evolution. *Am. Nat.* **108**:805-817.

Wilde, W. S. 1938. The role of Jacobson's organ in the feeding reaction of the common garter snake, *Thamnophis sirtalis sirtalis* (Linn.). *J. Exp. Zool.* **77**:445-464.

Wiley, F. H. and H. B. Lewis. 1927. The distribution of nitrogen in the blood and urine of the turtle. *Chrysemys picta. Am. J. Physiol.* **81**:692-695.

Will, L. 1893. Beiträge zur Entwicklungsgeschichte der Reptilien. 2. Die Anlage der Keimblätter bei der menorquinischen Sumpfschildkröte *(Cistudo lutaria* Gesn.). *Zool. Jahrb., Abt. Anat.* **6**:529-615.

Willemsens, N. M. C. 1974. Een Schildafwijking bij *Pseudemys scripta elegans. Lacerta.* **32**:165.

Williams, E. C., Jr. 1961. A study of the box turtle, *Terrapene carolina carolina* (L.), population in Allee Memorial Woods. *Proc. Indiana Acad. Sci.* **71**:399-406.

Williams, E. E. 1950. *Testudo cubensis* and the evolution of Western Hemisphere tortoises. *Bull. Am. Mus. Nat. Hist.* **95**:1-36.

Williams, E. E. 1960. Two species of tortoises in northern South America. *Brevoria.* No. **120**:1-13.

Williams, E. E. 1970. (Introduction to reprint edition of J. Sowerby and E. Lear: *Tortoises, terrapins and turtles.*) SSAR Facsimile Reprints in Herpetology, No. 28:iii–vi.

Williams, E. E. and S. B. McDowell. 1952. The plastron of soft-shelled turtles (Testudinata Trionychidae): A new interpretation. *J. Morphol.* **90**:263–280.

Williams, G. C. 1966. *Adaptation and natural selection*. Princeton, N.J.: Princeton University Press.

Williams, J. E. 1952. Homing behavior of the painted turtle and musk turtle in a lake. *Copeia.* **1952**:76–82.

Wilson, J. G. 1965. Embryological considerations in teratology. In J. G. Wilson and J. Warkany (Eds.), *Teratology: Principles and techniques,* pp. 251–261. Chicago: University of Chicago Press.

Wilson, V. J. 1968. The leopard tortoise, *Testudo pardalis babcocki*, in eastern Zambia. *Arnoldia* **40**(3):1–11.

Wise, L. M. and D. P. Gallagher. 1964. Partial reinforcement of a discriminative response in the turtle. *J. Comp. Physiol. Psychol.* **57**:311–313.

Witschi, E. 1972. Time variation in developmental stages: Mammals and birds. In P. L. Altman and D. S. Diltmer (Eds.), *Biology data book* 1, p. 173. Bethesda, Md.: Federation of American Societies of Experimental Biology.

Wojtusiak, R. J. 1933. Über den Farbensinn der Schildkröten. *Z. Vergl. Physiol.* **18**:393–436.

Wolach, A. H. and K. Latta. 1974. Reward magnitude shifts in turtles (*Pseudemys scripta elegans*). *Psychol. Rec.* **24**:237–241.

Wolf, S. 1933. Zur Kenntnis von Bau und Funktion der Reptilien lunge. *Zool. Jahrb. Abt. Anat. Ont.* **57**:139–190.

Wood, F. G., Jr. 1953. Mating behavior of captive loggerhead turtles, *Caretta caretta caretta. Copeia.* **1953**:184–186.

Woodbury, A. M. 1948. Marking reptiles with an electric tatooing outfit. *Copeia.* **1948**:127–128.

Woodbury, A. M. 1956. Uses of marking animals in ecological studies: Marking amphibians and reptiles. *Ecology.* **37**:670–674.

Woodbury, A. M. and R. Hardy. 1948. Studies of the desert tortoise, *Gopherus agassizii. Ecol. Monogr.* **18**:145–200.

Worrell, E. R. 1963. *Reptiles of Australia*. Sydney: Angus and Robertson.

Wright, A. H. 1918. Notes on *Clemmys. Proc. Biol. Soc., Wash.* **31**:51–58.

Wright, R. H. 1964. *The science of smell.* New York: Basic Books.

Wyssbrod, H. R. 1969. Slow cis-stimulation of sodium transport across isolated urinary bladders of the fresh-water turtle, *Pseudemys scripta. Biochem. Biophys. Acta.* **183**:577–590.

Yahner, R. 1974. Weight change, survival rate and home range change in the box turtle, *Terrapene carolina. Copeia.* **1974**:546–548.

Yerkes, R. M. 1901. Formation of habits in the turtle. *Pop. Sci. Mon.* **58**:519–525.

Yntema, C. L. 1960. Effects of various temperatures on the embryonic development of *Chelydra serpentina. Anat. Rec.* **136**:305–306.

Yntema, C. L. 1964. Procurement and use of turtle embryos for experimental procedures. *Anat. Rec.* **149**:577–586.

Yntema, C. L. 1968. A series of stages in the embryonic development of *Chelydra serpentina*. *J. Morphol.* **125**:219-252.

Yntema, C. L. 1970a. Extirpation experiments on embryonic rudiments of the carapace of *Chelydra serpentina*. *J. Morphol.* **132**:235-243.

Yntema, C. L. 1970b. Observations on females and eggs of the common snapping turtle, *Chelydra serpentina*. *Am. Midl. Nat.* **84**:69-76.

Yntema, C. L. 1970c. Survival of exogenic grafts of embryonic pigment and carapace rudiments in embryos of *Chelydra serpentina*. *J. Morphol.* **132**:353-359.

Yntema, C. L. 1974a. Incidence and progress of rejection of embryonic limb bud transplants in the turtle *Chelydra serpentina*. *J. Morphol.* **144**:453-460.

Yntema, C. L. 1974b. Survival of skin allografts following embryonic limb bud transplants in the turtle, *Chelydra serpentina*. *J. Morphol.* **144**:461-468.

Yntema, C. L. and M. Borysenko. 1971. Survival of embryonic limb bud transplants in snapping turtles. *Experientia.* **27**:567-569.

Young, J. D. 1950. The structure and some physical properties of the testudian eggshell. *Proc. Zool. Soc. London.* **120**:455-469.

Young, R. and H. M. Kaplan. 1960. Anesthesia of turtles with chlorpromazine and sodium pentobarbital. *Proc. Anim. Care Panel.* **10**:57-62.

Zagorulko, T. M. 1968. Effect of intensity and wavelength of photic stimulus on evoked responses of general cortex and optic tectum in turtles. *Fiziol. Zh. S.S.S.R.* **54**:436-446.

Zangerl, R. 1948. The vertebrate fauna of the Selma Formation of Alabama. *Fieldiana: Geol. Mem.* **3**:1-56.

Zangerl, R. 1953. Vertebrate fauna of the Selma Formation of Alabama. Part IV. The turtles of the family Toxochelyidae. *Fieldiana: Geol. Mem.* **3**:135-277.

Zangerl, R. 1959. Rudimentäre Carapaxbeschuppung bei jungen Exemplaren von Carettochelys und ihr morphogenetische Bedeutung. *Gestschr. Steiner. Vierteljahrschr. Naturforsch. Ges. Zürich.* **104**:138-147.

Zangerl, R. 1969. The turtle shell. In C. Gans, A. d'A. Bellairs, and T. S. Parsons (Eds.), *Biology of the Reptilia*, Vol. 1. Morphology A, pp. 311-339. New York: Academic Press.

Zangerl, R. and R. G. Johnson. 1957. The nature of shield abnormalities in the turtle shell. *Fieldiana: Geol. Mem.* **10**:341-362.

Zovickian, W. H. 1971. Captive nesting of bog turtles. *Int. Turtle Tortoise Soc. J.* **5**(4):14-15, 37.

Zovickian, W. H. 1973. [Cover photo]. *Int. Turtle Tortoise Soc. J.* **7**(2):cover, 2.

Zovickian, W. H. 1973. Captive reproduction of the radiated tortoise. *HISS News-J.* **1**(4):115-118.

Zug, G. R. 1966. The penial morphology and the relationships of cryptodiran turtles. *Occas. Pap. Mus. Zool. Univ. Mich.* (647):1-24.

Zug, G. R. 1971. Buoyancy, locomotion, morphology of the pelvic girdle and hindlimb and systematics of cryptodiran turtles. *Misc. Publ. Mus. Zool. Univ. Mich.* (142):1-98.

Zug, G. R. 1972a. A critique of the walk pattern analysis of symmetrical quadrupedal gaits. *Anim. Behav.* **20**:436-438.

Zug, G. A. 1972b. Walk pattern analysis of cryptodiran turtle gaits. *Anim. Behav.* **20**:439-443.

Zwick, H. and A. M. Granda. 1968. Behaviorally determined dark adaptation functions in the turtle, *Pseudemys. Psychon. Sci.* **11**:239-240.

Index

Accessory bladders, 162, 241
Accessory olfactory bulb, 268, 271, 272, 274, 275
Acclimation, 113, 115, 116, 218, 220. *See also* Acclimatization
Acclimatization, 110, 218, 220. *See also* Acclimation
Accommodation, 247, 249
Acherontemys, 8
Acinixys, 13, 39
Acinixys planicauda, 13
ACTH, 497, 498, 502, 503
Actinomyces, 399
Action potentials, 279, 280
Activity, adaptive significance of, 516-518
 controlling factors in, 510, 516
 crepuscular, 509, 510
 daily, 524, 525, 544, 546, 587-589
 and hormones, 479
 individual differences in, 486, 494
 in life history studies, 74, 91
 nocturnal, 509
 open field, 500
 and resource partitioning, 601
 seasonal, 510, 542, 544, 589, 590
 and social interaction, 477
 and tranquilizers, 496
Adaptation, sensory, 297, 298
Adocus, 11
Adrenal glands, 148, 149, 161, 498, 501, 504
Adrenal steroid hormones, 240
Aeromonas, 118
Age, 313, 320, 321. *See also* Longevity
Age determination, 75-79

Aggregations, 434, 475-478, 481, 483, 487, 548, 563, 564, 587, 591
Aggression, 235, 574, 576. *See also* Agonistic behavior
Agomphus, 11
Agonistic behavior, 475, 478-481, 483, 485-489, 491, 492. *See also* Aggression
Agrionemys, 17
Air, as positive reinforcer, 458, 466, 467
Alarm reaction, 494
Aldabrachelys, 15, 16
Aldosterone, 502, 503
Algae, 120, 218
Alimentary canal, 375
Alligators, 200, 420, 429, 434
Amblyomma, 121
Amphetamine, 496, 500
Amphibolurus barbatus, 386
Amphibolurus muricatus, 107
Amphichelydia, 2, 33, 41, 42
Anaerobic metabolism, 183-185
Analgesia, 136, 137, 151, 152
Anapsida, 1
Ancyclostoma, 123
Anesthesia, 127-136, 143, 148, 149, 151, 152, 496, 498
Annamemys, 23, 39
Annamemys annamensis, 23
Anosmia, 295-297
Anosteirinae, 30, 40
Anteromedial lesion, 500
Aperotemporalidae, 2, 41, 42
Apnea, 158, 171, 172, 179, 182, 183, 188, 189
Apparatus, availability of, 301

669

avoidance, 468
ballistocardiograph, 224
discrimination box, 461, 462, 464
filter spectacles, 431
guillotine, 143
head holder, 144
heat sensitive models, 214
jaw screw implant, 147, 148
mechanical respirator, 140-142
neutral density filter, 432
olfactory discrimination, 294, 295, 460
operant lever (bar, or key) press, 458-461, 464, 466, 467, 470
orientation arena, 432, 518, 519
for recording locomotion, climbing, 496
shuttle box, 494, 497
thermocouples, 218
Approach behavior, 480, 481, 483
Aquatic breathing, 184-186, 190
Aquatic species, 237, 239, 243.
　　See also Freshwater species; Marine turtles
Archelon ischyros, 34
Arizona, 119
Arousal, 500
Arribada, 32, 525, 529
Arteries, 158-162
Ascaris, 123
Aspergillus, 117, 399
Asterochelys, 15
Athecae, 2, 31
Atresia, 317
Audition, 431, 463, 464, 483
Avoidance, 263, 467, 468, 473, 497, 498.
　　See also Instrumental conditioning

Bacillus, 399
Bacterium, 399
Baenidae, 42
Baeninae, 41
Bartonella, 118
Basilemys, 11
Basking, 112, 212, 214-218, 475, 477, 525, 546, 587-589, 591
Batagur, 23, 39, 318
Batagur baska, 230, 316-318, 324, 326, 349, 366
Batagurinae, 3, 10, 18, 339
Batrachemys, 38
Behavior, filming of, 97-108.
　　See also Activity; Audition, Homing; Learning; Nesting; Orientation; Rhythms; Social behavior; Thermoregulation; Vision
Big-headed turtles, see *Platysternon*
Biotelemetry, 61-72, 213, 299, 433, 601
Bladder, 162, 169, 226, 237, 241-243, 502
Blood, 121, 149, 150, 170-174, 223, 240, 242, 249
Body size, 329. See also Carapace length
Bog turtle, see *Clemmys muhlenbergi*
Bowman's glands, 268, 270
Box turtles, see *Terrapene* species
Brain, 193-203
Brainstem, 193, 200, 202, 203
Breeding, 487
　　programs, 334
　　season, 480, 485, 542
　　see also Mating; Sexual behavior
Broilia, 25, 26
Bronchus, 160
Brumation, 222, 223, 308, 313, 475, 482.
　　See also hibernation
Buccal pumping, 186. See also Throat pumping
Buccopharyngeal respiration, 186, 293
Buoyancy control, 165, 167-170
Burrowing, 216, 217, 541, 542, 547

Calcarichelys, 33, 34
Callagur, 23, 39
Callagur borneoensis, 24, 339
Callopsis, see *Rhinoclemys*
Capture-release-recapture techniques, 88, 89, 91, 92
Carapace, blood supply of, 162
　　bones, labelled, 17
　　color, shape, and heat relationship, 214, 215
　　deformity, 412
　　and conductance, 224
　　embryonic development, 375
　　flexibility, 349
　　injury to, 125
　　length, of adults, 344
　　　　and body weight, 587
　　　　of hatchlings, 344-348
　　　　and egg size, 324, 325, 342
　　　　record size, 306
　　　　of terrestrial species, 556, 557
　　marginocostal artery, 160
　　and mating, 488

INDEX

measurement, 92
scutes, labelled, 17
sensory nerves, 227
surgery on, 148, 149
veins of, 163
Carcharinus spallanzani, 539
Cardiovascular system, 155-164
Caretta, 2, 39, 419
Caretta caretta, activity, 496
  buoyancy control, 168
  clutch size, 315, 325
  clutches per year, 318
  color preference, 465
  deformity, 402, 403
  diet, 523
  eggs, 353, 379, 380, 381, 383, 386
  energetics, 380
  female reproductive cycles, 312
  hatchling weight, 386
  incubation, 368, 372
  internesting intervals, 316, 372
  lacrimal salt excretion, 240
  migration, 32
  nesting, 32, 400, 420, 425, 525, 527
  olfaction, 293
  orientation, 430, 432
  pigmentation, 409
  predation, 543
  range, 32
  reproductive effort, 534
  sexual behavior, 485
  taxonomy, 30, 523
  visual discrimination, 462
Carettini, 31
Carettochelyidae, 2, 29, 39, 340, 352
Carettochelyinae, 40
Carettochelyoidea, 40
Carettochelys, 6, 27, 39, 444
Carettochelys insculpta, 29, 30, 324, 340, 352, 354, 450, 452
Carolinochelys, 32
Carteremys, 36
Caruncle, 378
Catecholamines, 197, 199
Census, 87, 88
Central nervous system, 193-203
Cerebellum, 194, 201, 202
Cerebrum, 195-201, 499, 500, 504
Cestodes, 124
Chelidae, 1, 36-42, 211, 319, 338, 344, 350, 356, 364, 378, 384, 406

Chelodina, 1, 39, 229, 390
Chelodina expansa, clutch size, 324
  description, 38
  eggs, 344, 350, 384, 398
  hatchling, 384
  incubation, 356, 364, 393, 425
  nesting, 394, 420, 425
Chelodina longicollis, body temperature, 211, 213, 221, 222
  clutch size, 324
  description, 38
  eggs, 338, 344, 348, 399
  incubation, 356, 364
  nesting, 394, 420, 421, 425
  nitrogen excretion, 239
Chelodina novaeguineae, 38, 344
Chelodina oblonga, 38, 344, 356, 364
Chelodina parkeri, 38
Chelodina rugosa, 38
Chelodina siebenrocki, 38
Chelodina steindachneri, 38
Chelonia, 2, 32, 39, 328, 419
Chelonia depressa, clutch size, 325
  clutches per year, 319
  coloration, 529
  diet, 523
  distribution, 32, 523
  egg, 386
  hatchling, 386
  migration, 32
  nesting, 32, 316, 419, 420, 525
  predation, 536
  taxonomy, 30, 523
Chelonia mydas, activity, 509, 510
  agonistic behavior, 488, 491, 507
  audition, 431
  basking, 214, 525
  body size, 323
  body temperature, 210, 525
  characteristics, 31
  choanal rakers, 230
  clutch size, 315, 322, 323, 325, 533, 534
  clutches per year, 319, 322
  collecting, 48
  color preference, 465
  color vision, 253-256, 261, 262
  cones, 254
  conservation of, 537
  deformity, 406, 407
  density spectra, 255
  diencephalon, 193

eggs, 341, 353, 386, 395, 396, 400
feeding, 233
female rejection of male, 488, 489
growth, 233, 530
hatchling, 386, 429, 516, 529
incubation, 363, 367, 368, 370, 372, 392
jaws, 230
lacrimal salt excretion, 240
locomotion, 436, 453
management, 537-540
marking, 59
mating, 485, 488, 507, 524, 525, 527
maximum sustainable yield, 538
migration, 526
mounting, 480, 481
nesting, 316, 322, 372, 391, 395, 396, 418, 419, 421, 423, 425, 426, 510, 525, 526
nitrogen excretion, 239
odor production, 290
oil droplets, 256-258
olfaction, 293-298, 431, 459, 464
operant behavior, 295, 299, 300, 459
orientation, 431-433, 518
photopigments, 253-256
pigmentation, 409
population density, 532, 533
predation on, 429, 430, 536, 539
radiotracking, 69
range, 31
reproductive cycle, 321
rods, 254
"sand smelling", 297
secondary sex characteristics, 83
sex ratio, 487
sexual maturity, 307, 531, 532
spectral sensitivity, 261, 262, 432
survival rate, 86
taxonomy, 30, 523
dosage, route for tubocurarine, 140
visual detectability, 464
winter dormancy, 526
Cheloniidae, 1, 31, 39, 40, 210, 253, 268, 318, 341, 353, 363, 367, 372, 381, 386, 406, 485, 491, 523
Chelonini, 31
Chelonioidea, 40
Chelonoidis, 15
Chelosphargis, 33
Chelospharginae, 33, 34, 40
Chelus, 27, 37, 39, 231
Chelus fimbriatus, 37, 172-179, 182, 188, 324, 338, 356
Chelycarapookidae, 42

Chelydra, 39, 47, 444
Chelydra serpentina, activity, 510, 588, 590
agonistic behavior, 491, 506, 574, 576
arterial ducts, 158
body size, 323, 345
body temperature, 209, 224
clutch size, 323, 324
clutches per year, 319
color preference, 465
as corpse detector, 289
"defensive posture", 489
deformity, 403, 407, 408, 412
description, 6, 7
dominance, 491
eggs, 335, 336, 339, 345, 348, 351, 381, 383, 396-398, 412
embryonic development, 370, 373, 376, 377, 391
energetics, 380
feeding, 231
female reproductive cycle, 312-314, 317
food competition, 486
food preference tests, 236
fossil history, 7
growth, 93, 585
habitat, 592
hatchling, 236, 384, 428
heat exchange ratio, 225
home range, 596
housing incompatability, 110
immunology, 388, 389
incubation, 335, 358, 364, 365, 369, 370, 392, 393
liver glycogen, 236
locomotion, 444
male reproductive cycles, 307, 308, 310
in mixed species populations, 513
nesting, 322, 336, 390, 398, 419, 421, 425, 588
niche, 593
odor production, 290
pigmentation, 409
population density, 586
radiotracking, 71
range, 6
respiration, 166, 167, 172, 173, 182, 191
sex ratio, 581
sexual behavior, 484
sexual dimorphism, 82, 574
sexual maturity, 306, 578
"social imprinting", 482
sperm storage, 320
trapping, 50, 51

INDEX

Chelydridae, 3, 6-8, 10, 39, 40, 209, 308, 319, 335, 339, 345, 351, 358, 364, 376, 378, 381, 384, 406, 436, 484, 491
Chelydrops, 8
Chelydropsis, 8
Chelyopsis, 32
Chelys fimbriata, see Chelus fimbriatus
Chemoreception, 234, 235, see also olfaction, taste
Chersina, 12, 39
Chersine angulata, 238, 324, 366, 484, 491, 556
Chicken turtle, see Deirochelys reticularia
Chinemys, 39
Chinemys kwangtungensis, 25, 324, 346
Chinemys megalocephala, 25
Chinemys reevesi, 25, 318, 324, 462
Chitra, 28, 39
Choana, 230, 268, 277
Chondrodendron, 137
Chrysemys, 4, 39
Chrysemys concinna, see Pseudemys concinna
Chrysemys floridana, see Pseudemys floridana
Chrysemys nelsoni, see Pseudemys nelsoni
Chrysemys picta, activity, 216, 510, 511, 517, 589
   age composition in populations, 575
   agonistic behavior, 235, 491, 506
   avoidance behavior, 468, 497, 507
   basking, 588, 589
   blood volume, 224
   body temperature, 209, 224
   brain, 193, 198, 200-202, 504
   breathing cycles, 171, 172
   brumation, 223
   climbing, 504
   clutch size, 315, 317, 324, 325
   clutches per year, 318, 323, 326
   color preference, 465
   corticosterone, 503
   courtship, 484, 492
   craniotomy, 131
   critical thermal maximum, 220, 517
   dark adaptation, 262
   deformity, 401, 402, 403, 405, 407, 408
   detour learning, 494-497
   ecdysis, 76
   eggs, 336, 337, 339, 346, 351, 354, 381, 383, 385, 394, 395, 397-399, 412
   embryology, 373, 376-378, 391, 400
   energetics, 380
   escape behavior, 494, 497
   excretory alkali metal level, 503-506
   extinction of runway response, 471
   feeding, 116, 230, 231, 233-235, 588
   female reproductive cycle, 312-315, 321
   forebrain, 196
   growth, 77-79, 93, 583-585
   habitat, 591
   habituation, 494, 497
   hatchling, 385
   heat exchange ratio, 225
   home range, 595, 598
   homing, 60, 598
   hypophysectomy, 130
   immunology, 388, 389
   incubation, 355, 360, 365, 366, 372, 392-394
   locomotion, 436, 437, 441, 444
   male reproductive cycle, 307, 308, 309-311
   maze learning, 457
   metobolic rate, 184
   migration, 588, 589
   in mixed species populations, 572, 573
   name dispute Pseudemys, 19, 20
   dosage, response, route for Nembutal, 130-132
   nesting, 316, 372, 394, 398, 419, 421, 422, 425, 472
   niche, 593
   olfaction, 464
   operant behavior, 459, 460, 496, 497
   orientation, 518
   pigmentation, 409
   population density, 586, 587
   population size, 89, 90
   position preference, 507
   probability learning, 471
   reproductive anatomy 80, 81, 83
   respiration, 179
   reversal learning, 470
   screening procedure, 493, 494
   SCUD infection, 118
   secondary sex characteristics, 82
   sex ratio, 580
   sexual behavior, 484, 589
   sexual maturity, 306, 577
   size, 346
   skeletal anatomy, 442
   social "imprinting", 482
   sodium reabsorption, 241
   spectral sensitivity, 262, 263
   stress, 503, 504, 506
   survival rate, 86, 87
   thermal relations, 214, 216, 219-221

thermal receptors, 227
threat behavior, 507
trailing, 60
trapping, 51
urogenital system, 80, 81, 83
visual discrimination, 461, 462, 464
water regulation, 505, 506
Chrysemys rubriventris, see *Pseudemys rubriventris*
Chrysemys scripta, see *Pseudemys scripta*
Cimochelys benstedi, 34
Circus movements, 432
Cistudo carolina, see *Terrapene carolina*
Citrobacter, 118-120
Classical conditioning, 456
Claudius, 39
Claudius angustatus, 9, 10, 345, 354, 357
Clemmydopsis, 26
Clemmys, 21, 22, 39, 457, 573
Clemmys caspica, see *Mauremys caspica*
Clemmys guttata, activity, 511, 517, 589, 590
 age composition in population, 575
 agonistic behavior, 491
 basking, 588, 590
 body temperatures, 209
 brumation, 223
 clutch size, 324
 courtship, 590
 description, 22
 eggs, 351, 358, 365, 384
 escape behavior, 495
 feeding, 588, 590
 growth, 93, 585
 habitat, 591
 hatchling, 384
 home range, 594-596
 homing, 598
 maze learning, 457, 495
 nesting, 349, 419, 425, 426, 590
 niche, 593
 population density, 586
 thermal relations, 218, 219
 sex ratio, 487, 580
 sexual behavior, 484
 sexual maturity, 306, 578
Clemmys insculpta, agonistic behavior, 491
 body color, 480
 body temperature, 209
 clutch size, 324
 clutches per year, 318
 collecting, 47

deformity, 401, 402
dominance, 486, 487, 491
eggs, 336, 346, 373, 381, 385
embryos, 378
escape, 480
female reproduction, 321
feeding, 233, 479
"flipping", 490
hatchling, 385
homing, 433, 434, 552
inactivity, 546
incubation, 358, 359, 365, 393
locomotion, 436, 444, 449
longevity, 556
mating, 480, 484
mirror image response, 481
nesting, 336
niche, 593
open-mouth response, 480
penis extension, 490
range, 21
secondary sex characteristics, 82
sex ratio, 559
size, 556
"subtle" social responses, 487, 488
as terrestrial emydid, 541
water pumping, 483
Clemmys japonica, see *Mauremys japonica*
Clemmys leprosa, see *Mauremys leprosa*
Clemmys marmorata, activity, 590
 age composition, 575
 agonistic behavior, 477, 491
 basking, 214, 477, 589
 biomass, 587
 body temperature, 71, 209
 eggs, 324, 346
 escape, 480
 foraging, 589
 habitat, 591
 home range, 595, 596
 movement, 599
 nesting, 477, 590
 open-mouth response, 480
 population density, 586, 587
 radio tracking, 70, 71
 range, 21
 sex ratio, 580
 sexual maturity, 578
 size, 346
Clemmys muhlenbergi, activity, 590
 eggs, 324, 336, 339, 385

INDEX

hatchlings, 385
niche, 593
range, 22
sexual behavior, 485
Climbing, 496, 498-500, 504
Cloaca, 162, 163, 168, 169, 213, 221, 292, 481, 483, 489
Clostridium, 118
Clutch, *see* Eggs
Cnemidophorous sexlineatus, 386
Cold storage, 113
Collecting, baited hooks, 49
  hand, 45-49
  "muddling," 47
  "noodling", 47
  regulations, 54, 55
  sampling, 54
  seining, 48
  shocking, 48
  "sounding," 46, 47
  techniques not recommended, 48
  trammel nets, 48
  "water goggling," 47, 48
Color, 98, 409-411, 480
Color vision, 252-265, 465, 466
Coluber constrictor, 62
Commercial breeders, 326, 333, 539, 540, 585, 602
Competition, 306, 322. *See also* Dominance; Resource partitioning
Conditioned emotional response, 504
Conditioning, 262, 263. *See also* Discrimination; Learning
Confinement, 494-496
Conservation, 45, 54, 55, 296, 524, 528, 537, 540, 568, 569, 601, 602
Cooters, see *Pseudemys* species
Cortex, 195, 274, 276, 286, 499, 500
Cortisone, 497, 498
Corticosterone, 502
Cotylosauria, 1, 5
Courtship, 105, 290, 291
Cranial nerves, 145, 194, 202
Crocodiles, 34, 386, 429
Crotalus atrox, 497
Crowding, 112, 235
Cryptodira, 2, 40, 42
Ctenochelys, 33
Ctenosaura pectinata, 69
Cuora, 23, 39, 328
Cuora amboinensis, 24, 324, 347

Cuora flavormarginata, 346
Cuora trifasciata, 324, 337, 347
Cuterebra, 124
Cyclanorbinae, 40
Cyclanorbis, 39
Cyclemys, 23, 39, 328
Cyclemys amboiensis, 462
Cyclemys dentata, 24, 324, 347, 436, 444, 446, 449
Cycloderma, 39
Cycloderma frenatum, 324
Cylindraspis, 12, 15

Dasypus novemcinctus, 565
Day length, captive conditions, 113, 114
DC shift, 278, 279
Decoys, 48, 481
Defensive behavior, 500
"Defense posture," 489, 495
Deirochelys, 4, 39
Deirochelys reticularia, activity, 510, 589
  body temperature, 210
  carapace marked, 56
  description, 21
  eggs, 318, 324, 346
  female reproductive cycle, 313, 315
  growth, 584
  in mixed species populations, 573
  nesting, 525
  niche, 593
  population density, 586
  secondary sex characteristics, 83
  sex ratio, 581
  sexual maturity, 578
  size, 346
Demography, 541-569. *See also* Population dynamics; Population structure
Dermatemyidae, 2, 10, 11, 338
Dermatemys, 3, 4, 39
Dermatemys mawi, 10, 11, 322, 324, 338, 390
Dermatemyidae, 39, 41
Dermochelyidae, 1, 31, 39, 40, 211, 318, 341, 353, 368, 372, 381, 386, 523
Dermochelyoidea, 40
Dermochelys, 2-4, 39, 532
Dermochelys coriacea, body temperature, 211, 234
  deformity, 402
  description, 30, 31
  eggs, 318, 325, 326, 328, 341, 353, 381, 383, 386, 398

"endothermy," 226
feeding, 230, 234, 523
incubation, 368, 372
nesting, 316, 322, 372, 424
predation, 537
subepidermal fat, 226
taxonomy, 523
thermal relations, 226, 234
Desert tortoise, see *Gopherus agassizi*
Desmatochelyidae, 42
Desmatochelyinae, 40
Desmemydinae, 6, 41
Desmemys, 6
Diencephalon, 193, 194, 197-199, 203
Digestion, 213, 218, 243
Discrimination, 262, 290-298, 301, 461-466, 494, 500
Disease, abscesses, 120
  alimentary disorders, 117
  amoebiasis, 117, 124
  bacterial infection, 118-120
  coccidian infection, 122, 123
  endoparasites, blood parasites, 121
    intestinal protozoans, 121-123
    worms, 123, 124
  fungal infections, 120
  gastroenteritis, 119
  human health implications, 119
  maladaption syndrome, 116, 117
  miasis, 124
  nematode infection, 123, 124
  pulmonary infection, 117
  respiratory, 117
  SCUD, 118, 119
  shell disorders, 120
  trichomoniasis, 122
  ulcerative stomatitis, 118
  viral infections, 117, 118
Disorientation, 142, 443, 494, 500
Display-action pattern, 107
Dithyrosternum, 25, 26
Diving, 182-189, 191
Dogania subplana, 409
Dominance, 480-483, 485-487, 491
Dorsal aorta, 161, 162
Drift fences, 92
Drinking, 237, 238
Drymarchon corais, 565
Duodenum, 158, 163
Dura mater, 146, 160, 195

Ear, 145-147
Echeneis, 48
Ecology, 220, 305, 307, 419. *See also*
  Population topics
Edwardsiella, 119
Eggs, albumin, 343, 373, 374, 379, 399
  allantois, 376, 377
  amnochorion, 376
  amnion, 376
  antimicrobial defenses of, 399
  chorion, 376
  clutch size, 84, 85, 322-331, 529, 533, 567, 574
  detection, 334
  development in nest, 390-400
  diameter, 306
  extraembryonic membranes, 343, 376-379
  hatching rate, 528
  immune system, 388, 389
  incubation, 334-336, 343, 348, 354-370, 371, 373, 412
  internesting interval, 371, 372
  jarring, 396-398
  laying of, 315, 423-425, 510, 516
  management of, 538
  morphogenesis, abnormal development, 400-412
    extraembryonic membranes, 376-379
    gastrulation, 371
    organogenesis, 375
    serial stages in, 373-379
    and temperature, 373, 376
  morphological variation, 400-412
  mortality, 399, 400
  nitrogen metabolism, 387
  oviducal, 333, 334, 392
  oxygen consumption, 377
  predation on, 420
  respiration, 377, 394, 395, 412
  retention by female, 314, 315
  and *Salmonella*, 399
  shape, 336, 337, 373-375
  shells, fibrous layer, 343, 348
    fibrous membrane, 354
    form of, 336
    interspecific variation of, 338-341
    mineral layer, 343, 348, 349
    minerals in, 354
    permeability of, 394, 395
    pores, 354

INDEX

sphaerulites, 349, 354
  whitening of, 374, 375
  size, 326, 327, 331, 336, 337, 338, 342-348, 533
  space limitations in, 380
  transportation of, 335
  turning, 396-398, 528
  viability, 334
  vitelline sac, 373-375
  vitellogenesis, 371
  weight, 381
  yolk, 379, 380, 396, 399
  yolk sac, 343, 376
Electric shock, 48, 294, 461, 462, 464, 467-469, 493, 497, 498, 502-504, 506
Electrolyte balance, 240-243, 501
Electromyography, 454
Electro-olfactogram, 278-280, 283
Electroretinogram, 258
Eleutherodactylus martinicensis, 371
Elochelys, 36
Elodea, 231
Elseya, 39, 42
Elseya dentata, 42, 350, 356, 375
Elseya latisternum, 42, 319, 344, 354, 384
Elseya novaeguineae, 42, 338, 344
Embryos, 333-413
Emigration, 87, 91, 560, 562, 564. *See also* Movement
Emotional reactivity, 493-507
Emydidae, 2, 3, 10, 17-26, 39, 41, 209, 214, 253, 268, 308, 318, 335, 337, 339, 346, 351, 358, 365, 372, 376, 378, 381, 384, 406, 436, 484, 491, 541
Emydinae, 3, 18, 339
Emydoidea, 4
Emydoidea blandingi, body temperature, 210
  collecting, 47
  clutch size, 324
  clutches per year, 318
  deformity, 403
  description, 21
  eggs, 346, 351, 361, 366, 381, 385
  hatchling, 385
  immunology, 388, 389
  locomotion, 436
  sexual behavior, 484
  size, 346
Emydura, 38, 39, 319

Emydura australis, 38
Emydura kreffti, 38
Emydura macquarri, 38, 324, 344, 356, 364, 369, 384, 420
Emydura subglossus, 344
Emys, 2, 39, 118
Emys orbicularis, arterial shunt, 158
  carapace, labelled, 17
  cloacal water flow, 168, 186
  clutch size, 324
  craniotomy, 130, 131
  description, 22, 23
  eggs, 339, 346, 351, 354, 359, 365
  eye, 250
  dosage for Nembutal, 130, 131
  olfactory bulb, 274
  plastron, labelled, 18
  range, 23
  sexual maturity, 578
  size, 346
  telencephalon, 274
  dosage, route for tubocurarine, 140
  visual discrimination, 462
Emys leprosa, see *Mauremys leprosa*
Encyclopedia Cinematographica, 101
Endocrine, 308, 497, 498, see also Leydig cells, Sertoli cells
Endolimax, 122
Energetics, 379-387
Energy exchange, 208, 212
Entamoeba, 117, 122
Enterobacter, 120
Eochelone, 32
Epididymides, 308, 312
Eretmochelys, 39
Eretmochelys imbricata, agonistic behavior, 491
  body temperature, 210
  description, 31
  diet, 523
  eggs, 319, 325, 341, 368, 386
  female reproductive cycle, 312, 313, 315
  hatchlings, 386, 429
  jaws, 230
  nesting, 31, 316, 322, 421, 423, 426, 505
  population differences, 523
  predation, 539
  range, 31
  "sand smelling", 419

sexual behavior, 485
sexual maturity, 532
taxonomy, 30, 523
Erquelinnesia, 32
Erymnochelys, 34, 39
Erymnochelys madagascariensis, 35
Escape behavior, 467, 468, 479, 488, 489, 499
Escherichia, 120
Esophagus, 159, 230, 237
Estivation, 222, 223, 475, 546, 547
Ethology, 417, 426, 427. *See also* Behavior
Eubaenidae, 42
Eubaeninae, 41
Euclastes, 32
Eunotosaurus africanus, 4
"*Eusarkia*" *rotundiformis*, 35
Euthanasia, 143
Evaporative heat loss, 190
Evolution, 4-6, 228, 305, 327, 330, 417, 422, 426, 427, 475, 490, 492, 574, 591
Excretion, 238-243, 481, 482
Excretory electrolytes, 501-506
Exploration, 466, 467, 496
Extant families, 39
Extinction of learned response, 471, 472, 497
Extradimensional shift learning, 470, 471
Eye, 247-262
  ciliary muscles, 248, 249
  cornea, 248, 249
  choroid, 247, 249
  deformity, 402
  electrophysiology, 258-262
  electroretinograms, 258
  eyeball, 247
  iris, 248, 249
  lens, 247-251
  movements, 456
  oil droplets in, 249-251, 256-258
  organogenesis, 375
  photochemistry of, 252, 253
  photopigments, 249, 250, 253-256
  pupil, 248, 249
  retina, 247-252
  sclera, 247-249
  spectral sensitivity of, 258-263

Familiarity, 482
Fat, 163, 226, 321
Feeding, anatomical adaptations, 229-231
  behaviors, 230-232, 475, 533
  calcium supplement in, 114, 116
  in captivity, 234, 235
  as daily activity, 587-589
  by force, 115-116
  grounds, 526
  habits, 531, 571
  by hatchlings, 116
  hierarchy, 235, 486
  inertial, 232
  role of Jacobson's organ in, 268
  neustophagia, 231, 232
  olfactory cues in, 289
  photographic analysis of, 97, 106
  and operant response rate, 458
  feeding preferences, 233-236
  recommended foods for, 114, 115
  and reproductive potential, 326
  schedule, 115
  and size, 326
  species-specific patterns of, 116
  and sympatry, 553
  vitamin supplements in, 115, see also food
Feet, 160, 422-426, 490
Fixed action pattern, 475
Flatback, see *Chelonia depressa*
Fishes, 298
Flukes, 123
Food, and aggregations, 478
  "chasing," 479
  habits, 74, 75, 94, 95, 542
  as reinforcing stimulus, 458-466, 468, 469, 473
  reserves, 236
  supply, 223
Forebrain, 193, 499
Freshwater species, 174, 571-602
  accessory bladders, 241
  collecting, 46-49
  color vision, 262, 263
  diving, 186
  eye, 248
  feeding aggregations, 478
  imprinting, 300
  literature synopsis, 572
  olfactory cues in feeding, 289
  orientation, 430, 431, 433
  population ecology, 571-602
  prey capture, 231
  salt loading, 240
  tagging, 60
  possible territoriality, 485
  throat pumping, 293. *See also families and species names*

INDEX

Gafsachelys, 32
Gaits, 423, 435
  asymmetric, 438
  classification, 439-441
  diagrams of, 437, 439, 441
  footfall patterns, 437-440
  formula, 439-441
  lateral sequences, 437, 444-448
  symmetric, 438, 441
  trot, 444, 447, 449
Galapágos tortoise, see *Geochelone elephantopus*
Gallbladder, 158, 163
Gaze, 487
Geiselmys, 26
Geochelone, eggs, 328, 370
  dominance in feeding, 486
  island populations, 14
  population limiting factors for, 15
  reproductive patterns, 328
  secondary sex characteristics, 558
  taxonomy, 39
Geochelone abrupta, 16
Geochelone carbonaria, agonistic behavior, 491
  aggregations, 563
  biotype, 543
  carapace length, 342
  eggs, 324, 342
  female reproductive cycle, 317
  head movements, 292, 480
  locomotion, 436, 444
  dosage, route for Nembutal for, 131
  olfaction, 292
  screw implant, 131
  sex ratio, 559
  sexual behavior, 292, 293, 484
  thermal relations, 224, 225
  zoogeography, 14
Geochelone chilensis, biotope, 543
  burrows, 547
  eggs, 318, 324, 342
  hibernation, 547
  nesting, 316
  sex ratio, 559
  sexual dimorphism, 555
  size, 342, 557
  zoogeography, 14
Geochelone denticulata, aggregation sites, 477
  beak, 230
  biotope, 543

  eggs, 324, 342
  head movements, 292, 480
  nesting, 420, 422
  olfaction, 292
  sex ratio, 559
  sexual behavior, 292, 293, 484
  size, 342
  zoogeography, 14
Geochelone elegans, biotope, 543
  description, 13, 14
  eggs, 318, 324, 342, 352, 354, 361, 373, 374, 381
  locomotion, 436
  sex ratio, 559
  size, 342, 557
  social behavior, 490
  zoogeography, 13
Geochelone elephantopus, amoebiasis treatment, 122
  biotelemetry, 69
  biotope, 543
  body temperature, 210
  color preference, 465
  dominance, 491
  dosage, response to Etorphine, 133
  eggs, 318, 324, 326, 340, 342, 348, 361, 367, 372, 385
  feeding trails, 550
  fertility, 568
  hatchlings, 385, 394
  locating in field, 46
  longevity, 557, 566
  migration, 422, 478, 551
  mortality, 565, 566
  nesting, 372, 419, 421, 422, 425
  orderly behavior, 478
  population dynamics, 561
  size, 342, 557
  thermal relations, 214, 217
  visual discrimination, 462
  zoo shelter behavior, 478, 482
  zoogeography, 14
Geochelone elongata, 14, 226, 324, 342, 436, 444, 543
Geochelone emys, 342, 543
Geochelone forsteni, 543
Geochelone gigantea, aggregations, 476
  biotope, 543
  eggs, 342
  longevity, 566
  marking, 58
  nesting, 426

population density, 561
sexual behavior, 484
size, 342
thermal relations, 210, 214, 217
Geochelone grandidieri, 16
Geochelone impressa, 543
Geochelone pardalis, biotope, 543
  eggs, 318, 324, 342, 362, 367, 372, 393, 567
  longevity, 557
  nesting, 372, 394, 421, 567
  sex ratio, 559
  sexual behavior, 484
  size, 342, 557
  zoogeography, 12
Geochelone platynota, 543
Geochelone radiata
  ancestors of, 16
  biotope, 543
  description, 15, 16
  eggs, 342, 352, 362, 567
  longevity, 557
  nesting, 422
  sex ratio, 559
  size, 342, 557
Geochelone sulcata, activity, 509, 510, 512, 513
  biotope, 543
  eggs, 342, 367, 385
  hatchling, 385
  sexual behavior, 484
  size, 342
  thermal relations, 214, 217, 226
  zoogeography, 12
Geochelone sumeirei, 566
Geochelone travancorica, 14, 342, 480, 484, 543, 558, 563
Geochelone yniphora, 12, 15, 16, 543
Geoclemys, 39
Geoclemys hamiltoni, 25, 347
Geoemyda, 23, 39. See also *Rhinoclemys*
Geoemyda spengleri, 24, 25, 347
Geomagnetism, 433
Geotaxis, 430, 431
Gerrhonotus coeruleus, 382
Glands, 290-293, 479
  thyroid
  see also Pineal; Pituitary; Thymus
Glarichelys, 32
Glomeruli, 272, 273, 284
Glomerulofiltration, 503
Glottis, 293

Glyptopsidae, 41
Gopher tortoise, see *Gopherus polyphemus*
Gopherus, anatomy in locomotion, 444
  chin glands, 291
  description, 16
  dominance in feeding, 486
  island populations, 14
  maze learning, 457
  olfaction, 292, 292
  species account, 561, 567
  taxonomy, 39
  zoogeography, 14
Gopherus agassizi, activity, 509, 546, 551
  aggregations, 481, 482, 564
  agonistic behavior, 491
  biotope, 543
  brumation, 481, 482, 547, 548
  burrow, 542
  collecting, 46
  eggs, 318, 324, 342, 362, 366, 367, 393
  growth, 77, 553-555
  home range, 548
  immigration, 563, 564
  jaws, 230
  response to lidocaine, 137
  longevity, 556
  marking, 57, 59
  migration, 510, 552
  nesting, 349, 422, 425, 564
  olfaction, 481, 482
  range, 16
  renal tubular permeability, 242
  salt loading, 240
  secondary sex characteristics, 83
  sex ratio, 559
  sexual behavior, 484
  sexual maturity, 306, 307
  size, 342, 556
  social behavior, 292, 490, 492
  soil "poking", 238
  thermal relations, 210, 214, 217, 223, 224
  urine, 240-243
Gopherus berlandieri, biotope, 543
  burrows, 547, 550
  color preference, 233, 234
  eggs, 318, 324, 349, 564
  estivation, 223
  feeding, 234, 551
  effects of hormone injections, 479
  nesting, 349, 420, 421
  nomadism, 550

olfaction, 291, 292
ornithine cycle enzymes, 239
orientation, 552
population density, 560, 561
range, 16
sex ratio, 559
sexual behavior, 484, 489
size, 556
social behavior, 481, 490-492
thermal relations, 210, 214, 224
Gopherus flavomarginatus, 16, 484, 491, 542, 543, 559
Gopherus polyphemus, action potentials, 280
activity, 477, 509, 511, 514-516, 542, 544, 545, 550
agonistic behavior, 491
biotope, 543
burrows, 542
collecting, 46
courtship, 291
drinking, 237
eggs, 324, 342, 352, 564
electro-olfactogram, 278
feeding, 550, 551
growth, 553-555
home range, 548, 549
locomotion, 436, 449
longevity, 557
mortality, 565
movement, 562-564
nasal capacity, 277
nesting, 390, 419, 420
olfaction, 280, 281, 285
orientation, 518, 519, 552
population density, 558-561
population dynamics, 562
range, 16
reproductive effort, 567, 568
sex ratio, 559
sexual behavior, 484
size, 342, 557
social behavior, 492
thermal relations, 210, 225
Granular cells, 273, 274, 277, 284, 285
Graptemys, 21, 39, 48, 50, 116, 231, 409, 573
Graptemys barbouri, deformity, 407
eggs, 318, 324, 346, 359, 385
embryology, 378
hatchling, 385
jaw musculature, 230

nesting, 316
niche, 593
range, 21
secondary sex characteristics, 82
sexual behavior, 484
sexual maturity, 306
size, 346
Graptemys caglei, 21
Graptemys flavimaculata, 21, 484, 492
graptemys geographica, deformity, 401, 402, 406
eggs, 318, 324, 336, 351, 359, 365, 373, 385
growth, 583, 584
hatchling, 385
nesting, 419, 426
niche, 593
range, 21
Graptemys kohni, 359, 385, 409-411, 484, 593
Graptemys nigrinoda, 21, 589, 591, 593
Graptemys oculifera, 21, 336, 359, 378, 385, 465
Graptemys pseudogeographica, body temperatures, 209
deformity, 403
eggs, 318, 324, 336, 337, 346, 359, 365, 385
embryology, 378
growth, 584, 585
nesting, 419, 593
pigmentation, 409, 411
range, 21
sex ratio, 581
sexual behavior, 484
sexual maturity, 578
social behavior, 492
speculation, 409-411
Graptemys pulchra, 21, 230, 318, 324, 385, 465, 578, 593
Graptemys versa, 21
Green turtle, *see* Chelonia mydas
Growth, 74, 553-555, 576, 583-585, 600
Growth determination techniques, 92-94
Gyremys, 25, 26

Habitat, 473
Habitat preferences, 591-593
Habituation, 102, 455, 456, 488, 494, 497, 501
Hadrianus, 14
Hardella, 23, 39
Hatchling, coloration, 529
deformities, 404, 407
early learning, 482

emergence, 354, 355, 428, 429
emigration, 562
energetics, 382, 383
growth, 530
hatching, 354, 377-379, 427-430
management, 538, 539
mortality, 394, 529, 530, 535, 536
nomadism, 562
overwintering, 400, 428
pipping, 355, 377, 379
predation, 420, 429, 430, 528-530, 539
and *Salmonella*, 399
size, 343-348, 387, 556, 557
social facilitation, 427, 428, 475
weight, 384-386
Hawksbill, see *Eretmochelys imbricata*
Hayemydinae, 41
Head, 159, 402
Head movements, 291-293, 456, 464, 494
Hearing, 431, 463, 464, 483
Heart, anatomy, 155-158
rate telemetry, 69-72
surgery, 149-150
Heliothermism, 214
Hemispheric lesions, 499
Heosemys, 1, 23, 39
Heosemys grandis, 25, 409
Heosemys spinosa, 25, 347, 349, 378
Heosemys sylvatica, 347
Hepatic portal system, 163
Hibernation, 184, 185, 190, 222, 223, 510, 516 525, 546-548. See also Brumation
Hieremys, 23, 39
Hieremys annandalei, 24, 324
Hindbrain, 194
Hippocampus, 274, 275, 286
Home range, 91, 485, 548-551, 594-593, 601
Homing, 91, 433, 434, 552, 598, 599
Homopus, 12, 39, 328, 367, 567
Homopus areolatus, biotope, 543
description, 12
drinking, 238
eggs, 324, 342, 349, 557
sex ratio, 559
sexual behavior, 484
size, 342, 557
Homopus boulengeri, 12, 342, 543
Homopus femoralis, 12, 324, 342, 543
Homopus signatus, 12
Hoplochelys, 8
Hormones, 243, 479, 506
Housing, 109-114
Hydrocortisone, 498

Hydromedusa, 37, 39
Hydromedusa maximiliani, 37
Hydromedusa tectifera, 37
Hypophysis, 146, 194, 197, 199, 503, 504
Hypothalamus, 197, 199, 227, 275, 286, 503

Iguana, 69, 382
Immigration, 88, 91, 567.
See also movement
Immobility, 500
Immune system, 388, 389
Imprinting, 236, 299, 300, 473, 482
Indotestudo, 14
Injury, 124, 125
Insect orientation, 298, 299
Instrumental conditioning, amount of reward in, 469
avoidance, 467, 468, 473, 497, 498
detour learning, 494-497
discrimination, 461-466
extinction, 471, 472
extradimensional shift, 470, 471
free operant, 457-461
maze learning, 457
motivation in, 466-469
probability learning, 471
reversal, 469-471
"Intention movement", 480
Islamichelys, 32

Jacobson's organ, 267, 268, 271
jaws, 229-232

Kachuga, 23, 39, 326
Kachuga dhongoka, 325
Kachuga smithi, 324, 347, 349
Kachuga tecta, 347
Kachuga trivittata, 324
Kallokibotiidae, 2, 42
Kallokibotiinae, 41
Kemp's ridley, see *Lepidochelys kempi*
Key, 75
Kidneys, 149, 161, 162, 164, 238, 241
Kinixys, 12, 13, 39
Kinixys belliana, 12, 14, 324, 328, 337, 342 543, 559
Kinixys erosa, 12, 342, 543
Kinixys homeana, 12, 238, 543
Kinosternidae, 3 8-10, 39, 41, 209, 233-235, 308, 319, 335, 337, 338, 345, 350, 357, 364, 378, 381, 384, 406, 436, 484, 490, 491
Kinosternon, activity, 587

collecting, 47
eggs, 328
feeding, 116
fossil history, 9
homing, 598, 599
nesting, 420, 422
prey capture, 230, 231
taxonomy, 2, 3, 39
Kinosternon abaxillare, 8, 324
Kinosternon acutum, 8
Kinosternon angustipons, 8, 324, 345
Kinosternon arizonense, 9
Kinosternon bauri, agonistic behavior, 491
body temperatures, 209
deformity, 408
eggs, 319, 323, 324, 336, 343, 357, 375, 381, 384, 398
habitat, 592
niche, 593
zoogeography, 8
Kinosternon creaseri, 9
Kinosternon dunni, 8, 320, 324, 339, 345
Kinosternon flavescens, activity, 590
agonistic behavior, 491
body temperatures, 209
brumation, 223
eggs, 319, 324, 348, 357, 384
estivation, 590
female reproductive cycle, 312-314
food preference test, 235
habitat, 592
hatchling, 384
home range, 596
male reproductive cycles, 307
neustophagia, 232
niche, 593
population density, 586
sex ratio, 582
sexual behavior, 484
sexual maturity, 306, 579
zoogeography, 8
Kinosternon herrerai, 8
Kinosternon hirtipes, 9, 324, 345
Kinosternon integrum, 9, 345
Kinosternon leucostomum, eggs, 319, 324, 330, 336, 345, 357, 364, 384
female reproduction, 312, 320
hatchling, 384
male reproduction, 309, 311
nesting, 336
sexual maturity, 306
size, 330
zoogeography, 8

Kinosternon scorpiodes, clutch size, 324
eggs, 336, 345, 350, 354, 357, 373, 374, 381, 384
hatchling, 384
nesting, 336
zoogeography
Kinosternon sonoriense, 8, 336, 345
Kinosternon subrubrum, activity, 510, 590
agonistic behavior, 491
body temperatures, 209
brumation, 22
eggs, 319, 324, 350, 357, 373, 374, 381, 384
embryo development, 377
female reproductive cycles, 312-314
food preference test, 235
growth, 585
habitat, 592
hatchling, 384
home range, 596
male reproductive cycles, 307
nesting, 421
niche, 593
population density, 586
sex ratio, 582
sexual behavior, 484
sexual maturity, 579
zoogeography, 8
Kurobechelys, 32

Large intestine, 159, 164
Larus novaehollandiae, 536
Learning, 262-263, 294, 455-474. *See also* Classical conditioning; Habituation; Instrumental conditioning; Operant conditioning
Leatherback, see *Dermochelys coriacea*
Leeches, 121
Leg, 162, 423-427, 456. *See also* Locomotion
Lepidochelys, 2, 39, 313, 419
Lepidochelys kempi, diet, 523
distribution, 32, 523
eggs, 319, 325, 368, 386
hatchling, 386
nesting, 316, 426, 529
sexual behavior, 485
species account, 537
taxonomy, 30, 523
Lepidochelys olivacea, body temperatures, 210
deformity, 406
diet, 523
eggs, 319, 325, 368, 374
migration, 32

nesting, 32, 316, 394
range, 32
taxonomy, 30, 523
Leydig cells, 308-312
Life history background information, 73, 74
Life history techniques, 73-95
Light, 215, 321, 430-432, 465, 476, 477, 496, 497, 510-516, 546
Lincoln index, 89-91
Lissemydinae, 27
Lissemys, 27, 39
Lissemys punctata, 161, 166, 307, 319, 324
Liver, 158, 162-164, 236, 239
Living turtles, 39
Lizards, 60, 98, 104, 107, 190, 200, 208, 227, 239, 382, 386, 511
Locomotion, 97, 435-454
  aquatic, 435, 450-454
  propulsion, 436, 438
  stance, 436
  stride, 438, 444, 447-451
  support, 436, 444-448
  terrestrial, 436-441
  see also Gait
Loggerhead, see *Caretta caretta*
Longevity, 75, 76, 556, 557, 566, 574
Lopochelyinae, 33, 40
Lophochelys, 33
Lungs, 165-173, 182-184, 188, 189, 191

Macrobaenidae, 41, 42
Macrocephalochelys pontica, 27
Macroclemys, 2, 39
Macroclemys auffenbergi, 8
Macroclemys schmidti, 8
Macroclemys temmincki, activity, 588
  agonistic behavior, 491
  beak, 230
  body temperatures, 209
  collecting, 47
  deformity, 401
  description, 8
  eggs, 319, 323, 324, 339, 351, 354, 358, 365, 381, 384, 394
  female reproductive cycle, 312, 313
  fossil history, 8
  growth, 79, 587
  habitat, 592
  hatchling, 384
  home range, 596
  male reproductive cycles, 307, 312

nesting, 390, 426
niche, 593
prey capture, 231
range, 7, 8
sexual maturity, 306, 578
Maintenance, 109-125
Malaclemys, 4, 39
Malaclemys terrapin, activity, 516
  age and reproductive potential, 323
  blood urea concentration, 239, 240
  body temperatures, 209
  breathing cycles, 171, 172
  brumation, 223
  commercial breeding, 326
  deformity, 402, 405
  description, 21
  eggs, 318, 323, 324, 346, 365, 393
  growth, 93, 585
  habitat, 516
  lacrimal salt excretion, 240
  nesting, 419, 420, 422
  orientation, 431
  pigmentation, 409
  range, 21
  respiratory muscles, 166
  saltwater housing, 112
  secondary sex characteristics, 82
  sex ratio, 576, 581
  sexual maturity, 306, 326, 578
  size, 346
  sperm storage, 320
  throat pumping, 293
  urine osmolality, 241
Malacochersus, 13, 39
Malacochersus tornieri, biotope, 543
  breeding program, 334
  defensive posture, 495
  description, 558
  eggs, 324, 342, 363, 385
  hatchling, 385
  longevity, 557
  sexual behavior, 484
  size, 342, 557
  social behavior, 490
  taxonomy, 13
Malayemys, 39, 116
Malayemys subtrijuga, 25, 347
Management, 528, 602
Manouria, 14
Map turtles, see *Graptemys* species
Marine Turtle Newsletter, 58

INDEX 685

Marine turtles, 30-34, 174, 268, 269, 523-540
  eye, 248, 251, 431
  extinct families, 32-34
  imprinting, 300
  locomotion, 444, 450
  orientation, 430-434
  prey capture, 231
  throat pumping, 293
  see also *Caretta; Chelonia; Dermochelys; Eretmochelys; Lepidochelys* species
Marking, 55-60, 88-89, 92, 102, 530, 531, 536
Mascarene tortoises, 14, 15
Matamata, see *Chelus fimbriatus*
Maternal behavior, 434
Mating, 308, 475, 483, 507, 532, 534, see also sexual behavior
Mauremys, 22, 23, 39
Mauremys caspica, conditioning, 262
  eggs, 324, 346, 366, 373
  female reproductive cycle, 312, 313
  lipid changes, 307
  male reproductive cycles, 307, 321
  sexual behavior, 485
  size, 346
  visual discrimination, 462
  zoogeography, 23
Mauremys japonica, 23, 312, 315, 316, 318, 324, 463
Mauremys leprosa, 23, 307
Mauremys mutica, 23, 346, 348, 361, 385, 436
Maze learning, 457, 462
Medication, see specific disease, injury, surgery
Medulla, 194
Megalochelys, 16
Meiolania oweni, 42
Meiolaniidae, 41, 42
Melanochelys, 23, 39
Melanochelys tricarinata, 23
Melanochelys trijuga, 23, 306, 318, 323, 324, 347, 348, 351, 354, 366, 373-375, 381
Mesencephalon, 193, 199, 201, 275
Mesenteries, 159
Mesochelydia, 3, 23
Mesoclemmys, 38
Mesodiencephalon, 199
Metachelydia, 2, 3
Microclimate, 228
Micrococcus, 399
Microhabitat, 54
Midbrain, 193

Migration, 31, 32, 91, 422, 434, 475, 510, 526, 527, 551, 552
Mineralocorticoid activity, 502
Mites, 121
Mitral cells, 272, 273, 284, 285
Moisture, 223
  and activity, 477, 512, 513, 516, 546
  and eggs, 334, 335, 420, 567
  and orientation, 433
  and reproductive cycles, 321, 322
  in growth, 585
  and nests, 393-395, 424
  and social behavior, 477
  see also Water
Morenia, 39
Morenia ocellata, 25, 347
Morenia petersi, 25
Morphogenesis, see egg
Morphology, 97
Mortality, 87, 117, 534-536, 564-566, 574, 583
Motivation, 466-469
Mouth, 402
Movement, 91, 92, 475, 499, 551, 552, 594-599
Movement sequence, 106
Mucor, 117
Mud turtles, see *Kinosternon* species
Muhlenberg's turtle, see *Clemmys muhlenbergi*
Musk turtle, see *Sternotherus* species
Mycobacterium, 118, 120, 399
Myopia, 431

Najas, 231
Natality, 87
Neck, 159
Nematodes, 123, 124
Neochelydia, 2, 3
Nerves, see Central nervous system; Cranial nerves; *individual nerves*
Nervus terminalis, 269
Nest(s), abnormal conditions in, 399, 400
  body pit, 421-423
  construction, 336
  covering, 425, 426
  destruction, 527
  egg cavity, 418, 422-424, 427
  emergence from, 427, 428, 475
  gas exchange in, 394, 395
  habitat, 369
  incubation period in, 369
  and microclimate, 390

moisture in, 393-395
site, density, 528, 529
  multiple species use of, 322, 528
  orientation, 430, 434
  preparation, 418, 421, 422
  selection, 418-420, 473, 477
tenacity, 527
spacing, 420
and temperature, 390-393
Nesting, 31, 32 and body temperature, 525
  digging egg cavity, 422-424
  and human interference, 528, 531
  and humidity, 420
  intervals, 371, 372, 526
  stages of, 418
  at time of day, 420, 509
  Neural transmitter, 517
Neurankylidae, 42
Neustophagia, 231, 232
Niolamia argentina, 6
Nitrogen excretion, 238-240
Nose, 267, 268, 375
Notochelys, 23, 39
Notochelys platnota, 24, 347

Ocadia, 39
Ocadia sinesis, 25, 324, 346
Ocypoda ceratophthalma, 536
Odor production, 290-292, 482
Olfaction, 267-301
  central nervous system connections, 274-277
  cortical responses, 286
  discrimination, 293-298, 481
  gross anatomy, 267-269
  in identification of nesting area, 419
  and imprinting, 299, 300, 482
  and orientation, 298, 299, 433, 552
  receptor anatomy, 269-274
  receptor physiology, 277-286
  and social behavior, 290-293, 481-483
Olfactory bulb, 194, 195, 197, 267-269, 271-276, 294, 296
Olfactory epithelium, 269-271, 277, 280, 297
Olfactory nerve, 268-272, 277, 280, 281, 285, 296
Olfactory tract, 274, 275, 284
Olive ridley, see *Lepidochelys olivacea*
Oogenesis, 312-315, 317
Open field activity, 500
Operant conditioning, and olfactory

  discrimination, 294-297
  learning research, 457-461
  and thermoregulation, 218
Operants, 476
Ophionyssus natricus, 121
Optic chiasm, 194
Optic tectum, 194, 203, 251, 375
Optomotor response, 456
Opuntia, 550
Orbit, 229
Organogenesis, 375
Orientation, 298, 299, 430-434, 478, 518, 551, 552
Orlitia, 23, 39
Orlitia borneoensis, 339
Ornithodoros, 121
Osteopyginae, 33, 40
Osteopygis, 33
Ovaries, 149, 164, 312-315, 317
Overwintering, 236, 400, 428
Oviducts, 334, 370, 371, 373, 392
Oviposition, 315, 335, 336

Pachychelys, 32
Pacific pond turtle, see *Clemmys marmorata*
Painted turtle, see *Chrysemys picta*
Palaeochelys, 25, 26
Paleotheca, 25, 26
Palatobaeninae, 41
Pancake tortoise, see *Malacochersus tornieri*
Pancreas, 158, 163, 164
Panting, 225, 226
Paralichelys, 36
Parasites, 120-124, 218
"Parental investment," 424, 425
Pectoral girdle, 441-443
Pelochelys, 39
Pelochelys bibroni, 27, 28, 319, 325
Pelomedusa, 34, 35, 39
Pelomedusa subrufa, 35, 344
Pelomedusidae, 2, 6, 34-36, 39, 41, 338, 344, 350, 356, 364, 378, 381, 384
Peltocephalus, 34, 39
Peltocephalus tracaxa, 1, 35, 325
Pelusios, 34, 35, 39
Pelusios adansoni, 324, 338
Pelusios gabonensis, 36, 344
Pelusios sinuatus, 336, 350, 354, 373, 381
Pelusios subniger, 324, 344, 356, 384
Pelvic girdle, 441-443
Penicillium, 399

INDEX

Penis, 4, 489, 490
Peptostreptococcus, 120
Perception, 433
Peritresius, 32
Photographic analysis, 97-108, 476
Photokinesis, 432
Photoperiod, 114, 218, 220, 223, 321
Phototaxis, 465
Phototropism, 432
Phrynops, 37, 39, 337
Phrynops dahli, 38, 319, 324, 369, 580
Phrynops geoffroanus, 38, 344, 384
Phrynops gibba, 38, 344, 356
Phrynops hilari, 38
Phrynops nasutus, 37, 38, 324
Phrynops rufipes, 38
Phrynops tuberculatus, 38
Phrynops vanderhaegei, 38
Phylogeny, 40, 41
Pineal, 194, 375
Piriform cortex, 274-276, 283-286
Pituitary, 146, 194, 197, 199, 503, 504
Planiplastron tatarinovi, 27
Plasma, 502
Plastomenus, 28
Plastron, bones, 18
  drag in locomotion, 444
  epigastric artery, 160
  flexibility, 349
  length and egg size, 329, 330
  length of hatchlings, 344-348
  length and limb anatomy, 444
  measurement, 92
  scutes, 18
  surgery, 149
  use in nest concealment, 426
Platemys, 37, 39
Platemys pallidipectoris, 37
Platemys platycephala, 37, 324, 337, 345, 419
Platemys radiolata, 37, 345
Platemys spixi, 37
Platycheloides, 36
Platysternidae, 1, 26, 27, 39, 40, 339
Platysternon, 3, 39
Platysternon megacephalum, 26, 324, 337, 339, 393, 436
Plesiochelyidae, 2, 33, 41, 42
Pleurodira, 2, 41
Pleurosternidae, 6
Pleurosterninae, 41
Podocnemis, 34, 35, 39, 328, 592

Podocnemis dumeriliana, see *Peltocephalus tracaxa*
Podocnemis cayennensis, 325
Podocnemis erythrocephala, 35
Podocnemis expansa, collecting, 48
  ecology, 231, 232
  eggs, 322, 325, 338, 349, 364, 369
  embryonic development, 370
  female reproductive cycle, 317
  nesting, 322, 391, 400
  taxonomy, 35
Podocnemis lewyana, 35
Podocnemis madagascariensis, see *Erymnochelys madagascariensis*
Podocnemis sextuberculata, 35, 324, 344
Podocnemis unifilis, 35, 231, 232, 325, 349, 364, 384, 391, 462
Podocnemis venezuelensis, 35
Podocnemis vogli, 35, 349, 384, 391
Polysternon, 36
Population density, 74, 420, 522, 533, 558-561, 587, 600
Population dynamics, 83, 523-601. *See also* Activity; Growth; Mortality; Movement; *and population entries*
Population estimation, 90, 91
Population integrity, 482
Population recruitment, 87, 532, 574
Population size, 87-91
Population structure, 79, 83, 571-587, 600, 601
Porthochelys, 33
Predation, 223, 228, 329, 420, 424-427, 429, 430, 516, 528-530, 534-536, 564-568
"Presentation posture," 489
Prionochelys, 33
Probability learning, 471
Procolpochelys, 32
Proganochelyidae, 2, 5, 41
Proganochelys quenstedi, 5
Proterochersidae, 2
Proterochersis robusta, 6
Proteus, 118, 120
Protostega, 34
Prostostegida, 6, 32-34
Protosteginae, 40
Psammobates, 12, 39, 567
Psammobates geometricus, 12, 324, 543
Psammobates oculifer, 12, 342, 543
Psammobates tentorius, 12, 237, 238, 324, 340, 342, 543, 557

Pseudemydura, 39
Pseudemydura umbrina, 1, 42, 350, 364
Pseudemys, feeding, 116
  maze learning, 457
  name dispute-*Chrysemys*, 19, 20
  nest site, 420
  prey capture, 231
  range, 18, 19
  SCUD infection, 118
  species groups, 18, 19
  taxonomy, 4, 39
Pseudemys alabamensis, 18-20
Pseudemys concinna, basking, 588
  blood properties, 173, 174
  body temperatures, 209
  breathing cycles, 171, 172
  courtship, 291
  description, 19
  dive length, 182
  eggs, 324, 334, 346, 351, 360, 381, 383, 385
  growth, 584
  jaws, 230
  nesting, 419
  niche, 593
  range, 19
  sex ratio, 580
  sexual behavior, 484
  sexual maturity, 577
  size, 346
Pseudemys decorata, 360
Pseudemys dorbignyi, 18, 318, 324
Pseudemys floridana, basking, 588
  body temperatures, 210
  courtship, 105
  description, 19
  eggs, 324, 346, 360, 366, 373, 375
  hatchling, 385
  home range, 595
  jaws, 230
  in mixed species populations, 573
  nesting, 424, 425, 315
  niche, 593
  operant lever pressing, 459
  size, 346
  terrestrial activity, 510
  thermal relations, 224, 225
  trapping, 52
Pseudemys malonei, 434
Pseudemys nelsoni, 18, 19, 105, 210, 360, 385, 588, 593
Pseudemys ornata, 469
Pseudemys rubriventris, body temperatures, 210

eggs, 323, 325, 360, 385
growth, 93, 584
jaws, 230
nesting, 426
range, 18, 19
sexual maturity, 577
Pseudemys scripta, activity, 510, 511, 516, 588
  age composition in population, 575
  agonistic behavior, 235, 506
  anaerobic metabolism, 183, 185
  apnea, 171, 172
  audition, 464
  avoidance, 468
  bacterial infections, 119
  basking, 588
  biotelemetry, 71, 72
  blood distribution, 157, 158
  brainstem, 193
  brumation, 223
  buoyancy control, 168, 169
  carotid body chemoreceptors, 189
  cerebral heat sensitivity, 227
  cerebrospinal fluid, 189
  cloacal water flow, 168, 169
  clutches, 318, 323-326
  collecting, 48
  color vision, 253-265
  commercial breeding, 326
  cones, 249, 254, 255
  corneal reflex, 128
  courtship, 484, 492
  cranial nerves, 145
  craniotomy, 128
  dark adaptation, 258, 259, 262, 264, 265
  deformity, 401-403, 407
  density spectra, 255
  ecdysis, 77
  eggs, 329, 336, 337, 346, 347, 373, 375, 385, 394, 396, 398, 399
  electrode implantation surgery, 72
  energetics, 380
  response to ether, 134
  extinction of learned response, 472
  eyeball, 251
  feeding, 231, 233-235, 588
  female rejecting male, 489
  female reproductive cycle, 312, 317
  food preferences, 233, 235, 236
  dosage, route for gallamine in, 140, 142, 143
  gross movement, 128
  growth, 93, 584
  growth rings, 76, 77

INDEX

habitat, 591, 592
habituation, 456, 494
hatchling, 385
heart rate, 69, 70, 128
home range, 594-596
homing, 598
incubation, 360, 361, 366, 392
jaw screw implant, 129
larynx, 141
lens, 251
lung gas volume, 169, 170
response to lidocaine, 137
male reproductive cycles, 307, 308, 311
marking, 59
in mixed species populations, 572, 573
metabolic rate, 183
mouth, 141
movement, 599
muscle tone, 128
response to Nembutal, 128-133
nesting, 322, 390, 391, 394, 419, 421
neural transmitter activity, 517
niche, 593
nitrogen excretion, 239
oil droplets, 249, 256-258
olfactory bulbs, 267, 271
olfactory discrimination, 294, 464
operant bar pressing, 458
orientation, 518
as popular research animal, 165
photperiod, 321
photopigments, 253-256
pigmentation, 409
plastron puncture surgery, 128
population density, 586
population size, 89
pupil, 128, 248
range, 18, 19
reproductive cycles, 320. See also Female; Male
reproductive potential, 84, 85
respiration, 141, 168-173, 175-182, 184-186, 188-190
retina, 252
reversal learning, 470
rods, 254, 255
sex ratio, 580, 583
sexual behavior, 484
sexual maturity, 326, 577
size, 346, 347
skull, 144, 145
"social imprinting", 482
sodium uptake, 241, 242

spectral sensitivity, 258-265
taste, 287
telencephalon, 267
thermal relations, 128, 179-182, 190, 210, 215, 225, 227, 234, 391
trapping, 50
dosage route for tubocurarine in, 140
ultrasound tracking, 70
uric acid secretion, 241
visual discrimination, 464
water "nibbling", 483
Pseudemys stejnegeri, 361
Pseudemys troosti, see *Pseudemys scripta troosti*
Pseudomonas, 118-120, 399
Pseudotestudo, 16
Ptychogaster, 26
Ptychogastridae, 26, 40
Punishment, 461, 468, 469
Pylorus, 158
Python, 226
Pyxidea, 6, 23, 39
Pyxidea mouhoti, 25, 347, 349
Pyxis, 13, 39
Pyxis arachnoides, 543

Radioactive tags, 60, 92
Radio telemetry, 223, 433, 601. See also Biotelemetry
Reinforcement, 295
Reinforcement schedules, 458-460
Reinforcers, see Stimulus, reinforcing
Releasers, 480, 483
Remora, 48
Renal portal system, 163
Reproduction, 305-331, asynchrony of male-female cycles, 320
factors influencing, 306, 307, 310, 320-323, 326-331
female cycle, 312-317, 320-322
male cycle, 307-312, 320-322
patterns of, 327-331
Reproductive effort, 566-568
Reproductive potential, 74, 75, 84-86, 322-326, 574
Resource partitioning, 322, 330, 476, 477, 572-574, 599-601
Respiration, 165-191
acid-base balance, 175, 178, 182
breathing control, 175-182, 189, 190
diving, 182-189, 191
embryonic, 376, 377
gas exchange, 170-174, 188-190

mechanical, 140-142
mechanics of, 165-168
Respondent conditioning, 456
Respondents, 475
Response rate, 458-460
Reticular formation, 200
Reversal learning, 469-471
Reward, see stimulus, reinforcing
Rhinolemys, 6, 39, 326, 328
Rhinoclemys annulata, 22, 320, 324, 347, 390, 485
Rhinoclemys areolata, 22, 336, 337, 343, 347, 349, 352, 354, 361, 373-375, 381, 383, 385
Rhinoclemys funerea, eggs, 318, 324, 326, 347, 349, 366
  female reproductive cycle, 312, 317
  male reproductive cycle, 311
  nesting, 322, 390
  sexual maturity, 306
  size, 347
  taxonomy, 22
Rhinoclemys nasuta, 22
Rhinoclemys pulcherrima, 22, 210, 334, 347, 349, 354, 373, 374
Rhinoclemys punctularia, 22, 318, 320, 324, 347
Rhinoclemys rubida, 22, 347
Rhythm, 223, 320, 509-520, see also activity
Ribs, 166
Ridley, see *Lepidochelys* species
Righting response, 220
Ringer's solution, 129, 139, 140, 142
Rodoterula, 399
"Rubner's hypothesis", 75, 76

Sacalia, 23, 39
Sacalia bealei, 324, 349
Sakya, 26
Sakyidae, 26
Salvation, 225, 226
Salmonella, 118-120, 399
Sampling biases, 54, 79, 87-89, 532, 583
Sampling estimate, 87-89
"Sand smelling," 297, 419, 423, 431
Sarcina, 399
Sarcophaga, 124
Satellites, 299
Sawbacks, see *Graptemys* species
Schnabel method, 90, 91
Sciurus niger, 565

Sea turtles, See Marine turtles
Secondary sex characteristics, 79-84, 479, 558, 576
Seminiferous tubules, 309-311
Senescence, 323
Sensory discrimination, 461, 466
Septum, 274, 275
Sergeev's formula, 93
Serratia, 118, 120
Sertoli cells, 308-311
Sex determination, 79-84, 482
Sex ratio, 532, 555, 559, 576, 580-583
Sexual behavior, 475, 479-481, 483-485, 487-490, 492
Sexual dimorphism, 82, 248, 483, 490, 555, 574, 476, 585
Sexual maturity, 306, 307, 554, 556, 557, 576-579, 585, 587
Shansiemys, 26
Shaping, 457
Shell, 92, 125, 166, 174, 349, 402, 406-409, 411, 412. See also Carapace; Plastron
Shweboemys, 36
Siebenrockiella, 39
Siebenrockiella crassicollis, 25, 348
Side necks, see Chelidae, Pelomedusidae
Sinaspideretes wimani, 28
Sinemyidae, 2, 41, 42
Skeletal anatomy, 442, 454
Skull, 145, 229, 375
Sliders, see *Pseudemys* species
Small intestine, 159, 163, 164
Snakes, 62, 118, 121, 122, 213, 221, 226, 235, 239, 386, 411, 429, 497, 499
Snapping, 456
Snapping turtles, see *Chelydra, Elseya, Macroclemys*
Social behavior, 290-293, 475-492, 506, 507
  dominance hierarchies, 486, 487
  factors influencing, 475-479
  during nesting, 434
  stimuli for, 479-483
  subtle responses, 487-490
  see also Agnostic behavior; Sexual behavior
Social facilitation, 427-429, 434, 475
Softshells, *see* Trionychidae
Sonic telemetry, 92
Spalerosophis cliffordi, 386
Spatial distribution, 476-479, 590-593. *See also* Aggregations

INDEX 691

Species identification, 75, 479, 482
Speckled turtle, see *Clemmys guttata*
Sperm storage, 312, 320, 534, 568
Spermatogenesis, 307-312
Sphenopalatine nerve, 269
Spinal cord, 160, 194, 195, 203
Spotted turtle, see Clemmys guttata
Staurotypus, 3, 9, 39
Staurotypus salvini, 9, 336, 348, 351, 357, 384
Staurotypus triporcatus, 9, 339
Stereogneys, 36
Stereotaxis atlas, 150, 151, 193
Sternotherus, activity, 587
   description, 9
   feeding habits, 230
   hunting, 231
   hybridization, 490
   jaw, 230
   in mixed species populations, 573
   nest construction, 422
   range, 9
   smelling, 231
   taxonomy, 2, 4, 39
Sternotherus carinatus, body temperature, 209
   brumation, 223
   eggs, 319, 324, 336, 345, 350, 357, 384
   female reproductive cycle, 312-314
   growth, 585
   habitat, 592
   hatchlings, 584
   home range, 596
   male reproductive cycles, 307
   niche, 593
   population density, 586
   sex ratio, 581
   sexual behavior, 484, 490
   sexual maturity, 307, 579
   shape and heating relationship, 214, 215
   size, 345
   taxonomy, 4
   zoogeography, 9
Sternotherus minor, agonistic behavior, 491
   anaerobic metabolism, 183, 185
   body temperatures, 209
   buoyancy control, 167, 168
   eggs, 319, 324, 336, 357, 364, 381, 384
   fasting, 167, 168
   growth, 585
   habitat, 591, 592
   hatchling, 384
   niche, 593
   respiration, 185
   sex ratio, 581
   sexual maturity, 306
   taxonomy, 4, 490
   water intake, 167, 168
   zoogeography, 9
Sternotherus odoratus, activity, 509-511, 516, 587
   age composition in population, 575
   body temperatures, 209
   burmation, 223
   clutches, 319, 323-326
   color preference, 465
   courtship, 290, 291, 484, 490
   deformity, 403-405, 408
   eggs, 336, 338, 350, 354, 373, 374, 381, 383, 384, 396
   embryonic development, 371, 378, 400
   female reproductive cycle, 312-315
   food preference test, 235
   growth, 93, 583-585
   habitat, 592
   hatchling, 384
   home range, 596
   homing, 598
   incubation, 355, 357, 358, 369
   male reproductive cycle, 307, 308, 312
   nesting, 390, 421
   niche, 593
   odor production, 290
   olfaction, 293
   population density, 586
   reproductive potential, 85
   secondary sex characteristics, 82
   sex ratio, 581, 583
   sexual behavior, 484, 490
   sexual maturity, 306, 579
   size, 323
   taxonomy, 4
   thermal relations, 214, 218-220
   throat pumping, 293
Stimulus, detectability, 461, 463, 464
   discriminative, 298, 461-464, 483
   eliciting, 456, 475, 478, 479-483
   as factor in rhythms, 510-516
   reinforcing, 295, 457-469, 475, 478
   setting occasion for social responding, 475-479
   stressful, 493
Stinkpots, see Sternotherus

Stomach, 158
Stomach analysis, 94, 95, 232
Streptomyces, 399
Stress, 290, 483, 493, 501, 502
Strongylids, 123
Stupendemys geographicus, 36
Subcortical structures, 500
Successive approximations, 457
Surgery, adrenalectomy, 148, 149
　cardiac puncture, 149, 150
　craniotomy, 144, 145
　ear surgery, 146, 147
　hypophysectomy, 146
　jaw screw implant, 147, 148
　thoracic and abdominal, 148
Survivorship, 74, 75, 86, 87
Syllomus, 32

Tail, 162, 402, 426, 488
Tameness, 531
Taste, 286, 287
Taxonomy, 1-4
Tectum opticum, 194, 203, 251
Tela choroidea, 194
Telencephalon, 193, 195, 197, 203, 267, 274
Temperature, and activity, 510-513, 546, 588-590
　for basking, 215, 588
　body, 207-227
　for captive turtles, 113
　effect on brumation and estivation, 223
　eccritic, 208
　and growth, 585
　and hatching emergence, 427-429
　and immune reactions, 388
　and incubation, 334, 336, 567
　and learning, 458
　limits, see thermal tolerance
　and mortality, 433
　in nests, 390-393
　and nitron excretion, 239
　normal activity range, 208
　optimum, 208, 213
　pigmentation, 409-411
　preferred, 208
　and reproductive cycles, 321
　and social behavior, 476, 477
　and terata, 412
　voluntary maximum, 208
　voluntary minimum, 208
Tenebrio molitor, 235

Teratology, 400-412
Terrapene, aggregations, 478
　collecting with dogs, 46
　drinking, 237
　evolution, 6
　feeding, 116, 479
　field studies, 476
　use of forms, 482
　home range, 548
　homing, 548
　inactivity, 546
　maze learning, 457
　mortality, 566
　reproductive physiology, 479
　spatial distribution, 479
　species account, 569
　taxonomy, 39
　terrestrial environment, 541
　trailing, 60
Terrapene carolina, activity, 477, 551
　aggregations, 477, 478
　agonistic behavior, 486, 491, 506
　body temperatures, 209
　brumation, 70, 223, 510, 547, 548
　courtship, 485
　critical thermal maximum, 220
　deformity, 405
　dominance, 491, 507
　eggs, 318, 324, 336, 351, 373, 381, 383, 385, 412
　embryonic development, 377
　feeding, 232, 478
　feeding order hierarchy, 235, 486
　female reproductive cycle, 312
　growth, 553
　habituation, 494
　home range, 551
　homing, 433, 434, 552
　incubation, 359, 365, 372
　locomotion, 436, 444, 446, 448, 449
　longevity, 76, 556
　male reproductive cycle, 307
　marking, 55
　mating, 483, 485
　dosage, route of Nembutal for, 130
　nesting, 372
　odor production, 290
　olfaction, 293, 294
　olfactory nerve, 293
　orientation, 518
　penis extension, 489, 490

INDEX

693

plastron puncture surgery, 130
radio tracking, 70
range, 22
respiration, 179, 190
screw implant, 130
secondary sex characteristics, 82, 248
sex ratio, 559
size, 556
social behavior, 480, 481, 490
sperm storage, 320
survival rate, 87
telencephalon, 193
thermal relations, 190, 225, 226
trailing, 60
visual discrimination, 462
Terrapene coahuila, activity, 552, 590
  eggs, 346
  home range, 595, 596
  population density, 586
  range, 22
  sex ratio, 559, 581
  sexual maturity, 578
  size, 346
Terrapene nelsoni, 22, 346, 559
Terrapene ornata, activity, 477, 486, 507
  agonistic behavior, 491
  biotelemetry, 68-70
  blood volume, 224
  body temperature, 209
  brumation, 69
  collecting, 46
  courtship, 485
  eggs, 315, 318, 324, 325, 335, 385
  embryonic development, 371
  feeding, 106, 478, 486
  female rejecting male, 489
  female reproductive cycle, 312-315
  growth rings, 76
  hatchlings, 385
  heart rate, 69-71
  homing, 552
  incubation, 359, 365
  longevity, 556
  male reproductive cycle, 307
  mating, 485, 507
  nesting, 421
  odor production, 290
  penis extension, 490
  range, 22
  reproductive potential, 323
  sex ratio, 559

sexual maturity, 306
size, 556
social behavior, 490
thermal relations, 224, 226
trapping, 52
Terrestrial species, collecting, 45, 46
  nitrogen excretion, 483, *See also* tortoise
    families and *Terrapene, Clemmys*
Territories, 483, 485
Testes, 149, 161, 164, 307-312
Testudinidae, 3, 4, 6, 11, 12, 14-17, 39, 40, 210,
  268, 290, 318, 339, 342, 352, 361, 366,
  372, 378, 381, 385, 406, 436, 484, 491,
  541
Testudinoidea, 40
Testudo, 2, 16, 39
Testudo elegans, see *Geochelone elegans*
Testudo europaea, see *Emys orbicularis*
Testudo gigantea, see *Geochelone gigantea*
Testudo graeca, biotope, 543
  eggs, 324, 342, 354, 361, 366
  response to ether, 134
  response to halothane, 135
  longevity, 557
  dosage, route, response for Nembutal, 131,
    132
  range, 16
  respiration, 166, 167, 171, 191
  sex ratio, 559
  sexual behavior, 484
  size, 342, 557
  sodium uptake, 242
  thermal relations, 226
  dosage, route for tubocurarine, 140
Testudo hermanni, activity, 511
  biotope, 543
  body temperatures, 210
  brainstem, 193
  breeding programs, 334
  eggs, 385
  incubation, 335, 361, 366, 369, 372
  nesting, 372, 425
  pigmentation, 409
  range, 16
  sex ratio, 559
  size, 557
  social behavior, 490
  dosage, route for tubocurarine, 140
Testudo horsfieldi, biotope, 543
  body temperatures, 210
  burrows, 547

craniotomy, 130, 131
eggs, 318, 324, 342, 366, 399
hibernation, 547
inactivity, 546
dosage, route for Nembutal, 130, 131
sex ratio, 559
size, 342
thermal relations, 214, 217
Testudo kleinmanni, 16, 237, 339, 543, 557
Testudo leithii, see *Testudo kleinmanni*
Testudo marginata, 16, 361, 543
Testudo pardalis, see *Geochelone pardalis*
Testudo sulcata, see *Geochelone sulcata*
Thalamus, 197-199, 251
Thalassemyidae, 6, 32, 33, 41, 42
Thalassia, 297
Thecophora, 2
Thermal tolerance, 218-222, 391-393, 517
Thermoregulation, 207-228
and behavior, 213-218
and body shape, 214, 215
and physiology, 224-228
Thinochelys, 33
Threat, 480
Thresholds, 262
Throat pumping, 293, 295, 480, 483
Thymus, 160, 375
Thyroid, 375
Ticks, 121
Time-motion projector, 103-108
Time out procedure, 469
Tissue transplantation, 388, 389
Titillation, 475
Tongue, 229, 231, 286, 287, 480
Tortoises, color preferences, 465
drinking, 237
social interactions, 481, 492
thermoregulation, 547, 548
see also Testudinidea; *Testudo*; *Geochelone*; *Kinixys*; *Malacochersus*; *Gopherus*; *Pyxis, Acinixys*
Tortual, 399
Touch, 483
Toxochelyidae, 6, 32, 33
Toxochelyinae, 40
Toxochelys, 33
Trachea, 159
Trachemys, 20
Trailers, 60, 92
Transmitters, 62-65
Trapping, 49-54, 87, 289

Tretosternon, 6
Triassochelys dux, 5
Trigeminal nerve, 269, 281, 282
Trionychidae, 2-4, 27, 28, 39, 56, 57, 156, 211, 230, 268, 308, 319, 335, 337, 341, 353, 363, 367, 372, 376, 378, 379, 381, 385, 436, 485, 491
Trionychinae, 27, 28, 40
Trionychoidea, 40
Trionyx, activity, 588, 590
anatomy, 444
collecting, 47
description, 27, 28
eggs, 328, 334, 348, 375, 397
fossil history, 28
in mixed species populations, 573
prey capture, 231
range, 28
rotenone susceptibility, 48
taxonomy, 34, 39
Trionyx crtilagineus, 324
Trionyx euphraticus, 353, 354
Trionyx ferox, activity range, 595
body temperatures, 211
eggs, 324, 353, 354, 363, 381, 385
embryonic development, 371
growth, 93
hatchlings, 385
marking, 57, 58
nesting, 419, 421
niche, 593
photograph of, 28
sex ratio, 582
thermal relations, 214, 215
Trionyx muticus, agonistic behavior, 491
deformity, 402-404
eggs, 319, 323-325, 336, 337, 381, 383, 385, 396
embryonic development, 371
hatchling, 385
incubation, 363, 367, 392
response to light, 465
male reproductive cycle, 308
marking, 57-60
nesting, 390, 391, 419, 425
niche, 593
ova, 379
sexual behavior, 485
sexual maturity, 306, 579
social behavior, 490
throat pumping, 293

INDEX

trapping, 49-51
Trionyx sinensis, eggs, 319, 323, 324, 326, 341
  incubation, 367, 372, 393
  nesting, 316, 372
  sexual behavior, 485
  sexual maturity, 578
Trionyx spiniferus, activity, 590
  agonistic behavior, 491
  aquatic breathing, 184
  body temperatures, 211
  corpora lutea, 317
  deformity, 403
  eggs, 319, 323-325, 353, 386
  embryos, 400
  growth, 587
  hatchlings, 386
  immunology, 389
  incubation, 363, 367, 392
  locomotion, 451, 452
  nesting, 419, 425, 426
  niche, 593
  pigmentation, 409
  secondary sex characteristics, 83
  sex ratio, 582
  sexual maturity, 579
  social behavior, 490
  sodium absorption, 241
  trapping, 51
Trionyx triunguis, 158, 161, 162, 184
Trinitychelyinae, 41
Tropotaxis, 432, 434
Tufted cells, 272, 273
"Turtle farming", 326, 539, 540
Tympanic membrane, 146, 147

Urea, 238-240
Urinary alkali metal level, 502, 503
Urination, 225, 226
  and nesting, 421, 422

Urine, 238, 241, 242, 292
Urogenital system, 80, 81, 83, 375

Varanus flavescens, 69
Varanus salvator, 386
Veins, 162-164
Vestibular mechanisms, 147
Videotape, 99, 100
Vision, 247, 266, 289, 301
  and orientation, 431-433, 552
  and social behavior, 479-481
  see also Eye
Visual discrimination. 461-467, 470, 471, 473
Visual thresholds, 262
Vocalization, 483
Vomeronasal nerve, 280-283
Vomeronasal organ, 267, 268, 271

Water, balance, 222
  and electrolyte balance, 240-243
  and orientation, 430, 431
  as positive reinforcer, 466
  regulation, 505, 506
  in stress research, 502
  in urine, 238
  see also moisture
Wind, 215, 216
Wood turtle, see *Clemmys insculpta*
Worms, 123, 124

Xenochelys, 3, 11
Xenochelys formosa, 9

Zangerlia testudinimorpha, 5, 6
Zeitgeber, 510
Zinc, 295-297
Zoogeography, 6-39, 42, 523, 541, 542, 571, 572
Zostera, 297